Regional Geology Reviews

The Geology of- series seeks to systematically present the geology of each country, region and continent on Earth. Each book aims to provide the reader with the state-of-the-art understanding of a regions geology with subsequent updated editions appearing every 5 to 10 years and accompanied by an online "must read" reference list, which will be updated each year. The books should form the basis of understanding that students, researchers and professional geologists require when beginning investigations in a particular area and are encouraged to include as much information as possible such as: Maps and Cross-sections, Past and current models, Geophysical investigations, Geochemical Datasets, Economic Geology, Geotourism (Geoparks etc), Geo-environmental/ecological concerns, etc.

More information about this series at http://www.springer.com/series/8643

Alfred Kröner • Axel Hofmann
Editors

The Archaean Geology of the Kaapvaal Craton, Southern Africa

Springer

Editors
Alfred Kröner
Institut für Geowissenschaften
University of Mainz
Mainz, Rheinland-Pfalz, Germany

Axel Hofmann
Department of Geology
University of Johannesburg
Auckland Park, South Africa

ISSN 2364-6438 ISSN 2364-6446 (electronic)
Regional Geology Reviews
ISBN 978-3-319-78651-3 ISBN 978-3-319-78652-0 (eBook)
https://doi.org/10.1007/978-3-319-78652-0

Library of Congress Control Number: 2018956600

This Springer imprint is published by the registered company Springer Nature Switzerland AG
The registered company address is: Gewerbestrasse 11, 6330 Cham, Switzerland

Preface

The evolution of the early part of Earth history has been a topic of continued interest. The Kaapvaal Craton of Southern Africa is featuring prominently in early Earth studies because of its excellent preservation of Archaean rocks as old as 3.6 billion years.

During the first half of its history, Earth underwent rapid and irreversible changes. Cooling of the Earth was associated with the evolution of a global tectonic regime, with a decrease in mantle-plume activity and the advent of plate tectonics. The evolution of life led to changes in atmospheric/oceanic redox state and the advent of biologically mediated processes. The distinct style of the early Earth is preserved in the Archaean geological record, and some of the best-preserved rocks are found in the Kaapvaal Craton (Fig. 1).

Following the plate tectonic revolution in the 1960s of the last century, many features seen in Archaean rocks of the Craton were interpreted in terms of actualistic geodynamic models. The widespread granitoid rocks of the tonalite–trondhjemite–granodiorite suite were seen as evidence of arc formation, the mafic–ultramafic rocks of greenstone belts were equated with oceanic crust, and many tectonic and/or metamorphic features were related to subduction, accretion and collision. As more research was undertaken globally on Archaean rocks, it became clear that many rock relationships did not fit simple plate tectonic models. Therefore, views emerged over the last 10–15 years or so that the Archaean may have been a transitional period in which crust formation and evolution evolved from a time of stagnant lid tectonics to a period resembling modern plate tectonic processes. This discussion is still ongoing, and the interpretations of Archaean Kaapvaal geology as documented in the contributions of this volume reflect current thinking on crustal evolution and surface processes in one of the best-preserved early crustal segments on Earth.

The last major overview of Kaapvaal Craton geology was published in 2006 in "The Geology of South Africa" (Geological Society of South Africa/Council for Geoscience). Review papers on the Archaean geology of the Kaapvaal Craton were also included in the global compilation of "The Earth's Oldest Rocks", published by Elsevier in 2007. Since then a wealth of new data and field observations have become available, leading to new or improved models for crust formation and the evolution of surface processes, some of which are contained in this volume. While the present volume was compiled, additional new review papers on the Archaean of the Kaapvaal Craton were prepared for an update of "The Earth's Oldest Rocks", to be published by Elsevier in 2018, and the reader is referred to this publication on aspects not discussed in this volume.

This volume contains 11 contributions covering aspects of the metamorphic basement of the Kaapvaal Craton and its volcano-sedimentary cover (Fig. 1). The oldest granitoid–greenstone basement rocks are Eo- to Palaeoarchaean in age and are found in the eastern part of the Craton. The chapter of Kröner et al. summarizes the Palaeoarchaean evolution of granitoid rocks and infolded supracrustal sequences in this part of the Craton, whereas Hofmann et al. discuss the Palaeoarchaean granitoid–greenstone geology in the southeastern part in southern Swaziland and KwaZulu-Natal Province of South Africa. C. Heubeck presents new data for sedimentary and volcanic rocks of the Moodies Group of the Barberton Greenstone Belt (BGB), and Agangi et al. discuss gold deposits in the BGB. As a result of the

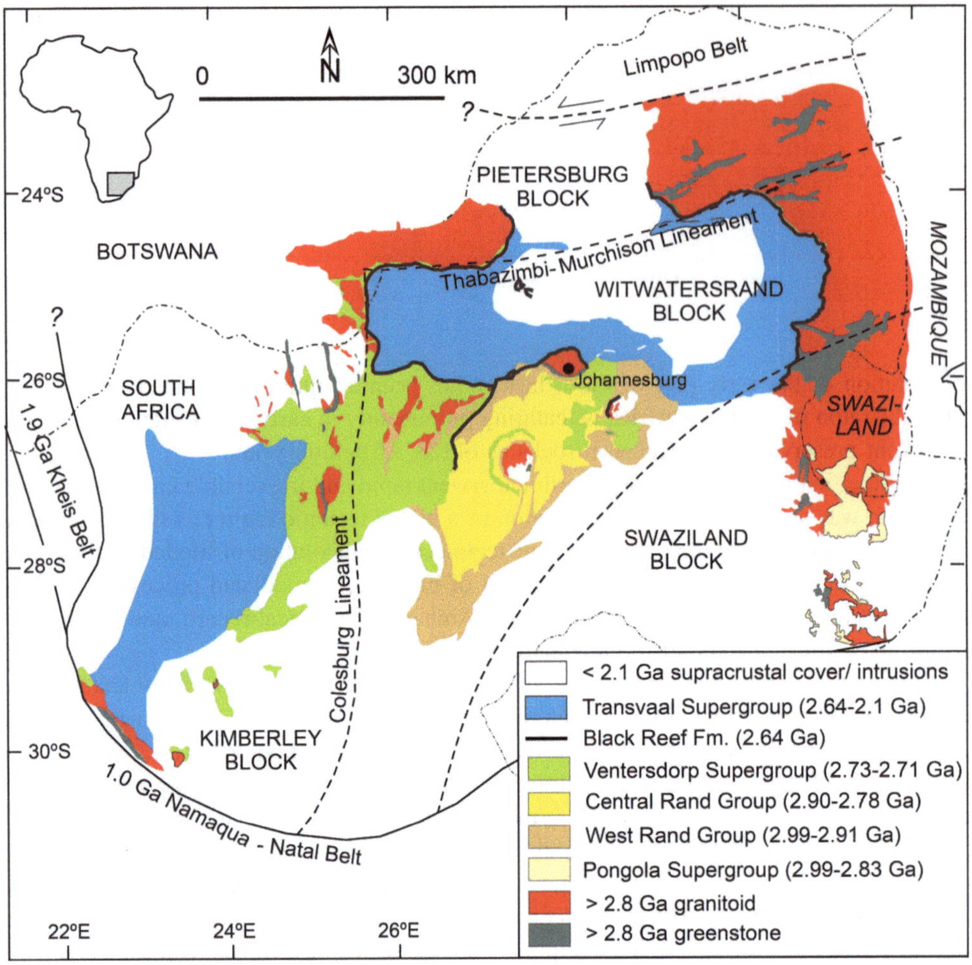

Fig. 1 Surface and subsurface distribution of the main Archaean stratigraphic units of the Kaapvaal Craton (modified after Frimmel 2014)

growth of the Kaapvaal Craton through time, northern and central parts of the Craton are dominated my basement rocks of Meso- to Neoarchaean age. C. R. Anhaeusser deals with the central and western parts, R. L. Gibson provides a summary of Mesoarchaean rocks in the Vredefort Dome, and Laurent et al. summarize the northeastern part up and into the Limpopo belt. Van Reenen et al. present a new tectonic model for the combined evolution of the northernmost part of the Kaapvaal Craton and the adjoining Limpopo belt. The remaining contributions discuss the Archaean cratonic cover sequences of the Pongola Supergroup (Luskin et al.), the Witwatersrand Supergroup and its gold deposits (H. E. Frimmel), and Neoarchaean to early Palaeoproterozoic within-plate volcanism of the Ventersdorp and Transvaal Supergroups (Humbert et al.).

We are grateful to the following reviewers who provided valuable comments and suggestions to improve individual manuscripts: A. Agangi, M. Andreoli, C. R. Anhaeusser, L. Y. Aranovich, R. Buick, A. Dziggel, B. Eglington, R. Ernst, C. Gauert, G. H. Grantham, N. Hicks, T. E. Johnson, M. Klausen, J. Kramers, D. R. Lowe, S. McCourt, D. Nelson, T. Pearton, J. Reinhardt, L. J. Robb, G. Stevens, D. van Reenen and R. P. Viljoen.

Mainz, Germany
Johannesburg, South Africa
July 2018

Alfred Kröner
Axel Hofmann

Reference

Frimmel HE (2014) A giant mesoarchean crustal gold-enrichment episode: possible causes and consequences for exploration. In: Kelley K, Golden, HC (eds) Building exploration capability for the 21st Century. Society of Economic Geologists, Special Publication 18, pp 209–234

Contents

Archaean Crystalline Rocks of the Eastern Kaapvaal Craton

1

Alfred Kröner, J. Elis Hoffmann, Jean M. Wong, Hong-Yan Geng, Kathrin P. Schneider, Hangqiang Xie, Jin-Hui Yang, and Noah Nhleko

Abstract

The eastern part of the Kaapvaal Craton represents a classical granitoid-greenstone terrain and contains the oldest rocks of the African continent, exhibiting about 1000 Ma of crustal evolution from 3.66 to 2.67 Ga. The granitoid rocks predominantly consist of the tonalite–trondhjemite–granodiorite (TTG) association with true granites becoming abundant at about 3 Ga. Greenstones are represented by the well-preserved and well-studied 3.54–3.2 Ga Barberton Greenstone Belt and smaller ca. 3.45 Ga greenstone belt remnants infolded in TTG gneisses around the BGB as well as in the Ancient Gneiss Complex in Swaziland. The origin of both the TTGs and greenstone units is still debated as strong deformation and medium- to high-grade metamorphism have obliterated most of the original rock relationships. This is largely due to an extensive tectono-magmato-metamorphic event at ca. 3.2 Ga that affected virtually the entire eastern part of the craton. Heat provided by mantle-derived melts during this event led to extensive intracrustal melting, with melt migration resulting in depletion of the lower crust in radioactive and other mobile elements. Long-lived extraction of granitoid magmas up to about 2.67 Ga increased the rigidity of the lower crust, causing tectonic stabilization of the Kaapvaal Craton.

Keywords

Ancient Gneiss Complex • Barberton granitoid-greenstone terrain • Crustal evolution Kaapvaal Craton • Swaziland • TTG • Zircon age

Electronic supplementary material
The online version of this chapter (https://doi.org/10.1007/978-3-319-78652-0_1) contains supplementary material, which is available to authorized users.

A. Kröner (✉)
Institut für Geowissenschaften, Universität Mainz, 55099 Mainz, Germany
e-mail: kroener@uni-mainz.de

A. Kröner · H. Xie
Beijing SHRIMP Centre, Chinese Academy of Geological Sciences, Beijing, China

J. E. Hoffmann · K. P. Schneider
Institut für Geologische Wissenschaften, Geochemie, Freie Universität Berlin, 12249 Berlin, Germany

J. M. Wong · H.-Y. Geng
Department of Earth Sciences, University of Hong Kong, Pokfulam Road, Hong Kong, China

H.-Y. Geng
University Research Facility in Chemical and Environmental Analysis, Hong Kong Polytechnic University, Hong Hum, Kowloon, Hong Kong, China

J.-H. Yang
State Key Laboratory of Lithospheric Evolution, Institute of Geology & Geophysics, Chinese Academy of Sciences, P.O. Box 9825 Beijing, China

N. Nhleko
Geological Survey and Mines Department, Mbabane, Swaziland

1.1 Introduction

The Kaapvaal Craton of southeastern Africa is one of the best preserved and best studied Archaean crustal segments worldwide and, together with the Pilbara Craton of Western Australia, has served as a global model for early Archaean granitoid-greenstone evolution (e.g. De Wit et al. 1992; Anhaeusser 2014; Van Kranendonk 2011). It preserves a history of granitoid magmatism, supracrustal deposition, deformation and metamorphism beginning in the Eoarchaean at ca. 3.66 Ga and extending into the Neoarchaean with the emplacement of post-tectonic granites in Swaziland and northern KwaZulu-Natal Province of South Africa at ca. 2850–2670 Ma (see Table 1 in Robb et al. 2006).

The oldest rocks occur within the Ancient Gneiss Complex of Swaziland and consist of foliated and often migmatized granitoid gneisses generally referred to as a

A. Kröner and A. Hofmann (eds.), *The Archaean Geology of the Kaapvaal Craton, Southern Africa*, Regional Geology Reviews, https://doi.org/10.1007/978-3-319-78652-0_1

tonalite–trondhjemite–granodiorite (TTG) suite (Hunter 1979). These rocks partly overlap in age with mafic-ultramafic to felsic metavolcanic assemblages generally referred to as greenstone belts such as exposed in the famous Barberton Greenstone Belt (BGB; e.g. Anhaeusser 2014; see also Byerly et al. 2018; Heubeck 2019; see Fig. 1.1) or in small greenstone remnants infolded in the TTG suite. There has been an ongoing discussion for almost 50 years on the original relationships between the granitoids and the greenstone sequences. Early investigations explained the greenstones as rift-related bimodal assemblages deposited on older granitoid crust (Anhaeusser et al. 1969) although field evidence for such a relationship could not be provided. Subsequently, plate tectonic interpretations envisaged the greenstones to have evolved in an intra-oceanic environment with no evidence for the older continental crust (Anhaeusser 1973; de Wit et al. 1987; Furnes et al. 2013; Drabon et al. 2017; de Wit et al. 2018). In such models, the mafic-ultramafic parts of the greenstones are seen as remnants of early Archaean oceanic crust, implying the operation of sea floor spreading and modern-style plate tectonics at least since about 3.5 Ga ago (e.g. Dilek and Furnes 2011). The alternative model is that the greenstones evolved on or near older continental crust, represented by the TTG granitoid suite (e.g. Kröner 1985; Kröner et al. 2016; Van Kranendonk 2011), and this model is based on indirect evidence such as the occurrence of old zircon xenocrysts in the greenstone volcanic rocks, the chemistry of felsic volcanic layers within the greenstone sequence, suggesting derivation from pre-greenstone felsic crust, as well as Hf-in-zircon isotopic data also implying the presence of older felsic crust (Kröner et al. 2016). This controversy remains unresolved.

Almost the entire eastern part of the craton, from northern Swaziland into central KwaZulu-Natal and including the southern BGB east of the town of Badplaas (now renamed to eManzana), was affected by a strong deformational, metamorphic and magmatic event at about 3.2 Ga. This resulted in ductile deformation and migmatization of lower crustal rocks of the TTG suite (Kröner 2007; Kröner et al. 2018) and in open to tight folding and low- to medium-grade metamorphism in the upper crustal greenstones, locally reaching uppermost amphibolite-facies (e.g. Dziggel et al. 2005; Nédélec et al. 2012; see Brown 2015 for discussion). This event led to extensive structural transposition and obliteration of the original granitoid-greenstone relationships as well as the emplacement of a widespread suite of ca. 3.2 Ga granitoid rocks. During this event, some of the greenstones were transferred to lower crustal levels, intensely deformed and metamorphosed together with the enclosing TTG gneisses and now occur as fragmented greenstone remnants such as in Swaziland and in the Piet Retief area of adjacent southern Mpumalanga Province of South Africa (see Hoffmann et al. 2018). Here again, tectonic models for the 3.2 Ga event vary from plate tectonic (i.e. mobile lid) scenarios (e.g. Moyen et al. 2006; Schoene and Bowring 2010) to models envisaging partial convective overturn (Van Kranendonk 2011) and extensive magmatic underplating (e.g. Kröner et al. 2018) in a stagnant lid tectonic environment (e.g. Bédard et al. 2003; François et al. 2014; Johnson et al. 2014; Moyen and Laurent 2018; Sizova et al. 2018).

The Nelspruit Migmatite Terrain (NMT) north of the BGB (Figs. 1.1 and 1.2), also known as Nelspruit Batholith or Nelspruit Suite (Robb 1983a; Robb et al. 2006), consists of variably migmatized potassic granitic gneisses and porphyritic granite, dated at 3105 ± 5 Ma (Kamo and Davis 1994), and distinct from the TTG terrain around and south of the BGB (Robb and Anhaeusser 1983; Robb et al. 2006). New isotopic data do not indicate derivation of these rocks from early Archaean crust (Zeh et al. 2009; Hao Wang, pers. comm. 2018), suggesting that there may be a terrane boundary between the BGB and NMT (Zeh et al. 2009).

The large, sheet-like Piggs Peak and Mpuluzi Batholiths that separate the BGB from the AGC in Swaziland (Fig. 1.1) are similar in age to the Nelspruit Batholith. Their chemistry, zircon geochronology and evolution were recently investigated by Murphy (2015), and these rocks will be discussed in Sect. 1.6.

The granitoid rocks and greenstones of the eastern Kaapvaal craton are overlain, unconformably, by the ca. 2985–2967 Ma Nsuze bimodal volcanic suite and overlying Mozaan metasediments of the Pongola Supergroup which are discussed in this volume by Luskin et al. (2019). In southern Swaziland, these rocks became involved in strong deformation metamorphism and local anatexis at about 2.72 Ga (Hofmann et al. 2015) as part of an event leading to emplacement of discrete, undeformed granite plutons into the older crust in Swaziland in the Neoarchaean between 2820 and 2690 Ma. During and shortly after deposition of the Pongola sequence, the 2990–2978 Ma mafic-ultramafic rocks of the Usushwana Complex (Gumsley et al. 2015) intruded into the Pongola and the older granitoid gneisses in southwestern Swaziland (Fig. 1.2) and in the neighbouring Piet Retief area in South Africa. Undeformed Neoarchaean granites terminate the magmatic evolution in the eastern Kaapvaal Craton.

In addition to providing a review on the early Archaean rocks of the eastern Kaapvaal Craton, this contribution provides new geochronological and geochemical data. All tables containing analytical data are in the Supplementary File 1 together with short descriptions of the analytical procedures. Sample locations and brief characterizations of dated samples are listed in Table 1.1.

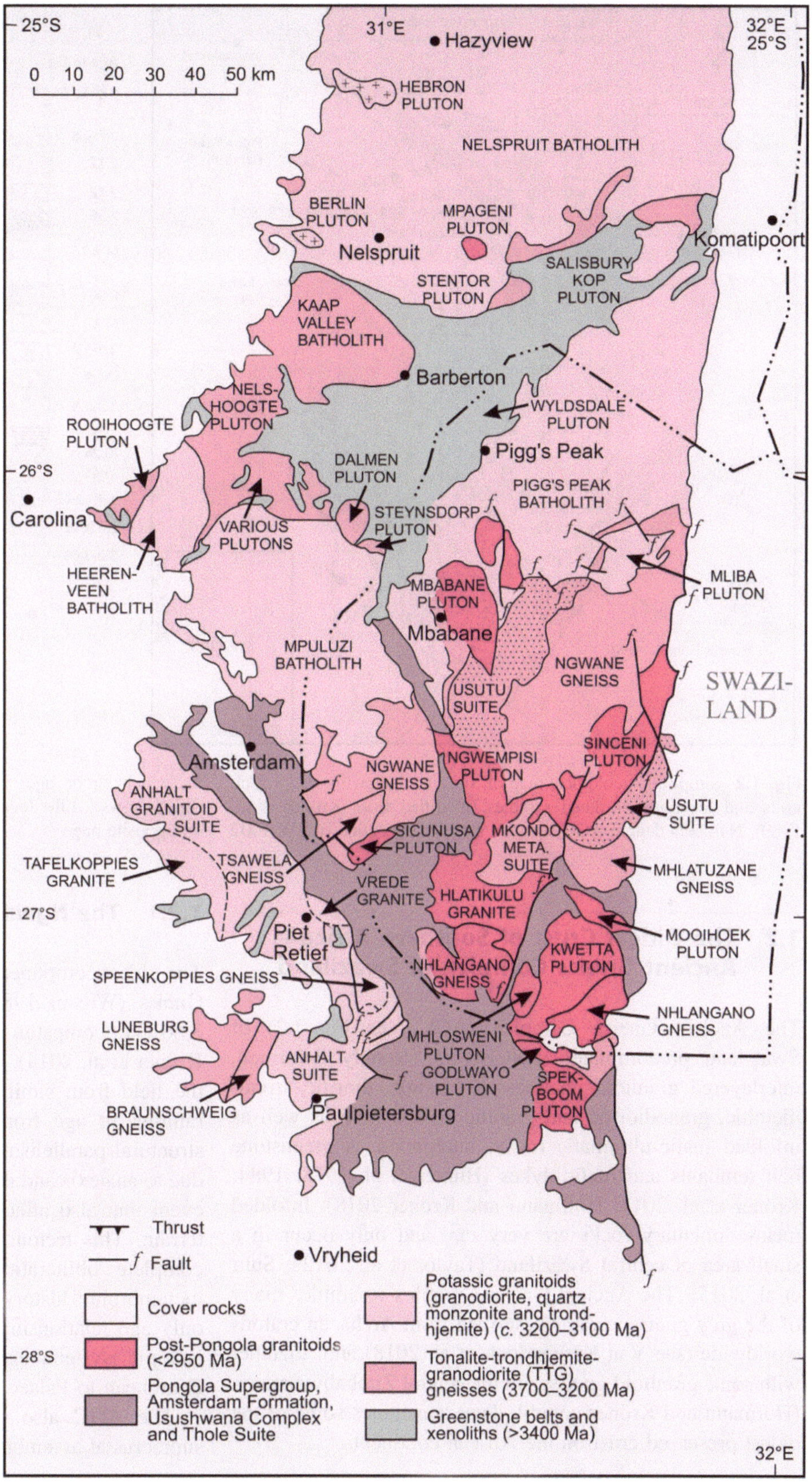

Fig. 1.1 Generalized map showing the main Archaean granitoid intrusions and greenstone occurrences in the eastern Kaapvaal Craton Modified from Robb et al. 2010

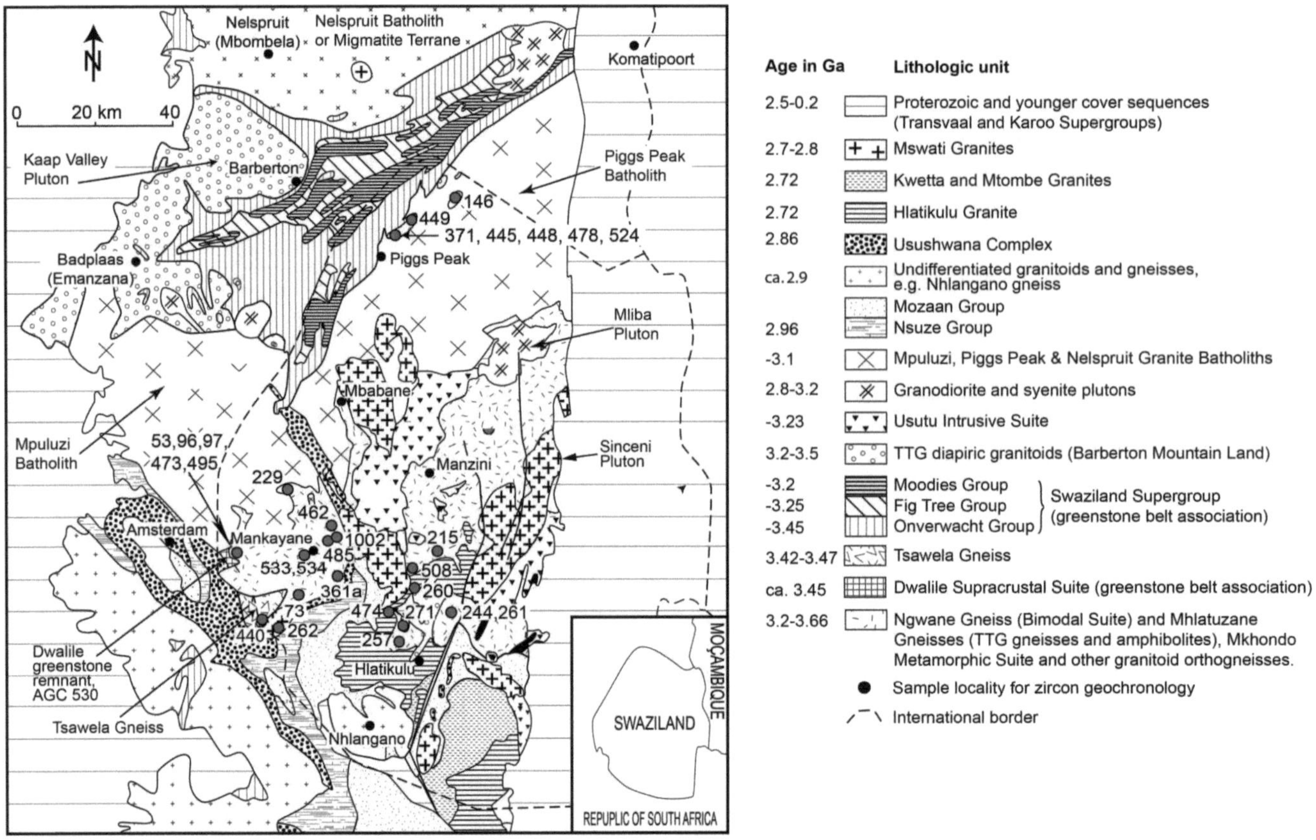

Fig. 1.2 Simplified geological map of Swaziland showing major rock units and locations of dated samples (modified from Kröner et al. 1989). Numbers denote AGC samples with the exception of SW 1002 for which zircon ages are reported in this contribution. Insets show (Inset shows ...) the location of Swaziland in southern Africa. Legend on opposite page

1.2 The Oldest Crust of Southern Africa—Ancient Gneiss Complex of Swaziland

The Ancient Gneiss Complex (AGC, see Fig. 1.2) of Swaziland predominantly consists of strongly deformed, interlayered granitoid gneisses of tonalitic, dioritic, trondhjemitic, granodioritic and granitic composition as well as infolded mafic-ultramafic rocks, interpreted as greenstone belt remnants and mafic dykes (Hunter et al. 1978, 1984; Kröner et al. 2014; Hoffmann and Kröner 2018). Infolded metasedimentary rocks are very rare and only occur in a small area of central Swaziland (Taylor et al. 2012a; Suhr et al. 2015). The Ancient Gneiss Complex resembles many of the grey gneiss terrains described from Archaean cratons worldwide (see Van Kranendonk et al. 2018) and, together with some granitoid gneisses in the central Zimbabwe craton (Hofmann and Kröner unpubl. data) constitutes some of the oldest preserved crust on the African continent.

1.2.1 The Ngwane Gneiss

The oldest component of the AGC has been named Ngwane Gneiss (Wilson 1982) and yielded zircon ages of 3.45–3.66 Ga (Compston and Kröner 1988; Schoene et al. 2008; Kröner et al. 2014). This gneiss is difficult to distinguish in the field from similar-looking, but younger TTG gneisses ranging in age from ca. 3.45 to 3.2 Ga because of the structural parallelism of all these units (Kröner et al. 2018) due to anatexis and deformation at ca. 3.2 Ga during a major event that also affected the Barberton granitoid-greenstone terrain. This tectono-metamorphic event led to the almost complete obliteration of the pre-3.2 Ga structural and metamorphic history in rocks of the AGC. Consequently, only age relationships based on zircon geochronology and isotopic systematics can be used to infer the Eo- to Eoarchean to Palaeoarchaean history of this complex.

The AGC also includes remnants of greenstone belt supracrustal assemblages that vary in size from xenoliths of

Table 1.1 Summary of sample localities, rock types, and zircon ages for ca. 3.2 Ga granitoid rocks and ca. 3.45 Ga amphibolites of the Ancient Gneiss Complex, Swaziland. For map location of dated samples see Fig. 1.2

Sample	S. Latitude	E. Longitude	Locality	Rock type	Age (Ma)
AGC53	26°45′35.7″	30°49′17.5″	Slope south of Dwalile village	Grey banded gneiss, partly migmatitic	3213.6 ± 2.5
AGC73	26°42′21.5″	31°03′56.6″	North bank of Ngwempisi River S of Mankayane	Light-coloured foliated grey gneiss	3221.2 ± 3.0
AGC96	26°41′23.5″	30°50′20.1″	East of road to Dwalile at gneiss–greenstone contact	Tonalitic gneiss of Ngwane type	3455.9 ± 4.4
AGC97	26°41′24.2″	30°50′22.1″	East of road to Dwalie at gneiss–greenstone contact	Trondhjemitic gneiss of Ngwane type	3444.1 ± 2.8
AGC146	25°58′31.0″	30°39′39.1″	Road cut on MR1 asphalt road near Hlohlo	Dark, banded trondhjemitic gneiss	3213 ± 6
AGC215	26°39′59.2″	31°23′46.7″	East bank of Ngwempisi River SW of Sidvokodvo	Dark, foliated bio- and hbl-bearing grey gneiss	3236.3 ± 2.0
AGC229	26°31′43.1″	31°09′21.8″	Roadcut west of Malkerns on road to Mhlambanyatsi	Fine-grained massive but foliated tonalite	3240.4 ± 3.7
AGC244	26°49′08.1″	31°25′19.9″	Matsanjeni River, Kubuta-Sishelweni farm	Granulite-facies metagreywacke	3541 − 3628
AGC257	26°53′48.8″	31°19′57.1″	Roadcut on MR9 asphalt road N of Hlatikulu	Heterogeneous banded grey gneiss	3224.6 ± 3.2
AGC260	26°47′19.6″	31°21′30.5″	Roadcut on N9 highway in Mkhondo Valley	Hlatikulu Granite	2726.8 ± 6.2
AGC261	26°49′12.5″	31°25′33.5″	Matsanjeni River, Kubuta area, central Swazild.	Tonalitic gneiss interlayered with metasediment	3556.2 ± 2.1
AGC262	26°51′54.8″	30°57′37.7″	Roadcut on road N13 near Sicunusa village	Sicunusa Granite	2724 ± 6
AGC271	26°53′12.2″	31°17′39.8″	Bank of Mkhondo River near bridge	Migmatitic granite-gneiss	2726.5 ± 3.1
AGC317	26°35′17.4″	31°11′09.3″	At bridge across Usutu River south of Malkerns	Massive, weakly foliated granodiorite of Usutu Suite	3226 ± 4
AGC361a	26°43′22.4″	31°08′22.9″	North bank of Ngwempisi River	Light grey, not layered trondhjemitic gneiss	3226.8 ± 4.5
AGC371	25°53′52.0″	31°17′48.2″	Phophonyane Waterfall	Strongly foliated dark grey homogeneous tonalite	3214.3 ± 2.3
AGC440	26°50′16.2″	30°55′08.7″	Farm just NE of Sicunusa Porder Post	Silicified felsic volcanic rock	3450.7 ± 4.2
AGC445	25°54′35.0″	31°18′11.1″	Blasted material from roadcut	Banded grey gneiss, similar to 3.66 Ga variety	3216.0 ± 0.9
AGC448	25°53′43.2″	31°18′32.4″	Phophonyane River NE of waterfall	Homogeneous dark grey banded gneiss	3216.8 ± 4.0
AGC449	25°53′24.3″	31°19′11.1″	Roadcut on MR1 asphalt road	Coarse-grained, layered, qtz-rich granitic gneiss	3229.3 ± 3.9
AGC473	26°40′56.1″	30°49′05.4″	East of Dwalile greenstone remnant	Very homogeneous layered trondhjemitic gneiss	3240.5 ± 2.9
ACG474	26°48′27.6″	31°19′21.5″	Roadcut along N9 in Mkhondo Valley	Well-foliated granodioritic gneiss	2712 ± 7
AGC478	25°54′35.0″	31°18′11.1″	Blasted material from roadcut	Grey, foliated trondhjemitic gneiss	3241.6 ± 1.1
AGC485	26°40′30.4″	31°07′37.1″	South bank of Mtimane River E of concrete drift	Diorite from large boudin in 3.5 Ga gneiss	3287.7 ± 2.0
AGC495	26°40′47.2″	30°49′06.4″	Forest SE of Dwalile village	Grey banded gneiss, partly migmatitic	3226.0 ± 1.9
AGC524	25°53′40.2″	31°18′07.7″	Phophonyane River above waterfall	Grey banded tonalitic gneiss	3223.0 ± 3.7

(continued)

Table 1.1 (continued)

Sample	S. Latitude	E. Longitude	Locality	Rock type	Age (Ma)
AGC530	26°40′44.6″	30°51′39.9″	Upper Mponono Riber, southwestern Swaziland	Felsic metavolcanic rock of Dwalile sequence	3458.7 ± 3.7
AGC533	26°41′08.8″	31°02′34.6″	Side stream of Mlatana River near Mankayane	Well-layered amphibolite	3452.2 ± 5.8
AGC534	26°41′17.3″	31°02′30.2″	Hill near Mlatana River SW of Mankayane	Massive calc–silicate rock	3447.9 ± 3.4
SW1002-1	26°40′31.6″	31°07′40.7″	Middle Mtimane River east of concrete drift	Strongly foliated and layered gneiss, light grey part	3273 ± 5
SW1002-2	26°40′31.6″	31°07′40.7″	Middle Mtimane River east of concrete drift	Strongly foliated and layered gneiss, dark grey part	3277 ± 4
KS-SW13	26°41′08.0″	31°02′34.5″	Small stream southwest of Mankayane	Amphibolite (metabasalt)	ca. 3450
KS-SW17	26°42′51.6″	31°02′39.6″	Small river near bridge of MR4 road S of Mankayane	Amphibolite (metabasalt)	ca. 3450
KS-SW23	26°45′24.4″	31°03′58.9″	Small tributary of Ngwempisi River S of Mankayane	Amphibolite (metabasalt)	ca. 3450
KS-SW24	26°45′12.9″	31°04′02.5″	Small tributary of Ngwempisi River S of Mankayane	Amphibolite (metabasalt)	ca. 3450
KS-SW25	26°45′13.6″	31°04′2.8″	Small tributary of Ngwempisi River S of Mankayane	Amphibolite (metabasalt)	ca. 3450
KS-SW26	26°47′13.0″	31°03′07.6″	Track crossing small stream south of Mankayane	Amphibolite (metabasalt)	ca. 3450

a few centimetres long to inliers several kilometres in width (Hunter 1970; Wilson 1982; Jackson 1984; Kröner and Tegtmeyer 1994). The largest of these are the Dwalile greenstone remnant in southwestern Swaziland (Fig. 1.2); Jackson 1984; Kröner and Tegtmeyer 1994; van Schijndel et al. 2017; Hoffmann et al., submitted), the Mankayane remnant near the small town of Mankayane in west-central Swaziland (Hunter 1970; Wilson 1982) and the Sicunusa remnant near the Sicunusa border post with South Africa in SW Swaziland (Wilson 1982).

The gneisses of the AGC are separated from the Barberton Greenstone Belt (BGB) by an extensive granitoid sheet-like pluton some 3.1 Ga in age and known as the Piggs Peak Batholith in Swaziland and the Mpuluzi Batholith in South Africa (Barton et al. 1983; Murphy 2015; see Figs. 1.1 and 1.2). In northwest Swaziland, a small inlier of 3.66 Ga AGC gneisses occurs in faulted and sheared contact with ca. 3.45 Ga BGB rocks (e.g. north of Piggs Peak, see Fig. 1.1; Schoene et al. 2008; Kröner et al. 2014). This suggests that the two units were in direct contact prior to about 3.1 Ga although the original relationships are not preserved.

The zircon geochronology, Hf-in-zircon isotopic systematics and geochemistry of the Ngwane Gneiss have been discussed by Kröner et al. (2014), and these data were summarized by Hoffmann and Kröner (2018). The ages of these compositionally heterogeneous, multicomponent, migmatitic gneisses range from 3.66 to ca. 3.50 Ga (Compston and Kröner 1988; Schoene et al. 2008; Kröner et al. 1996, 2014) and their major and trace element compositions show considerable compositional variability, probably reflecting heterogeneous crustal sources mixed with mantle-derived material (Hoffmann and Kröner 2018). Hf-in-zircon isotopic compositions are also highly variable, even in the oldest zircons dating back to 3.66 Ga, suggesting growth of zircon from melts representing a mix of juvenile and anatectic material of Eoarchean to late Hadean age (Kröner et al. 2014).

The Ngwane Gneiss was intruded by the predominantly tonalitic to dioritic 3.43–3.48 Ga Tsawela Gneiss in southwest Swaziland (Fig. 1.1 and 1.2; Jackson 1984; Hoffmann et al. 2016). This well foliated and generally coarse- to medium-grained suite is distinct from the compositionally layered Ngwane Gneiss, and rarely preserved intrusive contacts show that the Tsawela Gneiss cuts previously isoclinally folded Ngwane Gneiss (Kröner 2007). Such relationships demonstrate that the latter underwent strong ductile deformation prior to about 3.45 Ga. The Tsawela Gneiss is time equivalent to several 3.43–3.45 Ga banded grey gneisses of the Ngwane Gneiss (Hoffmann et al. 2016) and some of the TTG plutons intruding the southern margin of the BGB (Kröner et al. 1991; Kamo and Davis 1994; Kröner et al. 2016) as well as to felsic volcanic rocks within the BGB (Kröner et al. 2013).

1.2.2 Greenstone-Type Supracrustal Rocks and Mafic Dykes Infolded with the Ngwane Gneiss

A characteristic element of the AGC is tectonic interlayering of mafic-ultramafic rocks with the granitoid gneisses at highly variable scales. At many localities, the mafic layers (now predominantly amphibolite) are only several centimetres thick and constitute boudins of what in most cases have been interpreted as mafic dykes (Jackson 1984; Kröner 2007). This relationship had led Hunter (1974) to term the gneiss–amphibolite relationship "bimodal suite". Locally, some of these mafic boudins are well preserved, such as in the Mtimane River east of Mankayane town, and can be shown to represent strongly metamorphosed lenses of 3.45 Ga gabbroic rocks (Kröner et al. 2018).

Major and trace element data for amphibolites from the Mankanyane area are listed in Table 1.2, and primitive mantle-normalized trace element patterns are presented in Fig. 1.3. The trace element patterns show enrichment in LREE and Th, as well as negative Nb–Ta and, except for one sample, negative Ti anomalies (Fig. 1.3). These features are similar to amphibolites in the Dwalile greenstone remnant (Hoffmann et al. 2018) and are interpreted to reflect crustal contamination during magma ascent.

A larger xenolith of mafic-ultramafic rocks associated with partly metasomatized felsic metavolcanics and BIF is exposed just west of the town of Mankayane. It constitutes a tightly folded synform some 2 km in extent and surrounded by Ngwane Gneiss (see 1:50.000 Geological Map of Swaziland, sheet 2631CA). The mafic-ultramafic rocks are now amphibolite, locally garnet-bearing, and talc serpentine schist, whereas the massive rocks described as calc–silicate (Hunter 1970) grade into a well-foliated felsic metavolcanic layer that contains abundant magmatic zircon. Our sample AGC534 with a mean $^{207}Pb/^{206}Pb$ age of 3447.9 ± 3.4 Ma (Fig. 1.4a) is an example for this well-foliated metavolcanic unit. Zircons from an amphibole schist interlayered with the

Table 1.2 Major and trace element abundances of selected ca. 3.2 Ga granitoid gneisses and ca 3.45 Ga amphibolites of the Ancient Gneiss Complex, Swaziland

Lab.	Kiel	Kiel	Kiel	Durban	Kiel	Xi'an	Bremen	Durban	Kiel
Sample	AGC445	AGC448	AGC449	AGC368	AGC473	AGC371	AGC495	AGC361A	AGC361A
SiO_2	57.09	59.53	73.31	66.93	67.76	59.92	n.a.	72.00	n.a.
TiO_2	1.13	0.72	0.14	0.65	0.48	1.49	n.a.	0.40	n.a.
Al_2O_3	18.09	17.87	14.28	15.29	16.15	15.92	n.a.	14.52	n.a.
Fe_2O_3t	7.34	6.14	1.08	5.49	3.14	8.09	n.a.	3.05	n.a.
MnO	0.08	0.08	0.02	0.06	0.04	0.11	n.a.	0.04	n.a.
MgO	1.9	1.73	0.19	1.46	0.85	2.19	n.a.	0.97	n.a.
CaO	5.67	4.72	1.34	2.85	2.92	4.5	n.a.	1.44	n.a.
Na_2O	5.23	5.23	4.68	4.95	5.43	3.82	n.a.	5.16	n.a.
K_2O	1.57	1.64	3.45	1.17	1.74	2.42	n.a.	2.24	n.a.
P_2O_5	0.34	0.26	0.02	0.28	0.15	0.63	n.a.	0.14	n.a.
SO_3	0.22	0.04	n.a.	n.a.	n.a.	n.a.	n.a.	n.a.	n.a.
L.O.I.	0.27	0.83	0.49	0.51	0.51	0.87	n.a.	1.36	n.a.
Total	**98.93**	**98.79**	**99.00**	**99.64**	**99.17**	**99.96**		**101.23**	
Trace elements (in ppm)									
Li	178	83.7	40.3	n.a.	63.4	161	96.2	n.a.	13.7
Sc	3.27	6.00	0.317	5.42	2.10	10.1	2.11	2.58	1.78
V	104	75.4	3.23	n.a.	27.2	106	n.a.	n.a.	33.5
Cr	4.10	12.0	2.47	10.7	7.97	350	n.a.	7.34	7.34
Co	15.7	11.7	0.307	12.9	5.23	16.1	21	6.46	5.53
Ni	5.23	7.12	0.635	10.6	3.80	23.8	4.21	5.66	5.51
Cu	37.9	30.1	2.01	n.a.	10.5	10.9	n.a.	n.a.	3.43
Zn	128	99.1	12.6	n.a.	68.4	154	n.a.	n.a.	44.8
Ga	22.9	23.1	17.6	n.a.	21.0	23.6	n.a.	n.a.	19.0
Rb	50.6	63.7	86.8	7.13	28.8	161	44	58.2	43.9
Sr	499	826	135	635	367	875	523	504	482

(continued)

Table 1.2 (continued)

Lab.	Kiel	Kiel	Kiel	Durban	Kiel	Xi'an	Bremen	Durban	Kiel
Sample	AGC445	AGC448	AGC449	AGC368	AGC473	AGC371	AGC495	AGC361A	AGC361A
Y	11.7	15.8	3.654	10.0	11.2	40.6	10.7	7.35	7.98
Zr	244	270	98.9	175	199	551	197	94.0	196
Nb	9.48	10.0	6.66	8.12	9.76	31.6	11.20	6.04	6.60
Mo	0.565	0.437	0.413	n.a.	0.243	n.a.	0.159	n.a.	8.25
Cd	0.325	0.182	0.190	n.a.	0.157	n.a.	n.a.	n.a.	0.063
Sn	3.52	0.403	3.65	n.a.	1.77	n.a.	n.a.	n.a.	0.118
Sb	0.0442	0.116	0.0615	n.a.	0.0367	n.a.	n.a.	n.a.	0.128
Cs	13.8	7.9	11.5	n.a.	2.46	61.2	2.09	n.a.	0.398
Ba	446	613	545	518	384	899	464	765	688
La	37.3	43.0	11.1	37.8	28.8	122	25	35.7	31.4
Ce	47.6	76.5	15.7	81.7	37.8	241	57.5	66.8	52.6
Pr	9.21	9.07	2.09	7.77	6.20	28.7	5.96	6.85	6.12
Nd	36.9	34.2	7.39	30.6	23.2	96.0	21.8	25.0	21.6
Sm	6.82	6.04	1.38	4.53	4.02	15.0	4.56	3.41	3.32
Eu	1.90	1.73	0.327	1.50	1.07	3.42	1.26	1.35	1.33
Gd	5.26	4.92	1.09	3.58	3.14	12.3	3.69	2.61	2.56
Tb	0.626	0.642	0.142	0.422	0.391	1.51	0.468	0.308	0.320
Dy	2.70	3.23	0.715	2.29	1.92	7.96	2.29	1.69	1.59
Ho	0.390	0.560	0.121	0.388	0.337	1.40	0.362	0.279	0.286
Er	0.857	1.427	0.301	0.999	0.865	3.64	0.882	0.692	0.749
Tm	0.103	0.194	0.043	0.131	0.119	0.489	0.102	0.0875	0.097
Yb	0.611	1.206	0.264	0.828	0.725	2.97	0.619	0.518	0.597
Lu	0.088	0.183	0.0410	0.126	0.101	0.421	0.0785	0.0745	0.099
Hf	5.19	5.98	2.67	3.85	4.48	10.5	4.42	2.27	4.33
Ta	0.481	0.612	0.523	0.315	0.778	1.01	0.999	0.240	0.355
W	0.293	0.175	0.205	n.a.	0.120	n.a.	75.5	n.a.	8.23
Tl	0.631	0.510	0.681	n.a.	0.402	n.a.	n.a.	n.a.	0.315
Pb	9.69	11.7	13.37	7.04	9.87	8.95	9.12	7.97	6.45
Th	4.75	5.92	6.30	0.69	4.71	5.31	6.22	4.26	3.66
U	2.55	1.00	1.19	0.306	0.838	0.666	0.845	0.471	0.299
High precision HFSE, Lu, Sm–Nd concentrations (ppm)									
Zr	218	254	92.8	n.a.	183	n.a.	n.a.	n.a.	164
Hf	5.230	5.909	2.785	n.a.	4.616	n.a.	n.a.	n.a.	4.368
Nb	9.250	9.800	6.060	n.a.	8.850	n.a.	n.a.	n.a.	5.95
Ta	0.486	0.416	0.227	n.a.	0.786	n.a.	n.a.	n.a.	0.250
Lu	0.084	0.183	0.056	n.a.	0.111	n.a.	n.a.	n.a.	0.09763
Sm	6.800	5.640	2.980	5.165	4.930	15.31	n.a.	n.a.	3.51
Nd	37.51	31.98	17.55	33.19	30.46	100	n.a.	n.a.	23.13
Nb/Ta	23.6	25.9	15.3	25.8	17.3	31.2	11.2		23.8
Zr/Hf	41.68	42.99	33.32	45.45	39.64	52.65	44.57		36.8
Zr/Sm	32.1	45.0	31.1	38.6	37.1	36.8	43.2		46.7

(continued)

Table 1.2 (continued)

Lab.	Berlin	Berlin	Berlin	Berlin	Berlin	Berlin
Lab. Sample	Berlin KS-SW 13	Berlin KS-SW 17	Berlin KS-SW 23	Berlin KS-SW 24	Berlin KS-SW 25	Berlin KS-SW 26
SiO_2	54.3	50.7	56.0	53.6	54.0	46.8
TiO_2	0.975	1.29	0.592	0.768	0.666	2.64
Al_2O_3	13.2	13.2	14.5	9.10	8.26	12.4
Fe_2O_3t	10.4	14.7	9.14	11.5	10.6	19.6
MnO	0.149	0.193	0.151	0.163	0.164	0.209
MgO	5.98	5.23	4.29	9.91	10.9	4.37
CaO	7.92	8.90	7.68	8.95	10.2	8.89
Na_2O	2.11	2.01	5.12	3.08	2.35	2.24
K_2O	1.52	0.930	0.560	0.190	0.320	0.760
P_2O_5	0.107	0.244	0.129	0.112	0.089	0.087
H_2O	2.77	2.03	1.62	2.19	1.97	1.75
CO_2	0.125	n.a.	n.a.	0.112	0.0275	n.a.
Total	**99.6**	**99.5**	**99.8**	**99.7**	**99.6**	**99.6**
Li	n.a.	n.a.	n.a.	n.a.	n.a.	n.a.
Sc	n.a.	n.a.	29.0	46.5	42.8	37.7
V	247[a]	210[a]	174	791	391	220
Cr	207[a]	69[a]	773[a]	10[a]	84.1	242
Co	n.a.	n.a.	70.6	97.8	50.7	51.8
Ni	110[a]	76[a]	421[a]	6.21	101	112
Cu	n.a.	n.a.	107	43.4	n.a.	154
Zn	104[a]	117[a]	226	174	99[a]	370[a]
Ga	19[a]	16[a]	14.7	18.5	15.8	12.6
Rb	73[a]	47[a]	6.98	19.1	7.41	23[a]
Sr	200[a]	178[a]	231	202	107	248
Y	26[a]	30[a]	18.7	16.7	26.1	13.2
Zr	120[a]	136[a]	105	69.5	74.5	32.0
Nb	n.a.	n.a.	4.61	4.07	3.35	1.10
Mo	n.a.	n.a.	n.a.	0.181	n.a.	0.118
Cd	n.a.	n.a.	n.a.	n.a.	n.a.	n.a.
Sn	n.a.	n.a.	1.37	0.979	1.41	1.02
Sb	n.a.	n.a.	n.a.	0.0651	0.201	0.589
Cs	n.a.	n.a.	n.a.	n.a.	n.a.	n.a.
Ba	240[a]	217[a]	123	160	21.4	36.5
La	n.a.	n.a.	17.3	11.4	5.16	2.12
Ce	n.a.	n.a.	31.7	23.1	12.9	5.07
Pr	n.a.	n.a.	4.15	2.84	1.97	0.710
Nd	n.a.	n.a.	16.6	11.4	9.62	3.47
Sm	n.a.	n.a.	3.66	2.69	2.90	1.11
Eu	n.a.	n.a.	1.11	0.987	1.19	0.471
Gd	n.a.	n.a.	4.21	3.21	3.05	1.89
Tb	n.a.	n.a.	0.607	0.527	0.663	0.331
Dy	n.a.	n.a.	3.58	3.31	4.49	2.26

(continued)

Table 1.2 (continued)

Lab.	Berlin	Berlin	Berlin	Berlin	Berlin	Berlin
Ho	n.a.	n.a.	0.708	0.704	0.961	0.507
Er	n.a.	n.a.	1.90	1.96	2.69	1.48
Tm	n.a.	n.a.	0.271	0.288	0.407	0.220
Yb	n.a.	n.a.	1.67	1.85	2.55	1.42
Lu	n.a.	n.a.	0.248	0.284	0.396	0.211
Hf	n.a.	n.a.	2.75	2.19	2.04	0.734
Ta	n.a.	n.a.	0.314	0.342	0.242	0.088
W	n.a.	n.a.	0.336	0.488	n.a.	0.697
Tl	n.a.	n.a.	n.a.	n.a.	0.0449	n.a.
Pb	n.a.	n.a.	n.a.	6.19	0.653	1.57
Th	n.a.	n.a.	2.25	2.74	0.387	0.266
U	n.a.	n.a.	0.390	0.593	0.112	n.a.

[a]XRF data (Potsdam)

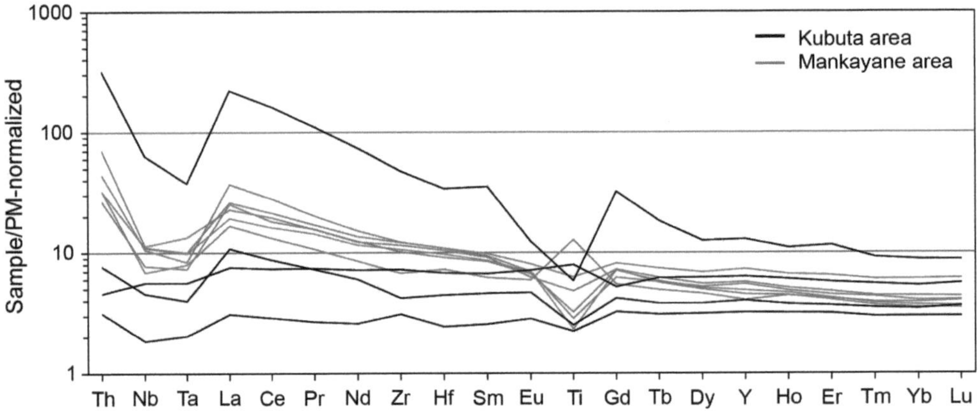

Fig. 1.3 Primitive mantle-normalized trace element abundances of amphibolites and one felsic metavolcanic rock from the Mankayane area of the AGC. Primitive mantle values are from Palme and O'Neill (2014). For analytical data see Table S1

felsic rock (sample AGC533) yielded a similar age of 3452.2 ± 5.8 Ma (Fig. 1.4b) and also shows overprinting at 3224.5 ± 7.0 Ma. We interpret these ages to reflect volcanic activity and consider the rock association exposed near Mankayane as a small greenstone remnant similar in age to the upper Onverwacht Group in the Barberton Greenstone Belt.

A slightly larger but poorly exposed greenstone remnant is exposed just northeast of the Swaziland–South Africa border post at Sicunusa (Fig. 1.1, see also 1:50.000 Geological Map of Swaziland, sheet 2630DD). No mafic-ultramafic rocks were found here, but a sample from a sequence of silicified felsic tuff contained zircons similar in morphology and CL images as found in the felsic rock near Mankayane, yielding a mean $^{207}Pb/^{206}Pb$ age of 3450.7 ± 4.2 Ma (Fig. 1.5a). We interpret this as reflecting felsic volcanism during greenstone deposition and again note the similarity of age with some of the BGB felsic volcanic rocks.

The largest greenstone remnant within the AGC is exposed near the village of Dwalile at the border between Swaziland and South Africa (Fig. 1.2). This sequence constitutes a boot-shaped structure consisting of amphibolite-facies mafic-ultramafic and minor felsic metavolcanic rocks as well as metagreywacke, chert and BIF (Jackson 1984; Kröner and Tegtmeyer 1994; Musese 2014; Hoffmann et al. 2018). It was mapped by Jackson (1984) who concluded that it is younger than the adjacent Ngwane Gneiss on structural grounds. The sequence consists of basal metakomatiite and metabasalt. Tegtmeyer (1989) suggested on the basis of REE data that some of the rocks reflect up to 10% contamination with the older continental crust. This conclusion is supported by Lu–Hf and Sm–Nd isotopic and trace element data (Musese 2014; Hoffmann et al. submitted).

New trace element analyses of Dwalile metavolcanic rocks show that the metabasalts have remarkably different trace element patterns compared to those of the similarly old Hoggenoeg Formation (upper Onverwacht Group) of the Barberton Greenstone Belt (Hoffmann et al. 2018; Fig. 1.6). They are characterized by enriched LREE and negative Nb anomalies as well as variable La/Yb ratios. This either

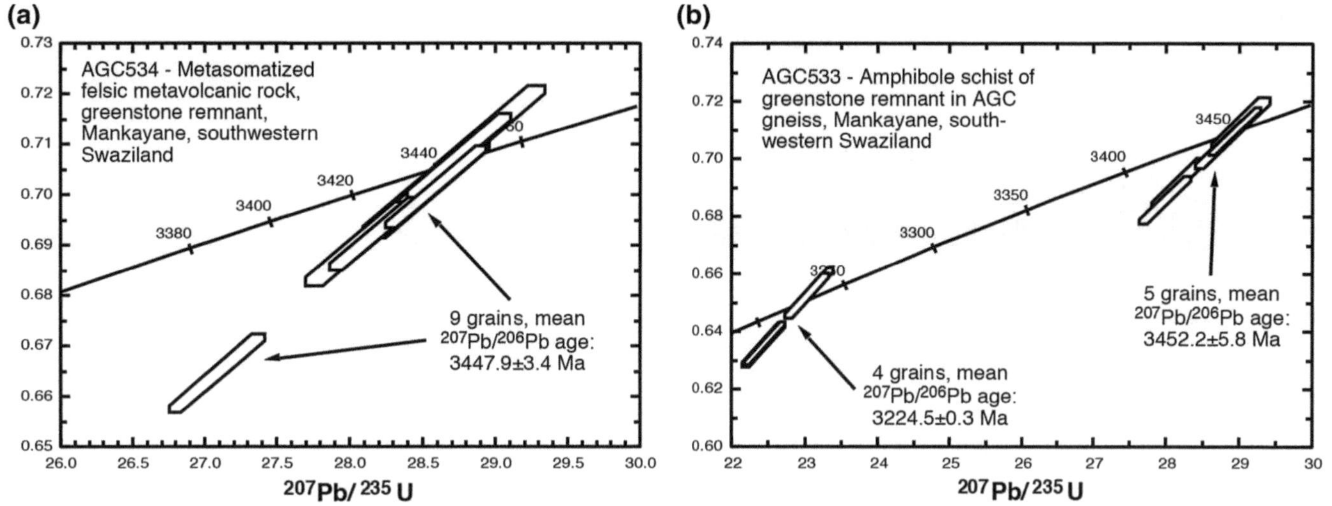

Fig. 1.4 Concordia diagrams showing analytical data for magmatic zircons from metavolcanic rocks of the Mankayane greenstone remnant, southwestern Swaziland. **a** Metasomatized felsic metavolcanic sample AGC534. **b** Amphibole schist sample AGC 533. Sample locations are given in Table 1.1 and Fig. 1.2. Data boxes are defined by standard errors (1σ) in $^{206}Pb/^{238}U$, $^{207}Pb/^{235}U$ and $^{207}Pb/^{206}Pb$. For analytical data see Table S2

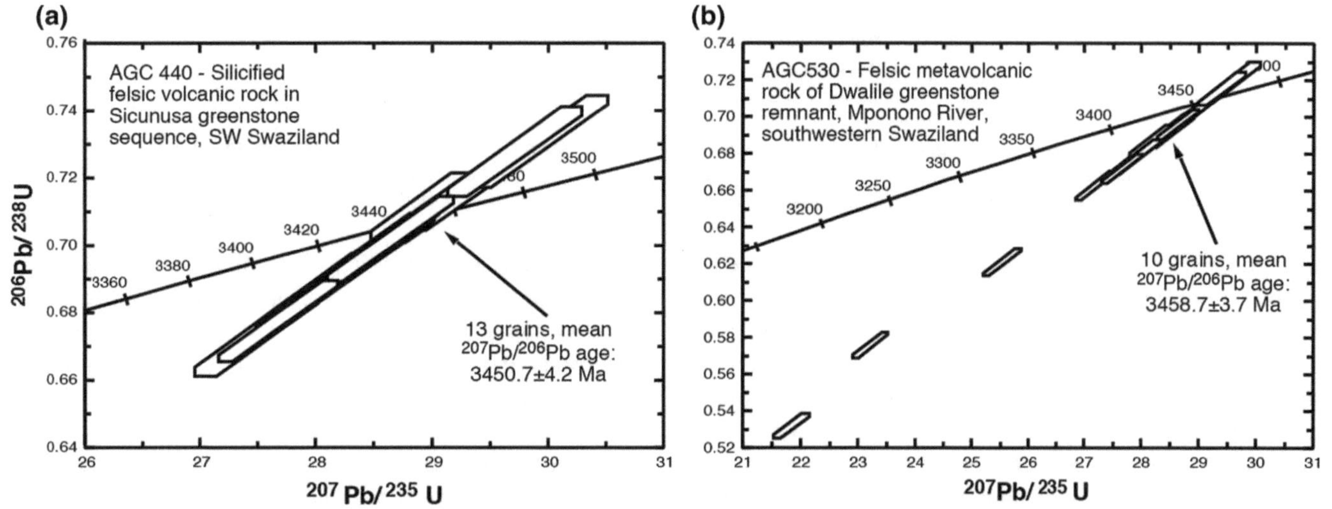

Fig. 1.5 Concordia diagrams showing analytical data for magmatic zircons from metavolcanic rocks of the Sicunusa and Dwalile greenstone remnants, southwestern Swaziland. **a** Silicified felsic metavolcanic sample AGC440. **b** Felsic metavolcanic sample AGC 530. Sample locations are given in Table 1.1 and Fig. 1.2. Data boxes are as in Fig. 1.4. For analytical data see Table S2

reflects crustal contamination or mixing with magmas from felsic volcanic rocks which are found interlayered with the mafic sequence in the field (Kröner and Tegtmeyer 1994; Hoffmann et al. 2018). Such magma mixing scenario has been proposed for the metavolcanic rocks of the Sandspruit and Theespruit Formations of the BGB (Schneider et al. submitted).

A layer of partly silicified felsic metavolcanic rocks associated with BIF seems to structurally overlie the mafic-ultramafic sequence and is well exposed in the northern part of the greenstone remnant as well as in the Mponono River to the east of the main sequence (Jackson 1984). Magmatic zircons from samples of the felsic unit yielded concordant to near-concordant results with mean $^{207}Pb/^{206}Pb$ ages of 3454.0 ± 1.2 and 3458.7 ± 3.7 Ma (Hoffmann et al. submitted; see Fig. 1.5b) that we interpret to reflect the time of felsic Dwalile volcanism. These data also agree with zircon ages of 3465 ± 2 and 3459 ± 2 Ma for two Dwalile felsic volcanic samples dated by van Schijndel et al. (2017).

The felsic rocks are interlayered with garnet–cordierite paragneiss, interpreted by Kröner and Tegtmeyer (1994) as

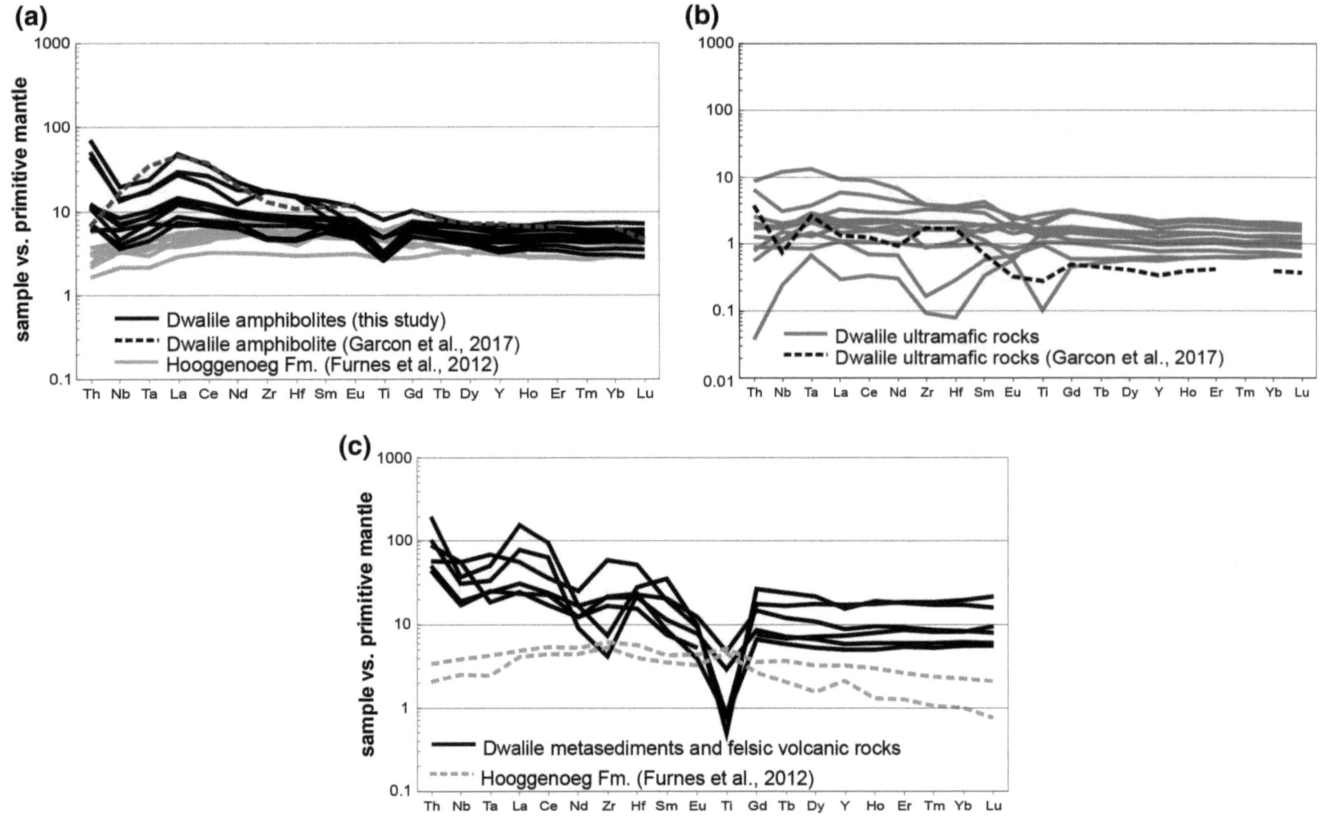

Fig. 1.6 Primitive mantle-normalized trace element diagrams for **a** amphibolite (metabasalt), **b** metakomatiite, **c** metasediments and felsic volcanic rocks from the Dwalile greenstone remnant. Literature data are from Garcon et al. (2017) and Hoffmann et al. (2018) and for the Hooggenoeg Formation from Furnes et al. (2012). Modified from Hoffmann et al. (submitted)

metagreywacke. The least discordant U–Pb data for detrital zircons from three metagreywacke samples yielded ages between 3423 ± 30 Ma and 3566 ± 33 Ma (Kröner and Tegtmeyer 1994) that probably reflect the original chronological heterogeneity of the source terrain, presumably consisting of old components of the Ngwane Gneiss but possibly also the Tsawela Gneiss.

Layers and boudins of amphibolite correlated with the main mafic-ultramafic Dwalile sequence are interlayered with granitoid gneisses at the eastern margin of the main greenstone remnant. Although most contacts are concordant, the granitoid foliation at two localities cuts across the foliation in the amphibolites; we interpret this relationship to reflect originally intrusive contacts. Sample AGC96 (for location see Fig. 1.2, Table 1.3) is a coarse-grained tonalitic gneiss from which U–Pb dating of magmatic zircons yielded a mean $^{207}Pb/^{206}Pb$ age of 3455.9 ± 4.4 Ma (Fig. 1.7a). Nearby sample AGC97, which also cross-cuts an amphibolite layer, is a biotite–hornblende trondhjemitic gneiss from which magmatic zircons yielded consistent concordant and somewhat discordant results with a mean $^{207}Pb/^{206}Pb$ age of 3444.1 ± 2.8 Ma (Fig. 1.7b).

We suggest that granitoid magmatism recorded in these two samples is broadly time equivalent with felsic volcanism in the Dwalile sequence. This is similar to the relationship between ca. 3445 Ma felsic volcanism in the upper Onverwacht Group of the BGB and ca. 3445 Ma granitoids in the Theespruit and Stolzburg Plutons of the southern BGGT (Kröner et al. 1991; Anhaeusser 2014; Kröner et al. 2016).

In summary, the ca. 3.45 Ga Dwalile, Mankayane and Sicunusa supracrustal suites infolded in granitoid gneisses of the AGC in Swaziland constitute remnants of early Archaean greenstone belts and may be correlated with parts of the upper Onverwacht Group of the BGB. Isotopic data suggest variable crustal contamination (Hoffmann et al. submitted), and it is possible that these sequences were originally deposited on an early phase of the Ngwane Gneiss as suggested by Jackson (1984).

1.2.3 Metasediments Infolded with the Ngwane Gneiss

Palaeoarchaean clastic metasedimentary rocks occur in a small inlier within TTG gneisses in south-central Swaziland and are assigned to the Mahamba Gneiss Complex (Wilson 1980; 1982). These gneisses are similar in appearance, style of deformation and metamorphism to the granitoid gneisses

Table 1.3 Estimated U-Pb ages for the major plutonic rocks in the southern Barberton granitoid-greenstone terrane, South Africa

Pluton	Rock type dated	Mineral phase(s)	Age (Ma)	Reference[a]
Steynsdorp	Trondhjemitic gneiss	Zircon	3509 + 8/−9	(a)
	Trondhjemite	Zircon	3509 ± 4	(d)
	Trondhjemite	Zircon xenocryst	3553 ± 2	(d)
	Trondhjemite	Zircon	3510 ± 2	(d)
	Tonalite	Zircon	3506 ± 3	(n)
	Tonalite	Zircon	3521 ± 6	(o)
	Tonalite	Zircon	3517.6 ± 0.7	(q)
Doornhoek	Trondhjemite	Zircon, monazite	3448 ± 4	(a)
Theespruit	Trondhjemite	Zircon, titanite	3443 + 4/−3	(a)
	Trondhjemite	Zircon	3437 ± 6	(b)
	Trondhjemite	Zircon	3452 ± 8	(n)
	Trondhjemite	Zircon	3448 ± 5	(n)
	Tonalitic gneiss	Zircon	3441 ± 3	(e)
	Tonalitic gneiss	Zircon inherited?	3432 ± 10	(f)
	Tonalitic gneiss	Zircon xenocrysts	3481 ± 21	(f)
	Tonalitic gneiss	Zircon	3450 ± 15	(n)
	Tonalitic gneiss	Zircon	3234 ± 12	(f)
Stolzburg	Trondhjemitic gneiss	Zircon	3459 + 35/−23	(a)
	Trondhjemitic gneiss	Zircon	3445 ± 4	(e)
	Trondhjemitic gneiss	Zircon	3457 ± 2	(g)
	Trondhjemitic gneiss	Zircon	3442 ± 5	(m)
	Trondhjemitic gneiss	Zircon	3454 ± 4	(n)
	Trondhjemite, undeformed	Zircon	3455.9 ± 0.5	(q)
	Trondhjemite, deformed	Zircon	3455.5 ± 0.6	(q)
	Granodiorite	Zircon	3225 ± 4	(m)
	Granodiorite	Zircon	3192 ± 15	(n)
Honingklip	Trondhjemite	Zircon	3222 ± 7	(m)
		Zircon	3225 ± 5	(n)
		Zircon xenocrysts	3473 and 3460	(m)
Badplaas	Trondhjemitic gneiss	Zircon	3238 ± 1	(h)
	Gneissic leuco-trondhjemite	Zircon	3256 ± 10	(i)
	Leuco-trondhjemite	Zircon	3230 ± 30	(i)
	Trondhjemitic gneiss	Zircon	3266 ± 6	(n)
	Trondhjemitic gneiss	Zircon	3269 ± 4	(n)
Rooihoogte	Trondhjemitic gneiss	Zircon	3257 ± 13	(i)
	Leuco-trondhjemitic gneiss	Zircon	3272 ± 9	(i)
Nelshoogte	Trondhjemitic gneiss	Zircon	3236 ± 1	(c)
	Trondhjemite	Zircon	3213 ± 1	(j)
	Tonalitic gneiss	Zircon	3236.0 ± 0.5	(q)
	Tonalitic gneiss	Zircon	3236.3 ± 0.4	(q)
Dalmein	Quartz monzonite	Titanite	3215 ± 2	(a)
	Quartz monzonite	Zircon	3203 ± 7	(g)
	Quartz monzonite	Zircon	3197 ± 27	(k)

(continued)

Table 1.3 (continued)

Pluton	Rock type dated	Mineral phase(s)	Age (Ma)	Reference[a]
Boesmanskop	Syenite	Zircon, titanite	3107 ± 2	(a)
Mpuluzi	Granodiorite	Zircon, titanite	3107 + 4/−2	(a)
	Granite, granodiorite	Zircon	3165–3090	(p)

[a](a) Kamo and Davis (1994); (b) Armstrong et al. (1990); (c) De Ronde and Kamo (2000); (d) Kröner et al. (1996); (e) Kröner et al. (1991); (f) Lana et al. (2010); (g) Kröner et al. (2016); (h) Van Kranendonk (2011); (i) Kisters et al. (2010); (j) Layer et al. (1998); (k) Zeh et al. (2009); (l) Layer et al. (1992); (m) Wang et al. (submitted); (n) Wang, unpubl.; (o) H. Xie, unpubl.; (p) Murphy (2015); (q) Schoene et al. (2008)

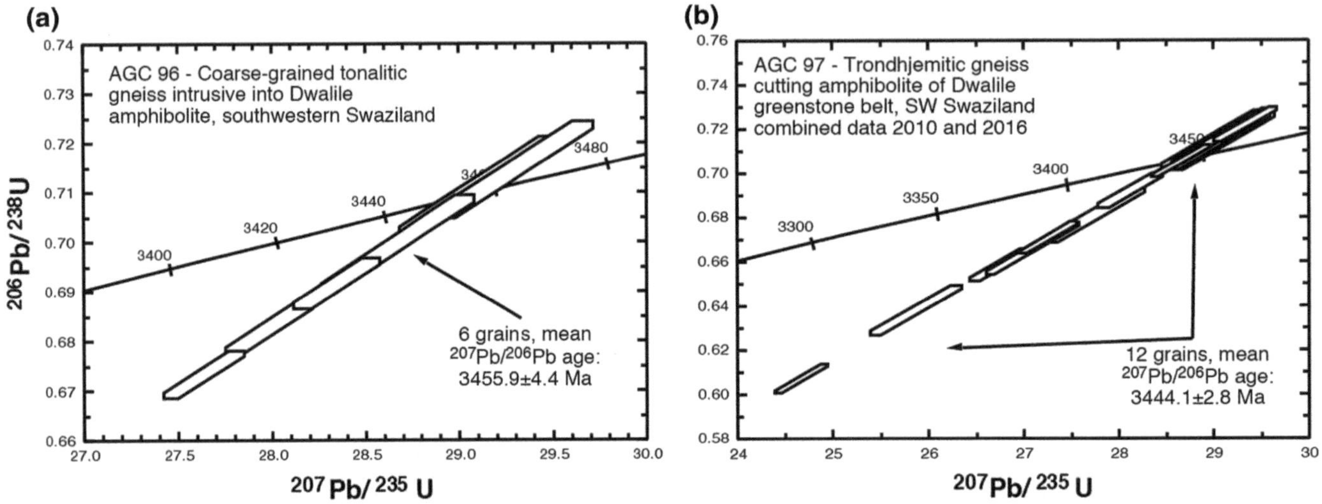

Fig. 1.7 Concordia diagrams showing analytical data for magmatic zircons from granitoid gneisses cutting amphibolite on the eastern margin of the Dwalile greenstone belt, southwestern Swaziland. **a** Tonalitic gneiss sample AGC96. **b** Trondhjemitic gneiss sample AGC97. Sample locations are given in Table 1.1 and Fig. 1.2. Data boxes are as in Fig. 1.4. For analytical data see Table S2

in the AGC, and there is no reason to assign them to a different rock unit. They occur to the west of the settlement of Kubuta and were previously described as Shishelweni gneiss (Kröner et al. 1993) or Kubuta gneiss (Taylor et al. 2012b). These gneisses are interlayered with rare bands (several cm to about 2 m thick) of amphibolite (metabasalt) and metakomatiite, the latter strongly weathered to talc-actinolite schist (Suhr et al. 2015). Infolded paragneisses are best exposed in a ca. 4 km long section in the Matsanjeni River on the Kubuta-Sishelweni cattle ranch and consist of garnetiferous metapelite, quartzite and biotite–plagioclase gneiss that may be a metagreywacke. Detailed descriptions of the metamorphic petrology were provided by Taylor et al. (2012b; 2017) and Suhr et al. (2015).

Petrological data (Taylor et al. 2012a; Suhr et al. 2015) indicate granulite-facies P-T conditions of 795–875 °C and >5.7 kbar for the peak metamorphic evolution, followed by decompression and cooling to 650 °C and 4.7 kbar. Nd–Hf whole-rock-garnet isotopic systematics are not conclusive but suggest an imprecise metamorphic age of ca. 3 Ga (Suhr et al. 2015), in agreement with zircon and titanite metamorphic ages reported by Taylor et al. (2012b).

The oldest of the predominantly trondhjemitic gneisses in the Matsanjeni River is interlayered with the garnetiferous metapelites. Magmatic zircons from a trondhjemitic gneiss yielded an age of 3558 ± 2 Ma (Suhr et al. 2015). Measurement of additional zircons here gave an identical age (3556.5 ± 2.1 Ma, Table S1, Fig. 1.8a). It cannot be established, based on field relationships, whether this gneiss originally constituted a basement to the metasedimentary sequence or was intrusive into it, as all contacts are parallel. Suhr et al. (2015) dated well-rounded detrital zircons from a garnetiferous metapelite in the Matsanjeni River and obtained laser ablation $^{207}Pb/^{206}Pb$ ages ranging from 3308 to 3502 Ma with one discordant analysis dating back to 3730 ± 43 Ma. Taylor et al. (2017) also dated detrital zircons of two metagreywacke samples from the upper Matsanjeni River. Their laser ablation ages range from 3430 to 3598 Ma with a pronounced peak at 3535 ± 5 Ma.

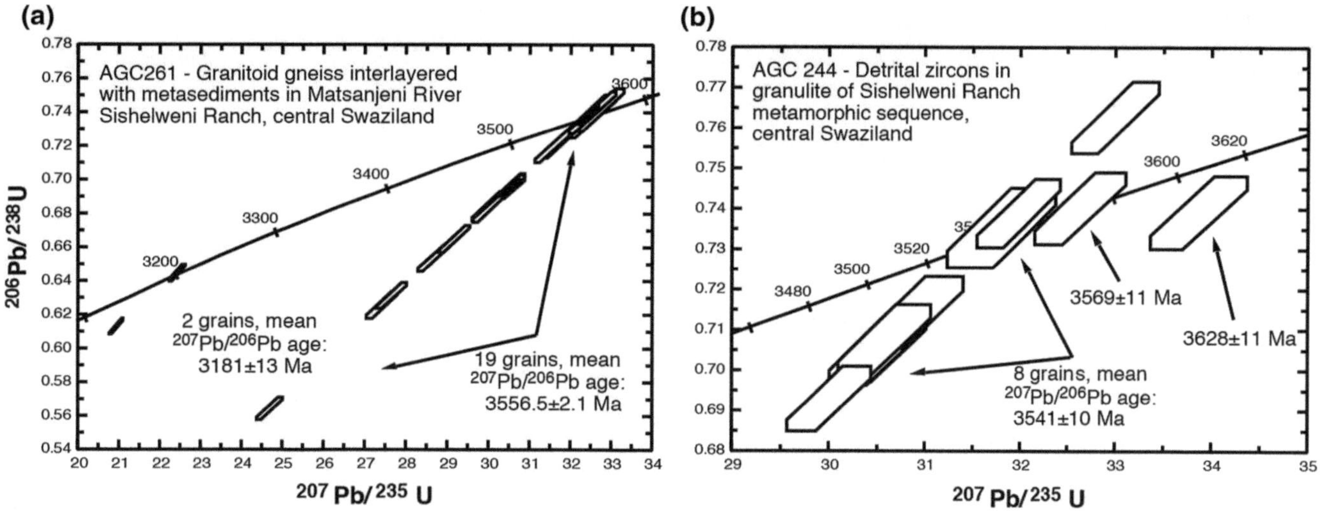

Fig. 1.8 Concordia diagrams showing analytical data for zircons from rocks in the Matsanjeni River, Kubuta area, central Swaziland. **a** Granitoid gneiss sample AGC261, interlayered with garnetiferous metapelite. **b** Detrital zircons from felsic granulite (metagreywacke). Sample locations are given in Table 1.1 and Fig. 1.2. Data boxes are as in Fig. 1.4. For analytical data see Table S2

We dated similar well-rounded zircons from a sample of granulite-facies, orthopyroxene-bearing metagreywacke in the middle Matsanjeni River and obtained SHRIMP $^{207}Pb/^{206}Pb$ ages ranging from 3541 to 3628 Ma (Fig. 1.8b). It is thus likely that the above 3557 Ma trondhjemitic gneiss constituted the basement and that the detrital zircons of approximately the same or older ages were derived from a granitoid domain that is similar to the Eo- to Palaeoarchaean gneisses in northwestern Swaziland. The above data suggest that early Precambrian crust extended into south-central Swaziland and was deformed and metamorphosed together with a supracrustal sequence of unknown original extent at about 3 Ga ago.

A further occurrence mapped as Mahamba Gneiss (Wilson 1982) occurs in southernmost Swaziland and is traversed by the Ncotshane River. This area exposes a complexly deformed and metamorphosed suite of Archaean amphibolite, gabbroic gneiss, quartzite, diorite, augen gneiss and granitic gneiss, all intimately associated with weakly foliated granite and dolerite (Dlamini et al. 2017). The supracrustal rocks are interpreted as metamorphic equivalents of the ca. 3.0 Ga Pongola Group. The term Mahamba Gneiss in central and southern Swaziland, introduced by Wilson (1982), lumps together metamorphosed rocks belonging to different litho-tectonic units (Taylor et al. 2012b; Dlamini et al. 2017) and should be abandoned.

1.3 Palaeoarchaean Rocks of the Southern BGGT and in NW Swaziland

The southern margin of the BGGT south of the Komati River predominantly consists of numerous granitoid plutons of the TTG suite with infolded remnants of greenstone metavolcanic rocks (Viljoen and Viljoen 1969; Anhaeusser and Robb 1980; Robb 1983a; Robb and Anhaeusser 1983; Meyer et al. 1994; Anhaeusser 2010; see Fig. 1.9). The granitoids, some of which are partly migmatized along their margins (Robb 1983b), range in age between 3521 and 3197 Ma (see Table 3) and were emplaced during four major periods of magmatism. Some have been considered to be genetically related to felsic volcanism in the lowermost Barberton greenstone sequence in the same area (De Wit et al. 1987; Agangi et al. 2018).

The greenstone remnants constitute a bimodal sequence of mafic-ultramafic (komatiite to basalt) and felsic (dacite to rhyolite) metavolcanic rocks originally assigned to the Sandspruit and Theespruit Formations of the lower Onverwacht Group (Viljoen and Viljoen 1969), later shown to constitute a single lithostratigraphic unit (Kröner et al. 2016). The felsic rocks contain abundant zircons and are partly metasomatized, and several geochronological studies suggested ages for volcanic activity ranging between 3552 and 3521 (Armstrong et al. 1990; Van Kranendonk et al. 2009; Kröner et al. 1996; 2013; 2016; Roerdink et al. 2016). An early phase of amphibolite-facies metamorphism at 3436 ± 18 Ma is recorded in monazite, and peak P-T conditions were ~ 4.5 kbar and 550 °C (Cutts et al. 2014). The main phase of widespread regional metamorphism occurred at ca. 3230 Ma, dated on zircon (Kröner et al., 2016) and titanite (Dziggel et al., 2005).

There has been some discussion in the literature on the petrogenesis of these Palaeoarchaean felsic metavolcanic rocks. Furnes et al. (2013) suggested derivation by partial melting of a basaltic source, whereas the occurrence of xenocrystic zircons as old as 3610 ± 33 Ma (Roerdink et al. 2016) and significantly negative $\varepsilon_{Hf(t)}$ values for many

Fig. 1.9 Geological sketch map of the southern Barberton granitoid-greenstone terrain showing major rock units. Modified from Kröner et al. 2016

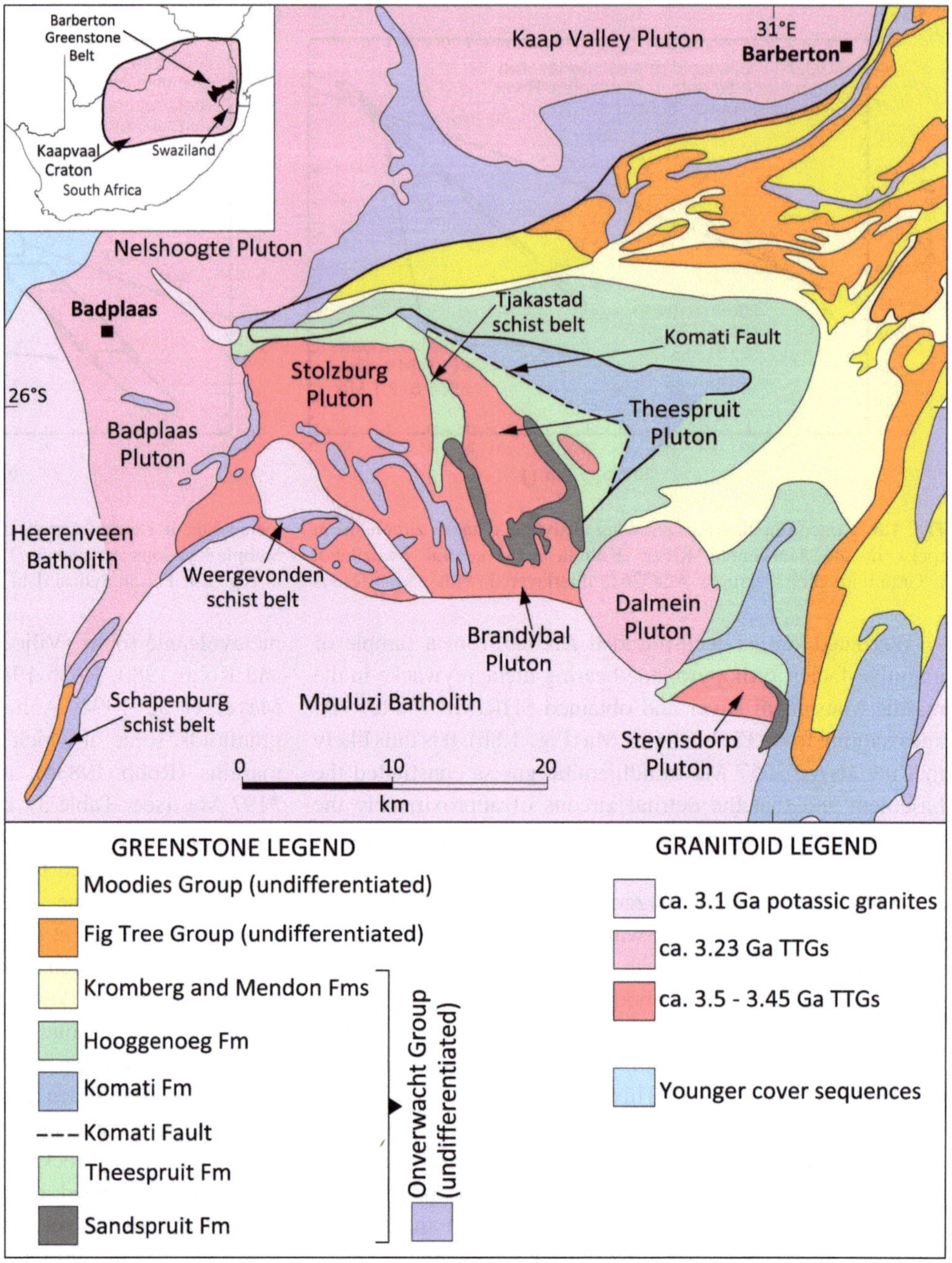

magmatic zircons of the Sandspruit and Theespruit Formations (Fig. 1.10) imply derivation from melting of an older crustal source, presumably components of the nearby Ancient Gneiss Complex, as well as mantle-derived material (Kröner et al. 2013; 2016).

There is considerable controversy on the structural evolution and tectonic setting of the Sandspruit–Theespruit sequence. De Wit et al. (1983, 2011, 2018) and Furnes et al. (2013) interpreted these rocks as a tectonic package, characterized by numerous thrusts and repetition of the lithology (Fig. 1.11a), and named this the Theespruit–Sandspruit Complex. Based on trace element discrimination diagrams, these authors suggested deposition of the original sequence as oceanic crust in a back-arc basin (e.g. De Wit et al., 2018, and references therein). Furnes et al. (2013) speculated that the stratigraphically lowest complex (Sandspruit–Theespruit) was tectonically emplaced onto an older gneissic basement, presumably the Ancient Gneiss Complex. Van Kranendonk et al. (2009) and Van Kranendonk (2011) also recognized narrow zones of intense deformation but interpreted most of these as extensional shears, originating from greenstones sinking as "cold drips" between rising granitoid plutons (Fig. 1.11b). Based on the bimodal composition of the Sandspruit–Theespruit sequence as well as the

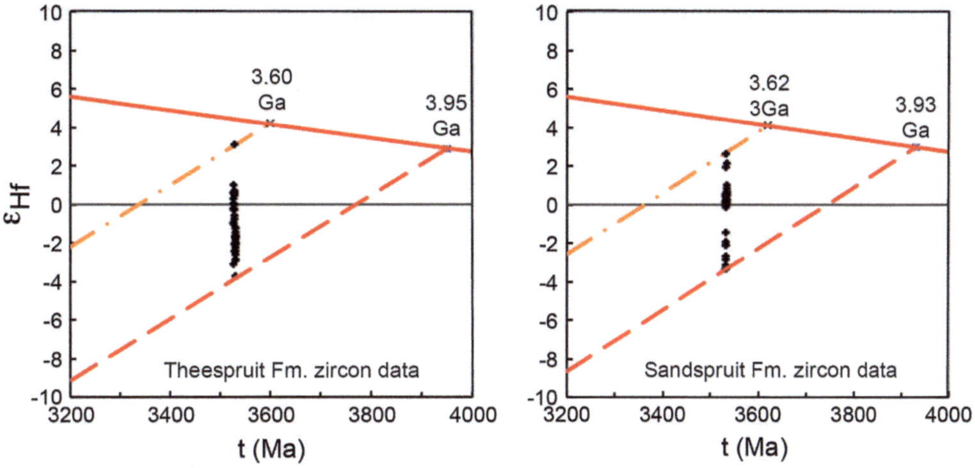

Fig. 1.10 Composite Hf-isotope evolution diagrams for zircons from felsic metavolcanic rocks of the Theespruit Formation (**a**) and Sandspruit Formation (**b**), lower Onverwacht Group, Barberton Greenstone Belt (from Kröner et al. 2014, 2016). The solid red line is the mantle evolution. The broken red Hf evolution lines are based on $^{176}Lu/^{177}Hf = 0.01$. For explanation see text and Supplementary File 1

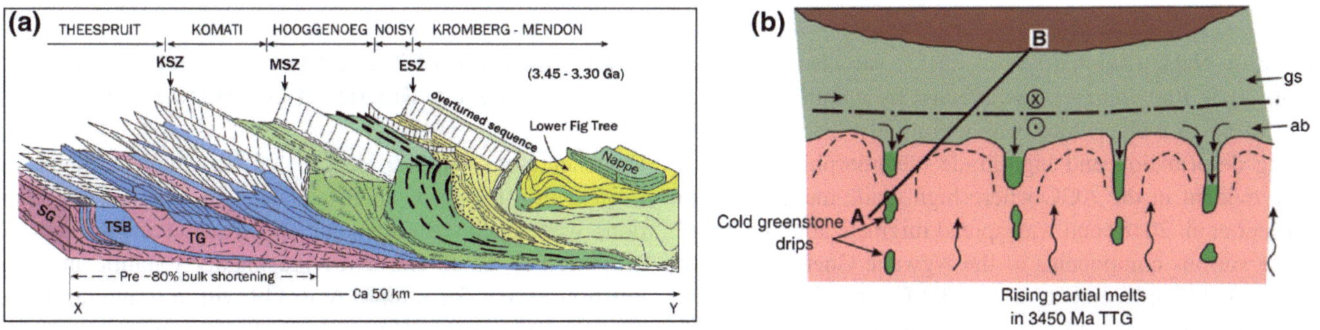

Fig. 1.11 Tectonic models for the Sandspruit–Theespruit sequence of the lower Onverwacht Group, southern Barberton Mountain Land. **a** Schematic N-S section across southern BGB South-directed thrusting is likely related to north-dipping subduction initiated about 3.45 Ga ago (from de Wit et al. 2018). **b** Schematic model for Barberton greenstone belt evolution, stage 1: initiation of greenstone sinking commenced at 3.26 Ga with cool greenstone drips into partially molten, ca. 3.45 Ga TTG. From Van Kranendonk 2011

occurrence of xenocrystic zircons in the felsic rocks and their distinctive Hf-in-zircon isotopic characteristics, Van Kranendonk (2011) and Kröner et al. (2016) interpreted these rocks as part of a thick volcanic plateau deposited on older continental crust.

Supracrustal rocks of the BGB are also exposed in northwestern Swaziland (Fig. 1.2) but were only mapped on a reconnaissance scale and subdivided into Onverwacht, Fig Tree and Moodies Groups (Wilson 1982). More detailed work on the stratigraphy and structures in the Fig Tree and Moodies Groups was undertaken by Lamb and Paris (1988) and Heubeck (2019). The Onverwacht Group has not been subdivided into formations and is mainly composed of strongly altered mafic-ultramafic metavolcanic rocks now occurring as talc-chlorite-serpentinite schist, serpentinite and amphibolite. Rare thin layers of felsic rocks consist of fine-grained silicified tuff and are exposed west of a major fault separating the BGB strata from the Phophonyane gneiss inlier. A sample from one such layer (BA169, see Table 1.1 and Fig. 1.2) contains short prismatic, near-euhedral zircons with a narrow range of magmatic ages between 3455 and 3479 Ma and a weighted mean $^{207}Pb/^{206}Pb$ age of 3469 ± 6 Ma (Fig. 1.12a). The Hf-in-zircon isotopes also display a narrow compositional range of $\varepsilon_{Hf(t)} = 0–1$ (Fig. 1.12b). The tuff zircon age and Hf-in-zircon data are similar to those of felsic volcanic rocks of the Hooggenoeg Formation in the southern BGB that were interpreted to be derived from melting of a TTG-type pre-Onverwacht basement (Kröner et al. 2013).

1.4 The ca. 3.2 Ga Event in the Eastern Kaapvaal Craton

Virtually the entire southeastern Kaapvaal Craton between the northern margin of the BGB and the gneissic and granitoid terrain to the south in the region around the town of Piet Retief (now Mkhondo; Fig. 1.1) and farther south into

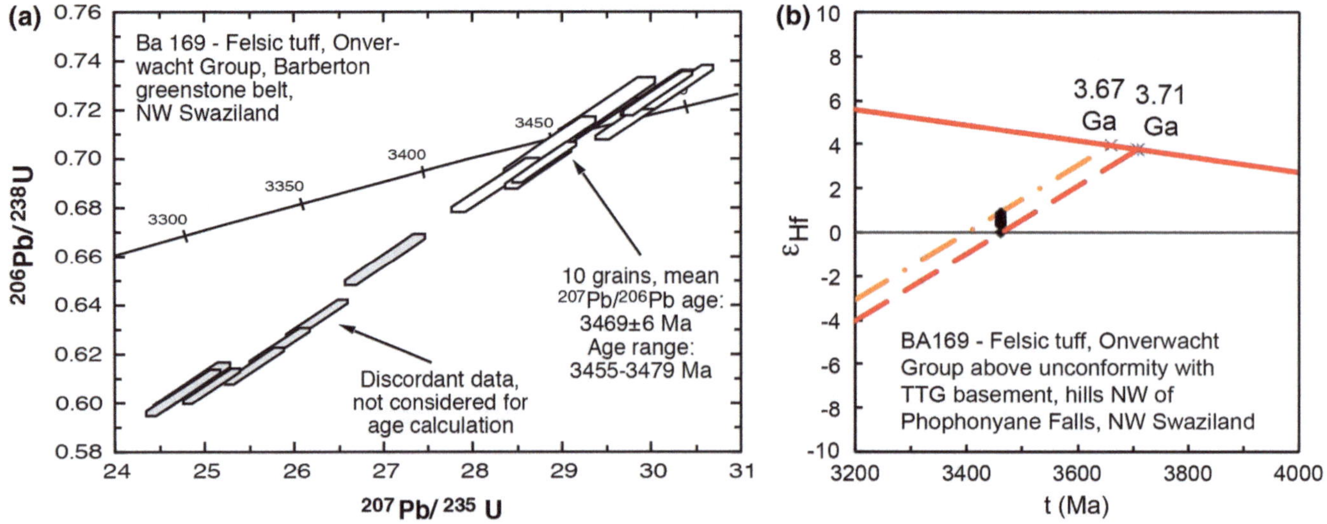

Fig. 1.12 Isotopic data for zircons from felsic tuff of the Onverwacht Group, Barberton Greenstone Belt in hills west of Phophonyane Lodge, northwest Swaziland. **a** Concordia diagram showing SHRIMP II zircon analyses with mean age of 3469 ± 6 Ma. Errors as in Fig. 1.4. **b** Hf evolution diagram for magmatic zircons from same sample. Explanation as in Fig. 1.10

KwaZulu-Natal (Reinhard et al. 2015; see also Hoffmann et al. 2019, this volume) was affected by granitoid magmatism at ca. 3.2 Ga, which was commonly associated with strong deformation and high-grade metamorphism. This is most evident in the AGC where high-grade metamorphism (Kröner et al. 2018) and widespread migmatization occurred in the various components of the Ngwane Gneiss and gave rise to the emplacement of ca. 3.2 Ga granitoids. These rocks were subsequently ductilely deformed together with the older gneisses and greenstone remnants, destroying almost all original contacts, and all granitoid rocks acquired a common parallel foliation (Kröner et al. 1989).

We now present and discuss new geochemical data, U–Pb zircon ages and Hf-in-zircon isotopic systematics for ca. 3.2 Ga granitoid gneisses in Swaziland and then compare the event that generated these rocks with similar data in the southern Barberton granitoid-greenstone terrain. This is of particular interest for the tectonic evolution of Archaean crust in the Kaapvaal craton and more generally because there is an ongoing debate on whether this event signifies the onset of modern-style plate tectonics (Moyen et al. 2006; Schoene and Bowring 2010) or reflects a widespread phase of magmatic underplating in a non-plate tectonic, stagnant lid-type tectonic scenario (Van Kranendonk 2011; Van Kranendonk et al. 2014; Kröner et al. 2018; Wang et al. submitted). The locations of samples discussed in this section are shown in Fig. 1.2 and Table 1.1, and presentation of the zircon ages, cathodoluminescence images and Hf-in-zircon data for selected samples can be found in Supplementary File 1.

1.4.1 Major and Trace Element Distribution of 3.2 Ga Granitoid Gneisses in Swaziland

Representative samples of ca. 3.2 Ga granitoid gneisses from within the AGC in Swaziland (see Fig. 1.2 for location) were analysed for major and trace elements and are presented in Table 1.2. They are mainly tonalitic in composition except for sample AGC449 with a granodioritic to granitic composition (Fig. 1.13a). Two samples overlap in their compositions with those of the Tsawela Gneisses and three with ca. 3.45 Ga Ngwane gneisses of trondhjemitic composition (Fig. 1.13b). The SiO_2 contents of the granitoids are 57–73 wt% (Table 1.2), showing a range similar to the Tsawela Gneisses. These data likely reflect either mixing processes of juvenile, mantle-derived melts and partial melts of older AGC crust or may be a consequence of fractional crystallization (Hoffmann et al. 2016).

The chondrite-normalized REE patterns of the granitoid samples are subparallel and show different degrees of overall REE abundances with enrichment in LREE and variable depletion of HREE (Fig. 1.14a). These steep REE patterns reflect partial melting of middle to lower crust in the garnet stability field. Eu-anomalies are either absent or small, implying minimal plagioclase-related fractionation. Samples AGC449, 445 and 473 show pronounced negative Ce anomalies likely resulting from rock alteration during weathering in the presence of oxidizing fluids (Cotton et al. 1995).

All samples show the characteristic negative Nb–Ta–Ti anomalies (Fig. 1.14b) reflecting residual Ti-bearing

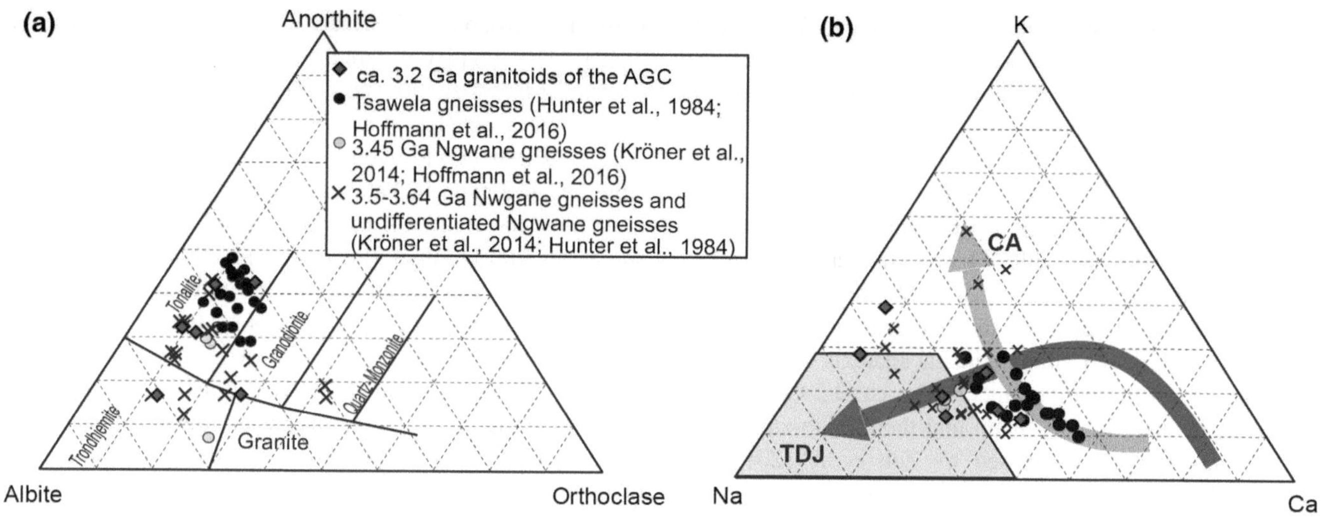

Fig. 1.13 **a** Feldspar diagram after O'Connor (1965), modified by Barker (1979), showing the compositional range of ca. 3.2 Ga granitoid gneisses of the AGC compared to the Ngwane and Tsawela Gneisses. **b** Na–Ca-K ternary diagram after Barker and Arth (1976) with highlighted trondhjemitic partial melting (TDJ) and fractional crystallization (CA) trends after Martin (1994). Data for the Ngwane Gneiss are from Hunter et al. (1984), Kröner et al. (2014a) and for the Tsawela Gneiss from Hunter et al. (1984) and Hoffmann et al. (2016)

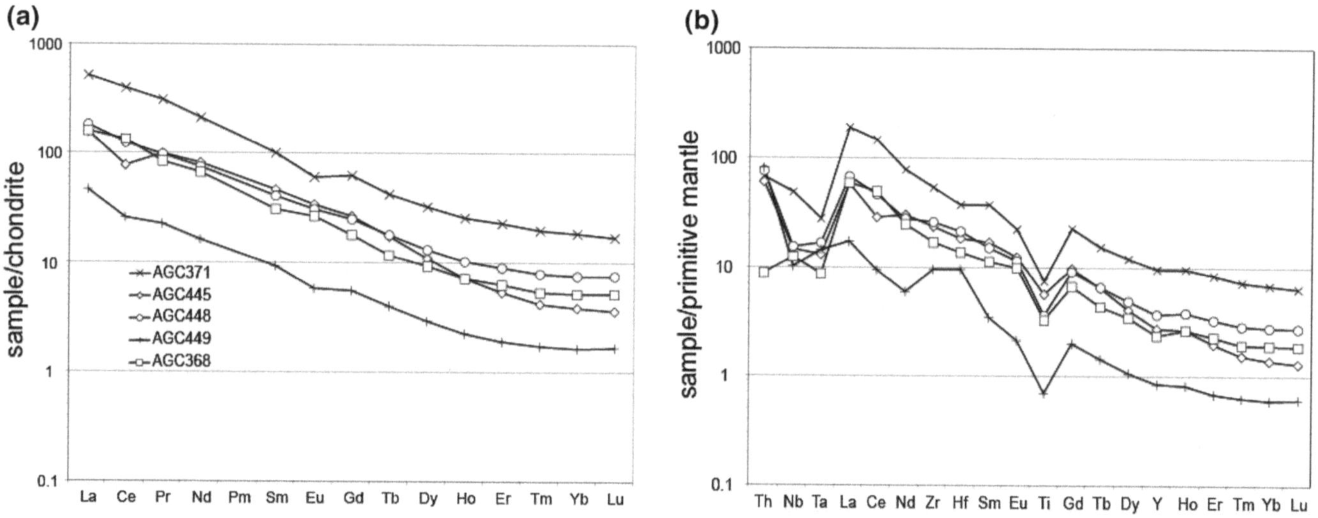

Fig. 1.14 Trace element characteristics of selected ca. 3.2 Ga granitoid gneisses of the Ancient Gneiss Complex, Swaziland. **a** Chondrite-normalized REE diagram. **b** Primitive mantle-normalized trace element variation diagram. Chondrite and primitive mantle compositions from McDonough and Sun (1995). For analytical data see Table S1

minerals such as rutile, ilmenite (e.g. Klemme et al. 2005, 2006), titanite (John et al. 2011) or low-Mg amphibole (Foley et al. 2002) during partial melting. Most samples have superchondritic Nb/Ta ratios of 24–31 (Table 1.2), suggesting that the Ti-phase was most likely rutile, because of its capacity to substantially fractionate Nb from Ta due to the higher partitioning of Ta over Nb in rutile rather than amphibole, which leads to subchondritic Nb/Ta ratios (see Hoffmann et al. 2011 for discussion). Sample AGC368 is depleted in Th compared to Nb (Fig. 1.14b) which may be explained with Th-loss or fractionation of a Th-bearing accessory phase. Positive Zr–Hf anomalies in samples AGC473 and AGC449 can be ascribed to amphibole breakdown in the source during partial melting.

1.4.2 Zircon Geochronology and Nd–Hf Isotopic Systematics of 3.2 Ga Granitoid Gneisses in the Ancient Gneiss Complex (for details see Supplementary File 1)

We determined U–Pb zircon ages and the Nd–Hf isotopic composition of whole-rock samples as well as Hf-in-zircon isotopes from late Palaeoarchaean ca. 3.2 Ga grey gneisses of the Ancient Gneiss Complex of Swaziland. These rocks reflect the ca. 3.2 Ga metamorphic–magmatic event (for locations see Fig. 1.2 and Table 1.1). Details on the analytical procedures are summarized in Supplementary File 2.

The zircon ages and Hf-in-zircon isotopic systematics of ca. 3.2 Ga granitoid rocks reflect magmatic activity throughout the AGC during a period of some 60 Ma from 3213 to 3277 Ma (see Table 1.1 and Supplementary File 1 for details), the rare occurrence of zircon xenocrysts as well as the Nd–Hf isotopic data suggest that these rocks predominantly originated through crustal anatexis with variable additions of mantle-derived melts. The ca. 3.2 Ga ages for granitoid gneisses in Swaziland and their Hf-in-zircon isotopic compositions as reported by Zeh et al. (2011) are similar to our data and reflect the same event.

The zircon ages and Nd–Hf isotopic systematics are summarized in Supplementary File 1 and document a regionally extensive tectono-thermal event at ca. 3.2 Ga in Swaziland that is characterized by medium-pressure and high-temperature metamorphism and strong deformation. In the lower crustal domain such as the AGC, this led to almost complete structural overprinting in older rocks as a result of ductile deformation, extensive migmatization, and the emplacement of syntectonic granitoids as reported above. The high temperature of this event is documented by in situ partial melting in the Mtimane gabbro at ca. 800 °C (Kröner et al. 2018) and the local occurrence of charnockites in the Usutu Suite. The variability of the whole-rock Hf and Nd isotopes as well as the Hf-in-zircon isotopic data in the 3.2 Ga Swaziland granitoids may reflect melting of most likely mafic lower crust as well as variable assimilation of older felsic crust with similar time-integrated Hf–Nd isotopic compositions, and/or deep crustal magma fractionation. However, it is also possible that the strong variation in $\varepsilon_{Hf(t)}$ values observed in magmatic zircons of some of our samples may not necessarily indicate magma mixing of multiple sources. The "zircon effect" during crustal anatexis may have resulted from disequilibrium melting of zircon at the source and may have led to the production of isotopically distinct batches of melts that do not represent the bulk isotope composition of the source (Tang et al. 2014).

Moyen and Laurent (2018) pointed out that using geochemical records to decipher Archaean tectonic scenarios is problematic, as one cannot rely on direct analogies with the modern Earth. Our study of 3.2 Ga granitoids in Swaziland supports the conclusion of these authors that Archaean granitoids predominantly formed by reworking of pre-existing Archaean crust.

The large area affected by the 3.2 Ga event lacks any of the typical linear plate tectonic features such as major shear belts as inferred by Moyen et al. (2006) and Schoene and Bowring (2010) or any other linear arrangement such as the strike-parallel metamorphic asymmetry that is typical of many Phanerozoic and Proterozoic orogenic belts (Brown, 2015). The relatively small size of the crustal blocks envisaged in subduction–collision models such as proposed by de Wit et al. (1992) and Schoene and Bowring (2010) (tens of kilometres to about 100 km) is considered implausible because even in Palaeoarchaean times the continental crust must have had a keel of subcrustal mantle lithosphere some 150–200 km thick (Pearson et al. 1995; Herzberg et al. 2010). These considerations together with our data do not support collision or accretion models for the evolution of the eastern Kaapvaal Craton at ca. 3.2 Ga. However, we do not preclude the possibility that the AGC crust at ca. 3.2 Ga ago was similar to the present middle to lower (Indian) crust beneath Tibet, and its pervasive ductile deformation may have resulted from extensive underthrusting due to collision far away at a distant plate boundary that is no longer preserved.

Kröner et al. (2018) suggested the following tectonic scenario for the 3.2 Ga thermal event in the eastern Kaapvaal Craton: (1) Plume-related magmatic under- and intraplating occurred, combined with delamination and reworking of the subcrustal lithosphere (c.f. Bird 1979; Kay and Kay 1993; Moyen et al. 2017). (2) Further delamination of part of the high-density lower mafic-ultramafic crust was followed by intrusion of hot asthenosphere to the base of the crust. This caused heating of the lower crust, mixing between juvenile and crustal material and generation of anatectic melts (Sinigoi et al. 2016). Delamination has commonly been inferred for Archaean tectonic scenarios (e.g. Bédard 2006; Moyen et al. 2017) because the generation of TTGs from lower crustal garnet amphibolite must have left super-dense residues in the lowermost crust thus facilitating gravitational instabilities and delamination (Johnson et al. 2014; Herzberg 2015).

1.5 The ca. 3.2 Ga Event in the Southern Barberton Granitoid-Greenstone Terrain

Whereas the core of the BGB was metamorphosed to very low grades (typically lower greenschist-facies) with maximum pressures in the range of 2.5–3 kbar (e.g. Grosch et al. 2011) and temperatures of ca. 320 °C, the gneissic terrain to the south of the Komati fault records mid- to upper amphibolite-facies grades and pressures of at least 8 kbar. This terrain has been termed the Stolzburg Block (Moyen et al. 2006) and comprises the Theespruit and Sandspruit Formations, in addition to the Theespruit, Stolzburg and Steynsdorp plutons. Rocks of the Stolzburg Block are polymetamorphic and have been metamorphosed to amphibolite-facies conditions during two separate events. The earlier of these, at ca. 3435 Ma, produced conditions of approximately 4.5 kbar and 550 °C (Cutts et al. 2014).

The extent of this event within the Stolzburg block is unknown due to comprehensive overprinting at ca. 3230 Ma by lower- to mid-amphibolite facies metamorphism with recorded peak pressures typically in excess of 7.5 kbar (Moyen et al. 2018). Details of the ca. 3230 Ma event have been documented in several different areas from within the Stolzburg block (see Table 1.4 and Fig. 1.15).

In most cases, pressure–temperature constraints are based on mineral assemblages that predate or define a steep extensional fabric. Thus, peak pressure estimates commonly represent minimum values for the depth of burial. Despite this, there is a pressure difference, at ca. 3230 Ma, of ~ 5 kbar between the amphibolite-facies rocks of the Theespruit Formation to the south of the Komati Fault and the Onverwacht Group rocks directly to the north of the fault. Thus, the Komati fault accommodated at least 18 km of uplift of the Stolzburg block relative to the greenstone belt to the north. Table 1.4 provides a summary of different metamorphic studies on the Stolzburg Block.

Metamorphic studies in the Stolzburg Block have been performed on both aluminous metapelites and metaluminous rocks of mafic to intermediate composition that contain clino-amphibole. The most reliable peak PT and PT loop constraints from the Stolzburg block are probably represented by the recent phase equilibrium study of the metapelites at Tjakastad town by Cutts et al. (2014; see Fig. 1.15).

Of particular interest in the southern Barberton granitoid-greenstone terrain is the postulated occurrence of a ca. 3.2 Ga suture zone in the southern extension of the Inyoni Shear Zone (ISZ), separating the Stolzburg block in the east from the Badplaas Block in the west (Fig. 1.16a). On the basis of metamorphic modelling (relatively high pressure and low geothermal gradient), together with the distinct nature and evolution of the granitoid domains on either side of the ISZ, Moyen et al. (2006) proposed that the NNE-striking Inyoni Shear Zone (ISZ) represents "…the mid- to lower crustal expression of a suture zone…" (Figure 1.16b).

However, detailed mapping across the ISZ revealed that high strain deformation is localized around tightly folded, kilometre-scale supracrustal rafts and occurred late in the structural history, clearly overprinting two older sets of fabric elements striking at right angles to the direction of proposed collision (Van Kranendonk et al. 2014). Panels of folded supracrustal rocks can be traced across and outside of the ISZ, on both sides of the zone, arguing against terrane accretion. The kinematic asymmetry of the shear foliations changes across opposing limbs of the tightly folded supracrustal rafts, indicative of simple shear related to buckle folding of more competent supracrustal rocks in a matrix of less competent granitoid gneisses. Late buckle folding within the ISZ was highly non-cylindrical and includes vertically plunging sheath folds along amphibolite/metagranite contacts. High-pressure (HP) garnet-bearing amphibolites are restricted to vertical late sheath folds and do not extend along strike. Results from detailed thermobarometry by Nédélec et al. (2012) and Van Kranendonk et al. (2014) show that the original P-T estimates (12–15 °C/km; Moyen et al. 2006) were too high. Peng et al. (2019) used thermobarometry to calculate peak P-T conditions of 670–690 °C/8.6–9.3 kbar for some garnet amphibolites. These data are well outside of conditions for subduction zone settings. Significantly, the high-temperature static recrystallization was shown by Van Kranendonk et al. (2014) to be late- to post-kinematic, indicating heating after vertical shearing, features that are consistent with thermal models of partial convective overturn (Collins and Van Kranendonk 1999; Van Kranendonk 2011).

Brown (2015) discussed the metamorphic evolution of the southern Barberton terrain in terms of the two contrasting models proposed by Moyen et al. (2006) and Van Kranendonk et al. (2014). He concluded that "there is no evidence in the Inyoni shear zone of the hallmark of subduction accretion, which is a unidirectional asymmetry in the thermal structure as has been registered by paired metamorphism in many Phanerozoic metamorphic belts".

Schoene et al. (2009) undertook Sm–Nd whole-rock isotopic mapping of the granitoid terrains surrounding the BGB and delineated lithospheric boundaries on the basis of significant differences in $\varepsilon_{Nd(t)}$ values. These authors speculated that ca. 3.23 Ga accretion of newly formed lithosphere north of the BGB occurred onto pre-existing ca. 3.66 Ga lithosphere south of the BGB along a doubly verging subduction margin. No structural evidence was provided for this model, but the isotopic data demonstrate that significant crustal growth occurred during the ca. 3.2 Ga event, whereas all younger granitoid rocks reflect intracrustal recycling.

1.6 The Post-3.2 Ga Potassic Granitoids

The Nelspruit, Heerenveen, Salisbury Kop, Mliba and Mpuluzi-Piggs Peak batholiths (Fig. 1.1) belong to a suite of post-3.2 Ga potassic granitoids that are believed to have been derived from partial melting of an older TTG basement (Robb et al. 2006; Schoene et al. 2009). The large Nelspruit Batholith comprises several texturally distinct granitoid varieties, some foliated and/or migmatized, and occurs north of the BGB. The various rock types were described in detail by Robb (1983a), and a summary can be found in Robb et al. (2006). The predominant rock is a coarse-grained porphyritic granodiorite to quartz monzonite with large variations in K_2O and Na_2O concentrations, and the contacts with gneissic varieties are gradational (Robb et al. 2006).

Table 1.4 Peak metamorphic conditions for different localities in the Stolzburg block at ca. 3230 Ma

	Important metamorphic assemblages	Location	Method to constrain PT conditions	Representative peak PT conditions
A.	Studies in the Theespruit Formation directly to the south of the Komati Fault or its extension in the Steynsdorp area			
Kisters et al. (2003)	Garnet + plagioclase + hornblende + biotite + quartz; Garnet + plagioclase + hornblende + biotite + quartz + epidote + calcite	No contact with Stolzburg Pluton	THERMOCALC average PT	500 ± 40°C and 6.0 ± 1 kbar
Diener et al. (2005)	Garnet + staurolite + chlorite + plagioclase + quartz; Garnet + hornblende + plagioclase + epidote + quartz; Kyanite + staurolite + biotite + muscovite + plagioclase + quartz	Tjakastad	THERMOCALC average PT	7.5 ± 1.0 kbar, 550 ± 30°C
Lana et al. (2010)	Garnet + hornblende + biotite +chlorite + quartz + epidote + calcite; Garnet + hornblende + clino-pyroxene + quartz + epidote + plagioclase	Northern margin of Steynsdorp Pluton	THERMOCALC average PT	12 ± 1.0 kbar and 600 to 700°C, followed by 8 ± 1.0 kbar and 600 to 740°C
Cutts et al. (2014)	Garnet + staurolite + biotite + kyanite + plagioclase + rutile + ilmenite + quartz ± muscovite; Garnet + staurolite + biotite + plagioclase + chlorite + quartz +rutile ± ilmenite	Tjakastad	Phase equilibrium study, Mn, Ti and excess O included.	Peak pressure of 8.5 kbar at 600°C, peak temperature of 640 ± 10°C over the pressure interval of 7 to 5 kbar
B.	Studies in the Inyoni shear zone and along the southern margin of the Stolzburg pluton			
Dziggel et al. (2002)	Garnet + hornblende + clino-pyroxene + plagioclase + quartz + epidote + titanite; Garnet + clino-pyroxene + plagioclase + quartz; Garnet + orthopyroxene + grunerite + magnetite + quartz	Inyoni shear zone and southern margin of Stolzburg pluton	A variety of exchange thermometers and barometers.	680 ± 40°C and 9.5 ± 1.5 kbar
Moyen et al. (2006)	Garnet + clino-pyroxene + hornblende + plagioclase + quartz; Retrograde epidote + hornblende + quartz simplectite formed as replacement of garnet	Inyoni shear zone	THERMOCALC average PT	12 to 15 kbar and 600 to 650°C, followed by 7 to 9 kbar and 650 to 750°C
Nédélec et al. (2012)	Garnet + hornblende + biotite +quartz + plagioclase + titanite + ilmenite; Garnet + hornblende + clino-pyroxene + biotite + quartz + plagioclase + titanite + melt	Inyoni shear zone	A combination of exchange thermometers and barometers and phase equilibrium modelling—1st generation amphibole as models	680 to 700°C and 9.5 to 10.5 kbar
Cutts et al. (2014)	Garnet + epidote + albite + quartz +hornblende (developed as replacement of trondhjemite veins in amphibolite)	Inyoni shear zone	Phase equilibrium modelling—1st generation amphibole as models	8-9 kbar at ~ 525°C

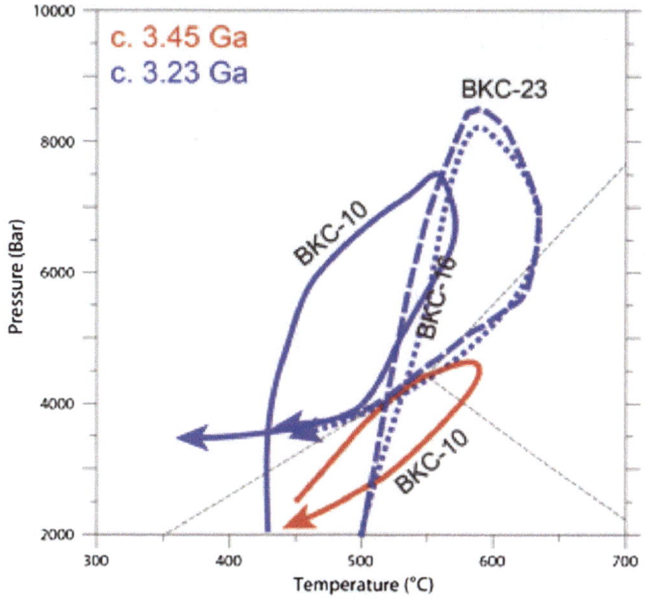

Fig. 1.15 PT loops reconstructed from garnet-bearing rocks of the Theesruit Formation in the Tjakastad area from Cutts et al. (2014). The red loop illustrates the path followed by rocks exposed in association with andalusite-bearing schists which were metamorphosed to amphibolite-facies conditions at ca. 3.45 Ga. The blue loops represent the PT evolution of garnet-, staurolite- and kyanite-bearing rocks at ca 3.23 Ga and probably represent the best constrained estimates available for the metamorphic evolution of the Stolzburg block at ca. 3.23 Ga

Dating of abraded zircons from the above two granitoid phases yielded identical $^{207}Pb/^{206}Pb$ ages of 3105 ± 3 and 3104 +3/−2 Ma (Kamo and Davis 1994), whereas whole-rock Nd isotopic systematics are highly variable with $\varepsilon_{Nd(t)}$ values between −2.76 and +6.39 (Schoene et al. 2009), suggesting crustal melting as well as mantle-derived crustal additions in the petrogenesis of these rocks. Kamo and Davis (1994) reported a zircon age of 3303 ± 6 Ma for a layered tonalitic gneiss xenolith in porphyritic granodiorite.

The Salisbury Kop Pluton is approximately oval in shape and is located in the extreme northwestern part of the BGB (Figs. 1.1 and 1.2). It constitutes a homogeneous granodioritic to quartz-dioritic intrusion and intrudes folded supracrustal rocks of the BGB (Heubeck et al. 1993; Robb et al. 2006). A U–Pb zircon age of 3109 +10/−8 Ma has been obtained by Kamo and Davis (1994), suggesting that this pluton is closely related to the nearby Nelspruit and Piggs Peak Batholiths (Robb et al. 2006; Murphy 2015).

The Mliba Pluton occurs in northeastern Swaziland (Fig. 1.2) and intrudes gneisses of the AGC (Hunter 1973; Wilson 1982). It is a coarse-grained slightly porphyritic granodiorite cut by pegmatite veins related to the nearby Piggs Peak Batholith. Maphalala and Kröner (1993) reported a $^{207}Pb/^{206}Pb$ zircon evaporation age of 3105 ± 5 Ma, identical to ages for the Salisbury Kop, Nelspruit and Piggs Peak plutons.

The Heerenveen Batholith (Anhaeusser and Robb 1983) occurs on the southwestern margin of the BGGT near Badplaas (now renamed Emananza) (Fig. 1.1). It consists of a core of porphyritic trondhjemite surrounded by granodiorite and quartz monzonite, locally migmatized along the pluton margins. There are no age determinations, but Robb et al. (2006) considered it a part of the ca. 3.1 Ga pluton generation.

The Mpuluzi Batholith, previously known as Lochiel Granite, is a large body, almost circular in shape, and occurs south of the BGGT (Fig. 1.1). It intrudes the older 3.51–3.2 Ga granitoid rocks of the southern Barberton terrain and caused migmatization on its margins and intrusion of innumerable K-rich granite and pegmatite veins in the adjacent rocks. It is separated from the identical Piggs Peak Batholith in Swaziland by a narrow belt of metamorphosed mafic-ultramafic rocks of the lower Onverwacht Group (Fig. 1.1). Both the Mpuluzi and Piggs Peak Batholiths constitute a kilometre-thick granitoid sheet overlying and intruding the older gneiss–greenstone terrain that is generally undeformed; and a weak foliation is developed only locally along its margins.

Murphy (2015) investigated the petrography, petrology and geochronology of the Mpuluzi and Piggs Peak Batholiths and concluded that the two plutons are not distinguishable in terms of their mineralogical, chemical or isotopic composition. Zircon ages range from ca. 3.08 to 3.15 Ga, suggesting that emplacement probably occurred over as much as 70 Ma. This is in contrast with previous data suggesting crystallization at ca. 3105 ± 3 Ma (Kamo and Davis 1994). Coarse-grained porphyritic quartz monzonite is the most dominant phase, in which large microcline megacrysts are a characteristic feature. Finer grained granodiorite occurs locally (see Murphy 2015).

Samples investigated by Murphy (2015) have a minor inherited zircon population at ~3.5 Ga, implying the involvement of older crustal material, presumably the Ngwane Gneiss of the Ancient Gneiss Complex. Zircon Hf-isotope data show considerable spread within individual samples but little variation between samples, and average Hf crustal model ages of the 3.1 zircons are ~3.5 Ga, corresponding to the age of the inherited population. These older grains commonly have Hf crustal model ages that date back to 4 Ga (Murphy 2015), similar to the oldest gneisses of the AGC (Kröner et al. 2014).

Whole-rock Sr and Nd isotopic data yielded consistent isochron ages of 3.11 Ga for each system, consistent with the U–Pb zircon data and suggest that the entire Mpuluzi-Piggs Peak granite mass formed and cooled as a cohesive entity (Murphy 2015). The isotopic evidence for mixing between an older crustal component and juvenile material suggests that the mantle may have both provided the heat for melting and contributed material to the magma

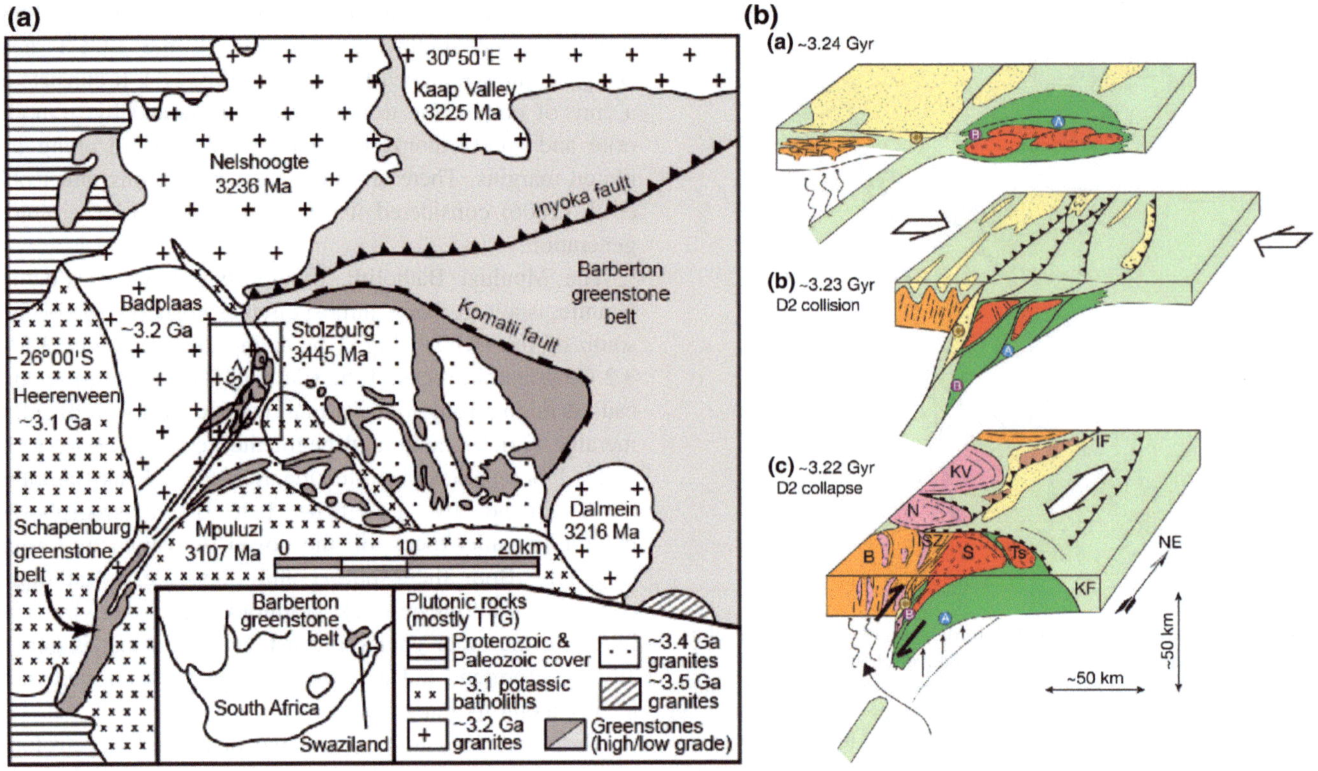

Fig. 1.16 **a** Regional map showing location of the Inyoni Shear Zone (ISZ). Upright rectangle shows area mapped by Moyen et al. (2006) and Van Kranendonk et al. (2014). **b** Geodynamic sketch summarizing the inferred tectonic evolution of the southern Barberton terrain. Reproduced from Moyen et al. 2006

itself. Moser et al. (2001) suggested that the regionally extensive ca. 3.1 Ga magmatic event was caused by partial delamination of mantle lithosphere.

Murphy (2015) proposed that the heat provided by these mantle-derived melts may have led to "draining" of fusible material from the lower crust, increasing its rigidity and limiting further tectonism. This is further discussed in Sect. 1.8.

1.7 Late Archaean Granites in Swaziland

All geological units described so far, including the ca. 3 Ga old cratonic supracrustal rocks of the Pongola Supergroup (Luskin et al. 2019), were intruded by what is generally termed anorogenic "post-Pongola" or "younger granites" (Hunter 1957; Robb 1983b). These undeformed granites fall into the age range between 2.99 and 2.67 Ma and terminate the Archaean crustal history in the eastern Kapvaal craton. On the Geological Map of Swaziland, these plutons are named the Mswati, Hlatikulu, Kwetta and Mtombe granites (Wilson 1982). Robb (1983b), Mukasa et al. (2013) and Hofmann et al. (2015) summarized the occurrence of these rocks in southern Swaziland which are shown in Figs. 1.1 and 1.2.

These granites are medium- to coarse-grained with well-preserved igneous textures and were classified as low-Ca and high-Ca granites by Meyer et al. (1994). The low-Ca variety has a chemical signature consistent with derivation from a metasedimentary source, whereas the high-Ca type was interpreted to derive from an igneous source (Meyer et al. 1994). Late pegmatoid phases of some plutons such as Sinceni and in the Mbabane area carry cassiterite, which has been worked from placer deposits, making them the oldest known tin-bearing pegmatites (Maphalala et al. 1989; Trumbull 1993) together with those in the Pilbara craton (Sweetapple and Collins 2002). Table 1.5 summarizes the published ages for these plutons, and our new data are presented and discussed below.

The Hlatikuli Granite is the largest Neoarchaean intrusion and occurs in several bodies in south-central and south-western Swaziland (Fig. 1.2). It is texturally similar to rocks of the Piggs Peak and Mpuluzi Batholiths and intrudes rocks of the older gneiss units and the Pongola Supergroup. Our sample AGC260 was collected from a road cut along the N9 road south of Sidvokodvo. The zircons are euhedral and long-prismatic with excellent concentric zoning in CL image, and SHRIMP dating of 6 grains yielded slightly discordant data with near-identical ^{207}Pb/^{207}Pb ratios (Table S1) corresponding to a weighted mean age of

2726.8 ± 6.2 Ma (Fig. 1.17a). This age is identical to previously published data (Table 1.5).

The coarse-grained megacrystic Sicunusa Granite forms a small pluton in southwestern Swaziland near the border post of Sicunusa and extends into South Africa. It has sharp intrusive contacts with the 3.45 Ga Tsawela Gneiss and ca. 3.0 Ga volcanic rocks of the Pongola Supergroup. Our sample AGC262 contains euhedral, prismatic zircons with well-developed concentric oscillatory zoning in CL image, and analysis of 6 grains yielded near-concordant and slightly discordant results with similar $^{207}Pb/^{206}Pb$ ratios (Table S1) and a weighted mean age of 2724 ± 6 Ma (Fig. 1.17b). This age is identical, within error, to results reported by Mukasa et al. (2013) (see Table 1.5).

Finally, we report on two gneissic or migmatitic granite-gneisses sampled in the Mkhondo Valley in close proximity to high-grade paragneisses of the Mkhondo Valley Metamorphic Suite (Wilson 1982; Condie et al. 1996). These granulite-facies rocks are exposed on Somhlolo Hill, a prominent oval-shaped hill at the Mkhondo River. Their metamorphic evolution was investigated by Taylor et al. (2012a) who reported an age of 2735 ± 11 Ma for presumed metamorphic rims of zircons in a tonalitic gneiss, considered to reflect the peak of high-grade metamorphism.

Table 1.5 Estimated ages for late Archaean granitoid plutons in Swaziland

Pluton	Rock type dated	Mineral phase(s)	Age (Ma)	Reference[a]
Mbabane	Granite	Zircon	2691 ± 4	(a)
	Granite	Hornblende	269	(a)
Ngwempisi	Granite	Zircon	2720 ± 4	(b)
Kwetta	Granite	Zircon	2722 ± 6	(b)
	Granite	Zircon	2671 ± 3	(b)
	Granite	Zircon	2911 ± 50	(b)
	Granite	Zircon	2720 ± 10	(d)
	Granite	Zircon cores	2911 ± 50	(d)
Sinceni	Granite	Zircon	3074 ± 4	(b)
Sicunusa	Granite	Zircon	2723 ± 7	(b)
	Granite	Zircon	2717 ± 11	(d)
	Granite	Zircon	2724 ± 6	(e)
Mhloasheni	Granite	Zircon	2822 ± 5	(b)
	Granite	Zircon	2837 ± 6	(d)
	Granite	Zircon	2838 ± 10	
Hlatikulu	Granite	Zircon	2722 ± 7	(b)
	Granite	Zircon rims	2742 ± 22	(d)
	Granite	Zircon cores	2960 ± 11	(d)
	Granite	Zircon	2728.9 ± 0.5	(c)
	Granite	Zircon	2729.8 ± 0.4	(c)
	Granite	Zircon	2726.8 ± 6.2	(e)
Mooihoek	Granite	Zircon	2824 ± 6	(b)
in Mkhondo Valley	Granite	Zircon	2734	(c)
	Granite	Zircon	2726.5 ± 3.1	(e)
		Zircon xenocryst	3250 ± 5	(e)
	Granite-gneiss	Zircon	2712 ± 7	(e)
Mahamba Gneiss	Tonalitic gneiss	Zircon	2735 ± 11	(f)
in Mkhondo Valley		Zircon core	2889 ± 13	(f)
Mahamba Gneiss	Granite	Zircon	2808 ± 10	(g)
in Ncotshane River				

[a](a) Layer et al. (1989); (b) Maphalala and Kröner (1993); (c) Schoene and Bowring (2010); (d) Mukasa et al. (2013); (e) Kröner et al. (2018); (f) Taylor et al. (2012a); (g) Dlamini et al. 2017)

Our sample AGC271 is a migmatitic granite-gneiss collected on the bank of the Mkhondo River and interlayered with the granulites. The field relationships suggest an anatectic origin for this gneiss, and our zircon dating of 6 grains (Table S1) showing strong rounding at their terminations yielded a weighted mean $^{207}Pb/^{206}Pb$ age of 2726.5 ± 3.1 Ma (Fig. 1.17c), identical to the age reported by Taylor et al. (2012a).

Sample AGC474 is a homogeneous granodioritic gneiss, taken from a heterogeneous suite of granitoid gneisses mapped as Bimodal Suite (Wilson 1982) and exposed in a roadcut along the N9 highway some 8 km northeast of Somhlolo Hill (Fig. 1.2, Table 1.1). Only three zircon grains with rounded terminations and striped or oscillatory zoning in CL image were analysed (Table S1) and provided a weighted mean $^{207}Pb/^{206}Pb$ age of 2712 ± 7 Ma (Fig. 1.17d) that likely reflects the time of protolith emplacement. It is possible that this rock is either of anatectic origin, similar to sample AGC271, or is a layer of the nearby Hlatikulu Granite that became involved in deformation along the pluton margin.

We suggest that the ca. 2.74 Ga high-grade event documented by Taylor et al. (2012a) from granulite-facies rocks in the Mkhondo Valley signifies a major thermal episode in central and southern Swaziland that led to melting in the lower crust and generated the distinct Neoarchaean high-level granitoid plutons described above.

1.8 Discussion and Conclusions

The eastern part of the Kaapvaal Craton contains the oldest rocks of the African continent and exhibits about 1000 Ma of crustal evolution. The early history is very difficult to reconstruct because of extensive tectonic and metamorphic overprinting, particularly during the pervasive ca. 3.2 Ga event. Therefore, tectonic models are largely based on

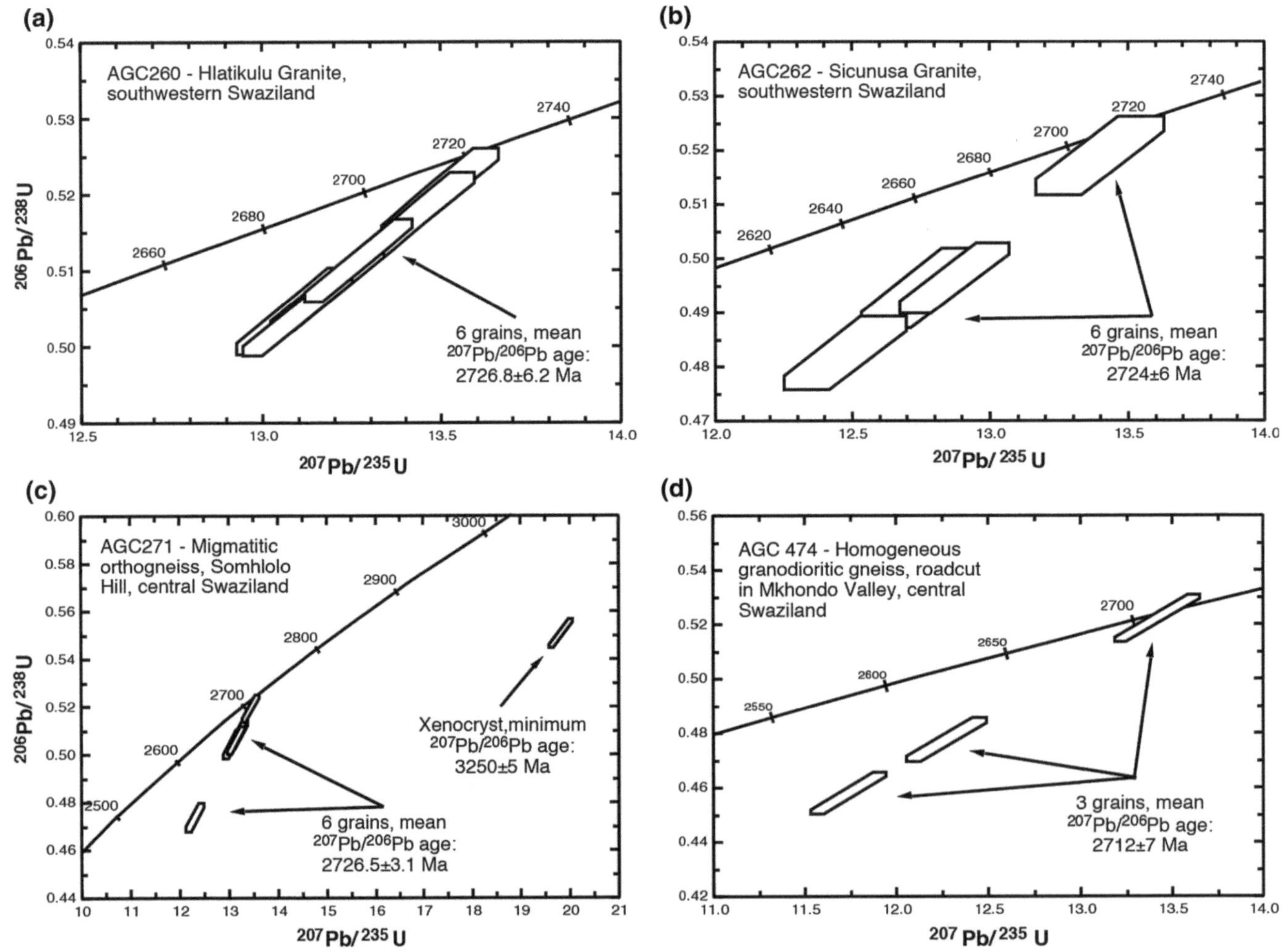

Fig. 1.17 Concordia diagrams showing SHRIMP II zircon analyses of Neoarchaean granites and granite-gneisses from southern Swaziland. Locations are given in Table 1.1 and Fig. 1.2, for analytical data see Table S2. **a** Hlatikulu Granite, roadcut on N9 road south of Sidvokodvo. **b** Sicunusa Granite, roadcut on N13 road near Sicunusa village. **c** Migmatitic gneiss AGC271, Mkhondo River near high-grade metasediments of Somhlolo Hill, Mkhondo Valley. **d** Granodioritic gneiss, roadcut along N9 road in Mkondo Valley. Errors as in Fig. 1.4

petrogenetic and geochemical/isotopic considerations. Both modern-style plate tectonic and stagnant lid tectonic models have been invoked for the granitoid-greenstone evolution in the Palaeoarchaean, and inconclusive field evidence will continue to cause disagreement between researchers on early evolutionary scenarios. For some of the currently discussed models, see references in Van Kranendonk et al. (2018).

Palaeoarchaean granitoids of TTG composition are the volumetrically most extensive rocks in the eastern part of the craton and exhibit strongly heterogeneous geochemical and isotopic signatures (Fig. 1.18). This is in agreement with mixtures of juvenile melts interacting with already differentiated and compositionally variable continental crust of Eoarchaean to late Hadean age. This pattern is similar to that observed in other cratons (e.g. Moyen 2011; Gardiner et al. 2017; Wan et al. 2018). Moyen and Laurent (2018) concluded from an assessment of Archaean TTG chemical compositions worldwide that subduction-like processes are essentially absent from the geological record before ca. 3.5 Ga and that early Archaean granitoids formed predominantly by reworking of pre-existing crust.

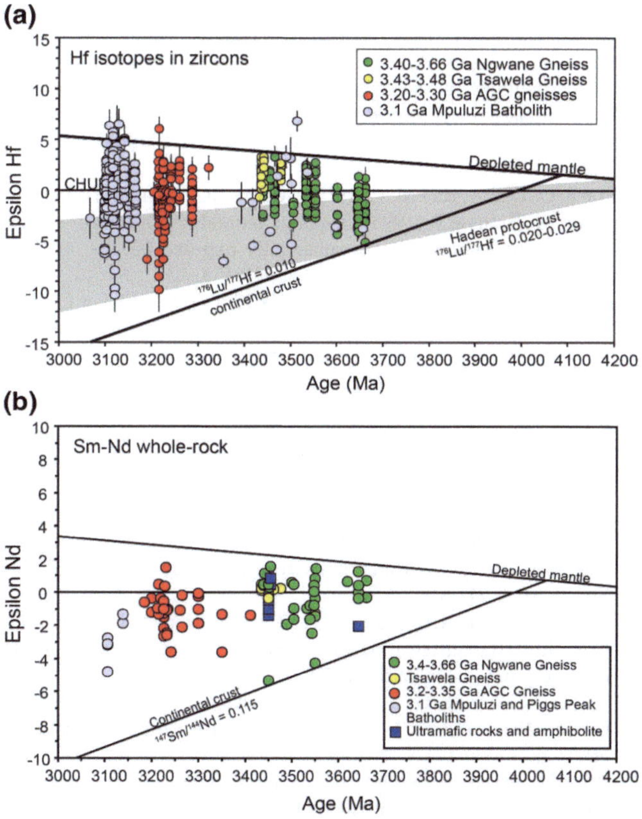

Fig. 1.18 Evolution diagrams for initial Hf-in-zircon isotopic compositions (**a**) and whole-rock initial Nd isotopic compositions (**b**) with time for granitoid rocks of the eastern Kaapvaal Craton. Data in **a** were compiled from Zeh et al. (2011), Kröner et al. (2014), Hoffmann et al. (2016) and Murphy (2015). The grey area marks Hadean protocrust as suggested by Kemp et al. (2010). Data in **b** were compiled from Kröner and Tegtmeyer (1994), Schoene et al. (2009), Kröner et al. (2014), Hoffmann et al. (2016) and Murphy (2015)

Hoffmann et al. (2018) reviewed the petrogenesis of early Archaean TTGs and concluded that melting of metamorphosed and hydrated high-Fe and Mg-tholeiites with low Al contents may constitute the precursor rocks of the early Archaean TTG crust. Later granitoids may have been derived by mixing with melts from an older TTG basement with mafic magmas, followed by fractional crystallization. Wang et al. (2017) proposed a similar model for TTG generation in the southern North China Craton. Martin et al. (2014) showed that the mafic source of TTGs cannot be MORB-type basalt but requires enrichment in LILE as found in plume-derived magmas. Nagel et al. (2012) and Johnson et al. (2014, 2017) favoured the formation of TTGs near the base of thick, plateau-like basaltic crust, and Zhang et al. (2013) showed that felsic melts can form at ca. 900 °C and 12 kbar from such crust. This rules out TTG generation during subduction of oceanic crust as has been proposed in numerous models for Archaean crust formation and growth (see reviews in Moyen and Martin 2012; Johnson et al. 2017), particularly in the eastern Kaapvaal craton (e.g. Kisters et al. 2010; Schoene and Bowring 2010).

Whether the Palaeoarchaean greenstone belts such as Barberton formed and evolved on the ancient oceanic crust or as bimodal volcanic plateau sequences on older continental crust remains one of the most contentious issues. Increasing evidence for contamination of mafic greenstone volcanic rocks with older continental material (e.g. Sossi et al. 2016) favours the view that at least some granitoid crust was already in existence prior to greenstone generation. Small-scale plate tectonic models for the Palaeoarchaean such as proposed by Moyen et al. (2006) and Schoene and Bowring (2010) seem to be incompatible with thick roots of ancient subcrustal mantle lithosphere (Carlson et al. 2000; Griffin et al. 2014). Such models are also incompatible with the observation that Archaean TTGs could not have been generated by melting of oceanic crust in subduction zones (Martin et al. 2014). At least for the widespread ca. 3.2 Ga tectono-metamorphic event a major phase of mantle lithosphere delamination, followed by extensive magmatic under- and intraplating, appears to be a more plausible scenario. This is compatible with recently proposed plume-lid tectonic models for the early Archaean (Fischer and Gerya 2016; Rozel et al. 2017; Sizova et al., in press).

Post-tectonic granitoids at ca. 3.1 and 2.8–2.6 Ga were predominantly generated by melting of older crust. The isotopic evidence for mixing between such an older crustal component and variable juvenile material suggests that the mantle may have provided the heat for melting and contributed material to the magma itself.

We suggest that continuous melt extraction since ca. 3.66 Ga ago from the subcrustal mantle lithosphere (SCML) beneath what would become the eastern Kaapvaal Craton caused this SCML to become depleted in basaltic constituents, thus became boyant and grew vertically. At the same time, melting of the lower crust induced by under- and intraplating of hot mantle-derived magma caused depletion in heat-producing elements, particularly during the ca. 3.2 Ga event. Thus, a combination of lower crustal reworking and juvenile mantle-derived additions during the Palaeo- and Mesoarchaean led to growth, and increased rigidity and tectonic stabilization of the eastern Kaapvaal Craton by about 3.1 Ga ago.

Acknowledgements We thank Martin Van Kranendonk, Axel Hofmann, Carl Anhaeusser, Gary Stevens and members of an international field workshop at Badplaas in 2016 for discussions. Guangshen Nie and Baoying Zheng of the Beijing SHRIMP Centre prepared most of the zircon concentrates, Liqin Zhou and Xiao-Chao Che provided the zircon CL images, Chun Yang prepared perfect zircon mounts, and Jianhui Liu and Zhiqing Yang made sure that SHRIMP II was in excellent operating conditions. Dieter Garbe-Schönberg, Anja Schleicher, Ernst Hegner and Axel Hofmann kindly provided some chemical or isotopic analyses listed in Tables S1 and S3. Jianfeng Gao of the Institute of Geochemistry, Chinese Academy of Sciences, provided an unpublished Excel-based program to construct the Hf evolution diagrams of Figs. 10, 12, S3 and S6. Gary Stevens and Alex Kisters kindly supplied Table 1.4. Comments of Tim Johnson, Laurence Robb and Axel Hofmann improved the manuscript. This study was supported by grants of the Deutsche Forschungsgemeinschaft to AK (KR590/92-1, 92-2, 94-1, 97-1, 98-1) and grant HO4794/1-1 to JEH.

References

Agangi A, Hofmann A, Elburg MA (2018) A review of Palaeoarchaean felsic volcanism in the eastern Kaapvaal craton: linking plutonic and volcanic records. Geosci Front 9:667–688

Anhaeusser CR (1973) The evolution of the early Precambrian crust of southern Africa. Philos Trans Roy Soc Lond A 273:359–388

Anhaeusser CR (2010) Magmatic and structural characteristics of the ca. 3443-3437 Ma Theespruit trondhjemite pluton, Barberton Mountain Land, South Africa. Am J Sci 310:1136–1167

Anhaeusser CR (2014) Archaean greenstone belts and associated granitic rocks—a review. J Afr Earth Sc 100:684–732

Anhaeusser CR, Robb LJ (1980) Regional and detailed field and geochemical studies of Archean trondhjemitic gneisses, migmatites and greenstone xenoliths in the southern part of the Barberton Mountain Land, South Africa. Precambr Res 11:373–397

Anhaeusser CR, Mason R, Viljoen MJ, Viljoen RP (1969) A reappraisal of some aspects of Precambrian shield geology. Geol Soc Am Bull 80:2175–2200

Armstrong RA, Compston W, de Wit MJ, Williams IS (1990) The stratigraphy of the 3.5–3.2 Ga Barberton Greenstone Belt revisited: a single zircon ion microprobe study. Earth Planetary Sci Lett 101:90–106

Barton JM Jr, Robb LJ, Anhaeusser CR, Van Nierop DA (1983) Geochronologic and Sr-isotopic studies of certain units in the Barberton granite-greenstone terrane, South Africa. Spec Publ Geol Soc S Afr 9:63–72

Bédard JH (2006) A catalytic delamination-driven model for coupled genesis of Archaean crust and subcontinental lithospheric mantle. Geochim Cosmochim Acta 70:1188–1214

Bédard JH, Brouillette P, Madore L, Berlaz A (2003) Archaean cratonization and deformation in the northern Superior Province, Canada: an evaluation of plate tectonic versus vertical tectonic models. Precambr Res 127(61):87

Bird P (1979) Continental delamination and the Colorado Plateau. J Geophys Res 84:7561–7571

Brown M (2015) Paleo- to Mesoarchean polymetamorphism in the Barberton granite-greenstone belt, South Africa: constraints from U-Pb monazite and Lu-Hf garnet geochronology on the tectonic processes that shaped the belt: discussion. Geol Soc Am Bull 127:1550–1557

Byerly G, Lowe DR, Heubeck C (2018) Geologic evolution of the Barberton greenstone belt—a unique record of crustal development, surface processes, and early life 3.55 to 3.20 Ga. In: Van Kranendonk MJ, Bennett VC, Hofmann JE (eds) Earth's oldest rocks, 2nd edn. Elsevier, Amsterdam (in press)

Carlson RW, Boyd FR, Shirey SB, Janney PE, Grove TL, Bowring SA, Schmitz MD, Dann JC, Bell DR, Gurney JJ, Richardson SH, Tredoux M, Menzies AH, Pearson DG, Hart RJ, Wilson AH, Moser D (2000) Continental growth, preservation, and modification in Southern Africa. GSA Today 10(2):1–7

Collins WJ, Van Kranendonk MJ (1999) Model for the development of kyanite during partial convective overturn of Archaean granite-greenstone terranes: the Pilbara Craton, Australia. J Metamorph Geol 17:145–156

Compston W, Kröner A (1988) Multiple zircon growth within early Archaean tonalitic gneiss from the Ancient Gneiss Complex, Swaziland. Earth Planet Sci Lett 87:13–28

Condie KC, Kröner A, Milisenda CC (1996) Geochemistry and geochronology of the Mkhondo suite, Swaziland: evidence for passive-margin deposition and granulite facies metamorphism in the Late Archean of Southern Africa. J Afr Earth Sci 21:483–506

Cotton J, Le Dez A, Bau M, Caroff M, Maury RC, Duski P, Fourcade S, Bohn M, Brousse R (1995) Origin of anomalous rare-Earth-element and yttrium enrichments in subaerially exposed basalts: evidence from French Polynesia. Chem Geol 119:115–138

Cutts KA, Stevens G, Hoffmann JE, Buick IS, Frei D, Münker C (2014) Paleo- to Mesoarchean poly-metamorphism in the Barberton Granite-Greenstone Belt, South Africa: Constraints from U-Pb monazite and Lu-Hf garnet geochronology on the tectonic processes that shaped the belt. Geol Soc Am Bull 126:251–270

de Ronde CEJ, Kamo SL (2000) An Archaean arc-arc collisional event: a short-lived (ca. 3 Myr) episode, Weltevreden area, Barberton greenstone belt, South Africa. J Afr Earth Sci 30:219-248

de Wit MJ, Fripp REP, Stanistreet IG (1983) Tectonic and stratigraphic implications of new field observations along the southern part of the Barberton Greenstone Belt. Spec Publ Geol Soc S Afr 9:21–30

de Wit MJ, Hart RA, Hart R (1987) The Jamestown Ophiolite Complex, Barberton mountain belt: a section through 3.5 Ga oceanic crust. J Afr Earth Sc 6:681–730

de Wit MJ, Roering C, Hart RJ, Armstrong RA, de Ronde CEJ, Green RWE, Tredoux M, Peberdy E, Hart RA (1992) Formation of an Archaean continent. Nature 357:553–562

de Wit MJ, Furnes H, Robins B (2011) Geology and tectonostratigraphy of the Onverwacht Suite, Barberton greenstone belt, South Africa. Precambr Res 186:1–27

de Wit MJ, Furnes H, MacLennan S, Doucouré M, Schoene B, Weckmann U, Martínez U, Bowring S (2018) Paleoarchean bedrock lithologies across the Makhonjwa Mountains of South Africa and Swaziland linked to geochemical, magnetic and tectonic data reveal early plate tectonic genes flanking subduction margins. Geosci Front 9:603–665

Diener JFA, Stevens G, Kisters AFM, Poujol M (2005) Metamorphism and exhumation of the basal parts of the Barberton greenstone belt, South Africa: Constraining the rates of Mesoarchaean tectonism. Precambr Res 143:87–112

Dilek Y, Furnes H (2011) Ophiolite genesis and global tectonics: Geochemical and tectonic fingerprinting of ancient oceanic lithosphere. Geol Soc Am Bull 123:387–411

Dlamini N, Hofmann A, Belyanin G, Xie H, Kröner A, Wilson A, Slabunov A (2017) Supracrustal gneisses in southern Swaziland: a basalt-sandstone assemblage of the upper Mozaan Group deformed in the Neoarchaean. S Afr J Geol 120:477–500

Drabon N, Lowe DR, Byerly GR, Harrington JA (2017) Detrital zircon geochronology of sandstones of the 3.6–3.2 Ga Barberton greenstone belt: no evidence for older continental crust. Geology 45:803–806

Dziggel A, Armstrong RA, Stevens G, Nasdala L (2005) Growth of zircon and titanite during metamorphism in the granitoid-gneiss terrane south of the Barberton greenstone belt, South Africa. Mineral Mag 69:1019–1036

Dziggel A, Stevens G, Poujol M, Anhaeusser CR, Armstrong RA (2002) Metamorphism of the granite-greenstone terrane south of the Barberton greenstone belt, South Africa: an insight into the tectono-thermal evolution of the lower portions of the Onverwacht Group. Precambr Res 114:221–247

Fischer R, Gerya T (2016) Early earth plume-lid tectonics: a high-resolution 3D numerical modeling approach. J Geodyn 100:198–214

Foley S, Tiepolo M, Vanucci R (2002) Growth of early continental crust controlled by melting of amphibolite in subduction zones. Nature 41:837–840

François C, Philippot P, Rey P, Rubatto D (2014) Burial and exhumation during Archean sagduction in the East Pilbara granite-greenstone terrane. Earth Planet Sci Lett 396:235–251

Furnes H, Robins B, de Wit MJ (2012) Geochemistry and petrology of lavas in the upper Onverwacht Suite, Barberton Mountain Land, South Africa. S Afr J Geol 115:171–210

Furnes H, de Wit MJ, Robins B (2013) A review of new interpretations of the tectonostratigraphy, geochemistry and evolution of the Onverwacht Suite, Barberton Greenstone Belt, South Africa. Gondwana Res 23:403–428

Garcon M, Carlsson RW, Shirey SB, Arndt NT, Horan MF, Mock TD (2017) Erosion of Archean continents: The Sm-Nd and Lu-Hf isotopic record of Barberton sedimentary rocks. Geochim Cosmochim Acta 206:216–235

Gardiner NJ, Hickman A, Kirkland CL, Lu Y, Johnson T, Zhao J-X (2017) Processes of crust formation in the early earth imaged through Hf isotopes from the East Pilbara Terrane. Precambr Res 297:56–76

Griffin WL, Belousova EA, O'Neill C, O'Reilly S, Malkovets V, Pearson NJ, Spetsius S, Wilde SA (2014) The world turns over: Hadean-Archean crust–mantle evolution. Lithos 189:2–15

Grosch EG, Kosler J, McLoughlin N, Drost K, Slama J, Pedersen B (2011) Paleoarchean detrital zircon ages from the earliest tectonic basin in the Barberton Greenstone Belt, South Africa. Precambr Res 191:85–99

Gumsley A, Olsson J, Söderlund U, de Kock M, Hofmann A, Klausen M (2015) Precise U-Pb baddeleyite age dating of the Usushwana Complex, Southern Africa—implications for the Mesoarchaean magmatic and sedimentological evolution of the Pongola Supergroup, Kaapvaal craton. Precambr Res 267:174–185

Herzberg C (2015) Archean drips. Nat Geosci 7:7–8

Herzberg C, Condie K, Korenaga J (2010) Thermal history of the earth and its petrological expression. Earth Planet Sci Lett 292:79–88

Heubeck C, Wendt JI, Toulkeridis T, Kröner A, Lowe DR (1993) Timing of deformation of the Archaean Barberton greenstone belt, South Africa: constraints from zircon dating of the Salisbury Kop pluton. S Afr J Geolol 96:1–8

Heubeck C (2019) The Moodies Group—A High-Resolution Archive of Archaean Surface and Basin-Forming Processes. In: Kröner A, Hofmann A (eds) The Archaean Kaapvaal Craton, Southern Africa. Springer Geosciences, Berlin, pp 133–169

Hoffmann JE, Kröner A, Hegner E, Viehmann S, Xie H, Iiacheri LM, Schneider KP, Hofmann A, Wong J, Geng H, Yang JH (2016) Source composition, fractional crystallisation and magma mixing processes of the 3.48–3.43 Ga Tsawela tonalite suite (Ancient Gneiss Complex, Swaziland)—implications for Palaeoarchean geodynamics. Precambr Res 276:43–66

Hoffmann JE, Musese E, Kröner A, Schneider KP, Hegner E, Hofmann A, Kasper H-U, Münker C (submitted). Hf-Nd isotope and trace element constraints on the mantle sources and contamination history of the ca. 3.46 Ga Dwalile Greenstone belt, Swaziland. Geoscience

Hoffmann JE, Kröner A (2018) Early Archaean crustal evolution in southern Africa—an updated record of the Ancient Gneiss Complex of Swaziland. In: Van Kranendonk MJ, Bennett VC, Hoffmann JE (eds) Earth's oldest rocks, 2nd edn. Elsevier, Amsterdam, pp 553–567

Hoffmann JE, Zhang C, Moyen J-F, Nagel TJ (2018) The formation of tonalites-trondjhemites-granodiorites in early continental crust. In: Van Kranendonk MJ, Bennett VC, Hoffmann JE (eds) Earth's oldest rocks, 2nd edn. Elsevier, Amsterdam, pp 133–168

Hofmann A, Kröner A, Xie H, Hegner E, Belyanin G, Kramers J, Bolhar R, Slabunov A, Reinhard J (2015) The Nhlangano gneiss dome in south-west Swaziland—a record of crustal destabilization of the eastern Kaapvaal craton in the Neoarchaean. Precambr Res 258:109–132

Hofmann A, Anhaeusser C, Dixon J, Kröner A, Saha L, Wilson A, Xie H (2019) Archaean Granitoid-Greenstone Geology of the Southeastern Part of the Kaapvaal Craton. In: Kröner A, Hofmann A (eds) The Archaean Geology of the Kaapvaal Craton, Southern Africa, Springer, pp 33–54

Hunter DR (1970) The Ancient Gneiss Complex in Swaziland. Trans Geol Soc S Afr 73:107–150

Hunter DR (1957) The geology, petrology and classification of the Swaziland granites and gneisses. Trans Geol Soc S Afr 60:85–125

Hunter DR (1973) The granitic rocks of the Precambrian in Swaziland. Spec Publ Geol Soc S Afr 3:131–147

Hunter DR (1979) The role of tonalitic to trondhjemitic rocks in the crustal development of Swaziland and the eastern Trasvaal, South Africa. In: Barker, F (ed) Trondhjemites, dacites and related rocks. Elsevier, Amsterdam, pp 301–322

Hunter DR, Barker F, Millard Jr, HT (1978) The geochemical nature of the Ancient Gneiss Complex and Granodiorite Suite, Swaziland: a preliminary study. Precambr Res 7:105–127

Hunter DR, Barker F, Millard Jr HT (1984) Geochemical investigation of Archaean bimodal and Dwalile metamorphic suites, Ancient Gneiss Complex, Swaziland. Precambr Res 24:131–155

Jackson MPA (1984) Archaean structural styles in the Ancient Gneiss Complex of Swaziland, southern Africa. In: Kröner A, Greiling R (eds), Precambrian Tectonics Illustrated. E. Schweizerbart`sche Verlagsbuchhandlung, Stuttgart, Germany, pp 1–18

John T, Klemd R, Klemme S, Pfänder JA, Hoffmann JE, Gao J (2011) Nb–Ta fractionation by partial melting at the titanite-rutile transition. Contrib Mineral Petrol 161:35–45

Johnson TE, Brown M, Gardiner NJ, Kirkland CL, Smithies RH (2017) Earth's first stable continents did not form by subduction. Nature 543:239–242

Johnson TE, Brown M, Kaus BJ, VanTongeren JA (2014) Delamination and recycling of Archaean crust caused by gravitational instabilities. Nat Geosci 7:47–52

Kamo S, Davis DW (1994) Reassessment of Archean crustal development in the Barberton Mountain Land, South Africa, based on U-Pb dating. Tectonics 13:167–192

Kay RW, Kay MS (1993) Delamination and delamination magmatism. Tectonophysics 219:177–189

Kemp AIS, Wilde SA, Hawkesworth CJ, Coath CD, Nemchin A, Pidgeon RT, Vervoort JD, DuFrane SA (2010) Hadean crustal evolution revisited: New constraints from Pb–Hf isotope systematics of the Jack Hills zircons. Earth Planet Sci Lett 296:45–56

Kisters AFM, Belcher RW, Poujol M, Dziggel A (2010) Continental growth and convergence-related arc plutonism in the Mesoarchaean: evidence from the Barberton granitoid-greenstone terrain, South Africa. Precambr Res 178:15–26

Kisters AFM, Stevens G, Dziggel A, Armstrong RA (2003) Extensional detachment faulting and core complex formation in the southern Barberton granite-greenstone terrain, South Africa: evidence for a 3.2 Ga orogenic collapse. Precambr Res 127:355–378

Klemme S, Prowatke S, Hametner K, Günther D (2005) Partitioning of trace elements between rutile and silicate melts: implications for subduction zones. Geochim Cosmochim Acta 49:2361–2371

Klemme S, Günther D, Hametner K, Prowatke S, Zack T (2006) The partitioning of trace elements between ilmenite, ulvospinel, armalcolite and silicate melts with implications for the early differentiation of the moon. Chem Geol 234:251–263

Kröner A (1985) Evolution of the Archean continental crust. Annu Rev Earth Planet Sci 13:49–74

Kröner A (2007) The Ancient Gneiss Complex of Swaziland and environs: record of early Archaean crustal evolution in southern Africa. In: Van Kranendonk MJ, Smithies RH, Bennett VC (eds), Earth's oldest rocks. Elsevier, Amsterdam, pp 465–480

Kröner A, Byerly GR, Lowe DR (1991) Chronology of early Archaean granite greenstone evolution in the Barberton Mountain land, South Africa, based on precise dating by single zircon evaporation. Earth Planet Sci Lett 103:41–54

Kröner A, Compston W, Williams IS (1989) Growth of early Archaean crust in the Ancient Gneiss Complex of Swaziland as revealed by single zircon dating. Tectonophysics 161:271–298

Kröner A, Hegner E, Wendt JI, Byerly GR (1996) The oldest part of the Barberton granitoid-greenstone terrain, South Africa: evidence for crust formation between 3.5 and 3.7 Ga. Precambr Res 78:105–124

Kröner A, Hoffmann JE, Xie H, Wu F, Münker C, Hegner E, Wong J, Wan Y, Liu D, (2013) Generation of early Archaean felsic greenstone volcanic rocks through crustal melting in the Kaapvaal, craton, southern Africa. Earth Planet Sci Lett 381:188–197

Kröner A, Tegtmeyer A (1994) Gneiss-greenstone relationships in the Ancient Gneiss Complex of southwestern Swaziland, Southern Africa, and implications for early crustal evolution. Precambr Res 67:109–139

Kröner A, Hoffmann JE, Xie H, Münker C, Hegner E, Wan Y, Hofmann A, Liu D, Yang J (2014) Generation of early Archaean grey gneisses through melting of older continental crust in the eastern Kaapvaal craton, Southern Africa. Precambr Res 255:823–846

Kröner A, Anhaeusser CR, Hoffmann JE, Wong J, Geng H, Hegner E, Xie H, Yang J, Liu D (2016) Chronology of the oldest supracrustal sequences in the Palaeoarchaean Barberton Greenstone Belt, South Africa and Swaziland. Precambr Res 279:123–143

Kröner A, Nagel TJ, Hoffmann JE, Liu X, Wong J, Hegner E, Xie H, Kasper U, Hofmann A, Liu D (2018) High-temperature metamorphism and crustal melting at ca. 3.2 Ga in the eastern Kaapvaal craton, southern Africa. Precambrian Research 317:101–116

Lamb S, Paris I (1988) Post-Onverwacht Group stratigraphy in the southeast part of the Archaean Barberton greenstone belt. J Afr Earth Sc 7:285–306

Lana C, Kisters AFM, Stevens G (2010). Exhumation of Mesoarchean TTG gneisses from the middle crust: insights from the Steynsdorp core complex, Barberton granitoid-greenstone terrain, South Africa. Geol Soc Am Bull 122:183–197

Layer PW, Kröner A, McWilliams M, York D (1989) Elements of the Archean thermal history and apparent polar wander of the eastern Kaapvaal Craton, Swaziland, from single grain dating and paleomagnetism. Earth Planet Sci Lett 93:23–34

Layer PW, Kröner A, York D (1992) Pre-3000 Ma thermal history of the Archean Kaap Valley pluton, South Africa. Geol 20:717–720

Layer PW, Lopez-Martinez M, Kröner A, York D, McWilliams M (1998) Thermochronometry and paleomagnetism of the Archean Nelshoogte Pluton, South Africa. Geophys J Inter 135:129–145

Luskin C, Wilson A, Gold D, Hofmann A (2019) The Pongola Supergroup: Mesoarchaean Deposition Following Kaapvaal Craton Stabilization. In: Kröner A, Hofmann A (eds) The Archaean Kaapvaal Craton, Southern Africa. Springer Geosciences, Berlin, pp 225–254

Maphalala R, Kröner A, Kramers JD (1989) Rb-Sr ages for Archaean granitoids and tin-bearing pegmatites in Swaziland, Southern Africa. J Afr Earth Sc 9:749–757

Maphalala R, Kröner A (1993) Pb-Pb singe zircon ages for the younger Archaean granitoids of Swaziland, southern Africa. Extended abstracts, vol II, 16th Colloquium of African Geology, Mbabane, Swaziland, pp 201–206

Martin H, Moyen J-F, Guitreau M, Blichert-Toft J (2014) Why Archaean TTG cannot be generated by MORB melting in subduction zones. Lithos 198–199:1–13

McDonough WL, Sun S-S (1995) The composition of the earth. Chem Geol 120:223–253

Meyer FM, Robb LJ, Reimold WU, Debruiyn H (1994) Contrasting low-Ca and high-Ca granites in the Archaean Barberton Mountain Land, Southern Africa. Lithos 32:63–76

Moser DE, Flowers RM, Hart RJ (2001) Birth of the Kaapvaal tectosphere 3.08 billion years Ago. Sci 291:465–468

Moyen J-F (2011) The composite Archaean grey gneisses: petrological significance, and evidence for a non-unique tectonic setting for Archaean crustal growth. Lithos 123:21–36

Moyen J-F, Martin H (2012) Forty years of TTG research. Lithos 148:312–336

Moyen JF, Laurent O, Chelle-Michou C, Couzinié S, Vanderhaeghe O, Zeh A, Villaros A, Gardien V (2017) Collision vs. subduction-related magmatism: two contrasting ways of granite formation and implications for crustal growth. Lithos 277:154–177

Moyen J-F, Laurent O (2018) Archaean tectonic systems: a view from igneous rocks. Lithos 302–303:99–125

Moyen J-F, Stevens G, Kisters AFM (2006) Record of mid-Archaean subduction from metamorphism in the Barberton terrain, South Africa. Nature 442:559–562

Moyen J-F, Stevens G, Kisters AFM, Belcher RW, Lamirre B (2018). TTG plutons of the Barberton granitoid-greenstone terrain, South Africa. In: Van Kranendonk MJ, Bennett VC, Hoffmann JE (eds) Earth's oldest rocks, 2nd Edition. Elsevier, Amsterdam, 293–327

Mukasa SB, Wilson AH, Young KR (2013) Geochronological constraints on the magmatic and tectonic development of the Pongola Supergroup (Central Region), South Africa. Precambr Res 224:268–286

Murphy RCL (2015) Stabilizing a craton: the origin and emplacement of the 3.1 Ga Mpuluzi Batholith. Unpublished PhD thesis, Macquarie University, Sydney, Australia, 224p and appendices

Musese E-M (2014) Geochemische Untersuchungen an paläoarchaischen Grünsteinen von Dwalile, Swaziland. Unpublished MSc thesis, University of Cologne, Germany, 87p

Nagel TJ, Hoffmann JE, Münker C (2012) Melting of Eoarchean tonalite-trondhjemite-granodiorite suite from thickened mafic arc crust. Geology 40:375–378

Nédélec A, Chevrel MO, Moyen JF, Ganne J, Fabre S (2012) TTGs in the making: natural evidence from the Inyoni shear zone (Barberton, South Africa). Lithos 153:25–38

Palme H, O'Neill HSC (2014) Cosmochemical estimates of mantle composition. In: Holland HD, Turekian KK (eds) Treatise on Geochemistry Volume 3: The Mantle and Core, 2nd edn. Elsevier, Amsterdam, pp 1–39

Pearson DG, Carlson RW, Shirey SB, Boyd FR, Nixon PH (1995) Stabilisation of Archaean lithospheric mantle: a Re-Os isotope study of peridotite xenoliths from the Kaapvaal craton. Earth Planet Sci Lett 134:341–357

Peng T, Wang GD, Wang YHC, Chen H-X, Wu C-M (2019). Metamorphic P-T path and geochronology of garnet-bearing amphibolite of the Inyoni Shear Zone, southwestern Barberton Greenstone Belt, South Africa. Precambrian Research 321:261–276

Reinhard J, Elburg MA, Andersen T (2015) Zircon U-Pb age data and Hf isotopic signature of Kaapvaal basement granitoids from the Archaean White Mfolozi Inlier, Northern Kwazulu-Natal. S Afr J Geol 118:473–488

Robb LJ (1983a) The nature, origin and significance of Archaean migmatites in the Barberton Mountain Land: a new approach in the assessment of early crustal evolution. In: Anhaeusser CR (ed) Contributions to the Geology of the Barberton Mountain Land. Geol Soc S Afr Spec Publ 9:81–101

Robb LJ (1983b) Geological and chemical characteristics of late granite plutons in the Barberton Region and Swaziland with emphasis on the Dalmein Pluton–a review. In: Anhaeusser CR (ed) Contributions to the Geology of the Barberton Mountain Land. Geol Soc S Afr Spec Publ 9: 153–168

Robb LJ, Anhaeusser CR (1983) Chemical and petrogenetic characteristics of Archaean tonalite-trondhjemite gneiss plutons in the Barberton Mountain Land. In: Anhaeusser CR (ed) Contributions to the Geology of the Barberton Mountain Land. Geol Soc S Afr Spec Publ 9: 81–101

Robb LJ, Brandl G, Anhaeusser CR, Poujol M (2006) Archaean granitoid intrusions. In: Johnson MR, Anhaeusser CR, Thomas RJ (eds) The geology of South Africa. Geological Society of South Africa and Council for Geoscience, Johannesburg and Pretoria, pp 57–94

Roerdink DL, Mason PRD, Whitehouse MJ, Brouwer FM (2016) Reworking of atmospheric sulfur in a Paleoarchean hydrothermal system at Londozi, Barberton Greenstone Belt, Swaziland. Precambr Res 280:195–204

Rozel AB, Golabek GJ, Jain C, Tackley PJ, Gerya T (2017) Continental crust formation on early earth controlled by intrusive magmatism. Nature 545:332–335

Schneider K, Hoffmann JE, Münker C, Patyniak M, Sprung P, Roerdink D, Garbe-Schönberg D, Kröner A (submitted). The origin of komatiites and basalts of the lower Onverwacht Group, Barberton Greenstone Belt (South Africa). Chemical Geology

Schoene B, de Wit M, Bowring SA (2008) Mesoarchean assembly and stabilization of the eastern Kaapvaal craton: a structural-thermochronological perspective. Tectonics 27: TC5010, https://doi.org/10.1029/2008tc002267

Schoene B, Bowring S (2010) Rates and mechanisms of Mesoarchean magmatic arc construction, eastern Kaapvaal craton, Swaziland. Geol Soc Am Bull 122:408–429

Schoene B, Dudas FOL, Bowring S, de Wit M (2009) Sm–Nd isotopic mapping of lithospheric growth and stabilization in the eastern Kaapvaal craton. Terra Nova 21:219–228

Sinigoi S, Quick JE, Demarchi G, Klötzli US (2016) Production of hybrid granitic magma at the advancing front of basaltic underplating: Inferences from the Sesia magmatic system (South-Western Alps, Italy). Lithos 252–253:109–122

Sizova E, Gerya T, Brown M, Stüwe K (2018) What drives metamorphism in early Archean greenstone belts? Insights from numerical modeling. Tectonophysics (in press). https://doi.org/10.1016/j.tecto.2017.07.020

Sossi PA, Eggins SA, Nesbitt RW, Nebel O, Hergt JM, Campbell IH, O'Neill H, St C, Van Kranendonk M, Davies DR (2016) Petrogenesis and geochemistry of Archean komatiites. J Petrol 57:147–184

Suhr N, Hoffmann JE, Kröner A, Schröder S (2015) Archaean granulitic Paragneisses from Central Swaziland: Inferences on Paleoarchaean crustal reworking and a complex metamorphic history. J Geol Soc London 172:139–152

Sweetapple MT, Collins PLF (2002) Genetic framework for the classification and distribution of Archean rare metal pegmatites in the North Pilbara Craton, Western Australia. Econ Geol 97:873–895

Tang M, Wang X-L, Shu, X-J, Wang D, Yang T, Gopon P (2014) Hafnium isotopic heterogeneity in zircons from granitic rocks: Geochemical evaluation and modeling of "zircon effect" in crustal anatexis. Earth Planet Sci Lett 389:188–199

Taylor J, Stevens G, Buick IS, Lana C (2012a) Successive midcrustal, high-grade metamorphic events provide insight into Mid-Archean mountain building along the SE margin of the proto-Kaapvaal craton. Geol Soc Am Bull 124:1191–1211

Taylor J, Stevens G, Lana C (2012b) Age and field relationshios of Mahamba orthogneisses and Mkhondo Valley Metamorphic Suite paragneisses from the Mkhondo River, Ancient Gneiss Complex, Swaziland. S Afr J Geol 115:369–384

Taylor J, Zeh A, Gerdes A (2017) U-Pb–Hf isotope systematics of detrital zircons in high-grade paragneisses of the Ancient Gneiss Complex, Swaziland: Evidence for two periods of juvenile crust formation, Paleo- and Mesoarchaean sediment deposition, and 3.23 Ga terrane accretion. Precambr Res 280:205–220

Trumbull RB (1993) A petrological and Rb-Sr isotopic study of an early Archeran fertile granite-pegmatite system: the Sinceni Pluton in Swaziland. Precambr Res 61:89–116

Van Kranendonk MJ (2011) Cool greenstone drips and the role of partial convective overturn in Barberton greenstone belt evolution. J Afr Earth Sc 60:346–352

Van Kranendonk MJ, Bennett VC, Hoffmann, JE (eds) (2018) Earth's oldest rocks, 2nd Edition. Elsevier, Amsterdam, in press

Van Kranendonk MJ, Kröner A, Hegner E, Connelly J (2009) Age, lithology and structural evolution of the c. 3.53 Ga Theespruit Formation in the Tjakastad area, southwestern Barberton Greenstone Belt, South Africa, with implications for Archean tectonics. Chem Geol 261:114–138

Van Kranendonk MJ, Kröner A, Hoffmann JE, Nagel T, Anhaeusser CR (2014) Just another drip: Re-analysis of a proposed Mesoarchean suture from the Barberton Mountain Land, South Africa. Precambr Res 254:19–35

van Schijndel V, Stevens G, Zeh A, Frei D, Lana C (2017) Zircon geochronology and Hf isotopes of the Dwalile supracrustal suite, Ancient Gneiss Complex, Swaziland: Insights into the diversity of Palaeoarchaean source rocks, depositional and metamorphic ages. Precambr Res 295:48–66

Viljoen MJ, Viljoen RP (1969) An introduction to the geology of the Barberton granite–greenstone terrain. Geol Soc S Afr Spec Publ 2:9–28

Wan Y, Xie H, Dong C, Kröner A, Bai W, Liu S, Xie S, Ma M, Li Y, Liu D (2018) Hadean to Paleoarchean rocks and zircons in China. In: Van Kranendonk MJ, Bennett VC, Hoffmann JE (eds) Earth's oldest rocks, 2nd Edition. Elsevier, Amsterdam, pp 293–327

Wang G-D, Wang HYC, Chen H-X, Zhang B, Zhang Q, Wu C-M (2017) Geochronology and geochemistry of the TTG and potassic granite of the Taihua complex, Mts. Huashan: implications for crustal evolution of the southern North China Craton. Precambr Res 288:72–90

Wilson AC (1980) 1:50.000 Geological Map of Swaziland, with Explanatory Notes. Swaziland Geological Survey and Mines Department, Mbabane, Swaziland

Wilson AC (Compiler) (1982) 1:250,000 Geological Map of Swaziland. Geological Survey and Mines Department, Mbabane, Swaziland

Zeh A, Gerdes A, Barton JM Jr (2009) Archean accretion and crustal evolution of the Kalahari craton—the zircon age and Hf isotope record of granitic rocks from Barberton/Swaziland to the Francistown arc. J Petrol 50:933–966

Zeh A, Gerdes A, Millonig L (2011) Hafnium isotope record of the Ancient Gneiss Complex, Swaziland, southern Africa; evidence for Archaean crust–mantle formation and crust reworking between 3.66 and 2.73 Ga. J Geol Soc London 168:1–11

Zhang C, Holtz F, Koepke J, Wolff PE, Ma C, Bédard JH (2013) Constraints from experimental melting of amphibolite on the depth of formation of garnet-rich restites, and implications for models of early Archean crustal growth. Precambr Res 231:206–217

Archaean Granitoid–Greenstone Geology of the Southeastern Part of the Kaapvaal Craton

2

Axel Hofmann, Carl R. Anhaeusser, John Dixon, Alfred Kröner, Lopamudra Saha, Allan Wilson, and Hangqiang Xie

Abstract

The southeastern Kaapvaal Craton is a Palaeoarchaean granitoid–greenstone terrain. Supracrustal rocks are dominated by metamorphosed mafic–ultramafic volcanic rocks intercalated with minor felsic volcanic and chemical sedimentary rocks, including carbonaceous chert and minor iron formation. Siliciclastic sedimentary rocks are rare. The greenstones occur in the Schapenburg and Dwalile fragments close to the Barberton greenstone belt, the Assegaai, De Kraalen, Witrivier and Commondale fragments in the vicinity of Piet Retief, and the Nondweni and Ilangwe greenstone belts together with several smaller fragments in the southern part of the craton. The greenstones are locally in tectonic contact with compositionally layered grey gneisses as old as ~3.5 Ga and, in the southern part, are intruded by the ~3.25 Ga Anhalt granitoid suite, providing a minimum age for the greenstones. The oldest felsic volcanic rocks so far identified are 3.53 Ga in age. The belts show evidence of polyphase deformation, with metamorphism at greenschist- to amphibolite-facies grade, the peak of which was at ~3.2 Ga. Ore deposits are restricted to sub-economic gold and rare massive sulphide Cu–Zn deposits.

Keywords

Palaeoarchaean • Greenstone belt • Granitoid–gneiss terrain

A. Hofmann (✉)
Department of Geology, University of Johannesburg, PO Box 524, Auckland Park, Johannesburg, 2006, South Africa
e-mail: ahofmann@uj.ac.za

C. R. Anhaeusser · A. Wilson
School of Geosciences, University of the Witwatersrand, Wits, 2050, South Africa

J. Dixon
School of Geological Sciences, University of KwaZulu-Natal, Private Bag X 54001, Durban, 4000, South Africa

A. Kröner · H. Xie
Beijing SHRIMP Centre, Chinese Academy of Geological Sciences, 26 Baiwanzhuang Road, Beijing, 100037, China

A. Kröner
Institut Für Geowissenschaften, Universität Mainz, 55099 Mainz, Germany

L. Saha
Department of Earth Sciences, Indian Institute of Technology Roorkee, Roorkee, 247667, India

2.1 Introduction

The southeastern Kaapvaal Craton is a classic granitoid–greenstone terrain that preserves some of the oldest crustal components on Earth. Palaeoarchaean granitoids and greenstones occupy an area extending from the Barberton greenstone belt (BGB) in Mpumalanga Province and Swaziland (now renamed to Eswatini) in the north to the Natal thrust front in KwaZulu-Natal Province in the south (Fig. 2.1; Brandl et al. 2006; Robb et al. 2006). The Barberton granitoid–greenstone terrain has been intensively studied with respect to the nature and processes of early crust formation (Viljoen and Viljoen 1969; Lowe and Byerly 2007). In contrast, the granitoid–greenstone terrain south of the Barberton belt has received much less scientific attention to date, mostly because of limited exposure and, on average, more intense deformation and higher metamorphic grade. Greenstones and granitoids of similar age to those of the Barberton granitoid–greenstone terrain are present, and new perspectives on regional processes leading to the earliest crust in southern Africa will be gained from a more rigorous study of these areas. The Barberton belt is a classic example of a Palaeoarchaean greenstone belt, consisting of a keel of tightly folded supracrustal rocks between granitoid domes stretching for more than 100 km. It consists predominantly of greenstones, lithostratigraphically mainly confined to the 3.55 to 3.28 Ga Onverwacht Group. Greenstones we define here as mafic/ultramafic volcanic rock-dominated

A. Kröner and A. Hofmann (eds.), *The Archaean Geology of the Kaapvaal Craton, Southern Africa*, Regional Geology Reviews, https://doi.org/10.1007/978-3-319-78652-0_2

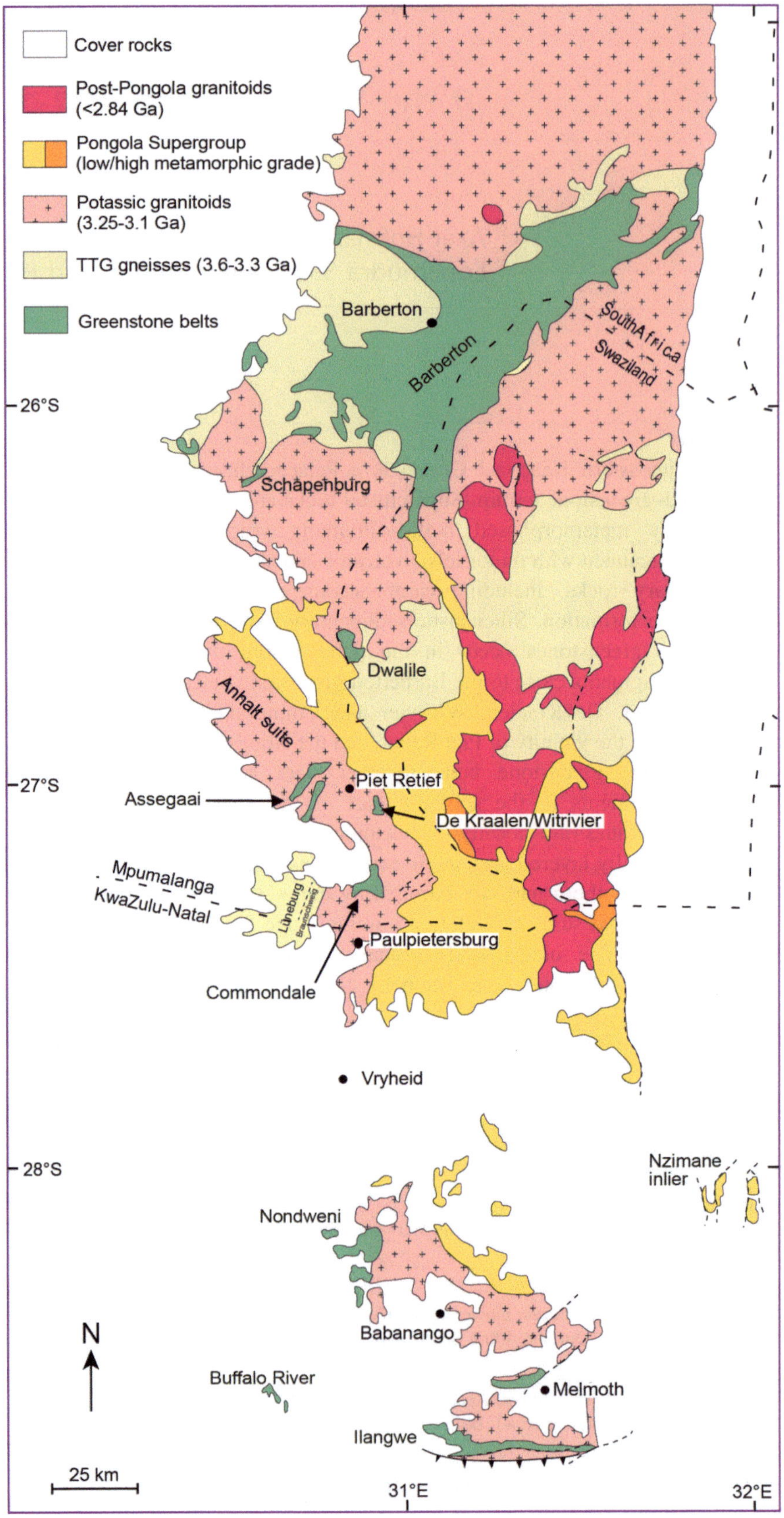

Fig. 2.1 Geological map of the southern part of the Kaapvaal Craton, showing the distribution of the major greenstone belts and remnants. Modified after Robb et al. (2006)

successions that extruded in a submarine setting and typically include pillow basalt and komatiite. Greenstones are, by definition, of greenschist-facies grade, although we do herein include rocks subjected to higher metamorphic grades. Important siliciclastic sedimentary successions (Fig Tree and Moodies groups) form the youngest part (3.26–3.22 Ga) of the supracrustal succession of the Barberton belt (see also Heubeck 2019, this volume), known as the Barberton Supergroup. Metamorphic grade is largely greenschist-facies.

Several occurrences of supracrustal rocks have been described south of Barberton. Frequently much smaller in size and metamorphosed at greenschist- to amphibolite-facies grade, they have been variably described as greenstone belts, schist belts and greenstone belt fragments or remnants, due to their disrupted nature as a result of deformation and granitoid intrusion as well as medium- to high-grade metamorphism in places. In addition, some greenstones crop out in inliers surrounded by younger sedimentary cover. The preserved Archaean greenstone successions exhibit a wide variety of rock types and include clastic and (bio-)chemical sedimentary rocks, pyroclastic rocks and volcanic rocks ranging from ultramafic to felsic in composition. The proportion of rock types between the various occurrences varies greatly, which precludes simple lithostratigraphic correlations. The successions are generally steeply dipping and have experienced varying degrees of deformation. The supracrustal fragments are tectonically interleaved with locally migmatitic tonalite–trondhjemite–granodiorite (TTG) gneisses and are intruded by a large variety of granitoids (Matthews 1985; Hunter and Wilson 1988; Hunter 1991; Brandl et al. 2006; Robb et al. 2006; Saha et al. 2010). The granitoid–greenstone terrain is unconformably overlain by the cratonic cover succession of the c. 3.0 Ga Pongola Supergroup (Luskin et al. 2019, this volume), and this contact marks a major transition in the Archaean history of the Kaapvaal Craton. Elsewhere, Palaeozoic and Mesozoic cover rocks rest directly on Archaean basement. Several localities exist where strongly deformed Pongola Supergroup strata of amphibolite-facies grade may be mistaken for greenstone successions (Fig. 2.1). Some of these units in southern Swaziland were mapped as Mahamba Gneiss (Wilson 1982), although a combination of field observations, trace element geochemistry and geochronology allows their differentiation (Wilson and Jackson 1988; Hofmann et al. 2015; Dlamini et al. 2017). The 'greenstone' succession of the Nzimane inlier in north-eastern KwaZulu-Natal (Fig. 2.1; Thomas et al. 1997) also belongs to this group of high-grade Pongola strata (unpublished U-Pb zircon ages of H. Xie). In the following, we provide a brief account of the Archaean granitoid–greenstone geology of the southeastern part of the Kaapvaal Craton. The focus will be on the greenstone belts for which more information is currently available than on the granitoid gneisses.

2.2 Granitoid–Greenstone Terrain South of the Barberton Belt

A wide variety of granitoid gneisses of Palaeoarchaean age is exposed immediately south of the Barberton belt and in north-eastern Swaziland. They are described in detail in Kröner et al. (2019, this volume). Greenstone belts include the Schapenburg and Dwalile remnants and several occurrences scattered within the gneiss terrain. Metamorphic grade is largely amphibolite-facies.

2.2.1 Schapenburg Greenstone Remnant

The Schapenburg remnant (Anhaeusser 1983) is situated ~25 km south of Badplaas (Fig. 2.2). It measures ~15 × 2.5 km, is surrounded and intruded by a variety of granitoid rocks, and can be traced intermittently in a northeasterly direction, where it ultimately merges with greenstone fragments linked to the Barberton belt, such as the Weergevonden fragment (Anhaeusser 1980; Anhaeusser and Robb 1980). The Schapenburg succession consists largely of cyclically repetitive, regularly layered volcano-sedimentary units comprising massive and spinifex-textured komatiite flows (Fig. 2.3a, b), overlain by basaltic komatiite lava flows that are often capped by banded iron formation (BIF, Fig. 2.3c; Anhaeusser 1983). Re–Os isotopic data define a regression line with an age of 3549 ± 99 Ma for the ultramafic lavas (Puchtel et al. 2009). In addition, the greenstone remnant contains a well-developed siliciclastic metasedimentary sequence (Fig. 2.3d). It contains detrital zircons as young as 3240 ± 4 Ma, suggesting a correlation with Fig Tree Group sediments in the BGB (Stevens et al. 2002). Peak metamorphic assemblages have yielded pressure–temperature conditions of 640 ± 40 °C and 4.8 ± 1.0 kbar, and the maximum age of the metamorphism is defined by the 3231 ± 5 Ma age of a syn-tectonic tonalite intrusion (Fig. 2.2; Stevens et al. 2002).

2.2.2 Dwalile Greenstone Remnant

Greenstone remnants in SW Swaziland were collectively referred to as the Dwalile Metamorphic Suite (Hunter 1970) and were renamed as the Dwalile Supracrustal Suite by Wilson (1982). These rocks are best preserved in the Mankayane and Dwalile areas of southwestern Swaziland (Jackson 1984), where they are infolded with the c. 3.5 Ga Ngwane gneiss suite (Kröner et al. 2019, this volume).

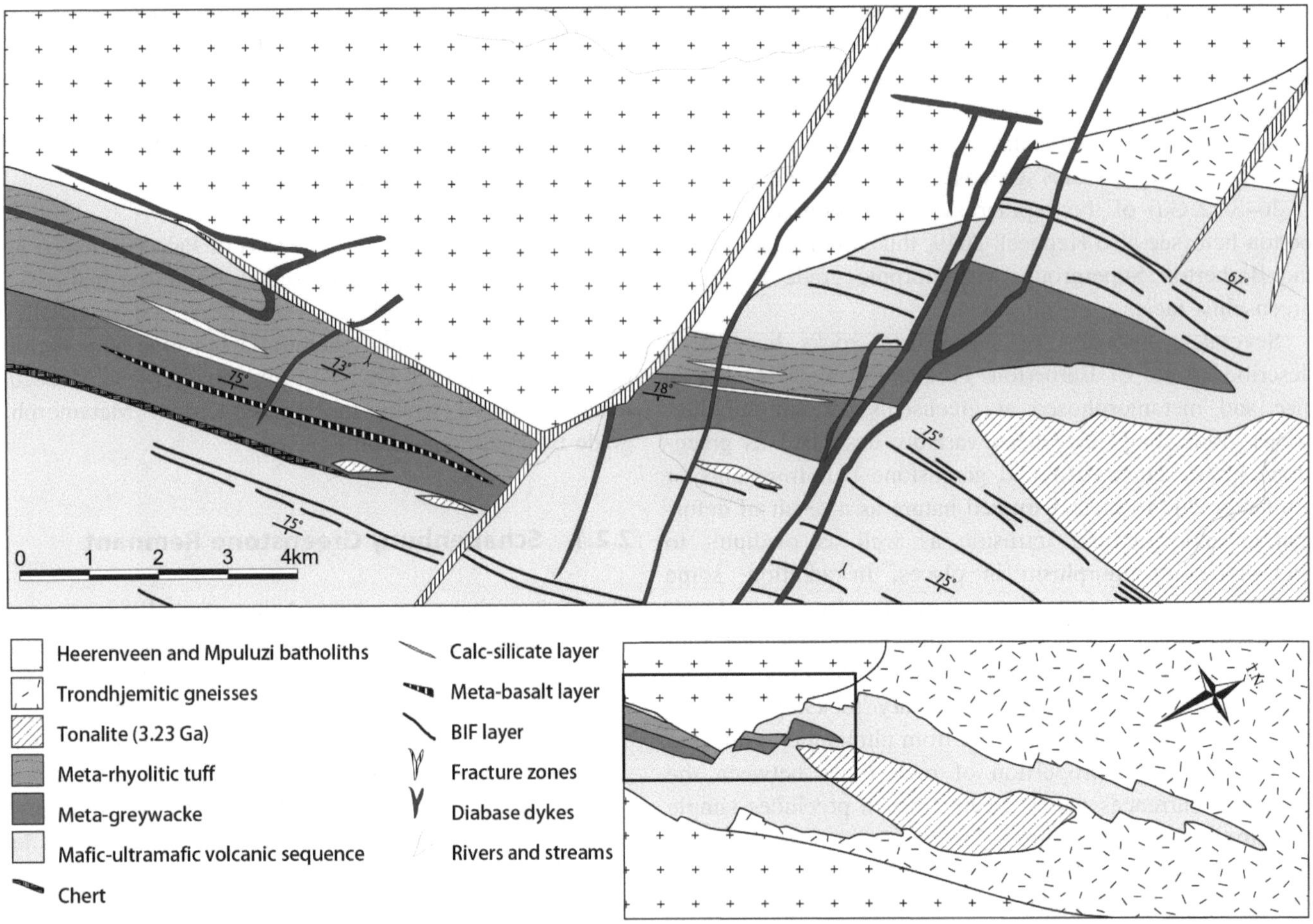

Fig. 2.2 The geology of the Schapenburg greenstone remnant (after Anhaeusser 1983; Stevens et al. 2002)

The boot-shaped, amphibolite- to lower granulite-facies Dwalile greenstone remnant occurs in western Swaziland near the border with South Africa (Fig. 2.4). It was mapped by Jackson (1984), who concluded from a structural study that contacts between rocks of the multiply deformed Dwalile and Ngwane suites are everywhere structurally conformable, but that the Dwalile suite structurally overlies the Ngwane Gneiss and may therefore be younger.

The structurally lowest rocks of the Dwalile suite are talc–tremolite schist and serpentinite, interlayered with amphibolite and actinolite schist (Fig. 2.4). Structurally overlying this mafic–ultramafic assemblage is metasedimentary rocks of greywacke, chert and BIF parentage. Minor felsic volcanic rocks also occur. Tegtmeyer (1989) estimated peak metamorphic conditions at 550–600 °C and ~4 kbar.

Structural data suggest that the Dwalile greenstones postdate the Ngwane Gneiss (Jackson 1984), and this was confirmed by U-Pb detrital zircon geochronology from Dwalile clastic metasediments, suggesting that these rocks were derived from a granitoid source some 3.42–3.57 Ga in age (Kröner and Tegtmeyer 1994). Nd whole-rock isotopic systematics of Dwalile metasediments yielded $\varepsilon_{Nd(t)}$ values of −0.5 to −2.4 and T_{DM} model ages of 3.7–3.9 Ga (Kröner and Tegtmeyer 1994). A sample of trondhjemitic Ngwane Gneiss structurally underlying the greenstones yielded a zircon age of 3521 ± 23 Ma and contained a 3683 ± 10 Ma old xenocryst (Kröner and Tegtmeyer 1994). This, together with old Nd model ages and a pronounced negative Eu anomaly for the metasedimentary rocks (Kröner and Tegtmeyer 1994), argues for the presence of old crustal components in these rocks. Siliceous schists intercalated with metasedimentary rocks and interpreted as felsic volcanic or volcaniclastic rocks yielded zircon ages of 3.46 Ga (van Schijndel et al. 2017; Kröner et al. 2019, this volume). These ages correspond to those reported for felsic volcanic and plutonic rocks from the Barberton greenstone belt (Agangi et al. 2019). Van Schijndel et al. (2017) also reported detrital zircon ages for Dwalile metasedimentary rocks and two groups were distinguished, one group containing zircons similar in age to the Hooggenoeg and Kromberg formations (3.46–3.40 Ga) of the BGB, and a second group containing zircons as young as 3.23 Ga and

Fig. 2.3 Lithologies of the Schapenburg greenstone remnant. **a** Parts of two komatiite flows younging to the left. A flow top (stippled line) from an underlying flow is visible to the right. In the centre is a cumulate basal zone of the overlying flow overlain by a random spinifex-textured zone (left). **b** Komatiite with random spinifex texture. **c** Grunerite-bearing silicate-facies BIF. **d** Planar-bedded clastic metasedimentary rock

thus possibly correlative with the Moodies Group (<3.23 Ga) or at least reflecting a similar provenance. Van Schijndel et al. (2017) reported evidence for a polymetamorphic history of the Dwalile belt, including an event at 3.23 Ga, pervasive amphibolite-facies metamorphism at 3.15 Ga and low-P granulite-facies metamorphism at 2.99 Ga. More detailed information on the Dwalile and related greenstone remnants can be found in Kröner et al. (2019, this volume).

2.3 Granitoid–Greenstone Terrain in the Vicinity of Piet Retief

In the vicinity of Piet Retief (now renamed as Mkhondo), situated ~ 100 km south of the Barberton belt (Fig. 2.1), the Assegaai, De Kraalen and Witrivier greenstone remnants (Fig. 2.5) preserve poorly exposed volcano-sedimentary sequences older than 3.25 Ga. These show evidence of multiple events of deformation and, locally, high-grade (upper amphibolite- to granulite-facies) metamorphism (Sleigh 1988; Verbeek 1991). The Commondale greenstone belt is a well-preserved supracrustal sequence in the Paulpietersburg (now renamed as eDumbe) area, ~ 150 km south of Barberton (Fig. 2.1).

Hunter et al. (1978, 1992) and Smith (1987) reported the presence of compositionally layered TTG gneisses and tectonically interleaved amphibolites, outcrops of which are preserved as inliers in younger granitoids (Fig. 2.6). The gneisses were equated with the Ngwane Gneiss of the Ancient Gneiss Complex in Swaziland and interpreted as a pre-greenstone basement. Contacts between supracrustal rocks and the layered gneisses are, however, blurred by later granitoid intrusions.

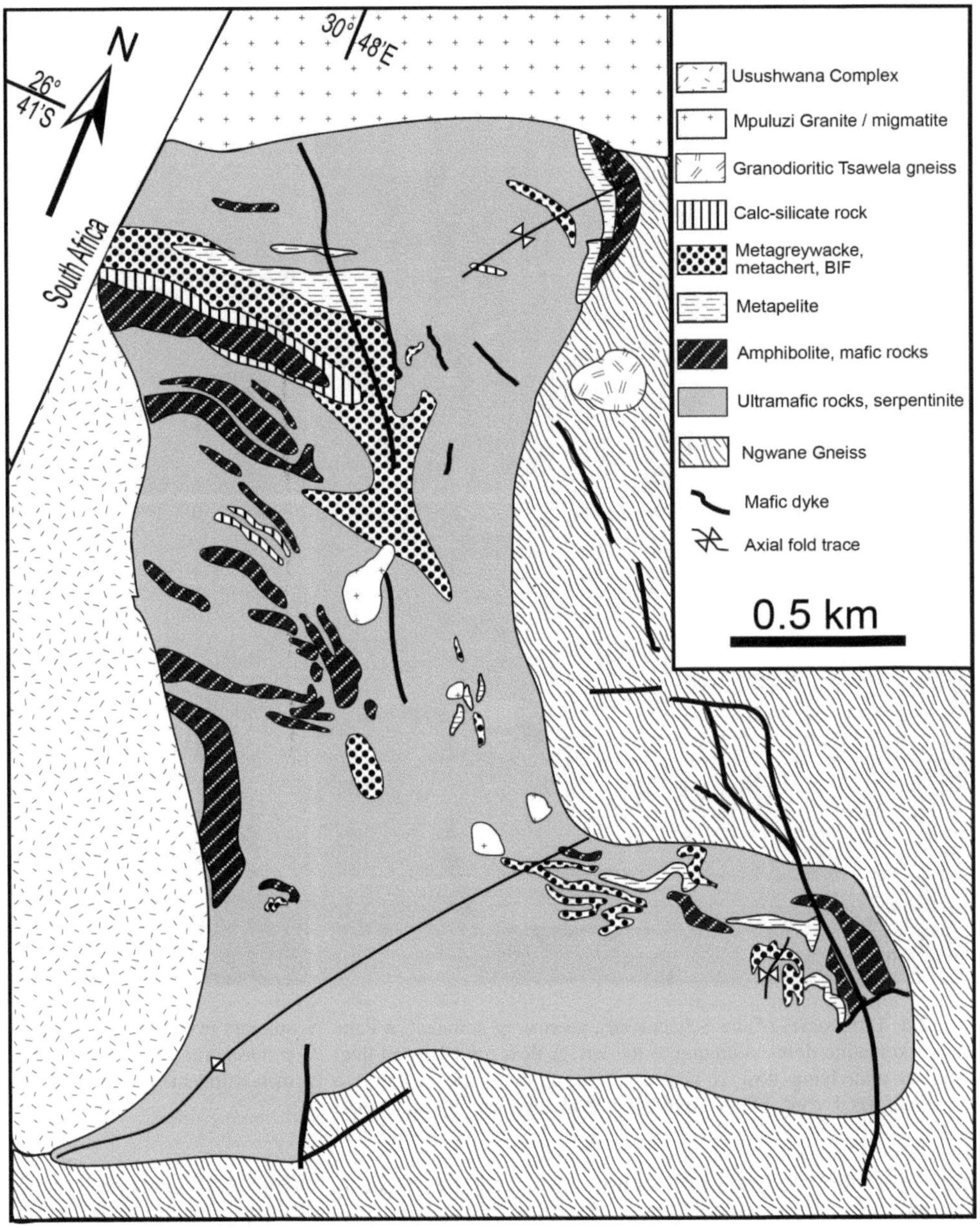

Fig. 2.4 Simplified geological map of the Dwalile greenstone remnant (after Jackson 1984; Kröner and Tegtmeyer 1994)

A suite of granitoid gneisses lacking amphibolite bands and regarded somewhat younger than those described above are widespread in the area. South and west of the Assegaai and Commondale belts (Fig. 2.1) are exposures of the Lüneburg Gneiss (Smith 1987; Hunter et al. 1992), which consists of compositionally layered TTG gneisses. Situated between the Lüneburg gneisses and the Commondale greenstone belt are rocks referred to as the Braunschweig Gneiss (Fig. 2.1), which comprises medium- to coarse-grained tonalite regarded by Hunter et al. (1992) as being chemically and mineralogically indistinguishable from the Tsawela Gneiss of southwestern Swaziland (Hoffmann et al. 2016; Kröner et al. 2019, this volume). A third variety of gneiss, referred to as the De Kraalen Gneiss (Sleigh 1988; Fig. 2.5), consists of layered tonalitic rocks and crops out southeast of the De Kraalen greenstone remnant in the valley of the Assegaai River. Lastly, a fourth gneiss variety, referred to by Hunter et al. (1992) as the Witrivier Gneiss, occurs approximately 35 km southeast of Piet Retief (Fig. 2.5). The Witrivier Gneiss consists of porphyroclastic, coarse-grained, strongly gneissic to mylonitic, tonalitic and trondhjemitic lithologies with supracrustal xenoliths, suggesting a post-Commondale greenstone age. No published radiometric ages are available for the various gneisses previously described, but they have been intruded by the c. 3250 Ma Anhalt granitoid suite described below. The gneisses were believed by Talbot et al. (1987) to predate the greenstone belt fragments present in the area.

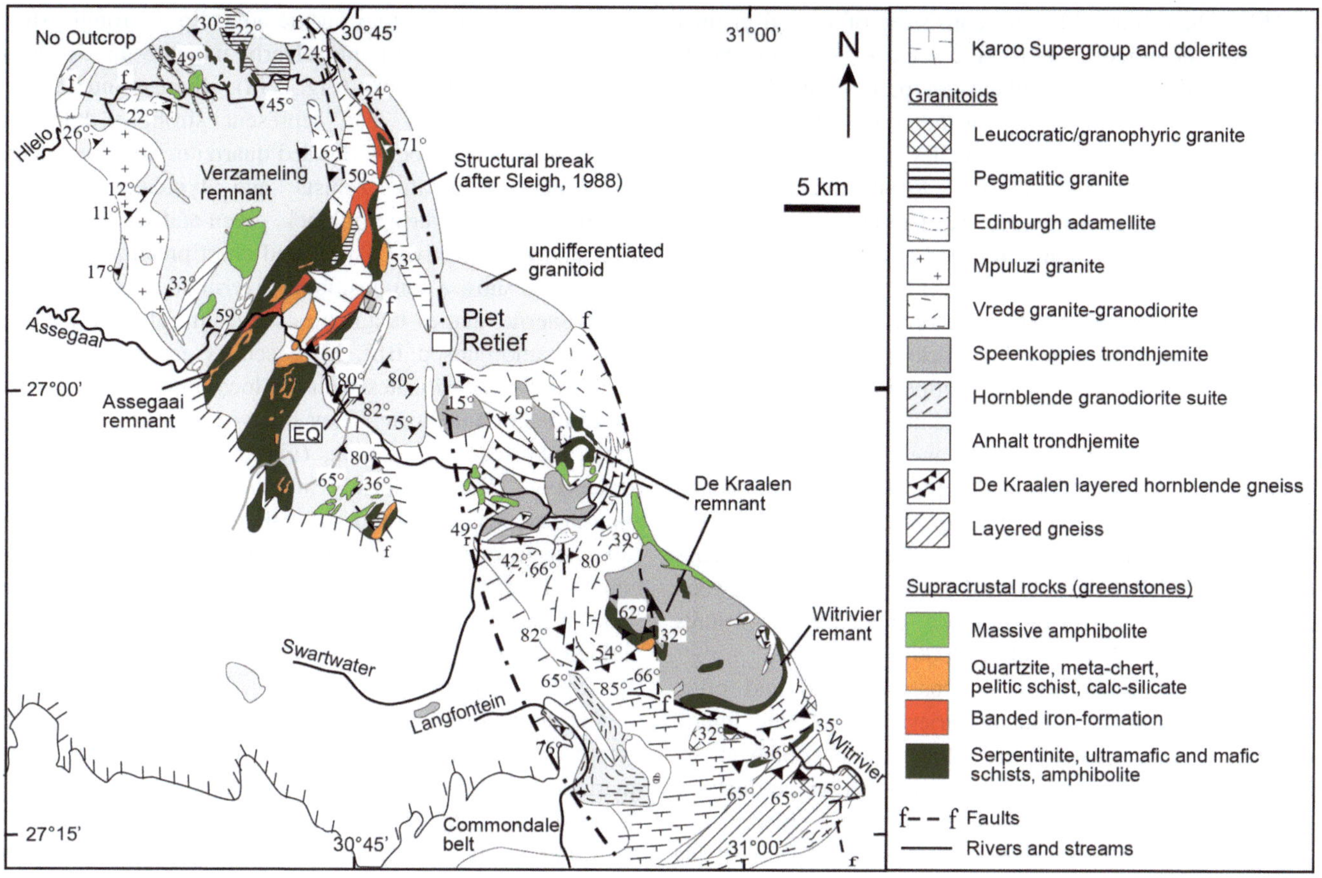

Fig. 2.5 Geological map of the granitoid–greenstone terrain of the Piet Retief area (after Sleigh 1988)

Fig. 2.6 Layered gneiss east of the Verzameling greenstone fragment consisting of different phases of TTG that are tectonically interleaved due to high strain

The Anhalt granitoid suite, consisting of several distinct batholiths, comprises the most extensive development of granitoid rocks in the southeastern part of the Kaapvaal Craton, and it underlies much of the country from north of Piet Retief to the Bivane River valley south of Paulpietersburg (Fig. 2.1; Hunter et al. 1992; Robb et al. 2006). The Anhalt batholiths generally occur as tabular bodies are up to 3 km thick, and commonly intruded at the base of the various large greenstone belt fragments. A well-developed fabric is always present, which becomes mylonitic towards the contact with greenstone belt rocks (Talbot et al. 1987). The Anhalt granitoids can be subdivided into several different phases, including trondhjemite, granodiorite and quartz monzonite (Anhalt and Bazane facies), trondhjemite, granodiorite and granite (Speenkoppies facies), a hornblende granodiorite facies intruding the north-eastern margin of the Commondale belt, and a tonalitic and trondhjemitic gneiss facies (Witrivier facies), located roughly midway between the De Kraalen and Commondale greenstone exposures (Hunter et al. 1992; Robb et al. 2006).

The trondhjemitic phase is coarse-grained and often porphyritic and is the dominant variety. The granitic, quartz monzonitic and granodioritic phases were believed by Hunter et al. (1992) to represent fractionation products of trondhjemite magma, derived from partial melting of amphibolite. An Rb–Sr whole-rock age of 3250 ± 39 Ma for Anhalt-suite rocks has been documented by Farrow et al. (1990). Single zircon SHRIMP ages of 3222 ± 8 and 3193 ± 5 Ma for two granitoids intrusive into the Assegaai greenstone remnant and probably part of the Anhalt granitoid suite were reported by Saha et al. (2010) from Eskay quarry (Fig. 2.5).

The Vrede granite–granodiorite postdates the Anhalt granitoid suite (Fig. 2.5). The latter is compositionally similar to the 3.1 Ga potassic Mpuluzi granite intruding the BGB (Hunter et al. 1983; Talbot et al. 1987; Hunter and Wilson 1988; Sleigh 1988).

2.3.1 Assegaai, de Kraalen and Witrivier Greenstone Remnants

The Assegaai greenstone remnant forms an NNE-elongated belt, ~28 by 8 km in size, that is separated from the De Kraalen greenstone remnant to the east by a linear belt of Anhalt and Speenkoppies trondhjemite, Vrede granite–granodiorite as well as Phanerozoic sedimentary rocks of the Karoo Supergroup (Fig. 2.5). Associated with the Assegaai belt, but separated from it by granitoids, are the Verzameling and other, smaller, fragments (Fig. 2.5). The supracrustal rocks include (a) talc–tremolite schist and amphibolite, (b) banded quartzite, (c) ferruginous rocks, (d) mica schist, and (e) calc–silicate rocks. Ultramafic schist and amphibolite are the dominant lithology in the area. Rare strongly stretched pillows in amphibolite clearly indicate an extrusive origin for the mafic rocks (Fig. 2.7a), whereas interlayered talc–tremolite schists may represent strongly deformed ultramafic volcanic rocks. Banded quartzites are intercalated with mafic–ultramafic schists and typically show a centimeter-scale layering of black, green and white quartzite (Fig. 2.7b), suggestive of a banded chert precursor. Ferruginous rocks show a compositional layering defined by quartz and grunerite–garnet–magnetite, suggesting that they represent metamorphosed BIF. Mica schists are dominated by quartz and phyllosilicates with the local presence of garnet, staurolite and andalusite. They likely represent fine-grained clastic metasedimentary rocks (Fig. 2.7c). Calc–silicate-like rocks (Fig. 2.7d) are locally intercalated with amphibolite and may represent protoliths of different parentages strongly subjected to metasomatism, including silicification and carbonate alteration (Hofmann and Harris 2008). One of these rocks, a metasomatized felsic volcanic or volcaniclastic rock, yielded a SHRIMP II $^{207}Pb/^{206}Pb$ age of 3457.8 ± 3.4 Ma (A. Kröner, unpubl. data).

The supracrustal rocks of the De Kraalen remnant predominantly consist of meta-BIF, calc-silicate gneisses, quartzite, amphibolite and mafic–ultramafic rocks. Most of the mafic–ultramafic rocks form an interlayered sequence of metamorphosed dunite, websterite and gabbro–norite. The presence of cumulate textures and the lack of pillow structures indicate that these rocks are likely intrusive into the supracrustal rocks (Verbeek 1991). Some of the ultramafic rocks are, however, devoid of any cumulate texture and, together with amphibolites, may represent part of an extrusive sequence associated with the metasedimentary rocks. The supracrustal sequence of the De Kraalen greenstone remnant is closely associated with, and forms lenses within, the De Kraalen layered tonalitic gneiss (Fig. 2.5). According to Verbeek (1991), this gneiss is interlayered with amphibolite and is characterized by concordant quartz veins and pegmatites. The supracrustal rocks of the De Kraalen remnant were deformed together with the tonalite gneiss and were metamorphosed to upper amphibolite-facies (Verbeek 1991). Verbeek (1991) reported the presence of boudins of garnet + clinopyroxene-bearing rocks in the layered ultramafic sequence, thus suggesting the possibility that the metamorphic grade reached granulite-facies.

South of the De Kraalen remnant, a greenstone succession of predominantly mafic–ultramafic schist and amphibolite with minor calc–silicate gneisses crops out in a small area that has been described as the Witrivier greenstone remnant (Fig. 2.5; Sleigh 1988; Verbeek 1991). This greenstone succession is closely associated with tonalitic to trondhjemitic Witrivier gneisses. The leucocratic layers in the gneisses are complexly folded, and within the melanocratic layers a strong gneissic foliation is defined by

Fig. 2.7 Photographs of field relationships in the Assegaai greenstone remnant. **a** Deformed pillows in amphibolite. **b** Quartzite with well-preserved lithological layering typical of a banded chert precursor. **c** Mica schist with large staurolite porphyroblasts. **d** Calc–silicate gneiss

hornblende + biotite + quartz. Amphibolites and other supracrustal lithologies are locally interleaved with the gneisses and typically show concordant relationships with the layering in the gneisses. These amphibolites are also closely associated with calc–silicates and are characterized by layers of quartzofeldspathic leucosome (Saha et al. 2010). The supracrustal successions of the Witrivier greenstone remnant and the associated gneisses are strongly deformed and have been affected by high-grade metamorphism.

The granitoid–greenstone terrain in the area around Piet Retief can be divided into a western and eastern structural domain, separated by an NNW-trending structural break (Fig. 2.5; Sleigh 1988). North of Piet Retief this structural break is represented by a major shear zone, whereas to the south it is partially covered by rocks of the Karoo Supergroup. Whereas the Assegaai belt and associated granitoids form the western structural domain, the eastern structural domain is represented by the De Kraalen and Witrivier greenstone remnants. Farther south, the structural break separates the Witrivier remnant in the east from the Commondale greenstone belt to the SW. The dominant regional structural trend in the western domain is NNE, whereas in the eastern structural domain it is NW (Fig. 2.5). Four major deformation events (D_1–D_4) have been postulated for both of these domains (Talbot et al. 1987; Sleigh 1988). According to these authors, pre-D_1 structures are restricted to the layered gneisses from the Assegaai and De Kraalen remnants. D_1 is recorded in the supracrustal rocks and the layered gneisses, whereas post-D_1 events also affected granitoids such as the Anhalt trondhjemite. Intrusion of the 3.25 Ga Anhalt trondhjemite has been related to pre- or syn-D_2 deformation that led to the development of a contact-parallel foliation (Talbot et al. 1987; Sleigh 1988; Verbeek 1991). Hunter et al. (1983) estimated peak metamorphic conditions at 550–600 °C and ~4 kbar during D_2. Post-D_2 events are characterized by the development of multiple events of shearing along discrete fault zones. Verbeek (1991) related the post-D_2 shearing events with the intrusion of the

Ushushwana Complex. Sleigh (1988) correlated the post-D_2 events with the successive intrusion of Vrede granite–granodiorite, extrusion and deposition of the 2.98 Ga (Hegner et al. 1994; Gumsley et al. 2015) volcano-sedimentary Nsuze Group and intrusion of the Ushushwana Complex.

2.3.2 Commondale Greenstone Belt

The Commondale greenstone belt comprises a suite of deformed and attenuated supracrustal rocks covering an area of ~10 by 8 km and located about 20 km north of Paulpietersburg (Fig. 2.1). The sequence consists of two lithostratigraphic units, namely, the Commondale Formation and the Commondale ultramafic suite (Fig. 2.8). The Commondale Formation is characterized by serpentinites, and mafic–ultramafic rocks metamorphosed at upper greenschist- to amphibolite-facies grade. In shear zones and in zones of high strain, these ultramafic rocks have been transformed into talc-carbonate, talc-chlorite and actinolite schist. Sedimentary units include fuchsitic chert, calc–silicate rocks, BIF and ferruginous quartzite (Hunter and Wilson 1988; Smith 1987; Wilson 2003; Wilson and Carlson 1989; Wilson et al. 2003).

The Commondale ultramafic suite is an important component of the supracrustal association and has been accurately dated by whole-rock Sm–Nd systematics at 3354 ± 18 Ma (Wilson and Carlson 1989). The suite comprises spinifex-bearing (olivine and orthopyroxene) komatiite lava flows, which are preserved in the core of the major synformal structure of the belt (Fig. 2.8). Flow thicknesses range from less than 1 m to greater than 25 m and comprise olivine cumulates and spinifex-textured components to the flows. Clinopyroxene is entirely absent. The unique chemical aspect of the lava flows is their highly depleted compositions (Wilson 2003) with exceptionally low LREE and incompatible element contents. Core compositions of olivine of Fo96.6 are the highest measured for any komatiite and attest to the high MgO content of the parental magma of 34.2%. At the calculated eruption temperature of 1638 °C, these are the highest temperature lavas encountered in Earth's history.

The origin of the Commondale komatiites was initially reported to be the result of hydrous melting of a depleted and highly refractory mantle wedge (Wilson 2003) and therefore could have provided evidence for subduction-related processes at this early stage of the Archaean. However, the Commondale komatiites do not have subduction (or boninitic) trace element signatures and may rather have formed by two-stage melting of a mantle plume at shallow depths (Arndt et al. 2008). Hoffmann and Wilson (2016) showed possible decoupling between the whole-rock isotopic systems of Nd–Sm and Hf–Lu, pointing to a heterogeneous mantle at that stage.

Some tonalitic gneiss and minor amphibolite layers, possibly related to the Ancient Gneiss Complex of Swaziland, occur in the area and were interpreted as a pre-greenstone basement. U-Pb dating of tonalitic gneiss yielded an age of 3419 ± 7 Ma for emplacement of the gneiss protolith (Xie et al. 2010). Dating of metamorphic zircon from tonalite and Commondale amphibolite yielded an age of 3.21 Ga (Xie et al. 2010), interpreted to reflect the age of peak metamorphism in the area.

The stratigraphy of the belt has to be understood on the basis of the deformation events. D_1 is the first recognizable event in the supracrustal suite and is characterized by recumbent folds and thrusts that extend into the granitoid gneisses. Horizontal tectonism and sodic granitoid underplating are interpreted to have caused thickening of the stratigraphy prior to the D_2 event. Concurrent with the D_2 event folding took place about NE-trending axes, causing folding of D_1 axial planes. This event resulted in considerable shortening. The sequence is therefore preserved as a series of sharply defined interference folds that form an F_2 sheath-fold structure (Smith 1987). Mylonites formed during the late D_2 event at the contact of some of the gneisses and the supracrustals and record a dextral shear sense. The D_3 event is characterized by sinistral shearing of the suite, whereas the last event (D_4) resulted in generally open refolding of previous structures about steeply inclined NW-trending axial directions (Smith 1987).

2.4 Granitoid–Greenstone Terrain of the Southern Margin of the Kaapvaal Craton

Several greenstone belts are present in the southernmost part of the Kaapvaal Craton, where they have been variably affected by the Namaqua–Natal orogeny in the Mesoproterozoic. The largest and least affected is the Nondweni greenstone belt, which has also been the most intensely studied. Others include the Mhlatuze and Melmoth greenstone inliers, the Ilangwe greenstone belt and three granitoid–greenstone inliers in the Buffalo River area (Fig. 2.9).

A massive, tabular batholith of commonly porphyritic granodiorite to granite extends from the Nondweni greenstone belt in the west to an area south of Ulundi (Fig. 2.9) and has been regarded to be part of the Anhalt granitoid suite (Robb et al. 2006). The granitoids include the weakly deformed Mvunyana granodiorite that is intrusive into the Nondweni greenstone belt. Matthews et al. (1989) reported an age of ~3.29 Ga on the basis of a combination of Rb–Sr, Pb–Pb and U–Pb zircon data. More recent, albeit unpublished, SHRIMP zircon dating provided a well-constrained age of the Mvunyana granodiorite at ~3.25 Ga (H. Xie, pers. comm., 2018). Zircons from three compositionally

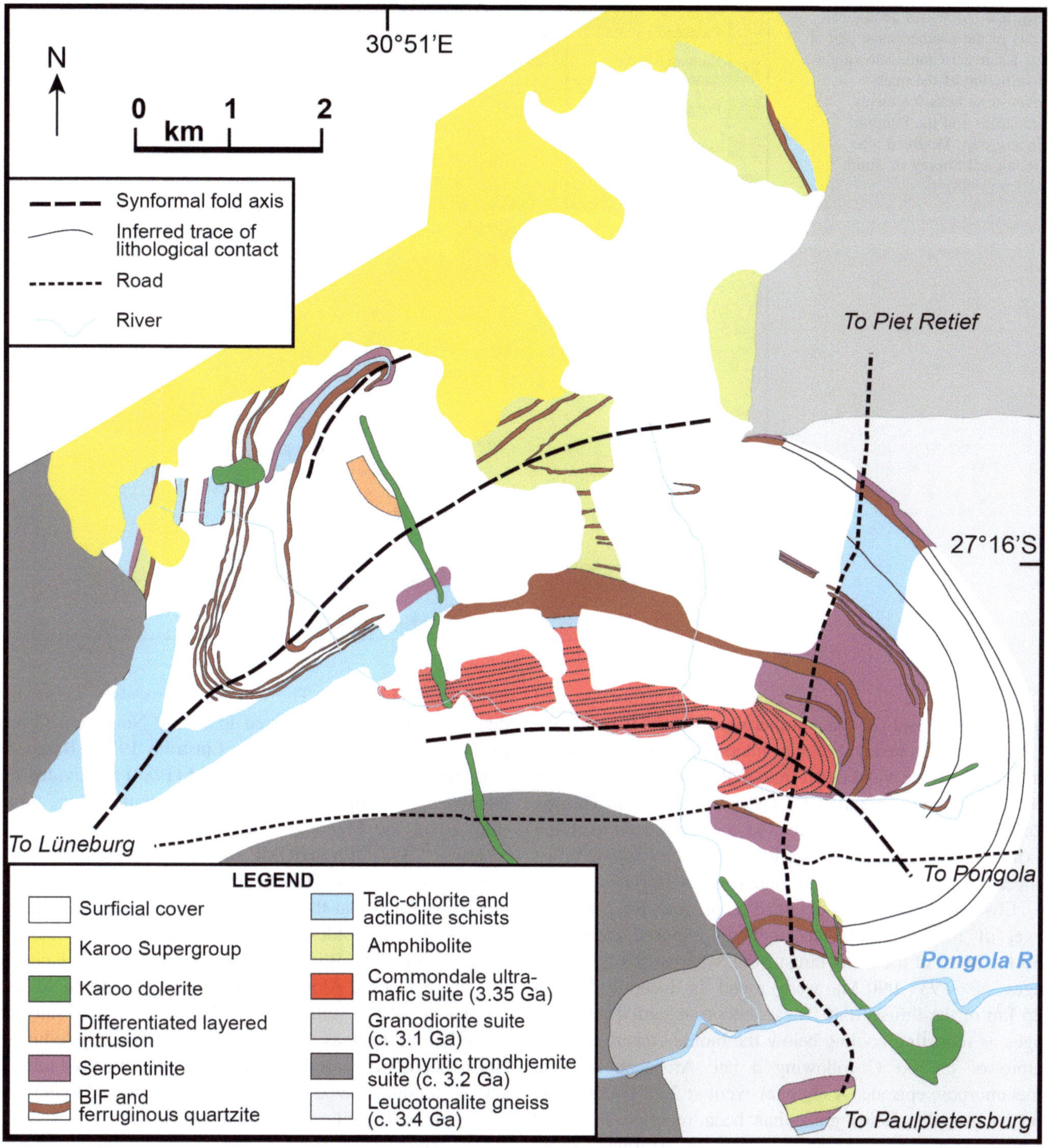

Fig. 2.8 Outcrop map of the Commondale greenstone belt (after Smith, 1987; Wilson, unpubl. data)

distinct samples of granite taken in the White Mfolozi inlier of Pongola strata (Fig. 2.9) were dated using LA-ICP-MS by Reinhardt et al. (2015) to have crystallized between 3.25 and 3.23 Ga. The same granite was investigated by Misra et al. (2017) who regarded it to be a peraluminous A-type granite that was generated by partial melting of a pre-existing TTG source under anhydrous conditions, possibly as a result of mafic underplating.

The contact of the Kaapvaal Craton with the Mesoproterozoic Namaqua-Natal mobile belt is referred to as the Natal thrust front along which the Tugela terrane was thrust to the north (Fig. 2.9; Thomas 1989). Jacobs et al. (1997), using

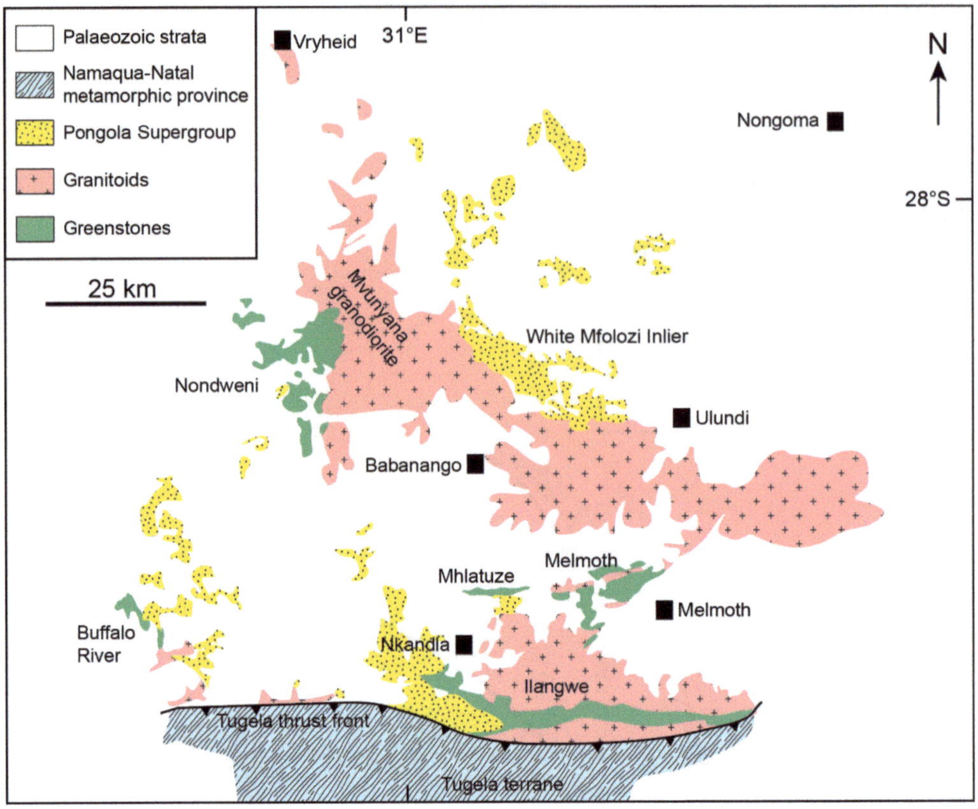

Fig. 2.9 Archaean geological units of the southernmost part of the Kaapvaal Craton showing the distribution of the main greenstone belts/fragments, granitoids and the Pongola Supergroup. Modified after Geological Survey of South Africa (1988)

$^{40}Ar/^{39}Ar$ thermochronology, showed that thrusting occurred at, or prior to, ~1135 Ma. Furthermore, titanite fission track data of samples taken at regular intervals along a ~250 km traverse from the Natal thrust front to the north into the craton revealed a concentration of 1.1 Ga ages (Jacobs and Thomas 2001). These ages were ascribed to cooling related to craton exhumation following tectonic loading of the crust by thrust sheets (Tugela nappes; Jacobs and Thomas 2001).

Elworthy et al. (2000) obtained whole-rock Rb–Sr biotite ages of 2614 ± 74 Ma for granitoids exposed more than 50 km north of the Natal thrust front, whereas Rb–Sr biotite ages of ~933–990 Ma were found in localities within 25 km of the thrust front. These authors regarded the older ages as reflecting cooling below the biotite closure temperature of c. 300 °C following a late Archaean regional metamorphic episode. A regional event at 2.72 Ga reaching up to granulite-facies grade has been reported from the southeastern part of the Kaapvaal Craton (Hofmann et al. 2015). The younger Rb–Sr biotite ages were ascribed to thermal overprinting during the Namaqua-Natal orogeny (Elworthy et al. 2000).

2.4.1 Nondweni Greenstone Belt

The Nondweni greenstone belt occurs near the southern margin of the Kaapvaal Craton ~50 km north of the Natal thrust front (Fig. 2.9). The volcano-sedimentary succession that forms the belt is referred to as the Nondweni Group (du Toit 1931; Matthews 1959; Linström 1987). Based on field evidence and lithology, Versfeld (1988) subdivided the Nondweni Group into three lithostratigraphic units (Fig. 2.10), the Magongolozi, Toggekry and Witkop formations. $^{207}Pb/^{206}Pb$ SHRIMP zircon ages obtained from the three formations by Tarduno et al. (2010) and Xie et al. (2012) revealed that the more strongly deformed Toggekry Formation is the oldest unit (up to 3.53 Ga), whereas the Magongolozi and Witkop formations are possibly coeval (~3.43–3.41 Ga). Along its eastern margin, the Nondweni greenstone belt is intruded by the ~3.25 Ga Mvunyana granodiorite, whereas to the west it is covered by sedimentary rocks of the Karoo Supergroup. Small-scale gold mining activity on mineralized quartz veins was restricted to the turn of the previous century (Versfeld 1988).

The Toggekry Formation consists of a bimodal succession of quartz–feldspar–sericite schist, representing rhyolite and rhyodacite flows and tuffs, and variably silicified mafic schist locally showing deformed pillow structures (Fig. 2.11a). The schists strike E-W and have a near-vertical dip (Versfeld 1988; Hofmann and Wilson 2007). Small Cu–Zn massive sulphide bodies are hosted by the felsic schists. The Toggekry Formation is conformably overlain by the mafic volcanic-dominated succession of the Witkop Formation.

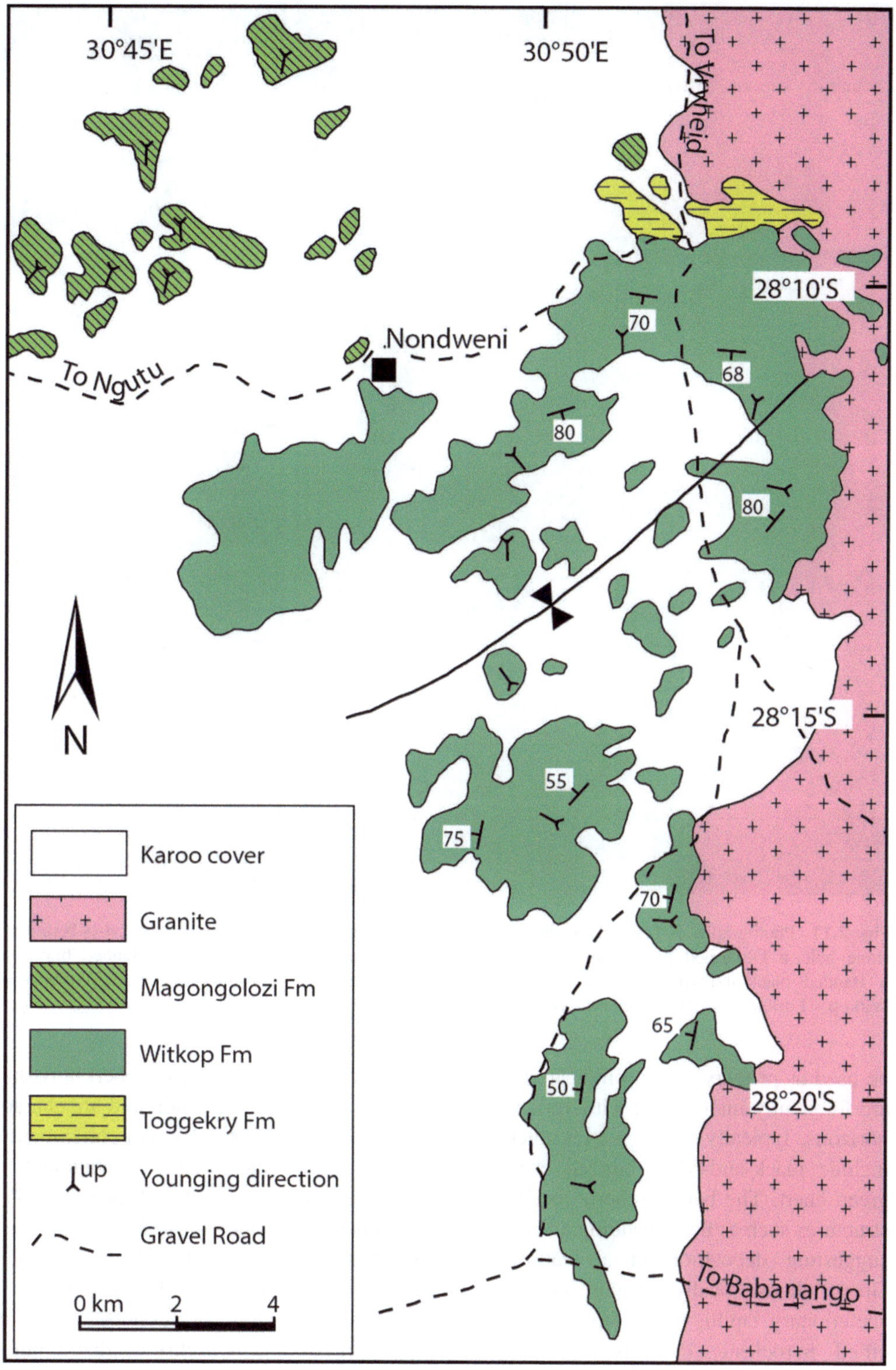

Fig. 2.10 Geological map of the Nondweni greenstone belt, showing the distribution of the Toggekry, Witkop and Magongolozi formations (after Versfeld, 1988)

The Witkop Formation is the most extensive among the stratigraphic units of the Nondweni Group. It is ~7000 m thick and mainly consists of pillowed or massive flows of basalt, komatiitic basalt and komatiite. The komatiites are distinct from those of the Barberton greenstone belt in that the MgO content does not exceed 22%, but they have a similar CaO/Al_2O_3 ratio >1 (Riganti and Wilson 1995). A unique feature is the development of pyroxene (now amphibole) bladed spinifex with crystals up to 5 m in length, commonly formed in cone-type structures (Wilson et al. 1989). These cones are always developed at the tops of flow units and fan downwards. Their consistency allows them to

Fig. 2.11 Photographs of field relationships in the Nondweni greenstone belt. **a** Deformed pillow basalt of the Toggekry Formation. **b** Banded black-and-white chert overlying silicified basalt of the Witkop Formation. Note chert vein cutting across basalt. **c** Herring-bone cross-bedding in pebbly sandstone of the Witkop Formation. **d** Banded black-and-white chert with silicified domical stromatolites of the Witkop Formation

be used as primary way-up indicators. Komatiite and basalt are frequently intercalated with variably recrystallized chert horizons, typically a few metres thick (Fig. 2.11b). These include black-and-white banded chert and more massive green chert. The banded cherts locally exhibit sedimentary structures such as flaser bedding and ripple cross-lamination, suggesting deposition in a shallow-water environment. Green cherts locally contain normally graded beds of accretionary lapilli, which suggest that they represent silicified pyroclastic deposits. An unusual association of carbon-bearing quartzite and subordinate chert occurs in the Patsoana area of the Witkop Formation as relatively thick bodies (10–100 m in size; Riganti 1996). These may have originated by chemical precipitation of silica emitted by fumaroles in a subaqueous environment, followed by later recrystallization (van den Kerkhof et al. 2004).

Below the chert horizons, alteration zones 20–50 m thick are common and are characterized by high SiO_2 contents up to 90 wt% and enrichment in K_2O (+Rb, Ba). Cross-cutting chert veins are abundant in these zones (Fig. 2.11b). Hofmann and Wilson (2007) concluded that these silicified zones formed by syn-depositional, low-temperature (<150 °C) sea-floor hydrothermal processes, a feature common to the Barberton greenstone belt and Palaeoarchaean greenstone successions worldwide (Hofmann and Harris 2008).

A highly localized, but well-preserved, sedimentary succession, 500 m thick, occurs at the top of the Witkop Formation and has been interpreted as a shallow subaqueous, or intermittently subaerial, succession (Wilson and Versfeld 1994). The basal sequence is characterized by banded and fragmented chert grading upwards into coarse-grained sandstone, which is commonly cross-bedded by coastal reworking (Fig. 2.11c). The upper conglomerate sequence is made up of normally graded mass-flow to sheet-flood deposits that contain a wide variety of both rounded and angular rock fragments (chert, rhyolite, silicified basalt). The uppermost part of the succession reflects a marked decrease of depositional energy and is made up of a stromatolite zone (Fig. 2.11d), overlain by silicified chemical precipitates and rhythmically layered accretionary lapilli tuff.

The Magongolozi Formation occurs as small inliers within Phanerozoic cover (Versfeld 1988). The strike of the Magongolozi strata is mostly E-W, and dips are nearly vertical. The estimated thickness of this formation is ~7500 m. It mainly consists of basalt, basaltic andesite and komatiitic andesite, the latter being characterized by high SiO_2 and MgO contents. The latter have similarities to arc-related volcanic rocks, such as boninites or crustally contaminated mantle melts (Riganti 1996). The volcanic units form an interlayered sequence with locally well-preserved pillow structures. Pyroxene spinifex is common in many of these rocks, and several occurrences are regarded as ancient lava lakes (Wilson and Riganti 1998). Small carbonate bodies associated with quartz veining and hosted by ultramafic rocks are present in both the Witkop and Magongolozi formations. They carry sub-economic gold mineralization of hydrothermal origin (Wilson and Versfeld 1994).

The Nondweni greenstone belt records two major episodes of deformation (Versfeld 1988; Matthews et al. 1989). An early deformation event (D_1) led to the development of thrust faults and associated small-scale isoclinals folds with strong penetrative axial planar cleavage, mostly recorded in the felsic schists of the Toggekry Formation. The second deformational event (D_2) led to the formation of a major synformal structure with ENE–WSW trending axial trace (Fig. 2.10). Small-scale D_2 folds show variable plunges (near-vertical to 30°), indicating a syn-tectonic progressive deformation event (Versfeld 1988). Whereas Matthews et al. (1989) considered that the D_2 folds accommodated strain related to the intrusion of the Mvunyana granodiorite, Versfeld (1988) considered the granodiorite to be post-tectonic with respect to D_2.

Versfeld (1988) suggested the regional metamorphism to be of greenschist-facies as recorded by tremolite–actinolite–plagioclase–chlorite–epidote-bearing assemblages in the basalts and attributed the metamorphism to D_1. Amphibolite-facies mineral assemblages (garnet–hornblende–plagioclase) are noted mainly in basalts of the Witkop Formation near the intrusive contact of the Mvunyana granodiorite with the supracrustal rocks, the latter containing sheets and dykes of granite and pegmatite. Metamorphism of the pillow basalts took place at P-T conditions of ~6–6.5 kbar and 600–650 °C, which marked syn-tectonic garnet growth in pillow rims. Post to peak metamorphism, the pillows underwent isobaric cooling to ~450 °C along an anticlockwise P-T path (Saha et al. 2012). The late syn-tectonic intrusion of the Mvunyana granodiorite into the supracrustal rocks and the presence of migmatites near the contact indicate that peak metamorphism of the Witkop basalts occurred at ~3.25 Ga.

2.4.2 Melmoth Granitoid–Greenstone Inlier

A granitoid–greenstone assemblage of the Melmoth inlier crops out over an area of about 360 km^2 and is situated 8 km WNW of Melmoth (Fig. 2.9). The greenstone sequence was mapped by Bullen (1990), who identified mafic metavolcanic rocks with lesser serpentinite, talc schist, quartz-muscovite schist, quartzite and calc–silicate rocks. Several phases of deformation have been recognized. Pillows are rarely present, whereas quartzite shows layering typical of recrystallized banded chert. Rocks of felsic composition likely represent silicified mafic volcanic rocks, although a thin unit of dacitic tuff has been described by Bullen (1990). The greenstones are intruded by syn-tectonic (relative to D_1) trondhjemitic gneisses, late-tectonic (relative to D_2) granodioritic gneisses and post-tectonic granite dykes. Metamorphic grade varies from greenschist- to amphibolite-facies close to intrusive granitoids. Contacts between lithologies strike east–west, parallel to a pervasive cleavage and gneissosity (attributed to D_1 and D_2), and dip steeply to the south. Small-scale gold mining took place at the end of the nineteenth century from deformed quartz veins (Bullen 1990).

2.4.3 Mhlatuze Greenstone Inlier

A small inlier of metavolcanic rocks intercalated with BIF and chert is present along the Mhlatuze River between Melmoth and Nkandla (Fig. 2.9). Metamorphic grade is greenschist-facies, and the rocks underwent three phases of deformation (Ovendale 1985), the youngest of which is exemplified by tight folding of Pongola Supergroup strata in the immediate south (Tainton 1977), potentially as a result of the Namaqua-Natal orogeny. The metavolcanic rocks mainly consist of chlorite–tremolite schist suggestive of a mafic precursor, but local quartz–sericite schist with quartz phenocrysts indicates the presence of felsic volcanic or volcaniclastic rocks. Zircons in a sample of such a rock yielded a SHRIMP II $^{207}Pb/^{206}Pb$ age of 3407 ± 5 Ma (H. Xie, unpubl. data). Sedimentary rocks are mainly characterized by ~50-m-thick continuous units of banded black-and-white chert and BIF. A serpentinized ultramafic intrusion occupies the contact between the Pongola and greenstone strata (Ovendale 1985).

2.4.4 Granitoid–Greenstone Inliers in the Buffalo River Area

Three granitoid–greenstone inliers are exposed over an area of approximately 10 km^2 in the valley of the Buffalo River and its tributaries (Fig. 2.12). The greenstones are intruded by porphyritic granite (Fig. 2.13a) and are unconformably overlain by rocks of the Nsuze Group. Dixon (2004) proposed the name Sifula Subgroup for the volcano-sedimentary succession and subdivided it into the Sibomvu and Mgunjane formations. From west to east the greenstone inliers are the Sifula (or western) inlier, which constitutes the type area for the subgroup, the Mgunjane (or central) inlier and the Buffalo (or eastern) inlier. Within each inlier, greenstones have a general N-S trend, dipping steeply to the east or west. Younging directions obtained from pillow basalts suggest that rocks of the eastern and western inliers form the limbs of a steeply plunging, upright synclinal structure with the greenstone formations of the Mgunjane inlier situated in the core of this fold and, by implication, being structurally higher.

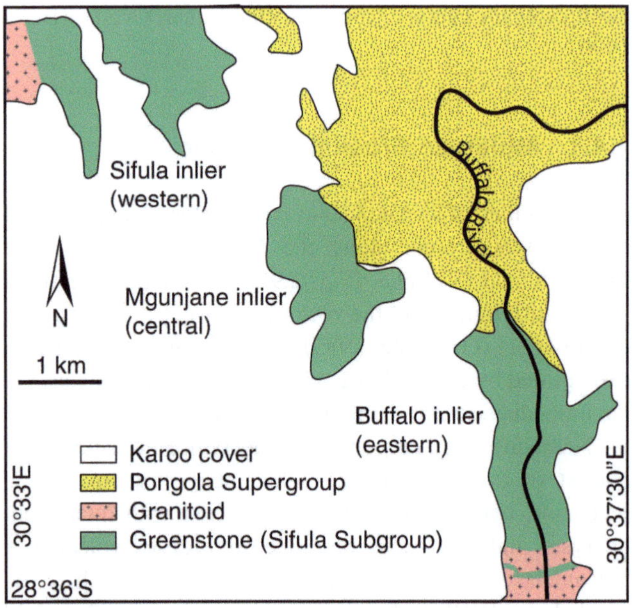

Fig. 2.12 Granitoid–greenstone inliers of the Buffalo River area (modified after Dixon, 2004). Greenstones are intruded by a porphyritic granite and are unconformably overlain by the northeasterly dipping Pongola Supergroup

Consistent with this structural interpretation, two lithological assemblages were observed and attributed to the Sibonvu and Mgunjane formations. The Sibonvu Formation is mainly found within the eastern and western inliers and consists mostly of locally pillowed mafic greenstones (Fig. 2.13b) and subordinate serpentinite, talc–tremolite–chlorite schist, quartz–mica schist, chert (Fig. 2.13c) and BIF. The Mgunjane Formation forms the core to the Sifula synform. It is a subvertically dipping faulted and folded assemblage of rocks comprising mostly serpentinite (komatiitic lava flows showing well-preserved spinifex textures; Fig. 2.13d) interlayered with lesser quartz–mica schist (felsic volcanic flows with locally well-preserved primary textures), chert and minor mafic greenstones.

The regional metamorphism of the Sifula greenstones is of lower greenschist-facies. Near intrusive granitoids, the presence of actinolitic hornblende, chlorite and almandine indicates an upper greenschist metamorphic grade.

Dixon (2004) proposed a polyphase deformational history for the Sifula greenstones including isoclinal folding related to granitoid intrusion, a possible rotation of the greenstones during folding and block faulting of the Pongola Supergroup and subsequent later faulting events. The orientation and vergence of structures (small-scale folds, foliation, lineation) within the greenstones and their parallelism with structures both within the granitoids and in the contact zones suggests cause and effect. The isoclinal folding, together with a transposed foliation and the repetition of rock assemblages with common facing directions, suggests structural repetition of the Sifula succession through transposition, thrusts being represented by zones of highly sheared rocks within otherwise slightly deformed greenstones. Later faulting events both pre- and post-dating depositions of the Pongola Supergroup are also recognized.

2.4.5 Ilangwe Greenstone Belt

The Ilangwe greenstone belt is situated along the southern margin of the Kaapvaal Craton immediately north of the Tugela thrust front (Fig. 2.9). It comprises steeply northerly dipping volcano-sedimentary rocks forming a narrow (~3 km), linear E-W-trending belt flanked by various granitoids and granitoid gneisses (Fig. 2.14; Mathe 1997). The rocks of the Ilangwe belt were included in the Nondweni Group by Linström (1987). He subdivided the sequence, of unknown thickness, into a lower unit comprising mafic to ultramafic igneous rocks with interbedded sedimentary rocks (Mhlatuze Formation) and an upper unit dominated by metasedimentary rocks (Nkandla Formation).

Mathe (1997) assigned subgroup status to Linström's formations and divided each into three tectonostratigraphic units or formations. The predominantly metavolcanic

Fig. 2.13 Photographs of field relationships in the granitoid–greenstone inliers of the Buffalo River area. **a** Sheets of granite (light grey) intrusive into greenstones in the Sifula inlier. Contact relationships have been overprinted by deformation. **b** Deformed pillow basalt, Sifula Formation. **c** Banded chert, Sifula formation. **d** Komatiite in the Mgunjane Formation showing platy olivine spinifex

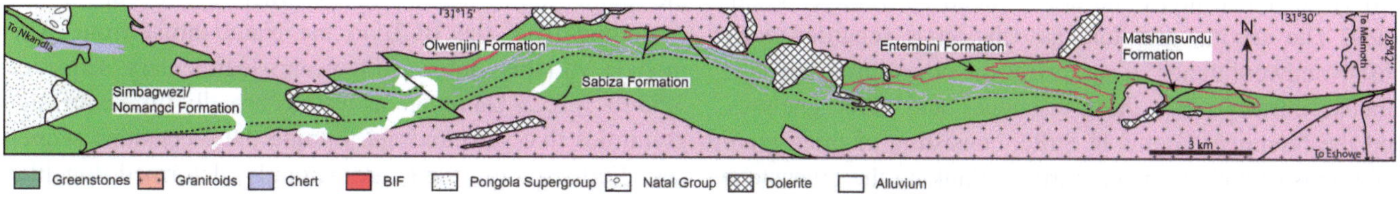

Fig. 2.14 Geological map of the Ilangwe greenstone belt (after Mathe, 1997). Stippled lines indicate formation boundaries

Mhlatuze Subgroup was subdivided into the Sabiza, Matshansundu and Olwenjini formations. Pillows, where present, indicate younging to the north. The Sabiza Formation occurs along the southern margin of the greenstone belt and consists mainly of amphibolite and massive and pillowed metabasalt with subordinate talc–tremolite schist and serpentinite. The Matshansundu Formation is represented by an amphibolite/BIF association that is poorly exposed in the eastern part of the belt. The Olwenjini Formation, situated to the north of the Sabiza Formation, is characterized by mafic rocks. These include amphibolite, actinolite schist and pillowed metabasalt and are intercalated with banded metachert and iron formation (Fig. 2.14).

The Nkandla Subgroup was subdivided by Mathe (1997) into the Entembini, Simbagwezi and Nomangci formations. The Entembini Formation, situated in the central and eastern parts of the belt (Fig. 2.14), is regarded to consist chiefly of phyllite (fine-grained quartz–sericite–chlorite schist) with intercalations of metamorphosed banded chert and iron formation, with minor pillowed metabasalt.

The Simbagwezi Formation comprises an association of phyllite, banded metachert, tremolite and actinolite schists, and minor pillowed metabasalt, occurring in the western part of the belt, southeast of Nkandla. The Nomangci Formation is a dominantly quartzite (metachert) and quartz–muscovite schist formation, also occurring southeast of Nkandla.

The contacts of the six tectonostratigraphic formations as defined by Mathe (1997) are all tectonic, and the greenstones are intruded by mafic–ultramafic bodies and calc–alkaline granitoids. The phyllites of the Nkandla Subgroup have been regarded metasedimentary in origin (Mathe 1997), but the reasoning for this interpretation is unclear. Preliminary fieldwork in the Ilangwe belt (A. Hofmann, unpubl. data) indicates an abundance of strongly sheared silicified mafic rocks that may superficially resemble sedimentary protoliths. Furthermore, the basis for stratigraphic subdivision into six formations remains unclear in the light of similar compositional make-up of several units and contacts drawn at high angles to the overall E-W structural grain. Keeping in mind the likely strong deformational overprint in the Mesoproterozoic due to its proximity to the Natal thrust front, the Ilangwe greenstone belt is the structurally most complex of all the belts near the southern margin of the Kaapvaal Craton.

A variety of granitoid rocks and gneisses occur north and south of the Ilangwe greenstone belt (Mathe 1997). These have been grouped together under the general term Ilangwe Granitoid/Gneiss Suite and subdivided into three associations on the basis of relative ages. The Amazulu Gneiss–Nkwalini Mylonitic Gneiss–Nkwalini Quartzofeldspathic Flaser Gneiss association, consisting of migmatitic paragneiss and mylonitic flaser gneiss, are regarded to be of similar age to the Nondweni Group. They occur in several widely separated areas and are intruded by younger granitoids. The second granitoid association is regarded early post-Nondweni in age. It comprises the Nkwalinye Tonalitic Gneiss, intrusive into the greenstones and older gneisses, and the Nsengeni Granitoid Suite, consisting of three granitoid units of batholithic proportions flanking the greenstone belt and intrusive into the greenstones, the older gneisses and the Nkwalinye Tonalitic Gneiss. Samples of Nsengeni Gneiss yielded whole-rock and biotite Rb-Sr ages clustering around 1.0 Ga, while one sample yielded a whole-rock and mineral (plagioclase, sphene) errorchron of 3475 ± 69 Ma (Reimold et al. 1989). The third suite is regarded as late post-Nondweni in age, comprising the Impisi-Umgabhi Granitoid Suite, a batholithic-scale, fine-grained to megacrystic association of five granitoid phases occurring north and south of the Ilangwe belt and also intrusive into greenstones, older gneisses and early post-Nondweni granitoids (Mathe 1997). One of these phases is the porphyritic Nkandla granite that is exposed northwest of the belt. Hicks et al. (2015) reported a zircon age of 3254 ± 8 Ma for a sample of this granite.

Three major episodes of deformation have been proposed by Mathe (1997), the first phase (D_1) being dominated by fold nappes and thrusts and prograde middle to upper amphibolite-facies metamorphism. During D_2 (a major compressional event, accompanied by greenschist-facies metamorphism and early post-Nondweni granitoid intrusions), sub-horizontal D_1 structures were refolded by large-scale structures with steeply inclined axial planes, interference folding, and doubly plunging periclinal folds with steeply inclined and overturned limbs. The D_3 deformation produced superimposed small-scale folding, late post-Nondweni granitoid intrusions and prograde as well as retrograde metamorphism. Late cross-cutting faulting and the intrusion of mafic–ultramafic bodies occurred in post-D_3 times. No age constraints were provided by Mathe (1997) for the timing of the various events. In addition, which of these events relates to deformation associated with the Namaqua-Natal orogeny was not specified. Judging from the recent 3.25 Ga age determination of a so-called late post-Nondweni granitoid associated with D_3 of Mathe (1997), deformation of the Ilangwe belt in the Mesoproterozoic was either minor or has not been fully recognized. The latter is consistent with the observation of andalusite overgrown by kyanite in the Nkandla Subgroup, which has been attributed to tectonic loading associated with emplacement of the Tugela nappes in the Mesoproterozoic (Morita et al. 2007).

2.5 Discussion and Conclusions

Palaeoarchaean granitoid–greenstone terrains are exposed in the southeastern Kaapvaal Craton where they occupy an area extending from the BGB to the Natal thrust front. They include variably deformed volcano-sedimentary sequences 3.53 to $\geq$3.25 Ga in age as well as broadly time-equivalent granitoids that surround and intrude the greenstone sequences. Furthermore, there are strongly deformed tonalitic–trondhjemitic gneisses, up to ~3.5 Ga in age, some of which may constitute a temporarily stable basement to the greenstone sequences. The BGB and environs have been an area of intensive geological research. It represents one of the best-studied greenstone terrains in the world from which unique information on the geological processes that operated on the early Earth has been obtained. Despite this, the tectonic setting and environment of emplacement remains a matter of debate. This may be explained by the limited lateral extent of the Barberton granitoid–greenstone terrain on which most studies have focused. However, the granitoid–greenstone terrain stretches from Barberton all the way to the Ilangwe greenstone belt some 300 km in the south. It thus represents a large tract of Archaean basement rocks, the study of which will shed more light on the formation of continental crust in the early Archaean.

Most greenstone belts and associated granitoids south of the Barberton belt and associated greenstone remnants have been mapped in some detail, largely by students of the University of Natal (Smith 1987; Sleigh 1988; Versfeld 1988; Verbeek 1991; Mathe 1997; Dixon 2004). With some exceptions (e.g., Dwalile, Nondweni), geochronological, geochemical and tectono-metamorphic data remain rudimentary or are lacking. However, detailed mapping has laid the foundation for follow-up geochronological–geochemical investigations. In order to increase our understanding of crustal processes that led to the formation of the earliest crustal blocks in southern Africa, studies of rock successions potentially correlative with the BGB may provide new perspectives on the topic. More specifically, future work needs (1) to clarify the depositional and tectonic setting in which the greenstone successions formed, (2) to evaluate the extent and age of continental basement that may have been in existence when the volcano-sedimentary successions were deposited and (3) to investigate the processes that led to the formation of this ancient segment of continental crust.

Sparse geochronological data have shown that the greenstone belts are at least 3.25 Ga old, whereas the ages of the gneisses are largely unknown. The Onverwacht and Nondweni groups, and the Sifula Subgroup, share many similar features. They consist predominantly of km-thick piles of submarine mafic–ultramafic volcanic rocks with thin intercalations of chemical sedimentary rocks, typically banded black-and-white chert and iron formation. Silicification of lava flow tops beneath chert horizons is characteristic (Hofmann and Wilson 2007). Felsic volcanic rocks occur in most belts, mainly in the form of pyroclastic rocks or massive quartz–feldspar–phyric rocks, the latter of which may include both thick massive lava flows or domes and sub-volcanic intrusive sills. Felsic volcanism at 3.53 Ga occurred in both the Barberton and Nondweni belts but may not be used for correlation, as felsic volcanism of that age is also observed in other Palaeoarchaean terrains such as the Pilbara and Singhbhum cratons (Buick et al. 1995; Mukhopadhyay et al. 2008). Felsic volcanism and plutonism have been widespread at 3.45 Ga, as recorded in the Barberton, Dwalile and Assegaai belts. In the southern part of the area, felsic volcanism seems to have taken place some 40 Ma later, as seen from the Nondweni and Mhlatuze areas.

A characteristic suite of c. 3.3 Ga old Al-enriched komatiites with high Mg contents and inferred eruption temperatures have been recorded from the Weltevreden Formation in the BGB (Robin et al. 2011) and in the Commondale belt. This may point to widespread, geochemically distinct plume-related magmatism at that time. However, the Weltevreden komatiites have $(La/Gd)_N = 0.8$ and are therefore far less depleted than those of Commondale, which have $(La/Gd)_N = 0.12$ (Wilson et al. 2003).

Siliciclastic sedimentary rocks form a prominent part of the BGB in the Fig Tree and Moodies groups. They contain extrabasinal material derived from the erosion of granitoid–greenstone terrains, indicative of tectonism and basement uplift. Rocks of a similar type and age have been described from the Schapenburg remnant (Stevens et al. 2002), but greenstone successions farther to the south are mainly made of volcanic and chemical sedimentary rocks. This may reflect non-deposition of clastic sediments in this area, or, more likely, their preferential erosion, assuming exposure of an overall deeper crustal level.

Despite the local presence of xenocrysts in felsic rocks and Hf crustal model ages of zircons (Kröner et al. 2016), Nd and Hf whole-rock isotopic data, as well as trace elements, indicate no widespread crustal contamination in rocks of the Onverwacht Group (Blichert-Toft et al. 2004; Chavagnac 2004). This has generally been interpreted to indicate an oceanic setting, at least for the basal part of the Barberton greenstone belt. However, the Barberton Supergroup evolved over a time interval of 300 million years, and its geological history records processes of crust formation, such as emplacement of a thick volcanic pile interspersed with TTG plutonism. Volcanic rocks deposited after 3.45 Ga were emplaced on a substrate that already had been intruded by large volumes of TTG's, thus already representing differentiated crust.

In the greenstone belts to the south of the BGB, there are reports of basaltic-komatiitic magma contamination by sialic crust, as suggested by trace elements, Nd isotope systematics and the presence of xenocrystic zircons, which suggests a continental setting for these belts or at least the presence of older crust nearby (Wilson and Carlson 1989; Kröner and Tegtmeyer 1994; Riganti and Wilson 1995; Xie et al. 2012). Banded TTG gneisses infolded with the Assegaai, De Kraalen and Commondale successions may well represent remnants of this early crust. However, most geochronological data of these gneisses indicate ages of ~3.40 Ga (Xie et al. 2010 and unpublished data) and are thus younger than some of the greenstone successions themselves.

A re-evaluation of elemental and isotopic (Nd, Hf) data and zircon U-Pb ages of felsic volcanic and plutonic rocks from the Kaapvaal Craton (Agangi et al. 2019) proposed the widespread presence of felsic crust as far back as ~3.7 Ga, stretching between the Barberton and Nondweni belts. This crustal domain was undergoing repeated addition of mantle-derived juvenile magmas and multiple episodes of crustal reworking, resulting in the formation of felsic plutonic rocks and their volcanic equivalents (Agangi et al. 2019). Deformation and metamorphism at ~3.2 Ga were accompanied by the emplacement of large amounts of granitoids, which eventually gave rise to the formation of stable crust. This process of stabilization may have already been largely completed at ~3.20 Ga with the emplacement of Anhalt-suite granitoids (Anhalt trondhjemite, White Mfolozi granite, Nkandla granite) in the southernmost part of the craton. In the Barberton area, stabilization was only completed with the emplacement of 3.1 Ga granites such as the Mpuluzi and Piggs Peak Batholiths and the Vrede granite in the Piet Retief area (Anhaeusser and Robb 1983; Hunter et al. 1992; Kamo and Davis 1994; Maphalala and Kröner 1993). The evidence for crustal contamination and earlier stabilization of the southernmost part of the craton may thus point to an, on average, northward growth of the Kaapvaal Craton at the end of the Palaeoarchaean.

Acknowledgements AH acknowledges support by the University of KwaZulu-Natal and the National Research Foundation of South Africa. We thank Gary Stevens for providing the Schapenburg map. Bruce Eglington and Geoff Grantham kindly reviewed this article.

References

Agangi A, Hofmann A, Elburg MA (2019) A review of Palaeoarchaean felsic volcanism in the eastern Kaapvaal craton: Linking plutonic and volcanic records. Geosci Front. https://doi.org/10.1016/j.gsf.2017.08.003

Anhaeusser CR (1980) A geological investigation of the Archaean granite-greenstone terrane south of the Boesmanskop syenite pluton, Barberton Mountain Land. Trans Geol Soc S Afr 83:93–106

Anhaeusser CR (1983) The geology of the Schapenburg greenstone remnant and surrounding Achaean granitic terrane south of Badplaas, Eastern Transvaal. Spec Publ Geol Soc S Afr 9:31–44

Anhaeusser CR, Robb LJ (1980) Regional and detailed field and geochemical studies of Archean trondhjemitic gneisses, migmatites and greenstone xenoliths in the southern part of the Barberton Mountain Land, South Africa. Precambr Res 11(3–4):373–397

Anhaeusser CR, Robb LJ (1983) Geological and geochemical characteristics of the Heerenveen and Mpuluzi batholiths south of the Barberton greenstone belt and preliminary thoughts on their petrogenesis. Geological Society of South Africa. Special publication 9, pp 189–219

Arndt NT, Lesher CM, Barnes SB (2008) Komatiite. Cambridge University Press, Cambridge, p 467

Blichert-Toft J, Arndt NT, Gruau G (2004) Hf isotopic measurements on Barberton komatiites: effects of incomplete sample dissolution and importance for primary and secondary magmatic signatures. Chem Geol 207(3–4):261–275

Brandl G, Cloete M, Anhaeusser CR (2006) Archaean greenstone belts. In: Johnson MR, Anhaeusser CR, Thomas RJ (eds) The geology of South Africa. Geological Society of South Africa and Council of Geoscience, Pretoria, pp 9–56

Buick R, Thornett JR, McNaughton NJ, Smith JB, Barley ME, Savage M (1995) Record of emergent continental crust ~3.5 billion years ago in the Pilbara Craton of Australia. Nature 375(6532):574

Bullen WD (1990) Gold mineralization in an Archaean granite-greenstone remnant west of Melmoth, Natal: Ore genesis and implications for exploration. MSc thesis (unpublished), Rhodes University, Grahamstown, 117 pp

Chavagnac V (2004) A geochemical and Nd isotopic study of Barberton komatiites (South Africa): implication for the Archean mantle. Lithos 75(3):253–281

Dixon JGP (2004) Archaean geology of the Buffalo River gorge, KwaZulu-Natal. Doctoral dissertation (unpublished), University of Natal, Durban, 274 pp

Dlamini N, Hofmann A, Belyanin G, Xie H, Kröner A, Wilson A, Slabunov A (2017) Supracrustal gneisses in southern Swaziland: a basalt-sandstone assemblage of the upper Mozaan Group deformed in the Neoarchaean. S Afr J Geol 120:477–498. https://doi.org/10.25131/gssajg.120.4.477

Du Toit AL (1931) The geology of the country surrounding Nkandla, Natal. Explanation Sheet 109 (Nkandla), Geological Survey of South Africa, 111 pp

Elworthy T, Eglington BM, Armstrong RA, Moyes AB (2000) Rb-Sr isotope constraints on the timing of late to post-Archaean tectonometamorphism affecting the southeastern Kaapvaal Craton. J Afr Earth Sc 30:641–650

Farrow DJ, Harmer RE, Hunter DR, Eglington BM (1990) Rb-Sr and Pb-Pb dating of the Anhalt Leucotonalite, northern Natal. S Afr J Geol 93:696–701

Geological Survey of South Africa, 1988, sheet 2830 Dundee, 1:250 000 Geological Series

Gumsley A, Olsson J, Söderlund U, de Kock M, Hofmann A, Klausen M (2015) Precise U-Pb baddeleyite age dating of the Usushwana Complex, southern Africa-Implications for the Mesoarchaean magmatic and sedimentological evolution of the Pongola Supergroup, Kaapvaal Craton. Precambr Res 267:174–185

Hegner E, Kröner A, Hunt P (1994) A precise U-Pb zircon age for the Archaean Pongola Supergroup volcanics in Swaziland. J Afr Earth Sc 18(4):339-341

Heubeck C (2019) The Moodies Group—A High-Resolution Archive of Archaean Surface and Basin-Forming Processes. In: Kröner A, Hofmann A (eds) The Archaean Kaapvaal Craton, Southern Africa. Springer Geosciences, Berlin

Hicks N, Elburg M, Andersen T (2015) U-Pb and Hf isotope constraints for emplacement of the Nkandla Granite, southeastern Kaapvaal Craton, South Africa. S Afr J Geol 118(2):119–128

Hoffmann JE, Kröner A, Hegner E, Viehmann S, Xie H, Iaccheri LM, Schneider KP, Hofmann A, Wong J, Geng H, Yang J (2016) Source composition, fractional crystallization and magma mixing processes in the 3.48–3.43 Ga Tsawela tonalite suite (Ancient Gneiss Complex, Swaziland)–Implications for Palaeoarchaean geodynamics. Precambr Res 276:43–66

Hoffmann JE, Wilson AH (2016) The origin of highly radiogenic Hf isotopic composition in 3.33 Ga Commondale komatiite lavas (South Africa). Chem Geol 449:162–178

Hofmann A, Wilson AH (2007) Silicified basalts, bedded cherts and other sea-floor alteration phenomena of the 3.4 Ga Nondweni Greenstone Belt, South Africa. In: Van Kranendonk MJ, Smithies RH, Bennett J (eds) Earth's oldest rocks. Elsevier, Amsterdam, pp 571–606

Hofmann A, Harris C (2008) Stratiform alteration zones in the Barberton greenstone belt: a window into sub seafloor processes 3.5–3.3 Ga ago. Chem Geol 257:224–242

Hofmann A, Kröner A, Xie H, Hegner E, Belyanin G, Kramers J, Bolhar R, Slabunov A, Reinhardt J, Horváth P (2015) The Nhlangano gneiss dome in south-west Swaziland—a record of crustal destabilization of the eastern Kaapvaal craton in the Neoarchaean. Precambr Res 258:109–132

Hunter DR (1970) The ancient Gneiss complex in Swaziland. Trans Geol Soc S Afr 73:107–150

Hunter DR (1991) Crustal processes during Archaean evolution of the southeastern Kaapvaal province. J Afr Earth Sc 13(1):13–25

Hunter DR, Wilson AH (1988) A continuous record of Archaean evolution from 3.5 Ga to 2.6 Ga in Swaziland and northern Natal. S Afr J Geol 91:57–74

Hunter DR, Allen AR, Millin P (1983) A preliminary note on Archaean supracrustal and granitoid rocks west of Piet Retief. Trans Geol Soc S Afr 86:301–306

Hunter DR, Barker F, Millard HT Jr (1978) The geochemical nature of the Archean Ancient Gneiss Complex and Granodiorite Suite, Swaziland: a preliminary study. Precambr Res 7(2):105–127

Hunter DR, Smith RG, Sleigh DWW (1992) Geochemical studies of Archaean granitoid rocks in the southeastern Kaapvaal Province: implications for crustal development. J Afr Earth Sc 15(1):127–151

Jackson MPA (1984) Archaean structural styles in the ancient gneiss complex of Swaziland, southern Africa. In: Kröner A, Greiling R (eds) Precambrian tectonics illustrated. E. Schweizerbart'sche Verlagsbuchhandlung, Stuttgart, pp 1–18

Jacobs J, Falter M, Thomas RJ, Kunz J, Jeßberger EK (1997) $^{40}Ar/^{39}Ar$ thermochronological constraints on the structural evolution of the Mesoproterozoic Natal Metamophic Province, SE Africa. Precambr Res 86:71–92

Jacobs J, Thomas RJ (2001) A titanite fission track profile across the southeastern Archaean Kaapvaal Craton and the Mesoproterozoic Natal Metamorphic Province, South Africa: evidence for differential cryptic Meso- to Neoproterozoic tectonism. J Afr Earth Sc 33:323–333

Kamo SL, Davis DW (1994) Reassessment of Archean crustal development in the Barberton Mountain Land, South Africa, based on U-Pb dating. Tectonics 13:167–192

Kröner A, Tegtmeyer A (1994) Gneiss-greenstone relationships in the Ancient Gneiss Complex of southwestern Swaziland, southern Africa, and implications for early crustal evolution. Precambr Res 67:109–139

Kröner A, Anhaeusser CR, Hoffmann JE, Wong J, Geng H, Hegner E, Xie H, Yang J, Liu D (2016) Chronology of the oldest supracrustal sequences in the Palaeoarchaean Barberton Greenstone Belt, South Africa and Swaziland. Precambr Res 279:123–143

Kröner A, Hoffmann JE, Wong JPM, Geng H-Y, Schneider KP, Xie H, Yang J, Nhleko N (2019) Archaean Crystalline Rocks of the Eastern Kaapvaal Craton. In: Kröner A, Hofmann A (eds) The Archaean Kaapvaal Craton, Southern Africa. Springer Nature Switzerland

Linström W (1987) The geology of the Dundee area. Explanation: Sheet 2730 (1:250 000), Geological Survey of South Africa, 52 pp

Lowe DR, Byerly GR (2007) An overview of the geology of the Barberton greenstone belt and vicinity: implications for early crustal development. In: Van Kranendonk MJ, Smithies RH, Bennett J (eds) Earth's oldest rocks. Elsevier, Amsterdam, pp 481–526

Luskin C, Wilson A, Gold D, Hofmann A (2019) The Pongola Supergroup. In: Kröner A, Hofmann A (eds) The Archaean Kaapvaal Craton, Southern Africa. Springer Geosciences, Berlin

Maphalala RM, Kröner A (1993) Pb-Pb single zircon ages for the younger Archaean granitoids of Swaziland, southern Africa. Abs 16th Colloquium Afr Geol, Mbabane, Swaziland 2:201–206

Mathe HLM (1997) Tectonostratigraphy, structure and metamorphism of the Archean Ilangwe granite-greenstone belt south of Melmoth, KwaZulu-Natal. PhD. thesis (unpublished), University of Natal, Durban, 365 pp

Matthews PE, Charlesworth EG, Eglington AA, Harmer RE (1989) A minimum 3.29 Ga age for the Nondweni greenstone complex in the south-eastern Kaapvaal Craton. S Afr J Geol 92:272–278

Matthews PE (1959) The metamorphism and tectonics of the pre-Cape Formations in the post-Ntingwe thrust-belt, S.W. Zululand. Natal Trans Geol Soc S Afr 62:257–322

Matthews PE (1985) Archaean post-Pongola granites in the southeastern Kaapvaal Craton. S Afr J Sci 81:479–484

Misra S, Reinhardt J, Wilson AH (2017) Petrochemical evolution of the White Mfolozi Granite pluton: evidence for a late Palaeoarchaean A-type granite from the SE Kaapvaal Craton, South Africa. Lithos 286:480–501

Morita H, Hiroi Y, Grantham GH (2007) Prograde andalusite in kyanite-bearing siliceous rocks in the Ilangwe Greenstone belt, southeastern margin of the Kaapvaal Craton, South Africa. S Afr J Geol 110(1):55–58

Mukhopadhyay J, Beukes NJ, Armstrong RA, Zimmermann U, Ghosh G, Medda RA (2008) Dating the oldest greenstone in India: a 3.51-Ga precise U-Pb SHRIMP zircon age for dacitic lava of the southern Iron Ore Group, Singhbhum craton. J Geol 116(5): 449–461

Ovendale B (1985) An interpretation of the geology of the northern Mhlatuzi gorge. BSc (Hons) thesis (unpublished), University of Natal, Pietermaritzburg, 63 pp

Puchtel IS, Walker RJ, Anhaeusser CR, Gruau G (2009) Re–Os isotope systematics and HSE abundances of the 3.5 Ga Schapenburg komatiites, South Africa: Hydrous melting or prolonged survival of primordial heterogeneities in the mantle? Chem Geol 262:355–369

Riganti A (1996) The northern segment of the early Archaean Nondweni greenstone belt, South Africa: field relations and petrogenetic constraints. Ph.D. thesis (unpublished), University of Natal, Pietermaritzburg, 384 pp

Riganti A, Wilson AH (1995) Geochemistry of the mafic/ultramafic volcanic associations of the Nondweni greenstone belt, South Africa, and constraints on their petrogenesis. Lithos 34:235–252

Reimold WU, Matthews PE, Meyer FM (1989) An isotopic and geochemical study of pre-and post-Pongola Archaean granitoids in the southeastern Kaapvaal craton. University of the Witwatersrand, unpublished report

Reinhardt J, Elburg MA, Andersen T (2015) Zircon U-Pb age data and Hf isotopic signature of Kaapvaal basement granitoids from the Archaean White Mfolozi Inlier, northern KwaZulu-Natal. S Afr J Geol 118(4):473–488

Robb LJ, Brandl G, Anhaeusser CR, Poujol M (2006) Archaean granitoid intrusions. In: Johnson MR, Anhaeusser CR, Thomas RJ (eds) The geology of South Africa. Geological Society of South Africa and Council of Geoscience, Pretoria, pp 57–94

Robin C, Arndt NT, Chauvel C, Byerly G, Sobolev A, Wilson A (2011) Barberton komatiites produced by deep critical melting with high melt retention. J Petrol 53:2191–2229

Saha L, Hofmann A, Xie H, Hegner E, Wilson A, Wan Y, Liu D, Kröner A (2010) Zircon ages and metamorphic evolution of the Archaean Assegaai—De Kraalen granitoid-greenstone terrain, southeastern Kaapvaal Craton. Am J Sci 310:1384–1420

Saha L, Hofmann A, Jele N, Harris C (2012) Palaeoarchaean sea-floor alteration and metamorphism of pillow basalts from the Nondweni Greenstone Belt, Southeastern Kaapvaal Craton, South Africa. S Afr J Geol 115:259–282

Sleigh DWW (1988) The Geology of the Archaean terrane in the Piet Retief District, Transvaal, M.Sc. thesis (unpublished), University of Natal, Pietermaritzburg, 208 pp

Smith RG (1987) Geochemistry and structure of the Archaean granitoid-supercrustal terrane, southeastern Transvaal and northern Natal. Ph.D. thesis (unpublished), University of Natal, Pietermaritzburg, 305 pp

Stevens G, Droop TR, Armstrong RA, Anhaeusser CR (2002) Amphibolite-facies metamorphism in the Schapenburg schist belt: a record of the mid-crustal response to ~3.23 Ga terrane accretion in the Barberton greenstone belt. S Afr J Geol 105:271–284

Tainton S (1977) An interpretation of the geology of the Mhlatuzi gorge, Nkandhla-Mtonjaneni district, Zululand. B.Sc. Hons. thesis (unpublished), University of Natal, Durban, 136 pp

Talbot CJ, Hunter DR, Allen AR (1987) Deformation of the Assegaai supracrustals and adjoining granitoids, Transvaal, South-Africa. J Struct Geol 9:1–12

Tarduno JA, Cottrell RD, Watkeys MK, Hofmann A, Doubrovine PV, Mamajek EE, Liu D, Sibeck DG, Neukirch LP, Usui Y (2010) Geodynamo, solar wind, and magnetopause 3.4 to 3.45 billion years ago. Science 327(5970):1238–1240

Tegtmeyer A (1989) Geochronologie und Geochemie im Präkambrium des südlichen Afrika. Unpublished doctoral dissertation, University of Mainz, Germany, 270 p

Thomas RJ (1989) A tale of two tectonic terranes. S Afr J Geol 92: 306–321

Thomas RJ, Gold DJC, Verbeek JA, Walraven F (1997) Geology of the Archaean Nzimane Inlier, Zululand. S Afr J Geol 100(2):123–136

van den Kerkhof AM, Kronz A, Simon K, Riganti A, Scherer T (2004) Origin and evolution of Archean quartzites from the Nondweni greenstone belt (South Africa): inferences from a multidisciplinary study. S Afr J Geol 107:559–576

van Schijndel V, Stevens G, Zeh A, Frei D, Lana C (2017) Zircon geochronology and Hf isotopes of the Dwalile Supracrustal Suite, ancient gneiss complex, Swaziland: Insights into the diversity of Palaeoarchaean source rocks, depositional and metamorphic ages. Precambr Res 295:48–66

Verbeek JA (1991) The geology and geochemistry of the Archaean granitoid-supracrustal terrane, southeast of Piet Retief, Transvaal. Ph.D. thesis (unpublished), University of Natal, Pietermaritzburg, 261 pp

Versfeld JA (1988) The geology of the Nondweni greenstone belt, Natal. Ph.D. thesis (unpublished), University of Natal, Pietermaritzburg, 298 pp

Viljoen MJ, Viljoen RP (1969) An introduction to the geology of the Barberton granite-greenstone terrain. Spec Publ Geol Soc S Afr 2:9–28

Wilson AC (compiler) (1982) 1:250,000 Geological Map of Swaziland. Geological Survey & Mines Department, Mbabane, Swaziland

Wilson AC, Jackson MPA (1988) Mantled gneiss domes in southern Swaziland and the concept of „stable" Pongola cratonic cover. S Afr J Geol 91:404–414

Wilson AH (2003) A new class of silica enriched, highly depleted komatiites in the southern Kaapvaal Craton, South Africa. Precambr Res 127:125–141

Wilson AH, Carlson RW (1989) A Sm-Nd and Pb-Pb isotopic study of Archaean greenstone belts in the southern Kaapvaal Craton, South Africa. Earth Planet Sci Lett 96:89–105

Wilson AH, Versfeld JA (1994) The early Archaean Nondweni greenstone belt, southern Kaapvaal Craton, South Africa, Part I: stratigraphy, sedimentology, mineralization and depositional environment. Precambr Res 67:243–276

Wilson AH, Riganti A (1998) A fractionated komatiitic basalt lava lake sequence in the Nondweni greenstone belt. In: International Volcanological Conference, IAVCEI, Cape Town, p 70

Wilson AH, Versfeld JA, Hunter DR (1989) Emplacement, crystallization and alteration of spinifex-textured komatiitic basalt flows in the Archaean Nondweni greenstone belt, southern Kaapvaal Craton, South Africa. Contrib Miner Petrol 101:301–317

Wilson AH, Shirey SB, Carlson RW (2003) Archaean ultra-depleted komatiites formed by hydrous melting of cratonic mantle. Nature 423:858–861

Xie H, Hofmann A, Wan YS, Saha L, Kröner A, Liu DY (2010) 3.2 Ga metamorphic event in the Paulpietersburg area, South Africa, and tectonic implications. In: Excursion guidebook and abstract volume, Field workshop, Early crustal evolution of the southeastern Kaapvaal craton in South Africa and Swaziland, 17–23 May 2010, Paulpietersburg/Badplaas, South Africa and Ezulwini, Swaziland, p. 98

Xie H, Hofmann A, Hegner E, Wilson A, Wan Y, Liu D (2012) Zircon SHRIMP dating confirms a Palaeo- to Mesoarchaean supracrustal terrain in the Kaapvaal craton, southern Africa. Gondwana Res 21:818–828

Palaeo- Meso- and Neoarchaean Granite-Greenstone Basement Geology and Related Rocks of the Central and Western Kaapvaal Craton, South Africa

3

Carl R. Anhaeusser

Abstract

The Kaapvaal Craton, one of the oldest well-preserved crustal fragments on Earth, enables an understanding of the development of the early Earth. The craton comprises an older eastern component (Witwatersrand Block) ranging in age from ca. 3660 to 2700 Ma and a ca. 3174 to 2782 Ma western domain (Kimberley Block). This contribution deals with the geological nature of the central and western sectors of the Kaapvaal Craton where Archaean basement granite-greenstone rocks occur as inliers surrounded by younger cover formations. Three main areas discussed include the Johannesburg and Makoppa domes and the Kraaipan terrane. Also described are the Marydale terrane, the Rand Anticline, the linear Colesberg magnetic anomaly, and the Modipe Gabbro and Gaborone Granite complexes. Geophysical and geochronological findings suggest that the Kimberley Block is a younger terrane accreted onto the older eastern or Witwatersrand Block along or near the Colesberg Lineament—this despite meagre physical evidence of a collisional zone. Geophysical, petrological and geochemical techniques have helped to understand the crustal structure and depth extent of the roots of the craton. Findings suggest the keel extends to depths of 250–300 km, important for the stability field of diamonds.

Keywords

Central and Western Kaapvaal Craton • Archaean greenstone belts • Archaean granitoid rocks Geochronology

C. R. Anhaeusser (✉)
Economic Geology Research Institute, School of Geosciences, University of the Witwatersrand, Private Bag 3, Wits 2050, Johannesburg, South Africa
e-mail: carl.anhaeusser@wits.ac.za

3.1 Introduction

The name 'Kaapvaal Craton' was first employed by Pretorius (1965) to describe a fragment of the old crust of the northeastern sector of South Africa—the name 'Kaapvaal' being derived from the Kaap Plateau in the south-western portion of the crustal fragment, the Vaal River in the central part, and the De Kaap Valley, adjacent to the Barberton Mountain Land, in the northeastern portion. The Kaapvaal Craton was later defined by Pretorius (1974) as being characterized by a well-preserved assemblage of Archaean granite-greenstone formations and Proterozoic sedimentary-volcanic strata, which experienced its last period of medium- to high-grade regional metamorphism approximately 3000 Ma ago. The craton was estimated by Brandl and de Wit (1997) to occupy an area of about 12,00,000 km^2, with an Archaean history covering a time span of more than 1000 Ma, extending from about 3660 to 2600 Ma.

The boundaries of the Kaapvaal Craton (Fig. 3.1) have been defined using geochronological, structural and geophysical data (Corner et al. 1990; Corner 1991). The eastern sector is bounded by the Lebombo monocline, and in the south and southwest by the Mesoproterozoic Namaqua-Natal Province consisting of igneous and metamorphic rocks formed or metamorphosed during the Namaqua Orogeny at ca. 1200–1000 Ma (Cornell et al. 2006, 2019; de Beer and Meyer 1984). The northern and northwestern limits of the Kaapvaal Craton are less well defined. In general, the boundary in the north is regarded as being where the broad zone of high-grade gneisses of the Southern Marginal Zone extends to the Central Zone of the Limpopo Belt (Kramers et al. 2006). In central Botswana, the northern limits of the Kaapvaal Craton are unclear due partly to poor exposure and the paucity of geophysical information (Key and Ayres 2000). These authors and Cornell et al. (2006) indicated that the western edge of the Kaapvaal Craton abutted against the Kheis Province (ca. 1750–1900 Ma Kheis orogenic belt, see

A. Kröner and A. Hofmann (eds.), *The Archaean Geology of the Kaapvaal Craton, Southern Africa*, Regional Geology Reviews, https://doi.org/10.1007/978-3-319-78652-0_3

Fig. 3.15b), which trends due north from South Africa into central Botswana.

It was noted by Brandl and de Wit (1997) that the Kaapvaal Craton may have extended beyond the present-day limits outlined above, as it has been suggested that parts of it may occur within the continent of Antarctica (de Wit et al. 1988; Groenewald et al. 1991). Similarly, seismic constraints on the evolution of the western margin of the Kaapvaal Craton suggested to Tinker et al. (2002) that the craton once formed part of a larger Mesoarchaean landmass. Other possible 'southern' cratons that may have been contiguous with the Kaapvaal Craton include the Pilbara, Zimbabwe, Dharwar, São Francisco, Amazon, and Congo cratons. However, reviews of their basic geology and

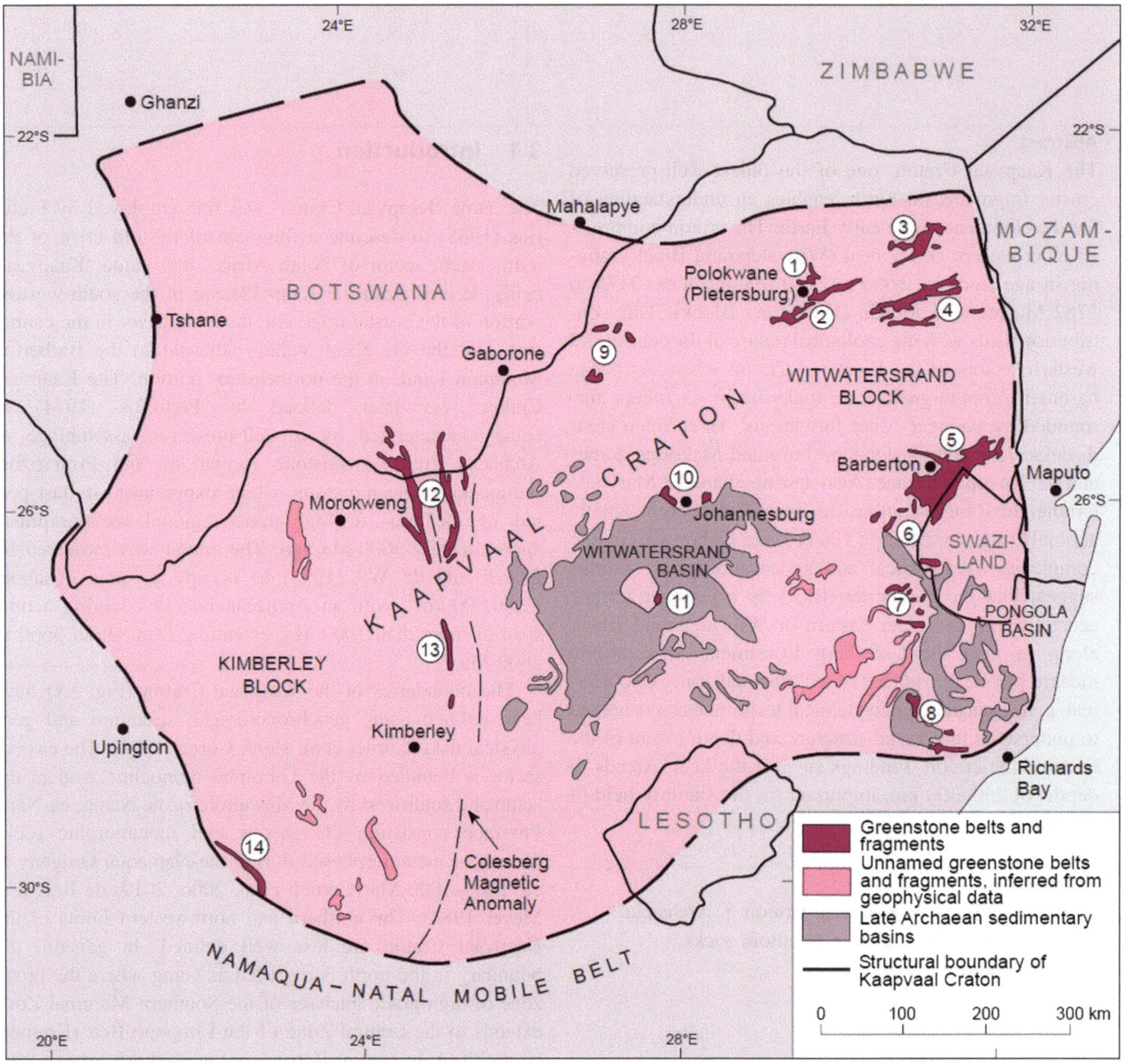

Fig. 3.1 Distribution of greenstone belts on the Kaapvaal Craton. *1* Rhenosterkoppies; *2* Pietersburg; *3* Giyani (Sutherland); *4* Murchison; *5* Barberton; *6* Weergevonden and Schapenburg; *7* Assegaai, De Kraalen, and Commondale; *8* Nondweni-Ilangwe; *9* Makoppa Dome; *10* Johannesburg Dome; *11* Vredefort Dome; *12* Kraaipan; *13* Amalia-Schweizer-Reneke; and *14* Marydale. Map modified after Brandl et al. (2006)

inferred evolution in syn-Witwatersrand basin times (ca. 3100–2800 Ma) show no obvious support for any such supercontinental amalgamations according to Eriksson et al. (2009, 2011). Despite these views chronostratigraphic, geochronological and palaeomagnetic evidence suggests that a strong link existed between the early phases of development of the Kaapvaal Craton and the Pilbara region in Western Australia (Trendall 1968; Button 1979; Cheney et al. 1988; Thorpe et al. 1992; Cheney 1996; Wingate 1998; Zegers et al. 1998; Nelson et al. 1999; de Kock et al. 2009).

The Kaapvaal Craton was divided into 12 Archaean crustal subdomains by de Wit et al. (1992). These authors conceded that it was not yet possible to distinguish whether these subdomains or terrains are allochthonous with respect to each other or if they were once part of a larger coherent geological province that has subsequently been tectonically disrupted. Anhaeusser (2006a) questioned the veracity of some of the proposed subdomains, believing they were not all adequately justified on the evidence provided. It was agreed that some of the suggested terrain boundaries were of a clearly defined tectonic nature (e.g. major shear displacements such as the Palala Shear Zone and the Thabazimbi-Murchison fault zone in the northern part of the craton), but that many others were less convincing. Eriksson et al. (2009, 2011) agreed that the 12 subdomains, as advocated by de Wit et al. (1992), were excessive and indicated that U–Pb and Lu–Hf isotope data from zircons rather suggest that the Kaapvaal Craton is composed of about four distinct terranes each of which underwent different crustal evolutions and were successively accreted between ca. 3230 Ma and about 2650 Ma. It was also suggested by Anhaeusser (2006a) that there was evidence that some terrain boundaries on the craton could be regarded as suture zones with accompanying ophiolite-like ultramafic rocks, playing a role similar to those reported by Burke et al. (1977) in major Proterozoic and Phanerozoic orogenic belts.

Much of the Archaean basement of the Kaapvaal Craton is covered by Neoarchaean, Proterozoic and Phanerozoic volcanic and sedimentary successions that blanket the high central southern African plateau (Table 3.1). In addition, vast areas are obscured by the intrusion of the Bushveld Complex, the largest and economically important layered igneous complex in the world, which has been estimated to occupy an area of ca. 65,000 km^2 in the north-central parts of the craton (Cawthorn et al. 2006). Exposures of the Archaean basement are best encountered east of the Great Escarpment of South Africa and areas south of the Limpopo Belt. These areas are described elsewhere in this volume.

Geophysical studies, summarized by Brandl and de Wit (1997), have been undertaken to determine the nature of the lower crust beneath the Kaapvaal Craton and adjacent high-grade metamorphic belts. These include heat flow studies (Jones 1987, 1988, 1993, 2017; Nyblade et al. 1990) that indicate a general heat flow pattern of increasing heat flow away from the centres of the Archaean cratons, suggesting a fundamental difference in lithospheric thermal structure between the Archaean cratons and the Proterozoic and early Palaeozoic mobile belts which surround them. Heat flow data reported from numerous sites in the Archaean Kaapvaal and Zimbabwe cratons and the Limpopo Belt (all part of the more extensive Kalahari Craton) indicate a mean heat flow of 47 ± 2 mW m^{-2} (Jones 1988). This value indicates that the cratonic lithosphere is considerably cooler and thicker than that below an equilibrium ocean (i.e. 84 to 251 mW/m^2; Sclater et al. 1980). Heat flow and heat production determinations in Archaean granite domes from northwest of the Witwatersrand Basin average 33 ± 2 mW m^{-2} and 2.3 ± 0.8 μW m^{-3}, respectively. The lower heat flow in the domes is thought to reflect a smaller radiogenic contribution from the crystalline crust below the domes (Jones 1988).

The deep structure of the Kaapvaal Craton has been studied by means of deep geoelectrical soundings as well as the analysis of gravity and magnetic data (de Beer and Stettler 1988). These authors found that in the low-grade granitoid terranes the upper 6–10 km of the Archaean crust is characterized by high resistivity rocks (40,000–100,000 Ω m) overlying a moderately resistive zone (~5000 Ω m) that extends to about the crust-mantle boundary. Below this is a conductive zone (1–100 Ω m) followed by a resistive zone in the upper mantle.

A generalized seismic model of the Kaapvaal Craton was proposed by Durrheim and Green (1992) who recorded the following features: (1) supracrustal strata 0–10 km thick; (2) upper crystalline basement with *P*-wave velocities of 6.0–6.2 km s^{-1}; (3) the boundary between upper and lower crust at a depth of 14–18 km; (4) a lower crust with a relatively uniform seismic velocity in the range 6.4–6.7 km s^{-1}; and (5) the crust/mantle transition over 1–3 km with the Moho at a depth of about 35 km (by contrast the depth to the Moho is about 42 km in the Namaqua-Natal belt (Green and Durrheim 1990; de Wit and Hart 1993; Nguuri et al. 2001). Durrheim and Green (1992) also found the lower crust to be electrically conductive and seismically attenuating with a Poisson's ratio of about 0.28, a value regarded by Zandt and Ammon (1995) as being consistent with tectonically stable shield or cratonic areas. In the south-western section of the craton the Moho depth was found by Stankiewicz et al. (2002) to vary between 37 and 40 km, except for one station that recorded a depth of 43 km. Farther north along the western block of the craton (into Botswana) they noted that the depth increases up to 43 km.

The tectospheric structure beneath southern Africa was determined by James et al. (2001) from P- and S-wave delay times from broadband data of the southern Africa seismic experiment. Three-dimensional images of velocity

Table 3.1 Stratigraphic framework, major volcanism, associated dyke events and age relationships of the geology of the Kaapvaal Craton

Stratigraphic unit	Volcanics	Associated dyke events	Age (Ma)
Karoo Supergroup	Mafic and felsic volcanics	Rooi Rand Dyke Swarm Okavango Dyke Swarm Olifants River Dyke Swarm Karoo Dolerite Suite	~ 180
Pilanesberg Complex	Various volcanics	Alkaline dyke swarm	~1300
Waterberg Group	None	Post-Waterberg dykes and sills NW-trending dykes	~1400–1700
Soutpansberg Group	Mafic volcanics	Dykes and sills	~1800
Phalaborwa Complex Bushveld Complex Rooiberg Group	Mafic and felsic volcanics	Bushveld Complex mafic sills; E–W-trending mafic dykes	~ 2050
Transvaal Supergroup	Mafic volcanics	None	
Ventersdorp Supergroup Wolkberg Group Godwan Group	Mafic volcanics	N-E-trending dykes	~2700
Usushwana Complex, Amsterdam Formation	Felsic volcanics	NW-trending dykes	~2800
Witwatersrand and Pongola Supergroups, Dominion Group	Mafic and felsic volcanics	NW-trending dykes	~3000
Granite-greenstone belts	Mafic and felsic volcanics	No subvolcanic dykes identified	~3500

Modified after Uken and Watkeys (1997)

perturbations in the mantle revealed high-velocity mantle roots that appear to extend to depths of at least 250 km, and locally to depths of 300 km, beneath both the Kaapvaal and Zimbabwe cratons as well as beneath the Limpopo Belt (Fouch et al. 2004). Lower seismic velocities were, however, found to affect cratonic mantle beneath and to the west of the Bushveld Complex.

It is the purpose of this chapter to deal specifically with the Archaean basement in the central and western parts of the Kaapvaal Craton where limited exposures exist in uplifted domes and ridges that occur as inliers on the central plateau region. Some of the key domical exposures include those of the Johannesburg Dome in the central craton area (Anhaeusser 1973), the Vredefort Dome, approximately 150 km south of the Johannesburg Dome (Gibson 2019, this volume), and the Makoppa Dome in the northwest, straddling the border with eastern Botswana (Anhaeusser and Poujol 2004; Robb et al. 2006). In the western sector of the craton Archaean basement rocks form the Kraaipan granite-greenstone terrane, which is exposed intermittently from southern Botswana towards Kimberley and beyond into the south-western part of the craton (Anhaeusser and Walraven 1999; Brandl et al. 2006). Similar rocks reappear sporadically in the region immediately north of the Namaqua-Natal metamorphic belt (Fig. 3.1). In this region, where discontinuous outcrops extend for about 100 km in the Marydale-Prieska region (Fig. 3.16), it is referred to as the Marydale greenstone belt with limited granitic rocks also present (Vajner 1974; McCourt et al. 2000; Brandl et al. 2006; Robb et al. 2006; Cornell et al. 2019).

3.2 Basement Granite Domes Surrounding the Witwatersrand Basin

A number of granitic basement domes surround the Witwatersrand Basin. With the exception of a few domical structures partially exposed in the northeast, north and west-southwest as well as the Vredefort Dome in the centre of the Witwatersrand Basin (Brock and Pretorius 1964; Minter et al. 1986; McCarthy 2006—see Figs. 2.2, 3.3), these domes are mainly subsurface features known only from exploration drilling and geophysics. These include the Wesselsbron Hub west of the Welkom Goldfield, and several domical granitoid masses reported along the southeastern margin of the Witwatersrand Basin by Brock and Pretorius (1964). Being elevated subsurface features, little is known of the geochemistry and age of these rocks, but they are thought to have influenced the deposition of sediments and associated gold mineralization into adjacent depressed areas or basins within the confines of the greater Witwatersrand Basin.

The exposed Archaean granite-greenstone basement, occurring as domical inliers in the central part of the Kaapvaal Craton, include the Johannesburg Dome, the Devon Dome approximately 70 km to the ESE, and the Vredefort Dome situated about 120 km southwest of Johannesburg. In addition, an elevated feature referred to as the Rand Anticline can be traced southwest of the Johannesburg Dome and extends around the northern arc of the Witwatersrand Basin (Fig. 3.2). This structure played an important role in the depositional, deformational and

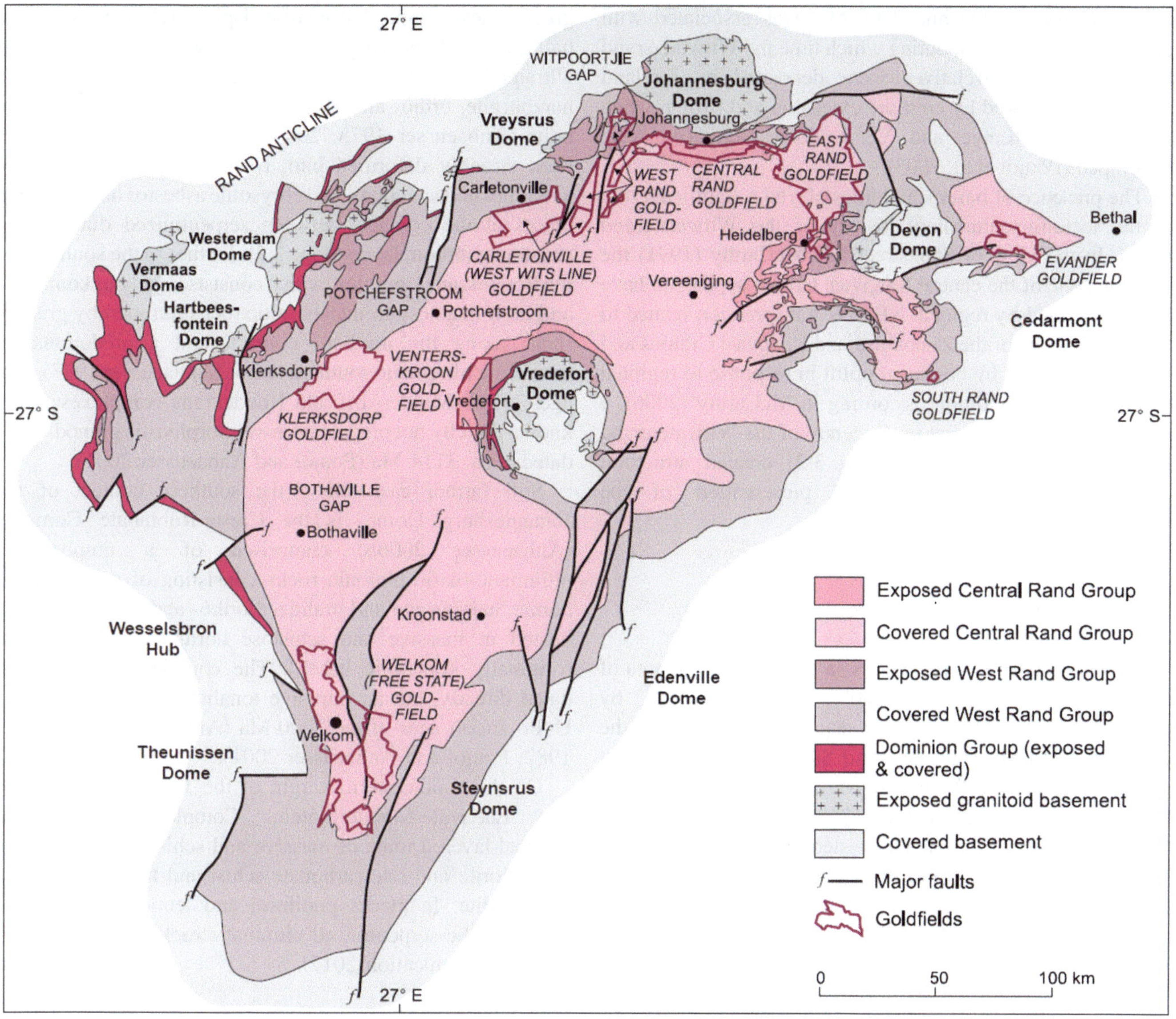

Fig. 3.2 Geological map of the Witwatersrand Basin, with younger cover sequences removed and showing the basement granitoid domes that occur around the margins of the basin. Modified after Brock and Pretorius (1964), McCarthy (2006)

exploration history of the Witwatersrand Basin (Pretorius 1979).

In the central Kaapvaal Craton the various granitoid rocks, including the youngest ca. 3000 Ma granodioritic and pegmatitic phases formed the basement onto which were deposited Mesoarchaean to Palaeoproterozoic (ca. 3074–2224 Ma) sedimentary and volcanic rocks of the Dominion Group and the Witwatersrand, Ventersdorp and Transvaal Supergroups.

The Witwatersrand Supergroup (Witwatersrand Basin, Fig. 3.2) is host to the world's largest goldfield (see Frimmel 2019). Despite more than 130 years of mining there still remains much controversy as to the source of the gold. Views range from placer-type gold having been derived from ancient granite-greenstone terranes on the Kaapvaal Craton (Viljoen et al. 1970), to suggestions by Robb et al. (1991, 1992) and Robb and Meyer (1987, 1990, 1995) that the Witwatersrand gold, uraninite and pyrite were derived from erosion of the footwall sequences and from remote provenance areas (greenstones and hydrothermally altered granites like those found along the Rand Anticline). Other suggestions included a hydrothermal origin involving pyritic auriferous exhalites forming along the tectonically active northwestern margin of the basin by shallow-marine hydrothermal discharge. This is thought to have been generated and accompanied very closely by andesitic-basaltic volcanism of the Dominion Group and lower Witwatersrand Supergroup (Hutchinson and Viljoen 1988). Still another suggestion is that the evolution of the craton, between 3000 and 2700 Ma, was associated with continent–arc collision, during which time the Witwatersrand Basin and its correlatives were deposited as foreland sequences, followed by episodic extension and rifting, when the Gaborone–Kanye and Ventersdorp sequences were developed (Poujol et al. 2003).

The presence of basement domes played a significant role in the tectono-sedimentary history of the Witwatersrand Basin. In the view of Stanistreet and McCarthy (1991) the tectonic style of the central Kaapvaal Craton appears to have been controlled by regional NE–SW transpression related to the convergence of the Zimbabwe and Kaapvaal Cratons and was characterized by basement uplift in response to regional compression. In addition, according to McCarthy (2006), a unique astrobleme impact in the centre of the Witwatersrand Basin (the Vredefort event—Fig. 3.2) created structural conditions which favoured the preservation of the depository.

3.2.1 Johannesburg Dome

The Johannesburg Dome (Figs. 3.3a and 3.4) has an area of approximately 700 km^2 and is surrounded by outward-dipping younger cover sequences that include the ca. 3074–2714 Ma Witwatersrand Supergroup in the south (Robb et al. 1991; Robb and Meyer 1995), and the ca. 2714 Ma Ventersdorp and ca. 2585–2224 Ma Transvaal Supergroup rocks rimming the western, northern and eastern sectors of the dome (Armstrong et al. 1991; Eriksson et al. 2006; Humbert et al. 2019, this volume).

The Johannesburg Dome displays a variety of granitoid magmatic events all of which postdate mafic and ultramafic greenstone remnants that are best displayed along the southern half of the dome (Fig. 3.4). In the west the Muldersdrif Ultramafic Complex comprises layered serpentinized dunite, harzburgite, ortho- and clinopyroxenite and minor gabbroic units (Anhaeusser 1978, 2006b). These assemblages have been strongly deformed into a series of doubly plunging anticlines and synclines, and chrysotile asbestos has formed in places at the contacts between serpentinized dunites and resistant gabbroic layers (Fig. 3.5). Farther to the southeast is the Roodekrans Complex, which consists largely of komatiitic and high-magnesium metabasaltic rocks intruded by granitic rocks along the northern contact, and a predominantly komatiitic ultramafic southern sector (Anhaeusser 1977). The ages of the Muldersdrif and Roodekrans complexes is not known directly but predate intrusive porphyritic granodiorites dated at ca. 3114 Ma (Poujol and Anhaeusser 2001).

Still farther east, along the southern contact of the Johannesburg Dome, is the Cresta-Robindale Complex (Anhaeusser 2006b) comprised of a number of ultramafic-mafic plutonic rocks consisting of serpentinized dunite, harzburgite and uralitized ortho- and clinopyroxenite hosted in massive and schistose hornblende amphibolite (originally komatiitic basalt). The complex has not been dated directly, but an intrusive tonalitic gneiss has yielded U–Pb zircon ages of ca. 3200 Ma (Anhaeusser and Burger 1982; Poujol and Anhaeusser 2001).

On the southeastern margin of the Johannesburg Dome the Edenvale-Modderfontein Complex consists of cyclical-layered units of massive and schistose serpentinite, talc-chlorite and talc-carbonate schist and tremolite-bearing amphibolite. In places podiform and lenticular chromitite occurs in the serpentinized ultramafic rocks (G. Guice, personal communication 2017).

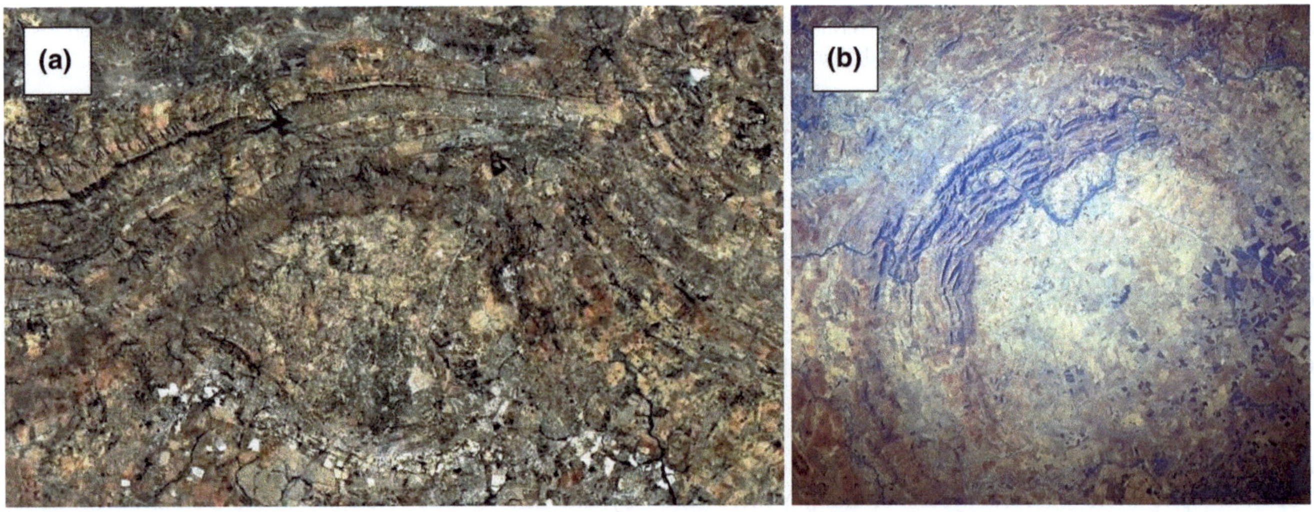

Fig. 3.3 Google Earth satellite images of the Johannesburg Dome **a** and the Vredefort Dome **b**. The domes comprise Archaean granitoids and greenstone remnants surrounded by Mesoarchaean to Eoproterozoic volcanic and sedimentary rocks. The Johanesburg Dome is approximately 44 km E–W and 34 km N–S and the core of the Vredefort Dome is 40 km in diameter

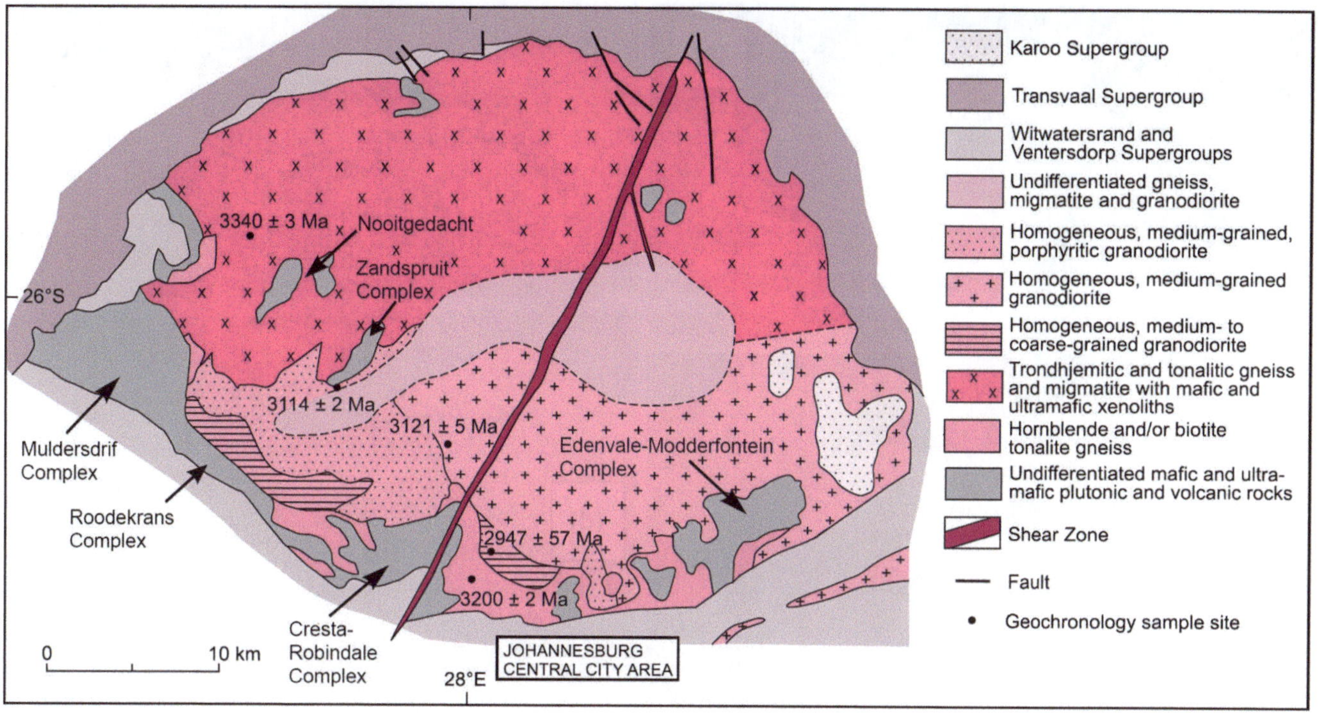

Fig. 3.4 Simplified map of the Johannesburg Dome showing the geology and the relative ages of the various granitoid rocks and the location of Archaean greenstone remnants. Modified after Anhaeusser (1973), Poujol and Anhaeusser (2001), Robb et al. (2006)

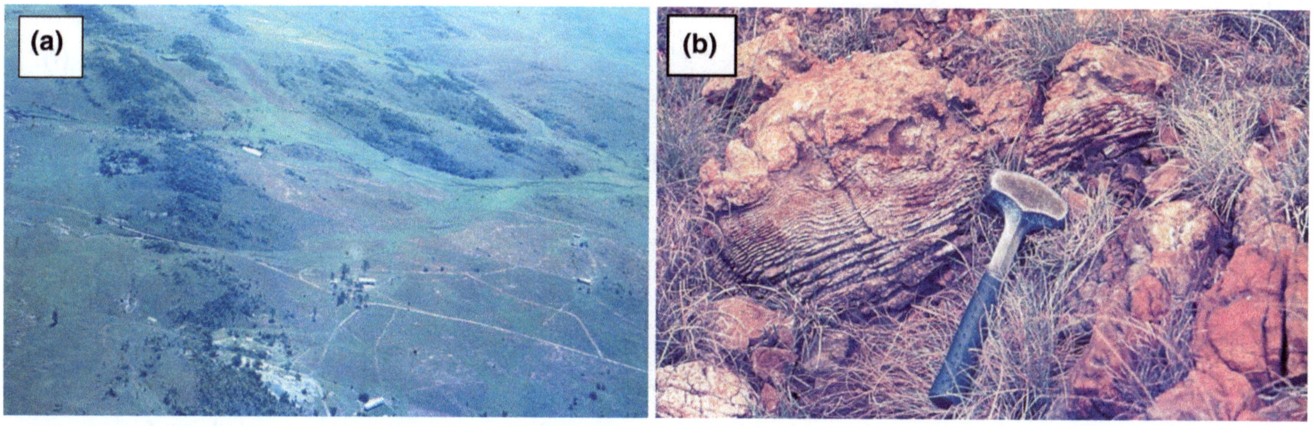

Fig. 3.5 **a** Aerial view showing the E–W layering of serpentiniized dunite, ortho- and clinopyroxenite and gabbro of the Muldersdrif Ultramafic Complex. Numerous small mine workings and prospect for chrysotile asbestos are also evident; **b** Asbestos layers formed at the contact between serpentinized dunite (left of hammer) and gabbro. After Anhaeusser (1978)

In the west-central part of the Johannesburg Dome is the Zandspruit Complex (Fig. 3.4), an Archaean greenstone remnant entirely surrounded by intrusive porphyritic granodiorites dated at ca. 3114 Ma (Poujol and Anhaeusser 2001). The greenstones constitute a layered ultramafic complex consisting of at least four alternating cyclical units of serpentinized and metamorphosed dunite, harzburgite and clinopyroxenite, the latter altered to tremolite amphibolite. Local assimilation, alteration and granitization of the greenstones has resulted in the development of hybridized dioritic rocks and a zone of K-metasomatized ultramafic rocks (Anhaeusser 1992, 2015; Fig. 3.6).

Several magmatic events involving granitoid rocks were identified on the Johannesburg Dome by Anhaeusser (1973). These included migmatitic gneisses and a TTG suite of tonalite, trondhjemite and granodiorite. Figure 3.4 shows the distribution of these rocks on the dome and their ages. Migmatites and biotite trondhjemite gneisses (Lanseria Gneiss)

Fig. 3.6 Outcrops of various rock types occurring on the Johannesburg Dome. **a** Layered harzburgites in the Zandspruit greenstone remnant; **b** Tectonized serpentinized dunite from the Muldersdrif Ultramafic Complex; **c** Blotchy patches of K-feldspar enveloped in a matrix comprising mainly biotite, together with quartz and carbonate; **d** Large, in places zoned, K-feldspar megacrysts (microcline and orthoclase) in a fine-grained matrix comprising mainly biotite, calcite, quartz and plagioclase. Examples **c** and **d** are from a zone of K-metasomatized ultramafic rocks of the Zandspruit greenstone remnant (Anhaeusser 1992, 2015); **e** Hybridized dioritic rock formed as a result of assimilation of ultramafic rocks by porphyritic granodiorite at Zandspruit (Anhaeusser 2015); **f** Partially altered mafic xenoliths and clots of amphibole in hybridized diorite near the contact with the Zandspruit greenstone remnant (after Anhaeusser 2015)

occupy most of the northern half of the dome. Dating of these rocks in the northwestern sector of the dome yielded a U–Pb zircon age of 3340 ± 3 Ma (Poujol and Anhaeusser 2001). In places spectacular migmatite exposures can be seen in some river platforms, particularly in the NW sector of the Dome, the most impressive of which is the Nooitgedacht river exposure (Figs. 3.4, 3.7 and 3.8a—Anhaeusser 1999). Lamprophyre dykes also intrude the granitic rocks at this locality (Prevec et al. 2004), suggesting a high depth of melting and low degrees of partial melting, which yields magmas rich in alkalis, particularly potassium (Rock 1991).

Hornblende and/or biotite tonalite gneisses, intruded into ultramafic rocks north of the Johannesburg central city area, occur along the southern margin of the dome (Figs. 3.4, 3.8b, c)

Fig. 3.7 Portion of the Nooitgedacht river platform in the northwestern part of the Johanneburg Dome (see Fig. 3.4), illustrating the complex, long-lived evolution (over 340 Ma) of the Archaean basement rocks (Anhaeusser 1999). Displayed is an amphibolite greenstone xenolith (top centre, extending towards front centre) which is intruded by ca. 3340 Ma trondhjemitic gneisses (light-grey foliated rock with white leuco-trondhjemite veins). A hornblende lamprophyre dyke (cutting obliquely from top right to lower left) intrudes the gneisses and greenstone xenolith, and is in turn cross-cut by a homogeneous granodiorite dyke of ca. 3121 Ma age. In the left foreground is a ca. 3000 Ma pegmatite, which cross-cuts trondhjemite gneiss (geochronology after Poujol and Anhaeusser 2001)

and yielded a U–Pb zircon age of 3170 ± 34 Ma (Anhaeusser and Burger 1982). More recently these tonalitic gneisses (Linden Gneiss) provided a single zircon emplacement age of 3199.9 ± 2 Ma (Poujol and Anhaeusser 2001).

Between the migmatitic and trondhjemitic gneiss terrane in the north and the tonalitic gneisses in the south are various potassic granitoid rocks that occupy most of the southern half of the dome (Fig. 3.4). In the central and eastern sectors the dominant rock types are homogeneous medium-grained, grey granodiorites. Locally some porphyritic granodioritic phases are present together with pegmatite dykes and veins. Granitic tors are developed in the central part of the Dome (Fig. 3.8d–f). The granodioritic rocks (Bryanston Granodiorite) have been dated at 3121 ± 5 Ma, and the porphyritic grey granodiorites to the west (Honeydew Granodiorite) gave an age of 3114 ± 2 Ma (Poujol and Anhaeusser 2001). The youngest granodiorite, a medium- to coarse-grained, pinkish variety (Victory Park Granodiorite) occurs on the southern edge of the dome and yielded a U–Pb zircon age of 2947 ± 57 Ma (Poujol and Anhaeusser 2001). Earlier Rb–Sr, Pb–Pb and Sm–Nd whole-rock, Rb–Sr biotite and Pb–Pb zircon evaporation age dating by Barton et al. (1999) suggested that the granodiorite, granite, and leucosomes from migmatites were emplaced ca. 3090 Ma ago, were genetically related, and were derived primarily from a source between ca. 3300 and 3500 Ma old.

3.2.2 Rand Anticline

Southwest of the Johannesburg Dome numerous Archaean basement 'windows' or inliers occur along an arched feature referred to as the Rand Anticline. Along this structure are numerous basement domes, which occur mainly along the northern flank of the Witwatersrand Basin. These domes (Vreysrus, Westerdam, Hartbeesfontein, Vermaas—Fig. 3.2) extend westwards from Randfontein to Ventersdorp, Coligny, Ottosdal and Schweizer-Reneke, a distance of over 260 km. Granitic rocks exposed vary in composition and texture but generally consist of coarse-grained to porphyritic monzonites and granites (sensu stricto), or gneisses and migmatites of tonalitic composition (Robb and Meyer 1987). Many of the granitic rocks exhibit an often intense hydrothermal alteration manifest as sericitization, chloritization, carbonate replacement, carbonate or quartz-carbonate vein-type alteration, pyritization and leucoxene development (Klemd and Hallbauer 1987; Robb and Meyer 1987; Robb et al. 1990).

U–Pb zircon and monazite dating by Robb et al. (1992) of some hydrothermally altered granitic domes, including the Westerdam Dome south of Ventersdorp, gave an age of 3086 ± 3 Ma (^{207}Pb/^{206}Pb zircon ages ranging between 3101.4 ± 2.7 and 3054 ± 3.6 Ma). To the west the Vermaas Dome, between Ottosdal and Coligny, gave ages of 3031 +11/−10 and 2927 ± 23/−6 Ma (^{207}Pb/^{206}Pb zircon ages ranging from 3029.6 ± 11 to 3004 ± 3.5 Ma). Also in this area peraluminous granodiorite-monzonite of the Hartbeesfontein Dome (Fig. 3.2) gave a zircon age of 3174 +9/−7 Ma (^{207}Pb/^{206}Pb zircon ages ranging from 3174.2 ± 3.3 to 3054.6 ± 3 Ma). Still farther west peraluminous monzonites of the Schweizer-Reneke Dome (Fig. 3.11) provided a monazite age of 2880 ± 2 Ma (^{207}Pb/^{206}Pb monzonite ages ranging from 2879.5 ± 3.2 to 2867.7 ± 2.2 Ma).

Again the ages obtained on the western half of the Kaapvaal Craton are consistently younger than granitoid rocks on the eastern side of the craton and fall into the age range of ca. 3100 to 2868 Ma.

Fig. 3.8 Exposures of granitoid rocks on the Johannesburg Dome. **a** Banded trondhjemitic gneiss ca. 3340 Ma intruded by ca. 3121 Ma homogeneous granodiorite and ca. 3000 Ma pegmatite dykes in the NW sector of the Dome; **b** foliated ca. 3200 Ma tonalitic gneiss (hornblende, biotite, albite and quartz) from the southern part of the Dome; **c** banded tonalitic gneiss, southwestern part of the Dome; **d** homogeneous porphyritic granodiorite (ca. 3121–3114 Ma) typical of the south-central part of the Dome (Fig. 3.4); **e** Homogeneous, massive, grey granodiorite (ca. 3121 Ma) with nebulitic or 'ghost-like' remnants of granitized, earlier mafic rock; **f** Homogeneous granodioritic tor or castle koppie boulders from the undifferentiated zone in the central part of the Dome (after Anhaeusser 1973). Granitic ages after Poujol and Anhaeusser 2001)

3.2.3 Vredefort Dome

Another dome of significance on the Kaapvaal Craton is the Vredefort Dome, which is discussed more fully by Gibson (2019, this volume). The Vredefort Dome is also the largest known impact structure on Earth. Located in the central part of the Kaapvaal Craton it is also a dominant feature in the centre of the Witwatersrand Basin (Fig. 3.2). Estimates of the original diameter of the impact structure vary from 250 to 300 km, and the currently accepted age of the impact event is 2023 ± 4 Ma (Gibson and Reimold 2008). The approximately 40 km diameter Archaean basement core (Fig. 3.3b) represents the central uplift of the impact structure and consists of a sequence of Archaean amphibolite- and granulite-facies granitoid gneisses with subsidiary mafic, ultramafic, pelitic and banded iron-formation xenoliths (Stepto 1979, 1990; Lana et al. 2004). These rocks range in age between ca. 3500 and 3084 Ma (Gibson and Reimold 2008), whereas the TTG granitoids in the core of the dome only formed between 3120 and 3100 Ma (Armstrong et al. 2006).

3.3 Basement Complexes in the North-Central Sector of the Kaapvaal Craton

Three Mesoarchaean complexes that include the Makoppa Dome, the Modipe Gabbro Complex and the Gaborone Granite Complex or Suite straddle the northwestern border of South Africa and the southeastern border of Botswana. Each of these complexes are described briefly in the following sections.

3.3.1 Makoppa Dome

In the far north-central part of the Kaapvaal Craton an ovoid-shaped domal structure (Makoppa Dome) occurs as an inlier exposing Archaean basement granitic rocks and sporadic greenstone remnants (Figs. 3.1 and 3.9). The Dome, with dimensions of approximately 120 km east-west and 85 km north-south and an area of about 7600 km^2, is largely rimmed in the southwest, south and southeast by outward-dipping Neoarchaean and Palaeoproterozoic (ca. 2800 to 2050 Ma) volcanic and sedimentary rocks of the Kanye Formation (Grobler 1996) and the Ventersdorp and Transvaal Supergroups (Anhaeusser and Poujol 2004). In the north and northeast subhorizontal Palaeoproterozoic (ca. 2060 to 1700 Ma) redbed rudites and arenites of the Waterberg Group overly the basement granitoids (Jansen et al. 1974).

The Makoppa Dome constitutes one of the least well-exposed Archaean granite-greenstone terranes in southern Africa. Nevertheless, Anhaeusser and Poujol (2004) identified several greenstone remnants consisting predominantly of metamorphosed tholeiitic basalt (now mainly amphibolite, some garnet-bearing), occurring mainly northeast of the Crocodile River between Makoppa and Sentrum, and sporadic exposures in the Marico River along the Botswana border and southwest of Thabazimbi on the southern edge of the dome (Fig. 3.9). Some chromitite deposits associated with mafic and ultramafic rocks have also been mapped and explored in certain greenstone remnants (R. P. Viljoen personal communication, 2018).

Despite the limited exposure three varieties of granitic rocks were recognized on the Makoppa Dome. These include: (1) grey, medium-to-coarse-grained migmatites showing well-developed stromatic (banded) structure, which is in the form of alternate leucosome and melanosome layers, and exhibiting a subhorizontal foliation (Vaalpenskraal trondhjemitic gneisses). Two samples of this rock type from northeast of the Crocodile River yielded U–Pb zircon ages of 3013 ± 11 Ma and 3034 ± 64 Ma respectively. Overgrowth rims on the zircons gave ages of 2629 ± 130 and 2842 ± 35 Ma respectively (Anhaeusser and Poujol 2004), suggesting that these rocks were influenced by the emplacement of the younger Makoppa or Rooibokvlei granodiorites discussed below; (2) also along the Crocodile River, near the village of Makoppa, a second granitic variety consisting of homogeneous, reddish, coarse-grained porphyritic granodioritic to monzogranitic rocks (Makoppa monzogranite) is intruded locally by late-stage, reddish, fine-grained aplitic dykes and coarse-grained pegmatites. This granitic variety yielded a U–Pb zircon age of 2886 +3/−2 Ma (Anhaeusser and Poujol 2004); (3) exposures of the third granitoid variety occur on the eastern side of the dome midway between Thabazimbi and Sentrum. These rocks consist of homogeneous, grey-to-pink, medium-grained granodiorites and monzogranites (Rooibokvlei granodiorite) and provided U–Pb zircon ages of 2777 ± 35 and 2797 ± 2 Ma respectively (Anhaeusser and Poujol 2004).

The various granitoid rocks of the Makoppa Dome, with a spread of ages from ca. 3030 to 2780 Ma, have counterparts described later that occur approximately 200 km to the southwest in the Kraaipan-Amalia area of the west-central part of the Kaapvaal Craton (Fig. 3.1).

3.3.2 Modipe Gabbro Complex

The Modipe Gabbro Complex is mainly situated in southeast Botswana, approximately 25 km east of Gaborone (Jones 1963; Fig. 3.9), but offshoots and stocks are found about 35 km to the ESE in the Marico District of the North West Province of South Africa (Jansen et al. 1974). The rocks of the complex in South Africa consist mainly of gabbro with associated norite, gabbro-norite, pyroxenite, wehrlite, dunite, serpentinite, anorthosite and bands of magnetite. The age of the body is believed to exceed the 2785–2830 Ma age range recorded by Sibiya (1988) and Moore et al. (1993) for the Gaborone Granite Complex/Suite that intrudes the Modipe gabbro (Jones 1963).

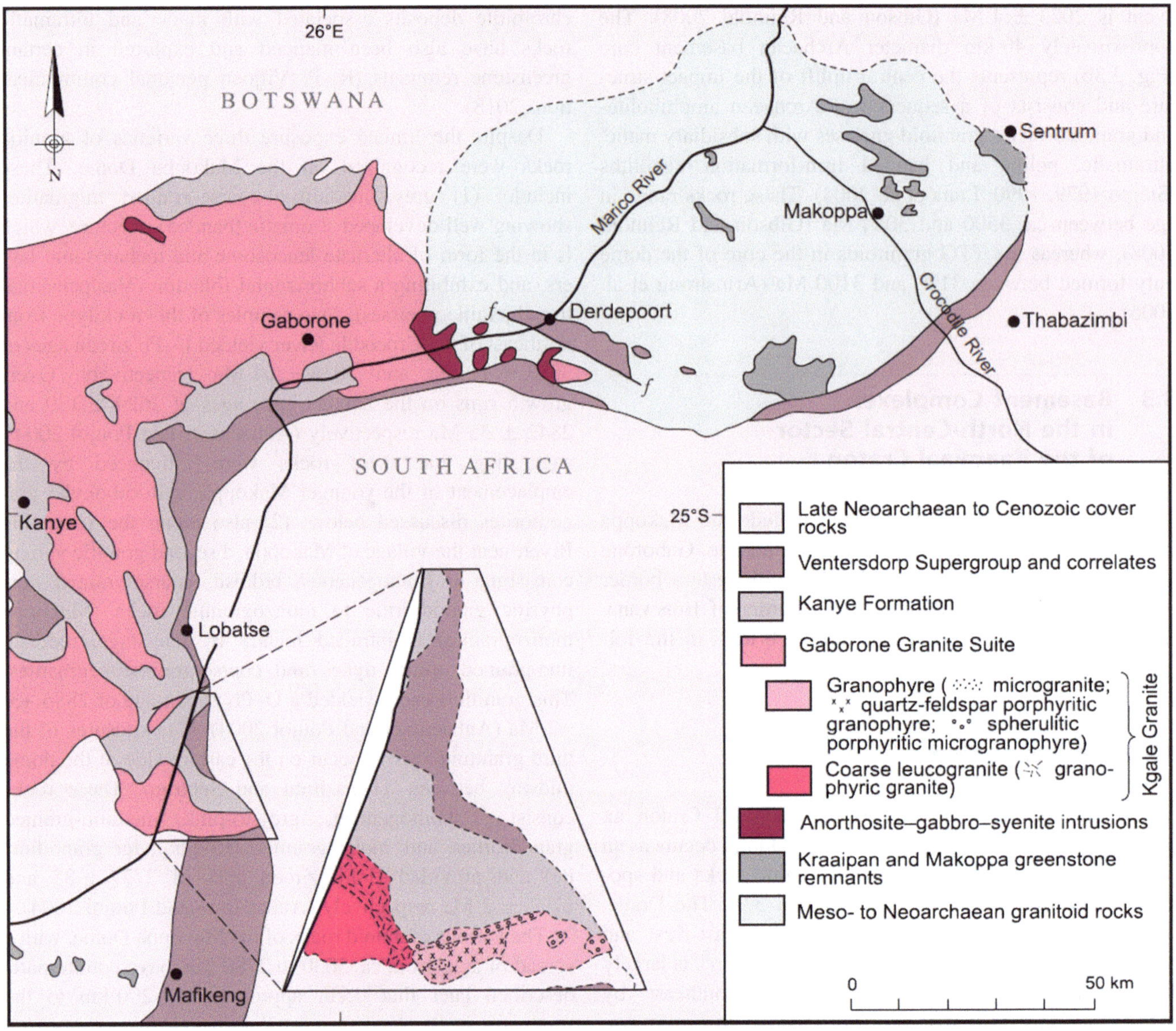

Fig. 3.9 Simplified geological map (modified after Robb et al. 2006) showing the ovoid-shaped Makoppa Dome (Anhaeusser and Poujol 2004), and the distribution of the Late-Archaean Gaborone Granite Suite and younger cover rocks exposed along the Botswana-South Africa border. The inset map shows details of the Gaborone Granite Suite exposed in the Botsalano Ring Complex north of Mafikeng (Grobler 1996)

3.3.3 Gaborone Granitic Complex/Suite

In the northwestern sector of the Kaapvaal Craton, mainly developed in southeast Botswana and straddling the border with South Africa in the Limpopo and North West Provinces, is the Gaborone-Kanye igneous terrane. Within this terrane is the Gaborone Granite Complex or Suite comprised of rapakivi feldspar-bearing granite, leucogranite, granophyric microgranite, and massif-type anorthosite exposed over an area of at least 6000 km^2 (Key and Wright 1982; Sibiya 1988; Moore et al. 1993). The granitic complex is spatially associated with pyroclastic and massive flow-banded rhyolites of the Kanye Formation. Together they represent a major magmatic episode characterized by an unusually old rapakivi granite-anorthosite-rhyolite suite that is atypical of the region and, indeed, of Archaean terranes worldwide.

The Gaborone Granite Suite consists of a central core of A-type rapakivi granite (Thamaga Granite), surrounded by leucogranite (Kgale Granite), microgranite and granophyric porphyry (Ntlhantlhe Microgranite) and minor gabbro-anorthosite (Sibiya 1988; Moore et al. 1993). The various components of the Gaborone Granite Suite and rocks of the Kanye Formation have yielded Neoarchaean ages ranging between ca. 2785 and 2782 Ma (Grobler and Walraven 1993; Moore et al. 1993).

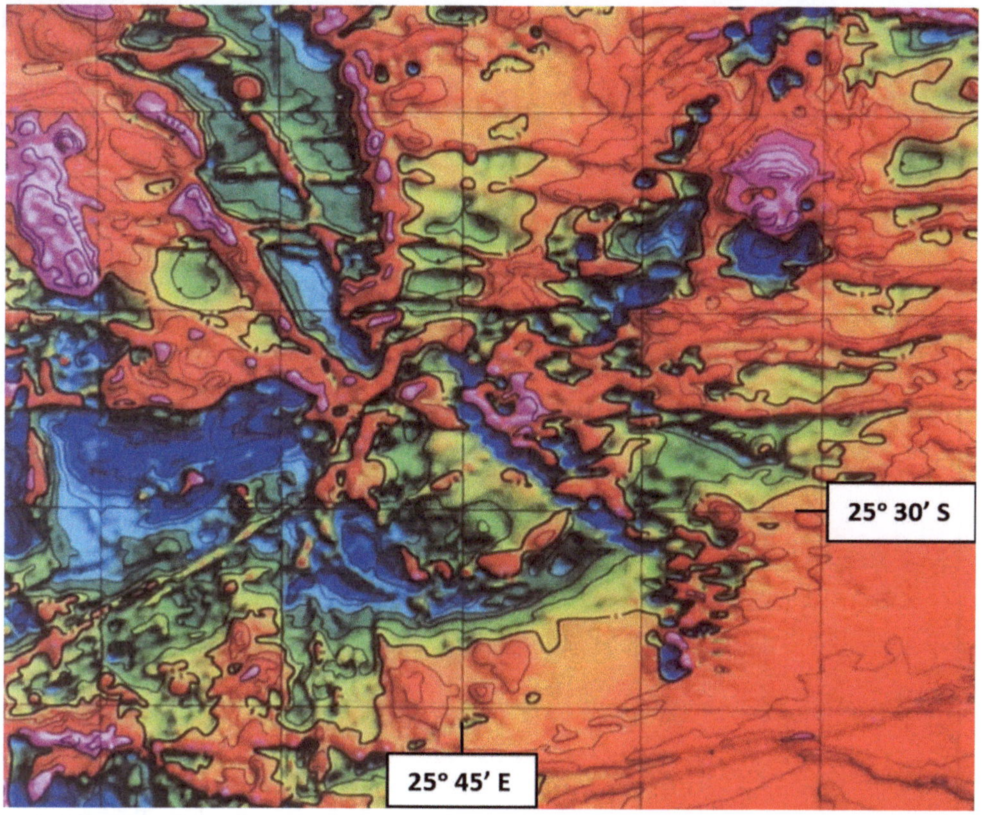

Fig. 3.10 Aeromagnetic image of the region shown in the inset map in Fig. 3.9. The western and southern part of the approximately 30 km diameter Botsalano Ring Complex is clearly evident, but the northeastern segment is obscured by Transvaal Supergroup sediments (after Grobler 1996)

In the south the Gaborone-Kanye rock assemblage between Mafikeng and Lobatse occurs as part of a large anorogenic multiple granitic ring complex (Botsalano Ring Complex; Figs. 3.9 and 3.10). According to Grobler (1996) the ring complex forms the eroded root of a caldera structure and at its present level has a diameter of about 30 km. Remnants of several other possible ring structures and intrusions were recognized by Key (1983), Sibiya (1988), Aldis et al. (1989), and Grobler (1996) in what was termed the Gaborone-Kanye anorogenic igneous terrane, which underlies an area of ca. 35,000 km^2. These structures, evident on aeromagnetic images, are regarded as typical of ring complexes and are further identified by the nature of the associated lithologies and their geochemical signatures (Grobler 1996).

3.4 Kraaipan Granite-Greenstone Terrane and Impact Structures

The Kraaipan granite-greenstone terrane in the Kimberley Block west of the Colesberg magnetic lineament (see below) extends from north of the Botswana border (Ramotoroko et al. 2016) southwards to Kraaipan, Amalia, Kimberley and Koffiefontein in the central part of the Kaapvaal Craton (du Toit 1906, 1908; van Eeden et al. 1963; Anhaeusser 1991; Bosch 1993; Keyser and du Plessis 1993). Rocks of the Kraaipan Group occur in the north as three narrow, discontinuous, NNW-trending successions (Fig. 3.11) spaced at 30–40 km intervals with intervening granitic areas covered by Cenozoic and older sedimentary and volcanic rocks. Gravity and magnetic intensity data show generally steep-sided greenstones with a maximum depth of 4.7 and 4.4 km for the younger granite plutons (Ramotoroko et al. 2016).

In the east is the Madibe Belt with the Kraaipan and Stella belts to the west providing the most continuous stratigraphic successions of the Gold Ridge, Ferndale and Khunwana Formations of the Kraaipan Group. These rocks, together with those in the Amalia greenstone belt, are gold-bearing in places (Ward and Wilson 1998; Hirner 2001; Kiefer 2004; Hammond and Morishita 2009) and consist mostly of tholeiitic metabasalt with interlayered banded iron formation, jaspilite, serpentinized dunite, phyllitic, chloritic and calcareous schists and clastic sediments (Zimmermann and Anhaeusser 1991; Zimmermann 1994; Brandl et al. 2006). A 12 km long by 1 km-wide sub-vertical Pt–Pd-bearing sill, comprising variably layered magnetite-bearing gabbros, leucogabbros and anorthosites (Stella Layered Intrusion—Fig. 3.11; Andrews 2003; Maier et al. 2003), is hosted by greenstones of the Stella Belt and intruded by meta-diabase dykes and granite veins and has yielded a zircon age of 3033.5 ± 0.3 Ma (Schmitz et al. 2004).

In the south the Amalia greenstone belt (Kiefer 2004) is flanked on the east by the Schweizer-Reneke Dome

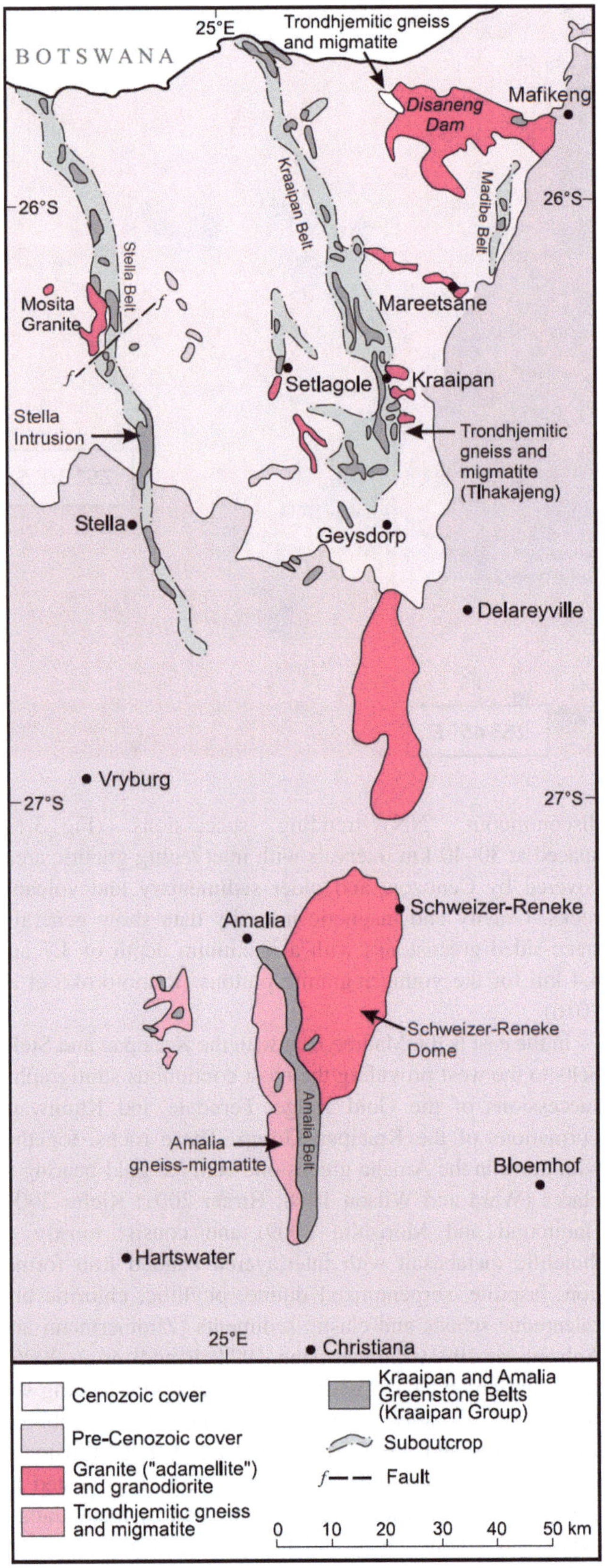

Fig. 3.11 Simplified geological map of the Archaean Kraaipan-Amalia terrane on the west-central side of the Kaapvaal Craton. The main areas of granite-greenstone exposure occur southwest of Schweizer-Reneke and in the north between Geysdorp and the Botswana border. Modified after Anhaeusser and Walraven (1999), Brandl et al. (2006), Robb et al. (2006)

(Fig. 3.11) that has yielded a U–Pb ID-TIMS monazite age of 2880 ± 2 Ma (Robb et al. 1992). Tonalitic and trondhjemitic gneisses and migmatites occur on the western side of the approximately north–south-trending Amalia greenstone belt (Fig. 3.12c) and yielded zircon ages of 3008 ± 4 to 2939 ± 9.6 Ma (Poujol et al. 2002).

The age of the Kraaipan Group of rocks is not well constrained. Anhaeusser and Walraven (1999) reported a whole-rock Pb–Pb age of ca. 3410 Ma, and Hirner (2001) recorded SHRIMP zircon ages as old as ca. 3407 and 3344 Ma from phyllites in the Madibe Belt. Ages in excess of ca. 3180 Ma were interpreted as being derived from xenocrystic zircons. Robb (1991) attempted to date silicic felsic schists from north and south of Kraaipan Station in the central belt, using the U–Pb zircon technique, and reported a mean age of ca. 3075 Ma and a xenocrystic age of ca. 3157 Ma.

Age data for the Amalia greenstone belt gave a Neoarchaean age of 2750.1 ± 4.6 Ma (Poujol et al. 2005). However, this greenstone belt is surrounded and intruded by older granitoid rocks, suggesting that this age is incorrect. The samples dated were collected near accretionary lapilli thought to have been associated with Kraaipan banded iron formations (Jones and Anhaeusser 1993). The most likely interpretation is that the accretionary lapilli are part of the Ventersdorp Supergroup rocks which fall within the age range 2782 ± 5 and 2714 ± 16 Ma (Wingate 1998). The best estimate for emplacement of the Kraaipan volcanic rocks in the Amalia belt is now thought to range between ca. 3098 and 3082 Ma.

Kraaipan granitic rocks occupy extensive areas south of Botswana and west and southwest of Mafikeng, but exposures are widely spaced and covered by Kalahari sediments. The Bouguer anomaly gravity map of the region shows gravity highs associated with the greenstone belts and major gravity lows between the belts, which are dominated by low-density granitic rocks (Stettler et al. 1990, 1997). Aeromagnetic maps clearly show this arrangement with, in places, the low-magnetic intensity areas occupied by granitic domes flanked by higher magnetic intensity greenstones, many of which contain banded iron formations and jaspilite (Figs. 3.12a and 3.13).

A threefold subdivision of the granites, similar to those of the Makoppa Dome discussed earlier, was reported by Anhaeusser and Walraven (1999), and yielded single-grain Pb-evaporation ages on zircons ranging from ca. 3162 to 2846 Ma. The episodic emplacement of the Kraaipan granitic rocks was confirmed later by single zircon U–Pb dating carried out by Poujol et al. (2002). Foliated tonalitic and trondhjemitic leucogneisses and migmatites (Disaneng, Tlhakajeng and Amalia gneisses; Figs. 3.11 and 3.12b, c), containing xenoliths of Kraaipan amphibolite and banded iron formation, gave ages of ca. 3008 Ma, whereas massive, fine-to medium-grained, grey-to-pink, homogeneous

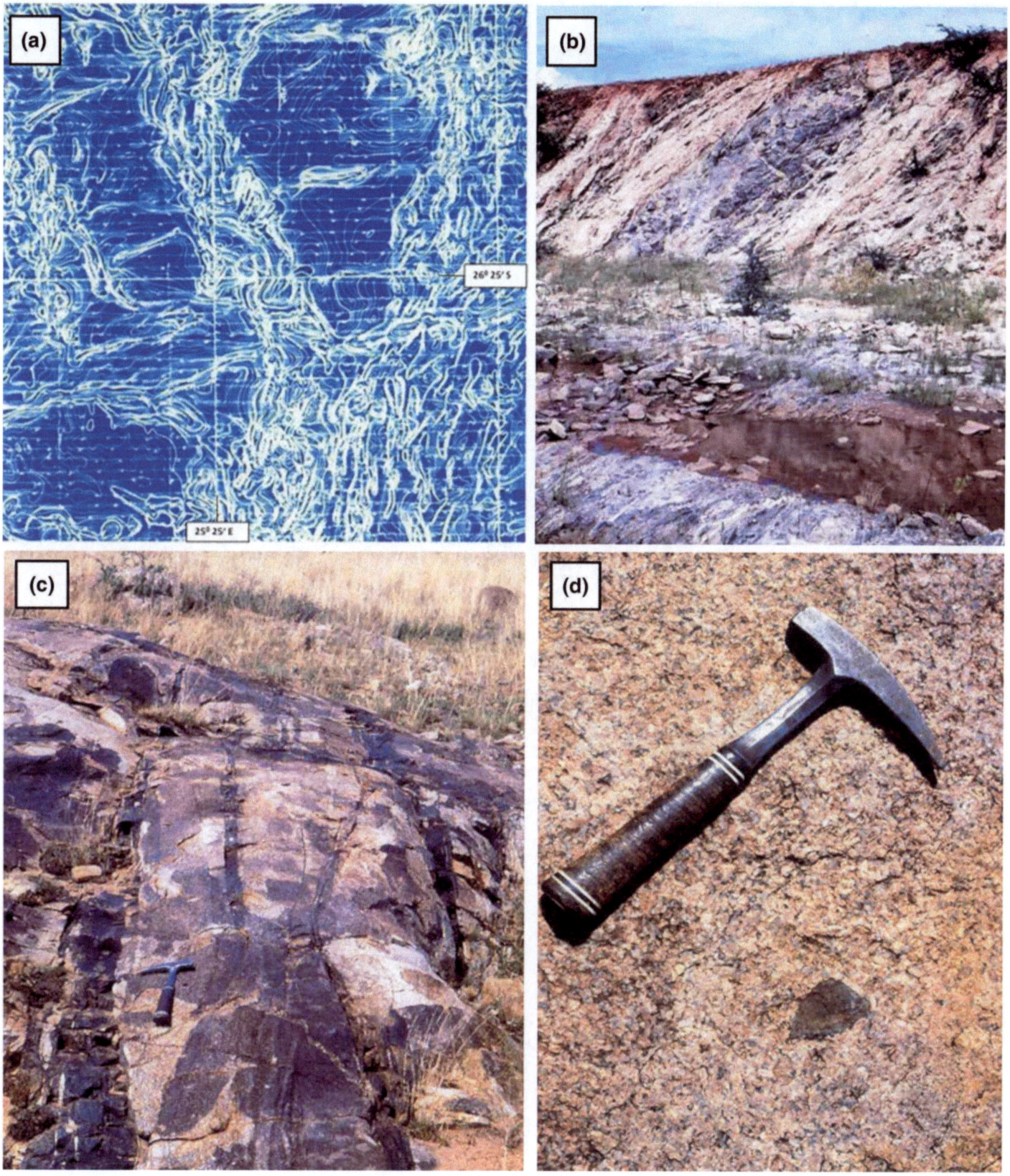

Fig. 3.12 Kraaipan granitoid rocks. **a** Aeromagnetic map showing the low-magnetic intensity areas occupied by granitic domes flanked by higher magnetic intensity greenstone belts containing banded iron formations. This pattern also conforms with Bouguer anomaly gravity data which shows gravity highs associated with the greenstone belts and gravity lows dominated by low-density granitic rocks between the belts (Stettler et al. 1990, 1997); **b** Trondhjemitic gneisses, migmatites and Kraaipan amphibolite xenoliths in spillway cutting at the Disaneng Dam 30 km west of Mafikeng. The gneisses gave ages of 3008 Ma and the cross-cutting granodiorites and pegmatites ages of ca. 2915–2879 Ma (Anhaeusser et al. 2002); **c** amphibolite xenoliths in ca. 3008 Ma Amalia trondhjemitic gneisses west of the Amalia greenstone belt (Fig. 3.11); **d** coarse-grained, homogeneous, pink granite with mafic xenolith. The Mosita granite is the youngest intrusive in the Kraaipan terrane dated at ca. 2791 Ma (Poujol et al. 2002)

Fig. 3.13 Kraaipan lithologies. **a** Undeformed tholeiitic pillow basalts in the Molopo River at the Botswana-South African border (Makgobistad), 60 km west of Mafikeng; **b** deformed and metamorphosed amphibolitic pillow basalts near Setlagole; **c** Jaspilite iron formation from the western Kraaipan belt (Logageng Dam); **d** banded ferruginous cherty iron formation, Amalia greenstone belt; **e** Intensely folded banded magnetite quartzite approximately 13 km NNE of Geysdorp (Fig. 3.11); **f** Folded banded iron formation in Amalia greenstone belt (Goudplaats gold mine)

granitoids (Schweizer-Reneke, Kraaipan and Mareetsane granodiorites and quartz monzonites) provided ages ranging from ca. 2915 to 2879 Ma. To the south of the Kraaipan greenstone belt a foliated pink granitoid from Koppiespan, 18 km WSW of Delareyville, gave a U–Pb zircon age of 3061 ± 9 Ma (Cornell et al. 2011). The youngest age of 2791 ± 8 Ma was derived from the Mosita Granite (Fig. 3.12d) consisting of massive, coarse-grained, homogeneous, pink, granite-monzonite (Poujol et al. 2002).

SHRIMP U–Pb zircon dating by Mapeo et al. (2004) of granitoids in southeast Botswana yielded ages of 2939 ± 9 Ma for a biotite granite gneiss, 2943 ± 9 Ma for a pink granite and 2782 ± 6 Ma for a red granite. An age of ca. 2780 Ma corresponds to emplacement of the Gaborone Igneous Complex of southeast Botswana.

In the Kimberley area (Fig. 3.1) Archaean basement gneisses, migmatites and pegmatites are exposed in underground workings of diamond mines in these areas (Drennan 1988; Lundell 2015). U–Pb ion microprobe dating of zircons from the gneisses gave older cores of ca. 3250 Ma with rims of ca. 2940 Ma (Drennan et al. 1990). Lundell (2015) reported a U–Pb zircon age of 3019 ± 47 Ma for granitic rocks considered to belong to the TTG group. Greenstone schists and amphibolites were also found in the Kimberley mines by Fritiofsson (2014) who obtained an age of 2908 ± 18 Ma for a schist sample, an age considered to coincide with a postulated subduction and collision event amalgamating the Witwatersrand and Kimberley crustal blocks.

Also described from the Kimberley mines together with the migmatites and gneisses are garnet-bearing amphibolite, mantle peridotite, eclogite and eclogite-hosted diamonds. These rocks have provided Re–Os, Sm–Nd and Rb–Sr ages ranging from ca. 3246 to 2800 Ma (Schmitz et al. 2004). These authors claimed furthermore that the Re–Os age distributions of Kimberley Block cratonic mantle xenoliths, at around 2800 to 2900 Ma, supported their view that coupling of continental crust and lithospheric mantle through arc magmatism, subduction and collisional thickening resulted in the formation of the tectosphere beneath the western Kaapvaal Craton at ca. 2900 Ma.

In summary, the nature and ages of the granitic rocks in the central and western half of the Kaapvaal Craton are shown in Table 3.2. These are consistently younger than the Palaeoarchaean granitoids found on the eastern half of the Kaapvaal Craton and appear to constitute a period of Mesoarchaean to Neoarchaean cratonic magmatism marked by the development of a major, crescent-shaped juvenile arc that was accreted onto the northern and western sectors of the evolving Kaapvaal Craton (Poujol et al. 2003). Cratonization, both in the eastern and western parts of the craton, was also accomplished by the emplacement of major potassic granitoid batholiths at around 3100–3000 Ma (Barberton-Swaziland, Johannesburg, Makoppa, and the Kraaipan granitic terranes), which resulted in thickening and stabilization of the continental crust.

The Kraaipan granite-greenstone terrane south of Botswana hosts two meteorite impact localities. One is located near Setlagole, southwest of Mafikeng (Fig. 3.11), and regarded by Anhaeusser et al. (2010) to be Mesoarchaean in age (ca. 2790–2710 Ma). The impact structure, which to date remains unconfirmed, was initially identified following the discovery of megabreccia exposures and the aeromagnetic recognition of a ring structure approximately 25–30 km in diameter (Fig. 3.14a).

One hundred and thirty kilometres to the west of Setlagole is the Morokweng impact structure (Andreoli et al. 1995; 2007; Figs. 3.1 and 3.15a, b). The rim diameter and morphology of this impact crater has long been an issue of debate (Wela et al. 2016; Wela 2017). Corner et al. (1997) suggested that geophysical evidence indicated an original size of this impact structure of at least 70 km but that several annular anomalies outside of the study region could be interpreted as indicating a diameter of 300–340 km. Andreoli et al. (2007) recognized multiple rings centred on the impact structure and concluded, by integrating available geophysical, remote sensing and borehole data, that the diameter of the crater may have been approximately 240 km. In contrast Henkel et al. (2002) and Reimold et al. (2002) estimated, from gravity and magnetic data, that the crater diameter was 65–70 km. If the 240 km diameter proves to be correct Morokweng would represent the third largest known terrestrial impact structure.

The age of the Morokweng structure has been reported to be 145 ± 2 Ma (Hart et al. 1997; Koeberl et al. 1997; Reimold and Koeberl 2014), it having impacted with surrounding Neoarchaean orthopyroxene-bearing felsic granophyric and micropegmatitic granitoids that yielded a SHRIMP zircon $^{207}Pb/^{206}Pb$ age of 2878 ± 8 Ma. Interpreted as the emplacement age of the granitoid rocks (Reimold et al. 2002) these rocks, together with granitic exposures dated by Cornell et al. (2011) in the range ca. 2882–2854 Ma (at Kgokgole and Ditshiping in the Ganyesa Dome area which surrounds the Morokweng crater), are the most westerly ages recorded on the Kaapvaal Craton and reflect the widespread occurrence of Neoarchaean rocks in the far western sector of the Kimberley Block. A zircon age of ca. 3300 Ma has been recorded in banded pyroxene gneisses from borehole core drilled into the central uplift of the Morokweng crater (M. Andreoli personal communication, 2018).

3.5 South-western Kaapvaal Craton

The Archaean basement south of Kimberley is covered by sedimentary rocks of the Karoo Supergroup and only reappears in the far south-western sector of the Kaapvaal Craton

Table 3.2 Emplacement ages for granitoid intrusions in the central and western sectors of the Kaapvaal Craton (after Robb et al. 2006)

Unit/locality	Sample nature	Age (method)	Ref.
Makoppa Dome			
Vaalpenskraal Gneiss	Trondhjemite gneiss	3034 ± 64 (D)	1
Vaalpenskraal Gneiss	Trondhjemite gneiss	3013 ± 11 (D)	1
Makoppa Granitoid	Porphyritic granodiorite	2887 +3/−2 (A)	1
Rooibokvlei Granitoid	Granodiorite	2797 ± 2 (A)	1
Rooibokvlei Granitoid	Granodiorite	2777 ± 35 (D)	1
Johannesburg Dome			
Lanseria Gneiss	Trondhjemite gneiss	3340 ± 3.3 (A)	2
Linden Gneiss	Tonalitic gneiss	3199.9 ± 2 (A)	2
Linden Gneiss	Tonalitic gneiss	3,170,134 (A)	3
Bryanston Granodiorite	Granodiorite	3121.2 ± 5 (A)	2
Honeydew Granodiorite	Porphyritic granodiorite	3114.2 ± 2.4 (A)	2
Vredefort Dome			
Vredefort Dome	'Outer Granitic Gneiss'	3100 (E + G)	4
Vredefort Dome	Granite pegmatite	3094 ± 7 (A)	5
Vredefort Dome	Tonalite gneiss	3092 ± 7 (A)	5
Westerdam and Coligny Domes			
Hartbeesfontein Dome	Granodiorite- adamellite	3174 ± +9/−7 (A)	6
Hartbeesfontein Dome	Granitoid	3120 ± 5 (B)	7
South of the Vredefort Dome	Low-Ca granite	3101 ± 2 (A(M))	6
Westerdam Dome	Low-Ca granite	3086 ± 3 (A (M))	6
Coligny Dome	Low-Ca leucogranite	3031 +11/−10 (A)	6
Amalia-Kraaipan area			
Kraaipan: Setlagole area	Migmatite (borehole)	3162 ± 8(C)	8
Kraaipan: Setlagole area	Migmatite (borehole)	3070 ± 7 (C)	8
Amalia area	Biotite trondhjemite gneiss	3008 ± 4 (A)	9
Amalia area	Leuco-trondhjemite gneiss	2939 ± 9.6 (B)	9
Kraaipan: Madibogo area	Granodiorite	2915 ± 9 (B)	9
Kraaipan area	Granodiorite	2913 ± 17(B)	9
Schweizer-Reneke area	Adamellite	2880 ± 2 (A (Mo))	6
Kraaipan: Disaneng Dam area	Granodiorite	2879 ± 9 (B)	9
Kraaipan: Khunwana area	Granodiorite	2879 ± 11 (B)	9
Kraaipan: Setlagole area	Migmatite (borehole)	2866 ± 25 (C)	8
Kraaipan: Mareetsane area	Granodiorite	2846 ± 22 (C)	8
Mosita Granite	Adamellite/Qz mon- zonite	2791 ± 8 (B)	9
Gaborone Granite Suite			
Gaborone Suite	Granophyre	2784.9 ± 1.9 (A)	10
Gaborone Suite	Kgale Granite	2783.2 ± 1.8 (C)	11
Gaborone Suite	Leucocratic Granite	2783.1 ± 2 (A)	10
Gaborone Suite	Kgale Granite	2782 ± 5 (C)	11

In addition to the granitoid intrusions listed there are several mafic and felsic granulites (enderbites and charnockites) that have been encountered in various areas on the Kaapvaal Craton (e.g. Vredefort Dome; Gibson and Reimold 2008; Morokweng Dome; Wela 2017) and as eclogite xenoliths in various kimberlite pipes on the craton (e.g. Lace, Orapa, Bellsbank, Jwaneng, Kimberley, Roberts Victor and Premier;—Aulbach and Viljoen 2015)

Method A = U–Pb ID-TIMS; B = SHRIMP; C = Pb–Pb zircon evaporation; D = LAM-ICP-MS; E = Whole-rock Rb/Sr; G = Whole-rock U–Pb; (M) = Muscovite; (Mo) = Monazite. Mineral analysed is zircon, unless otherwise mentioned. All errors are at 2σ

Refs: 1. Anhaeusser and Poujol (2004); 2. Poujol and Anhaeusser (2001); 3. Anhaeusser and Burger (1982); 4. Reimold et al. (1988); 5. Moser et al. (2001); 6. Robb et al. (1992); 7. Armstrong et al. (1991); 8. Anhaeusser and Walraven (1999); 9. Poujol et al. (2002); 10. Moore et al. (1993); 11. Grobler and Walraven (1993)

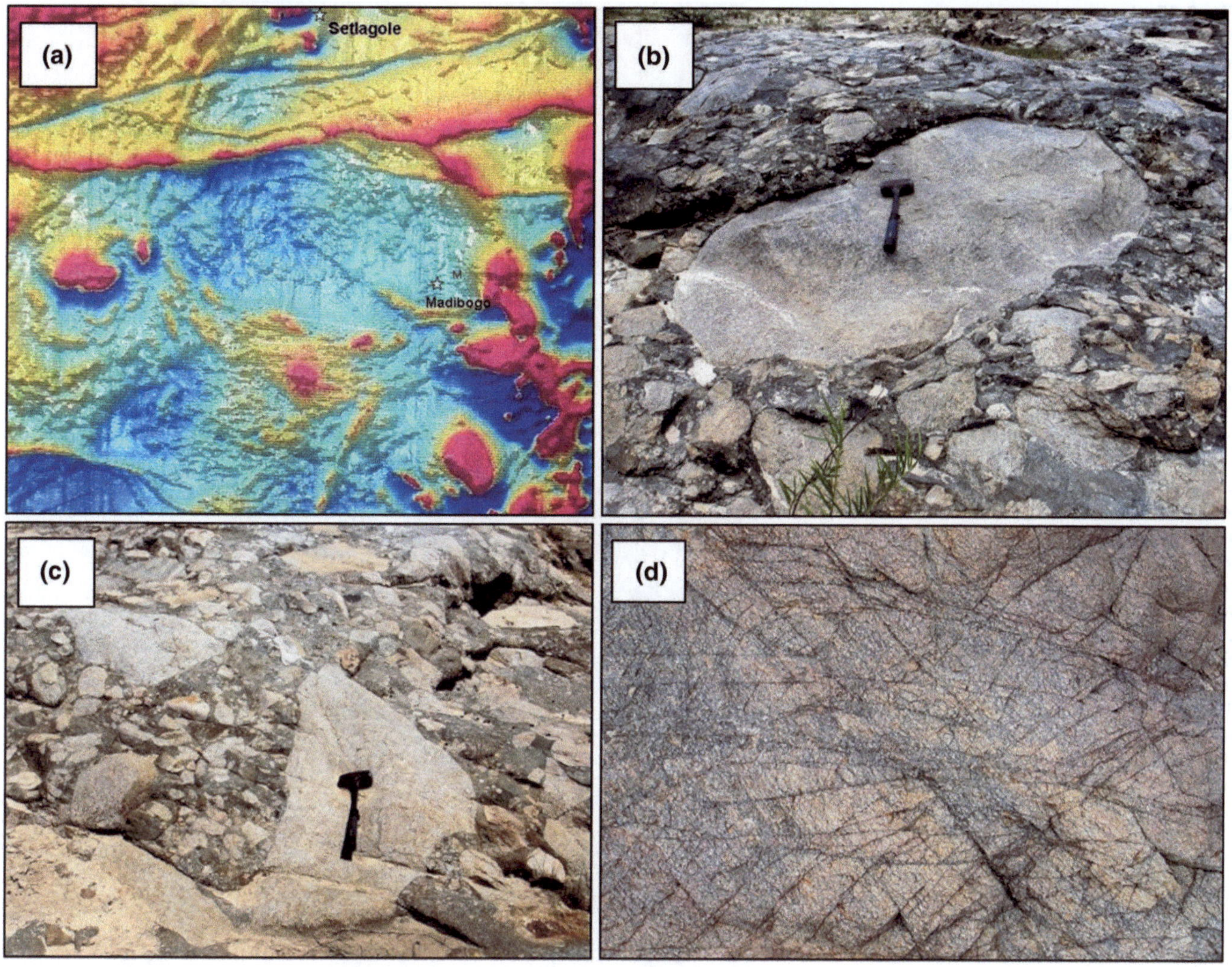

Fig. 3.14 Setlagole impact structure (locality Fig. 3.11). **a** Aeromagnetic image of the Setlagole-Madibogo area showing a circular anomaly and cross-cutting lineaments interpreted as dykes. The warm colours reflect magnetite-rich rocks, including the Archaean Kraaipan iron formations. The terrane dominated by the cooler colours is underlain mainly by Archaean granitoid rocks. Megabreccia outcrops (shown in **b** and **c**) lie just off the image immediately north of Setlagole. The distance between Setlagole and Madibogo is 18 km. **b** Setlagole megabreccia illustrating the highly variable clast size. The large block shown measures 3.2 × 1.5 m; **c** granitic blocks surrounded by smaller breccia fragments encased in matrix breccia. The largest megabreccia block measured 6 × 3 m near Setlagole; **d** Closely spaced intersecting impact fracture sets in granitic gneiss between Setlagole and Madibogo. After Anhaeusser et al. (2010)

adjacent to the high-grade rocks of the Namaqua-Natal Belt. Exposures of what has been termed the Marydale Greenstone Belt are of limited extent. They were described by Vajner (1974), Cornell (1975), Barton et al. (1986) and McCourt et al. (2000) and occur mainly near Marydale and southeast of Prieska. Cornell et al. (2019) have shown that the Marydale terrane lies between the Doornberg Fault and the Brakbos Shear Zone (Fig. 3.16).

Brandl et al. (2006) assigned the Marydale rocks to the Marydale Group, in which five formations are recognized. The Steenkop Formation at the base is comprised of metamorphosed clastic sediments, including arkose, greywacke, quartzite, conglomerate and phyllite with minor volcanoclastic rocks and lavas of felsic to intermediate composition. This unit is overlain conformably by the Perdeput Formation composed of amygdaloidal metabasites that are altered mainly to texturally schistose to massive amphibolites. The Steenkop and Perdeput formations are assigned to the Prieskapoort Subgroup (Fig. 3.16).

The overlying Doornfontein Subgroup comprises the Modderfontein Formation in which two horizons of banded iron formation are intercalated with metabasalt, chlorite-muscovite schist, dolomite and chert. The banded iron-formation units are several hundred metres apart, the lower one being more prominent with a maximum thickness of 5 m. Higher in the succession is the Uitzigt Formation, composed of metamorphosed tuffs, lavas and quartzo-feldspathic sediments. Barton et al. (1986) obtained a total

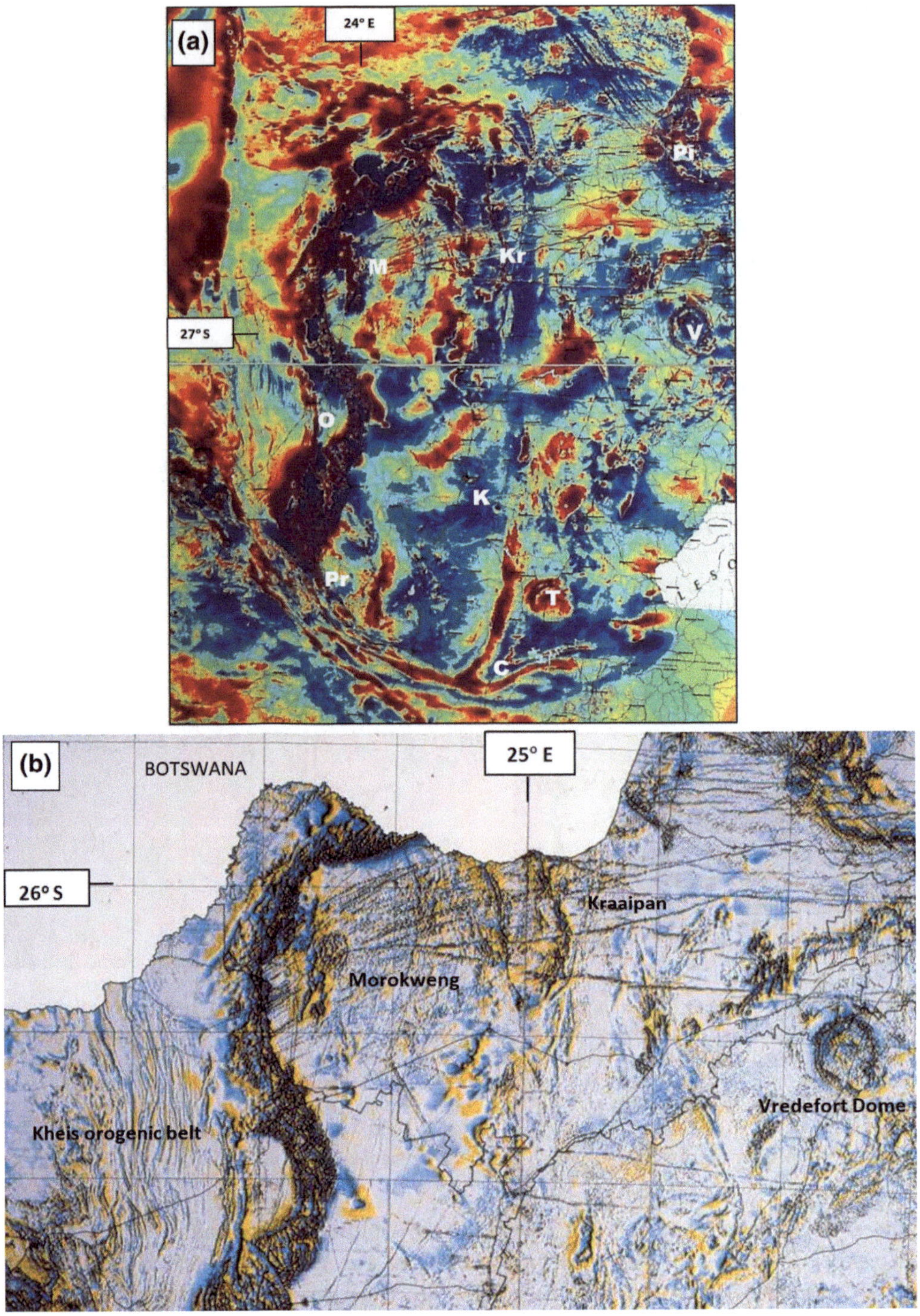

Fig. 3.15 Aeromagnetic maps relating to the western or Kimberley Block of the Kaapvaal Craton. **a** Total field aeromagnetic grid of the western half of the Kaapvaal Craton (after Reed 1996) showing the main magnetic features, including the N-S-trending Colesberg magnetic anomaly. Places shown on map include: Pi—Pilanesberg; M—Morokweng; Kr—Kraaipan terrane; V—Vredefort Dome; O—Olifantshoek; K—Kimberley; Pr—Prieska; T—Trompsberg; and C—Colesberg. Width of map 700 km; **b** portion of the Magnetic Fabric Map of South Africa (Council for Geosciences 1997) showing part of the region west of the Vredefort Dome. The Kraaipan terrane is followed westwards by the Morokweng impact locality, the strongly magnetic N–S-trending zone of ca. 2500–2420 Ma Transvaal Supergroup iron formations and the N–S-trending ca. 1750–1900 Ma Kheis orogenic belt marking the western boundary of the Kaapvaal Craton. Width of map 675 km

rock Pb–Pb age of ca. 3010 Ma for the greenschist-facies banded iron formations and a Pb–Pb age of ca. 2915 Ma for a Perdeput Formation metabasalt. At the top of the sequence is the Spioenkop Formation comprised of less than 100 m of bluish and greenish quartzites and quartz-mica schists with thin intercalations of amphibolitic mafic lavas and gabbroic sills. In general the mafic rocks of the Marydale Group are predominantly basaltic in composition with tholeiitic affinities. Ultramafic rocks are rare and, where present, generally contain less than 20% MgO.

Granitoid rocks of Mesoarchaean to Neoarchaean age occur in the Marydale-Prieska region, and prior to a recent compilation by Cornell et al. (2019) little was known of these rocks. The oldest granitoid is the Draghoender Granite which is generally tonalitic to trondhjemitic in composition and occurs in the region NW and SE of Marydale and again in the Franzenhof area adjacent to the Doornberg Fault (Fig. 3.16). A discordant zircon $^{207}Pb/^{206}Pb$ age of 2853 ± 4 Ma was reported for this granitoid by McCourt et al. (2000), but more recent precise U–Pb zircon dating yielded ages ranging from

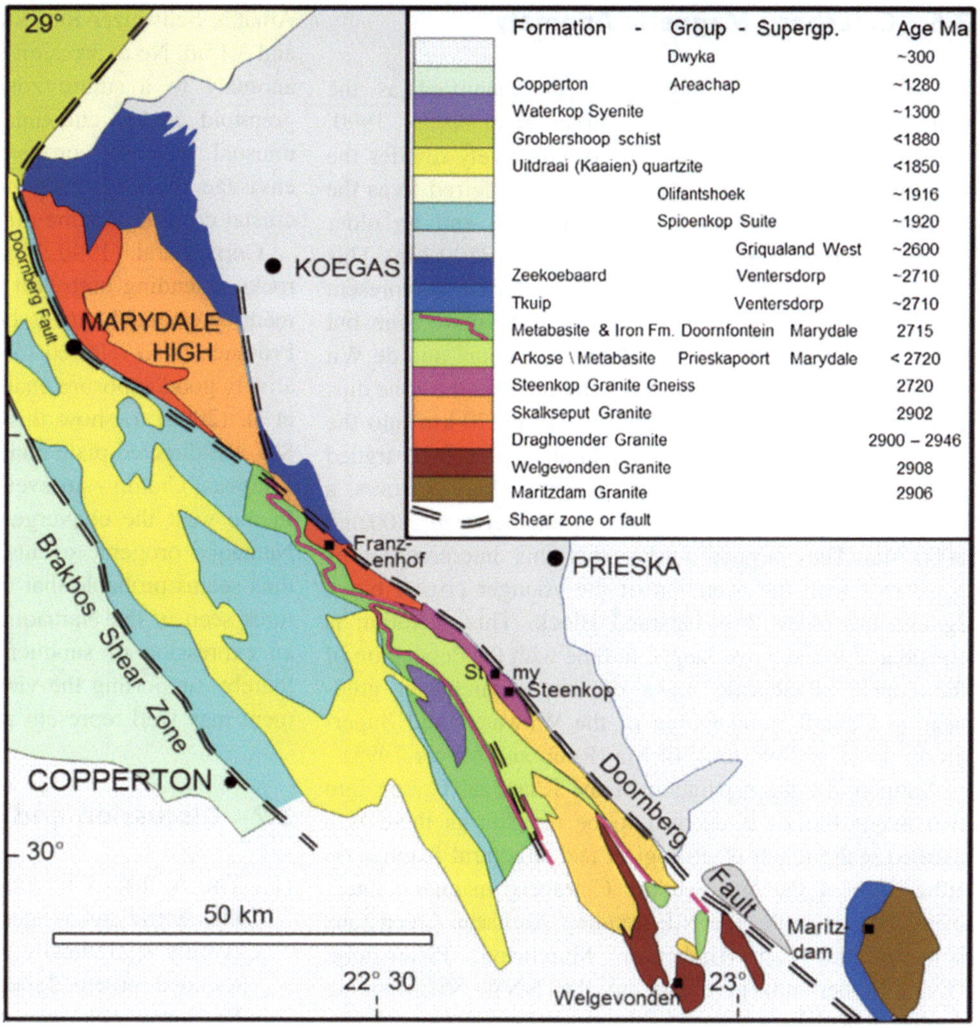

Fig. 3.16 Geological map of the Marydale-Prieska granite-greenstone terrane exposed in the southwestern sector of the Kaapvaal Craton (after Cornell et al. 2019). Granitoid rocks in this region range in age from ca. 2946 to ca. 2718 Ma (McCourt et al. 2000; Cornell et al. 2019)

2946 ± 9 to 2892 ± 6 Ma (Cornell et al. 2019). The narrow sliver of Draghoender Granite near Franzenhof was dated at 2931 ± 9 Ma.

The granodioritic to monzogranitic Skalkseput Granite, which intrudes the Draghoender Granite, occurs northeast, east and southeast of Marydale and comprises leucocratic biotite-muscovite granitoids that are locally porphyritic. SHRIMP zircon ages ranging between 2938 ± 5 and 2718 ± 8 Ma were reported for this granitoid type by McCourt et al. (2000). Core analyses on zircons from one sample yielded a spread of $^{207}Pb/^{206}Pb$ ages in the range ca. 3111–2930 Ma. A second sample of the Skalkseput Granite from the southern part of the Marydale Terrane gave a younger age of 2718 ± 8 Ma (McCourt et al. 2000). According to Cornell et al. (2019) the Draghoender and Skalkseput granitic rocks overlap in time and space. These authors contend that the more evolved Skalkseput granitoid, which yielded a coherent U–Pb zircon age of 2901 ± 14 Ma, was possibly derived by melting of the Draghoender-type granitic rocks during assembly of the Kaapvaal Craton at the time of the postulated collision of the Kimberley and Witwatersrand terranes discussed in the following section.

A third granite type is exemplified by the 2721 ± 6 Ma monzogranitic Steenkop Granite Gneiss, corresponding in age to a 2718 ± 8 Ma monzogranite dyke, which intrudes a ca. 2871 Ma Skalkseputs monzogranite.

Additional granitic exposures occur approximately 50 km SSE of Prieska, north and south of the NW–SE trending Doornberg Fault. According to Cornell et al. (2019) the 2907 ± 4 Ma Welgevonden Granite south of the fault underlies volcanic rocks of the Marydale Group and indicates that the Kaapvaal Craton extends across the fault zone. The 2905 ± 7 Ma monzogranitic Maritzdam Granite north of the fault is basement to Ventersdorp Supergroup volcanic rocks and corresponds to Skalkseput-type granitoids.

3.6 Colesberg Magnetic Anomaly

A prominent N-S-trending lineament, identified as the Colesberg magnetic anomaly (Corner et al. 1986, 1990; Corner 1991—Figs. 3.1 and 3.15a), effectively divides the Kaapvaal Craton into two major domains referred to as the Kimberley Block in the west (<3260 Ma) and an older Witwatersrand Block in the east (ca. 3700 to 3100 Ma). This prominent magnetic anomaly has been inferred to represent the near-surface expression of a Neoarchean suture zone, but there is no consensus on the polarity. Doucouré and de Wit (2002) maintained on geophysical grounds that the zone dips towards the east and may extend for some 70 km into the lithospheric mantle. By contrast Schmitz et al. (2004) argued that the surface geology and geochronology favours a west-dipping suture and a time of suturing between 2800 and 2000 Ma. They argued further that this interpretation is consistent with the accretion of the younger crustal block against the older Witwatersrand Block. This accretion is considered to have overlapped in time with the deposition of the coarse siliciclastic rocks of the Neoarchaean gold-bearing Central Rand Group of the Witwatersrand Supergroup dated at 2894 to 2714 Ma (Robb and Meyer 1995).

Support for the separation of the Kaapvaal Craton into two major blocks is deemed to be realistic as there is a marked realignment of geological and structural features on either side of the N-S-trending Colesberg magnetic lineament. For example, the N-E-trending Archaean greenstone belts in the east (Barberton, Murchison, Pietersburg, Giyani/Sutherland) give way to the NNW–SSE-trending Archaean Kraaipan terrane in the west (Figs. 3.1 and 3.11). Furthermore the available geochronological data suggests an older eastern terrane (Poujol et al. 2003; Eglington and Armstrong 2004).

Boreholes drilled directly into the magnetic anomaly were examined by Drennan (1988) who found gneisses that are not unlike the strongly foliated granodioritic to tonalitic granitoids found west of the anomaly. However the gneisses directly in the anomaly display phenocryst development, some brecciation, and intense hematitization as well as up to 5% magnetite. East of the anomaly the granitoids comprise more massive and homogeneous granitic to monzonitic rock types.

The magnetic anomaly has to date not been adequately explained, but speculation abounds that it represents a period of subduction and terrane collision or suturing between 2930 and 2880 Ma (de Wit et al. 1992; Doucouré and de Wit 2002; Tinker et al. 2002; Schmitz et al. 2004). Drennan (1988) on the other hand showed from borehole data that the magnetic anomaly became discontinuous to the north between Kimberley and Bloemfontein and appears to split into several N-S-trending splays in the vicinity of the Amalia, Schweizer-Reneke and Kraaipan regions (Figs. 3.1 and 3.15a). No expression of any surface features linking the anomaly to a suture zone have ever been found—only granitoid rocks, encountered in boreholes, that showed unusual magnetic properties. It is therefore difficult to envisage the Colesberg magnetic anomaly to represent a crustal collisional zone on the available evidence.

Corner et al. (1990, Fig. 4), showed anomalous magnetic rocks extending both east and west of the Colesberg lineament, presumably in association with the Namaqua-Natal Province (also referred to as a Mobile Belt, Fig. 3.1). Relatively good exposure enabled Matthews (1972) and Cornell et al. (2006) to show this belt in the east to be a zone of SW-NE-directed plate convergence and subduction with the Kaapvaal Craton—an event that occurred ca. 1300 Ma ago. In the west the convergent phase culminated in the main Namaqua orogenic events between 1220 and 1150 Ma. It thus seems probable that the high-intensity magnetic signatures seen in the Namaqua-Natal Province (Fig. 3.15a) are an expression of subduction and island arc development, thereby supporting the view that the older Colesberg lineament may well represent a Neoarchaean suture zone.

3.7 Discussion and Conclusions

(1) The Archaean basement granite-greenstone terrane in the central and western sectors of the Kaapvaal Craton is only sporadically preserved in domical inliers surrounded or enveloped by younger formations. Some basement rocks can also be seen in gold exploration boreholes drilled to locate extensions to the Witwatersrand Goldfield.
(2) In general the ages of the granite-greenstone terrane become progressively younger from east (Witwatersrand Block) to west (Kimberley Block). The older eastern basement terrane has ages in the range ca. 3660–2700 Ma, whereas the granitoids and greenstones west of the Colesberg magnetic anomaly, which separates the Kaapvaal Craton into roughly two halves, give ages generally ranging between ca. 3174 to 2782 Ma (Table 3.2). Ages exceeding about 3200 Ma are thought to reflect the presence of xenocrystic zircons.
(3) Geological, geochemical, geophysical and geochronological investigations appear to favour a close genetic link between the Kaapvaal Craton of southern Africa and the Pilbara terrane of Western Australia, as first proposed by Cheyney et al. (1988). Indications suggest a close spatial association of these two terranes prior to about 2700 Ma (Nelson et al. 1999).
(4) The nature of the upper crust of the Kaapvaal Craton has been studied, where surface exposures permit, using a

variety of techniques including geological mapping, geochronology, petrology and geochemical analysis. Various granitoid events have been identified, but no consensus has been reached on the origin and tectonic evolution of the generally older granitoids, viz. the TTG gneisses and migmatites. Nor have the genetic and emplacement histories been satisfactorily explained of the vast potassic batholiths present in the Barberton-Swaziland region, as well as those in the central part of the craton (e.g. Johannesburg Dome), the north-central region (Makoppa Dome) and the Kraaipan terrane in the western half of the craton. The depth extent of the Archaean greenstone belts that are general enveloped by granitoid rocks has also not been satisfactorily resolved.

(5) The nature and evolution of the lower crust of the Kaapvaal Craton has presented more of a challenge due to lack of exposure, but geophysical advances involving heat flow measurements, deep geoelectrical soundings, vibroseis seismic profiles, as well as gravity, magnetic and seismic studies have contributed to better estimates of the subsurface geology. Despite these efforts the nature of the Archaean crust of the craton still remains enigmatic. There remains considerable controversy as to whether the Archaean lower crust is comprised dominantly of rocks of felsic to intermediate or mafic composition or even whether it is feasible to ascribe a 'characteristic' composition to the Archaean lower crust in a global sense. Seismic array data reported by Niu and James (2002) from the Kimberley region provided clues as to the nature of the crust and crust-mantle boundary beneath the array. The findings suggested (1) the crust is relatively thin (35.4 km) with a low average crustal Poisson ratio of ~ 0.254; (2) the density contrast across the Moho is $\sim 15.4\%$, indicating a lowermost crustal density around 2.86 g/cm^3, (3) the Moho is very sharp, with a crust-mantle transition zone less than 0.5 km thick; (4) the Moho is surprisingly flat with topographic relief less than 1 km over a region of ~ 2400 km^2; and (5) there appears to be an absence of a mafic lower crust.

(6) Based on Palaeoarchaean granitic evolutionary cycles (e.g. in the Barberton-Swaziland region; Robb et al. 2006) whereby granitic terranes evolve from older sodium-rich granitoids (tonalite-trondhjemite-granodiorite—the so-called TTG granitoids) to younger more potassic granitoids (granodiorite-monzonite-granite sensu stricto), this relationship has also been shown to apply to the Mesoarchaean and Neoarchaean granitoid rocks found in the central and western sectors of the Kaapvaal Craton (Anhaeusser 1973; Anhaeusser and Walraven 1999; Poujol and Anhaeusser 2001; Poujol et al. 2002).

(7) Archaean volcanic rocks also show a trend from mainly high-magnesium komatiitic and basaltic komatiitic rocks in the east (viz., the Barberton greenstone terrane) to mainly tholeiitic basalts with few high-magnesium basalts and rare occurrences of komatiites in the west. Geochemical data from metavolcanic rocks in the Makoppa Dome and the Kraaipan terrane (also borehole BS1 drilled through the Ventersdorp Supergroup into Kraaipan rocks between Delareyville and Geysdorp—Fig. 3.11) show similar trends and no high-magnesium basalts (C.R. Anhaeusser unpublished data).

(8) Granitoid-greenstone geochemistry and ages appear to support the suggestions by Doucoure and de Wit (2002); Schmitz et al. (2004) and Tinker et al. (2002) that the Kimberley Block may represent an allochthonous crustal fragment that was accreted or sutured onto the Witwatersrand Block along the Colesberg lineament by some form of arc magmatism or subduction at about 2900 Ma. However, this proposal is not universally accepted as there are still some questions that remain unanswered.

(9) The presence of domical basement structures in the areas surrounding the Witwatersrand Basin played a significant role in the progressive sedimentation and concentration of gold mineralization in the intervening synclinal areas and along basin edges (as pre-, concurrent-, and post-depositional structures) as was suggested first by Brock and Pretorius (1964).

Acknowledgements The author wishes to thank reviewers Marco Andreoli and Richard Viljoen for their constructive comments, which helped to improve the manuscript. Sonja Van Eck of the Council for Geoscience, Pretoria, granted the author permission to use diagrams sourced from *The Geology of South Africa* textbook, which was produced in 2006 jointly by the Geological Society of South Africa and the Council for Geoscience. Sara Burness, at the University of the Witwatersrand, Johannesburg, helped to modify the diagrams used in the review and assisted with the photographic layout. Finally, the author thanks Alfred Kröner for having initiated this contribution on the Kaapvaal Craton and for inviting my participation.

References

Aldis DT, Tombale AR, Mapeo RBM, Chiepe M (1989) The geology of the Kanye Area. Bull Geol Surv Botswana 33:170

Andrews G (2003) Platinum-group element mineralization in the Stella layered intrusion, Kraaipan greenstone belt. MSc thesis (unpubl) Rhodes Univ Grahamstown, p 115

Andreoli MAG, Ashwal LD, Hart RJ, Smith CB, Webb SJ, Tredoux M, Gabrielli F, Cox RM, Hambleton-Jones BB (1995) The impact origin of the Morokweng Ring Structure, southern Kalahari, South Africa. Abstr Cent Geocongr Geol Soc S Afr Johannesburg, pp 541–554

Andreoli MAG, Hart RJ, Cooper GRJ, Stengel I, Webb S, Haddon I, Skála R, Viola G (2007) The 144 Ma Morokweng impact crater, South Africa: geophysical and borehole evidence for a ~240 km structure. In: Short pap 10th SAGA Bien Tech Meet Exhib, Wild Coast 24–26 Oct 2007, p 5

Anhaeusser CR (1973) The geology and geochemistry of the Archaean granites and gneisses of the Johannesburg-Pretoria Dome. In: Lister LA (ed) Symposium on granites, gneisses and related rocks, vol 3. Spec Publ Geol Soc S Afr, pp 361–385

Anhaeusser CR (1977) Geological and geochemical investigations of the Roodekrans ultramafic complex and surrounding Archaean volcanic rocks, Krugersdorp District. Trans Geol Soc S Afr 80:17–28

Anhaeusser CR (1978) The geology and geochemistry of the Muldersdrif Ultramafic Complex, Krugersdorp District. Trans Geol Soc S Afr 81:193–203

Anhaeusser CR (ed) (1991) The Archaean Kraaipan Group volcano-sedimentary rocks and associated granites and gneisses of the southwestern Transvaal, northwestern Cape Province and Bophuthatswana—Excurs Guidebk Inform Circ, Econ Geol Res Unit, Univ Witwatersrand, Johannesburg, vol 45, p 244

Anhaeusser CR (1992) Archaean granite-greenstone relationships on the farm Zandspruit 191 IQ, North Riding area, Johannesburg Dome. S Afr J Geol 94:94–101

Anhaeusser CR (1999) Archaean crustal evolution of the central Kaapvaal Craton, South Africa: evidence from the Johannesburg Dome. S Afr J Geol 102:303–322

Anhaeusser CR (2006a) A re-evaluation of Archean intracratonic terrane boundaries on the Kaapvaal Craton, South Africa: Collisional suture zones? In: Reimold, WU, Gibson, RL (eds) Processes on the early Earth, vol 405. Spec Pap Geol Soc Am, pp 193–210

Anhaeusser CR (2006b) Ultramafic and mafic intrusions of the Kaapvaal Craton. In: Johnson MR, Anhaeusser CR, Thomas RJ (eds) The geology of South Africa. Geol Soc S Afr Johannesburg/Counc Geosci Pretoria, pp 95–134

Anhaeusser CR (2015) Metasomatized and hybrid rocks associated with a Palaeoarchaean layered ultramafic intrusion on the Johannesburg Dome, South Africa. J Afr Earth Sci 102:203–217

Anhaeusser CR, Burger AJ (1982) An interpretation of U–Pb zircon ages for Archaean tonalitic gneisses from the Johannesburg-Pretoria granite dome. Trans Geol Soc S Afr 85:111–116

Anhaeusser CR, Walraven F (1999) Episodic granitoid emplacement in the western Kaapvaal Craton: evidence from the Archaean Kraaipan granite-greenstone terrane, South Africa. J Afr Earth Sci 28:289–309

Anhaeusser CR, Poujol M (2004) Petrological, geochemical and U–Pb isotopic studies of Archaean granitoid rocks of the Makoppa Dome, northwest Limpopo Province, South Africa. S Afr J Geol 107:521–544

Anhaeusser CR, Stettler E, Gibson RL, Cooper GRJ (2010) A possible Mesoarchaean impact structure at Setlagole, North West Province, South Africa: aeromagnetic and field evidence. S Afr J Geol 113:413–436

Armstrong RA, Compston W, Retief EA, Williams IS, Welke HJ (1991) Zircon ion microprobe studies bearing on the age and evolution of the Witwatersrand Triad. Precambrian Res 53:243–266

Armstrong RA, Lana C, Reimold WU, Gibson RL (2006) SHRIMP zircon age constraints on Mesoarcahaen crustal development in the Vredefort Dome, central Kaapvaal Craton, South Africa. In: Reimold WU, Gibson RL (eds) Processes on the early Earth, vol 405. Spec Pap Geol Soc Am, pp 233–253

Aulbach S, Viljoen KS (2015) Eclogite xenoliths from the Lace kimberlite, Kaapvaal craton: From convecting mantle source to palaeo-ocean floor and back. Earth Planet Sci Lett 431:274–286

Barton JM Jr, Barton ES, Kröner A (1999) Age and isotopic evidence for the origin of the Archaean granitoid intrusives of the Johannesburg Dome, South Africa. J Afr Earth Sci 28:693–702

Barton ES, Armstrong RA, Cornell DH, Welke HI (1986) Feasibility of total-rock Pb–Pb dating of metamorphosed banded iron formation: The Marydale Group, South Africa. Geochem Geol Isotope Geosci Sect 59:255–271

Bosch PJA (1993) Die geologie van die gebied Kimberley. Explanation Sheet 2824 (1:250 000), Geol Surv S Afr, p 60

Brandl G, de Wit MJ (1997) The Kaapvaal Craton. In: de Wit MJ, Ashwal LD (eds) Greenstone belts, vol 35. Oxford Mono Geol Geophys, Oxford, pp 581–607

Brandl G, Cloete M, Anhaeusser CR (2006) Archaean greenstone belts. In: Johnson MR, Anhaeusser CR, Thomas RJ (eds) The geology of South Africa. Geol Soc S Afr, Johannesburg/Counc Geosci Pretoria, pp 9–56

Brock BB, Pretorius DA (1964) An introduction to the stratigraphy and structure of the Rand Goldfield. In: Haughton SH (ed) The geology of some ore deposits in southern Africa, vol. I. Geol Soc S Afr, p 25–61

Burke K, Dewey JF, Kidd WSF (1977) World distribution of sutures: the sites of former oceans. Tectonophys 40:69–99

Button A (1979) Transvaal and Hamersley basins—review of basin development and mineral deposits. Miner Sci Engr 8:262–290

Cawthorn RG, Eales HV, Walraven F, Uken R, Watkeys MK (2006) The Bushveld Complex. In: Johnson MR, Anhaeusser CR, Thomas RJ (eds) The geology of South Africa. Geol Soc S Afr, Johannesburg/Counc Geosci, Pretoria, pp 261–281

Cheney ES (1996) Sequence stratigraphy and plate tectonic significance of the Transvaal succession of southern Africa and its equivalent in Western Australia. Precambrian Res 79:3–24

Cheney ES, Roering C, Stettler EH (1988) Vaalbara. Ext Abstr Geocongr '88, Geol Soc S Afr Durban, pp 85–88

Cornell DH (1975) Petrology of the Marydale metabasites. PhD thesis (unpublished), Cambridge Univ p 216

Cornell DH, Thomas RJ, Moen HFG, Reid DL, Moore JM, Gibson RL (2006) The Namaqua-Natal Province. In: Johnson MR, Anhaeusser CR, Thomas RJ (eds) The geology of South Africa. Geol Soc S Afr, Johannesburg/Counc Geosci, Pretoria, pp 325–379

Cornell DH, Van Schijndel V, Ingolfsson O, Schersten A, Karlsson L, Wojtyla J, Karlsson K (2011) Evidence from Dwyka tillite cobbles of Archaean basement beneath the Kalahari sands of southern Africa. Lithos 125:482–502

Cornell DH, Minnaar H, Frei D, Kristoffersen M (2019) Precise microbeam dating defines three Archaean granitoid suites at the southwestern margin of the Kaapvaal Craton. Precambrian Res 304:21–38

Corner B (1991) Crustal architecture of the Precambrian of the Kaapvaal Province from geophysical data. Abstr Precambrian sedimentary basins of southern Africa. In: Eriksson PG (compiler) Terra Nova 3:5 Pretoria

Corner B, Durrheim RJ, Nicolaysen LO (1986) The structural framework of the Witwatersrand basin as revealed by gravity and aeromagnetic data. In: Ext Abstr Geocongr '86 Geol Soc S Afr, pp 27–30

Corner B, Durrheim RJ, Nicolaysen LO (1990) Relationship between the Vredefort structure and the Witwatersrand Basin within the tectonic framework of the Kaapvaal Craton as interpreted from regional gravity and aeromagnetic data. Tectonophys 171:49–61

Corner B, Reimold WU, Brandt D, Koeberl C (1997) Morokweng impact structure, Northwest Province, South Africa: geophysical imaging and shock petrographic studies. Earth Planet Sci Lett 146:351–364

Council for Geosciences (1997) Magnetic fabric map of the Republic of South Africa. Reg Aeromag Ser, Geophys Div

de Beer JH, Meyer R (1984) Geophysical characteristics of the Namaqua-Natal Belt and its boundaries, South Africa. J Geodyn 1:473–494

de Beer JH, Stettler EH (1988) Geophysical characteristics of the southern African crust. J Petrol Spec Lithosph Vol:1–22

de Kock MO, Evans DAD, Beukes NJ (2009) Validating the existence of Vaalbara in the Neoarchean. Precambrian Res 174:145–154

de Wit MJ, Hart RA (1993) Earth's earliest continental lithosphere, hydrothermal flux and crustal recycling. Lithos 30:309–335

de Wit MJ, Jeffery M, Bergh H, Nicolayson LO (1988) The geological map of Gondwana, with explanatory notes. Am Assoc Petrol Geol, Tulsa, OK, USA

de Wit MJ, Roering C, Hart RJ, Armstrong RA, de Ronde CEJ, Green RWE, Tredoux M, Peberdy E, Hart RA (1992) Formation of an Archean continent. Nature 357:553–562

Doucoure CM, de Wit MJ (2002) Temporal variation in rigidity and mechanical behavior of old thick continental lithosphere. S Afr J Geol 107:13–32

Drennan GR (1988) The nature of the Archaean basement in the hinterland to the Welkom Goldfield. MSc thesis (unpubl), Univ Witwatersrand, Johannesburg, p 187

Drennan GR, Robb LJ, Meyer FM, Armstrong RA, de Bruiyn H (1990) The nature of the Archaean basement in the hinterland of the Witwatersrand Basin: II. A crustal profile west of the Welkom Goldfield and comparisons with the Vredefort crustal profile. S Afr J Geol 93:41–53

Durrheim RJ, Green RWE (1992) A seismic refraction investigation of the Archaean Kaapvaal Craton, South Africa, using mine tremors as the energy source. Geophys J Int 108:812–832

du Toit AL (1906) Geological survey of portions of the division of Vryburg and Mafeking. Ann Rep Geol Comm Cape of Good Hope 1905:205–258

du Toit AL (1908) Geological survey of portion Mafeking and Vryburg. Ann Rep Geol Comm Cape of Good Hope 1907:123–157

Eglington BM, Armstrong RA (2004) The Kaapvaal Craton and adjacent orogens, southern Africa: a geochronological database and overview of the geological development of the craton. S Afr J Geol 107:13–32

Eriksson PG, Altermann W, Hartzer FJ (2006) The Transvaal supergroup and its precursors. In: Johnson MR, Anhaeusser CR, Thomas RJ (eds) The geology of South Africa. Geol Soc S Afr, Johannesburg/Counc Geosci, Pretoria, pp 95–134

Eriksson PG, Banerjee S, Nelson DR, Rigby MJ, Catuneanu O, Sarkar S, Roberts RJ, Ruban D, Mtimkulu MN, Sunder Raju PV (2009) A Kaapvaal Craton debate: nucleus of an early small supercontinent or affected by an enhanced accretion event? Gond Res 15:354–372

Eriksson PG, Rigby MJ, Bandopadhyay PC, Steenkamp NC (2011) The Kaapvaal Craton, South Africa: no evidence for a supercontinental affinity prior to 2.0 Ga? Int Geol Rev 53:1312–1330

Fouch MJ, James DE, VanDecar JC, van der Lee S, the Kaapvaal Seismic Group (2004) Mantle seismic structure beneath the Kaapvaal and Zimbabwe Cratons. S Afr J Geol 107:33–44

Frimmel H (2019) The Witwatersrand Basin and Its Gold Deposits. In: Kröner A, Hofmann A (eds) The Archaean Geology of the Kaapvaal Craton, Southern Africa. Springer Nature Switzerland AG

Fritiofsson S (2014) Kaapvaal Craton basement exposed in diamond mines at Kimberley, South Africa: Greenstone belt schists and amphibolites. BSc thesis (unpubl), Univ Gothenburg, Sweden, p 56

Gibson RL (2019) The Mesoarchaean Basement Complex of the Vredefort Dome—A Mid-Crustal Section Through the Central Kaapvaal Craton Exposed by Impact. In: Kröner A, Hofmann A (eds) The Archaean Geology of the Kaapvaal Craton, Southern Africa. Springer Nature Switzerland AG

Gibson RL, Reimold WU (2008) Geology of the Vredefort impact structure—A guide to sites of interest. Mem Counc Geosci Pretoria 97:181

Green RWE, Durrheim RJ (1990) A seismic refraction investigation of the Namaqualand metamorphic complex, South Africa. J Geophys Res 95:19927–19932

Grobler DF (1996) The geology, geochemistry and geochronology of the Gaborone Granite Suite and Kanye Formation north of Mafeking, South Africa. PhD thesis (unpubl), Univ Witwatersrand, Johannesburg, p 459

Grobler DF, Walraven F (1993) Geochronology of the Gaborone Granite Complex extensions in the area north of Mafikeng, South Africa. Chem Geol (Isotope Geosci Sect) 105:319–337

Groenewald PB, Grantham GH, Watkeys MK (1991) Geologic evidence for a Proterozoic to Mesozoic link between southeastern Africa and Dronning Maud Land, Antarctica. J Geol Soc Lond 148:1115–1123

Hammond NQ, Morishita Y (2009) Source of ore fluids at the Kalahari Goldridge deposit, Kraaipan greenstone belt, South Africa: evidence from Sr, C and O isotope signatures in carbonates. Geofluids 9 (4):356–364

Hart RJ, Andreoli MAG, Tredoux M, Moser D, Ashwal LD, Eide EA, Webb SJ, Brandt D (1997) Late Jurassic age for the Morokweng impact structure, southern Africa. Earth Planet Sci Lett 147:25–35

Henkel H, Reimold WU, Koeberl C (2002) Magnetic and gravity model of the Morokweng impact structure. J Appl Geophys 49(3):129–147

Hirner AJ (2001) Geology and gold mineralization of the Madibe greenstone belt, eastern part of the Kraaipan terrain. PhD thesis (unpubl), Univ Witwatersrand, Johannesburg, p 453

Humbert F, de Kock M, Lenhardt N, Altermann W (2019) Neoarchaean to Early Palaeoproterozoic Within-Plate Volcanism of the Kaapvaal Craton: Comparing the Ventersdorp Supergroup and the Ongeluk and Hekpoort Formations (Transvaal Supergroup). In: Kröner A, Hofmann A (eds) The Archaean Geology of the Kaapvaal Craton, Southern Africa. Springer Nature Switzerland AG

Hutchinson RW, Viljoen RP (1988) Re-evaluation of gold source in Witwatersrand ores. S Afr J Geol 9:157–173

James DE, Fouch MJ, VanDecar JC, van der Lee S, Kaapvaal Seismic Group (2001) Tectospheric structure beneath southern Africa. Geophys Res Lett 28(13):2485–2488

Jansen H, Schifano G, Schutte IC, (Compilers) (1974) Sheet 2426 Thabazimbi (geological map with explanatory notes, 1:250 000 Geol Series), Geol Surv S Afr

Jones MQW (1987) Heat flow and heat production in the Namaqua Mobile Belt, South Africa. J Geophys Res 92(B7):6273–6289

Jones MQW (1988) Heat-flow in the Witwatersrand basin and environs, and its significance for the South African shield geotherm and lithosphere thickness. J Geophys Res 93:3243–3260

Jones MQW (1993) Heatflow in South Africa. Geol Surv S Afr Handbk 14:p109

Jones MQW (2017) Heat flow in the Bushveld Complex, South Africa: implications for upper mantle structure. S Afr J Geol 120(3):351–370

Jones MT (1963) The geology of the area to the east of Gaborones. An explanation of the quarter-degree sheet 2426C. Geol Surv Botswana Rec 1959/1960, pp 17–22

Jones IM, Anhaeusser CR (1993) Accretionary lapilli associated with Archaean banded iron formations of the Kraaipan Group, Amalia greenstone belt, South Africa. Precambrian Res 61:117–136

Key RM (1983) The geology of the area around Gaborone and Lobatse, Kweng, Kgatleng, Southern and Southeast Districts. Distr Mem Geol Surv Botswana 5:230

Key RM, Ayres N (2000) The 1998 edition of the National Geological Map of Botswana. J Afr Earth Sci 30(3):427–451
Key RM, Wright EP (1982) The genesis of the Gaborone rapakivi granite complex in southern Africa. J Geol Soc London 139:109–126
Keyser N, du Plessis CP (1993) The geology of the Vryburg area. Expl Sheet 2624 (1:250 000), Geol Surv S Afr, p 28
Kiefer RD (2004) Regional geology, tectonic evolution and controls of gold in the Archaean Amalia greenstone belt, Kraaipan terrane, South Africa. PhD thesis (unpubl) Univ Witwatersrand, Johannesburg, p 515
Klemd R, Hallbauer DK (1987) Hydrothermally altered peraluminous Archaean granites as a provenance model for Witwatersrand sediments. Miner Deposita 22:227–235
Koeberl C, Armstrong RA, Reimold WU (1997) Morokweng South Africa: a large impact structure of Jurassic-Cretaceous boundary age. Geology 25:731–734
Kramers JD, McCourt S, van Reenen DD (2006) The Limpopo Belt. In: Johnson MR, Anhaeusser CR, Thomas RJ (eds) The geology of South Africa. Geol Soc S Afr Johannesburg/Counc Geosci, Pretoria, pp 209–236
Lana C, Reimold WU, Gibson RL, Koeberl C, Siegund S (2004) Nature of the Archaean mid-crust in the central Kaapvaal craton, South Africa—evidence from the Vredefort dome. Geochim Cosmochim Acta 68:623–642
Lundel C (2015) Kaapvaal Craton basement exposed in the Dutoitspan diamond mine and the Bultfontein diamond mine, South Africa—Granitoids. BSc thesis (unpubl), Univ Gothenburg, Sweden, p 52
Maier WD, Barnes S-J, Gartz V, Andrews G (2003) Pt-Pd reefs in magnetitites of the Stella layered intrusion, South Africa: A world of new exploration opportunities for platinum group elements. Geology 31(10):885–888
Mapeo RBM, Armstrong RA, Kampunzu AB, Ramokate LV (2004) SHRIMP U-Pb zircon ages of granitoids from the Western Domain of the Kaapvaal Craton, southeastern Botswana: implications for crustal evolution. S Afr J Geol 107(1–2):159–172
Matthews PE (1972) Possible Precambrian obduction and plate tectonics in southeastern Africa. Nature 240:37–39
McCarthy TS (2006) The Witwatersrand Supergroup. In: Johnson MR, Anhaeusser CR, Thomas RJ (eds) The geology of South Africa. Geol Soc S Afr, Johannesburg/Counc Geosci, Pretoria, pp 155–186
McCourt S, Hilliard P, Armstrong RA (2000) SHRIMP U-Pb zircon geochronology of granitoids from the western margin of the Kaapvaal Craton: Implications for crustal evolution in the Neoarchaean. In: Kisters AFM, Thomas RJ (eds) 27th Earth Science Congress of the Geological Society of South Africa, Stellenbosch, South Africa. Abstr J Afr Earth Sci 31(1A), p 48
Minter WEL, Hill WCN, Kidger RJ, Kingsley CS, Snowden PA (1986) The Welkom Goldfield. In: Anhaeusser CR, Maske S (eds) Mineral deposits of Southern Africa, vol. I. Geol Soc S Afr, p 497–539
Moore M, Davis DW, Robb LJ, Jackson MC, Grobler DF (1993) Archaean rapakivi granite-anorthosite-rhyolite complex in the Witwatersrand hinterland, southern Africa. Geology 21:1031–1034
Moser D, Flowers RM, Hart RJ (2001) Birth of the Kaapvaal tectosphere 3.08 billion years ago. Science 291:465–468
Nelson DR, Trendall AF, Altermann W (1999) Chronological correlations between the Pilbara and Kaapvaal cratons. Precambrian Res 97:165–189
Nguuri TK, Gore J, James DE, Webb SJ, Wright C, Zengeni TG, Gwavava O, Snoke JA, Kaapvaal Seismic Group (2001) Crustal structure beneath southern Africa and its implications for the formation and evolution of the Kaapvaal and Zimbabwe cratons. Geophys Res Lett 28(13):2501–2504
Niu F, James DE (2002) Fine structure of the lowermost crust beneath the Kaapvaal Craton and its implications for crustal formation and evolution. Earth Planet Sci Lett 200:121–130
Nyblade AA, Pollack HN, Jones DL, Podmore F, Mushayandebvu M (1990) Terrestrial heat flow in east and southern Africa. J Geophys Res 95(B11):17371–17384
Poujol M, Anhaeusser CR (2001) The Johannesburg Dome, South Africa: new single zircon U–Pb isotopic evidence for early Archaean granite-greenstone development within the central Kaapvaal Craton. Precambrian Res 108:139–157
Poujol M, Anhaeusser CR, Armstrong RA (2002) Episodic granitoid emplacement in the Archaean Amalia-Kraaipan terrane, South Africa: confirmation from single zircon U–Pb geochronology. J Afr Earth Sci 35:147–161
Poujol M, Robb LJ, Anhaeusser CR, Gericke B (2003) A review of the geochronological constraints on the evolution of the Kaapvaal Craton, South Africa. Precambrian Res 127:181–213
Poujol M, Kiefer R, Robb LJ, Anhaeusser CR, Armstrong RA (2005) New U–Pb data on zircons from the Amalia greenstone belt, southern Africa: insights into the Neoarchaean evolution of the Kaapvaal Craton. S Afr J Geol 108:317–332
Pretorius DA (1965) Towards a regional synthesis of the results of the research work. Ann Rep 1964. Econ Geol Res Unit Univ Witwatersrand, Johannesburg, pp 25–27
Pretorius DA (1974) The structural boundary between the Kaapvaal and Sonama crustal provinces. Inform Circ Econ Geol Res Unit Univ Witwatersrand Johannesburg 88:27
Pretorius DA (1979) The crustal architecture of southern Africa. Alex L. du Toit Memorial Lecture No. 13, Annex vol LXXVI (1973) Trans Geol Soc S Afr, p 60
Prevec SA, Anhaeusser CR, Poujol M (2004) Evidence for Archaean lamprophyre from the Kaapvaal Craton, South Africa. S Afr J Sci 100:549–555
Ramotoroko CD, Ranganai RT, Nyabeze P (2016) Extension of the Archaean Madibe-Kraaipan granite-greenstone terrane in southeast Botswana: constraints from gravity and magnetic data. J Afr Earth Sci 123:39–56
Reed MDP (1996) 1:1 000 000 scale total field aeromagnetic grid of the southern Africa region. Kaapvaal Craton Study, Southern Africa Seismic Experiment, Anglo American Corp Geophys Serv Depart, Johannesburg
Reimold WU, Koeberl C (2014) Impact structures in Africa: a review. J Afr Earth Sci 93:57–175
Reimold WU, Colliston WP, Robertson AS (1988) Chronological and structural work in the Vredefort Dome. Abstr Geocongr '88, Geol Soc S Afr, pp 501–504
Reimold WU, Armstrong RA, Koeberl C (2002) A deep drill core from the Morokweng impact structure, South Africa: petrography, geochemistry, and constraints on the crater size. Earth Planet Sci Lett 201:221–232
Robb LJ (1991) U–Pb isotope determinations—felsic volcanics, Kraaipan Group. In: Anhaeusser CR (ed) The Archaean Kraaipan Group volcano-sedimentary rocks and associated granites and gneisses of the southwestern Transvaal, northwestern Cape Province and Bophuthatswana—Excurs Guidebk, vol 244. Inform Circ Econ Geol Res Unit, Univ Witwatersrand, Johannesburg, pp 28–30
Robb LJ, Meyer FM (1987) The nature of the Archaean basement in the hinterland of the Witwatersrand Basin: I, The Rand Anticline between Randfontein and Rysmierbult. S Afr J Geol 90:44–63
Robb LJ, Meyer FM (1990) The nature of the Witwatersrand hinterland: conjectures on the source area problem. Econ Geol 85:511–536
Robb LJ, Meyer FM (1995) The Witwatersrand Basin, South Africa: geological framework and mineralization processes. Ore Geol Rev 10:67–94
Robb LJ, Davis DW, Kamo SL (1991) Chronological framework for the Witwatersrand basin and environs: towards a time-constrained depositional model. S Afr J Geol 94:86–95

Robb LJ, Davis DW, Kamo SL, Meyer FM (1992) Ages of altered granites adjoining the Witwatersrand Basin with implications for the origin of gold and uranium. Nature 357:677–680

Robb LJ, Brandl G, Anhaeusser CR, Poujol M (2006) Archaean granitoid intrusions. In: Johnson MR, Anhaeusser CR, Thomas RJ (eds) The geology of South Africa. Geol Soc S Afr Johannesburg/Counc Geosci, Pretoria, pp 57–94

Robb LJ, Meyer FM, Ferraz MF, Drennan GR (1990) The distribution of radio-elements in Archaean granites of the Kaapvaal Craton, with implications for the source of uranium in the Witwatersrand Basin. S Afr J Geol 93:5–40

Rock NMS (1991) Lamprophyres. Blackie, Glasgow, p 285

Schmitz MD, Bowring SA, de Wit MJ, Gartz V (2004) Subduction and terrane collision stabilize the western Kaapvaal craton tectoshere 2.9 billion years ago. Earth Planet Sci Lett 222(2):363–376

Sclater JG, Jaupart C, Galson D (1980) The heat flow through oceanic and continental crust and the heat loss of the Earth. Rev Geophys Space Phys 18(1):269–311

Sibiya VB (1988) The Gaborone Granite Complex: an atypical rapakivi granite-massif anorthosite association. Free Univ Press, Amsterdam, p 317

Stanistreet IG, McCarthy TS (1991) Changing tectono-sedimentary scenarios relevant to the development of the late Archaean Witwatersrand Basin. J Afr Earth Sci 13:65–82

Stankiewicz J, Chevrot S, van der Hilst RD, de Wit MJ (2002) Crustal thickness, discontinuity depth, and upper mantle structure beneath southern Africa: constraints from body wave conversions. Phys Earth Planet Int 130:235–251

Stepto D (1979) A geological and geophysical study of the central portion of the Vredefort Dome Structure. Ph D thesis (unpubl), Univ Witwatersrand, Johannesburg, p 378

Stepto D (1990) The geology and gravity field in the central core of the Vredefort structure. Tectonophys 171:75–103

Stettler EH, de Beer JH, Eberle D, Ludden J, Mareschal M (1997) Geophysics and deep structures. In: de Wit MJ, Ashwal LD (eds) Greenstone belts, vol 35. Oxford Mono Geol Geophys, Oxford, pp 339–375

Stettler EH, du Plessis JG, Share FG (1990) Preliminary results of a geophysical investigation into the Kraaipan rocks. Ext Abstr Geocong '90, Geol Soc S Afr, Cape Town (late abstr p 25–28)

Thorpe RI, Hickman AH, Davis DW, Mortensen JK, Trendall AF (1992) U-Pb zircon geochronology of Archaean felsic units in the Marble Bar region, Pilbara Craton, Western Australia. Precambrian Res 56:169–189

Tinker J, de Wit MJ, Grotzinger J (2002) Seismic stratigraphic constraints on Neoarchean-Paleoproterozoic evolution of the western margin of the Kaapvaal Craton, South Africa. S Afr J Geol 105:107–134

Trendall AF (1968) Three great basins of Precambrian iron formation deposition: a systematic comparison. Geol Soc Am Bull 79:1527–1533

Uken R, Watkeys MK (1997) An interpretation of mafic dyke swarms and their relationship with major magmatic events on the Kaapvaal Craton and Limpopo Belt. S Afr J Geol 100:341–348

Vajner V (1974) The tectonic development of the Namaqua Mobile Belt and its foreland in parts of the northern Cape. Bull Precambrian Res Unit Univ Cape Town 14:201

van Eeden OR, de Wet NP, Strauss CA (1963) The geology of the area around Schweizer-Reneke. Expl Sheets 2724B (Pudimoe) and 2725A (Schweizer-Reneke), Geol Surv S Afr, p 76

Viljoen RP, Saager R, Viljoen MJ (1970) Some thoughts on the origin and processes responsible for the concentration of gold in the early Precambrian of southern Africa. Miner Deposita 5:164–180

Ward JHW, Wilson MGC (1998) Gold outside the Witwatersrand Basin. In: Wilson MGC, Anhaeusser CR (eds) The mineral resources of South Africa, vol 6. Handbk Counc Geosci, pp 350–386

Wela SS (2017) A lithological, petrographic and geochemical investigation of the M4 borehole core, Morokweng impact structure, South Africa. PhD thesis (unpubl), Univ Witwatersrand, Johannesburg, p 254

Wela SS, Gibson RL, Andreoli MAG (2016) Shock petrography of the target rocks and impactites from the M4 drillcore, Morokweng impact structure, South Africa. Abstr 35th Int Geol Congr, Cape Town

Wingate MTD (1998) A palaeomagnetic test of the Kaapvaal-Pilbara (Vaalbara) connection at 2.78 Ga. S Afr J Geol 101:257–274

Zandt G, Ammon CJ (1995) Continental crust composition constrained by measurements of crustal Poisson's ratio. Nature 374:152–154

Zegers TE, de Wit MJ, White SH (1998) Vaalbara, Earth's oldest assembled continent? A combined structural, geochronological, and palaeomagnetic test (PDF). Terra Nova 10(5):250–259

Zimmermann OT (1994) Aspects of the geology of the Kraaipan Group in the northern Cape Province and the Republic of Bophuthatswana. MSc thesis (unpubl) Univ Witwatersrand, Johannesburg, p 145

Zimmermann OT, Anhaeusser CR (1991) The northern Kraaipan granite-greenstone terrane. In: Anhaeusser CR (ed) The Archaean Kraaipan Group volcano-sedimentary rocks and associated granites and gneisses of the southwestern Transvaal, northwestern Cape Province and Bophuthatswana—Excurs Guidebk Inform Circ Econ Geol Res Unit, Univ Witwatersrand, Johannesburg 244, p 26–42

4 Granitoids and Greenstone Belts of the Pietersburg Block—Witnesses of an Archaean Accretionary Orogen Along the Northern Edge of the Kaapvaal Craton

Oscar Laurent, Armin Zeh, Günther Brandl, Adrien Vezinet, and Allan Wilson

Abstract

The Pietersburg Block (PB), the northernmost terrane of the Kaapvaal Craton, can be subdivided into four Archaean grey gneiss units: Makhutswi-Murchison South (MMS), Groot Letaba-Duivelskloof (GDL), Goudplaats-Hout River (GHR) and Southern Marginal Zone (SMZ). These domains are limited by greenstone belts (Pietersburg, Murchison, Rhenosterkoppies and Giyani) and major shear zones. These lithologies were intruded by a mafic-layered intrusion (the Rooiwater Complex) and voluminous syn- to post-tectonic granitoids forming either small bodies in grey gneisses or voluminous plutons/batholiths. The grey gneisses are made up of Mesoarchaean (3.12–2.85 Ga) and seldom Palaeoarchaean (3.46–3.17 Ga) TTGs formed by melting of metabasaltic rocks at >45 km depth. The greenstone belts consist of Palaeoarchaean (3.46–3.20 Ga) mafic metabasalts, metakomatiites (and intrusive counterparts) and rare Mesoarchaean (2.95–2.84 Ga) felsic volcanics, intercalated with BIF, chert and siliciclastic rocks deposited at 3.05–2.92 Ga in the Murchison belt; 2.88–2.80 Ga in the Pietersburg and Giyani belts; and 2.76–2.71 Ga in the Rhenosterkoppies belt and SMZ. All lithologies were affected by greenschist—to upper amphibolite-facies metamorphism, mainly during Mesoarchaean N–S-directed compression. The 2.97–2.96 Ga Rooiwater Complex and coeval felsic volcanics of the Murchison belt formed through fractionation of mantle-derived magma. Granitoids of the PB include Bt-(Ms-) granites formed at 2.84–2.75 Ga by melting of TTGs (± sediments), and Hbl-Bt-(Cpx-)sanukitoids and Bt-Ep-granitoids at 2.70–2.67 Ga by melting of enriched lithospheric mantle and interactions with the crust during magma ascent. After formation of a Palaeoarchaean (3.46–3.18 Ga) micro-terrane(s) in an island arc and/or oceanic plateau setting, the geodynamic evolution of the PB features a Meso- to Neoarchaean (3.15–2.65 Ga) accretionary orogen formed during protracted southward subduction. The Mesoarchaean evolution began at 3.15–2.97 Ga with the formation of an arc/back-arc system along the northern margin of the proto-Kaapvaal Craton. Subsequently, it was controlled by the docking of the Palaeoarchaean micro-terrane at 2.97–2.84 Ga, causing intense crustal reworking in an Andean-type continental margin setting (2.84–2.73 Ga) and by Neoarchaean (2.73–2.65 Ga) continent-continent collision with a crustal block located farther North, possibly the precursor (>2.7 Ga) rocks of the Central Zone of the Limpopo Belt.

Keywords

Pietersburg block • Granitoids • Greenstone belts • Geochemistry • Zircon U–Pb–Hf isotopes • Archaean accretionary orogen

Electronic supplementary material
The online version of this chapter (https://doi.org/10.1007/978-3-319-78652-0_4) contains supplementary material, which is available to authorized users.

O. Laurent (✉)
ETH Zürich, Institute for Geochemistry and Petrology, Clausiusstrasse 25, 8092 Zürich, Switzerland
e-mail: oscar.laurent@erdw.ethz.ch

A. Zeh
Mineralogie und Petrologie, Karlsruher Institut für Technologie, Campus Süd, Institut für Angewandte Geowissenschaften, Adenauerring 20b, Geb. 50.4, 76131 Karlsruhe, Germany

G. Brandl
Council for Geoscience, Limpopo Unit, Polokwane, South Africa

A. Vezinet
Department of Earth and Atmospheric Sciences, University of Alberta, T6G 2E3 Edmonton, Alberta, Canada

A. Wilson
School of Geosciences, University of the Witwatersrand, Johannesburg, 2050, South Africa

A. Kröner and A. Hofmann (eds.), *The Archaean Geology of the Kaapvaal Craton, Southern Africa*, Regional Geology Reviews, https://doi.org/10.1007/978-3-319-78652-0_4

4.1 Introduction

The Kaapvaal Craton of Southern Africa consists of several crustal domains characterized by the distinct timing of crust formation and evolution (Eglington and Armstrong 2004; Poujol et al. 2003; Zeh et al. 2009) and bounded by major intra-cratonic lineaments (Anhaeusser 2006) (Fig. 4.1a). The Pietersburg Block (PB) (Eglington and Armstrong 2004) is the northernmost of these terranes and the focus of this chapter. Between 3.1 and 2.7 Ga, rocks of the PB became accreted to the proto-Kaapvaal craton which, by that time, consisted of the Swaziland and Witwatersrand blocks formed and stabilized between 3.7 and 3.1 Ga (de Wit et al. 1992b; Poujol et al. 2003; Zeh et al. 2009, 2013; Wilson and Zeh 2018). Subsequently, collision between the Kaapvaal and Zimbabwe cratons resulted in the formation of the Limpopo mobile belt, which records two major tectono-meetamorphic/magmatic events at 2.7–2.5 Ga and ca. 2.0 Ga (Barton et al. 2006; de Wit et al. 1992b; Eriksson et al. 2009; Kramers and Mouri 2011; Kramers et al. 2006; Kröner et al. 2018; Roering et al. 1992; van Reenen et al. 1987; Zeh et al. 2009).

The PB represents a >600 km-long and >150 km-wide crustal domain, bounded by the Palala/Tshipise Shear Zones to the North, separating it from the Central Zone of the Limpopo Belt (Mason 1973; van Reenen et al. 1992), and the Murchison-Thabazimbi Lineament (MTL) to the South (Good and de Wit 1997), separating it from the Witwatersrand block (Fig. 4.1a). The MTL is, however, poorly defined at the level of exposure, and some authors place the southern limit of the PB between the Murchison greenstone belt and the Ofcolaco-Mica supracrustal suite (Anhaeusser 2006). This area, comprising the Makhutswi gneiss suite, intrusive granitoids (Brandl and Kröner 1993; Poujol and Robb 1999; Robb et al. 2006) and ultramafic bodies (Anhaeusser 2006), is considered here as part of the PB (Fig. 4.1b). Most of the published data on the PB stem from the relatively restricted (ca. 30,000 km^2) eastern part of the block (Fig. 4.1b; hereafter, the 'eastern PB'), and only minor data are from the central and western parts, comprising the Makoppa Dome exposed in the western Limpopo Province and Botswana (Fig. 4.1a). In this chapter, we mainly focus on the eastern PB, but for the sake of comparison also compile published (Anhaeusser and Poujol 2004) and new data from the Makoppa Dome.

The scientific interest in the northern Kaapvaal Craton has long been motivated by the study of Neoarchaean (ca. 2.7 Ga) granulite-facies metamorphism in the Southern Marginal Zone (SMZ) of the Limpopo Belt (Barton et al. 1992, 2006; Bohlender et al. 1992; Gerya et al. 2000;

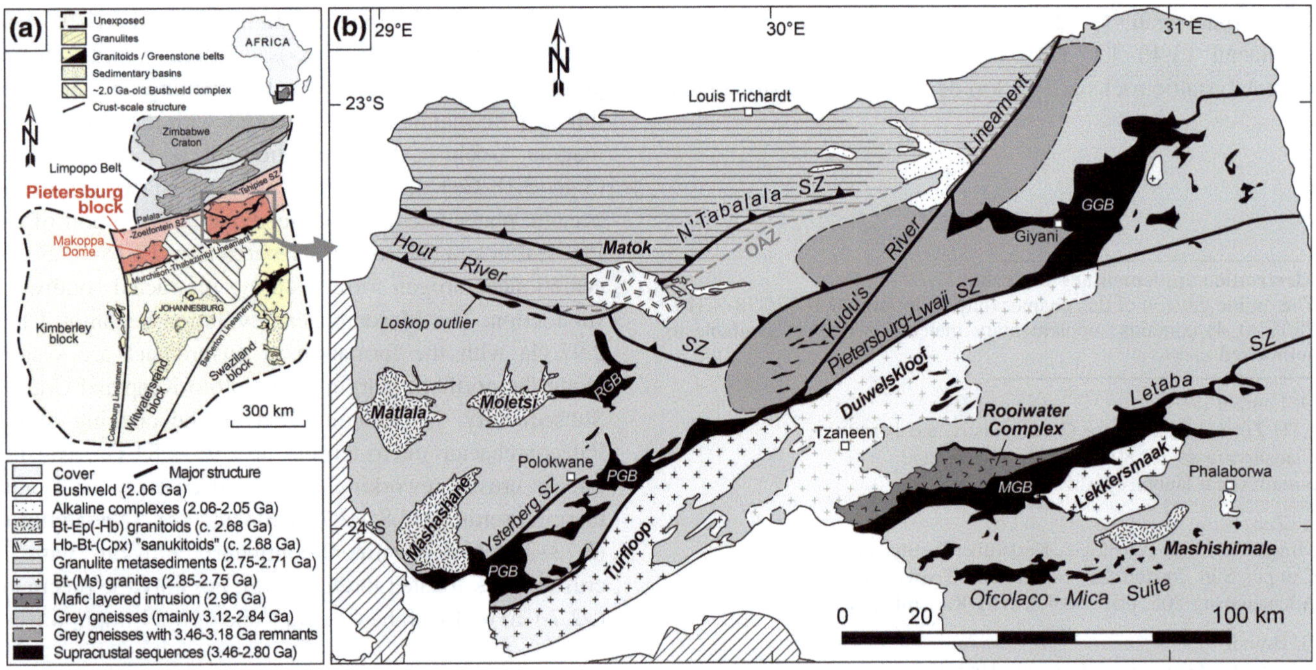

Fig. 4.1 a General structural map of the Kaapvaal Craton showing the position of major lithotectonic blocks and intervening crustal lineaments (after Anhaeusser 2006; Eglington and Armstrong 2004); b sketch geological map of the eastern Pietersburg Block (after Brandl et al. 2006; Robb et al. 2006 and the interpretations of the crustal architecture of the PB by Kramers et al. 2014; Laurent and Zeh 2015 and Vezinet et al. 2018). MGB = Murchison greenstone belt; GGB = Giyani greenstone belts; PGB = Pietersburg greenstone belt; RGB = Rhenosterkoppies greenstone belt; SZ = shear zone. The shaded, horizontally hatched area corresponds to the possible extension of the granulite-facies domain of the SMZ, i.e. to the 'orthoamhibole isograd' (van Reenen 1986) illustrated by a grey dashed line ('OAI'); see text (Sect. 4.2) for more details

Kreissig et al. 2000, 2001; Perchuk et al. 2000; Smit et al. 1992, 2001; Stevens 1997; Stevens and van Reenen 1992a, b; Stevens et al. 1997; van Reenen et al. 1992). This motivation reached an additional burst by the identification of Ultra-High-Temperature (UHT) rocks in the SMZ (Tsunogae et al. 2004; Belyanin et al. 2012), although the existence of such UHT metamorphism was disputed by Nicoli et al. (2014). These studies were accompanied by new geological, geochronological and geochemical works (Belyanin et al. 2014; Nicoli et al. 2015; Rajesh et al. 2014; Taylor et al. 2014; Tsunogae and van Reenen 2014), and provided new insights into the processes that shaped the SMZ during the Neoarchaean (van Reenen et al. 2019).

Nevertheless, preliminary dating on grey gneisses and greenstones to the South of the Hout River Shear Zone has revealed a long and complex geological history prior to Neoarchaean granulite-facies metamorphism in the SMZ. Such dating indeed showed that some rocks of the PB are as old as 3.45–3.33 Ga (Brandl and Kröner 1993; Byron and Barton 1990); and that the terrane was affected by several tectono-magmatic episodes between ca. 3.2 and 2.7 Ga (Brandl and Kröner 1993; Brandl et al. 1996; de Wit et al. 1993; Kröner et al. 2000; Poujol 2001; Poujol and Robb 1999; Poujol et al. 1996). The Palaeo- and Mesoarchaean geodynamic evolution of the PB yet remained poorly understood until the last decade, during which extensive new research has been carried out on grey gneisses and granitoids of the eastern PB (Laurent et al. 2013, 2014a, c; Laurent and Zeh 2015; Vezinet et al. 2018; Zeh et al. 2009), as well as on the greenstone belts (Block et al. 2013; Jaguin et al. 2012; Kramers et al. 2014; Zeh and Gerdes 2012; Zeh et al. 2013). The results of these studies provided new geological, geochemical and geochronological constraints to unravel the crustal architecture and geodynamic evolution of the PB.

The aim of this chapter is to (1) review the geological, petrographic and isotope geochemical characteristics of the main lithological units of the PB (greenstone belts, grey gneisses, intrusive granitoids and the Rooiwater Complex); (2) summarize the interpretation of these data in terms of petrogenesis; and (3) discuss a possible geodynamic model for crust formation and evolution in the PB in line with the compiled datasets. We provide along with this chapter compilation of literature and new whole-rock geochemistry on various PB lithologies (ESM Table 1) as well as U–Pb ages (ESM Table 2).

4.2 Crustal Architecture of the Eastern Pietersburg Block

The Archaean rocks in the eastern PB can be classified into four main lithotectonic groups:

(1) **Greenstone Belts**: Rocks of this group consist of supracrustal sequences of the Murchison, Pietersburg, Giyani and Rhenosterkoppies greenstone belts, as well as the Bandelierkop Formation in the SMZ (Fig. 4.1b). Furthermore, they form sparse and small enclaves in most grey gneiss units.
(2) **Grey gneiss Units**: These units consist mainly of orthogneisses of TTG affinity, representing the most widespread lithology in the PB (Fig. 4.1b). Here, we distinguish four grey gneiss units in the eastern PB, separated from each other by major shear zones and/or greenstone belts (Figs. 4.1b and 4.2). These are from South to North: (i) the Makhutswi-Murchison South (MMS) unit bounded by the Ofcolaco-Mica suite to the South and the Murchison Belt to the North; (ii) Groot Letaba-Duivelskloof (GLD) unit between the Murchison Belt and the Pietersburg-Lwaji Shear Zone; (iii) Goudplaats-Hout River (GHR) unit between the Pietersburg-Giyani Belts and the Hout River Shear Zone; (iv) Southern Marginal Zone (SMZ; North of the Hout River Shear Zone).
(3) **Intrusive Granitoids**: These rocks form batholiths, plutons and small bodies, which clearly show cross-cutting relationships and intrude the grey gneisses and greenstone belt lithologies.
(4) **Rooiwater Complex**: This complex comprises a mafic-layered intrusion and associated felsic plutonic rocks flanking the northern edge of the Murchison Greenstone Belt (Fig. 4.1b).

Importantly, the SMZ comprises two domains: (i) a southeastern, low-grade unit where the orthogneisses and supracrustal rocks show amphibolite-facies metamorphic assemblages; and (ii) a north-western high-grade domain, including granulite-facies orthogneisses (Bavianskloof gneiss) and supracrustal rocks (Bandelierkop Formation) (van Reenen et al. 2019). The boundary between these two domains would be represented either by the N'Tabalala Shear Zone (Vezinet et al. 2018) or the 'orthoamphibole isograd' (van Reenen 1986) (Fig. 4.1b). The low-grade domain would represent either a unit of former high-grade (granulite-facies) rocks that were retrogressed in the amphibolite-facies (see van Reenen et al. 2019), or a distinct grey gneiss unit that never underwent granulite-facies metamorphism (Vezinet et al. 2018).

In this chapter, we follow the latter revision of the lithotectonic model of the SMZ by Vezinet et al. (2018) (Fig. 4.1b). The resulting interpretations are nevertheless not univocal, and the reader is referred to van Reenen et al. (2019) (and references therein) for a different view. We importantly note that either of these two possibilities can be reconciled with the general geodynamic model presented here (Sect. 4.7) for crust formation and evolution in the PB.

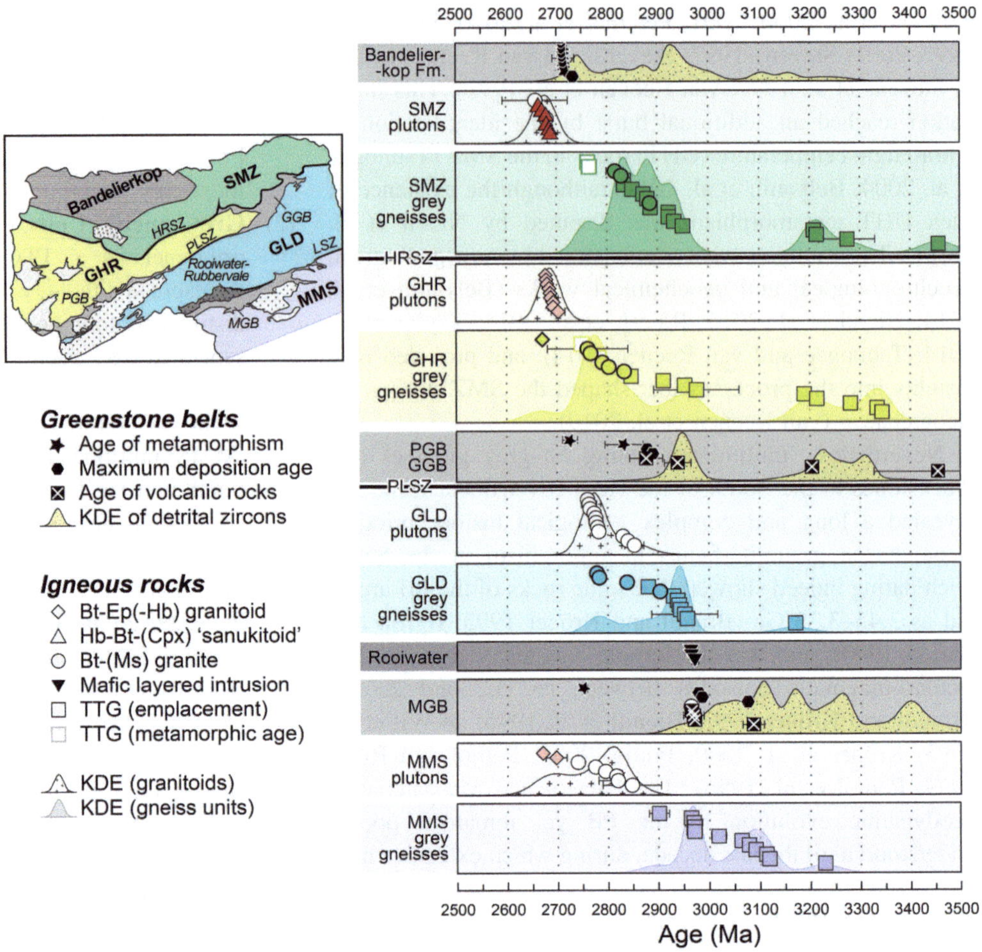

Fig. 4.2 Compilation of geochronological data from the eastern PB (Poujol et al. 2003 with new data from Kramers et al. 2014; Laurent and Zeh 2015; Laurent et al. 2013; Nicoli et al. 2015; Rajesh et al. 2014; Vezinet et al. 2018; Zeh and Gerdes 2012; Zeh et al. 2009; 2013—see ESM Table 2). The data are presented according to the geographic distribution of the dated samples in the main lithotectonic units of the eastern PB (shown on the inset map). The probability curves are Kernel Density Estimates generated using the DensityPlotter programme of Vermeesch (2012); using emplacement ages for grey gneiss units and plutons, and individual detrital zircon ages for the greenstone belts

4.3 Greenstone Belts

This section intentionally excludes the metasedimentary rocks of the Bandelierkop Formation because these are the focus of another chapter in this volume (van Reenen et al. 2019). All mineral abbreviations in the chapter are after Whitney and Evans (2010).

4.3.1 Murchison Greenstone Belt

The Murchison greenstone belt is about 140 km-long and at most 20 km-wide (Fig. 4.1b). It has a pronounced ENE orientation, considered to define the trace of the crustal-scale Murchison-Thabazimbi Lineament at its easternmost section (Good and de Wit 1997) (Fig. 4.1). Maximum depth of the belt as revealed by geophysical studies is between 8.8 and 12.3 km, but generally, it is less than 4.5 km (de Beer et al. 1984). Vearncombe et al. (1992) subdivided the belt into four domains: Silwana Amphibolite, Rubbervale and La France Formations and the prevalent Murchison Unit. The contacts between the different units are almost nowhere exposed but are likely of a tectonic nature (Block et al. 2013).

The **Silwana Amphibolite** forms a 300 m-thick succession along the NE flank of the belt. It comprises intensely deformed amphibolite, Hbl-schist and along the southern contact, a thin band of Bt-schist (possibly representing a strongly tectonized zone). The **Rubbervale Formation** occurs along the north-western margin of the greenstone belt, in close vicinity to the Rooiwater Complex (see Sect. 4.5). It comprises a northern unit of deformed Qz-feldspar porphyries, made up of bluish Qz clasts and sericitised Pl in a fine matrix of Qz, Ab and Ms (Fig. 4.3a). The southern unit consists of felsic lavas, well-bedded tuff and volcanic breccia. The two units are separated by a prominent tectonic break. Ages of ca. 2.97 Ga were obtained for the felsic lavas (Brandl et al. 1996; Poujol et al. 1996; Zeh et al. 2013).

The **La France Formation** comprises fuchsitic quartzite and Qz-mica schists. The former is pervasively deformed with no primary sedimentary structures preserved. The schists contain either Bt, Chl or white mica, and locally Grt, St and Ky overgrowing the foliation. Detrital zircon grains in

Fig. 4.3 Selected field pictures showing the main characteristics of greenstone belts in the PB. **a** Felsic, volcanic quartz-bearing porphyry (bluish quartz phenocrysts are circled) from the Rubbervale Formation in the Murchison greenstone belt, coeval with the Rooiwater Complex (ca. 2.96 Ga; Brandl et al. 1996; Poujol et al. 1996; Zeh et al. 2013) and genetically linked with it (see Sect. 4.5 for details). **b** Greenschist-facies meta-volcanic, mafic rock (Qz + Ab + Chl ± Tlc) of the Murchison Unit (Murchison Greenstone Belt) displaying a sub-vertical foliation. **c** Landscape picture showing the stack of thrust sheets in the Mount Maré area SW of Polokwane (Pietersburg Greenstone Belt); the thrust sheets (D2 event) are NE–SW-striking, dip to the SE and are NW-verging, leading to tectonic imbrication of the $\leq$2.88 Ga-old Uitkyk Formation quartz-rich metasediments (small hill to the left) and the $\geq$2.95 Ga-old 'simatic basement' (prominent hill to the right). **d** Medium-grade amphibolite of the Giyani Greenstone Belt displaying a series of well-preserved pillow structures. **e** Thinly laminated and strongly foliated Banded Iron Formation of the Giyani Greenstone Belt. **f** Metapelite from the Giyani Greenstone Belt (Qz + Ser + Chl + Ab) containing plurimillimetric phenoblasts of andalusite (And). **g** Medium-grained amphibolite cropping out as a large enclave in the Hout River grey gneiss unit close to the western boundary of the Rhenosterkoppies Greenstone Belt

the La France Formation suggest a maximum deposition age of ca. 2.99 Ga (Zeh et al. 2013; Fig. 4.2).

The **Murchison Unit** includes the MacKop, Weigel, Leydsdorp and Mulati Formations of SACS (1980). It consists of (ultra)mafic volcanic rocks, sedimentary and volcano-sedimentary rocks. The ultramafic rocks (serpentinite, Chl ± Tlc-bearing schists) occur along the southern flank of the belt as discontinuous lenses. The mafic rocks include mainly Ab-Chl-Act schists and Qz-Chl-schists (Fig. 4.3b), representing strongly altered basalts as shown by the preservation of pillow structures in the less deformed zones. The Murchison Unit also contains up to 200 m-wide layers of rocks ranging from iron-poor chert to BIF, the latter being commonly interlayered with mafic lavas or clastic sediments. The latter form steep ridges ('bars') and range from rare massive quartzite to Qz-mica schist (containing locally fuchsite and/or Chl) and rare, thin conglomerate horizons (Vearncombe et al. 1992). The U–Pb ages of detrital zircon grains in the siliciclastic sediments of the Murchison unit yielded a similar maximum deposition age as in the La France Formation, i.e. 2.98 Ga (Zeh et al. 2013; Fig. 4.2).

The rocks of the Murchison belt are strongly deformed, showing ENE-trending isoclinal folds and vertical foliations (Fig. 4.3b) with steeply plunging lineations (Jaguin et al. 2012). The kinematic criteria indicate tectonic movements ranging from strike-slip, left-lateral shearing to top-to-the-S thrusting, e.g. along the Antimony Line (Jaguin et al. 2012; Vearncombe et al. 1992). The latter is a 35 km-long, up to 250 m-wide steeply N-dipping (Sb-)mineralised shear zone within the Murchison Unit, developed early in the deformation history. In addition, the different domains record contrasted metamorphic conditions, from lower greenschist-facies in the Murchison Unit (4–6 kbar, ca. 550 °C) to middle and upper amphibolite-facies in the La France Formation (6–8 kbar, 550–650 °C) and Silwana amphibolites (9 kbar, 680 °C) respectively (Block et al. 2013). Combined, pressure–temperature–deformation paths suggest that juxtaposition of the domains was accommodated by S-verging thrusting (Block et al. 2013). This structural-metamorphic pattern resulted from an N–S compressional stress field, which affected the Murchison Belt between 2.97 and 2.92 Ga, indicated by U–Pb dating of zircons in syntectonic granitoids and monazite in a Qz-mica schist of the La France formation (Block et al. 2013; Jaguin 2012).

The Murchison greenstone belt is most famous for hosting the world's largest Sb-single deposit in Archaean rocks. Most of the Sb (and by-product Au) occurs along the Antimony Line as stibnite-bearing Qz-Cal veins which occupy tension gashes (Vearncombe et al. 1992). Prominent Cu–Zn mineralisation is restricted to the narrow Cu–Zn line developed at the top of the Rubbervale Formation. Mineralisation consists of VMS-like, zoned stratiform lenses of massive volcanogenic sulphides (Py + Po) containing Ccp at the bottom and Sph at the top (Schwarz-Schampera et al. 2010; Taylor 1981).

4.3.2 Pietersburg Greenstone Belt

The Pietersburg Greenstone Belt is an elongated, NE-trending feature with a length of 120 km and a maximum width of 20 km (Fig. 4.1b). It comprises two major tectono-stratigraphic sequences which are separated by a well-defined unconformity:

(1) The '***simatic basement***', i.e. a lower, mainly (ultra)-mafic sequence (Jones 1989; de Wit et al. 1992a; Brandl and de Wit 1997; Brandl et al. 2006; Zeh and Gerdes 2012) separated into the Mothiba, Eersteling, Vrischgewaagd and Ysterberg Formations with no particular stratigraphy implied.
(2) The **Uitkyk Formation** (SACS 1980), an upper sequence dominated by siliciclastic rocks.

The **Mothiba Formation** consists mostly of massive to foliated serpentinite (mainly Atg with minor Tlc, Tr, Chl, relict Ol (Fo) and Chr). The infiltration of high-CO_2 fluids led to the local development of ophicarbonate, talc-carbonate schist and quartz-carbonate rocks. The protoliths were most probably lherzolites, komatiites or komatiitic basalts, emplaced either as sills (de Wit et al. 1992a) or flows (Grobler 1972). The Mothiba formation also includes a layered complex of altered peridotite and pyroxenite (Rooikop Complex). BIF occur as prominent horizons (up to 50 m-thick), discontinuous lenses or mega-boudins up to 20 m long. Both oxide-facies (Mag + Gth) and silicate-facies (Gru + Fe-Act ± Mag, sulphides) varieties are present, consisting of alternating microcrystalline chert/jasper and iron-rich minerals (Franey 1987).

The **Eersteling Formation**, the most widespread unit of the belt, comprises mainly amphibolites representing basic, tholeiitic extrusive rocks. They contain minor intercalations of BIF, metagabbro and ultramafic rocks. The amphibolites range from massive to well-foliated, medium- to fine-grained and are composed of Amp (Act or Hbl), Qz, Ep, Pl and minor Ttn, Grt and Cpx. Prominent pillow structures and lava tubes (up to 4 m long and 0.7 m across) are commonly well developed. They locally consist of thin concentric zones of spherulites at the margins and Qz-filled, former elongated gas cavities. Rare, coarse-grained metagabbro represents intrusive bodies or concordant sills (Franey 1987; de Wit et al. 1992a). An imprecise Pb–Pb isochron age of 3455 +120/−128 Ma has been reported for the amphibolites of the Eersteling Formation by Byron and Barton (1990). The **Vrischgewaagd Formation** comprises fine-grained, schistose mafic rocks (Qz + Chl + Cal/Mgs ± Ser, Ms, Pl) displaying a sub-vertical foliation. Their close spatial relationship with the amphibolites of the Eersteling Formation suggests that they represent strongly sheared varieties of the latter. The **Ysterberg Formation** is made up of felsic volcanics, i.e. massive flows, tuffs and Qz-porphyries, dated at 2.95 Ga (Kröner et al. 2000) (Fig. 4.2). They are rare and interspersed throughout the belt, but locally reach up to 50–300 m in thickness and form concordant to semi-concordant horizons of limited extent within the amphibolites of the Eersteling Formation.

The **Uitkyk Formation** comprises an up to 1000 m-thick sequence of steeply south-dipping beds of siliciclastic rocks (conglomerate, quartzite, greywacke and minor shale). Its upper contact with the greenstones is considered to be a thrust, whereas, the lower contact defines an angular unconformity (de Wit et al. 1993). The conglomerates are matrix-supported, poorly sorted and polymict with clasts of greenstones, granitoid gneisses and earlier quartzite. The presence of flyspeck carbon and buckshot pyrite is noteworthy (Saager and Muff 1980). The quartzites and greywackes are massive to bedded, with local cross-bedding accentuated by fuchsitic mica. The maximum age of deposition of the Uitkyk sediments is ca. 2.88 Ga, as indicated by detrital zircon grains (Zeh and Gerdes 2012) (Fig. 4.2).

The Pietersburg belt was affected by three main deformational events D1, D2 and D3/D4. During D1 (unknown age), open to isoclinal folds with E–W trending axial planes were produced, together with a weak to strong schistosity. The D2 event marks a period of major, top-to-the-NW thrusting characterized by a large fan thrust complex comprising >10 individual thrusts (Jones 1989; de Wit et al. 1992a) (Fig. 4.3c). This event was responsible for tectonic stacking of the 'simatic basement' and Uitkyk Formation (Fig. 4.3c) and thrust-related stratigraphic repetition (de Wit et al. 1992a). In the NE part of the belt, NE-trending D2 shear zones deformed granitoid bodies emplaced in the belt, which yielded a zircon U–Pb crystallization age of 2853 +19/−18 Ma (Kröner et al. 2000). This sets a maximum age for the D2 deformation. Several left-lateral, NE-striking strike-slip shear zones characterize the late D3/D4 event and also affected the adjacent granitoids. The most prominent is the Ysterberg shear zone, a transpressional system forming a positive flower structure (de Wit et al. 1992a). The belt seems to have been affected by only one major metamorphic event. In the SW part, both the 'simatic basement' and the Uitkyk Formation underwent greenschist-facies conditions whereas farther NE, the grade of metamorphism changes to lower to upper amphibolite-facies. The age of metamorphism is unknown, but by analogy to the adjacent (and possibly correlated) Giyani belt, it could have taken place between 2.87 and 2.84 Ga (Kramers et al. 2014) (Fig. 4.2), i.e. simultaneous with D2 thrusting.

The Pietersburg belt can thus be interpreted as a sequence of mafic to ultramafic rocks developed in an oceanic environment (the 'simatic basement') between 3.46 and >2.95 Ga, and overlain by molasse-filled, interlinked deposits (the Uitkyk Formation) developed in a terrestrial foreland basin at <2.88 Ga (de Wit et al. 1992a, 1993; Zeh and Gerdes 2012). Subsequently, a main tectono-metamorphic event at 2.87–2.84 Ga, related to N–S directed horizontal shortening, resulted in thrust-related stacking of these lithologies.

Gold is the principal mineralisation in the Pietersburg belt, hosted in several E–W-trending sub-vertical shear zones ('reefs'). It was at Eersteling that the first reef gold in South Africa was discovered in 1871 (Franey 1987; Byron and Barton 1990). The mineralizing fluids were guided by the D2 thrusts (Kalbskopf and Barton 2003), setting the age of the gold mineralisation to be younger than 2.85 Ga.

4.3.3 Giyani Greenstone Belt

The Giyani greenstone belt is about 15 km-wide and 70 km-long (Fig. 4.1b). Towards the SW, it splits into a northern Khavagari arm and a southern Lwaji arm. The latter is bounded to the South by the Pietersburg-Lwaji Shear Zone (Fig. 4.1b), a left-lateral, strike-slip feature (Kramers et al. 2014; Laurent et al. 2013). Geophysical data suggests a mean depth extent of only 1.5 km, but reaching up to 3 km in the SW (Kleywegt et al. 1987; de Beer and Stettler 1992).

The greenstone belt is poorly exposed, so that earlier workers have made no attempt to define a stratigraphic sequence, and all lithologies are assigned to the Giyani Group (SACS 1980). The rare outcrops show the predominance of (ultra)mafic rocks, locally associated with BIF, felsic schists and metasediments (Henzen and Steidle 2000; Kramers et al. 2014). Amphibolites (Act or Hbl + Pl + Qz) are dominant and comprise banded and spotted varieties. The banded variety commonly occurs close to shear zones and in association with BIF, suggesting that it was most probably basalt. Moreover, lensoid, coarse-grained and less deformed domains are often associated with pillow structures (Fig. 4.3d). The spotted amphibolite has a more homogeneous texture and occasionally retains relict magmatic Cpx mantled by Hbl, suggesting a gabbro protolith (Kramers et al. 2014). Minor occurrences of metabasaltic andesite (Amp + Qz + Pl) intercalated with amphibolite occur at Giyani town and yielded a zircon U–Pb age of 3.20 Ga (Kröner et al. 2000).

The ultramafic rocks occur as strongly foliated Act-Chl (+ Mag ± Rt) and Act-Tlc schists. The latter contain relict Pl and textural evidence for replacement of Ol and Opx by Tlc, suggesting that the original rock was a peridotite or pyroxenite. Their close association with BIF suggests that they were sills. In the Khavagari arm, altered dunites of the Leonde intrusion (serpentinised Ol, Tlc and minor Ath) are considered to represent the basal cumulate of a layered ultramafic complex. In the northernmost part of the belt, a spotted ultramafic rock is present, comprising large Ol within a matrix of Tr (Pretorius et al. 1988). At short distance East of the belt (Busizi Hills), the surrounding grey gneisses contain a raft of serpentinite body flanked by ultramafic schist and pyroxenite (Aug + Hyp ± Pl, Mag) (Vorster 1979; Anhaeusser 2006).

BIF horizons (Fig. 4.3e) up to several metres across and comprising Qz, Act and Mag (commonly weathered to Gth and Lm) are intercalated in the (ultra)mafic rocks. Locally, they grade at the top into a ferruginous dolomite. Clastic metasediments, comprising mainly metagreywacke and metapelite, are quite common and locally show graded bedding (Kramers et al. 2014). The metapelites consist mainly of Qz, Ser, Chl and Ab (Fig. 4.3f), with the metagreywackes containing additional Kfs, Bt, Ep and Ttn. Porphyroblasts of pre-tectonic, sericitized And were locally observed (Fig. 4.3f) and may have resulted from contact metamorphism related to the intrusion of mafic sills (Kramers et al. 2014).

According to Kramers et al. (2014), the Giyani belt consists of two domains with contrasted tectono-metamorphic histories. The main domain in the central and SE parts of the belt is characterized by a single generation of early structures (D1), consisting of moderately to steeply NW-dipping S1 schistosity and axial planes of open to isoclinal folds, having shallowly NE-plunging lineation and axes, respectively (Kramers et al. 2014). The rocks show prograde greenschist to lower amphibolite metamorphic grade, dated at 2.87–2.80 Ga by Pb–Pb stepwise leaching on Grt from a metapelite (Kramers et al. 2014), which probably corresponds to the age of D1 deformation as well. This is supported by a Pb–Pb zircon evaporation age of 2.87 Ga obtained for a weakly foliated feldspar porphyry in the central part of the belt (Kröner et al. 2000) (Fig. 4.2). In contrast, in the NW part of the belt, D1 structures are overprinted by a number of steeply N- and NW-dipping, discontinuous D2 shear zones with an oblique to reverse, SW-verging sense of movement (de Wit et al. 1992c; Kramers et al. 2014). Close to D2 shear zones, metapelites carry

Grt and St porphyroblasts and contain thin laminated horizons of Ky-bearing quartzitic schists (Qz + Ky + Ms + Rt; McCourt and van Reenen 1992; de Wit et al. 1992c; Henzen and Steidle 2000) with local replacement of Ky by Sil (Kramers et al. 2014). This indicates upper amphibolite-facies peak metamorphism of >4 kbar and >500 °C, with post-kinematic retrogression to greenschist-facies (Henzen and Steidle 2000; Kramers et al. 2014). This event was dated at 2.74–2.71 Ga by Pb–Pb stepwise leaching on Grt and St (Kreissig et al. 2001; Kramers et al. 2014).

From this dataset, the Giyani belt is interpreted as a dominantly mafic sequence with minor siliciclastic rocks, which underwent NW-vergent thrusting at greenschist- to lower amphibolite-facies conditions at 2.87–2.80 Ga (Kramers et al. 2014). This is quite similar to the lithologies and tectono-metamorphic evolution observed in the Pietersburg Greenstone Belt, suggesting that there is a correlation between the latter and the Giyani belt (Kramers et al. 2014). The D2 shear zones are considered as subsidiaries of the SW-verging Hout River Shear Zone (Fig. 4.1b) (de Wit et al. 1992c; McCourt and van Reenen 1992).

The Giyani belt contains several mineral deposits, mainly gold, magnesite and various pegmatite-related minerals. The more significant gold occurrences (4 major mines, 10 t of Au extracted) are all concentrated along the NW margin of the belt and related to the D2 shear zones. The mineralisation occurs preferentially in BIF and mafic meta-volcanic rocks (Pretorius et al. 1988; McCourt and van Reenen 1992) and is related to the percolation of CO_2-rich brines along the D2 structures (van Reenen et al. 1994).

4.3.4 Rhenosterkoppies Greenstone Belt

The small Rhenosterkoppies belt occurs close to the Hout River Shear Zone and includes the Loskop outlier situated along the latter (Fig. 4.1b). The belt is up to 10 km-wide, with two narrow limbs extending west- and northwestwards.

The belt comprises a thin (300–500 m) tectono-stratigraphic succession, originally thought to be part of the Pietersburg greenstone belt (SACS 1980). It is dominated by amphibolite and BIF, with intercalated ultramafic schist, serpentinite, quartzose rocks, metapelite, calc–silicate rocks and Ky-bearing quartzose schist (Passeraub 1998; Passeraub et al. 1999; Brandl et al. 2006). De Beer and Stettler (1992) reported that the belt has no gravity signature and consequently must have a very limited depth extent.

The amphibolites are characterized by banded varieties (leuco- to melanocratic) in the lower part and spotted variety (Grt porphyroclasts or small boudins) in the upper part of the succession. They consist mainly of Hbl, Pl and Qz (Fig. 4.3g), with additional Act, Tr, Mag, Bt and Grt, and locally relict Cpx. The latter phase may indicate the presence of minor metagabbro, but not granulite-facies conditions. The BIF succession consists of six distinct individual layers, the basal one occurring throughout the belt. The rock displays alternating laminae of Qz and Mag (ca. 40%) + Act/Gru, Ap and Cal (Passeraub, 1998; Passeraub et al. 1999).

The ultramafic schists comprise Tr-Act, Chl, Mgs, Ath, Tlc and opaques. Tr and Ath may post-tectonically overgrow an earlier fabric defined by Chl and a first generation of Tr. Mgs is a further retrograde, post-tectonic phase replacing Tr. Undeformed serpentinite (Atg, Mag, Chl, Mgs and relict Ol) has been observed in the north-eastern part of the belt. Quartzose rocks (Qz ± Pl, Kfs, Bt, Grt, Ky and Cum) occur as concordant, decametric intercalations. Metapelite occurs as rare, thin horizons in layered amphibolite and contains mostly Grt and Bt with minor Ser, Chl, St, Ab and Qz. In the latter two rock types, Grt, Ky and/or St occur as cogenetic, pre-kinematic phases whereas Bt, Chl and Ser defines the matrix. This suggests upper amphibolite-facies peak metamorphism followed by deformation under retrograde, upper greenschist-facies conditions and precludes that Ky growth was caused by S-verging thrusting ('hot iron effect' of Van Reenen et al. 2018) of the Southern Marginal Zone over the Rhenosterkoppies belt (Passeraub 1998; Passeraub et al. 1999; Kramers et al. 2014).

A banded layer of calc–silicate rocks is sporadically present at the top of the succession and contains the metamorphic assemblage Di, Hbl, Cal, Qz and Ttn. The protolith was probably a calcareous shale. Finally, a Ky-bearing quartzose schist crops out in the north-western arm as a discordant, 50 m-wide and 1 km-long horizon containing Qz, Ky, Sil and Tur together with fuchsite. Ky appears to be pre-tectonic, whereas fuchsite and Tur define the matrix and are both syntectonic. This paragenesis may be related to hydrothermal activity in the northernmost part of the belt, close to the Hout River Shear Zone. The presence of kyanite suggests temperatures exceeding 450 °C and pressures >5 kbar (Passeraub 1998; Passeraub et al. 1999).

The Rhenosterkoppies belt is characterized by three generations of structures. D1 corresponds to intense isoclinal folding, responsible for local tectonic thickening of the BIF horizons. The main deformational episode D2 involves N-verging thrusting and shearing under upper greenschist- to lower amphibolite-facies conditions. Dating of syn- and post-tectonic titanite and epidote constrains the D2 event between 2.76 and 2.71 Ga (Passeraub et al. 1999). The final D3 episode corresponds to large-scale, open folding responsible for the broad antiformal structure of the belt, with a shallowly SE-plunging axis (Passeraub 1998; Passeraub et al. 1999).

4.4 Grey Gneiss Units and Granitoids

4.4.1 Grey Gneiss Units in the Eastern PB

The grey gneiss units cover the widest area in the eastern PB. They are grey gneisses in the sense of Moyen (2011), i.e. an intricate mixture of various rock types including granitoid orthogneisses, as well as meta-mafic and metasedimentary rocks overprinted mostly under amphibolite-facies conditions. The rock textures range from macroscopically isotropic to strongly foliated (Robb et al. 2006; Fig. 4.4a–d). The foliation generally strikes ENE–WSW with sub-vertical dip, but can vary significantly on a local scale. In the southern part of the SMZ the foliation is mostly sub-horizontal (Kramers et al. 2014; Smit et al. 2014).

The orthogneisses belong to the Na-rich Tonalite–Trondhjemite–Granodiorite (TTG) suite and represent the prevailing (and locally exclusive) rock type of the gneiss units (Anhaeusser 1992; Kramers et al. 2006; Robb et al. 2006; Laurent et al. 2013, 2014b; Laurent and Zeh 2015; Vezinet et al. 2018). Trondhjemitic gneisses are predominant. They are leuco- to mesocratic, greyish and locally show banded structures. The least deformed gneisses are equigranular, medium- to fine-grained (Fig. 4.4c), but some varieties contain Pl phenocrysts. The trondhjemites locally form discordant layers and bodies in fine-grained tonalitic gneisses (Fig. 4.4b). The latter occurs mostly in the eastern GHR unit and SMZ. Granodioritic varieties are scarce. The typical mineral association of the TTGs is Pl (mostly oligoclase) + Qz + Bt ± Hbl ± Ep and rare (<5%) Kfs, with accessory Ap and Zrn (Laurent et al. 2014a; Vezinet et al. 2018). Some rare TTGs associated with the Bandelierkop Formation in the SMZ contain Opx rather than Bt or Hbl, indicating granulite-facies conditions (Bohlender et al. 1992; Vezinet et al. 2018) (Fig. 4.1b).

Zircon U–Pb dating revealed two periods of TTG intrusion in the PB: (i) a minor Palaeoarchaean period between 3.46 and 3.18 Ga; and (ii) a major Mesoarchaean period between 3.12 and 2.85 Ga (Kröner et al. 2000; Laurent and Zeh 2015; Vezinet et al. 2018; Zeh et al. 2009) (Fig. 4.2). Palaeoarchaean TTGs with ages ≥ 3.18 Ga are scarce and occur almost exclusively in a NE–SW-trending strip in the GHR and SMZ (9 out of 11 samples) (Figs. 4.1b and 4.2). U–Pb zircon ages also reveal that the Mesoarchaean TTGs become systematically younger from South to North. TTGs with ages of 3.12–3.02 Ga are restricted to the MMS unit, whereas TTGs in the MMS and GLD units intruded mainly at 2.97–2.94 Ga (10 out of 12 samples) and those in the GHR unit and SMZ were emplaced continuously over a period ranging from 2.95 to 2.85 Ga (Fig. 4.2). Zircon grains in Opx-bearing TTGs of the Banderlierskop Formation of the SMZ yielded significantly younger ages of ca. 2.71 Ga (Vezinet et al. 2018). This age corresponds to the high-grade metamorphic event identified within the Bandelierkop Formation (Rajesh et al. 2014; Taylor et al. 2014; Nicoli et al. 2015), but has never been reported for TTG emplacement elsewhere in the eastern PB (Vezinet et al. 2018).

Geochemically, the orthogneisses are typical members of the TTG suite (Moyen and Martin 2012). They are slightly peraluminous, low-K (generally $K_2O < 2$ wt%) and Na-rich rocks with distinctively low K_2O/Na_2O ratios (<0.5), characterized by a moderate range of SiO_2 (65–75 wt%), low contents of ferromagnesian oxides ($FeO_t + MgO < 5$ wt%) and variable Mg# (0.2–0.6) (Fig. 4.5). Tonalitic varieties are slightly more mafic with SiO_2 contents below 70 wt% and $FeO_t + MgO$ around 5 wt% (Fig. 4.5). Most TTGs show fractionated REE patterns due to moderate LREE and low HREE contents, elevated Sr concentrations, and Nb–Ta and Ti negative anomalies (Fig. 4.5). The compositions of the TTGs from all grey gneiss units (MMS, GLD, GHR and SMZ, see Fig. 4.2) overlap (Fig. 4.5), suggesting a similar origin.

The major element composition of the PB TTGs is consistent with melting of relatively low-K mafic rocks (Fig. 4.6), as is suggested for TTGs in general (Martin et al. 2014; Moyen and Martin 2012). Trace element systematics reveal that they mostly belong to the 'high-pressure' TTG group of Moyen (2011), characterized by low Y and HREE contents and conversely high Sr/Y, La/Yb and Ce/Sr ratios (Fig. 4.7). This entails that garnet is a major phase of the residual assemblage, requiring melting pressures in excess of 1.5 GPa (Johnson et al. 2013; Moyen and Stevens 2006; Nagel et al. 2012). Notable exceptions are the Palaeoarchaean TTGs within the GHR and SMZ units. These are characterized by high Y, HREE and low Sr contents that makes them akin to the 'low-pressure' group (Fig. 4.7).

In addition to TTGs, the grey gneiss units frequently contain amphibolites (Fig. 4.3g), consisting of Hbl + Pl with minor Qz, Kfs, accessory Ap, Zrn, Ttn and secondary Chl, Act and Ser. They form concordant bodies (Fig. 4.4a) and remnants ranging in size from a few decimetres to several hundreds of metres. Given that these seem to be most abundant in the vicinity of the greenstone belts, they could represent remnants thereof (Robb et al. 2006), possibly incorporated in the original TTG magmas upon their emplacement. Geochemical data of these mafic remnants (Kreissig et al. 2000; Laurent et al. 2014a; Vezinet 2016) show typical compositions for Archaean mafic rocks (Condie 1981), i.e. tholeiitic basalts (ca. 50–55 wt% SiO_2) with mostly very low K_2O (<0.2 wt%) and K_2O/Na_2O ratios (<0.5), high FeOt + MgO contents (10–30 wt%) and high Mg# in the range 0.60–0.75 (Fig. 4.5). These characteristics clearly point to a mantle origin. The trace element

Fig. 4.4 Selected field pictures showing the main characteristics of grey gneisses and intrusive granitoids in the PB. **a** Grey gneisses of the GLD unit characterized by sub-vertical foliation and lithological banding, consisting dominantly of migmatitic, 2.95 Ga-old TTG and amphibolite [Amp] with (undated) granite [g] bodies. **b** River bed pavement exposing grey gneisses of the GHR unit (Goudplaats locality), comprising 3.34–3.28 Ga dark, tonalitic gneisses and 2.85 Ga trondhjemitic layers, cross-cut by 2.77 Ga coarse-grained granitic patches and a dyke of 2.67 Ga Bt-Ep-granite (zircon U–Pb dates from Laurent and Zeh 2015). **c** Undeformed, 2.92 Ga trondhjemite (Pl + Qz + Bt ± Ep, Ap, Zrn) in the SMZ (U–Pb date from Vezinet et al. 2018). **d** Foliated and folded, 2.92 Ga trondhjemitic gneiss from the SMZ, intruded by pinkish, ≤2.84 Ga-old coarse-grained granite [g] (U–Pb dates from Vezinet et al. 2018). **e** Weakly deformed, equigranular Bt-(Ms) granite of the Turfloop batholith (ca. 2.77 Ga; Henderson et al. 2000; Laurent and Zeh 2015; Zeh et al. 2009) containing polycrystalline Bt patches. **f** Vaalpenskraal trondhjemitic orthogneiss (ca. 3.06 Ga; Anhaeusser and Poujol 2004) from the Makoppa Dome, western PB. **g** Homogeneous Makoppa granodiorite (ca. 2.84 Ga; Anhaeusser and Poujol 2004) from the Makoppa Dome, western PB. **h** Coarse-grained, Kfs-porphyritic pinkish granite of the Mashashane pluton (ca. 2.68 Ga; Laurent et al. 2013) containing typical, centimetric Bt–Ep clots. **i** Figures of magma mingling in the Matok pluton where diorite [d] interacts with granitic liquid to form a hybrid, Hbl–Bt granodiorite [gd] dated at 2.68 Ga (U–Pb date from Laurent et al. 2013). **j** Kfs-porphyritic Hbl–Bt granodiorite of the Matok pluton (ca. 2.68 Ga; Laurent et al. 2013; Zeh et al. 2009) containing a small enclave of diorite [d]

systematics depict a non-depleted to slightly enriched mantle source, similar to other Archaean mafic rocks (Bédard 2018; Moyen and Laurent 2018). They are characterized by flat incompatible element patterns at 3–4 times the primitive mantle compositions (Fig. 4.5) and consistent with a primitive to slightly enriched mantle source (Pearce 2008; Fig. 4.8). Interaction with the surrounding TTGs and/or granites could account for the elevated concentrations in LILE (Cs, Rb, Ba, Pb and Sr) and U as well as some outliers characterized by higher K_2O, low Mg# and low FeO_t + MgO (Fig. 4.5).

4.4.2 Intrusive Granitoids

This group encompasses all granitoids showing intrusive contacts with the grey gneisses and greenstone belt lithologies. They mostly consist of K-rich granites and

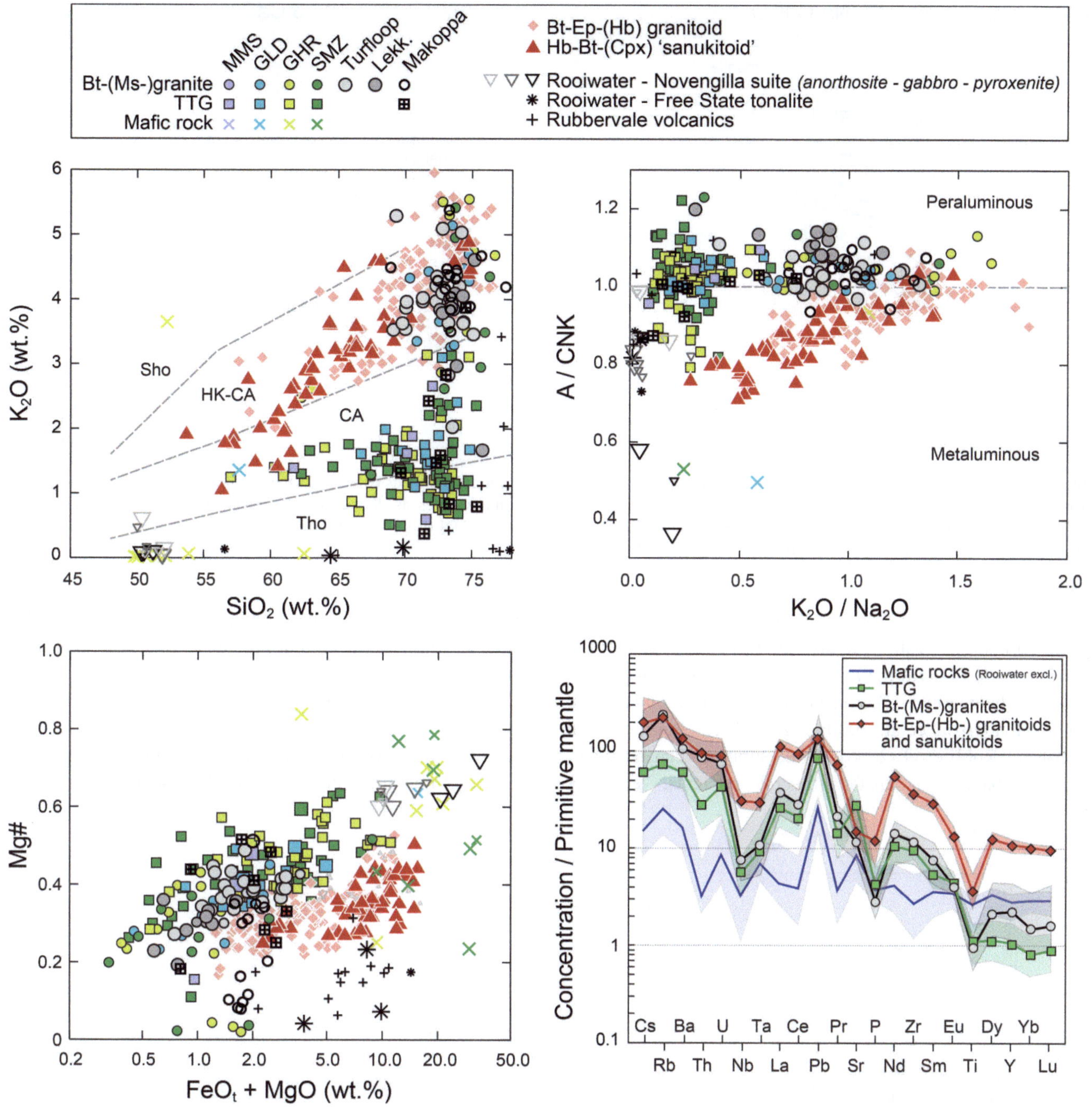

Fig. 4.5 Diagrams showing major- and trace element compositional characteristics of igneous rocks of the eastern PB, including those from the grey gneiss units (TTG, Bt-(Ms-) granites and mafic rocks—colour-coded by unit), intrusive granitoids (Turfloop, Lekkersmaak Bt-(Ms-) granite plutons; Bt-Ep-granitoid and Hbl-Bt-(Cpx-)sanukitoid plutons) the Rooiwater Complex and associated Rubbervale Formation of the Murchison greenstone belt (data from Jaguin 2012; Henderson et al. 2000; Kreissig et al. 2000; Kröner et al. 2000; Laurent et al. 2014a, c; Schwarz-Schampera et al. 2010; Vearncombe et al. (1987); Vezinet (2016); Vezinet et al. 2018 and this study—see ESM Table 1). Larger symbols for the Rooiwater Complex corresponds to data from this study; smaller symbols are literature values. Data from the Makoppa Dome (Anhaeusser and Poujol 2004) and this study—see ESM Table 1) are shown for comparison. In the bottom right panel, the multi-element patterns show the average composition of the various granitoid groups (Bt–Ep granitoids and sanukitoids are pooled together) and the shaded patterns indicate the first and third quartiles of the data; concentrations are normalized to Primitive Mantle values of McDonough and Sun (1995)

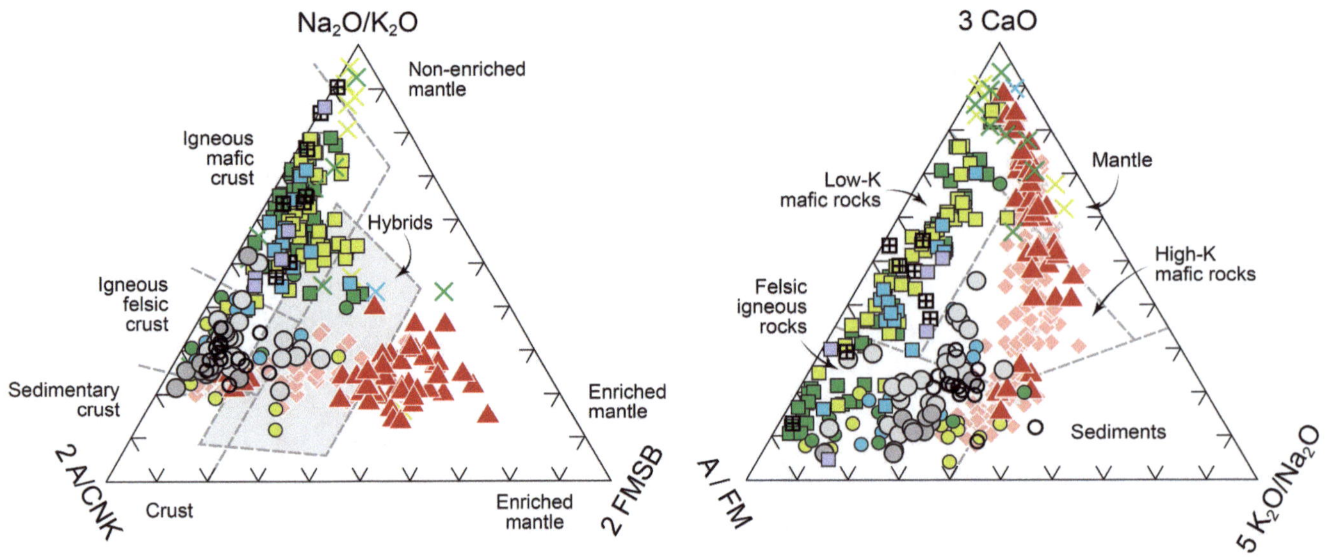

Fig. 4.6 Chemical composition of the PB granitoids and mafic rocks from the grey gneiss units plotted in the ternary 2 A/CNK (molar Al_2O_3/[$CaO+Na_2O+K_2O$])—Na_2O/K_2O—2 FMSB ([FeO_t+MgO] · [Sr +Ba]$_{wt\%}$) and A/FM (Al_2O_3/[FeO_t+MgO])—3 CaO—5 K_2O/Na_2O diagrams. The fields (modified after Laurent et al. 2014a) correspond to magma sources as defined on the basis of natural (left triangle) and experimental (right triangle) data. The symbols are as in Fig. 4.5

granodiorites, distinct from the dominant TTGs of the grey gneiss units by greater amounts of Kfs (Pl ≥ Kfs ≈ Qz) and the lack of deformation, except along the margins or discrete shear zones (Barton et al. 1992; Bohlender et al. 1992; Henderson et al. 2000; Jaguin 2012; Laurent et al. 2014b, c; Robb et al. 2006). Three subtypes can be distinguished based on mafic mineral contents, enclaves, geographic distribution and emplacement ages: (i) Bt-(Ms-) granites; (ii) Bt-Ep-granitoids; and (iii) Hbl-Bt-(Cpx-) sanukitoids.

Bt-(Ms-)bearing granites and granodiorites are coarse- to medium-grained and generally equigranular rocks. They have Zrn and Ap as main accessory phases, lack any mafic/intermediate rocks and locally contain enclaves of

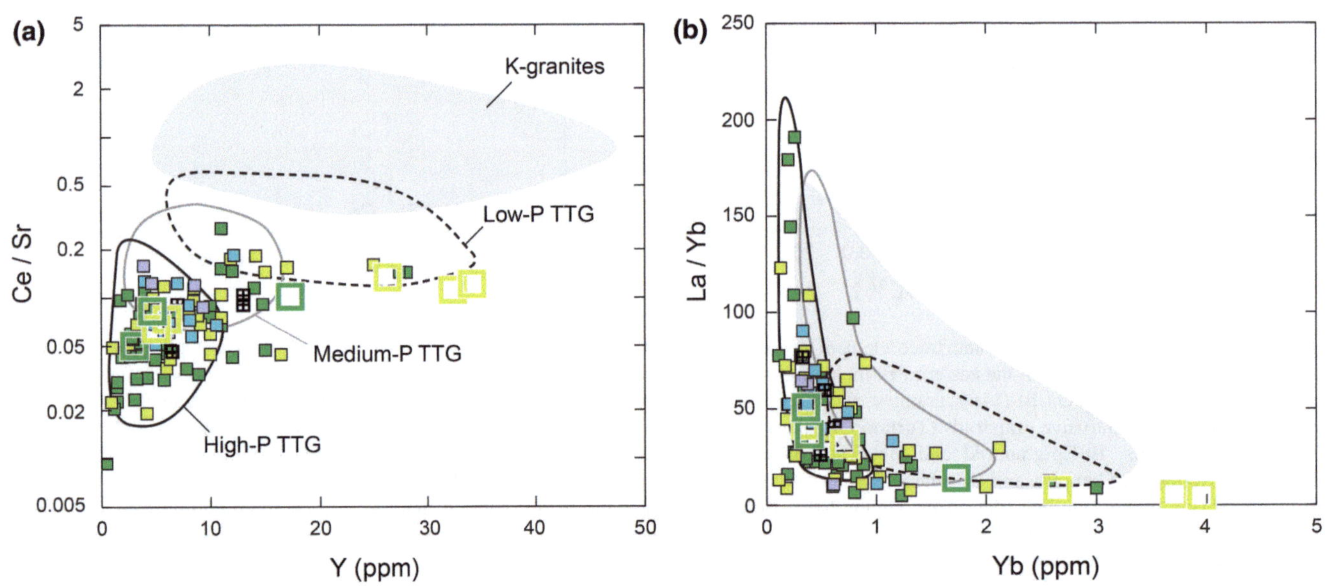

Fig. 4.7 Selected discrimination plots for TTG rocks: **a** Ce/Sr versus Y and **b** La/Yb versus Yb diagrams showing the composition of TTG samples from the grey gneiss units of the eastern PB and Makoppa Dome (symbols and data sources as in Fig. 4.5) compared to the fields of low-, medium- and high-pressure (P) TTGs and K-rich granites defined by Moyen (2011). The larger, empty symbols correspond to samples dated at ≥ 3.18 Ga

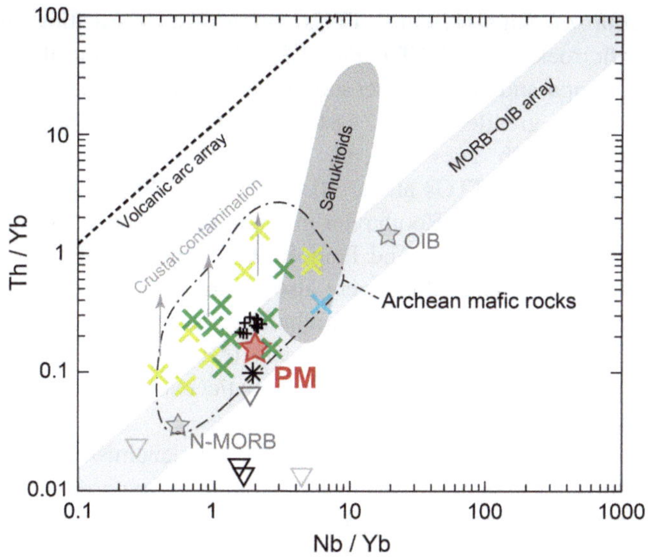

Fig. 4.8 Th/Yb versus Nb/Yb diagram of Pearce (2008) showing the composition of the mafic rocks embedded in grey gneiss units of the eastern PB; the Rooiwater Complex (Vearncombe et al. (1987); this study—see ESM Table 1) and Rubbervale Formation volcanics (Schwarz-Schampera et al. 2010) (symbols are as in Fig. 4.5). The field of sanukitoids is after Laurent et al. (2014a) and the field of Archean mafic rocks after Bédard (2018) and Moyen and Laurent (2018); the composition of the Primitive Mantle (PM) is after McDonough and Sun (1995). The 'crustal contamination' arrows depict possible increase of the Th/Yb ratio owing to interactions with felsic gneisses of TTG composition

grey gneisses. The occurrence of Ms is occasional and always subordinate relative to Bt, except in the Lekkersmaak pluton where it is a typical, ubiquitous phase (Jaguin 2012). The granites occur either as decametric to kilometric, concordant to cross-cutting dykes and bodies in grey gneisses with a weakly to moderately deformed texture (Fig. 4.4d, e) (Robb et al. 2006; Laurent et al. 2013; 2014a), or as the main, generally undeformed and locally Kfs-porphyritic phase of vast batholiths in the MMS (Lekkersmaak) and GLD (Turfloop) units (Fig. 4.1b and 4.2) (Henderson et al. 2000; Robb et al. 2006; Jaguin 2012; Laurent et al. 2013; 2014a; Laurent and Zeh 2015). The Duivelskloof 'batholith' in the GLD unit (Fig. 4.1b) represents an intermediate situation: it is characterized by clearly identifiable grey gneisses, cross-cut by very abundant bodies of Bt-(Ms-) granite (Laurent et al. 2013). The small Bt-(Ms-) granites bodies in the grey gneiss units crystallized mostly between 2.84 and 2.77 Ga (n = 10) and rarely earlier (2.91–2.88 Ga; n = 2), whereas larger plutons and batholiths crystallized between 2.85 and 2.74 Ga (n = 17) (Fig. 4.2).

The Bt-(Ms-) granites show the same mildly peraluminous character as the TTGs, albeit being generally more silicic (SiO_2 = 70–75 wt%). They clearly differ from TTGs by their much more potassic signature (K_2O > 3 wt%) and conversely higher K_2O/Na_2O ratios (0.5–1.5) (Fig. 4.5). They show the same low ferromagnesian contents and variable Mg# like the TTGs (Fig. 4.5) and comparable trace element patterns. However, they have higher contents in the most incompatible elements (Cs, Rb, Ba, Th and U) and HREE, and lower Sr concentrations (Fig. 4.5). These features generally apply to both the small granite bodies and the larger plutons and batholiths (Turfloop, Lekkersmaak), pointing to a common origin via melting of felsic crust (Fig. 4.6). Pre-existing TTGs are generally considered as the most likely source (Henderson et al. 2000; Laurent et al. 2014a), backed by the close association of the granites with migmatitic TTGs in the grey gneiss units (Laurent et al. 2013; 2014a; Vezinet et al. 2018). However, the more peraluminous character and lower $FeO_t + MgO$ contents of some of the granites, especially from the Lekkersmaak Ms-bearing pluton, additionally point to the involvement of metasedimentary rocks in their origin (Jaguin 2012).

Bt-Ep-bearing granites and granodiorites form large plutons in the MMS (Mashishimale) and GHR (Mashashane, Matlala and Moletsi) units (Figs. 4.1b and 4.2), in addition to dykes cross-cutting the surrounding grey gneisses (Fig. 4.4b). All those plutons were emplaced within a narrow timespan between 2.70 and 2.67 Ga (Fig. 4.2), and contain subordinate diorite enclaves and bodies, and locally show intermingling with more mafic magmas (Doucet et al. 2017; Laurent et al. 2014a). At the margins of the plutons, xenoliths of the surrounding grey gneisses and greenstone belts are common. Intrusive contacts between granites and country rocks are locally exposed such as at the SE margin of the Mashashane pluton where it cross-cuts the Pietersburg greenstone belt (de Wit et al. 1993). The typical rocks are medium- to coarse-grained, equigranular to Kfs-porphyritic pinkish granites (Fig. 4.4h), containing Bt and Ep as ubiquitous mafic phases and associated together with accessory Zrn, Ap, Ttn and Mag. The granites typically contain centimetric mafic clots consisting of a very fine-grained (<0.1 mm) intergrowth of Bt + Ep (Fig. 4.4h), with Qz and minor oxides. These clots locally contain relict Hbl crystals, suggesting that they represent Hbl pseudomorphs (Laurent 2012). The Mashashane granites frequently contain centimetric phenocrysts of bluish quartz (Fig. 4.4h) and locally show a hypovolcanic texture, suggesting that they represent high-level intrusions.

Hbl-Bt-(Cpx-)sanukitoids are restricted to the Matok pluton in the SMZ (Figs. 4.1b and 4.2), emplaced at 2.69–2.66 Ga (Fig. 4.2). Mafic rocks are volumetrically significant and notably much more common than in the former type. These consist of fine-grained, Cpx-bearing monzodiorites (Pl + Kfs ± Hbl, Cpx, Bt, Qz) forming small stocks and enclaves of decametric to kilometric size in the more felsic rocks, commonly showing mingling relationships (Fig. 4.4i) (Laurent et al. 2014a, c; 2017). Previously, they have been referred to as 'enderbites' or 'charnockites'

(Barton et al. 1992; Bohlender et al. 1992; Rajesh et al. 2014) because of the occasional presence of Opx, even though Cpx is by far the most common pyroxene and also forms cores in Hbl in the more felsic rocks (Laurent et al. 2014b, c; Rapopo 2010). The felsic rocks are mostly Kfs-porphyritic, medium- to coarse-grained Hbl-Bt-bearing granodiorites with a typical spotted appearance (Fig. 4.4j), and subordinate Bt-Ep-bearing granodiorites and granites. All rocks contain Zrn, Ap, Ttn and Mag as accessory phases.

The Bt-Ep-granitoids and Hbl-Bt-(Cpx-)sanukitoids share some common geochemical features. They are both metaluminous rocks, in contrast to the TTGs and Bt-(Ms-) granites, and straddling a larger compositional range in SiO_2 from ca. 50 to 75 wt% SiO_2 (mafic to felsic). Furthermore, they show higher FeOt + MgO contents (up to 15 wt%) (Fig. 4.5). The K_2O content (1.5–6.0 wt%) and K_2O/Na_2O ratio (0.5–1.5) increase from the mafic to the felsic rocks (Fig. 4.5). The most felsic varieties show the same contents in LILE (Rb, Ba, Sr and Pb), Th and U like the Bt-(Ms-) granites but much higher contents in REE, Nb, Ta, Zr and Y (Fig. 4.5).

The richness of the Hbl-Bt-(Cpx-)sanukitoids in both FeOt + MgO and incompatible elements (K_2O, Rb, Ba, REE, HFSE…) points to their formation from an enriched mantle source (Laurent et al. 2014a, c) (Fig. 4.6). Isotopic data (whole-rock Sr, Nd, Pb and zircon Hf) indicate that the metasomatic agent at the origin of mantle enrichment corresponds to local crustal material from the eastern PB (Barton et al. 1992; Laurent et al. 2017; Laurent and Zeh 2015; Zeh et al. 2009) (see Sect. 4.6.1). The position of the Bt-Ep-granitoids on the same compositional trend like Hbl-Bt-(Cpx-)sanukitoids suggests that they formed by differentiation of the sanukitoid magmas, possibly coupled to interactions with melts formed by melting of the surrounding grey gneisses (Doucet et al. 2017; Laurent et al. 2014a). This is further suggested by the intermediate composition of the Bt-Ep-granites between Hb-Bt-(Cpx-) sanukitoids and Bt-(Ms-) granites in many diagrams (Figs. 4.5 and 4.6).

4.4.3 Grey Gneisses and Granitoids of the Makoppa Dome

The Makoppa Dome hosts grey gneisses, mostly trondhjemitic orthogneisses more or less migmatized (i.e. the Vaalpenskraal gneisses of Anhaeusser and Poujol 2004) (Fig. 4.4f), intruded by relatively homogeneous granites and granodiorites (i.e. the Makoppa and Rooibokvlei granitoids of Anhaeusser and Poujol 2004) (Fig. 4.4g). The Vaalpenskraal gneisses are typical TTGs with major and trace element contents very similar to the TTGs exposed in the eastern PB (Figs. 4.5 and 4.6). They also belong to the same group of 'high-pressure' TTGs formed by the melting of mafic rocks at >1.5 GPa (Fig. 4.7). The composition of the Makoppa and Rooibokvlei granitoids is very similar to the Bt-(Ms-) granites in the eastern PB (Figs. 4.5 and 4.6). Zircon U–Pb ages indicate emplacement ages for the Vaalpenskraal TTGs at ca. 3.03, 3.01 and 2.84 Ga and for the Makoppa and Rooibovlei granitoids between 2.88 and 2.78 Ga (Anhaeusser and Poujol 2004). Zircon U–Pb ages of ca. 2.94 and 2.78 Ga were also obtained from Archaean granitoids sampled from a borehole at the western edge of the Makkoppa Dome in Botswana (Mapeo et al. 2004), both corresponding to major magmatic events in the PB (Fig. 4.2).

Overlapping lithological types, geochemistry and emplacement ages suggest that the lithologies exposed in the eastern PB extend westward under the Waterberg basin to the Makoppa Dome and possibly even farther West. It is, therefore, reasonable to infer that all domains of the PB formed through comparable geodynamic processes during the Meso- and Neoarchaean.

4.5 The Rooiwater Complex

The Rooiwater Complex is located to the North of the Murchison greenstone belt, separated from it by a tectonic contact (Fig. 4.1b). It has a thickness of about 7.5 km and is genetically linked with the felsic volcanic rocks of the Rubbervale Formation in the adjacent Murchison greenstone belt (Vearncombe et al. 1992; Zeh et al. 2013).

The complex has a sub-vertical attitude, presumably resulting from southward tilting, as indicated by southward polarity and younging (Vearncombe et al. (1987); Vearncombe 1991). The complex consists of a northerly (lowermost) mafic Novengilla Suite and a southerly (uppermost) felsic Free State (Tonalite) Suite (Vearncombe et al. (1987)). These two units are locally separated from each other by small bodies of the Quagga Qz-diorite (or amphibolite) (Vearncombe et al. 1992).

The Novengilla Suite comprises altered, coarse-grained equigranular gabbro (Fig. 4.9a) with subordinate anorthosite and pyroxenite. Rhythmic magmatic layering is well developed and highlighted by abrupt to gradational changes in modal proportions of igneous Pl and Cpx (Fig. 4.9b). In many places, these igneous minerals are pervasively altered and/or replaced (e.g. Hbl ± Act at the expense of Cpx; Pl by Ser). The Novengilla Suite contains a sulphide-bearing horizon (up to 100 m-thick) in the middle of the stratigraphy. Above this horizon, discontinuous layers of Ti-V-bearing magnetite are common (Vearncombe et al. (1987); Vearncombe 1991) (Fig. 4.9c). The Free State Suite consists mostly of a leucocratic, coarse-grained rock consisting of Ab + Qz + Hbl with minor (secondary) Ms and

Chl (Vearncombe et al. (1987)). The Quagga Qz-diorite (Fig. 4.9d) shows a distinct mineral layering and consists of Pl + Hbl ± Qz with minor Grt.

U–Pb dating of zircon from the Free State Tonalite yielded a crystallization age of 2964 ± 7 Ma (Zeh et al. 2013) (Fig. 4.2), which is, within uncertainties, identical to ages obtained from the Rubbervale volcanics in the adjacent Murchison greenstone belt (see Sect. 4.3.1), ranging from 2975 to 2963 Ma (Brandl et al. 1996; Poujol et al. 1996; Schwarz-Schampera et al. 2010; Zeh et al. 2013). Both units also show identical zircon Hf isotopic compositions (Zeh et al. 2013). Both observations support the early interpretation of Vearncombe et al. (1987) that the Rubbervale volcanics represent effusive counterparts of the Free State Suite.

The mafic rocks of the Novengilla Suite show a similar, low-K tholeiitic signature as other mafic rocks found in grey gneiss units. Whole-rock geochemical data reveal low SiO_2 (49–52 wt%), high Mg# (0.60–0.65) and a clear compositional trend controlled by the variable proportions of igneous Pl and Cpx, from aluminous, (FeO_t + MgO)-poor anorthosites to metaluminous, (FeO_t + MgO)-rich pyroxenites—the gabbros plotting in between (Fig. 4.5). The Free State Tonalite and the Rubbervale volcanics have both higher SiO_2 contents (from 57 up to 78 wt%), but low K_2O and K_2O/Na_2O ratios, which are similar to those of the Novengilla Suite. The felsic rocks are additionally characterized by relatively high FeO_t + MgO (4–10 wt%), at low Mg# (<0.3) (Fig. 4.5). All petrographic and geochemical features suggest that the Novengilla Suite and the Free State Suite + Rubbervale volcanics, respectively, represent the cumulates and residual liquids after fractionation of a common parental, mantle-derived mafic magma (Vearncombe et al. 1987; Vearncombe 1991; Zeh et al. 2013).

This interpretation is also supported by the results of semi-quantitative trace element modelling presented in Fig. 4.10. The results show that the composition of the Free State Tonalite can be explained by about 95% of fractionation of 60% Pl + 40% Cpx (±Opx, Fe–Ti oxides) from a mafic parental magma (Fig. 4.10). The modelling results are also in agreement with the observed predominance of mafic over felsic rocks in the Rooiwater Complex and the modal composition of gabbros in the Novengilla Suite (Vearncombe et al. 1987, 1992; Vearncombe 1991). Furthermore, they can account for the trace element composition of the Novengilla cumulate rocks (Fig. 4.10). The calculated parental liquid has a flat trace element pattern with concentrations of about 3–6 times those of the primitive mantle (Fig. 4.10), comparable to the composition of the mafic rocks found within the grey gneiss units (Fig. 4.5). This, together with geochronological data suggest that the rocks of the Rooiwater Complex and the amphibolites within the grey gneiss units resulted from melting of comparable mantle sources, yet underwent different extents of differentiation at shallow levels.

4.6 Hf- and Nd-Isotope Systematics of the Igneous and (Meta)Sedimentary Rocks

4.6.1 Igneous Rocks

Age-Hf isotopic data obtained from zircon grains in magmatic rocks of the eastern PB reveal two distinct crustal evolutionary trends (Zeh et al. 2009; Laurent and Zeh 2015; Vezinet et al. 2018): (i) a poorly defined Palaeoarchaean trend (n = 3) starting from chondritic mantle sources at 3.35–3.20 Ga and (ii) a well-defined Mesoarchaean trend (n = 42), starting from depleted mantle sources at 3.1–3.0 Ga and that can be traced until 2.0 Ga (Fig. 4.11a).

Fig. 4.9 Selected field pictures showing the main characteristics of the Rooiwater Complex. **a** Coarse-grained gabbro from the Novengilla Suite. **b** Layering in a gabbro from the Novengilla Suite, showing the alternance of, and gradational contact between plagioclase-rich (Pl) and pyroxene-rich (Px) layers (now largely transformed to amphibole) at the scale of a few tens of cm. **c** Metric layer of Ti- and V-bearing magnetite interlayered with the gabbros of the Novengilla Suite. **d** Medium-grained quartz-diorite from the so-called 'Quagga amphibolite' unit

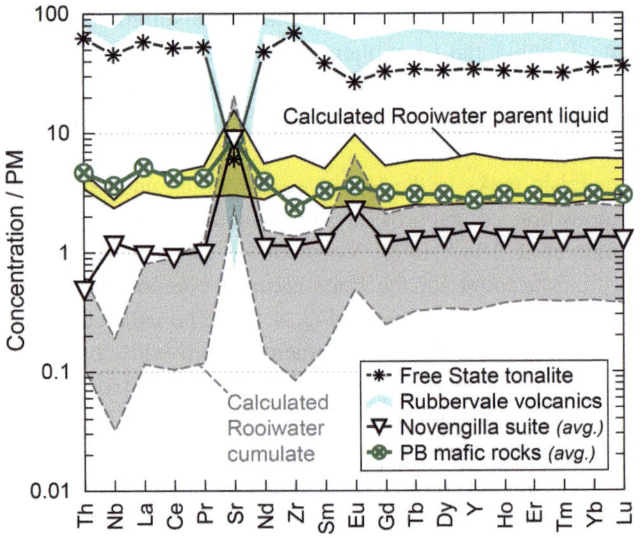

Fig. 4.10 Multi-element diagram with concentrations normalized to Primitive Mantle (PM) values of McDonough and Sun (1995), and showing the composition of main rock types in the Rooiwater Complex and the Rubbervale Formation of the Murchison Greenstone Belt. Data from the Free State Tonalite are from this study (ESM Table 1), the Rubbervale volcanics from Schwarz-Schampera et al. (2010) and the average concentrations in the Novengilla Suite from this study (ESM Table 1) and Vearncombe et al. (1987). The range of possible compositions for the parent liquid of the Rooiwater Complex (yellow field) was calculated considering that the Free State Tonalite represents 5% of residual liquid after 95% crystallization of a cumulate consisting of 60% Pl + 40% Cpx and using a range of partition coefficients appropriate for mafic liquids (compiled by Laurent et al. 2012). The grey field represents the range of modelled compositions for the cumulate. The green pattern shows the average trace element composition of mafic rocks embedded in the grey gneisses of the eastern PB; data are from Laurent et al. (2014a), Vezinet (2016) and this study (ESM Table 1)

The three TTGs orthogneisses defining the Palaeoarchaean trend are restricted to a few outcrops within the GHR grey gneiss unit. Their near-chondritic composition suggests formation by partial melting of mafic rocks, extracted from a chondritic mantle source (with respect to the Lu–Hf isotope system). Alternatively, they may have resulted from melting of Palaeoarchaean mafic rocks, derived from a depleted mantle source about 150 Ma prior to felsic magma formation. The occurrence of abundant detrital zircon grains with ages between 3.35 and 3.19 Ga (and near-chondritic εHf_t) in clastic sedimentary rocks of the Pietersburg Greenstone Belt indicate that the Palaeoarchaean felsic crust was reworked until <2.88 Ga (Zeh and Gerdes 2012). Nonetheless, it probably represents a small fraction of the present crustal volume, as suggested by the observations that (i) inherited zircon grains with Palaeoarchaean ages are rare in Meso- and Neoarchaean grey gneisses and granitoids (Laurent and Zeh 2015; Laurent et al. 2013) and (ii) that most of the grey gneisses and granitoids ($n = 42$) plot along the very consistent Mesoarchaean evolutionary trend (±0.5 εHf-unit). A few Mesoarchaean TTGs and Bt-(Ms-) granites (mainly from the GHR and SMZ) plot about 1–2 εHf-units below (Laurent and Zeh 2015; Zeh et al. 2009), possibly indicating minor contribution of Palaeoarchaean crust in their origin (Fig. 4.11a).

The Mesoarchaean trend is characterized by two important features. First, it is defined by magmatic rocks that were emplaced during a nearly continuous period of magmatic activity between 2.97 ($\varepsilon Hf_t = +5.0$) and 2.67 Ga ($\varepsilon Hf_t = -3.6$), and further by a websterite of the Schiel Alkaline Complex formed at 2.05 Ga ($\varepsilon Hf_t = -17$). This linear trend requires a very low $^{176}Lu/^{177}Hf$ ratio of ca. 0.0022 (Fig. 4.11a), which is typical of TTG crust (Kamber et al. 2002; Guitreau et al. 2012). Second, the trend involves a wide range of magmatic rocks of very different composition and origin (Fig. 4.12): (i) mantle-derived mafic and related felsic rocks of the Rooiwater Complex and Rubbervale Formation at 2.97 Ga, and amphibolites exposed in the grey gneiss units; (ii) TTGs abundantly formed between 2.96 and 2.84 Ga in all grey gneiss units by partial melting of juvenile mafic rocks; (iii) Bt-(Ms-) granites with crystallization ages between 2.91–2.88 Ga (minor phase) and 2.84–2.75 Ga (major phase); and (iv) Bt-Ep-granitoids and Hb-Bt-(Cpx-) sanukitoids emplaced at 2.70–2.67 Ga.

The linear εHf_t-age trend suggests that most of the volume of the PB crust was formed at 3.15–2.88 Ga, first by mantle melting (3.15–2.97 Ga), shortly followed by transformation of the resulting mafic crust into a TTG crust (2.97–2.88 Ga) (Fig. 4.12). Subsequent magmatism entailed massive reworking of this crust (Bt-(Ms-) granites, Bt-Ep-granitoids) and minor addition from an enriched mantle source (Hb-Bt-(Cpx-)sanukitoids, websterite) (Fig. 4.12). The data additionally require that crustal reworking and crust–mantle interactions occurred in an isotopically closed system, i.e. without significant input of either radiogenic Hf from the convective mantle or non-radiogenic Hf from a crust significantly older than 3.0 Ga. This excludes any implication of the adjacent proto-Kaapvaal Craton (Fig. 4.11a). The position of the 2.70–2.67 Ga PB sanukitoids on the Mesoarchaean trend further implies that the Lu–Hf budget of their lithospheric mantle source (Laurent et al. 2014a; Fig. 4.6) was controlled by enrichment processes, which resulted from the subduction of sedimentary rocks derived solely from the PB (Laurent and Zeh 2015). The enriched lithospheric mantle persisted at least until the Palaeoproterozoic at ca. 2.05–2.03 Ga, as indicated by Hf isotopic data obtained from websterites and felsic alkaline rocks in the SMZ (Schiel Complex, Entabeni Granite; Laurent and Zeh 2015; Zeh et al. 2009).

The whole-rock Sm–Nd systematics depicts a very similar evolution like the zircon Hf isotopic data (Fig. 4.11b). The Meso-/Neoarchaean granitoids plot on a trend which

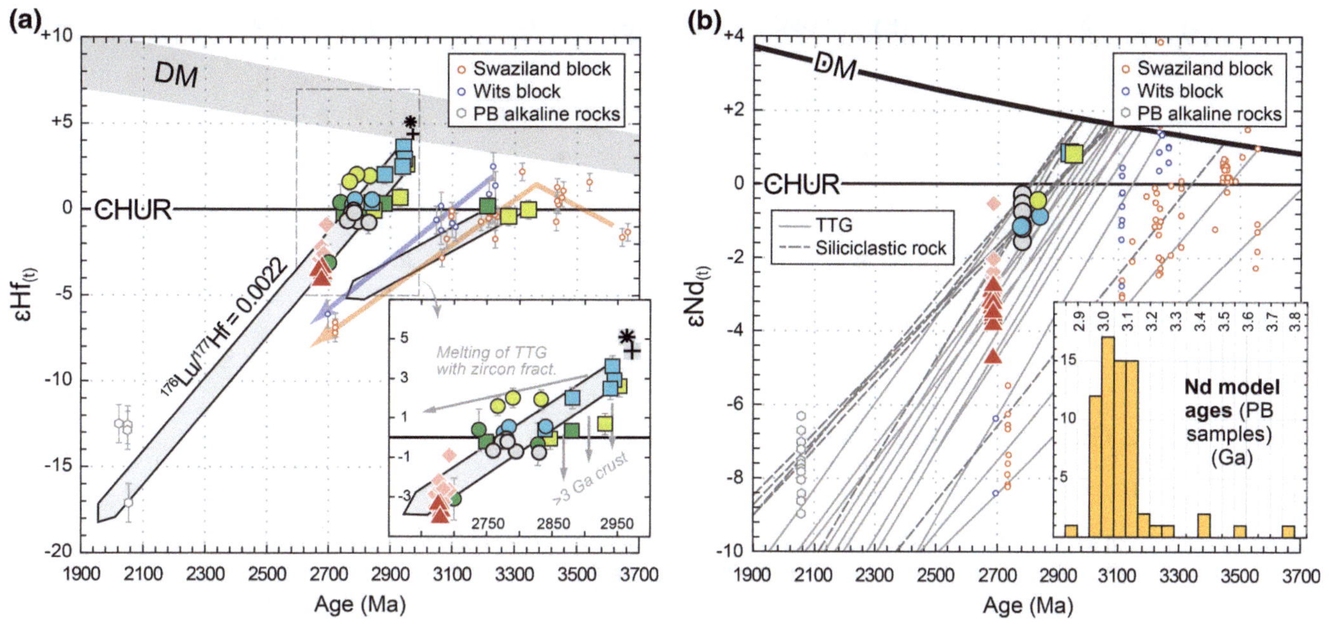

Fig. 4.11 Hf–Nd systematics of the eastern PB rocks. **a** Plot of average zircon $\varepsilon Hf_{(t)}$ versus emplacement age for igneous samples (symbols are as in Fig. 4.5; data from Laurent and Zeh 2015; Vezinet 2016; Zeh et al. 2009; 2013). One spot corresponds to several individual zircon analyses and Hf isotopic composition calculated at the average U–Pb crystallization age. The grey arrows correspond to the isotopic evolution trend of the Paleoarchean TTGs ($^{176}Lu/^{177}Hf$ of 0.018 corresponding to the average, elemental Lu/Hf composition of those rocks) and Mesoarchean TTG ($^{176}Lu/^{177}Hf$ of 0.0022 as determined by Laurent and Zeh 2015). In the inset, possible deviations from the Mesoarchean TTG trend are highlighted, including involvement of Paleoarchean (>3 Ga) crust and disequilibrium melting of TTGs with zircon remaining partly in the residue (Laurent and Zeh 2015). The diagram also shows the age, zircon $\varepsilon Hf_{(t)}$ and isotopic evolution trends (coloured arrows) of granitoids in the Swaziland and Witwatersrand block for comparison (data from Frimmel et al. 2009; Zeh et al. 2009; 2011). The CHUR model is after Bouvier et al. (2008), the DM field is bounded by the regression from present-day MORB (Griffin et al. 2002) and the 'Archean DM' model (Næraa et al. 2012). **b** Plot of whole-rock $\varepsilon Nd_{(t)}$ versus emplacement age. Igneous samples on which a zircon U–Pb crystallization age is available are plotted as symbols (as in Fig. 4.5), the undated ones as lines (based on measured $^{143}Nd/^{144}Nd$ and $^{147}Sm/^{144}Nd$) (data from Kreissig et al. 2000; Laurent et al. 2014a, c). Data from the alkaline rocks of the PB (Schiel Complex; Graupner et al. 2018) and the granitoids in the Witwatersrand and Swaziland blocks (Schoene et al. 2008) are also shown for comparison. The inset shows the distribution of Nd model ages for the PB samples

starts from the depleted mantle source at about 3.0 Ga (range 2.95–3.15 Ga; $n = 59$; 87% of the data) and is characterized by a systematic decrease in $\varepsilon Nd_{(t)}$ towards younger ages (Fig. 4.11b). The ca. 2.95 Ga TTGs show positive $\varepsilon Nd_{(t)}$ (ca. +1), the 2.84–2.75 Ga Bt-(Ms-) granites $\varepsilon Nd_{(t)}$ between −0.3 and −1.6, the 2.70–2.67 Ga Bt-Ep granitoids and Hb-Bt-(Cpx-) sanukitoids are characterized by more negative values (between −2 and −5). Interestingly, Nd isotopic data from the 2.05 Ga alkaline rocks of the Schiel Complex ($\varepsilon Nd_{(t)}$ between −6 and −9; Graupner et al. 2018) also fall on the same trend, as observed for zircon Hf isotopes (Fig. 4.11b). This trend is also supported by Nd-isotope data of undated TTGs obtained by Kreissig et al. (2000) (see Fig. 4.11b). As a consequence, the Nd model ages of most igneous rocks of the eastern PB point to a main even of crust formation between 2.95 and 3.15 Ga (Fig. 4.11b). Only a few TTGs ($n = 3$) from the SMZ and GHR units display Palaeoarchaean Nd model ages >3.2 Ga (Fig. 4.11b), suggesting that they represent remnants of the early Palaeoarchaean crust.

4.6.2 Sedimentary Rocks

Detrital zircons in ca. 2.99 Ga quartzites of the Murchison greenstone belt (La France Formation and Murchison Unit) display the same U–Pb–Hf isotopic signatures (ages between 3.45 and 3.12 Ga and chondritic to subchondritic εHf_t) like magmatic rocks exposed in the Witwatersrand and Swaziland blocks to the south of the PB (Fig. 4.13). This hints at a distal, southern provenance for the detritus within the proto-Kaapvaal Craton (Zeh et al. 2013). By contrast, detrital zircon grains in <2.88 Ga conglomerates of the Uitkyk Formation in the Pietersburg greenstone belt (Zeh and Gerdes 2012) show distinct characteristics. They reveal a bimodal age distribution with peaks at 3.40–3.17 Ga and 2.96–2.91 Ga and mostly chondritic to supra-chondritic $\varepsilon Hf_{(t)}$ from −2 to +4 (Fig. 4.13). The U–Pb–Hf isotopic signature of the older zircon population overlaps with that obtained for the Palaeoarchaean TTGs of the PB, while the younger population has the same Hf-age characteristics as zircons from the Mesoarchaean TTGs within the GLD and

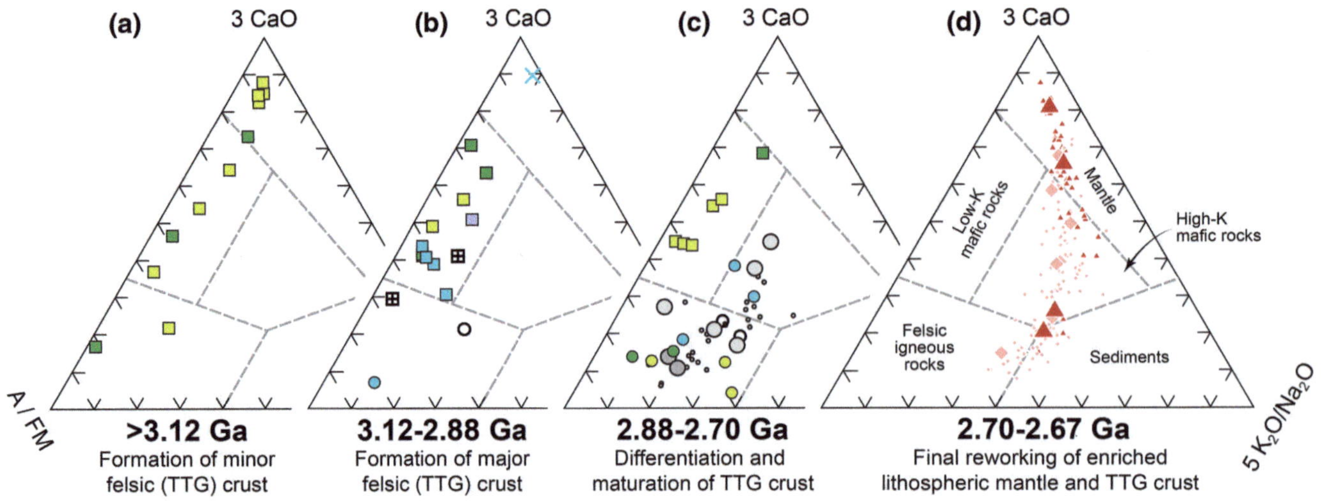

Fig. 4.12 Ternary A/FM—3 CaO—5 K_2O/Na_2O diagram (as in Fig. 4.6) where the chemical compositions of granitoids from the PB are reported as a function of their crystallization age. Modified after Vezinet (2016). The symbols are as in Fig. 4.5. The large symbols correspond to samples for which a zircon U–Pb crystallization age is available; the small symbols are from homogeneous plutons and batholiths from which other samples have been dated, and can be attributed with confidence to the corresponding time span

GHR grey gneiss units (Fig. 4.13). This overlap suggests proximal sources within the PB. Proximal sources are also required for the <2.76–2.71 Ga metasedimentary rocks within the high-grade Bandelierkop Formation of the SMZ. These show a similar detrital zircon age spectrum as the Uitkyk conglomerates (between 3.4 and 3.0 Ga) and additionally contain abundant detrital zircon grains having the same ages (2.95–2.76 Ga) like the Mesoarchaean TTGs and Bt-(Ms-) granites of the PB (Nicoli et al. 2015; Rajesh et al. 2014).

Supply of sediments from proximal sources is further supported by whole-rock Sm-Nd isotopic data obtained from (meta-)sedimentary rocks of the Pietersburg, Giyani and Rhenosterkoppies Greenstone Belts (Kreissig et al. 2000). Most of these rocks yielded Nd model ages of ca. 3.1–2.9 Ga, identical to local Mesoarchaean TTGs (Fig. 4.11b). The only exception is provided by a Nd model age of 3.44 Ga obtained from a sedimentary rock from the Giyani belt (sample 96/234 of Kreissig et al. 2000) (Fig. 4.11b). This age suggests an origin by erosion of Palaeoarchaean TTGs.

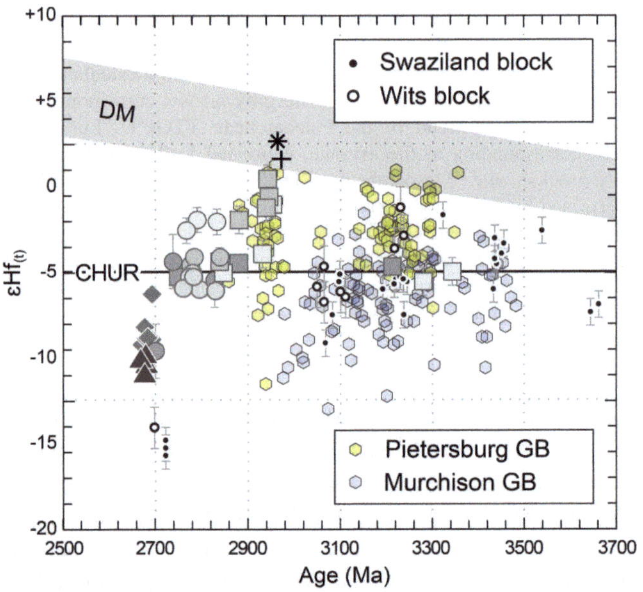

Fig. 4.13 Plot of $\varepsilon Hf_{(t)}$ versus age for individual detrital zircons from siliciclastic sediments of the Pietersburg greenstone belt (Uitkyk Formation; Zeh and Gerdes 2012) and Murchison greenstone belt (Murchison Unit and La France Formation; Zeh et al. 2013) compared to average age and zircon $\varepsilon Hf_{(t)}$ from granitoid rocks in the eastern PB (symbols as in Fig. 4.5, in shades of grey) as well as the Swaziland and Witwatersrand blocks (data sources, CHUR and DM models as in Fig. 4.11)

4.7 Geodynamic Model for the Formation of the PB

The geodynamic model for the PB presented below integrates but also extends previous models presented by Zeh et al. (2009, 2013), Laurent et al. (2014a), Laurent and Zeh (2015), Vezinet (2016), and Vezinet et al. (2018). The combined geochronological, geochemical, structural and metamorphic dataset presented above suggests that the PB hosts rocks that formed during (i) a minor Palaeoarchaean phase reflected by minor TTGs and greenstones formed between 3.46 and 3.18 Ga and (ii) a predominant Meso- to Neoarchaean phase between 3.11 and 2.67 Ga.

4.7.1 Palaeoarchaean Evolution

The close spatial association of Palaeoarchaean TTGs (Fig. 4.1b), and coeval mafic sequences in the Pietersburg and Giyani greenstone belts (3.46–3.18 Ga; Fig. 4.2) suggests that these rocks represent remnants of a coherent micro-terrane (Fig. 4.14a). This terrane was of limited size and became intensely structurally (sheared and northward thrusted), but barely magmatically reworked during the Meso- to Neoarchaean (de Wit et al. 1992c; Laurent and Zeh 2015). The chondritic Hf isotopic composition of the Palaeoarchaean igneous rocks (and related detrital zircon grains) suggests formation without the involvement of any older felsic crust (Laurent and Zeh 2015). The Palaeoarchaean TTGs generally show a tonalitic composition and a 'low-pressure' signature (Fig. 4.7), suggesting their formation either within an oceanic plateau, thick enough (maximum 30 km) to undergo modest differentiation and formation of felsic magmas, and/or an island arc regime with fractionation of hydrous basaltic melts to form tonalites. In any case, the coeval mafic rocks from the greenstone belts may represent the remnants of the oceanic plateau or arc.

4.7.2 Meso- to Neoarchaean Evolution

This phase was mainly controlled by south-verging subduction and docking of distinct terranes along the northern margin of the proto-Kaapvaal Craton, resulting in the formation of an accretionary orogenic system that developed successively from South to North (Fig. 4.14). This interpretation is supported by:

(1) a general ENE–WSW-trending orientation of greenstone belts and shear zones, separating the different grey gneiss units of the eastern PB and comprising the crustal-scale Murchison-Thabazimbi Lineament, which separates the PB from the proto-Kaapvaal Craton (Fig. 4.1a);
(2) progressive northward younging of sediment deposition ages and metamorphism in the Greenstone Belts (Murchison: 2.99–2.92 Ga; Pietersburg and Giyani: 2.88–2.80 Ga; Rhenosterkoppies and Bandelierkop: 2.76–2.71 Ga) and TTG intrusion in the grey gneiss units (MMS: 3.12–2.95 Ga; GLD: mostly ca. 2.95 Ga; GHR + SMZ: 2.95–2.84 Ga) (Fig. 4.2);
(3) the presence of reverse faults, thrusts and transpressive shear zones in the greenstone belts (Fig. 4.3c), indicative of an overall N–S compressive tectonic regime (de Wit et al. 1992a, c, 1993; Jaguin et al. 2012; Block et al. 2013; Kramers et al. 2014);
(4) the nature, geochemistry and temporal evolution of the Meso- to Neoarchaean granitoids (Figs. 4.5, 4.6, 4.7 and 4.12). Nearly all TTGs suggest melting of mafic crust at high pressure (>1.5 GPa), and the younger sanukitoids and alkaline rocks indicate coeval enrichment of the mantle beneath the PB due to subduction of surface material (Laurent et al. 2014a, c: Laurent and Zeh 2015);
(5) the Hf–Nd isotopic data requiring the formation of voluminous juvenile crust at 3.15–2.95 Ga, subsequently reworked in an isotopically closed system until 2.05 Ga (Fig. 4.11).

4.7.2.1 Initiation of Subduction Beneath the Proto-Kaapvaal Craton (3.15–2.98 Ga) (Fig. 4.14a)

This phase is only reflected by TTGs magmatism within the MMS unit and by some volcanic and clastic sedimentary rocks of the Murchison Greenstone Belt (Brandl and Kröner 1993; Jaguin 2012; Poujol 2001; Poujol and Robb 1999; Poujol et al. 1996). The TTGs were emplaced mostly between 3.12 and 3.02 Ga and belong to the high-pressure group formed at >45 km depth, potentially representing a continental magmatic arc. These rocks partly fed a fore-arc to intra-arc sedimentary system (see Zeh et al. 2013) represented by the clastic sedimentary rocks, containing 3.12–2.99 Ga detrital zircons (Fig. 4.2) together with distal supply from the proto-Kaapvaal Craton (3.45–3.22 Ga detrital zircons).

4.7.2.2 Formation and Accretion of an Oceanic Arc-Back System and Docking of the Palaeoarchaean Micro-Terrane (2.97–2.84 Ga) (Fig. 4.14b)

This phase began with the formation of the Rooiwater Complex and the genetically linked Rubbervale volcanics at 2.97–2.95 Ga (Brandl et al. 1996; Poujol et al. 1996; Zeh et al. 2013). Field relationships, geochemistry (Fig. 4.5), zircon characteristics (Fig. 4.11a) and the occurrence of VMS-type deposits (Schwarz-Schampera et al. 2010) most likely indicate a back-arc setting with an oceanic arc located to the North of the Rooiwater Complex (Zeh et al. 2013). The ca. 2.95 Ga TTGs within the GLD unit (Fig. 4.2) reveal high-pressure melting (>50 km; Fig. 4.7), and juvenile εHf_t and εNd_t (Fig. 4.11) indicate that they likely represent such oceanic arc. Syntectonic TTGs with ages between 2.97 and 2.96 Ga within the Murchison Unit (Minnitt and Anhaeusser 1992; Jaguin et al. 2012) suggest contemporaneous subduction and melting beneath the Murchison Greenstone Belt. The belt became affected slightly later by an intense structural-metamorphic overprint (2.95–2.92 Ga) imposed by the juxtaposition of various supracrustal units in a N–S-directed compressional regime (Block et al. 2013; Jaguin et al. 2012; Zeh et al. 2013). This event was related to the successive southward migration of the GLD arc followed by

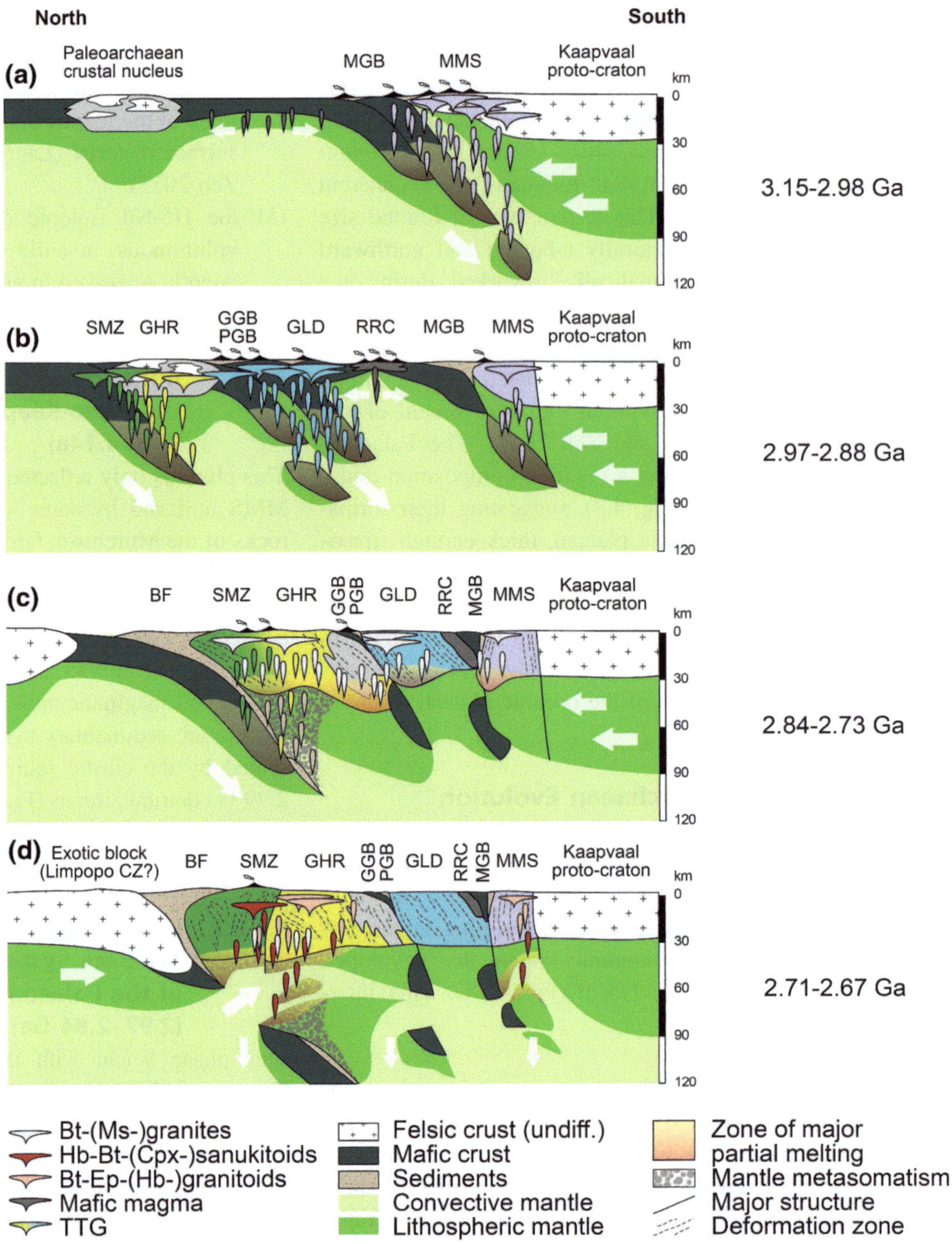

Fig. 4.14 Sketch cartoon depicting the proposed geodynamic evolution of the PB based on the presented dataset. The geodynamic context of formation for the Paleoarchean crustal nucleus being speculative, it has not been shown. The Paleoarchean micro-terrane is represented in light grey in the whole cartoon. **a** 3.15–2.98 Ga: incipient S-verging subduction below the northern margin of the Kaapvaal proto-continent; **b** 2.97–2.88 Ga: onset of tectonic switching characterized by formation of the Rooiwater-Rubbervale complex (RRC) as a back-arc basin and subsequent accretion of the Paleoarchean micro-terrane and intervening arcs; **c** 2.84–2.73 Ga: maturation of the crust in an Andean-type, continental margin setting characterized by crustal melting, subduction of sediments derived from the local crust and enrichment of the mantle by those sediments; **d** 2.71–2.67 Ga: collision with a terrane located to the North (precursor of the Central Zone (CZ) of the Limpopo Belt?), triggering delamination of the lithospheric mantle over-thickened during previous accretion events. The letters labelling the cross sections refer to the unit acronyms defined in the text (see Sect. 4.2 and Fig. 4.2); BF = Bandelierkop Formation. See text (Sect. 4.7.2) for details

docking to the proto-Kaapvaal Craton. This caused closure of the Rooiwater-Rubbervale back-arc (located in between) and southward thrusting of the Murchison supracrustals and MMS rocks onto the proto-Kaapvaal craton.

Continuous southward subduction led to the collision of the Palaeoarchaean micro-terrane with the previously formed GLD arc at 2.88–2.84 Ga. This collision (the 'PGB-Lwaji Orogeny' of Kramers et al. 2014) led to north-verging thrusting, metamorphism and minor volcanism in the Pietersburg and Giyani greenstone belts (de Wit et al. 1993; Kramers et al. 2014; Kröner et al. 2000) and formation of a foreland basin filled with sedimentary rocks of the Uitkyk Formation, containing zircon detritus supplied from both the GLD arc and the Palaeoarchaean micro-terrane (Zeh and Gerdes 2012) (Fig. 4.13). Voluminous Mesoarchaean TTGs with ages between 2.95 and 2.84 Ga, high-pressure characteristics, and supra-chondritic $\varepsilon Hf_{(t)}$ and $\varepsilon Nd_{(t)}$ in the GHR and SMZ (and volcanic counterparts in the Giyni and Pietersburg Belts) suggest ongoing subduction-related magmatism, most likely sourced by a subduction zone limiting the northern edge of the accreted micro-terrane.

4.7.2.3 Crustal Reworking in an Andean-Type Setting (2.84–2.73 Ga) (Fig. 4.14c)

The collision between the GLD and the Palaeoarchaean micro-terrane marks the starting point for intense crustal reworking, which affected the entire PB between 2.84 and 2.73 Ma and is reflected by the extensive formation of Bt-(Ms-) granites (Laurent and Zeh 2015). During the same period, TTG formation terminates (Fig. 4.2), except in the SMZ (Vezinet et al. 2018). The widespread occurrence of Bt-(Ms-) granites from the MMS unit to the SMZ (together with the Makoppa Dome) suggest that the PB already existed as a coherent crustal segment during the late Mesoarchaean/early Neoarchaean. This crustal segment became additionally deformed and heated, presumably through significant crustal thickening that promoted crustal anatexis (Laurent and Zeh 2015). Deformation still occurred in an N–S convergent tectonic regime, indicated by conspicuous north-verging reverse structures and folds within the Giyani, Pietersburg and Rhenosterkoppies greenstone belts (de Wit et al. 1992a, c, 1993; Kramers et al. 2014; Kröner et al. 2000; Passeraub et al. 1999).

Isotope geochemical signatures of the Hb-Bt-(Cpx-) sanukitoids and genetically related Bt–Ep granitoids (2.70–2.67 Ga) additionally require enrichment of the lithospheric mantle beneath the PB during the same period, as is reflected by combined U–Pb–Hf isotopic data (Laurent and Zeh 2015; Laurent et al. 2014a, c). Enrichment can be explained by subduction of sedimentary rocks, similar to those exposed in the Pietersburg and Rhenosterkoppies greenstone belts and/or in the Bandelierkop Formation, consisting all of proximal detritus derived from the PB (Fig. 4.2; Sect. 4.6.2).

In summary, thickening and reworking of the PB crust, burial of sedimentary rocks to mantle depth and minor TTG magmatism suggest that the PB was affected by ongoing southward subduction between 2.84 and 2.73 Ga, in an 'Andean-type' continental margin setting (Fig. 4.14c).

4.7.2.4 Continental Collision and Lithosphere Reworking (2.73–2.65 Ga) (Fig. 4.14d)

The sedimentary rocks of the Bandelierkop Formation in the northernmost PB were deposited in a fore-arc basin at 2.75–2.73 Ga (Fig. 4.2), followed by rapidly buried to a depth of ≥ 30 km (10–12 kbar) and affected by granulite-facies metamorphism in excess of 750 °C at 2.71 Ga (Nicoli et al. 2015; Rajesh et al. 2014; Taylor et al. 2014). Burial was accommodated by N-directed thrusting, and exhumation of the granulite-facies rocks by SW-verging, 'back-thrusting' along the Hout River Shear Zone (Perchuk et al. 2000; Roering et al. 1992; Smit et al. 1991; 2001, 2014; van Reenen et al. 1992; Vezinet et al. 2018) at 2.69–2.64 Ga (Kramers et al. 2014; Kreissig et al. 2001).

This tectono-metamorphic episode has collectively been attributed to 2.71–2.67 Ga collision of the PB with an exotic terrane located to the North of it (Barton et al. 2006; Kramers and Mouri 2011; Kramers et al. 2006; Laurent and Zeh 2015; Laurent et al. 2014c; Zeh et al. 2009). The Central Zone of the Limpopo Belt, which is located to the North of the PB, was largely structured at 2.7–2.6 Ga (Barton et al. 1994; Jaeckel et al. 1997; Kröner et al. 1999; Kramers et al. 2006, 2011). Nevertheless, it hosts >2.7 Ga-old rocks, such as parts of the Sand River TTG gneisses emplaced at ca. 3.31–3.28 Ga (Kröner et al. 1999; Zeh et al. 2007). Likewise, some of the Beit Bridge Complex metasediments host 3.95–3.18 Ga detrital zircons with <2.7 Ga-old metamorphic rims (Zeh et al. 2014), pointing to sediment deposition between 3.18 and 2.7 Ga. These 'precursor' rocks of the Central Zone therefore plausibly represented the crustal entity that collided with the PB at ca. 2.71–2.67 Ga (Brandt et al. 2018; Kröner et al. 2018; Laurent and Zeh 2015). This has to be confirmed by further work on the Central Zone of the Limpopo Belt and its relationship with the PB.

In any case, the collision was likely accompanied by delamination of the over-thickened lithospheric mantle formed during the protracted convergent margin setting of the PB. Delamination was accompanied by asthenosphere upwelling and, consequently, intense heating of the deep crust, resulting in the formation of the Banderlierkop Formation granulites. In addition, it caused extensive melting of the previously enriched lithospheric mantle (see Sect. 4.7.2.3), ending up in the formation of sanukitoid magmas. The latter interacted with the crust and/or melts derived from it during their ascent, evolving to Bt-Ep-granitoids (Doucet et al. 2017; Laurent et al. 2014a).

4.8 Concluding Remarks

The presented evolutionary model is based on the compilation of a comprehensive set of data, suggesting that the PB represents a 3.12–2.73 Ga Archaean accretionary orogen, which was generally driven by southward-directed subduction and terminated by the collision of the PB units with a crustal terrane located farther North (possibly the precursor of the Limpopo Central Zone) at 2.71–2.65 Ga. Amalgamation generally evolved from South to North and began with continental arc formation in a compressive regime at 3.12–2.98 Ga, followed by back-arc extension at 2.98–2.96 Ga and accretion of the entire Mesoarchaean arc/back-arc system at 2.95–2.92 Ga. This period of 'switching tectonics' was accompanied by the formation of large volumes of high-pressure TTGs with supra-chondritic Hf–Nd isotope signatures, forming the bulk of the PB crust.

Further on, the evolution of the PB was controlled by the collision with a Palaeoarchaean micro-terrane, hosting an oceanic plateau and/or island arc formed at 3.46–3.18 Ga.

This collision caused thickening and intense partial melting (reworking) of the previously formed TTG crust between 2.84 and 2.73 Ga, and ongoing subduction and extensive enrichment of the mantle lithosphere beneath the PB. The metasomatized mantle became the source of sanukitoids at 2.70–2.67 Ga, triggered by delamination of the lithospheric mantle after collision.

The extensive and collective evidence from the PB provides a strong case for the operation of horizontal plate movements, subduction and collision zone processes during the Meso- and Neoarchaean, akin to modern-style plate tectonics.

Acknowledgements We thank the editors for inviting us to contribute to the present book and for their patience regarding ever-delayed submission. Reviews by Jan Kramers, Dirk van Reenen and Andre Smit were appreciated and contributed to improve the original manuscript. We also gratefully thank the colleagues who contributed to the ideas presented here, either through stimulating discussion or/and their valuable contributions to understand the complex geology of the PB: S. Block, A. Chauvet, S. Couzinié, R. Doucelance, L.-S. Doucet, C. Dusséaux, A. Gerdes, J. Jaguin, N. Madlakana, H. Martin, J.-F. Moyen, G. Nicoli, J.-L. Paquette, M. Poujol, M. Rapopo, G. Stevens, J. Taylor and many others.

References

Anhaeusser CR (1992) Structures in granitoid gneisses and associated migmatites close to the granulite boundary of the Limpopo Belt, South Africa. Precambr Res 55:81–92

Anhaeusser CR (2002) Archaean granitoid rocks of the Makoppa Dome, Limpopo Province, South Africa: preliminary petrological and geochemical results. Extended Abstract, 11th Quadrennial IAGOD Symposium and Geocongress 2002, Windhoek, Namibia. CD-ROM Geological Survey of Namibia

Anhaeusser CR (2006) A reevaluation of Archean intracratonic terrane boundaries on the Kaapvaal Craton, South Africa: collisional suture zones? In: Reimold WU, Gibbson R (eds) Processes of the early earth. Geological Society of America Special Publication, vol. 405, pp 315–332

Anhaeusser CR, Poujol M (2004) Petrological, geochemical and U–Pb isotopic studies of Archaean granitoid rocks of the Makoppa Dome, northwest Limpopo Province, South Africa. S Afr J Geol 107:521–544

Barton JM Jr, Doig R, Smith CB, Bohlender F, Van Reenen DD (1992) Isotopic and REE characteristics of the intrusive charnoenderbite and enderbite geographically associated with the Matok Pluton, Limpopo Belt, southern Africa. Precambr Res 55:451–467

Barton JM Jr, Holzer L, Kamber B, Doig R, Kramers JD, Nyfeler D (1994) Discrete metamorphic events in the Limpopo belt, southern Africa: implications for the application of P-T paths in complex metamorphic terranes. Geology 22:1035–1038

Barton JM Jr, Klemd R, Zeh A (2006) The Limpopo belt: a result of Archean to Proterozoic, Turkic-type orogenesis? Geol Soc Am Spec Pap 405:315–332

Bédard JH (2018) Stagnant lids and mantle overturns: implications for Archaean tectonics, magma genesis, crustal growth, mantle evolution, and the start of plate tectonics. Geosci Front 9:19–49

Belyanin GA, Kramers JD, Vorster C, Knoper MW (2014) The timing of successive fluid events in the Southern Marginal Zone ofthe Limpopo Complex, South Africa: constraints from ^{40}Ar–^{39}Ar geochronology. Precambr Res 254:169–193

Belyanin GA, Rajesh HM, Sajeev K, Van Reenen DD (2012) Ultrahigh-temperature metamorphism from an unusual corundum +orthopyroxene intergrowth bearing Al–Mg granulite from the Southern Marginal Zone, Limpopo Complex, South Africa. Contrib Miner Petrol 164:457–475

Block S, Moyen J-F, Zeh A, Poujol M, Jaguin J, Paquette J-L (2013) The Murchison Greenstone Belt, South Africa: accreted slivers with contrasting metamorphic conditions. Precambr Res 227: 77–98

Bohlender F, van Reenen DD, Barton JM Jr (1992) Evidence for metamorphic and igneous charnockites in the Southern Marginal Zone of the Limpopo Belt. Precambrian Res. 55:429–449

Bouvier A, Vervoort JD, Patchett PJ (2008) The Lu–Hf and Sm–Nd isotopic composition of CHUR: constraints from unequilibrated chondrites and implications for the bulk composition of terrestrial planets. Earth Planet Sci Lett 273:48–57

Brandl G, Cloete M, Anhaeusser CR (2006) Archean greenstone belts. In: Johnson MR, Anhaeusser CR, Thomas RJ (eds) The geology of South Africa. Geological Society of South Africa, Johannesburg/ Council for Geoscience, Pretoria, pp 9–56

Brandl G, Jaeckel P, Kröner A (1996) Single zircon age for the felsic Rubbervale Formation, Murchison greenstone belt, South Africa. S Afr J Geol 99:229–234

Brandl G, Kröner A (1993) Preliminary results of single zircon studies from various Archaean rocks of the northeastern Transvaal. Extended Abstract, 16th International Colloquium of African Geology. Mbabane, Swaziland, pp 54–56

Brandl G, de Wit MJ (1997) The Kaapvaal Craton. In: De Wit MJ, Ashwal LD (eds) Greenstone belts. Oxford University Press, Oxford, pp 581–607

Brandt S, Klemd R, Li Q, Kröner A, Brand G, Fisher A, Bobek P, Zhou T (2018) Pressure-temperature evolution during two granulite-facies metamorphic events (2.62 and 2.02 Ga) in rocks from the Central Zone of the Limpopo Belt, South Africa. Precambr Res 310:471–506

Byron CL, Barton JM Jr (1990) The setting of mineralization in a portion of the Eersteling goldfield, Pietersburg granite–greenstone terrane, South Africa. S Afr J Geol 93:463–472

Condie KC (1981) Archaean greenstone belts. In: Windley BF (ed) Developments in precambrian geology, vol 3. 434 p

de Beer JH, Stettler EH (1992) The deep structure of the Limpopo Belt from geophysical studies. Precambr Res 55:173–186

de Beer JH, Stettler EH, Duvenhage AWA, Joubert SJ, de W Raath CJ (1984) Gravity and geoelectrical studies of the Murchison greenstone belt, South Africa. S Afr J Geol 91:184–197

de Wit MJ, Armstrong RA, Kamo SL, Erlank AJ (1993) Gold-bearing sediments in the Pietersburg greenstone belt: age equivalents of the Witwatersrand Supergroup sediments, South Africa. Econ Geol 88:1242–1252

de Wit MJ, Jones MG, Buchanan DL (1992a) The geology and tectonic evolution of the Pietersburg Greenstone Belt, South Africa. Precambr Res 55:123–153

de Wit MJ, Roering C, Hart RJ, Armstrong RA, de Ronde REJ, Green RW, Tredoux M et al (1992b) Formation of an Archaean continent. Nature 357:553–562

de Wit MJ, van Reenen DD, Roering C (1992c) Geologic observations across a tectono-metamorphic boundary in the Babangu area, Giyani (Sutherland) Greenstone Belt, South Africa. Precambr Res 55:111–122

Doucet LS, Laurent O, Mattielli N, Debouge W (2017) Zn isotope heterogeneity in the continental lithosphere: new evidence from Archean granitoids of the northern Kaapvaal craton, South Africa. Chem Geol 476:260–271

Eglington BM, Armstrong RA (2004) The Kaapvaal Craton and adjacent orogens, southern Africa: a geochronological database and overview of the geological development of the craton. S Afr J Geol 107:13–32

Eriksson PG, Banerjee S, Nelson DR, Rigby MJ, Catuneanu O, Sarkar S, Roberts RJ et al (2009) A Kaapvaal craton debate: nucleus of an early small supercontinent or affected by an enhanced accretion event? Gondwana Res 15:354–372

Franey NJ (1987) A geological model of the shear zone gold deposits in the Pietersburg greenstone belt, South Africa. Unpublished M.Sc. thesis, Rhodes University, Grahamstown, 114 p

Frimmel HE, Zeh A, Lehrmann B, Hallbauer D, Frank W (2009) Geochemical and geochronological constraints on the nature of the immediate basement next to the Mesoarchaean auriferous Witwatersrand Basin, South Africa. J Petrol 50:2187–2220

Gerya T, Perchuk LL, van Reenen DD, Smit CA (2000) Two-dimensional numerical modeling of pressure–temperature-time paths for the exhumation of some granulite facies terrains in the Precambrian. J Geodyn 30:17–35

Good N, de Wit MJ (1997) The Thabazimbi-Murchison lineament of the Kaapvaal Craton, South Africa: 2700 Ma of episodic deformation. J Geol Soc London 154:93–97

Graupner T, Klemd R, Henjes-Kunst F, Goldmann S, Behnsen H, Gerdes A, Dohrmann R, Barton JM, Opperman R (2018) Formation conditions and REY enrichment of the 2060 Ma phosphorus mineralization at Schiel (South Africa): geochemical and geochronological constraints. Miner Deposita, in press. https://doi.org/10.1007/s00126-018-0791-7

Griffin WL, Wang X, Jackson SE, Pearson NJ, O'Reilly SY, Xu X, Zhou X (2002) Zircon chemistry and magma mixing, SE China: in-situ analysis of Hf isotopes, Tonglu and Pingtan igneous complexes. Lithos 61:237–269

Grobler NJ (1972) The geology of the Pietersburg belt. Unpublished Ph.D. thesis, University Orange Free State, Bloemfontein, 156 p

Guitreau M, Blichert-Toft J, Martin H, Mojzsis S, Albarède F (2012) Hafnium isotope evidence from Archean granitic rocks for deep-mantle origin of continental crust. Earth Planet Sci Lett 337–338:211–223

Henderson DR, Long LE, Barton JM Jr (2000) Isotopic ages and chemical and isotopic composition of the Archaean Turfloop Batholith, Pietersburg granite greenstone terrane, Kaapvaal Craton, South Africa. S Afr J Geol 103:38–46

Henzen M, Steidle L (2000) The Giyani Greenstone Belt, Northern Province, South Africa. Unpublished Diploma thesis, Bern University, Switzerland; 73 p

Jaeckel P, Kröner A, Kamo SL, Brandl G, Wendt JI (1997) Late Archaean to early Proterozoic granitoid magmatism and high-grade metamorphism in the central Limpopo Belt, South Africa. J Geol Soc London 154:25–44

Jaguin J (2012) Datation et caractérisation de processus minéralisateurs à l'Archéen : Application à l'Antimony Line, Ceinture de Roches Vertes de Murchison, Afrique du Sud. Unpublished doctoral thesis, University of Rennes, France, 351 p

Jaguin J, Gapais D, Poujol M, Boulvais P, Moyen J-F (2012) The Murchison greenstone belt (South Africa): a general tectonic framework. South Afr J Geol 115:65–76

Johnson T, Brown M, Kaus BJP, van Tongeren JA (2013) Delamination and recycling of Archaean crust caused by gravitational instabilities. Nat Geosci 7:47–52

Jones MG (1989) The geology and associated mineralization of part of the Pietersburg greenstone belt, South Africa. Unpublished Ph.D. thesis, Imperial College, London, UK, 336 pp

Kalbskopf SP, Barton JM (2003) The Zandrivier Deposit, Pietersburg Greenstone Belt, South Africa: an auriferous tourmalinite. S Af J Geol 106:361–374

Kamber BS, Ewart A, Collerson KD, Bruce MC, McDonald GD (2002) Fluid-mobile trace element constraints on the role of slab melting and implications for Archaean crustal growth models. Contrib Miner Petrol 144:38–56

Kleywegt RJ, De Beer JH, Stettler EH, Brandl G, Duvenhage AWA, Day RW (1987) The structure of the Giyani greenstone belt, as derived from geophysical studies. S Afr J Geol 90:282–295

Kramers JD, Mouri H (2011) The geochronology of the Limpopo Complex: a controversy solved. In: van Reenen DD, Kramers JD, McCourt S, Perchuk LL (eds) Origin and evolution of Precambrian high-grade gneiss terranes, with special emphasis on the Limpopo Complex of Southern Africa. Geological Society of America, Memoir, vol. 207, pp 85–106

Kramers JD, Henzen M, Steidle L (2014) Greenstone belts at the northernmost edge of the Kaapvaal Craton: timing of tectonic events and a possible crustal fluid source. Precambrian Res. 253: 96–113

Kramers JD, McCourt S, Roering C, Smit CA, van Reenen DD (2011) Tectonic models proposed for the Limpopo Complex: mutual compatibilities and constraints. In: van Reenen DD, Kramers J., McCourt S, Perchuk LL (eds) Origin and evolution of Precambrian high grade gneiss terranes, with special emphasis on the Limpopo Complex of Southern Africa. Geological Society of America Memoir, vol 207, pp 311–324

Kramers JD, McCourt S, van Reenen DD (2006) The Limpopo Belt. In: Johnson MR, Anhaeusser CR, Thomas RJ (eds) The Geology of South Africa. Geological Society of South Africa, Johannesburg, South Africa, pp 209–236

Kreissig K, Holzer L, Frei IM, Villa JD, Kramers JD, Kröner A, Smit CA, van Reenen DD (2001) Geochronology of the Hout River shear zone and the metamorphism in the Southern marginal zone of the Limpopo belt, Southern Africa. Precambr Res 109:145–173

Kreissig K, Nagler TF, Kramers JD, Van Reenen DD, Smit CA (2000) An isotopic and geochemical study of the Kaapvaal Craton and the Southern Marginal Zone of the Limpopo belt: are they juxtaposed terranes? Lithos 50:1–25

Kröner A, Brandl G, Brandt S, Klemd R, Xie H (2018) Geochronological evidence for Archaean and Palaeoproterozoic polymetamorphism in the Central Zone of the Limpopo Belt, South Africa. Precambr Res 310:320–347

Kröner A, Jaeckel P, Brandl G (2000) Single zircon ages for felsic to intermediate rocks from the Pietersburg and Giyani greenstone belts and bordering granitoid orthogneisses, northern Kaapvaal Craton, South Africa. J Afr Earth Sci 30:773–793

Kröner A, Jaeckel P, Brandl G, Nemchin AA, Pidgeon RT (1999) Single zircon ages for granitoid gneisses in the Central Zone of the Limpopo Belt, Southern Africa and geodynamic significance. Precambr Res 93:299–337

Laurent O (2012) Les changements géodynamiques à la transition Archéen-Protérozoïque: étude des granitoïdes de la marge Nord du craton du Kaapvaal (Afrique du Sud). Unpublished doctoral thesis, University Blaise Pascal, Clermont-Ferrand II, France, 512 p

Laurent O, Zeh A (2015) A linear Hf isotope-age array despite different granitoid sources and complex Archean geodynamics: example from the Pietersburg block (South Africa). Earth Planet Sci Lett 430:326–338

Laurent O, Martin H, Moyen J-F, Doucelance R (2014a) The diversity and evolution of late-Archean granitoids: evidence for the onset of "modern-style" plate tectonics between 3.0 and 2.5 Ga. Lithos 205:208–235

Laurent O, Nicoli G, Zeh A, Stevens G, Moyen J-F, Vezinet A (2014b) Comment on "Ultrahigh temperature granulites and magnesian charnockites: evidence for the Neoarchean accretion along the northern margin of the Kaapvaal craton". Precambr Res 255: 455–458

Laurent O, Rapopo M, Stevens G, Moyen J-F, Martin H, Doucelance R, Bosq C (2014c) Contrasting petrogenesis of Mg–K and Fe–K granitoids and implications for post-collisional magmatism: case study from the late-Archean Matok pluton (Pietersburg block, South Africa). Lithos 196:131–149

Laurent O, Doucelance R, Martin H, Moyen J-F (2012) Differentiation of the late-Archaean sanukitoid series and some implications for crusta l growth: insightsfrom geochemical modelling on the Bulai pluton, Central Limpopo Belt, SouthAfrica. Precambr Res 227:188–203

Laurent O, Paquette J-L, Martin H, Doucelance R, Moyen J-F (2013) LA-ICP-MS dating of zircons from Meso- and Neoarchean granitoids of the Pietersburg block (South Africa): crustal evolution at the northern margin of the Kaapvaal craton. Precambr Res 230:209–226

Laurent O, Zeh A, Gerdes A, Villaros A, Gros K, Słaby E (2017) How do granitoid magmas mix with each other? Insights from textures, trace element and Sr–Nd isotopic composition of apatite and titanite from the Matok pluton (South Africa) Contrib Mineral Petrol 172 (9):80

Mapeo RBM, Armstrong RA, Kampunzu AB, Ramokate LV (2004) SHRIMP U–Pb zircon ages of granitoids from the Western Domain of the Kaapvaal Craton, southeastern Botswana: implications for crustal evolution. S Afr J Geol 107:55–68

Martin H, Moyen J-F, Guitreau M, Blichert-Toft J, Le Pennec J-L (2014) Why Archaean TTG cannot be generated by MORB melting in subduction zones. Lithos 198–199:1–13

Mason R (1973) The Limpopo mobile belt—Southern Africa. Philos Trans Royal Soc London Ser A 273:463–485

McCourt S, van Reenen DD (1992) Structural geology and tectonic setting of the Sutherland greenstone belt, Kaapvaal Craton, South Africa. Precambr Res 55:93–110

McDonough WF, Sun SS (1995) The composition of the Earth. Chem Geol 120:223–253

Moyen J-F (2011) The composite Archaean grey gneisses: petrological significance, and evidence for a non-unique tectonic setting for Archaean crustal growth. Lithos 123:21–36

Moyen J-F, Laurent O (2018) Archaean tectonic systems: a view from igneous rocks. Lithos 302–303:99–125

Moyen J-F, Martin H (2012) Forty years of TTG research. Lithos 148:312–336

Moyen J-F, Stevens, G (2006) Experimental constraints on TTG petrogenesis: implications for Archean geodynamics. In: Benn K, Mareschal J-C, Condie KC (eds) Archean geodynamics and environments. American Geophysical Union, Monographs vol 164, pp 149–178

Næraa T, Scherstén A, Rosing MT, Kemp AIS, Hoffmann JE, Kokfelt TF, Whitehouse MJ (2012) Hafnium isotope evidence for a transition in the dynamics of continental growth 3.2 Gyr ago. Nature 485:627–630

Nagel TJ, Hoffmann JE, Münker C (2012) Generation of Eoarchean tonalite–trondhjemite–granodiorite series from thickened mafic arc crust. Geology 40:375–378

Nicoli G, Stevens G, Buick IS, Moyen J-F (2014) A comment on ultrahigh-temperature metamorphism from an unusual corundum + orthopyroxene intergrowth bearing Al–Mg granulite from the Southern Marginal Zone, Limpopo Complex. South Africa. Contrib Mineral Petrol 167:1022

Nicoli G, Stevens G, Moyen J-F, Frei D (2015) Rapid evolution from sediment to anatectic granulite in an Archean continental collision zone: the example of the Bandelierkop Formation metapelites, South Marginal Zone, Limpopo Belt, South Africa. J Metamorph Geol 33:177–202

Passeraub M (1998) The Rhenosterkoppies Greenstone Belt, Northern Transvaal, South Africa. Unpublished diploma thesis, University of Bern, Switzerland, 96 p

Passeraub M, Wüst T, Kreissig K, Smit CA, Kramers JD (1999) Structure, metamorphism and geochronology of the Rhenosterkoppies greenstone belt, Northern Province, South Africa. S Afr J Geol 102:323–334

Pearce JA (2008) Geochemical fingerprinting of oceanic basalts with applications to ophiolite classification and the search for Archean oceanic crust. Lithos 100:14–48

Perchuk LL, Gerya TV, van Reenen D, Smit CA, Krotov AV (2000) P-T paths and tectonic evolution of shear zones separating high-grade terrains from cratons: examples from Kola Peninsula (Russia) and Limpopo Region (South Africa). Miner Petrol 69: 109–142

Poujol M (2001) U–Pb isotopic evidence for episodic granitoid emplacement in the Murchison greenstone belt, South Africa. J Afr Earth Sc 33:155–163

Poujol M, Robb LJ (1999) New U–Pb zircon ages on gneisses and pegmatite from south of the Murchison greenstone belt, South Africa. S Afr J Geol 102:93–97

Poujol M, Robb LJ, Anhaeusser CR, Gericke B (2003) A review of the geochronological constraints on the evolution of the Kaapvaal Craton, South Africa. Precambr Res 127:181–213

Poujol M, Robb LJ, Respaut J-P, Anhaeusser CR (1996) 3.07–2.97 Ga greenstone belt formation in the northeastern Kaapvaal Craton: implications for the origin of the Witwatersrand Basin. Econ Geol 91:1455–1461

Pretorius AI, van Reenen DD, Barton JM (1988) BIF-hosted gold mineralization at the Fumani Mine, Sutherland greenstone belt, South Africa. S Afr J Geol 91:429–438

Rajesh HM, Santosh M, Wan Y, Liu D, Liu SJ, Belyanin GA (2014) Ultrahigh temperature granulites and magnesian charnockites: evidence for Neoarchean accretion along the northern margin of the Kaapvaal Craton. Precambr Res. 246:150–159

Rapopo M (2010) Petrogenesis of the Matok pluton, South Africa: implications on the heat source that induced regional metamorphism in the Southern Marginal Zone of the Limpopo belt. Unpublished M.Sc. thesis, University of Stellenbosch, South Africa

Robb LJ, Brandl G, Anhaeusser CR, Poujol M (2006) Archean granitoid intrusions. In: Johnson, MR, Anhaeusser, Thomas RJ (eds) The Geology of South Africa. Geological Society of South Africa: Johannesburg/Council of Geosciences, Pretoria, pp 57–94

Roering C, van Reenen DD, Smit CA, Barton JM, de Beer JH, de Wit MJ, Stettler EH, van Schalkwyk JF, Stevens G, Pretorius S (1992) Tectonic model for the evolution of the Limpopo Belt. Precambr Res 55:539–552

SACS, South African Committee for Stratigraphy, 1980. Stratigraphy of South Africa. Part1, Kent, LE (comp). Lithostratigraphy of the Republic of South Africa, South West Africa/Namibia, and the Republics of Bophuthatswana, Transkei and Venda. Handbook Geol Surv S Afr, vol 8, 690 p

Saager R, Muff R (1980) A new discovery of possible primitive life-forms in conglomerate of the Archaean Pietersburg greenstone belt, South Africa. Geol Rundsch 69:179–185

Schwarz-Schampera U, Terblanche H, Oberthür T (2010) Volcanic-hosted massive sulfide deposits in the Murchison greenstone belt, South Africa. Miner Deposita 45:113–145

Schoene B, Dudas FOL, Bowring SA, de Wit M (2008) Sm–Nd isotopic mapping of lithospheric growth and stabilization in the eastern Kaapvaal craton. Terra Nova 21:219–228

Smit CA, Roering C, Van Reenen DD (1992) The structural framework of the southern margin of the Limpopo Belt, South Africa. Precambr Res 55:51–67

Smit CA, van Reenen DD, Gerya TV, Perchuk LL (2001) P–T conditions of decompression of the Limpopo high-grade terrain: record from shear zones. J Metam Geol 19:249–268

Smit CA, van Reenen DD, Roering C (2014) Role of fluids in the exhumation of the Southern Marginal Zone of the Limpopo Complex, South Africa. Precambr Res 253:81–95

Stevens G (1997) Melting, carbonic fluids and water recycling in the deep crust: an example from the Limpopo Belt, South Africa. J Metamorph Geol 15:141–154

Stevens G, van Reenen DD (1992a) Constraints on the form of the P-T loop in the Southern Marginal Zone of the Limpopo Belt. South Africa. Precambrian Res. 55:279–296

Stevens G, van Reenen DD (1992b) Partial melting and the origin of metapelitic granulites in the Southern Marginal Zone of the Limpopo Belt, South Africa. Precambr Res 55:303–319

Stevens G, Clemens JD, Droop GTR (1997) Melt production during granulite–facies anatexis: experimental data from primitive metasedimentary protoliths. Contrib Miner Petrol 128:352–370

Taylor J, Nicoli G, Stevens G, Frei D, Moyen J-F (2014) The processes that control leucosome compositions in metasedimentary granulites: perspectives from the Southern Marginal Zone migmatites, Limpopo Belt, South Africa. J Metam Geol 32:713–742

Taylor RS (1981) Volcanogenic copper-zinc sulphide deposits of the Murchinson greenstone belt, northeastern Transvaal, South Africa. Unpublished Ph.D. thesis, Durham University, UK, 361 p

Tsunogae T, van Reenen DD (2014) High- to ultrahigh-temperature metasomatism related to brine infiltration in the Neoarchean Limpopo Complex: petrology and phase equilibrium modeling. Precambr Res 253:157–170

Tsunogae T, Miyano T, van Reenen DD, Smit CA (2004) Ultrahigh-temperature metamorphism of the Southern Marginal Zone of the Archean Limpopo Belt, South Africa. J Mineral Petrol Sci 99:213–224

van Reenen DD (1986) Hydration of cordierite and hypersthene and a description of the retrograde orthoamphibole isograd in the Limpopo Belt, South Africa. Am Mineral 71:900–915

van Reenen DD, Barton JM Jr, Roering C, Smit CA, van Schalkwyk JF (1987) Deep crustal response to continental collision: the Limpopo Belt of South Africa. Geology 15:11–14

van Reenen DD, Roering C, Ashwal LD, de Wit MJ (1992) Regional geological setting of the Limpopo Belt. Precambrian Res. 55:1–5

van Reenen DD, Pretorius AI, Roering C (1994) Characterization of fluids associated with gold mineralization and with regional high-temperature retrogression of granulites in the Limpopo belt, South Africa. Geochim Cosmochim Acta 58(3): 1147–1159

van Reenen DD, Smit CA, Perchuk AL, Huizenga JM,Safonov OG, Gerya TV (2019) The Neoarchaean Limpopo Orogeny: Exhumation and Regional-Scale Gravitational Crustal Overturn Driven by a Granulite Diapir. In: Kröner A, Hofmann A (eds) The Archaean Geology of the Kaapvaal Craton, Southern Africa. Regional geology reviews. Springer International Publishing AG

Vearncombe JR (1991) A possible Archaean island arc in the Murchison Belt, Kaap-vaal Craton, South Africa. J Afr Earth Sc 13:299–304

Vearncombe JR, Barton JM, Walsh KL (1987) The Rooiwater Complex and associated rocks, Murchison granitoid–greenstone terrane, Kaapvaal Craton. S Afr J Geol 90:361–377

Vearncombe JR, Barton JM, Cheshire PE. De Beer JH, Stettler EH, Brandl G (1992) Geology, geophysics and mineralisation of the Murchison schist belt, Rooiwater Complex and surrounding granitoids. Geological Survey of South Africa Memoirs, vol 81, 139 p

Vermeesch P (2012) On the visualization of detrital age distributions. Chem Geol 312–313:190–194

Vezinet A (2016) Différenciation et stabilization de la croûte continentale archéenne, l'exemple de la marge Nord du craton du Kaapvaal, Afrique du Sud. Doctoral thesis, University of Saint-Etienne, France, 321 p

Vezinet A, Moyen J-F, Stevens G, Nicoli G, Laurent O, Couzinié S, Frei D (2018) A record of 0.5 Ga of evolution of the continental crust along the northern edge of the Kaapvaal Craton, South Africa: consequences for the understanding of Archean geodynamic processes. Precambr Res 305:310–326

Vorster CJ (1979) Die geologie van die Klein-Letabagebied, noordoos-Transvaal met spesiale verwysing na die granitiese gesteentes. Unpublished M.Sc. thesis, Rand Afrikaans University of Johannesburg, 138 p

Whitney DL, Evans BW (2010) Abbreviations for names of rock-forming minerals. Am Mineral 95:185–187

Wilson AH, Zeh A (2018) U–Pb and Hf isotopes of detrital zircons from the Pongola Supergroup: constraints on deposition ages, provenance and Archaean evolution of the Kaapvaal Craton. Precambr Res 305:117–196

Zeh A, Gerdes A (2012) U–Pb and Hf isotope record of detrital zircons from gold-bearing sediments of the Pietersburg Greenstone belt (South Africa)—is there a common provenance with the Witwatersrand Basin? Precambr Res 204–205:46–56

Zeh A, Gerdes A, Barton JM Jr (2009) Archean accretion and crustal evolution of the Kalahari Craton—the zircon age and Hf isotope record of granitic rocks from Barberton/Swaziland to the Francistown Arc. J Petrol 50:933–966

Zeh A, Gerdes A, Klemd R, Barton JM Jr (2007) Archean to Proterozoic crustal evolution in the Central Zone of the Limpopo Belt (South Africa/Botswana): constraints from combined U–Pb and Lu–Hf isotope analyses of zircon. J Petrol 48:1605–1639

Zeh A, Gerdes A, Millonig L (2011) Hafnium isotope record of the Ancient Gneiss Complex, Swaziland, southern Africa: evidence for Archaean crust–mantle for-mation and crust reworking between 3.66 and 2.73 Ga. J Geol Soc Lond 168:953–963

Zeh A, Jaguin J, Poujol M, Boulvais P, Block S, Paquette J-L (2013) Juvenile crust formation in the northeastern Kaapvaal Craton at 2.97 Ga—Implications for Archean terrane accretion, and the source of the Pietersburg gold. Precambr Res 233:20–43

Zeh A, Stern RA, Gerdes A (2014) The oldest zircons of Africa—their U–Pb–Hf–O isotope and trace element systematics, and implications for Hadean to Archean crust–mantle evolution. Precambr Res 241:203–230

5 The Mesoarchaean Basement Complex of the Vredefort Dome—A Mid-Crustal Section Through the Central Kaapvaal Craton Exposed by Impact

Roger L. Gibson

Abstract

Mesoarchaean Basement Complex rocks exposed in the core of the Vredefort Dome represent a composite terrane in which polydeformed high-grade TTG gneisses and granites with subsidiary greenstone remnants were tectonically juxtaposed against low-grade mafic–ultramafic metavolcanic greenschists along a major NE-trending shear zone prior to 3.07 Ga. The high-grade portion exposes a ≥ 10 km section through the mid-crust as a result of extreme rotation and differential uplift related to the 2.02 Ga Vredefort impact event. Syntectonic, 3.13–3.10 Ga, TTG gneisses and late-tectonic, 3.10–3.08 Ga, granites intruded and partially assimilated 3.3–3.5 Ga old supracrustal gneiss protoliths. Prior to impact rotational effects, the gneisses contained subhorizontal S2 and subvertical, NW-trending, S3 foliations. D3 straddled the metamorphic peak, which ranged from ~680 °C, 0.5 GPa to ~885 °C, 0.71–0.77 GPa. Together with the lithological and structural evidence, the tight clockwise P-T path derived from migmatitic-restitic metapelitic gneisses is interpreted to suggest the formation and docking of a volcanic arc with the southwestern edge of the Kaapvaal protocraton at 3.13–3.08 Ga. The D4 shear zone that juxtaposes the gneisses against greenschist facies (400 °C) komatiite to komatiitic metabasalts to the southeast is interpreted as an originally shallow NW-dipping, retrograde, crustal-scale extensional feature that aided rapid exhumation of the high-grade terrane prior to deposition of the 3.07 Ga Dominion Group. In addition to differential uplift of the Basement Complex, the impact caused significant post-shock heating that generated metamorphic reaction coronas, melting and annealing of Archaean features to produce granofelses under ultrahigh-T/low-P (>1050 °C, 0.2–0.3 GPa) conditions in the centre of the dome at 2.02 Ga.

R. L. Gibson (✉)
School of Geosciences, University of the Witwatersrand,
P O WITS, Johannesburg, South Africa
e-mail: roger.gibson@wits.ac.za

Keywords
Granulite–amphibolite transition • Impact • Juvenile arc • Kaapvaal craton • Komatiite • Mid-crust • Mesoarchaean • Vredefort Dome

5.1 Introduction

The Archaean Basement Complex rocks of the Vredefort Dome occupy a crucial location and context within the Kaapvaal craton. Situated broadly centrally within the craton, the exposures also lie within the limits of the world's largest gold deposit, the Witwatersrand Basin (Frimmel 2019, this volume, Figs. 5.1 and 5.2). The Basement Complex rocks underlie the core of a 90-km-wide domal uplift that exposes not only almost the entire Mesoarchaean-to-Palaeoproterozoic sedimentary and volcanic supracrustal sequence that overlies the Archaean Basement Complex over much of the craton (estimated to be between 14 and 16 km thick), but also a substantial section through the underlying mid-crust. The unusual geological circumstances leading to the exposure of the Basement Complex in the core of the Vredefort Dome, namely, extreme near-instantaneous uplift triggered by a giant impact event, also produced an array of exceptional structural and metamorphic features that must be recognised if the Archaean history of the rocks is to be deciphered.

5.2 Geological Setting

The Vredefort Dome lies approximately 120 km southwest of Johannesburg (Anhaeusser, this volume, Fig. 5.2). It comprises a 40-km-wide core that exposes crystalline Archaean Basement Complex rocks and a 20–25 km-wide collar consisting of younger (3.07 to ~2.10 Ga) metamorphosed sedimentary and volcanic rocks of the Dominion Group and Witwatersrand, Ventersdorp and Transvaal

A. Kröner and A. Hofmann (eds.), *The Archaean Geology of the Kaapvaal Craton, Southern Africa*,
Regional Geology Reviews, https://doi.org/10.1007/978-3-319-78652-0_5

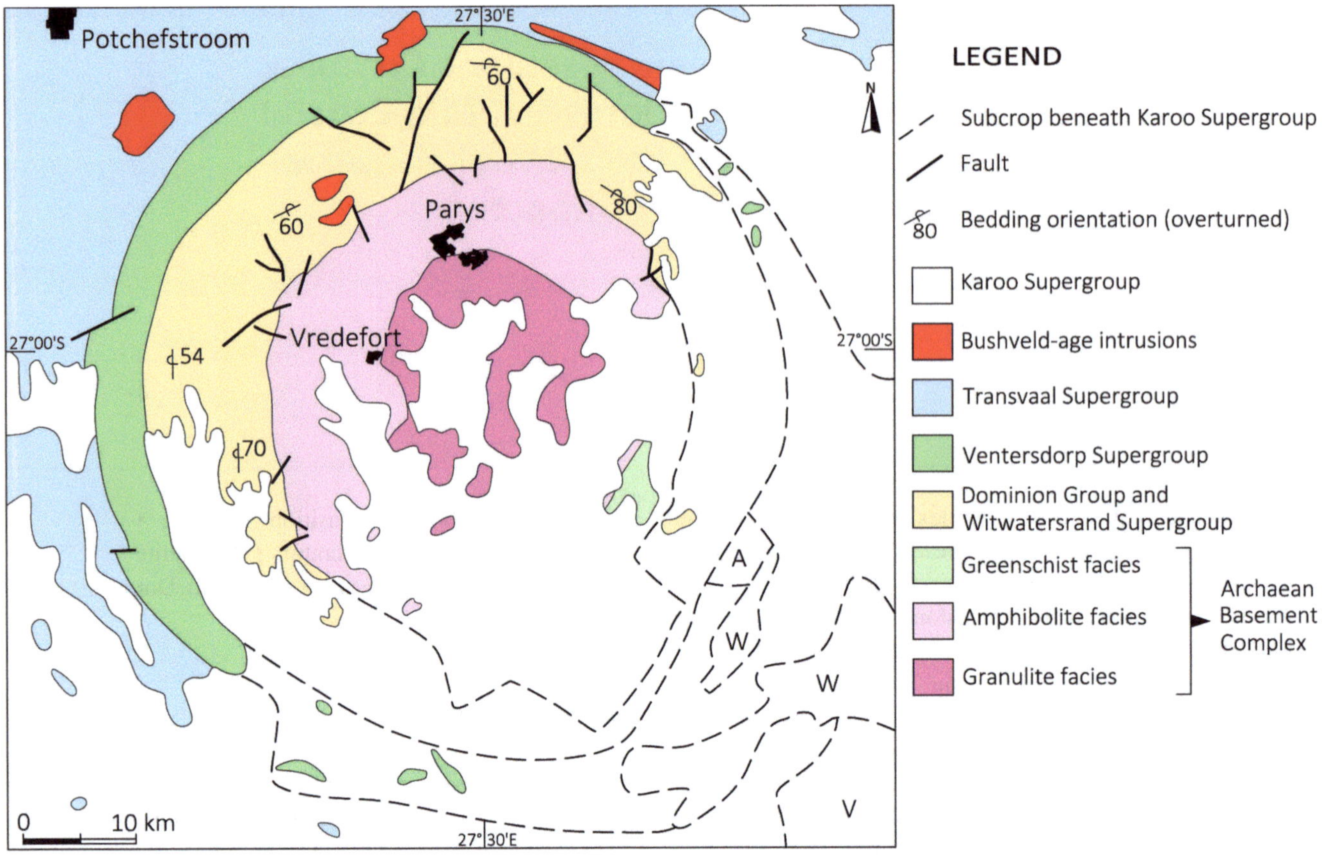

Fig. 5.1 Simplified geological map of the Vredefort Dome (modified from Martini 1992) showing the distribution of granulite, amphibolite and greenschist facies rocks in the Archaean Basement Complex, the collar of overturned Mesoarchaean-to-Palaeoproterozoic supracrustal strata, the main intrusions and the subcrop beneath the Karoo Supergroup unconformity (A—Archaean Basement Complex; W—Witwatersrand Supergroup; V—Ventersdorp Supergroup)

supergroups that display steep to overturned dips (Fig. 5.1). Both the core and collar rocks contain mafic intrusions that have been attributed, respectively, to the 2.71 Ga Ventersdorp, 2.06 Ga Bushveld and 1.1 Ga Umkondo events; additionally, several small ultramafic to peralkaline granite intrusions have been correlated with the Bushveld event (Figs. 5.1 and 5.2; see review in Gibson and Reimold 2008). Based on an array of shock deformation phenomena (shatter cones, high-pressure silica polymorphs, planar microdeformation features in quartz and zircon), unusually voluminous and pervasive pseudotachylitic breccias and kilometres-long dykes of an impact melt rock (Fig. 5.2), the dome is interpreted as the deeply eroded central uplift of a 2023 ± 4 Ma (Kamo et al. 1996) impact structure that had an original estimated diameter of between 250 and 300 km (see review in Gibson and Reimold 2008). Current post-impact erosion levels are estimated at 8–10 km (Gibson et al. 1998).

The central and southeastern portions of the dome are covered by a thin veneer of Phanerozoic Karoo Supergroup shales and dolerite sills, with only a few, poorly exposed inliers of the basement rocks (Fig. 5.1). Geophysical and drillcore evidence suggests that the broadly annular core-collar contact is disrupted in the southeast beneath this Phanerozoic cover by pre-Ventersdorp Supergroup block faulting (Fig. 5.1; Martini 1992).

The steeply overturned orientation of the collar strata (Fig. 5.1) provides an exceptional cross section through the Mesoarchaean-to-Palaeoproterozoic supracrustal successions of the Kaapvaal craton. Based on geochemical profiles through Ventersdorp-age metadolerite sheets intruding the Basement Complex, Bisschoff (1972) demonstrated that the rocks in the outermost 3 km of the core of the dome are also overturned. Lana et al. (2003c) further extended the zone of rotation up to 6 km into the core based on foliation patterns in the gneisses (Fig. 5.2; see Sect. 5.6). They also established that the transition to relatively unrotated rocks in the central core ($\leq$ 12 km radius) is relatively sharp. When this is added to the estimated cumulative, 14–16 km, stratigraphic thickness of the collar rocks, the evidence suggests that the Vredefort Dome exposes an ~25–27 km crustal section, of which at least 10 km is through the crystalline basement.

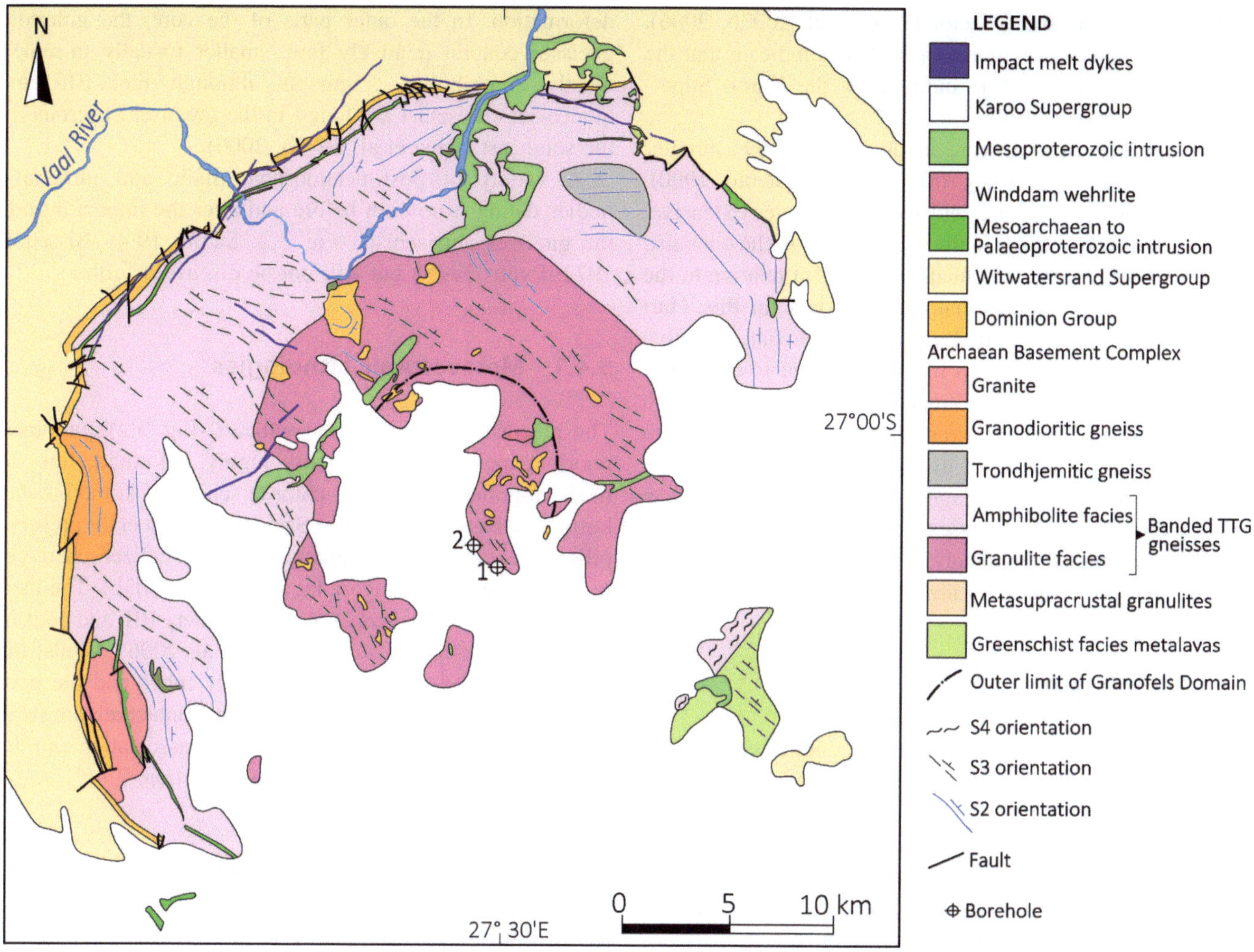

Fig. 5.2 Simplified geological map of the Archaean Basement Complex in the core of the Vredefort Dome, modified from Bisschoff (2000) and Lana et al. (2004). The map includes several Neoarchaean-to-Mesoproterozoic intrusions, including impact melt rock dykes, and the central impact-induced Granofels Domain (see text for details). Structural data from Stepto (1979) and Lana et al. (2003a, b, c, 2006). The high-grade gneisses are separated from low-grade metavolcanics to the southeast by a D4 shear zone. Boreholes are referred to in the text: 1—Beta-1 core; 2—Inlandsee core

Gibson et al. (1998, 2002) and Gibson and Reimold (2005) demonstrated from metamorphic parageneses that the rocks in the dome experienced progressively higher post-impact temperatures towards the centre of the dome that are interpreted to reflect a combination of centripetally increasing shock heating and uplift of progressively deeper crustal levels as a result of the impact. Numerical modelling (Ivanov 2005) confirmed that temperatures of >1000 °C were possible in the central core, sufficient to cause significant recrystallization, reaction and even localised melting of the Basement Complex gneisses (Gibson 2002; Gibson et al. 2002). The limit of this inner domain, called the 'Inlandsee Leucogranofels' by Stepto (1979, 1990) and the 'Granofels Domain' in this chapter, is shown in Fig. 5.2. The modelling also predicted that ~25 km of structural uplift would have occurred, consistent with field estimates.

5.3 Overview of the Archaean Basement Complex in the Vredefort Dome

The Archaean Basement Complex in the core of the Vredefort Dome comprises two terranes separated by a major structural discontinuity (Fig. 5.2). The bulk of the core comprises polydeformed, polymetamorphic, predominantly trondhjemitic, granodioritic and granitic (TTG) gneisses that contain increasing proportions of metre-to-kilometre-scale mafic, ultramafic, pelitic-greywacke and banded ironstone (BIF) gneiss fragments in the more deeply exhumed central parts, and which are metamorphosed to upper amphibolite to granulite facies grades. These gneisses are juxtaposed along a subvertical, NE-trending, mylonite zone against greenschist facies ultramafic metalavas with minor metasedimentary units in the southeastern sector (Lana et al. 2003b, 2006). Geophysics and coal exploration drillcores indicate that the greenstones extend further south below the Karoo Supergroup cover (Martini 1992).

The central parts of the core are characterised by a circular +30 mGal gravity anomaly (Maree 1947; Stepto 1990). Although outcrop is sparse and the Karoo cover extensive, several shallow boreholes in the central 5 km of the core also support a higher mafic–ultramafic component relative to the rocks in the outer annulus that could account for this. Hart et al. (1990a, 1995) reported a prominent arcuate negative aeromagnetic anomaly of up to −800 nT that is most pronounced between Parys and Vredefort, and which appears to broadly coincide with the transition from amphibolite to granulite facies (Flowers et al. 2003; Lana et al. 2003a). Although several studies sought to relate this to Archaean tectonics or impact processes, Carporzen et al. (2012) demonstrated that the anomaly is related to high magnetic remanence values caused by intense lightning strike activity.

5.4 Lithologies

The Archaean Basement Complex is dominated by predominantly syntectonic trondhjemitic and granodioritic–granitic gneisses and weakly deformed late-tectonic granodiorites and granites; tonalitic gneisses are relatively rare and appear to be confined to the granulite zone (Lana et al. 2004). Most outcrops are lithologically heterogeneous and reveal a composite intrusive character on a centimetre-to-metre scale, with cross-cutting relationships and degree of deformation establishing the more leucocratic–potassic lithologies as the younger phases (e.g. Fig. 5.3a, b). This small-scale heterogeneity is also exemplified in the Inlandsee drillcore (Fig. 5.2). Lana et al. (2004) described this banded gneiss as 'migmatitic gneiss' whilst noting that evidence that the 'leucosome' represents locally derived anatectic melt is lacking in both the amphibolite and granulite facies zones. It is referred to here as 'banded TTG gneiss' to distinguish it from more homogeneous bodies that are found locally (Fig. 5.2).

The next most abundant lithologies after the granitoid gneisses are mafic and ultramafic rocks. The southeastern greenstone terrane comprises almost exclusively komatiite to komatiitic basalt metalavas of greenschist grade, whereas the TTG gneisses in the central core contain relatively abundant mafic ± ultramafic granulite gneiss fragments, as well as metapelitic/greywacke and meta-BIF granulite fragments up to 1–2 km in size (Fig. 5.2). Contact relations between the fragments and the TTG gneisses are not well exposed, but some adjacent fragments show structural continuity, suggesting that they represent an original greenstone succession that was disrupted by granitoid intrusion and strong deformation. In the outer parts of the core, the granitoid gneisses contain relatively few, smaller (usually m-scale), mostly amphibolite, fragments although meta-BIF and quartz–sericite schist are found in the greenschist terrane in the southeast (Lana et al. 2003b, 2006).

A variety of post-metamorphic mafic and ultramafic bodies dating from both before and after the impact intrude the gneisses and schists (Fig. 5.2; Stepto 1990; Bisschoff 1972; Pybus 1995), but will not be discussed further.

5.4.1 Metavolcanic Lithologies

The greenstone inlier in the southeast (Fig. 5.2) comprises poorly exposed, predominantly highly deformed, fine- to medium-grained, actinolite–chlorite schists of komatiite and komatiitic basalt affinity (see Sect. 5.5) that locally preserve pillows, vesicles and variolites, and recrystallized chert–cemented hyaloclastite Fig. 5.4a; Lana et al. 2003b). The metalavas contain thin layers of metachert, meta-BIF and pods of sericite–quartz–biotite schist (Lana et al. 2006). Whilst the highly deformed nature of most of the rocks and the poor exposure make it impossible to establish contact relationships between these lithologies, the chert–BIF association and the pillow structures support subaqueous eruption.

5.4.2 Metasedimentary Lithologies

Metasedimentary fragments, up to 1.5 × 0.5 km in size and comprising interlayered metapelite, metagreywacke and meta-BIF lithologies, occur in the granulite facies zone. The metapelite and metagreywacke comprise medium- to coarse-grained stromatic migmatite and restite with centimetre-to-decimetre-scale banding (Fig. 5.4b). Their mineralogy comprises garnet–biotite–quartz–plagioclase–ilmenite, with either K-feldspar, cordierite, spinel and relict sillimanite (metapelite) or orthopyroxene (metagreywacke; see Sect. 5.7). Leucosomes in the metapelite are dominated by quartz–K-feldspar–plagioclase, as opposed to quartz–plagioclase in the metagreywacke. They occur as millimetre-to-centimetre-scale layers and lenses generally concordant to the dominant fabric in the rocks and are weakly foliated, but locally form discordant veins. Metagreywacke may also host more diffuse, locally nebulitic, leucosomes. Stepto (1979) reported garnet- and cordierite-bearing granitoid gneisses in the vicinity of the metasedimentary fragments, suggesting some mobility of the anatectic melts.

Banded iron formation occurs as a minor component associated with the metalavas in the greenstone inlier where it comprises millimetre-to-centimetre-scale siliceous metachert bands alternating with amphibole–quartz–magnetite bands (Lana et al. 2003b). In the high-grade terrane, meta-BIF

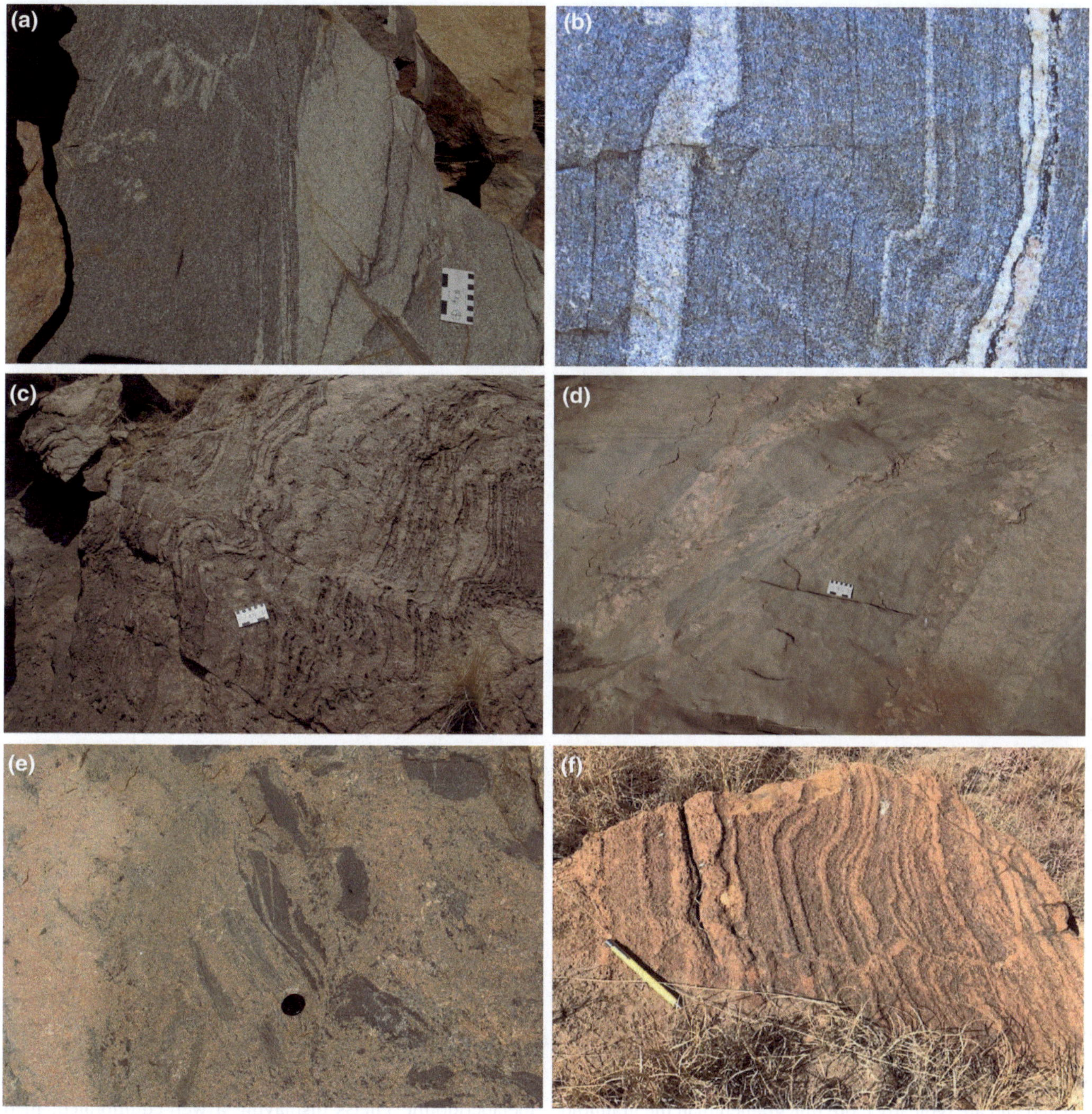

Fig. 5.3 Archaean Basement Complex granitoid lithologies in the Vredefort Dome. **a** Banded TTG gneiss showing grey biotite trondhjemite gneiss and slightly boudinaged and locally discordant, folded, quartz–plagioclase leucocratic veins. **b** Mesocratic trondhjemite gneiss intruded by more leucocratic trondhjemite gneiss and irregular quartz–plagioclase vein (right). **c** Granulite facies banded gneiss showing finer grained, dark grey, enderbitic gneiss hosting charnockitic veins subparallel to S2 foliation. The disharmonic D3 folds suggest high ductility, consistent with peri-anatectic conditions. Dark spots (bottom, centre) are weathered centimetre-scale pyroxene–mica aggregates. **d** Pink granodioritic gneiss showing grey trondhjemitic gneiss xenolith (centre left) and multiple pegmatite generations. The foliation is marked by variable mafic content and leucocratic veins. **e** Partial assimilation of mafic granulite xenoliths in felsic enderbite. Note the ghost outlines of assimilated clasts (left of lens cap), coarse hornblende (black specks) and the indistinct contact with charnockite (left). **f** 'Inlandsee leucogranofels' showing strong banding of enderbitic gneiss and coarser grained charnockite that show D3 folding and pinch-and-swell features. The discordant irregular vein is a granite leucosome formed as a result of impact-induced anatexis

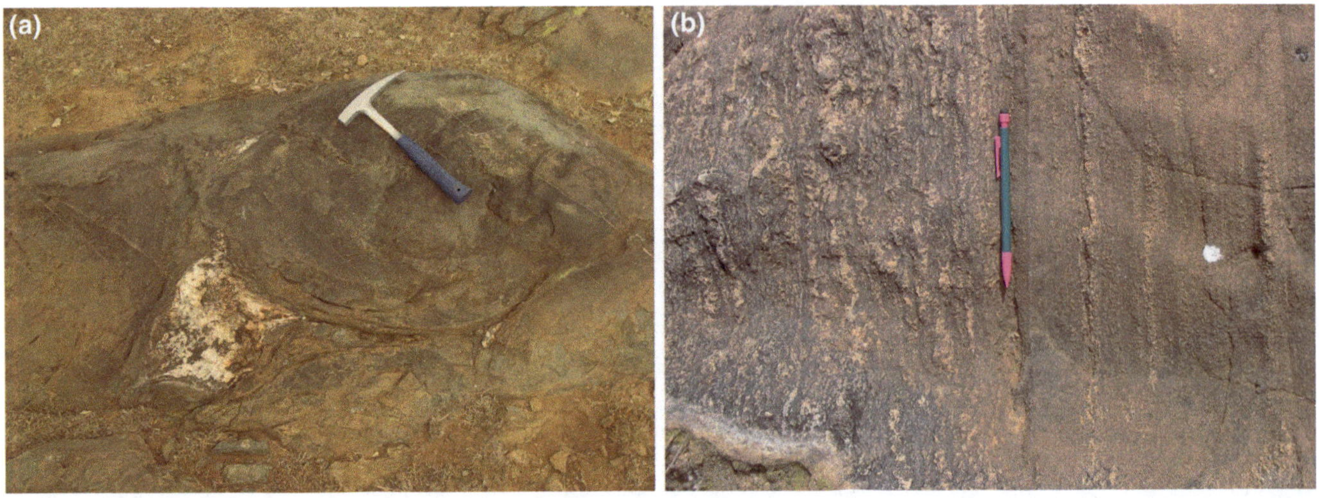

Fig. 5.4 Supracrustal rocks in the Archaean Basement Complex of the Vredefort Dome. **a** Weakly deformed pillows with interstitial hyaloclast-bearing chert (now quartz) in greenschist-grade komatiitic metabasalt. **b** Interlayered pelitic (left) and greywacke (right) granulite showing stromatic leucosome parallel to vertical S2/S3 fabric. The pelite contains more leucosome and coarse peritectic garnet

occurs intercalated with or close to the metapelitic–metagreywacke lithologies, except for an isolated xenolith within the major shear zone in the southeastern sector.

5.4.3 Metamafic Lithologies

Metamafic fragments occur in the granitoid gneisses throughout the high-grade terrane and appear to increase in abundance towards the centre of the dome. Although they show geochemical differences relative to the low-grade metalavas, (see Sect. 5.5), Stepto (1990) argued that their intimate association with the metasediments favours a volcanic origin. These fine- to medium-grained rocks weather preferentially relative to the granitoid gneisses, which may mean that their abundance is underestimated.

In the amphibolite facies zone, amphibolite fragments are comparatively rare and small (decimetre-to-metre-scale, rarely tens of metres) and occur as lensoid-to-angular boudinaged pods and irregular, variably disaggregated, agmatitic bodies hosted in granitoid gneisses. The amphibolites are foliated and are typically more fine grained than the granitoid gneisses. They lack obvious mineral banding but are intruded by foliation parallel to slightly discordant, centimetre-scale plagioclase + quartz $\pm$ biotite veins. Some banding, related to variable hornblende and plagioclase contents, is found in the largest body, which occurs within the retrograde shear zone in the southeast (Lana et al. 2003b).

Mafic fragments in the granulite zone reach up to hundreds of metres in length. Stepto (1979) noted that they are aligned parallel to the foliation in the enclosing granitoid gneisses and that foliation intensity increases from their centres towards their edges. Smaller fragments show disaggregation and variable amounts of assimilation by the enclosing granitoid, to the extent that only ghost outlines defined by higher mafic mineral content may remain (Fig. 5.3e). The mafic granulites are medium grained, and display a granoblastic microtexture. Locally, orthopyroxene forms large subidioblastic oikocrysts enclosing other phases. Stepto (1979) also described rare, predominantly monomineralic, amphibole gneiss in the southwest.

5.4.4 Meta-ultramafic Lithologies

Meta-ultramafic rocks are found only within the Granofels Domain (Fig. 5.2), which complicates their petrographic analysis owing to the strong impact-induced thermal overprint. These ultramafic gneisses crop out poorly, usually as highly serpentinised rubble, but have also been intersected in several shallow drill cores. With the exception of an undeformed wehrlite body (Fig. 5.2) that is linked by Stepto (1990) to the Bushveld magmatic event, all display similar fabrics to the enclosing granitoid gneisses. The best-studied occurrence involves the Beta-1 core (Fig. 5.2) that is dominated by metaperidotitic gneiss displaying a well-equilibrated, medium- to coarse-grained, granoblastic olivine–orthopyroxene–hornblende–clinopyroxene assemblage with minor biotite and magnetite. Subsidiary micaceous metapyroxenite, mica-rich bands and magnetite pods define a centimetre-to-decimetre-scale banding parallel to a strong foliation that is defined by large (up to 5 mm) poikiloblastic hornblende grains. Hart et al. (1990b) classified the dominant lithology as hornblende harzburgite, although Broccardo (2009) argued on the basis of normative mineralogy for a plagioclase lherzolite precursor.

Meta-ultramafic lithologies lacking olivine and dominated by orthopyroxene with variable amounts of plagioclase, clinopyroxene and/or amphibole occur elsewhere in the granulite zone (Stepto 1979, 1990; Lekgothoane 2017) and can only be distinguished from the mafic granulites by their higher MgO content (see Sect. 5.5).

5.4.5 Granitoid Lithologies

The Basement Complex is dominated by banded TTG gneisses that comprise centimetre-to-metre-scale, white to pink-grey, leucocratic veins in a finer grained, more highly deformed and more mafic, medium-grey, trondhjemitic host (Fig. 5.3a–c). The leucocratic component may exceed 50% of an outcrop and is locally pegmatitic. Leucocratic veins are broadly parallel to the dominant (S2 or S3) foliation in the host (Fig. 5.3a–c) and may show an internal magmatic-to-subsolidus fabric defined by aligned subhedral, centimetre-scale, K-feldspar phenocrysts and millimetre-scale mafic mineral aggregates in an undeformed matrix. Locally, veins transgress the host rock foliation and may either be slightly folded or boudinaged (Fig. 5.3a). They may also grade laterally into irregular to lensoid granodioritic to granitic bodies up to several metres to tens of metres in size. In the amphibolite facies zone, the veins contain <5% biotite with accessory ilmenite and vary from plagioclase–quartz-dominated to plagioclase–K-feldspar–quartz, with the more granitic veins being slightly younger based on outcrop relationships. Plagioclase in the veins displays similar compositional ranges to the host (An_{10-20}), further supporting an intrusive, rather than in situ anatectic, origin (Lana et al. 2004).

The banded TTG gneisses in the granulite zone display similar macroscopic features to those in the amphibolite facies zone, except that orthopyroxene ± clinopyroxene are the dominant mafic minerals, with biotite ± hornblende being secondary. The foliation is defined by highly flattened, millimetre-to-centimetre-scale, blue-grey quartz lenses and by mafic mineral aggregates and banding (Fig. 5.3c). Lana et al. (2004) distinguished three interbanded lithologies on a centimetre-to-metre-scale—dark grey enderbite, medium-grey felsic enderbite and light grey to pink charnockite (Fig. 5.3d). The charnockite has <5 vol.% pyroxene + biotite, and is more coarse grained and less deformed than, and intrusive into, the other lithologies, locally transgressing the foliation and leading to its disruption. A complete compositional spectrum exists from enderbitic gneiss to felsic enderbitic gneiss, with the former containing 10–20 vol.% pyroxene + biotite and more calcic plagioclase (An_{35-55}), whereas the latter contains 5–10 vol.% mafic minerals and plagioclase with An_{10-35} (Lana et al. 2004). The outcrop relationships between the various lithologies are largely indistinguishable from those seen in the amphibolite facies zone, and compositional analysis (see Sect. 5.5) confirms a similar range of protoliths as well (Lana et al. 2004).

Homogeneous gneiss bodies lacking banding are relatively rare in the Vredefort Dome. The largest body, comprising mesocratic biotite (5–10 vol.%) trondhjemitic gneiss, occurs northeast of Parys (Fig. 5.2). This strongly foliated gneiss itself contains decimetre-to-metre-scale boudinaged remnants of a more melanocratic, dark grey, biotite trondhjemitic gneiss, although the ages of both gneisses are indistinguishable (see Sect. 5.8). In the granulite zone, Stepto (1990) described discrete 'hypersthene tonalite gneiss' bodies up to 1 km in size; however, Lana et al. (2004) only noted smaller bodies of foliated pink quartz diorite gneiss up to a few tens of metres in size. The latter contains more K-feldspar and plagioclase (An_{15-25}) and less quartz (<10 vol.%) and ferromagnesian minerals (orthopyroxene + clinopyroxene, <5 vol.%) than the trondhjemitic gneisses, and displays a centimetre-to-decimetre-scale banding defined by modal variations in pyroxene content, rather than veins (Lana et al. 2004). Its contacts with the banded TTG gneiss are not exposed, but its strong fabric and more calcic nature suggest that it is an early intrusive phase, although whether its composition is unmodified by anatexis is debatable (see Sect. 5.5).

Pink-grey, coarse-grained, granodioritic gneiss (Fig. 5.3 d) has been extensively mined for dimension stone in several quarries west and northwest of Parys where it occurs as irregular to lensoid bodies within the banded TTG gneiss. It displays a weaker S3 foliation than the trondhjemitic gneiss and is clearly intrusive, but the contact between the two is not always sharp, and the granodiorite may display continuity with veins in the banded gneiss. The granodiorite contains euhedral K-feldspar megacrysts in a plagioclase–K-feldspar–quartz–biotite–magnetite–ilmenite matrix. Its fabric is best defined by grey gneiss schlieren in various stages of disaggregation (Fig. 5.3d), or by millimetre-long biotite aggregates that appear to survive this disaggregation and assimilation.

Along the western contact with the collar rocks, Lana et al. (2004) noted a 6 × 3 km body of porphyritic granodiorite with sharp to locally diffuse contacts that contain euhedral to subhedral perthitic K-feldspar phenocrysts and biotite clots that define a strong magmatic flow fabric, parallel to the S2 foliation in the enclosing TTG gneiss. A large body of coarse-grained granite occurs south of this granodiorite (Fig. 5.2). It is largely homogeneous and displays only a weak S3 foliation, defined by small biotite clots and aligned stringers of K-feldspar phenocrysts in a weakly foliated to undeformed plagioclase–quartz–K-feldspar matrix, and rare schlieren (Lana et al. 2004). Irregular, slightly discordant, intrusive veins of the granite occur in the surrounding TTG gneiss. Smaller bodies of schlieric granite with a weak S3 foliation are found locally in the amphibolite facies zone, particularly west of Parys (Lana et al. 2004).

5.4.6 Granofels Domain

Although largely covered by Karoo sediments and dolerite (Fig. 5.2), lithologies within ~5 km of the centre of the dome have attracted considerable attention because they are both the most deeply exhumed rocks in the dome and the most intensely shocked (Gibson and Reimold 2005, 2008). Despite retaining their gross banded gneissic character (Fig. 5.3f, d), they were described by Stepto (1990) as granofelses owing to their microtexture, which is attributed to annealing of feldspar and quartz shock glasses caused by the extreme post-shock temperatures (Schreyer 1983; Gibson 2002; Gibson and Reimold 2005). The exception is highly aluminous cordierite–K-feldspar–spinel–corundum–rutile granofelses that are interpreted as peritectic crystallisation products from almost complete partial melting of metapelitic granulite formed under ultrahigh-T/low-P metamorphic conditions (>1050 °C, 0.2–0.3 GPa; Gibson 2002; Ogilvie 2010). These temperatures are consistent with the numerical modelling predictions of impact-induced heating (Ivanov 2005). Additional impact-induced thermal effects include very fine-grained localised reaction coronas and partial replacement textures of the Archaean metamorphic assemblages in mafic and ultramafic rocks (Gibson and Reimold 2005, 2008), and small, millimetre-to-centimetre-scale, autochthonous to parautochthonous leucogranite melt veins in the felsic gneisses (Schreyer 1983; Gibson 2002). Although Moser (1997) and Cupelli et al. (2015) have suggested that 'norite' in the Granofels Domain represents either a mantle decompression melt or impact melt, field and petrographic analysis suggests that it represents ultrametamorphosed and annealed Archaean mafic granulite (Gibson and Reimold 2008).

5.5 Geochemistry

Initial geochemical studies of the Archaean Basement Complex lithologies in the Vredefort Dome were directed mainly at examination of radial profiles as a proxy for crustal depth (Slawson 1976; Hart et al. 1981, 1990a), particularly as some workers, such as Hart et al. (1990b), interpreted ultramafic rocks exposed in the central parts of the core as representing upper mantle. More recently, Moser et al. (2001) suggested that the radial transect comprises a section from melt-depleted, restitic, lower crust into a mid-crustal granitic batholith derived from the lower crust. However, more recent work (Lana et al. 2003c, 2004) suggests that such gross models do not take into account the lithological or structural complexity of the Archaean Basement Complex in the dome and that a more limited crustal section is exposed.

5.5.1 Metalavas

The greenschist facies metalavas show two distinct compositional groupings (Fig. 5.5)—komatiites (MgO 32–33 wt%) and komatiitic basalts (MgO 10–19 wt%). The latter are depleted in Ni, and to a lesser extent Cr, relative to the meta-komatiites but both show similar Co, Ba, Zn, Y and Zr values (Lana et al. 2003b). Overall, the metalavas display strong similarities to komatiites and komatiitic basalts in the Johannesburg Dome and Barberton greenstone belt; however, they contain lower alkalis than the Johannesburg Dome schists and higher Al_2O_3, and slightly lower CaO, TiO_2 and Fe_2O_3, than the Barberton lavas (Lana et al. 2003b).

5.5.2 Metasediments

Based on major element composition, the garnet-bearing granulites can be divided into subaluminous metagreywackes and more aluminous and potassic metapelites. Given their migmatitic-to-restitic character, Ogilvie (2010) calculated the original bulk compositions by reintegrating melt and found that although most major oxides approach the average Archaean shale composition of McLennan (1982), the Vredefort samples are still relatively enriched in FeO and MgO and are strongly depleted in CaO. The latter, with strong depletion in Na_2O, would be consistent with aggressive chemical weathering owing to high atmospheric CO_2 in the Archaean (Fedo et al. 1995) but, together with the elevated Ni and MgO contents, could also reflect a primitive, CaO-poor, ultramafic provenance, as proposed by Hart et al. (2004) from PGE data. The metapelites display similar K_2O values to average Archaean shale, but the metagreywackes are relatively K_2O-depleted, suggesting a varying contribution of a felsic component in the sediment source area. They are nonetheless more potassic, and slightly more ferromagnesian, than the greywackes described by Dziggel et al. (2006) from the Barberton region.

5.5.3 Mafic Gneisses

The amphibolites and mafic granulites show only limited overlap with the greenschist metalavas and display a predominantly tholeiitic basalt composition, but with some komatiitic basalt affinities (Fig. 5.5b, c; Stepto 1979, 1990; Hart et al. 2004; Ogilvie 2010). Overall they follow a tholeiitic trend (Fig. 5.5a). In general, the mafic granulites show flat HREE profiles lacking a Eu anomaly and with only limited LREE enrichment occurring except in amphibole-rich gneisses that are more strongly LREE enriched (Lekgothoane 2017). Hart et al. (2004) argued that the mafic granulites

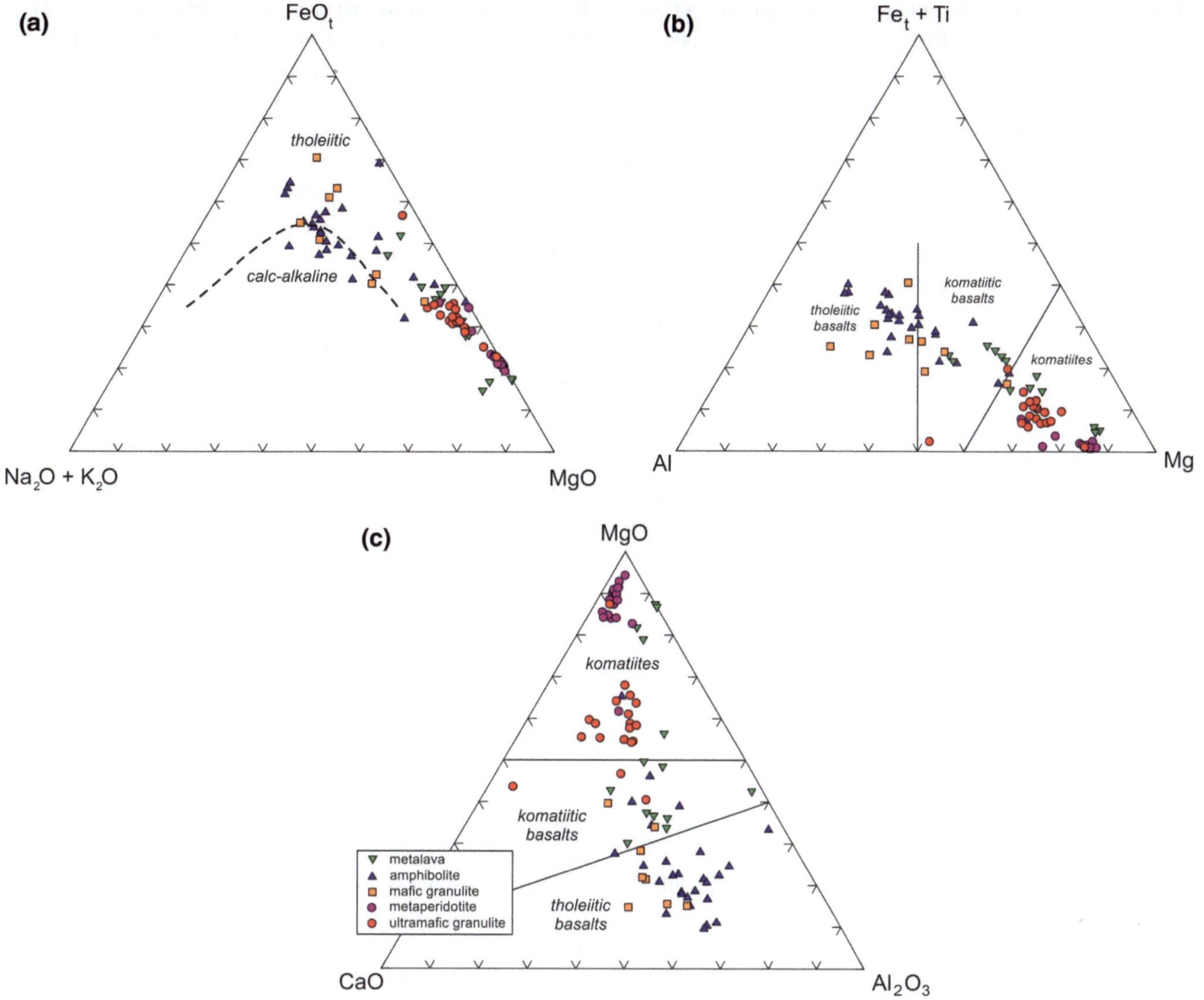

Fig. 5.5 Major oxide geochemical variation diagrams for metalavas and mafic and ultramafic gneisses in the Vredefort Dome. **a** AFM ternary (Irvine and Baragar 1971). **b** (Fe_{Total} + Ti)–Al–Mg ternary (Jensen 1976). **c** CaO–MgO–Al_2O_3 ternary of Viljoen et al. (1982). Data from Stepto (1979, 1990), Hart (1978), Hart et al. (1990a, b), Lana et al. (2003b, 2004), Broccardo (2009), Lieger et al. (2011) Ogilvie (2010) and Lekgothoane (2017)

show a similar PGE pattern (depleted Ir, Ru, enriched Pt, Pd) to basalts from the Barberton greenstone belt, including the slightly flatter profile for komatiitic compositions.

5.5.4 Ultramafic Gneisses

With the exception of the Beta-1 core metaperidotites (MgO = 33–38 wt%), the ultramafic granulite gneisses are only slightly more magnesian (MgO = 20–25 wt%) than the mafic gneisses (Fig. 5.5). Both the latter lack discernible layering, and their intimate association with the metasedimentary gneisses supports Stepto's (1979) proposal that they both represent volcanic units. Whilst ultramafic gneisses appear to be absent in the amphibolite facies zone, they are more common in the granulite zone. It is possible that compositions in the granulite zone have been modified by anatectic melt loss; however, Ogilvie (2010) showed that, even in the fertile pelitic-greywacke compositions, melt loss did not exceed 21 vol.%.

The Beta-1 metaperidotites are enriched in Fe_2O_{3Total} and depleted in Al_2O_3 relative to the komatiitic metalavas (Fig. 5.5b, c) and are enriched in SiO_2, Fe_2O_{3Total} and CaO relative to similar lithologies in the Johannesburg Dome (Broccardo 2009). Metapyroxenite bands within the metaperidotite are compositionally distinct, being relatively Al_2O_3-depleted (Fig. 5.5c). The metaperidotites show very flat, essentially chondritic, REE profiles with very limited enrichment and no Eu anomaly, although biotite–amphibole–orthopyroxene bands in the Beta-1 core do show significant LREE enrichment and a slight negative Eu anomaly (Broccardo 2009).

Hart et al. (1990b) suggested that the major- and trace-element geochemistry of the metaperidotite, and specifically HREE enrichment with lower LREE abundances compared to depleted mantle nodules of shallower lithospheric origin, is consistent with a fertile mantle precursor. However, based on its very unradiogenic Os, low Re/Os and very low $^{187}Os/^{188}Os$ and $^{187}Re/^{188}Os$ ratios, Tredoux et al. (1999) suggested that it is derived from a restitic subcontinental upper mantle precursor. The REE pattern argues against both a cumulate origin and significant melt abstraction; however, Hart et al. (1990b) suggested that this might be reconciled with an originally hydrated, chemically altered, fertile oceanic lithosphere. The absence of a significant Eu anomaly is also consistent with a lack of plagioclase fractionation. Hart et al. (2004) noted depleted Pt and Pd and enriched Ir and Ru, consistent with the Re depletion noted by Tredoux et al. (1999). Whilst not ruling out a depleted Archaean upper mantle origin based on Os, Ir and Ru contents, they did note that the very low Os, Ir ($\sim$1.5 ppb) and Cr (800-1200 ppm; Hart et al., 1990b) contents would be unusual for mantle restite, and that a more plausible precursor might be an intrusive oceanic crustal cumulate that later lost Pt, Pd and Re.

5.5.5 Granitoid Lithologies

Compositional analysis indicates that the granitoid lithologies in the Vredefort Dome together form a largely continuous, calc-alkaline, trend from trondhjemite to granite, with a very limited tonalitic component (Fig. 5.6). Overall, the granitoids are metaluminous to slightly peraluminous (Al/[Al–(Na + K + Ca)] < 1.1). The trondhjemitic gneisses and their granulite-grade equivalents (enderbite, felsic enderbite) are slightly more sodic, and slightly less aluminous and cafemic, than the average Archaean grey gneiss, and correspond with the field for juvenile Archaean TTG suites (Barker 1979; Fig. 5.6a). They possess relatively low K_2O/Na_2O and display relative depletion in LREE (relatively high HREE/LREE; Fig. 5.7) and typical characteristics of arc-related magmas (high Y, Yb_N; low $(La/Yb)_N$ and Sr/Y; negative Nb, Ta, Ti and P anomalies (Lana et al. 2004).

The trondhjemitic gneisses and banded TTG gneisses in the amphibolite facies zone are chemically similar (Lana et al. 2004) and display low mafic oxide content (<5 wt%), K_2O/Na_2O ($\leq$0.5) and Rb/Sr (0.3–0.5), and a fractionated REE pattern with a small negative Eu anomaly (Fig. 5.7b). In the granulite zone, tonalitic gneisses and quartz diorite gneisses display enrichment in cafemic components with respect to the more common gneisses. The tonalites are depleted in SiO_2, K_2O, Ba, Th and U (see also Hart et al. 1981), enriched in Sr and have a low Rb/Sr ratio (0.07), as befits their higher modal plagioclase content. They are slightly enriched in REE relative to the trondhjemitic gneiss, and display a flatter REE pattern but a similar slightly negative Eu anomaly (Fig. 5.7b). However, the quartz diorite gneiss shows extreme enrichment in Ba, Rb, Sr and REE, and stronger REE fractionation (Fig. 5.7b; Lana et al. 2004). Enderbitic lithologies in the granulite zone show a slight positive Eu anomaly (Fig. 5.7a).

Charnockites in the granulite zone display similar major and rare-earth element compositions to the granitic veins in the amphibolite facies banded gneiss; however, both show considerable REE content variation (Fig. 5.7a, c). They are enriched in SiO_2 and K_2O and depleted in Th and U relative to the trondhjemitic gneisses and have higher K_2O/Na_2O and Rb/Sr ratios and a steeper REE pattern than the granitic veins, with a stronger depletion in HREE, and a pronounced positive Eu anomaly (Fig. 5.7a). In contrast, granitic veins in the amphibolite facies zone with relatively depleted REE have a slightly positive Eu anomaly, whereas those with flatter, more enriched, profiles display no anomaly (Lana et al. 2004). Lana et al. (2004) interpreted HREE enrichment as representing apatite and zircon crystallisation.

Among the discrete granitoid bodies distinguished by Lana et al. (2004), the porphyritic granodiorite shows no Eu anomaly and displays only slight REE enrichment relative to the granitic veins, and is indistinguishable from the slightly younger magnetite-rich granodiorite (Fig. 5.7c). The schlieric granite also strongly resembles the granitic veins in the banded gneiss and shows a variable Eu anomaly, which may reflect variable assimilation of the host TTG. The homogeneous granite is relatively REE-enriched and has a strong negative Eu anomaly, and high Th and U contents (Lana et al. 2004). This contrasts with K-feldspar-rich granite that is relatively REE-depleted but has a positive Eu anomaly.

In summary, whilst Lana et al. (2004) noted some slightly more cafemic compositions in the granulite zone rocks, they found no geochemical evidence to suggest that the deepest exposed rocks are a melt-depleted refractory lower crust as proposed by Hart et al. (1990a) and Moser et al. (2001). Given that peak metamorphic temperatures in the granulite zone exceeded 870 °C (Ogilvie 2010; see Sect. 5.8), dehydration melting of biotite and hornblende in trondhjemitic gneiss, leaving a tonalitic restite, is possible (Vielzeuf et al. 1990). However, the tonalitic gneiss is intimately interlayered with granitic bands with no evidence of palaeosome, so an alternative origin for the tonalites by chemical assimilation of xenoliths (Fig. 5.3e) linked to intrusion and/or structural intercalation of more mafic bands cannot be ruled out. Relatively uniform whole-rock $\delta^{18}O$ values of 8–10 per mil (normalised to SMOW) also support the absence of large-scale crustal differentiation in the exposed section (Fagereng et al. 2008).

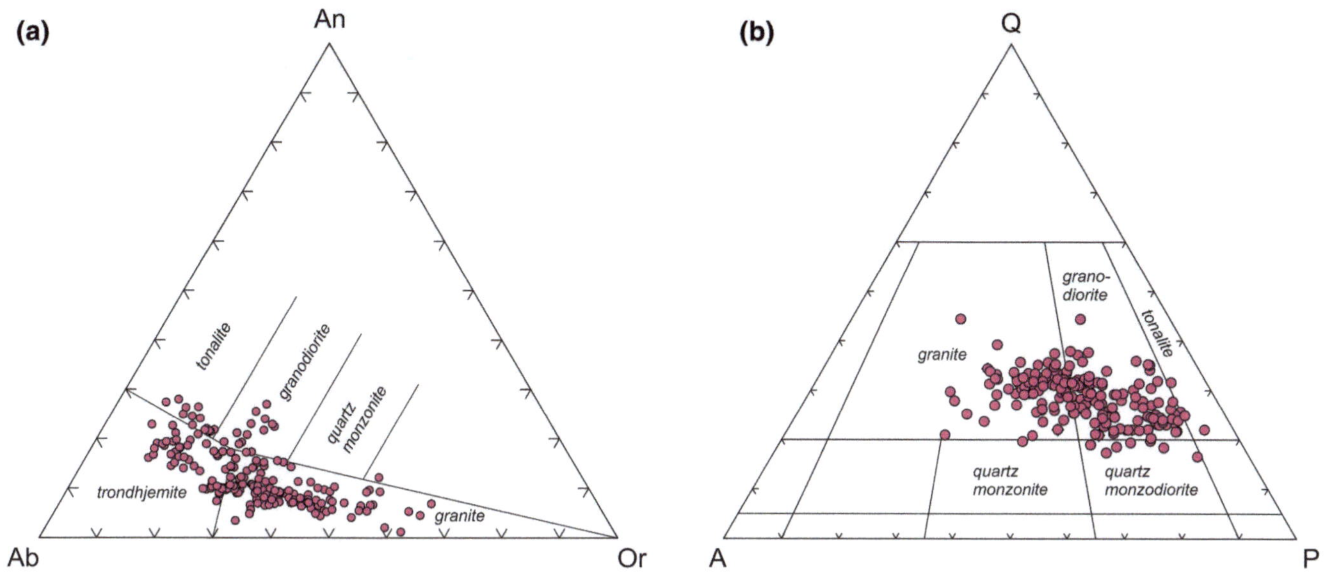

Fig. 5.6 Major element compositional variation diagrams for trondhjemitic and granitoid rocks in the Vredefort Dome. Data from Hart (1978), Hart et al. (1990a), Qoza (2001), Lana et al. (2004) and Lieger et al. (2011). **a** An-Ab-Or plot after Barker (1979). **b** QAP plot after Streckeisen (1976)

Fagereng et al. (2008) showed that the $\delta^{18}O$ values of the granitoid gneisses are consistent with an I-type source, but with a significant S-type component. Although cordierite- and garnet-bearing leucosomes are found in the gneisses surrounding the largest metasedimentary granulite fragments (Stepto 1990), Fagereng et al. (2008) demonstrated that the exposed metapelites cannot be the contributing source for the leucosomes in the rest of the terrane as they have $\delta^{18}O$ values that are, on average, 1.5 per mil higher than those of the TTG gneisses and granitoids.

Notwithstanding the lack of unambiguous large-scale crustal differentiation in the exposed section, the geochemical variation within the granitic and charnockitic lithologies supports more localised fractionation processes in their formation (Lana et al. 2004). The charnockites contain the strongest positive K and Eu anomalies among the granitoids, whereas the schlieric and homogeneous granites have smaller K anomalies and negative or no Eu anomalies. This suggests that the plagioclase fractionation may have occurred between the charnockites in the granulite zone and the larger granitoid bodies in the amphibolite facies zone. In the granulite zone, Lana et al. (2004) suggested that enrichment in MgO, Fe_2O_3, TiO_2, CaO, Zr and REE, and depletion in some LILE (Ba, Rb, K) in the enderbite relative to the felsic enderbite could, in theory, be balanced by the granitoid leucosomes (charnockite), some of which are depleted in REE and cafemic elements (Fig. 5.7). However, there is considerable overlap between all three lithologies (Fig. 5.7a), which makes it also possible that this variability may reflect original heterogeneity in the TTG suite. In this case, the granitic component within the banded TTG gneisses would have had to have been derived from a deeper, presently unexposed, source.

5.6 Structure

At the map scale, the Archaean Basement Complex comprises four main structural domains (Fig. 5.2): (a) a zone of broadly tangential, steeply dipping, foliations within a few kilometres of the contact with the supracrustal sequence in the northern, northeastern and western parts; (b) a 10–15-km-wide, NW-trending, domain of steep foliation with low-strain lenses that dominate the central and northwestern outer parts of the core; (c) an ~1-km-wide, vertical, NE-trending, mylonite zone in the southeastern sector that separates the high-grade gneisses from lower grade schists to the southeast and (d) a steep, NW-trending schistosity anastomosing around low-strain lenses in the greenstone inlier in the southeastern sector. All structures are truncated by the Dominion Group unconformity, indicating that they are Mesoarchaean in age. Lana et al. (2003b, 2006) identified four main deformation events based on outcrop-scale structural fabric analysis, the relative chronology of granitoid intrusions and analysis of the timing of metamorphic mineral growth relative to foliation development. The D1 features have been largely transposed by prograde D2 and peak metamorphic D3 structures, whereas the D4 event is retrograde. All structures are cut by impact-formed pseudotachylite.

5.6.1 Mesoarchaean Structures

The most widespread gneissic foliation in the Archaean Basement Complex in the dome is attributed to D2 by Lana et al. (2003a), as it contains relic centimetre-to-metre-scale

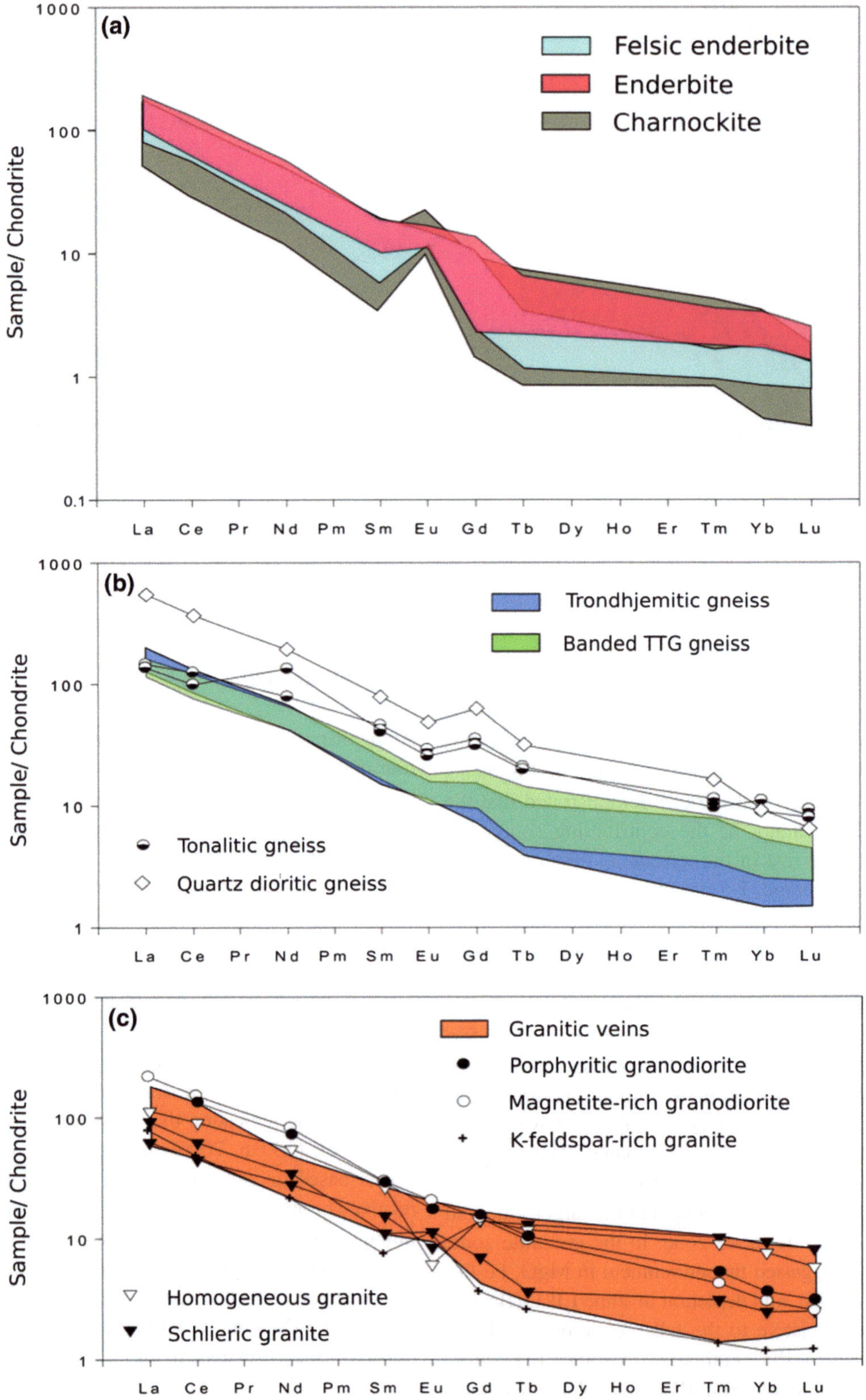

Fig. 5.7 Chondrite-normalised REE plots for the Archaean Basement Complex granitoids in the Vredefort Dome, using the classification scheme of Lana et al. (2004). Data from Hart et al. (1990a) and Lana et al. (2004)

intrafolial folds of an earlier, S1, foliation (Fig. 5.8c), particularly in the trondhjemitic gneisses northeast of Parys. S1 has not been identified in the metasedimentary granulites, possibly owing to the strong D2 and D3 overprints and syn-D3 anatexis in these rocks, but compositional banding (transposed bedding) is parallel to S2.

The steep, tangential, orientation of S2 in the outer parts of the core of the Dome, only slightly oblique to the Dominion Group unconformity (Fig. 5.2), suggests that it had a broadly subhorizontal to shallow dip at 3.07 Ga. The S2 and S3 foliations also show local rotations in the vicinity of competent mafic and metasedimentary fragments.

The S2 foliation intensity in the granitoids decreases in more leucocratic rocks, consistent with both field relations and geochronological data, which support a trend of decreasing mafic content in the younger magmas. It is absent

Fig. 5.8 **a** Upright-plunging D3 fold in pelitic granulite, showing folding of stromatic leucosomes with more irregular leucosome parallel to the axial plane (above pen). **b** Weakly deformed granodiorite hosting melanocratic biotite trondhjemitic gneiss with strong S2 foliation that is defined by plagioclase augen. Note the thin mafic streaks (centre left) formed by disruption and assimilation of the gneiss defining a weak foliation in the granodiorite. **c** Interfering D2 and D3 folds in enderbitic gneiss. S2 is enhanced by leucocratic veins. **d** Granulite banded gneiss with strong S3 foliation defined by leucosomes that show evidence of folding and boudinage. Late-stage charnockite vein occurs along an extensional shear band with sinistral shear sense. **e** Late-D3 mylonite (foreground) in charnockite with mafic schlieren that shows sinistral drag folding along the mylonite margins. **f** Vertical D4 mylonite in granitoid gneiss close to contact with greenschist metalavas. Feldspar augen and steeply SE-dipping (to the right) extensional shear bands indicate SE-side-up shear displacement

in the youngest (late-D3) granites. In the granitoid gneisses in the amphibolite facies zone, S2 comprises a strong gneissosity defined by aligned biotite ± hornblende and highly attenuated quartzofeldspathic aggregates and, locally, feldspar augen (Fig. 5.8b). It may be locally enhanced by intrusive leucocratic veins that are typically less well foliated (Figs. 5.3a, b and 5.8c). A late- (rather than post-) D2 timing for at least some of these veins is confirmed by evidence of folding where they are discordant to S2, and by pinch-and-swell features (Fig. 5.3a). In the porphyritic granodiorite along the western core-collar contact, the alignment of euhedral K-feldspar phenocrysts parallel to S2 in the adjacent migmatitic gneisses and absence of evidence of subsolidus deformation of the matrix suggests syn-D2 crystallisation. Where post-D2 granite veining is particularly intense, S2 may be differentially rotated in agmatitic blocks (Fig. 5.8b).

The S2 foliation in the mafic granulites is defined by generally concordant, weakly foliated veins and by elongate, highly flattened, millimetre-to-centimetre-scale, quartz and/or orthopyroxene + clinopyroxene ± biotite/hornblende aggregates. It is not clear if this well-equilibrated granoblastic texture is a mimetic feature formed after D2, or whether it indicates that D2 occurred under granulite facies conditions in the central parts of the dome.

The central and northwestern parts of the core of the dome are dominated by an upright, broadly NW-trending, S3 foliation that is developed in a series of high-strain zones up to several hundred metres wide that separate lower strain lenses characterised by gently to tightly folded S2 and gneissic banding (Figs. 5.3c and 5.8c). Kilometre-scale upright-vertical D3 folds are defined by the supracrustal granulites south of Parys (Fig. 5.2), but smaller scale folds (Fig. 5.8a) display highly variable hinge plunges and asymmetries. Chocolate tablet boudinage of mafic sheets and centimetre-to-metre-scale conjugate extensional shear bands suggest a large component of flattening strain during D3, but asymmetric extensional shear bands and asymmetric boudinage locally reflect sinistral non-coaxial strain (Lana et al. 2003b; Fig. 5.8d).

The general parallelism of leucocratic veins with S3, the development of stringer or pod-like granite bodies parallel or slightly discordant to S3 and the development of peak metamorphic parageneses in the axial planar foliation to F3 folds in metapelitic and mafic granulites indicate that D3 coincides with the peak of high-grade metamorphism. However, between Parys and Vredefort, blastomylonites up to several metres wide that are essentially parallel to S3 display a shallow northwest-plunging mineral rodding lineation and consistent sinistral shear-sense indicators (S-C fabrics, asymmetric extensional shear bands, foliation asymmetry; Lana et al. 2003a) (Fig. 5.8e). Lana et al. (2003a) attributed these mylonites to strain partitioning during the final stages of D3 under declining temperature conditions.

Whilst S2 also appears to have formed under relatively high-grade conditions and synchronous with granitoid magmatism (e.g. the syn-D2 porphyritic granodiorite), it is cut and locally disrupted by the more leucocratic intrusions. The granodioritic and granitic veins, stringers and pods are enclosed in gneisses with a strong S3 fabric but they contain a less-well-developed S3 fabric that is defined by, inter alia, aligned K-feldspar phenocrysts. Late-D3 crystallisation is supported by locally discordant veins of granodiorite, granite and charnockite. Granitic veins oblique to S3 may occupy extensional shear band sites (Fig. 5.8d), but are deformed by the late-D3 mylonites (Lana et al. 2003a).

In the amphibolite facies zone granitoid gneisses, S3 is defined by biotite ± hornblende stringers and millimetre-to-centimetre-scale mafic aggregates. In the granulite zone granitoid gneisses, it is defined by flattened quartz discs up to a few mm wide and up to 1–2 cm in length, and by stringers and flattened millimetre-to-centimetre-scale lensoid orthopyroxene–biotite ± hornblende aggregates. In the metapelitic and metagreywacke granulites, S3 is defined by elongate aligned peak metamorphic biotite and cordierite or orthopyroxene bands that wrap around garnet porphyroblasts that locally contain fibrolitic sillimanite defining micro-crenulations. Leucosomes are generally parallel to the composite S2/S3 foliation (Fig. 5.4b) but in low-strain lenses may be concordant to slightly discordant to S2 and the compositional banding that define the D3 folds. In larger scale D3 fold hinges, the leucosomes in the metapelites themselves may be folded or crenulated and display an axial planar S3 foliation.

In the greenschist terrane, the metavolcanics are dominated by a subvertical, predominantly NW-trending, schistosity and shallow SE-plunging mineral lineation defined by peak metamorphic chlorite and actinolite (Lana et al. 2003b). This schistosity is axial planar to upright, shallow-plunging, m-scale open to tight folds and millimetre-scale tight upright intrafolial folds with moderately steep-plunging hinges defined by an older schistosity that is locally accentuated by millimetre-to-centimetre-scale quartz veins. Despite the lower metamorphic grade and intervening D4 shear zone (Fig. 5.2), based on these relationships, Lana et al. (2003b) proposed that this schistosity is the low-grade equivalent of S3 in the gneisses, and that the folded foliation is S2. Pillows, vesicles and variolites in a low-strain domain display NE–SW coaxial flattening strain (Lana et al. 2003b).

The D4 event is represented by a ~1–km-wide, NE-trending, subvertical, mylonite zone along the southern contact of the gneisses with the greenschist metalavas (Fig. 5.2), and in the high-grade terrane by decimetre-scale blastomylonites that lack clear shear-sense indicators.

The main mylonite zone is exclusively developed in high-grade gneisses and displays a strong dip-parallel lineation (Lana et al. 2006). The shear fabric ranges from proto- to ultramylonitic and is highly heterogeneous and anastomoses on an outcrop scale, with low-strain lenses containing migmatitic banding that defines decimetre-to-metre-scale, upright-plunging to vertical folds, including rare vertically plunging sheath folds (Lana et al. 2006). The foliation and lineation in the mylonite are defined by elongate quartz and feldspar aggregates in the gneisses and by neoblastic hornblende and biotite in amphibolites. Although some feldspar porphyroclasts show a slight monoclinic asymmetry (Fig. 5.8f), most show orthorhombic symmetry in their mantles. Together with evidence of strain-related myrmekites, the stability of hornblende and biotite and the generally coarse-grained recrystallization of feldspars and quartz ribbons, this suggests a high-T blastomylonite fabric with a strong flattening component. Lana et al. (2006) identified extensional shear bands cutting the mylonitic foliation with a SE-side-up sense of movement towards the greenschist contact (Fig. 5.8f). Microscopically, these shear bands contain a fine-grained chlorite–sericite–epidote paragenesis indicative of greenschist-grade rehydration of the higher temperature mylonitic assemblage, suggesting that the shear zone evolved from amphibolite to greenschist facies conditions and, thus, that D4 followed shortly after the metamorphic peak in the gneiss terrane.

Back rotation of 100°–120° applied to the structural elements of this shear zone to account for impact-induced rotation suggests that it originated with a shallow northwesterly dip and, therefore, a normal dip-slip shear sense (Lana et al. 2006). Given its temporal link to high-to-moderate-T retrograde cooling of the gneiss terrane, this geometry suggests that it could have contributed to the ca 17 km of unroofing of the mid- to lower crust that exposed upper amphibolite facies rocks at surface prior to deposition of the Dominion Group lavas. Based on the difference in peak metamorphic temperatures between the two terranes, Lana et al. (2006) estimated that it may have accommodated more than 10 km of slip and at least 5 km of throw.

5.6.2 Neoarchaean-to-Palaeoproterozoic Structures

The Mesoproterozoic-to-Palaeoproterozoic supracrustal strata and intrusions in the collar of the dome contain abundant evidence of post-D4 structural events. Most obvious on a map scale are faults that displace the core-collar contact by up to several kilometres (Fig. 5.2). Faults with strike separations up to hundreds of metres have been linked by Wieland et al. (2005) and Mashabela (2015) to initial tangential shortening and subsequent collapse-related extensional inversion related to the impact-induced formation of the dome. Such faults are typically pseudotachylite-bearing and differ from rare larger faults in the inner collar that are attributed to Ventersdorp-age rifting (Mashabela 2015). Geophysical analysis and drillcores suggest that the largest faults in the dome are related to the development of a basement horst in the southeast of the dome that is pre-Ventersdorp in age (Martini 1992; Fig. 5.1).

The faults are generally very poorly exposed and, although Moser et al. (2001) and Flowers et al. (2003) proposed pseudotachylite-bearing 'impactogenic faults' with kilometre-scale displacements in the gneisses southwest of Parys, Lana et al. (2003a) found no evidence for them or the major brittle tectonic contact between the amphibolite and granulite facies zones hypothesised by Hart et al. (1990a). Impact-related pseudotachylite dyke networks that range up to hundreds of metres in length and tens of metres width are common in the gneisses but show negligible displacements.

Lana et al. (2003a) concluded that the subvertical concentric fabric in the outer 4–6 km of the core (Fig. 5.2) represents originally subhorizontal S2 that was rotated by ~90° during dome formation; however, Wieland et al. (2005) preferred a larger rotation of bedding (~120°) in the adjacent collar. Toward the centre of the core, impact-related rotation decreases quite sharply and S2 in the central core has a broadly shallow orientation. This 'piston-like' model, in which the volume of rock closest to the point of impact shows strong vertical uplift, but limited or no rotation, is supported by Ivanov's (2005) numerical modelling.

5.7 Metamorphism

5.7.1 Mesoarchaean Metamorphic Effects

All pre-impact lithologies in the core of the dome, with the exception of the metalavas in the southeast and Neoarchaean-to-Mesoproterozoic mafic intrusions, contain well-equilibrated upper amphibolite to granulite facies metamorphic mineral assemblages, with progressively higher grade rocks occurring towards the centre of the core (Fig. 5.2). These assemblages are cut by ubiquitous pseudotachylitic breccias and contain shock deformation effects (planar deformation features in quartz, feldspar and zircon, annealed diaplectic glasses and possible mineral shock melts after quartz, feldspar, biotite and amphibole; planar features, including twinning and shock-induced cleavage in pyroxenes, amphiboles and garnet (Fig. 5.9); kink banding and oxidation in biotite; Gibson and Reimold 2005), indicating that the metamorphism pre-dated the impact event. The overall intensity of these shock effects increases towards the centre of the dome (Gibson et al. 2002; Gibson and Reimold 2005). Post-impact metamorphic parageneses overprint pseudotachylite and occur in heterogeneous corona textures that can be distinguished from the pre-impact assemblages by their extremely fine grain size (e.g. Fig. 5.9b) and lack of mineral shock deformation features (Gibson et al. 1998, 2002; Gibson 2002). This post-shock thermal metamorphism also increases in grade from amphibolite to ultrahigh-T-low-P granulite facies towards the centre of the core (Gibson et al. 1998; Gibson 2002; Ogilvie 2010).

Mafic rocks are characterised by hornblende–biotite–plagioclase–quartz–ilmenite parageneses in the amphibolite facies zone and orthopyroxene–clinopyroxene–hornblende–plagioclase–quartz–ilmenite ± biotite ± olivine parageneses in the granulite zone. In the latter, hornblende replaces clinopyroxene and biotite replaces orthopyroxene and/or ilmenite, indicating some retrograde rehydration. Lana et al. (2006) estimated peak metamorphic conditions of 670–690 °C, 0.5 ± 0.1 GPa for an amphibolite in close proximity to the unconformity with the supracrustal sequence. Chlorite thermometry applied to the metalavas (chlorite–hornblende/actinolite–magnetite ± oligoclase) in the southeast of the dome indicates mid-greenschist facies conditions (~400 °C; Lana et al. 2003b), confirming that the D4 shear zone represents a significant terrane boundary.

The meta-BIF in the southeastern greenschist terrane contains a low-grade grunerite–ferrohornblende–quartz–magnetite ± ilmenite paragenesis (Lana et al. 2003b). Stepto (1979) identified a paragenesis of ferrohypersthene–hedenbergite–quartz–magnetite, with accessory plagioclase, biotite, apatite and garnet in the granulite facies meta-BIF. According to him, garnet is rare and occurs only as inclusions in orthopyroxene.

The TTG gneisses and granitoids in the amphibolite facies zone typically contain biotite, with only subsidiary hornblende. Thin, discontinuous, biotite ± hornblende selvedges adjacent to leucocratic veins are interpreted as rehydration effects linked to crystallisation of the veins, rather than melt-depleted restite. In the granulite zone, the gneisses and granitoids contain perthitic alkali feldspars and orthopyroxene + clinopyroxene ± biotite/hornblende, with biotite also occurring as replacement after orthopyroxene, and hornblende enclosing clinopyroxene.

The garnet-bearing leucosomes in the metapelitic and metagreywacke granulites (Figs. 5.4b and 5.9a) indicate anatexis. Schreyer and Abraham (1978) and Perchuk et al. (2002) proposed P-T conditions of 700 °C, 0.5 GPa, whereas Stevens et al. (1997) inferred peak conditions between 850 and 920 °C at 0.5–0.55 GPa, based on experimental constraints on biotite stability (Fig. 5.10a). More recent THERMOCALC modelling of the assemblages by Ogilvie (2010) constrained peak P-T conditions to 870–885 °C, 0.71–0.77 GPa (Fig. 5.10a, b). These estimates correspond to depths of burial of 25–27 km at the time of metamorphism, which fits better with Lana et al.'s (2006) P estimate for the amphibolite facies zone.

Although Stevens et al. (1997) proposed an anticlockwise P-T path for the granulite facies metamorphism in the metapelites (Fig. 5.10a), an intensive search by Ogilvie (2010) for the reported relict spinel–quartz association revealed only zincian spinel inclusions in cordierite (Fig. 5.9b), compatible with spinel stabilisation during prograde staurolite breakdown. Based on the presence of

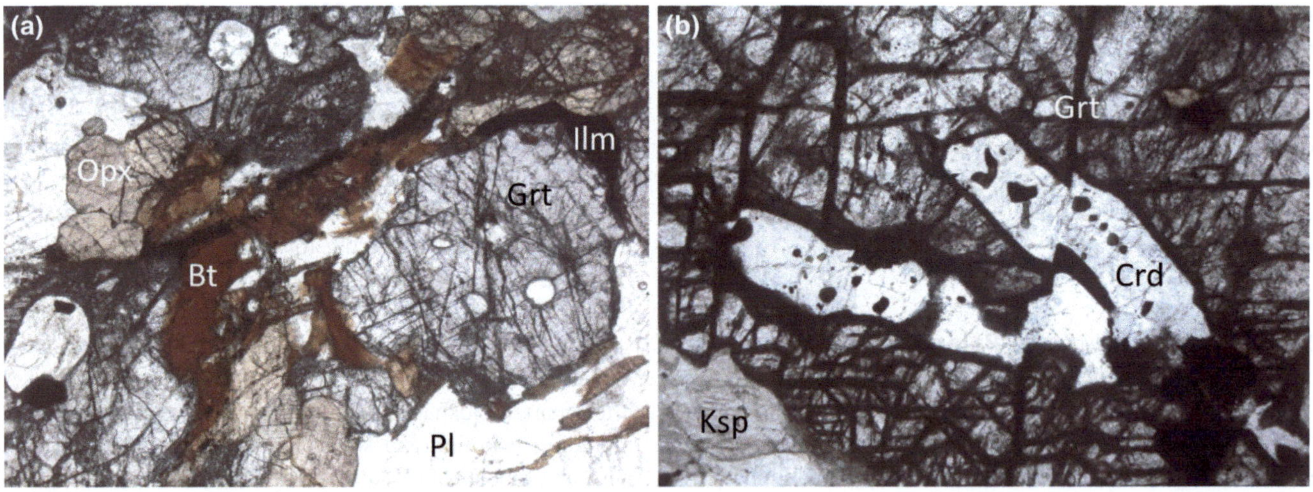

Fig. 5.9 **a** Photomicrograph of metagreywacke granulite sample from south of Parys. Note the intense planar shock fractures in the garnet. Field of view is 4 mm across. **b** Photomicrograph of metapelitic granulite from south of Parys showing green zincian spinel inclusions in a cordierite inclusion in garnet, which provides constraints on the prograde P-T path (see text). The opaque fill rimming the cordierite and in the planar shock fractures in the garnet is an extremely fine-grained cordierite + orthopyroxene ± spinel symplectite produced by the low-P post-shock metamorphic overprint. One vertically oriented fracture displaces the cordierite inclusion. Field of view is 1.8 mm across. Opx—orthopyroxene; Grt—garnet; Bt—biotite; Crd—cordierite; Pl—plagioclase; Ksp—mesoperthite; Ilm—ilmenite

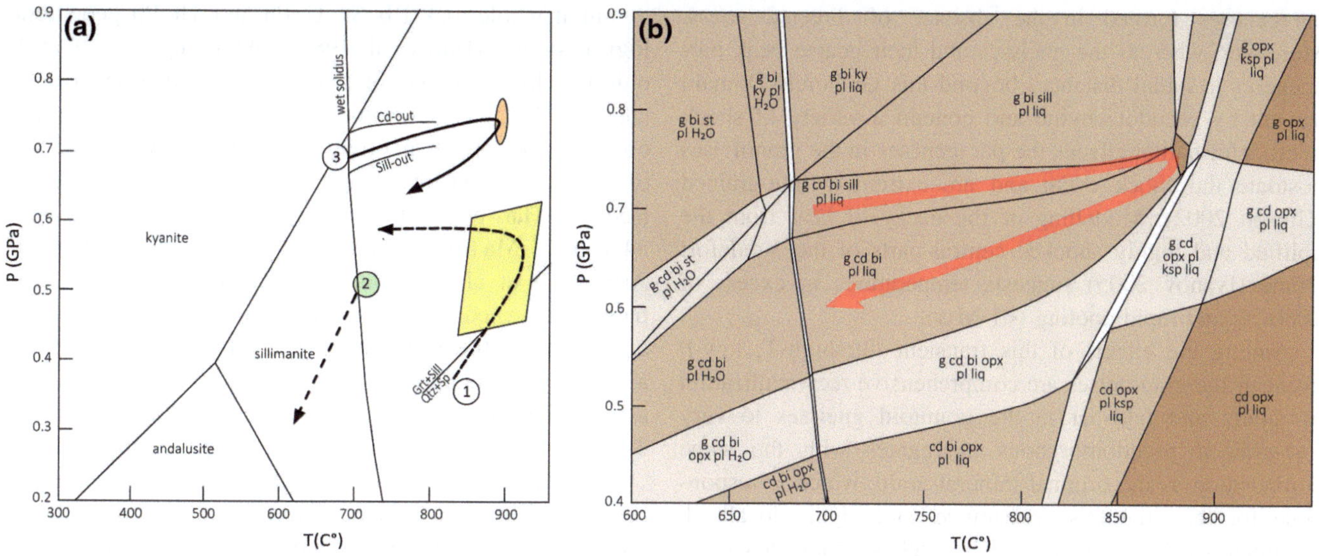

Fig. 5.10 **a** P-T graph summarising the different P-T paths and peak P-T estimates inferred from metapelitic granulites in the Vredefort Dome (modified from Ogilvie 2010). (1) Stevens et al. (1997) inferred peak P-T conditions based on biotite breakdown reactions (yellow box), and an anticlockwise P-T path based on their identification of spinel + quartz inclusions in garnet. This was challenged by Ogilvie (2010), who concluded that the inclusions comprise zincian spinel + cordierite (see Fig. 5.9b). (2) Peak estimate of Schreyer (1983) and Perchuk et al. (2002). Perchuk et al. (2002) suggested that Grt breakdown to symplectitic cordierite–orthopyroxene coronas indicates a decompression cooling path during the same event. (3) Ogilvie (2010) argued for a clockwise P-T path and significantly higher peak P based on THERMOCALC modelling (see b). **b** NCKFMASHTO (+ Qtz, Ilm) P-T pseudosection for a pelitic granulite from south of Parys (modified from Ogilvie 2010). The prograde P-T path is constrained by cordierite and sillimanite inclusions in garnet. Peak and retrograde conditions are constrained by cordierite stability

sillimanite inclusions in garnet and cordierite, Ogilvie (2010) concluded from THERMOCALC modelling that the rocks followed a prograde P-T trajectory from 0.65 GPa, 700 °C to 0.75 GPa, 850 °C (Fig. 5.10). Furthermore, the persistence of cordierite under retrograde cooling in the metapelites and its absence in the metagreywackes specifies a clockwise path (Fig. 5.10b). Although these features are ascribed to pre-3.07 Ga metamorphism based on field and microtextural relationships (particularly the syn-D3 timing of the metamorphic peak), radiometric dating of the granulites has yet to unambiguously confirm this (see Sect. 5.8). Consequently, further work is needed to unravel the polymetamorphic history of the terrane.

5.7.2 Neoarchaean and Younger Metamorphic Events

Most of the gneisses show evidence of low-temperature rehydration postdating the 2.02 Ga impact event. Feldspars show some retrogression to sericite and epidote, whereas biotite, hornblende and pyroxene alter to chlorite and, in the case of biotite, fine-grained ilmenite or magnetite. Given the widespread evidence of Pb loss from zircons at ca. 1.1 Ga (Kamo et al. 1996; Armstrong et al. 2006; see Sect. 5.8), this retrogression may be largely related to the subhorizontal Umkondo gabbro intrusion of this age that underlies much of the core of the dome (Fig. 5.2; Pybus 1995); however, coarser grained, higher-T, alteration of feldspars (to epidote and white mica) and ilmenite (to titanite) may imply other, higher-T, events, with likely candidates being related to the Ventersdorp or Bushveld magmatic events and/or the impact event. The distinctive mineralogical shock deformation effects associated with the impact (Gibson and Reimold 2005) are useful in discriminating impact-related metamorphism from older events.

The metapelitic and metagreywacke granulites located 5–8 km from the centre of the dome show complex coronas of cordierite + orthopyroxene ± spinel ± plagioclase ± biotite ± K-feldspar after garnet that have variously been attributed to decompression cooling following the pre-impact peak metamorphism (Schreyer 1983; Perchuk et al. 2002; Fig. 5.10a) and an impact-induced rapid heating–cooling event (Stevens et al. 1997; Gibson 2002; Ogilvie 2010). The latter proposed low-P (0.2–0.3 GPa) isobaric cooling following the impact uplift event. This is consistent with the ~0.2 GPa pressure estimates from fluid inclusions trapped along planar shock features in quartz (Fricke et al. 1990). In contrast, the intercalated less hydrous meta-BIF, mafic and ultramafic granulites show no impact-related corona textures.

Within 5 km of the centre of the dome (Granofels Domain, Fig. 5.2), however, metamorphic parageneses developed under dynamothermal conditions are more widely overprinted by unusually fine-grained high-grade parageneses formed in the absence of directed stress. Moreover, whereas the gneisses and their coarse peak parageneses at radial distances beyond this Granofels Domain are cut by pseudotachylite and contain a variety of shock microdeformation effects, the parageneses in the granofelses postdate the shock event and are extremely fine-grained (Gibson 2002). Modelling of the decay of heat from the uplifted and highly shocked central parts of the Vredefort Dome (Ivanov 2005) suggests temperatures in excess of 1000 °C and rapid cooling (<1 Myr).

Among the effects of this transient ultrahigh-T, low-P event in the granofelses are comprehensive recrystallization of quartz and feldspar in the granitoid gneisses to very fine-grained pseudomorphous aggregates with the same composition as the original mineral grain, which is responsible for the granofels appearance, and highly localised, millimetre-scale, interstitial melt patches characterised by graphic quartz–feldspar intergrowths and hypersolvus alkali feldspars (Schreyer 1983; Gibson 2002; Gibson and Reimold 2005). Meta-BIF, mafic and ultramafic rocks in the Granofels Domain also show unusual reaction corona textures, but still retain evidence of their dynamothermal metamorphic parageneses. The mafic and ultramafic rocks contain coronas and fine-grained polygonal aggregates replacing grain margins and hydrous minerals (Schreyer 1983; Gibson and Reimold 2005). Proximal to these lithologies near the Beta-1 core site (Fig. 5.2) are aluminous granofelses (K-feldspar–cordierite–corundum/sillimanite–spinel–plagioclase–rutile/ilmenite ± biotite) that lack any evidence of dynamothermal fabric. These are interpreted as the products of 2.02 Ga peritectic crystallisation from a high-proportion melt of a likely pelitic/greywacke garnet granulite precursor (Gibson 2002).

5.8 Geochronology

Outcrop-scale, petrographic and geochemical observations from the Archaean Basement Complex in the Vredefort Dome indicate that a polyphase suite of granitoid magmas intruded precursor mafic and ultramafic volcanic ± intrusive and sedimentary rocks, and that the later granitoid intrusions occurred contemporaneous with polyphase regional deformation and high-grade metamorphism prior to deposition of the 3.07 Ga Dominion Group lavas. However, studies of the overlying Neoarchaean-to-Palaeoproterozoic supracrustal sequences in the collar indicate that the central craton experienced multiple further magmatothermal and tectonic events leading up to the 2.02 Ga impact event, which was itself associated with significant thermal effects. As all of these events had the potential to disturb radiogenic isotopic systems, it is therefore not surprising that significant debate remains about the significance of radiometric ages from the Archaean Basement Complex.

Initial whole-rock Rb–Sr, U–Pb and Th–Pb geochronological studies (Hart et al. 1981, 1990a) suggested that the oldest rocks in the Basement Complex formed between 3.3 and 3.5 Ga ago, but some ages of ~3.1 Ga were also obtained. Tredoux et al. (1999) calculated Re–Os model ages of 3.3 and 3.5 Ga for metaperidotites from the Beta-1 core (see also Hart et al. 2004). Single-zircon U–Pb ages up to 3410 ± 14 Ma from an enderbitic gneiss were interpreted by Armstrong et al. (2006) as representing xenocrysts from partially digested mafic xenoliths, and a 3200 ± 6 Ma age as a possible tonalite intrusion age. Hart et al. (1999) interpreted a 3425 ± 10 Ma upper intercept age from a pelitic granulite as a detrital grain age and Flowers et al. (2003) obtained a 3.31 Ga upper intercept age from a 'quartz syenite'.

In general, zircon data from the basement gneisses are complex, falling on splayed discordia consistent with multistage Pb loss during one or more events at ~2.6–2.7, 2.2, 2.02, 1.1 and 0.6 Ga (Kamo et al. 1996; Hart et al. 1999; Flowers et al. 2003; Armstrong et al. 2006). This requires careful interpretation of upper intercept ages using discordant data (Armstrong et al. 2006). Although the 2.02 Ga impact shock produced crystallographically controlled planar fractures, twins, internal lattice strain and subgrain development in zircon, all of which may contribute to Pb loss (e.g. Kamo et al. 1996), Armstrong et al. (2006) pointed out that Pb loss related to the impact is by no means ubiquitous, and cannot be assumed.

Zircons in the granitoid gneisses are generally subhedral and show signs of marginal resorption consistent with metamorphic modification. Cathodolumniscence imaging reveals up to three distinct domains—a central dark luminescence core (not developed in all grains), an intermediate, oscillatory zoned, magmatic domain that may be continuous with the core zoning or truncate it, and a discontinuous, dark luminescence (high U, low Th/U) rim that may embay the magmatic core and that Armstrong et al. (2006) interpreted as metamorphic. Despite field evidence of two phases of intrusion of the homogeneous trondhjemitic gneisses northeast of Parys, Armstrong et al. (2006) obtained an identical intrusion age (3114 ± 6 Ma) for both, although they noted that the Pb loss related to the penecontemporaneous high-grade metamorphism may have caused discordance and underestimation of the crystallisation age. Zircons from the banded TTG gneiss provide a regressed upper intercept magmatic crystallisation age of 3107 ± 8 Ma but also contain inherited core ages up to 3154 ± 7 Ma (Armstrong et al. 2006). The latter ages are consistent with upper intercept TIMS ages between 3127 and 3162 Ma obtained by Flowers et al. (2003) from discordant data. The age resembles the weighted mean crystallisation age of 3105 ± 5 Ma for the syn-D2 porphyritic granodiorite west of Vredefort (Fig. 5.2) from which concordant zircon core data show a 3125 Ma mean age (Armstrong et al. 2006).

Armstrong et al. (2006) obtained a concordia age of 3084 ± 9 Ma for zircons from a late-D3 magnetite-bearing granodiorite, similar to the upper intercept TIMS age of 3084 ± 1 Ma from 'quartz syenite' obtained by Flowers et al. (2003) However, they pointed out that cores (3092 ± 6 Ma) and rims (3074 ± 10 Ma), which are either metamorphic overgrowths or related to marginal Pb loss during metamorphism, form distinct, concordant, clusters, highlighting the problem in characterising zircon textures unambiguously in such high-grade rocks. Limited concordant to near-concordant data from zircon rims in the homogeneous trondhjemitic gneisses produce consistent ages for the high-grade metamorphism between 3093 ± 10 and 3077 ± 7 Ma, and zircon rims from a granulite facies charnockite give a weighted mean age of 3095 ± 9 Ma (Armstrong et al. 2006). These are slightly younger that the 3107 Ma age inferred for the metamorphism by Hart et al. (1999) from a single concordant monazite analysis from a metapelitic granulite.

A weighted mean $^{207}Pb/^{206}Pb$ age of 3068 ± 6 Ma obtained from zircons from an undeformed aplite dyke intruded into a D4 mylonite constrains the end of deformation and magmatism in the Basement Complex (Armstrong et al. 2006).

In summary, despite the discordant nature of the analyses in many studies, single-grain U–Pb zircon and monazite geochronology suggest that granitoid magmatism, polyphase deformation and high-grade metamorphism were largely restricted between 3150 and 3080 Ma, and that data indicate a mafic crustal component possibly as old as ~3420 Ma. Less certain is whether an older, 3200 Ma, tonalitic component exists. The data are consistent with the metamorphic peak occurring between 3100 and 3080 Ma and the field-based intrusion chronology of early trondhjemite (>3110 Ma) and later granodiorite (3105 Ma) to granite (3092–3084 Ma). The D2 event is constrained to ~3105 Ma, the D3 event to 3095–3084 Ma and D4 must predate 3068 Ma. This is consistent with Lana et al.'s (2006) assertion that D4 is linked to the Dominion rifting event.

5.9 Discussion

General consensus exists that the Archaean Basement Complex rocks in the core of the Vredefort Dome expose both a significant terrane boundary and a profile through mid-to-deep crustal levels that include the transition between upper amphibolite and granulite facies. However, there is considerable debate not only about the vertical extent of the profile and its maximum depth, but also about the nature and age of the terrane boundary and its geodynamic significance and, therefore, that of the entire Basement Complex. Furthermore, uncertainty persists regarding the exact number, ages and grades of metamorphic events that have affected these rocks.

5.9.1 Nature of the Archaean Crust in the Vredefort Dome

The Archaean Basement Complex in the Vredefort Dome comprises one or more greenstone sequences with ages between 3.3 and 3.5 Ga and younger, ca. 3.1 Ga, intrusive TTG lithologies. The greenstones appear similar in age to those in the Barberton and Johannesburg regions, but the trondhjemitic lithologies appear to be considerably younger (by between 50 and 200 Myr) than their eastern equivalents (although rare xenocrystic zircon cores may indicate an older, ~3.2 Ga, TTG component). The composition of the greenschist mafic–ultramafic lavas, preservation of pillows and association of both the low- and high-grade units with BIF ± pelitic-greywacke sediments supports formation in a marine setting. In contrast to the greenstones southeast of the D4 terrane boundary, the high-grade metasupracrustals occur only as smaller fragments, albeit that some structural continuity is retained between fragments. Unusually, these fragments are more abundant in the deeper, higher grade, parts of the terrane and are virtually absent in the amphibolite facies zone, suggesting significant intrusive disruption. This may be a feature of the keels of other greenstone belts in the craton; however, none have been exhumed to the same extent as in the Vredefort Dome.

The granitoid rocks are syn- to late-tectonic, I-type, calc-alkaline intrusions that show characteristics consistent with juvenile arc-related granitoids derived from hydrated mafic rocks and subsidiary sediments (Fagereng et al. 2008) that Martin (1995) and Drummond and Defant (1990) suggested would reflect relatively shallow subduction, characteristic of hot (new) oceanic crust. The main pulse of magmatism appears to have occurred between 3130 and 3080 Ma, but intrusions may be as old as ~3150 Ma. The limited difference in ages between trondhjemites and granites is distinct from other parts of the craton, and an outstanding question is whether the granites might be the final product of a fractional crystallisation series with the trondhjemites, rather than being partial melts from older TTG gneisses, as has been proposed in other parts of the craton.

Widespread evidence of Pb loss owing to penecontemporaneous high-grade metamorphism and multiple subsequent events significantly complicates zircon age data analysis. Given the lithological complexity of the terrane, more detailed geochronological sampling is required to elucidate these questions. Finding more evidence of an older TTG component could, for instance, explain the apparent anomaly of greenstone crust that is 150–400 Myr older than the oldest granitoid magmatism.

Despite broadly similar whole-rock ages having been obtained from the greenschist facies metalavas in the southeastern part of the dome (Lana et al. 2006) and the high-grade mafic–ultramafic gneisses, no genetic relationship can be inferred; however, Stepto (1990) suggested that the high-grade mafic lithologies also formed as komatiitic basalt to tholeiitic lavas. In contrast, the layered nature of the metaperidotite suggests an intrusive origin; however, its geodynamic setting remains unresolved. The greenschist metalavas and subsidiary metasediments appear to be part of a larger terrane to the south (Martini 1992) that was tectonically juxtaposed against the gneiss terrane following the metamorphic peak, contrary to the suggestion by Hart et al. (2004) that the metalavas form part of the same crustal block as the metaperidotites that was then thrust northwards prior to the peak of metamorphism.

Compositional analysis of the metagreywacke and metapelitic granulites suggests a mixed ultramafic and granitoid provenance, with the limited detrital zircon age evidence suggesting that the source may have been as old as 3.43 Ga. More detailed analysis is needed to ascertain if this includes a TTG component of this age; however, if this exists it is unlikely to have been the source of the 3125–3100 Ma magmas, which have a juvenile arc geochemical signature (Lana et al. 2004).

5.9.2 Depth and Extent of the Vredefort Crustal Section

The size of the Vredefort Dome, together with the structural evidence that doming involved overturning of the ca. 15 km thick Mesoarchaean-to-Palaeoproterozoic supracrustal sequence, and the evidence of increasing metamorphic grade in the Archaean Basement Complex towards its centre, has engendered considerable debate about the maximum depth of exhumation found in the dome. Early geochemical studies of the granitoid gneisses (Slawson 1976; Hart et al. 1981, 1990a) were used in support of centripetally increasing depths of crustal exposure, however, actual depths were not inferred. Hart et al. (1990b) were the first to propose that the dome exposes a full crustal section, based on their inference that the metaperidotite in the Beta-1 core is geochemically similar to fertile upper mantle lithologies. This interpretation has been used by, inter alia, Moser et al. (2001) as the basis for proposing that the core of the dome represents an exceptional profile from restitic (>40% anatexis), mafitised, lower crust into an overlying mid-crust comprising batholithic granite plutons derived from the lower crust.

The interpretation of exposure of the Moho and, thus, a full crustal section, has been challenged on structural, metamorphic and geophysical grounds (Gibson et al. 2005), and is contradicted by modelling of the structural response of the Kaapvaal crust to the impact event (Ivanov 2005). Specifically, Lana et al. (2003a) showed that the strong rotation of the crust seen in the collar of the dome is limited to the outer 4–6 km of the core. In this context, the depth estimate of 25–27 km obtained from peak metamorphic parageneses in the metasedimentary granulites located 5–8 km from the centre of the dome is interpreted as being close to the maximum level of exhumation. The ~0.5 GPa pressure estimate from an amphibolite close to the collar unconformity (Lana et al. 2006), which would correspond to an ~17 km depth of burial under peak metamorphic conditions, is also consistent with Ivanov's (2005) numerical modelling. Thus, the basement crustal section exposed in the core of the Vredefort Dome, whilst significant (~10 km), does not appear to be as extensive, nor as deep, as proposed by Hart et al. (1990b) and Moser et al. (2001). A somewhat shallower mid-crustal profile is consistent with the absence in the dome of evidence of ~1 GPa ultrahigh-T (1100 °C) lower crustal metamorphic parageneses similar to those seen in xenoliths in the Lace and Voorspoed kimberlites that lie only 60 km southwest of the dome (Schmitz and Bowring 2003).

The restricted nature of the Vredefort depth profile through the Archaean Basement Complex suggests that caution should be exercised in inferring gross-scale crustal layering from the exposures in the dome. Specifically, the crustal model of undeformed batholithic granites passing downwards into more strongly deformed gneisses that represent the melt-depleted source of these granites (e.g. Hart et al. 1981, 1990a; Moser et al. 2001) cannot be reconciled with the field and geochemical evidence of Lana et al. (2003a, 2004) that indicates extreme small-scale lithological heterogeneity yet strong overall compositional similarities between major trondhjemitic and granitoid lithologies across the profile. A more likely scenario is that the Vredefort profile straddles the upper levels of the lower crustal source of granite melts and the immediately overlying feeder system through which these melts escaped (Lana et al. 2004).

Whilst the profile exposed in the core of the dome undoubtedly spans the mid-crust, integrated mapping and geochemical and geochronological analysis (Lana et al. 2003a, 2004; Armstrong et al. 2006) have confirmed that it does not span a major tectonic boundary between distinct high-grade terranes (Hart et al. 1990a; De Wit et al. 1991) or a crustal-scale intrusive contact (Moser et al. 2001). However, it does expose the boundary between upper amphibolite and granulite facies gneisses (Lana et al. 2003a; Flowers et al. 2003), and a major, ~3.07–3.08 Ga, terrane boundary does separate the high-grade gneisses from the greenschist facies greenstones to the southeast (Lana et al. 2006).

5.9.3 Geodynamic Model

Geodynamic models for the formation of the Kaapvaal craton (De Wit et al. 1992; Lowe 1999; Poujol et al. 2003) have implicitly favoured a two-stage evolution involving initial building of a continental core ('Kaapvaal Shield', comprising the present southeastern-to-central portion of the craton) between 3.6 and 3.1 Ga, and then lateral growth between 3.0 and 2.7 Ga via further collisions of magmatic arcs along its northern and western margins. Such a model, whilst biased towards the well-exposed eastern and northern parts of the craton where younger supracrustal cover is lacking, seeks as much to explain the emergence of the exceptional ~15 km thickness of Neoarchaean-to-Palaeoproterozoic supracrustal rocks that are distributed broadly across the craton, as the different granite–greenstone terranes. Given its geographic location within the Witwatersrand Basin and demonstrable exposure of this full supracrustal section, the Vredefort Dome is generally considered by these authors to be located within the 'stable Kaapvaal Shield'. However, in several respects the Archaean Basement Complex rocks in the dome are anomalous with regard to exposures in the Barberton and Johannesburg Dome regions, suggesting a more complex Mesoarchaean geodynamic model is required for the craton.

De Wit et al. (1992) and Lowe (1999) proposed that the Kaapvaal Shield itself formed in two stages—initial tectonic amalgamation of small protocontinental blocks comprising early-formed mafic–ultramafic volcanics and sediments and ca. 3.2 Ga, syntectonic, subduction-related, calc-alkaline magmatic arc TTGs, was followed by a 'cratonic magmatism' phase of widespread voluminous intrusions of post-tectonic granitoid batholiths at 3.1 Ga. This latter event, to which is attributed stabilisation of the craton and its lithospheric root (e.g. Moser et al. 2001), was postulated by Poujol et al. (2003) to have been linked to the formation of a major SE-dipping subduction zone along the northern and western margins of the Kaapvaal Shield. They proposed that magmatism accompanied crustal thickening and subsequent erosion, although Belcher and Kisters (2006) argued for transtensional tectonics synchronous with this granite emplacement.

The contradiction to this model presented by the Vredefort rocks is that the arc-related trondhjemite–granodiorite magmatism in the dome is ~100 Myr younger than that seen in the Barberton and Johannesburg Dome regions and that it thus overlaps with the model's 'regional batholith phase'. Given the exceptional (>17 km) levels of exhumation of the Basement Complex in the Vredefort Dome, it is not possible to confirm whether 3.1 Ga upper crustal granitoid batholiths were originally present in the Vredefort section, but evidence of anatexis in the granulite terrane and widespread evidence of granite intrusions in the dome make this plausible. A further contradiction is the evidence of NE–SW-directed D3 compressional tectonics, which is difficult to reconcile with both Poujol et al.'s (2003) tectonic vergence and Belcher and Kister's (2006) regional stress field.

Based on this evidence, including the Vredefort Dome within the Kaapvaal Shield domain appears premature, as does linking events on the craton at 3.1 Ga to a single large subduction system along the Murchison trend. It also raises the question about whether the Vredefort Dome rocks could have formed part of a separate microcontinental block that only docked with the Shield at 3.1 Ga. Inherent within this is the possibility that the Vredefort arc may have formed initially separately from the Shield above a SW-dipping subduction zone (although the absence of structural vergence in the D2 and D3 features make this only speculation at this point).

Poujol et al. (2003) proposed that the 3074 Ma Dominion rifting could have been triggered by NW-directed slab rollback along the northwestern margin of the Kaapvaal Shield. This stress geometry is orthogonal to that associated with the peak metamorphic D3 event in the Vredefort Dome; however, the kinematics of the retrograde D4 mylonite forming the terrane boundary between the high-grade gneisses and greenschist metalavas to the southeast is consistent with such NW–SE extension. Given that the metamorphic peak in the Archaean Basement Complex is statistically indistinguishable from the age of 3074 ± 6 Ma obtained by Armstrong et al. (1991) for the Dominion lavas, this change in far-field stresses is significant. However, the complexity of modern subduction–collisional settings means that such a scenario is not impossible.

Based on the available evidence it is thus suggested that the Vredefort terrane comprises both low-grade and high-grade remnants of submarine mafic–ultramafic lavas and sediments that likely formed in one or more oceanic plateau or extensional back-arc settings between 3.3 and 3.5 Ga, possibly overlying existing (extended) sialic crust. These were then incorporated into a 3.13–3.08 Ga magmatic arc dominated by trondhjemitic–granodioritic magmas that docked with the Kaapvaal Shield as a result of NE–SW convergence. Tectonic and magmatic thickening of the arc crust was associated with high-grade metamorphism and concomitant partial exhumation. Changing far-field stresses in the growing craton, in turn, triggered significant, rapid erosional and tectonic exhumation that continued into the development of the Dominion rift system, with the latter process likely responsible for strongly variable exhumation levels. The high-grade portion of the Vredefort terrane may offer a glimpse of what exists below the current levels of exposure in other granite–greenstone terranes in the craton. If it were not for the unusual, and extreme, exhumation that occurred during the 2.02 Ga impact, these levels would have remained hidden.

5.9.4 Younger Events

Given the exceptional levels of exhumation in the Vredefort Dome, any interpretation of the Archaean Basement Complex must acknowledge the possible influence of post-Dominion Group events in the south-central parts of the Kaapvaal craton. In addition to the 2.02 Ga impact, two major magmatic events affected the central Kaapvaal craton —the 2.71 Ga Ventersdorp volcanism and associated rifting, and the 2.06 Ga Bushveld magmatism. Mafic sills related to the Ventersdorp event occur abundantly in the Neoarchaean supracrustal succession in the collar and in the outer core of the dome, and mafic sills and small ultramafic to alkaline granite plutons linked to the Bushveld Complex intrude mainly in the collar (Fig. 5.1) but also include the Winddam wehrlite in the central core (Fig. 5.2; Stepto 1990). Despite evidence of ultrahigh-T assemblages of Ventersdorp-age in kimberlite xenoliths ~60 km southwest of the dome (Schmitz and Bowring 2003), no Ventersdorp-age metamorphic overprint has been identified in the silicate assemblages in the Vredefort Dome. This may reflect the fact that the Archaean Basement Complex rocks currently exposed in the dome were likely to have lain at depths of between 7 and 20 km at the time; consequently, even an extreme geotherm of >30 °C/km would have raised the gneisses only into the mid-amphibolite facies, at most. Nonetheless, the thermal perturbation at this time appears to have been capable of inducing at least some Pb loss in zircons in these rocks.

Gibson and Wallmach (1995) attributed the lower greenschist to mid-amphibolite facies metamorphism in the collar strata to enhanced heat flow during the Bushveld event; however, despite predicted temperatures of ≥ 600 °C through much of the core of the dome, no unambiguous evidence of metamorphic effects in the gneisses has been found. Radiogenic isotope systems also show only limited evidence of resetting, with a single Bushveld $^{40}Ar/^{39}Ar$ age having been obtained by Allsopp et al. (1991), and possible partial Pb loss in zircon (Kamo et al. 1996).

The impact-related shock metamorphism, high-strain-rate deformation and subsequent short-lived static metamorphism, reaching ultrahigh-T conditions in the Granofels Domain and exceeding 550 °C in much of the Basement Complex (Gibson et al. 1998), also appears to have had only limited effect on silicate mineral assemblages (Fig. 5.9b) and on zircon isotopic systems, except possibly where the latter were strongly shocked (Kamo et al. 1996). Based on the evidence, it appears that resetting of isotopic systematics may be highly localised and highly variable, depending primarily on the availability of hydrothermal fluids rather than host rock temperature.

5.10 Conclusions

The Archaean Basement Complex rocks in the Vredefort Dome straddle an important period in the geodynamic evolution of the Kaapvaal craton in which consolidation of the cratonic core as a result of extensive partial melting of the lower crust and high-level granite emplacement occurred. However, the Vredefort rocks appear to represent a new, predominantly juvenile, arc terrane that was being accreted during NE–SW convergence to the cratonic core at the time. Recent work suggests that they represent a 10–12 km deep crustal section through upper amphibolite and granulite facies mid-crust with a maximum depth of exposure of 25–27 km. Peak metamorphic conditions are consistent with the onset of partial melting in the granulite zone and the complex nature of the lithologies suggests that the terrane represents a section through the mid-crustal plumbing system between partially melted lower crust and upper crustal batholiths. A major retrograde shear zone, interpreted as a 3.07 Ga low-angle normal fault, juxtaposes high-grade rocks against greenschist-grade greenstones to the southeast. The rocks have been strongly structurally disrupted, and display a range of shock features and an unusual ultrahigh-T static metamorphic overprint in the central domain that are related to the 2.02 Ga impact event.

Acknowledgements The ideas expressed in this chapter have grown out of fruitful discussions with colleagues and postgraduate students over the past two decades, particularly Uwe Reimold, Cristiano Lana, Paula Ogilvie, Allan Wilson, Luisa Broccardo and Richard Armstrong. Comments from Dirk van Reenen, Gary Stevens and Axel Hofmann helped improve the manuscript. Lyn Whitfield is thanked for assisting with drafting and Kimberley Beaton, Grant Bybee, Lew Ashwal and S'lindile Wela with the geochemical plots. The financial support of the NRF Rated Researcher Programme is gratefully acknowledged.

References

Allsopp HL, Fitch FJ, Miller JA, Reimold WU (1991) $^{40}Ar/^{39}Ar$ stepheating age determinations relevant to the formation of the Vredefort Dome, South Africa. S Afr J Sci 87:431–442

Armstrong RA, Compston W, Retief EA, Williams LS, Welke HJ (1991) Zircon ion microprobe studies bearing on the age and evolution of the Witwatersrand Basin. Precambr Res 53:243–266

Armstrong RA, Lana C, Reimold WU, Gibson RL (2006) SHRIMP zircon age constraints on Mesoarchean crustal development in the Vredefort dome, central Kaapvaal craton, South Africa. In: Reimold WU, Gibson RL (eds) Processes on the early earth, vol 405. Geological Society of America Special Paper, pp 233–253

Barker F (1979) Trondhjemites: definition, environment and hypotheses of origin. In: Barker F (ed) Trondhjemites. Dacites and Related Rocks, Elsevier, Berlin, pp 1–12

Belcher RW, Kisters AFM (2006) Syntectonic emplacement and deformation of the Heerenveen batholith: Conjectures on the structural setting of the 3.1 Ga granite magmatism in the Barberton granite-greenstone terrain, South Africa. In: Reimold WU, Gibson RL (eds) Processes on the Early Earth, vol 405. Geological Society of America Special Paper, pp 211–231

Bisschoff AA (1972) The dioritic rocks of the Vredefort Dome. Transactions of the Geological Society of South Africa 75:31–45

Bisschoff AA (2000) The geology of the Vredefort Dome (Explanation of Geological Sheets 2627CA, CB, CC, CD, DA, DC, and 2727AA, AB, BA, 1:50 000 scale). Council for Geoscience, Pretoria, 49 p. and map 1:50 000 scale

Broccardo L (2009) Petrography and Geochemistry of ultramafic rocks from the central parts of the Vredefort Dome, South Africa. Unpubl Honours Report: 47p

Carporzen L, Weiss BP, Gilder SA, Hart RJ (2012) Lightning remagnetization of the Vredefort impact crater: no evidence for impact-generated magnetic fields. J Geophys Res Atmos 117 (E1):1007. https://doi.org/10.1029/2011JE003919

Cupelli CL, Moser DE, Barker IR, Darling JR, Bowman JR, Dhuime B (2015) Discovery of mafic impact melt in the center of the Vredefort dome: archetype for continental residua of early Earth cratering? Geology 42:403–406

De Wit MJ, Roering R, Hart RJ, Armstrong RA, De Ronde CEJ, Green RWE, Tredoux M, Peberdy E, Hart RA (1992) Formation of an archaean continent. Nature 357:553–562

Drummond MS, Defant MJ (1990) A model for trondhjemite-tonalite-dacite genesis and crustal growth via slab melting: archean to modern comparisons. J Geophys Res 95:21503–21521

Dziggel A, Stevens G, Poujol M, Armstrong RA (2006) Contrasting source components of clastic sedimentary rocks in the lowermost Formations of the Barberton greenstone belt. In: Reimold WU, Gibson RL (eds) Processes on the Early Earth, vol 405. Geological Society of America Special Paper, pp 157–172

Fagereng A, Harris C, La Grange M, Stevens G (2008) Stable isotope study of the Archaean rocks of the Vredefort impact structure, central Kaapvaal Craton, South Africa. Contibutions Miner Pet 155:63–78

Fedo CM, Nesbitt HW, Young GM (1995) Unraveling the effects of potassium metasomatism in sedimentary rocks and paleosols, with implications for paleoweathering conditions and provenance. Geology 23:921–924

Flowers RM, Moser DE, Hart RJ (2003) Evolution of the amphibolite-granulite facies transition exposed by the Vredefort impact structure, Kaapvaal craton, South Africa. J Geol 111:455–470

Fricke A, Medenbach O, Schreyer W (1990) Fluid inclusions, planar elements and pseudotachylites in the basement rocks of the Vredefort structure, South Africa. Tectonophysics 171:169–183

Frimmel H (2019) The Witwatersrand Basin and Its Gold Deposits. In: Kröner A, Hofmann A (eds) The Archaean Geology of the Kaapvaal Craton, Southern Africa. Springer Nature Switzerland AG

Gibson RL (2002) Impact-induced melting of Archaean granulites in the Vredefort Dome, South Africa. I: Anatexis of metapelitic granulites. J Metamorph Geol 20:57–70

Gibson RL, Reimold WU (2005) Shock pressure distribution in the Vredefort impact structure, South Africa. In: Kenkmann T, Hörz F, Deutsch A (eds) Large Meteorite Impacts and Planetary Evolution III, vol 384. Geological Society of America Special Paper, pp 329–349

Gibson RL, Wallmach T (1995) Low pressure-high temperature metamorphism in the Vredefort Dome, South Africa: anticlockwise pressure-temperature path followed by rapid decompression. Geol J 30:319–331

Gibson RL, Reimold WU (2008) Geology of the Vredefort impact structure; a guide to sites of interest. Counc Geosci Pretoria Memoir 97:181p

Gibson RL, Reimold WU, Lana C (2005) Comment on: new PGE and Re/Os-isotope data from lower crustal sections of the Vredefort dome and a reinterpretation of its "crust-on-edge" profile. S Afr J Geol 108:309–313

Gibson RL, Reimold WU, Stevens G (1998) Thermal-metamorphic signature of an impact event in the Vredefort Dome, South Africa. Geology 26:787–790

Gibson RL, Reimold WU, Ashley AJ, Koeberl C (2002) Metamorphism on the Moon: a terrestrial analogue in the Vredefort Dome, South Africa? Geology 30:475–478

Hart RJ (1978) A study of the isotopic and geochemical gradients in the old granite of the Vredefort structure, with implications for continental heat flow. University of the Witwatersrand, Thesis, p 172

Hart RJ, Nicolaysen LO, Gale NH (1981) Radioelement concentrations in the deep profile through Precambrian basement of the Vredefort structure. J Geophys Res 86:10639–10652

Hart RJ, Andreoli MAG, Tredoux M, de Wit MJ (1990a) Geochemistry across an exposed section of Archaean crust at Vredefort: with implications for mid-crustal discontinuities. Chem Geol 82:21–50

Hart RJ, Andreoli MAG, Smith CB, Otter ML, Durrheim R (1990b) Ultramafic rocks in the centre of the Vredefort structure: possible exposure of the upper mantle. Chem Geol 83:233–248

Hart RJ, Hargraves RB, Andreoli MAG, Tredoux M, Doucouré CM (1995) Magnetic anomaly near the center of the Vredefort structure: implications for impact-related magnetic signatures. Geology 23:277–280

Hart RJ, Moser D, Andreoli MAG (1999) Archaean age for the granulite metamorphism near the center of the Vredefort Structure, South Africa. Geology 27:1091–1094

Hart RJ, McDonald I, Tredoux M, de Wit MJ, Carlson RW, Andreoli MAG, Ashwal LD (2004) New PGE and Re/Os-isotope data from lower crustal sections of the Vredefort Dome and a reinterpretation of its "crust on edge" profile. S Afr J Geol 107:83–94

Irvine TN, Baragar WRA (1971) A guide to the chemical classification of the common volcanic rocks. Can J Earth Sci 8:523–548

Ivanov BA (2005) Numerical modeling of the largest terrestrial meteorite craters. Sol Syst Res 39:381–409

Jensen LS (1976) A new cation plot for classifying subalkaline volcanic rocks. Ontario Division Mines, Miscellaneous Paper 66: 22p

Kamo SL, Reimold WU, Krogh TE, Colliston WP (1996) A 2.023 Ga age for the Vredefort impact event and a first report of shock metamorphosed zircons in pseudotachylitic breccias and granophyre. Earth Planet Sci Lett 144:369–388

Lana C, Gibson RL, Reimold WU (2003a) Impact tectonics in the core of the Vredefort dome, South Africa: implications for central uplift formation in very large impact structures. Meteorit Planet Sci 38:1093–1107

Lana C, Gibson RL, Reimold WU (2006) The Broodkop shear zone, southeast Vredefort dome (South Africa): a dominion-related extensional feature? S Afr J Geol 109:265–278

Lana C, Gibson RL, Kisters AFM, Reimold WU (2003b) Archean crustal structure of the Kaapvaal craton, South Africa—evidence from the Vredefort dome. Earth Planet Sci Lett 206:133–144

Lana C, Gibson RL, Reimold WU, Minnitt RCA (2003c) Geology and geochemistry of a granite-greenstone association in the southeastern Vredefort dome, South Africa. S Afr J Geol 107:291–314

Lana C, Reimold WU, Gibson RL, Koeberl C, Siegesmund S (2004) Nature of the Archaean mid-crust in the central Kaapvaal craton, South Africa—evidence from the Vredefort dome. Geochim Cosmochim Acta 68:623–642

Lekgothoane D (2017) The petrographic and geochemical analysis of the Archaean metabasites and meta-ultramafics in the Vredefort Dome. Honours report, University of the Witwatersrand, 40p

Lieger D, Riller U, Gibson RL (2011) Petrographic and geochemical evidence for an allochthonous, possibly impact melt, origin of pseudotachylite from the Vredefort Dome, South Africa. Geochim Cosmochim Acta 75:4490–4514

Lowe DR (1999) Geological evolution of the Barberton greenstone belt and vincity. In: Lowe DR, Byerly GR (eds) Geological evolution of the Barberton greenstone belt, South Africa, vol 329, Geological Society of America Special Paper, pp 287–312

Maree BD (1947) The Vredefort Structure as revealed by a gravimetric survey. Trans Geol Soc S Afr 47:183–196

Martin H (1995) The Archean grey gneisses and the genesis of the continental crust. In: Condie KC (ed) The Archean Crustal Evlution. Elsevier, New York, pp 205–259

Martini JEJ (1992) The metamorphic history of the Vredefort dome at approximately 2 Ga as revealed by coesite-stishovite-bearing pseudotachylites. J Metamorph Geol 10:517–527

Mashabela S (2015) Analysis of pre-impact and impact-induced geological structures in the northern collar of the Vredefort Dome, South Africa. Dissertation, University of the Witwatersrand, 305p

McLennan SM (1982) On the geochemical evolution of sedimentary rocks. Chem Geol 37:335–350

Moser DE (1997) Dating the shock wave and thermal imprint of the giant Vredefort impact, South Africa. Geology 25:7–10

Moser D, Flowers RM, Hart RJ (2001) Birth of the Kaapvaal tectosphere 3.08 billion years ago. Science 291:465–468

Ogilvie P (2010) Metamorphic studies in the Vredefort dome, South Africa. Thesis, University of the Witwatersrand, 737p

Perchuk LL, Tokarev DA, van Reenen DD, Varlamov DA, Gerya TV, Sazonova LV, Fel'dman VI, Smit CA, Brink MC, Bisschoff AA (2002) Dynamic and thermal history of the Vredefort explosion structure in the Kaapvaal craton, South Africa. Petrology 10:395–432

Poujol M, Robb L, Anhaeusser CR, Gericke B (2003) Geochronological constraints on the evolution of the Kaapvaal craton, South Africa. Precambr Res 127:181–213

Pybus GQJ (1995) Geological and mineralogical analysis of some mafic intrusions in the Vredefort Dome, Central Witwatersrand basin. Dissertation, University of the Witwatersrand, 376p

Qoza ZR (2001) Petrographic and geochemical study of Inlandsee Leucogranofels, central Vredefort Dome. Unpubl Honours Report: 41p

Schmitz MD, Bowring SA (2003) Ultrahigh-temperature metamorphism in the lower crust during Neoarchean Ventersdorp rifting and magmatism, Kaapvaal craton, southern Africa. GSA Bull 115:533–548

Schreyer W (1983) Metamorphism and fluid inclusions in the basement of the Vredefort Dome, South Africa: guidelines to the origin of the structure. J Petrol 24:26–47

Schreyer W, Abraham K (1978) Symplectitic cordierite-orthopyroxene-garnet assemblages as pro-ducts of contact metamorphism of pre-existing basement granulites in the Vredefort structure, South Africa. Contrib Miner Petrol 68:53–62

Slawson WF (1976) Vredefort core: a cross-section of the upper crust? Geochim Cosmochim Acta 40:117–121

Stepto D (1979) A geological and geophysical study of the central portion of the Vredefort Dome Structure. Thesis, University of the Witwatersrand, 378p

Stepto D (1990) The geology and gravity field in the central core of the Vredefort structure. Tectonophysics 171:75–103

Stevens G, Gibson RL, Droop GTR (1997) Mid-crustal granulite facies metamorphism in the central Kaapvaal craton: the Bushveld Complex connection. Precambr Res 82:113–132

Streckeisen A (1976) To each plutonic rock its proper name. Earth-Sci Rev 12:1–33

Tredoux M, Hart RJ, Carlson RW, Shirey SB (1999) Ultramafic rocks at the center of the Vredefort structure: further evidence for the crust-on-edge model. Geology 27:923–926

Wieland F, Gibson RL, Reimold WU (2005) Structural analysis of the collar of the Vredefort dome, South Africa—significance for impact-related deformation and central uplift formation. Meteorit Planet Sci 40:1537–1554

Vielzeuf D, Clemens JD, Pin C, Moinet E (1990) Granites, granulites and crustal differentiation. In: Vielzeuf D, Vidal P (eds) Granulites and Crustal Evolution, Kluwer: 25–36

Viljoen MJ, Viljoen RP, Pearton TN (1982) The nature and distribution of Archaean komatiitic volcanics in South Africa. In: Arndt NT, Nisbet EG (eds) Komatiites. George Allen and Unwin Publications, London, pp 53–80

The Moodies Group—a High-Resolution Archive of Archaean Surface Processes and Basin-Forming Mechanisms

6

Christoph Heubeck

Abstract

The Moodies Group is the uppermost and youngest unit of the Barberton Supergroup, the volcanic-sedimentary sequence forming the Barberton Greenstone Belt (BGB), the largest and best-preserved greenstone belt in the basement of the Kaapvaal Craton. The Moodies Group consists predominantly of fine- to coarse-grained, compositionally immature to mature, quartzose sandstones up to 3.6 km thick, with significant units of conglomerates and siltstones and minor volcanic rocks and ferruginous sediments. Strata were deposited, in part syndeformationally, in alluvial, fluvial, deltaic, tidal and shallow-marine environments. They represent a unique early Archaean record of marine-terrestrial transition. Abundant tidal microbial mats provide insights into the early shallow-water biosphere. The incompletely investigated stratigraphic architecture likely records uniquely Archaean "sag basin" dynamics between rising plutons. The quartz-dominated Moodies sandstones mark long-term, large-scale access of surface systems to crust-stabilizing, high-level granitoid igneous rocks.

Keywords

Archaean • Sandstone • Microbial mats • Evaporites BIF • Tectonics • Basin formation

6.1 Introduction

The Moodies Group crops out over much of the BGB's 110 km strike length and 40 km width. It is bounded at its base by the top of the underlying Fig Tree Group or older rocks of the Onverwacht Group; its top is not exposed (Fig. 6.1; Byerly et al. 2019). Due to the high resistance of the quartz-rich sandstones to weathering, many Moodies units form prominent rocky ridges with steep slopes. They contribute greatly to the >1700 m of topographic relief of the Barberton-Makhonjwa Mountains and form their highest elevation, Maid-of-the-Mists, at 1903 m.

Hall (1918) originally defined the Moodies Series, named after a local surveyor, to include all sedimentary units of the Swaziland System of the BGB. The modern definition of the Fig Tree and Moodies Groups goes back to pre-WW II mapping by South African government geologists (Visser et al. 1956). Anhaeusser (1976) provided stratigraphic detail and subdivided the Moodies Group into three lithostratigraphic formations: Clutha, Joe's Luck, and Baviaanskop, and several informal lithostratigraphic members. His terminology will be used in this paper (Fig. 6.1). Eriksson (1977, 1978, 1979, 1980) was the first to focus on the sedimentary aspects of the Moodies Group, based on observations in the central BGB, and suggested an alternative subdivision into five fining-upward sequences (MD1 through MD5; Eriksson 1977).

Although the overall grade of metamorphism in the Moodies Group is only lower greenschist-facies in most areas, multiple regional alteration events have affected the greenstone belt in the Archaean and Palaeoproterozoic during burial, including aqueous alteration and thermal events. Mineral alteration, strain and metamorphic grade differ regionally; they are highest along the margins of the greenstone belt (Anhaeusser 1972, 1976) and near major fault zones within the belt (Heubeck and Lowe 1994b). Ductile deformation appears to increase regionally towards the southeast where Moodies conglomerates are commonly highly strained. Selective silicification during early diagenesis likely played a significant role in preserving extensive microbial mats near the Lomati River Sill, part of the mafic to intermediate Moodies-age Lomati River Complex intrusion, stockwork and lava in the central part of the BGB (Fugmann 2017). Following Late Cretaceous exhumation of the BGB, prolonged subtropical weathering has affected the near-surface composition of the rocks, most importantly by oxidative weathering.

C. Heubeck (✉)
Institut für Geowissenschaften, Friedrich-Schiller-Universität Jena, Burgweg 11, 07749 Jena, Germany
e-mail: christoph.heubeck@uni-jena.de

A. Kröner and A. Hofmann (eds.), *The Archaean Geology of the Kaapvaal Craton, Southern Africa*, Regional Geology Reviews, https://doi.org/10.1007/978-3-319-78652-0_6

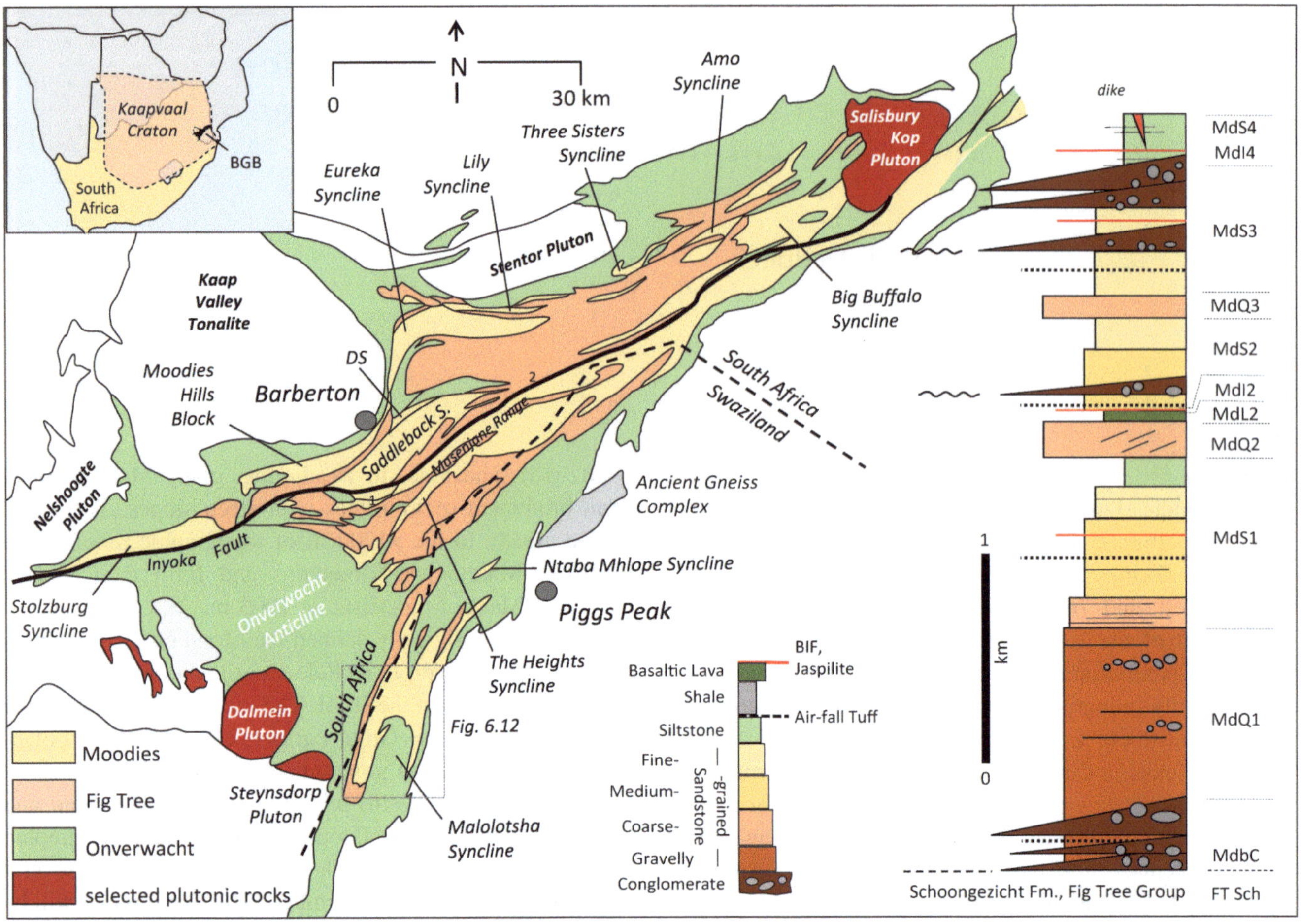

Fig. 6.1 Geologic map of the BGB, emphasizing outcrop area of the Moodies Group. Inset (top left) shows location within the Kaapvaal Craton of southern Africa. Inset right shows generalized stratigraphic column north of the Inyoka Fault; unit abbreviations follow Anhaeusser (1976). The lower part of the section, approximately 2 km thick, represents a gradually fining- and deepening-upward sequence; the upper part, ca. 1.5 km thick, above the prominent basaltic lava MdL2, is heterogeneous and overall coarsening- and shallowing-upward. At least one progressive unconformity above the MdL2 lava indicates the onset of syndepositional deformation. Moodies stratigraphy south of the Inyoka Fault is poorly differentiated. 1 = Maid-of-the-Mists (1903 m), highest point in the BGB; 2 = Shiyalongubo Dam. DS = Dycedale Syncline

6.2 Stratigraphy and Lithology; Depositional Environments

Moodies Group strata can be divided into two principal tectonostratigraphic facies (Heubeck and Lowe 1999): The northern facies, north of the BGB-axial Inyoka Fault, reaches up to 3.7 km thick and is locally feldspar-rich. It is stratigraphically and petrographically subdivided into two parts: An overall deepening-upward, fining-upward section from braided-fluvial to subtidal facies below a basaltic lava complex which occurs approximately mid-section in the Moodies Group; and a conglomerate-rich upper section with rapid lateral and vertical facies changes. Heubeck and Lowe (1994a) interpreted the former trend as intramontane or extensional while they related the composition and architecture of the upper Moodies Group to syndepositional up-to-the-north faulting along the northern margin of the BGB, basin margin uplift and inversion in the context of overall NW-to-SE greenstone belt shortening. Up to four jaspilitic units are known in the northern facies; they serve as useful regional correlation markers (Fig. 6.2).

The southern facies of the Moodies Group in South Africa and Swaziland (now renamed Eswatini) south of the Inyoka Fault is not well studied (Urie 1970, 1971; Lamb 1984a; Lamb and Paris 1988; Heubeck and Lowe 1999; Lowe et al. 2012). It appears to be thinner (<2 km), lacks feldspar, has a higher metamorphic grade, and consists, as currently known, mostly of thick, dominantly horizontally stratified and cross-bedded quartzose sandstones and conglomerates. Ongoing geologic mapping is tracing marker units in Eswatini and South Africa. Stratigraphic correlation

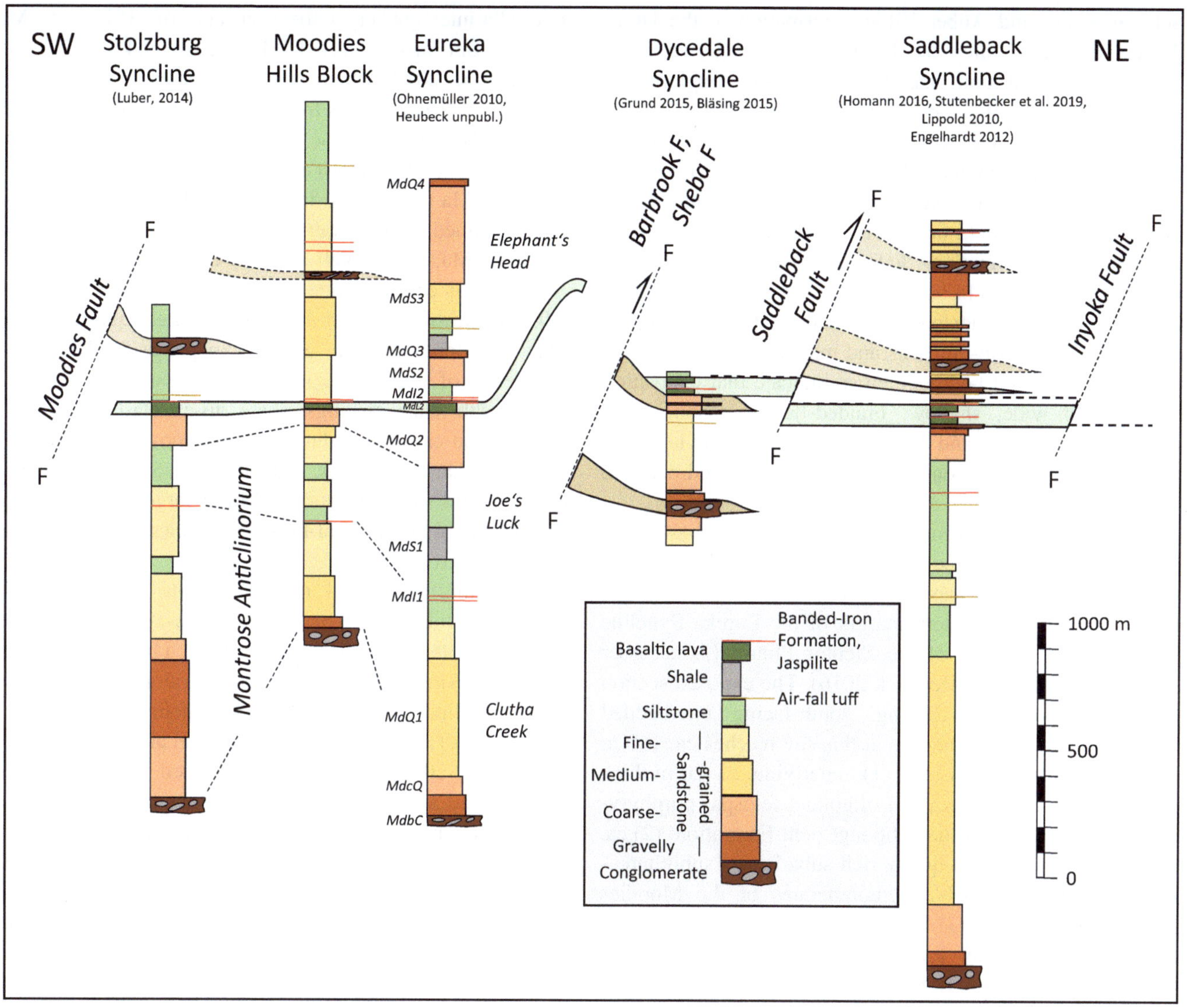

Fig. 6.2 Stratigraphic correlation diagram of Moodies Group lithologies among the major tectonic synclines in the BGB north of the Inyoka Fault (Heubeck et al. 2016a). The top of the Moodies basaltic lava MdL2 serves as stratigraphic datum but offsets by major faults are recognized. The stratigraphic subdivision south of the Inyoka Fault is problematic and not shown here

and limited dating of detrital zircons and interbedded tuffs indicate that sedimentation was contemporaneous with the northern Moodies facies (Wiechert 2014).

6.2.1 Moodies Strata North of the Inyoka Fault

Lithologic and facies variability define first-order lithostratigraphic subdivisions that can be correlated throughout five adjacent tectonic units north of the Inyoka Fault: They include the Eureka Syncline, the Moodies Hills Block, the Saddleback Syncline, the Dycedale Syncline, and the Stolzburg Syncline (Visser et al. 1956; Fig. 6.1). Other tectonic elements comprising Moodies strata north of the Inyoka Fault, such as the Lily, Three Sisters, Amo and Big Buffalo Synclines have not been thoroughly stratigraphically investigated (Kohler and Anhaeusser 2002; Fig. 6.1). In the Eureka Syncline, Anhaeusser (1976) formally recognized (from base to top) three fining-upwards formations: Clutha, Joe's Luck, and Baviaanskop Formations. Each of these formations begins with a basal conglomerate or quartzite and includes several members. The following stratigraphic description follows the terminology of Anhaeusser (1976) but expands and modifies his informal lithostratigraphic units.

The base of the Moodies Group is conformable with, and gradational from, dacitic volcanics and commonly coarse-grained volcaniclastic sediments of the underlying

Schoongezicht and Auber Villiers Formation in the Umuduha and Kaap Valley blocks, respectively (Lowe et al. 1999). The basal unit, the Moodies basal conglomerate (MdbC), consists of interbedded thick-bedded cobble conglomerate and gravelly sandstone (Fig. 6.3). It is defined by a matrix rich in monocrystalline quartz and a polymict clast composition; both properties distinguish this unit from underlying volcaniclastics and conglomerates of the Schoongezicht Formation. Sedimentary textures in MdbC include imbricated clasts, abundant horizontally stratified and cross-bedded sandstone, and shale coatings on rippled sandstones. Shale rip-up clasts and mudcracks are common. The lithologies and bed geometries indicate that deposition occurred in wide, shallow, braided-fluvial channels, on sandy fluvial plains and locally perhaps in alluvial environments. Erosional contacts to underlying bedrock are nowhere observed.

An easily accessible outcrop of the Moodies basal conglomerate above Fig Tree strata is well exposed in a roadcut along the Barberton-Kaapmuiden road (R38) at Ezzy's Pass, located on the northeastern margin of the Eureka Syncline and ca. 1.4 km south of the Stentor Pluton (Anhaeusser 1976; Zeh et al. 2013; Heubeck 2016). The exposed section of steeply southward-dipping, south-facing interbedded sandstones and conglomerates at this site reaches ca. 160 m thick (Fig. 6.3). It consists of (1) underlying, ca. 20 m thick basal volcanic litharenites and oligomict feldspar-porphyry-clast conglomerates of the Schoongezicht Formation; (2) ca. 80 m thick interbedded quartz-rich subarkoses, sublitharenites and polymict cobble conglomerates of the Moodies Group (MdbC); and (3) ca. 60 m thick, horizontally and cross-bedded gravelly sandstones of the lower Moodies Group (MdQ1) (Fig. 6.3). Sedimentation at this site probably reflects deposition in a proximal fluvial plain by rapid and shallow subaerial flows, as shown by the multiple, moderately erosive gravel-filled channels, the paucity of grading, common clast imbrication, single-clast-thick conglomerate stringers and, in the upper part of the section, abundant parallel stratification in thin- and wavy-bedded sandstones, shale lamination, and possibly some poorly preserved subaerial exposure surfaces.

Granitoid clasts in the basal conglomerate at this site have attracted much interest as possible components of old continental crust. Increasingly detailed zircon ages from cobbles (van Niekerk and Burger 1978; Tegtmeyer and Kröner 1987; Kröner and Compston 1988; Hessler and Lowe 2006; Sanchez-Garrido et al. 2011; Agangi et al. 2019; Kröner et al. 2019) and from the sandy conglomerate matrix (A. Kröner, pers. comm. 2016) yielded a heterogeneous age population whose peaks can be correlated with major peaks of magmatism throughout the stratigraphic record of the BGB. The oldest clasts were identified by Kröner and Compston (1988). Clast MD6 yielded a SHRIMP concordia-intercept crystallization age of 3570 ± 6 Ma (n = 10; confirmed by a SHRIMP age of 3564.5 ± 2.2 Ma; Kröner et al. 2019) which is slightly older than the oldest zircon crystallization ages from the lowest stratigraphic units of the BGB, felsic volcanic schists of the basal Onverwacht Group near the Steynsdorp anticline in the southern BGB, at 3552 ± 1 Ma (n = 10; Kröner et al. 2016). That same clast also contained two xenocrystic zircons of 3602 ± 3 and 3634 ± 3 Ma, respectively (Kröner et al. 2019).

The basal conglomerate is overlain everywhere by ca. 300–1100 m of poorly to moderately sorted, quartzose and lithic sandstones that are gravelly in places (MdQ1). The occurrence of mudcracks in thin shale partings, common ripples with shale coatings, shale rip-up clasts, evaporitic nodules, and calcareous duricrusts in exposures of this unit in the Eureka, Saddleback and Stolzburg Synclines associated with aridisols suggests that MdQ1 largely developed in braided-fluvial and coastal-floodplain facies (Fig. 6.4; Nabhan et al. 2016a, b).

Abundant mm-spaced microbial mats in medium- to high-energy tidal-facies sandstones are best preserved and in part spectacularly exposed over about 1100 m of stratigraphic thickness and for about 14 km along strike on the overturned limb of the Saddleback Syncline in the central BGB (Fig. 6.1; Heubeck 2009; Gamper et al. 2012; Homann et al. 2015, 2016). They are interspersed with large-scale cross-bedded sandstones that reflect subtidal and estuarine conditions. These microbial mats consist of wavy kerogenous laminae arranged in densely spaced, mm-high tufts, domes, and crinkly laminations which coat bedding planes, foresets, and the tops of conglomerate clasts; they also occur locally as microbial-chip conglomerates in wide and shallow tidal channels (Fig. 6.5). Microbial mats apparently retarded dewatering and gave rise to fluid-escape structures up to 6.8 m high; surficially, wet sand laced with microbial filaments bulged upward to form fluid-filled "blisters" which were internally colonized by microbial mats (Homann et al. 2016). Detailed stratigraphic mapping by Homann et al. (2015) with focus on the relationship between sedimentary facies and microbial-mat morphology showed that the microbial mats were supple, mechanically strong structures that were adapted to the environmental stresses of medium- to high-energy inter- and supratidal environments. SEM images document filaments and coccoidal structures. Based on the mat morphology, the inferred shallow, photic-zone setting, and the filamentous morphologies, Homann et al. (2015) reasoned that the microbial mats strongly resembled extant cyanobacterial microbial mats in recent shallow-water settings and argued that Moodies microbial mats likely consisted of oxygenic photosynthesizers.

The overlying fine-grained, commonly tuffaceous sandstones and siltstones of MdS1 reach 400–1000 m thick but are generally poorly exposed. In the Eureka and Stolzburg

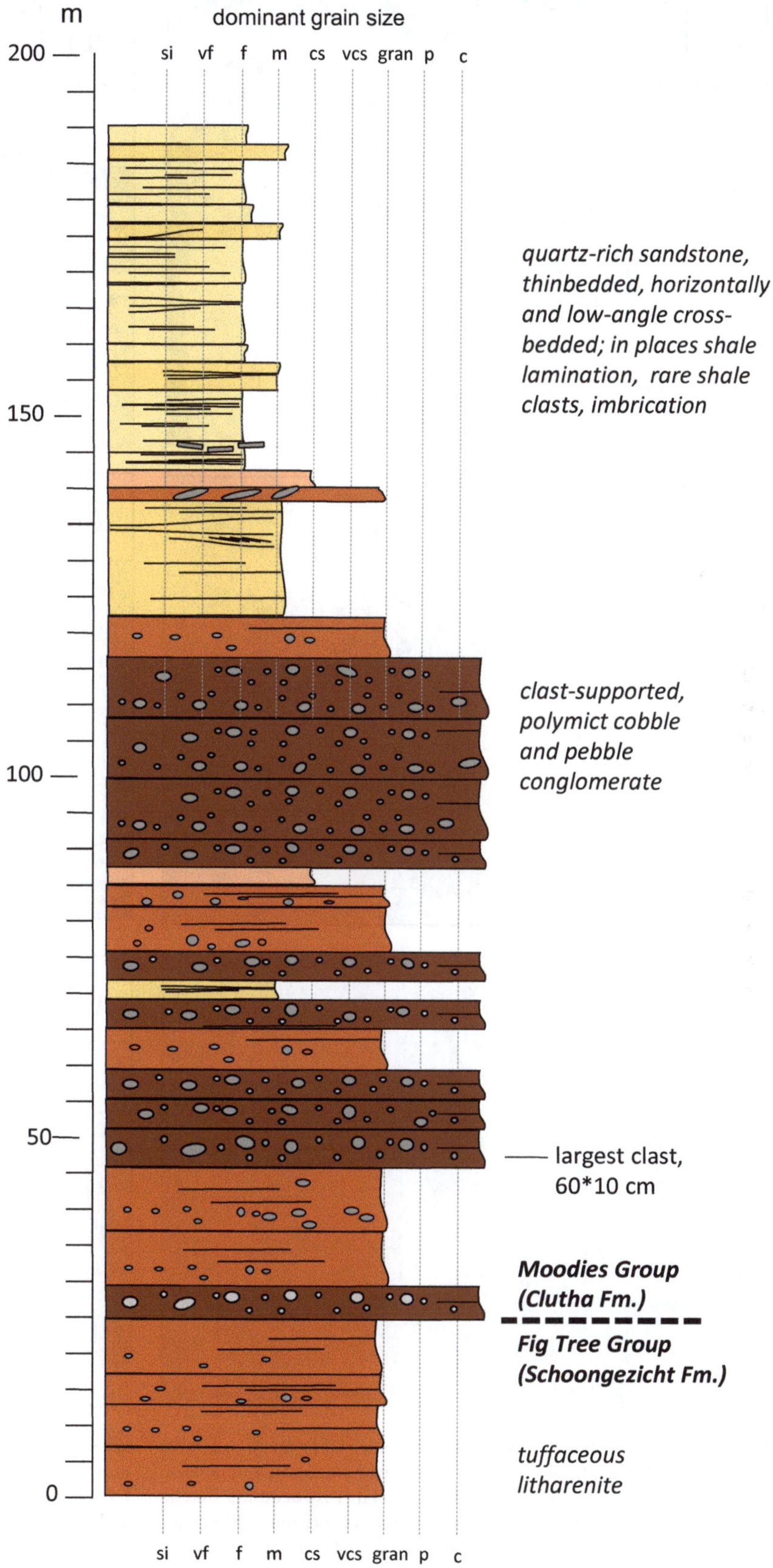

Fig. 6.3 Stratigraphic column of upper Schoongezicht and lowermost Moodies strata at the Ezzy's Pass roadcut. Legend as in Fig. 6.2

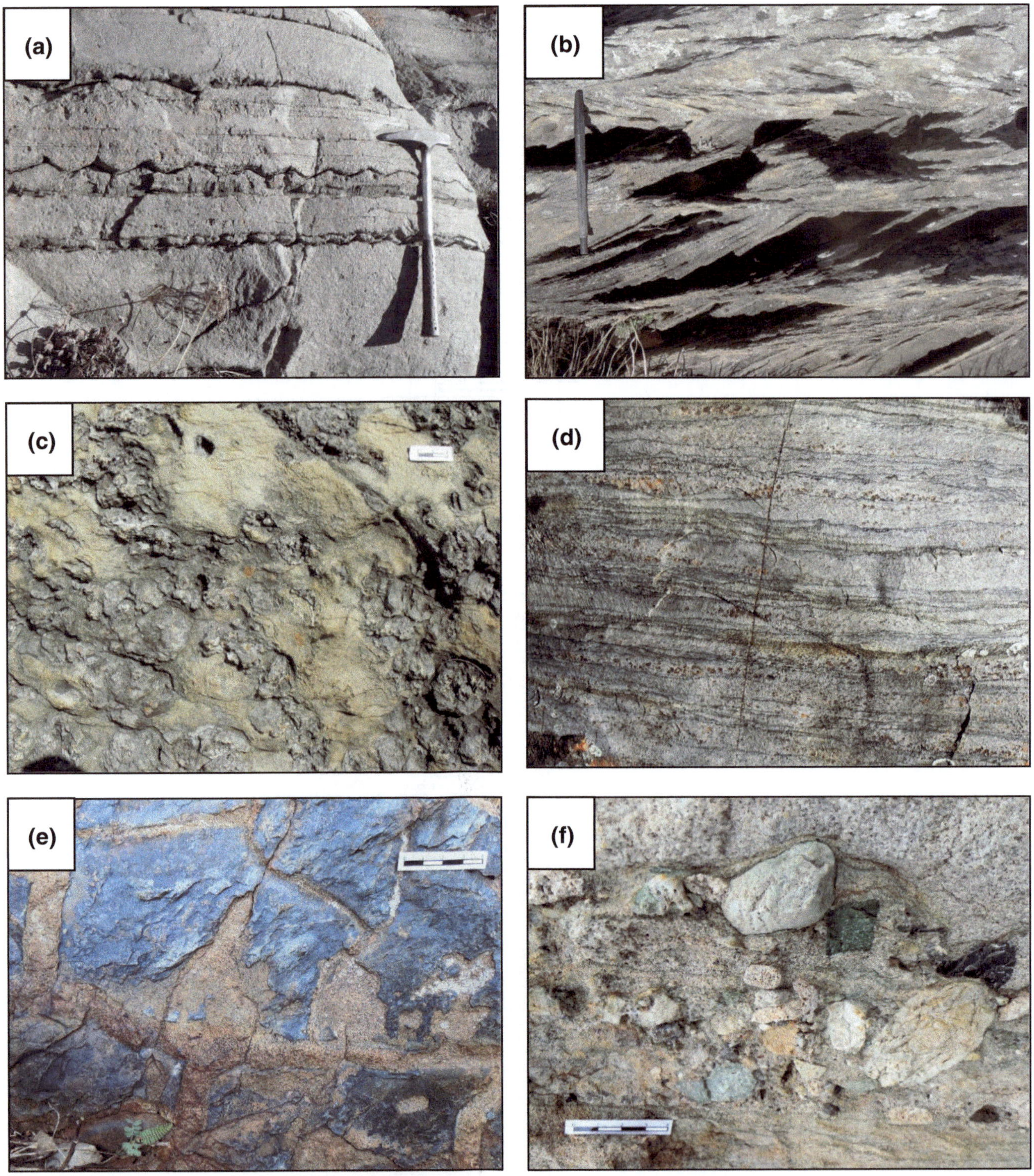

Fig. 6.4 Outcrop photographs of characteristic sedimentary structures in unit MdQ1. **a** Ripple lamination in medium-grained quartzose sandstone, Saddleback Syncline; **b** Bidirectional cross-bedding, Eureka Syncline; **c** Evaporitic vadose-zone nodular concretions, Stolzburg Syncline; **d** Rhythmically bedded microbial mats in coarse-grained arkosic sandstone, Saddleback Syncline; **e** Mudcracked shale (dark grey), coating rippled bedding plane, Saddleback Syncline; **f** Cross-sectional view of microbial mats (pale green) below, within and on top of polymict conglomerate overlain by coarse-grained sandstone, Saddleback Syncline

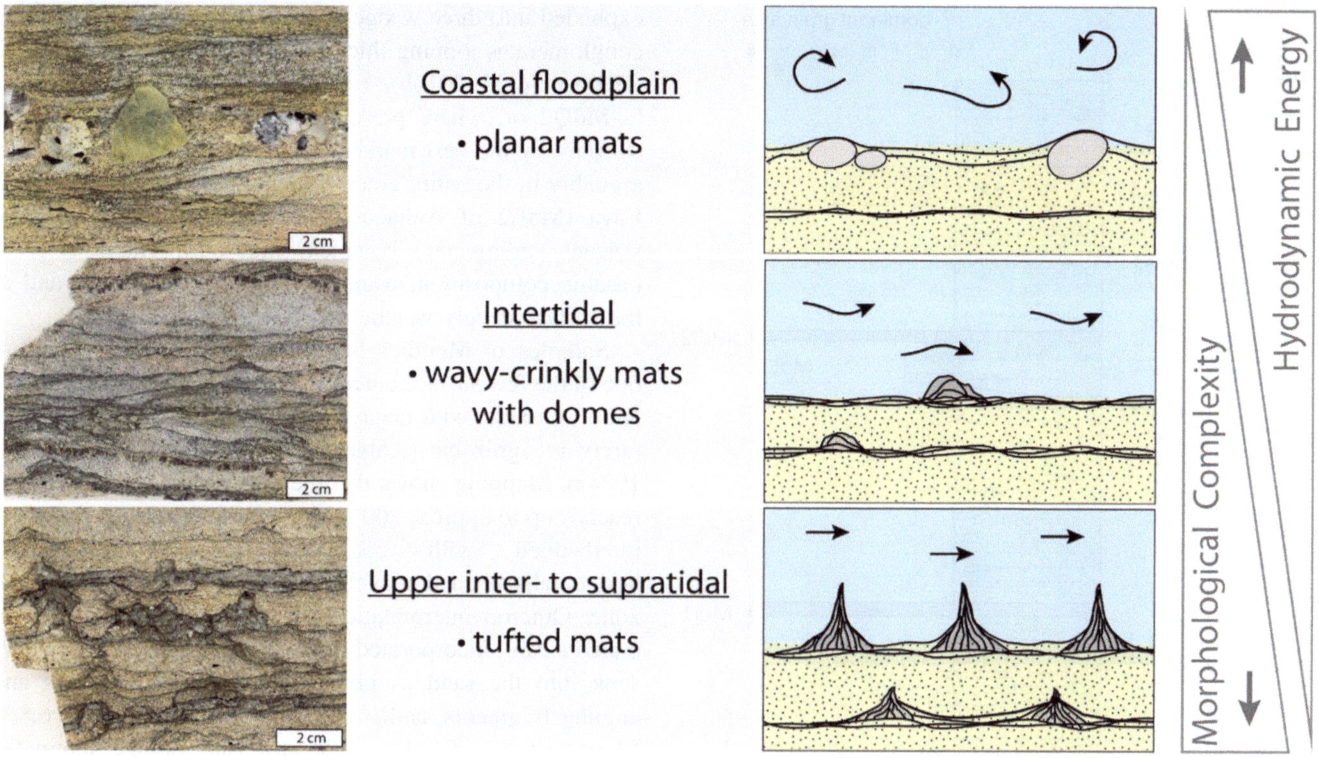

Fig. 6.5 Morphological classification of Moodies microbial mats and their environmental interpretation (Homann et al. 2015)

Synclines, this unit includes well-developed Bouma T_{cde} sequences, commonly with slump folds, rafts of semiconsolidated siltstone and sandstone slabs and other soft-sedimentary deformation structures in very-fine-grained sandstones, siltstones and minor shales, indicating below-wavebase deposition on prodelta slopes (Luber 2014). In the Saddleback Syncline, a 5 km-long wedge-shaped deltaic sandstone complex progrades SW-to-NE over ca. 8 km strike length within this unit (Stutenbecker et al. 2019). This complex includes spectacular shale coatings and thick beds of rip-up clasts; the world's oldest raindrop imprints were found there on shale-coated bedding planes (Altermann et al. 2017). MdS1 also includes several thin, regionally traceable tuff beds and a BIF-jaspilite unit (MdI1) composed of thin (1–30 cm) individual beds.

Overall, the up to 2000 m thick fining- and thinning-upward section between the base of the Moodies basal conglomerate to the top of MdS1 represents a transgressive systems tract and gradual increase in water depth. Facies range from subaerial gravelly fluvial plain (MdbC, MdQ1) through extensive coastal (shoreline, tidal, estuarine; MdQ1) to below-wavebase prodelta (MdS1, MdI1) settings. One of the best exposed and thickest sections, uninterrupted by faulting and without any recognizable stratigraphic breaks, is located in the eastern Stolzburg Syncline (Reimer 1967; Luber 2014; Fig. 6.6).

The fine-grained sandstones and siltstones of MdS1 are gradationally (Eureka Syncline) or abruptly (Stolzburg Syncline) overlain by up to 150 m of coarse- to granuly-grained and quartz-cemented quartzarenite of MdQ2, which is morphologically prominent in all Moodies tectonic units north of the Inyoka Fault (Fig. 6.7). Stacked planar foresets up to 5 m thick in the central Saddleback Syncline fill erosional channels into laterally adjacent low-angle cross-bedded sandstones. MdQ2 appears gravel-free and lacks mudchips or any other textural indications of subaerial exposure. Eriksson (1980) postulated deposition in a prograding barrier-beach and tidal-inlet setting; Heubeck and Lowe (1994a) favoured an inner-shelf dune field. Recent mapping and section measurements by Heubeck and students (unpublished field notes) suggests that MdQ2 in the central BGB may represent stacked delta-mouth bars prograding NE-to-SW along the present-day structural trend of the greenstone belt. In the Dycedale Syncline, this MdQ2 facies is replaced by a well-exposed estuarine complex (Heubeck et al. 2016a).

In the Saddleback Syncline, MdQ2 is locally abruptly overlain by an erosive and discontinuous cobble conglomerate that is up to a few m thick. This is the first occurrence of conglomerate above the top of MdQ1, about 1000 m lower in the section. It may represent a sedimentary response to a rapid base-level drop, abruptly juxtaposing a fluvial

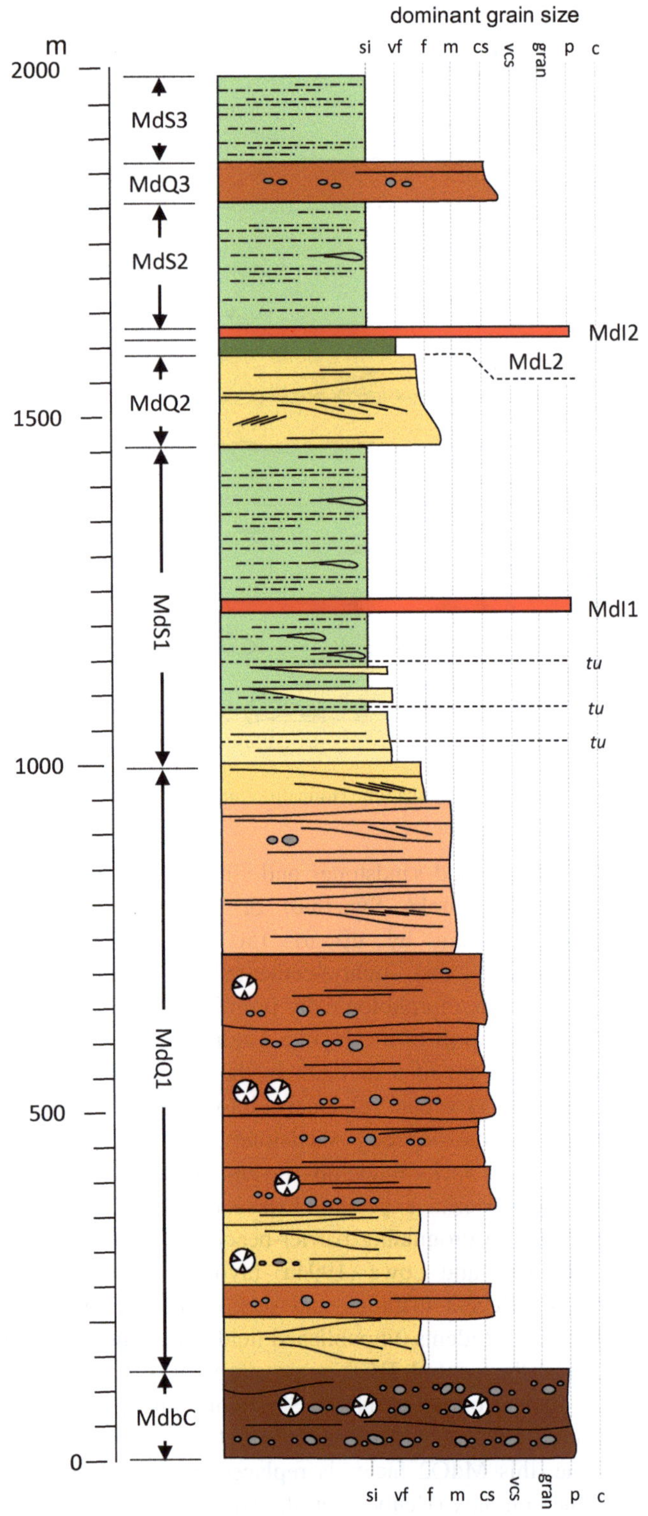

Fig. 6.6 Stratigraphic column of Moodies strata in the eastern Stolzburg Syncline (modified from Luber 2014)

gravelly-channel facies over MdQ2 barrier-beach or delta-mouth-bar sandstones, or could mark a subtle local angular unconformity. In the Dycedale Syncline, this unit is expanded into three wedge-shaped, probably alluvial cobble conglomerates topping thick estuarine sandstones of MdQ2 (Heubeck et al. 2016b).

MdQ2 or, where present, its topping conglomerate, is overlain by the best marker unit in the Moodies Group and arguably in the entire greenstone belt, the Moodies Basaltic Lava (MdL2 of Anhaeusser 1976), which is a regional volcanic-sedimentary complex (Fig. 6.8). The lava is of basaltic composition, usually amygdaloidal, altered and at the surface deeply weathered.

Samples of Moodies basalt from underground or from one of the few surface outcrops consist of chlorite, carbonate and iron oxide, with metasomatized plagioclase phenocrysts rarely recognizable (Anhaeusser 1976; Hose 1990; Lamb 1984a). Mapping shows that the unit is in places absent but reaches up to approx. 200 m thick. Individual lava flows are interbedded with coarse-grained, low-angle- and horizontally-bedded gravelly sandstone of a coastal swash zone. Outcrop interpretation suggests that lava flows generated steam, incorporated trains of sand and gravel, locally sank into the sand as pillows, broke up in bulbous and angular fragments, and were partially reworked (Fig. 6.8c). Elsewhere, interbedded dark grey and orange laminated siltstones and magnetic shales at the base, the top or within this unit indicate that multiple lava flows occurred in low-energy coastal environments.

MdL2 is overlain by fine-grained, ferruginous, thinly bedded banded-iron formation, thick-bedded jaspilite, and magnetic shale interbedded with siltstones (MdI2). This unit reaches a thickness of up to 22 m in the Moodies Hills Block and the Eureka Syncline.

The upper part of the Moodies Group, up to 1500 m thick and corresponding to the upper part of the Joe's Luck and the Baviaanskop formations of Anhaeusser (1976), is a heterogeneous siliciclastic sequence which is petrographically and lithologically variable, in particular in the Saddleback Syncline. Regional lithologic correlation above MdI2 is difficult and suggests, in contrast to the regionally correlatable units below MdI2, a highly dynamic and differentiated environment responding to local controls on composition, facies and thickness.

Thick gravelly and cross-bedded sandstone with mud-cracked shale coatings, conglomerate and minor tuffaceous siltstone (MdS2) overlie MdI2 in the Saddleback Syncline. At least one intraformational angular unconformity occurs near the base of this unit on the steeply south-dipping northern limb of the Saddleback Syncline near Saddleback Pass (Heubeck and Lowe 1994a; Fig. 6.9). Discontinuous dacitic ash-fall tuffs truncated at this unconformity by overlying conglomerate yield ages around 3220 ± 3 Ma (Heubeck et al. 2013). Shallow subvolcanic and hydrothermal clastic dykes locally crosscut the sandstones.

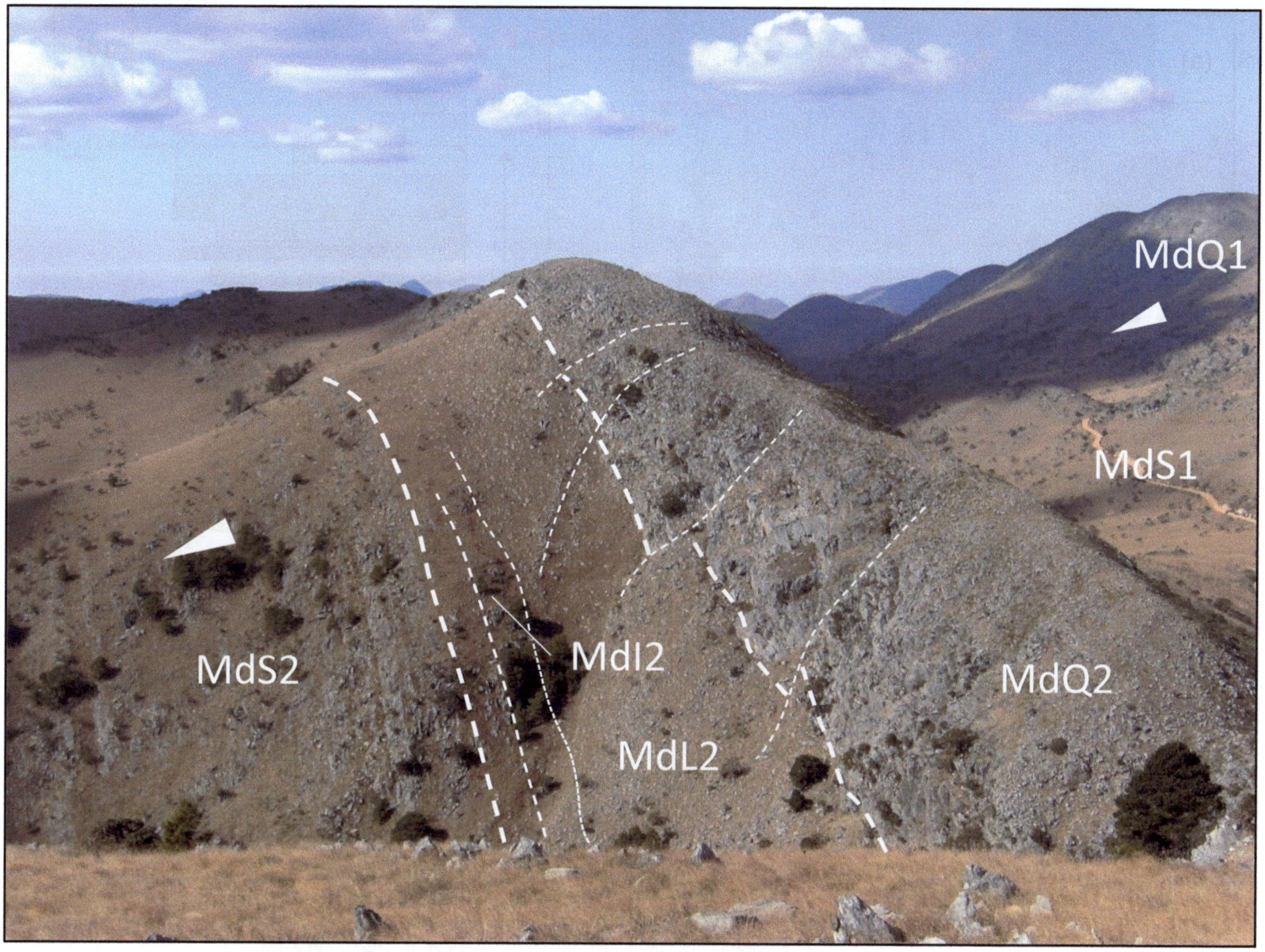

Fig. 6.7 Panoramic photograph of the stratigraphic succession in the western Saddleback Syncline. View is along strike toward the east. All units dip subvertically and are slightly overturned; stratigraphic-up is to the left. The quartz-cemented quartzose sandstone units of MdQ1 (right) and MdQ2 (center) make up craggy, resistant ridges. Silty-tuffaceous sandstones (MdS1 and MdS2) and the basaltic lava MdL2, along with the overlying jaspilite unit MdI2, in contrast, weather recessively and make grassy slopes

The uppermost Moodies units are represented in the Saddleback Syncline by thick quartzarenites (MdQ3) and a cobble conglomerate (MdQ3cgl), up to 100 m thick and locally rich in red jaspilite clasts, which extends over >5 km along strike. This conglomerate probably represents best a provenance at right angle to the structural trend; it is, however, unclear whether this unit reflects uplift and erosion to the north, related to up-to-the-north displacement along the Lily, Moodies, Sheba, Barbrook and/or Saddleback Faults, or uplift up-to-the-south and erosion along the Inyoka Fault. MdQ3 and MdQ3cgl form a prominent hill, Elephant's Head (1638 m), just south and above Barberton (Fig. 6.10). It is not clear whether the base of this conglomerate is unconformable.

Possibly correlative chert-cobble conglomerates only ca. 5–10 m thick occur in the Stolzburg Syncline and the Moodies Hills Block but not in the Eureka Syncline where MdQ3 is dominated by large-scale, in places gravelly cross-bedded quartzarenite in prograding-delta facies (Anhaeusser 1976).

The overlying unit (MdS3) consists of sandstones, siltstones and shales that are well exposed along the Eureka City Road track on both limbs of the Eureka Syncline. There, they include two thick regressive cycles representing prograding delta complexes (Fig. 6.11).

In the Moodies Hills Block and the Stolzburg Syncline, stratigraphic equivalents of MdS3 are generally finer-grained and more distal. They consist largely of rhythmically laminated and soft-sedimentary deformed siltstones in prodelta facies that are dark green when fresh and weather orange; they reach 600 m thick in the northern Moodies Hills Block. Of all Moodies units, MdS3 may best record a NE-to-SW depositional trend: lithology, palaeocurrent orientation, subunit thicknesses and facies patterns suggest alluvial fan

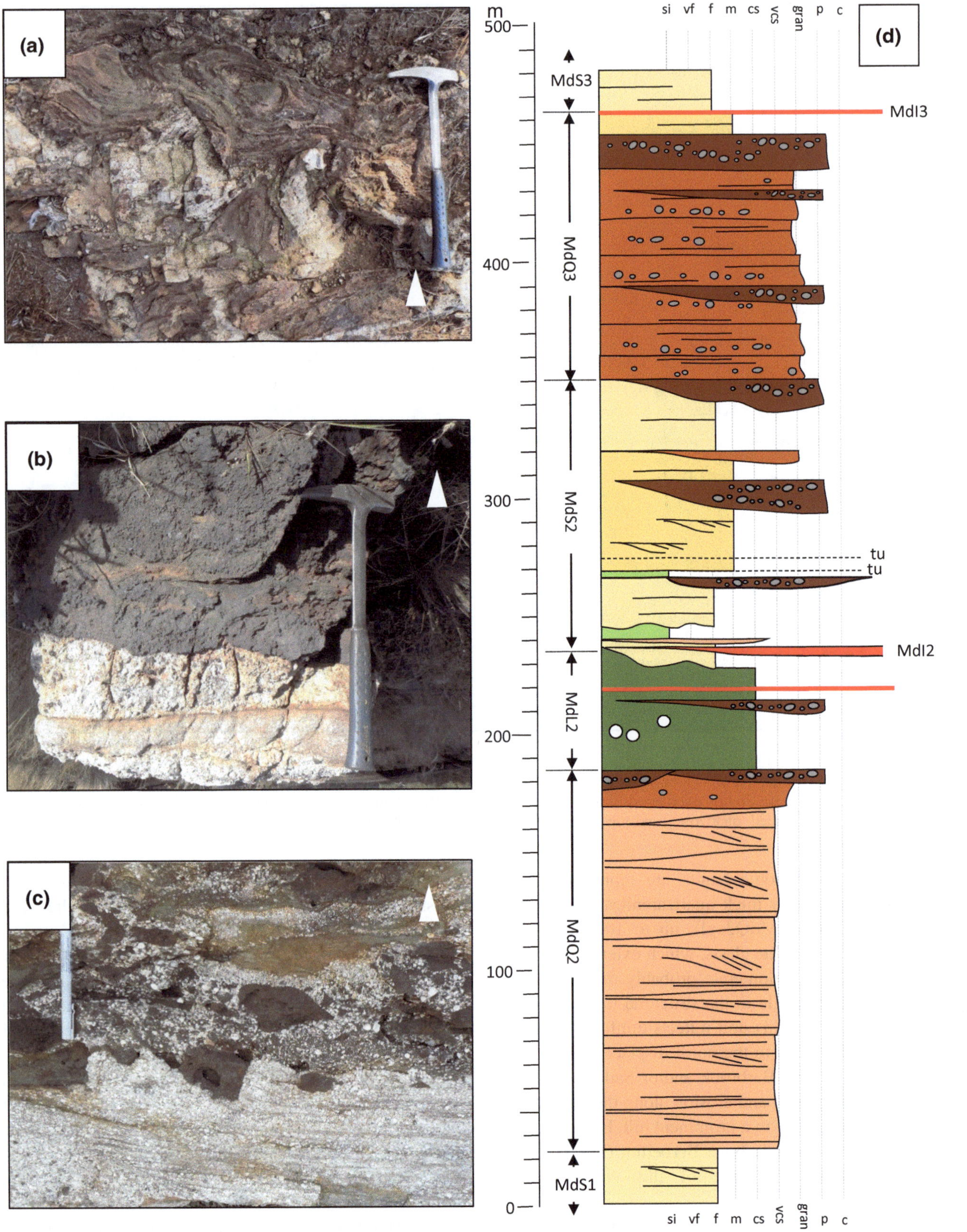

Fig. 6.8 Geological setting of the Moodies Basaltic Lava Complex (MdL2), a coastal or terrestrial basaltic flow complex. Photographs from outcrops in the Saddleback Syncline (**a**) and from Eswatini (**b**, **c**). **a** Amygdaloidal texture of the lava. **b** Sharp contact of lava to underlying, thermally affected volcanic litharenites. **c** Reworked angular lava clasts in low-angle, cross-bedded, coarse-grained sandstones; possibly a shoreline deposit. **d** Generalized stratigraphic column through the Moodies Lava Complex and over- and underlying sedimentary units in the northwestern Saddleback Syncline (Engelhardt 2012)

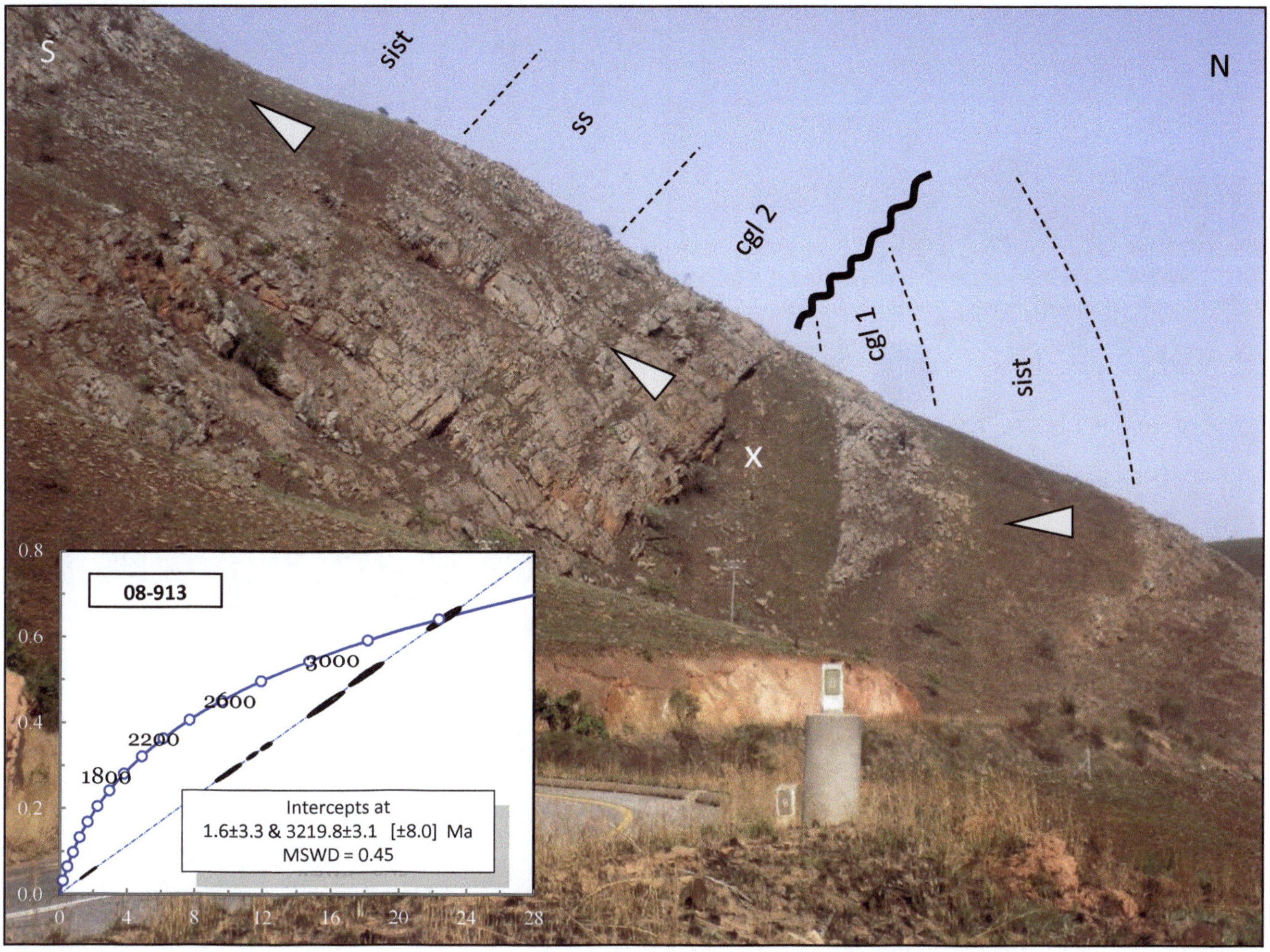

Fig. 6.9 View westward along the north slope of Saddleback Hill near Saddleback Pass. Moodies strata on the northern limb of Saddleback Syncline dip moderately to the south. An alluvial-facies conglomerate wedge (cgl 1) and overlying fine-grained strata are unconformably overlain and erosionally truncated by an orange-weathering cobble conglomerate (cgl 2). The angular unconformity dies out to the south. Field relationships, clast composition, paleocurrents and zircon dating of an air-fall tuff (sample 08-913; Heubeck and Lowe 1994a; Heubeck et al. 2013; inset) at this site demonstrate up-to-the north uplift of the Moodies basin margin, concurrent with erosion and southward progradation of alluvial fans into a subsiding, deforming basin. White "X" marks sample location for dating

sedimentation in the northeastern BGB near the Three Sisters Syncline, grading into deltaic complexes in the Eureka Syncline and prodelta slope facies in the Moodies Hills Block and the Stolzburg Syncline.

The uppermost units of the Moodies Group are cross-bedded gravelly sandstones in the cores of the Saddleback and Eureka Synclines (MdQ4); in the former, a thin, poorly developed BIF/jaspilite along with some ferruginous magnetic shales (MdI4) also occurs.

The stratigraphic top of the Moodies Group is nowhere preserved. Younger units of the basal Kaapvaal craton cover sequence unconformably overlie Onverwacht and Fig Tree units, but nowhere overlie Moodies strata.

6.2.2 Moodies Stratigraphy, Alteration and Age South of the Inyoka Fault

Detailed geological mapping in the Moodies Group south of the Inyoka Fault is in its early stage. Sandstones assigned to the Moodies Group south of the Inyoka Fault superficially resemble Moodies strata north of that fault because they are both dominated by a medium-grained quartzose composition and shallow-water sedimentary structures. They differ, however, strongly in petrographic detail, stratigraphic order, thickness, degree of alteration and possibly metamorphic grade. Tight folding is apparent but geologic mapping is currently insufficient to construct meaningful regional

Fig. 6.10 Field photograph of Elephant's Head, looking west. Southward-facing conglomerates overlain by quartzose sandstones of MdQ3 and a thick cobble conglomerate (MdQ3 cgl) form the resistant hill, prominently located about 2 km south of Barberton. White triangles indicate stratigraphic-up direction. The existence of an angular unconformity at the base of the conglomerate cannot be confirmed due to the steep topography

geologic profiles or stratigraphic columns. The maximum stratigraphic thickness may occur in the Masenjane Range where a homoclinal sequence of steeply southward-dipping quartzose sandstones reaches approximately 1600 m thickness.

Stratigraphy and Facies. Lamb and Paris (1988) erected a lithostratigraphic scheme in Eswatini which was semi-independent from the stratigraphy across the border in South Africa: A thick basal stack of mafic and ultramafic lavas (the Onverwacht Group), overlain by heterogeneous siltstones, shales, BIF, sandstones and conglomerates (the Diepgezet Group), in turn overlain by coarse clastic metasediments (the Intermediate Group) and quartz-rich sandstones and conglomerates (the Malolotsha Group). Because type localities are not identified, this scheme cannot be easily reconciled with the SACS (1980) system even though, for all practical purposes, the Diepgezet and Malolotsha Groups can be considered equivalents to the Fig Tree and the Moodies Groups, respectively. The stratigraphic position of the Intermediate Group is uncertain.

Lamb (1987), based on structural mapping in the Eswatini part of the BGB, mapped progressive unconformities at the base of the Moodies (his Malolotsha) Group. Reconnaissance mapping by the writer and M. Schmitz (2016), however, could not confirm this relationship.

In the southwestern BGB of Eswatini, quartzose and inward-facing Malolotsha/Moodies sandstones in the Malolotsha Syncline are interleaved with several bedding-plane-parallel talcose units several tens of m thick (Fig. 6.12). It is uncertain whether this is a structural or a stratigraphic relationship (Hunter and Jones 1968, 1969; Urie 1970, 1971; Lamb 1984b; our unpublished mapping). This sequence is regionally overlain, parallel to bedding, by a several-hundred-m-thick unit of highly altered and talcose

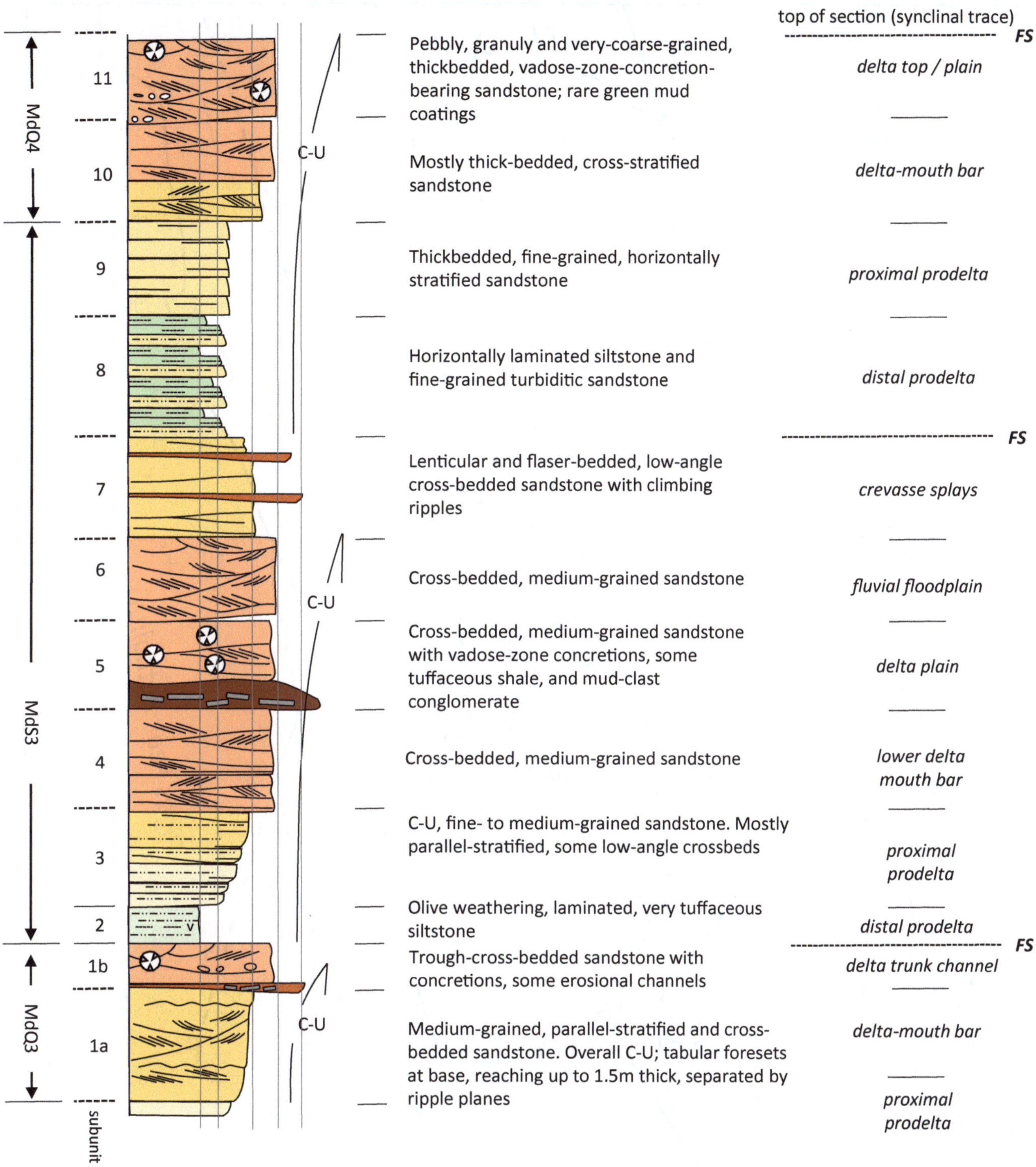

Fig. 6.11 Stratigraphic column along the Eureka City Road (a rough track), Eureka Syncline. This succession, comprising two complete prograding deltaic complexes, is exposed in identical facies and in similar thicknesses on both limbs of the syncline. The overall thickness of the column shown is 800 m on the western (normal) limb, 900 m on the eastern (overturned) synclinal limb. Topmost unit MdQ4 is the Bickenhall Member of Anhaeusser (1976). *C-U* coarsening-upward trend, *FS* flooding surface

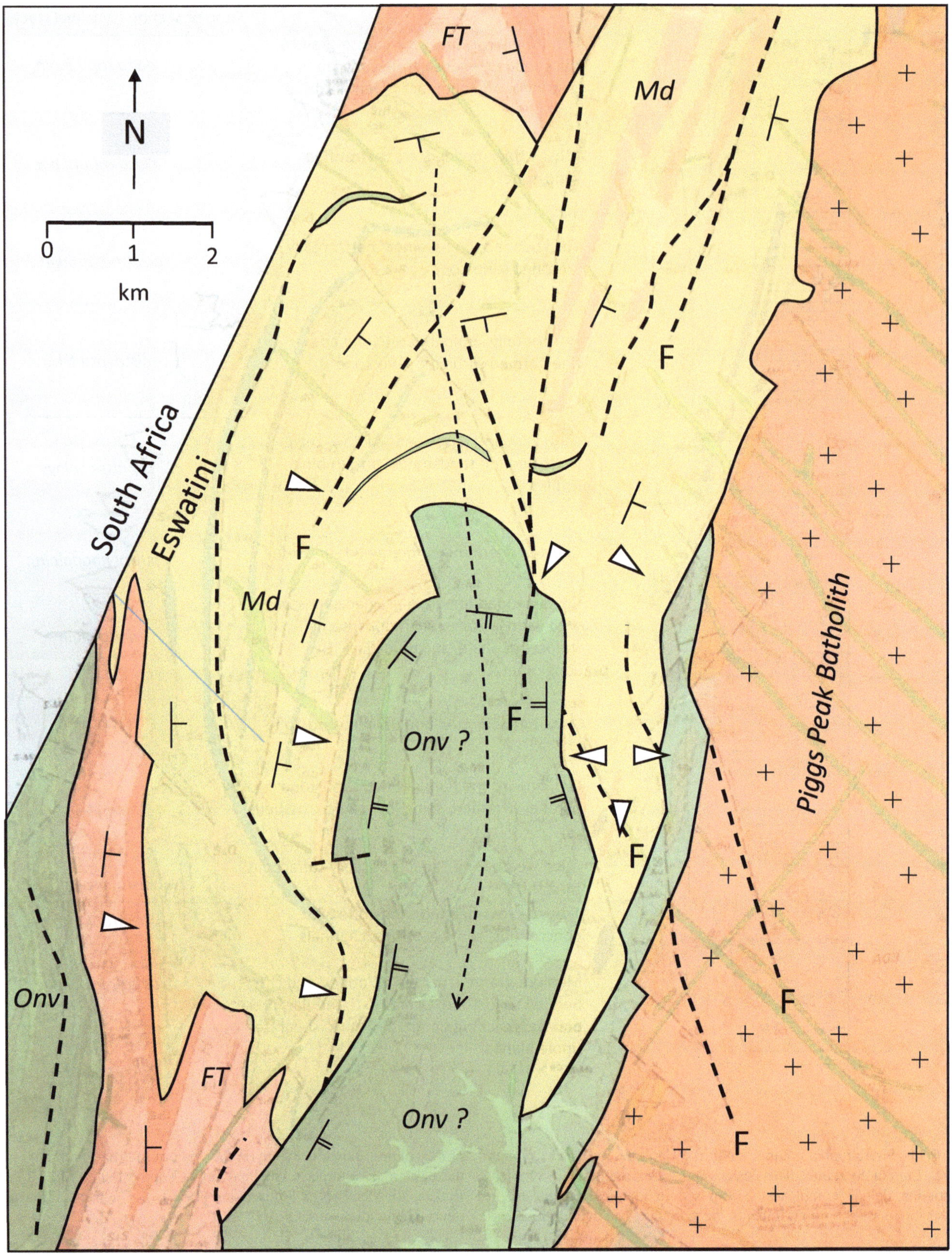

Fig. 6.12 Simplified geologic map of the Malolotsha Syncline, Eswatini, modified from Urie (1970). Location and colour code in Fig. 6.1. Upward-facing Moodies/Malolotsha sandstones are overlain by talcose komatiites with few chert beds in the core of the southward-plunging Malolotsha Syncline. *Onv* Onverwacht group, *FT* Fig Tree group, *Md* Moodies group, *F* fault

komatiites with minor chert and ignimbritic flows which occupies the core of that syncline (Fig. 6.12; Urie 1970). Lamb (1984b) interpreted this relationship as a tectonic nappe of altered Onverwacht lavas above a bedding-parallel base which had been overthrust for >10 km towards the west prior to regional-scale folding. If, however, the contact at the base of this komatiitic unit was stratigraphic, Moodies quartzose sandstones would be overlain by an unnamed younger komatiitic unit.

6.3 Petrography and Provenance

General. Moodies provenance carries special interest because it may represent (except for a sandstone unit at the base of the ca. 3.4 Ga Strelley Pool Formation; Wacey et al. 2006) the first-time access of surface systems to voluminous quartz-rich felsic igneous rocks. This resulted in the production of large quantities of coarse-grained monocrystalline quartz and K-feldspar grains. Provenance can be constrained by palaeocurrent and facies analysis, regional geologic mapping, conglomerate clast composition, sandstone petrography, shale geochemistry, and zircon age spectra but is incompletely understood.

Conglomerate composition. Eriksson (1978), Reimer et al. (1985), Hose (1990), Heubeck and Lowe (1999) and Hessler and Lowe (2006) quantitatively analysed Moodies conglomerate clast compositions. Overall, the clast population is dominated by several varieties of chert (ca. 60%) and feldspar porphyry (ca. 30%), and by sandstone/quartzite clasts (ca. 6%). Altered and silicified green ultramafic rocks, silicified lapilli tuff and jaspilite make up conspicuous but volumetrically minor contributions. Virtually all clast types can be matched to intra-BGB units. Together with the laterally rapidly changing clast proportions and unit thicknesses, the data strongly suggest the dominance of local sources. Except for the few granitoid plutonic clasts from the Ezzy's Pass locality of the Moodies basal conglomerate, no extrabasinal clasts are known with certainty; no TTG clasts from adjacent plutons have been documented in Moodies conglomerates. The clast composition may thus represent a mixture of those greenstone belt rock types most resistant to intensive chemical weathering, long-distance transport and abrasive reworking (quartzites, chert), mixed with a locally derived clast population eroded from underlying felsic igneous rocks and presumably having overlain the 3.23-Ga TTG plutons adjacent to the BGB (feldspar porphyries, very few granites).

Sandstone petrofacies and provenance. The petrographic composition of Moodies sandstones is diverse and includes quartzarenite, litharenite and arkose (Fig. 6.13). A systematic petrographic study (Heubeck and Lowe 1999), principally based on samples from the Saddleback Syncline, distinguished two petrofacies north of the Inyoka Fault: (1) The Oosterbeek petrofacies, making up the lower two-thirds of the Moodies Group, characterized by a high proportion of monocrystalline quartz, common K-feldspar, a low proportion of unstable lithic grains and chert, and a high percentage of framework grains (Fig. 6.14). This petrofacies corresponds stratigraphically to the upward-fining and -deepening sequence between the basal conglomerate (MdbC) and the top of MdQ2. (2) The overlying Elephant's Head petrofacies has a low abundance of monocrystalline quartz, a variable but high proportion of chert, a near-absence of feldspar, and a low abundance of unstable lithic grains. This petrofacies corresponds to the overall upward-coarsening trend, syndepositional deformation and possibly a southwestward progradation of units in the approx. upper one-third of the Moodies Group in the Saddleback Syncline.

Hessler and Lowe (2006) quantified the estimates of Heubeck and Lowe (1999) by averaging sandstone composition values from literature and accounting for different degrees of weathering resistance. They concluded that the inferred sources of Moodies sandstones consisted of approximately 54% (granitic) plutonic rock, 30% felsic volcanic rock, 11% chert, and less than 1% each of sedimentary, ultramafic-mafic volcanic, and other volcanic rock. Hessler and Lowe (2006) may have somewhat overstated the granitic contribution because they assigned all monocrystalline quartz without clearly recognizable volcanic features to a plutonic origin; they may also have overestimated the Moodies mean feldspar content of 9–15%. Future work may be able to better differentiate stratigraphically and regionally individual source rock contributions. However, surface exposures of true granites are not required to produce Moodies sandstones: Abundant large monocrystalline quartz grains can be liberated from the readily weathering and widespread but microcline-free quartz-feldspar porphyries of the Schoongezicht Formation. Microcline, in turn, is common in the quartz-poor dacites of the upper Hooggenoeg formation (H6 unit) and occurs in the tonalite-trondhjemite-granodiorite (TTG) plutons adjacent to the BGB (Robb et al. 1986; Kohler 1994; Moyen et al. 2007). It is, however, uncertain to what degree, if at all, the TTGs had been exposed at the surface at Moodies time. Overall, the degree of contribution of plutonic rocks, including TTGs and true granites, to Moodies sandstones is poorly constrained because plutonic clasts may have easily disintegrated in a chemically aggressive Archaean weathering regime. The few conglomerate clasts of granite, known only from the Ezzy's Pass outcrop, suggest that plutonic contributions were generally minor. One could also argue that some Ezzy's Pass "granites" actually represented a particularly coarse-grained and densely packed variety of feldspar porphyry, not true plutonic rocks. Nearby true granites, i.e. the Dalmein (3203 ± 7 Ma) and Salisbury Kop (3109 + 10/−8 Ma) plutons, are slightly younger than

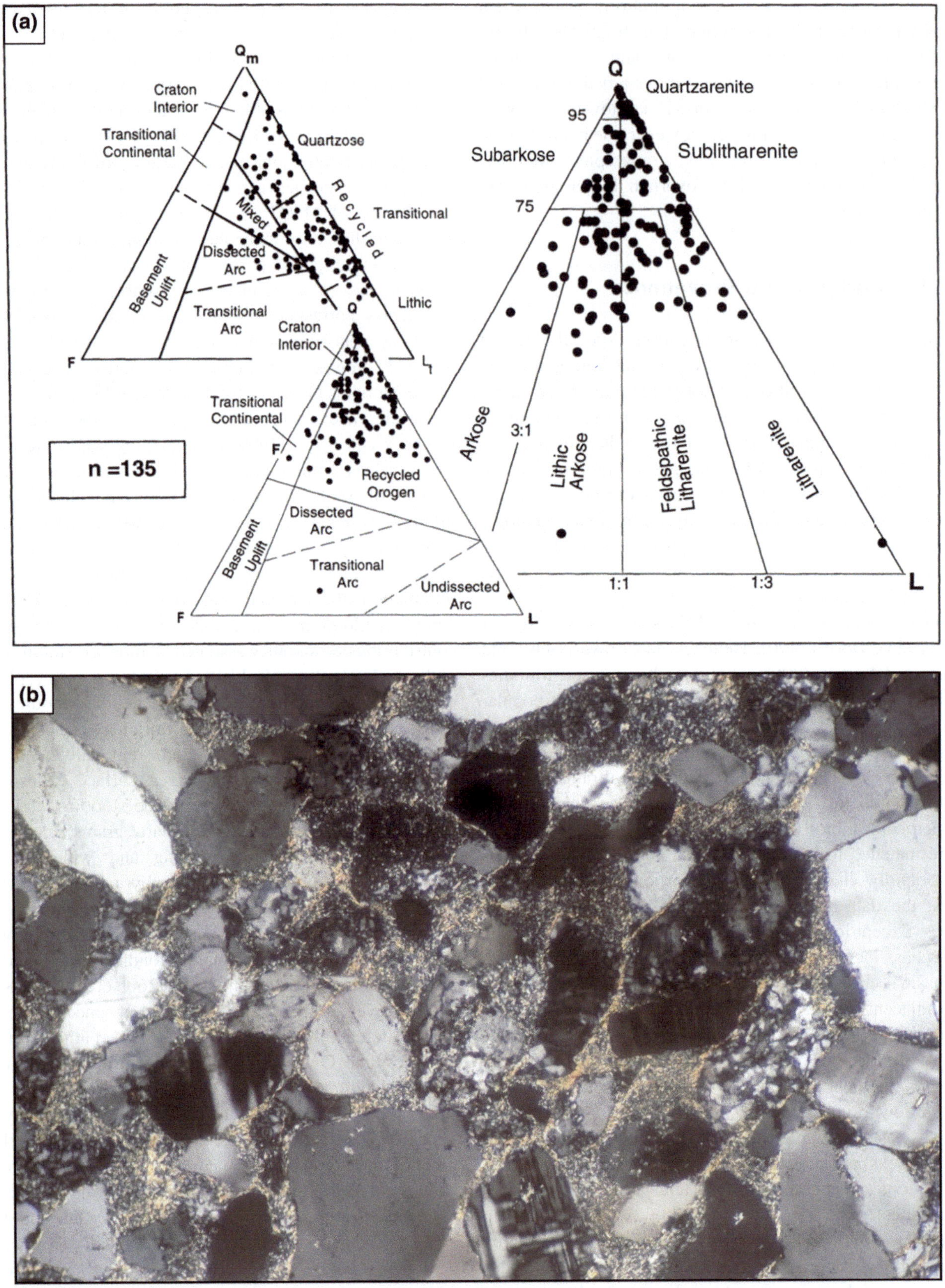

Fig. 6.13 Compositional features of Moodies Group sandstones. **a** Ternary sandstone classification (Folk 1965) and compositional diagrams (QFL, Q_mFL_t) of Moodies Group sandstones in the Saddleback Syncline (Heubeck and Lowe 1999). Sandstones are generally quartzose but span a wide range of compositions. Most plot in the "recycled orogen" field of Dickinson et al. (1983) because of their high chert content. **b** Thin-section photomicrograph from unit MdQ1, Saddleback Syncline. Note fine-grained interstitial quartzose-sericitic (pseudo-)matrix and well-preserved microcline grains

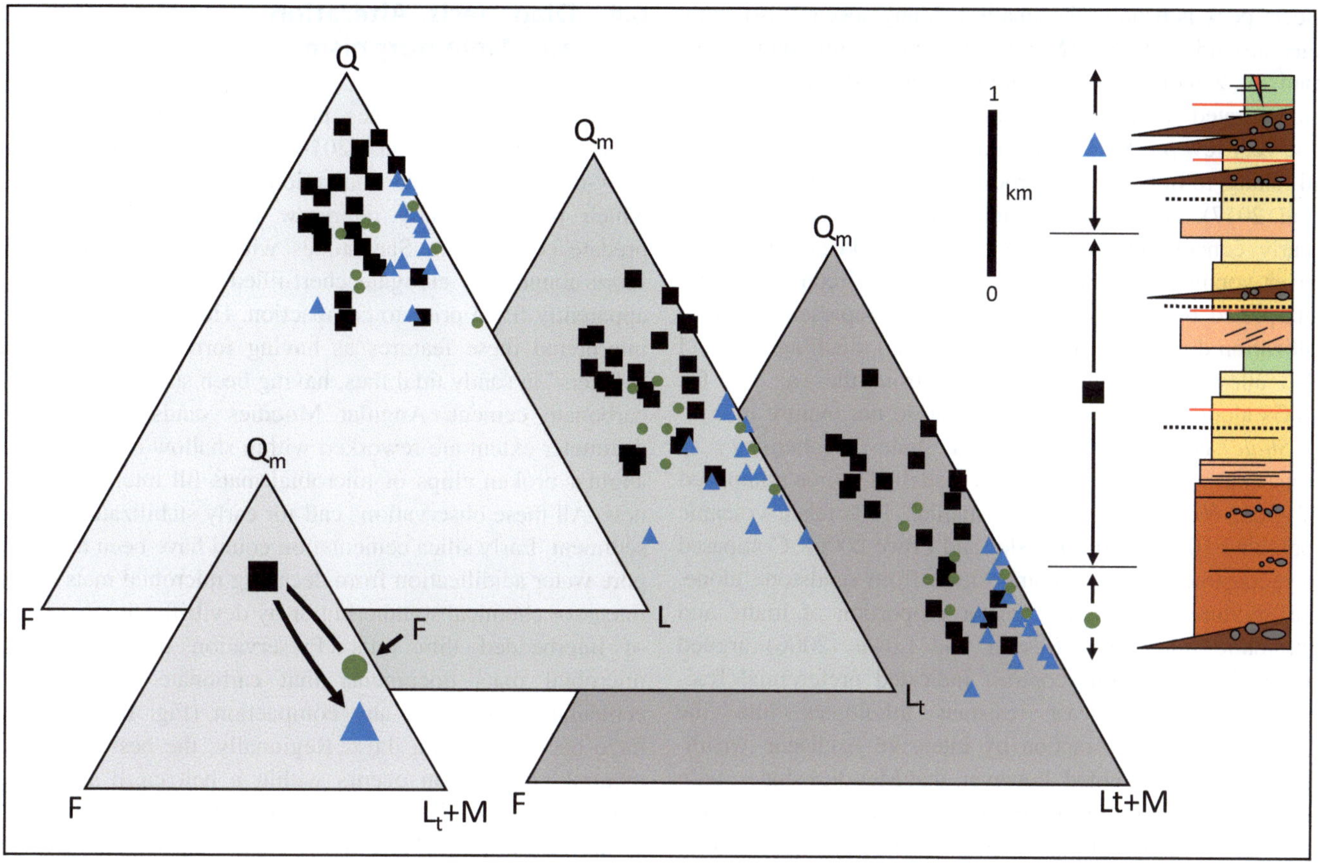

Fig. 6.14 Petrographic trends in Moodies sandstones with stratigraphic height in the eastern Saddleback Syncline, shown in three ternary diagrams and a generalized evolution diagram (lower left; modified from Heubeck and Lowe 1999). A trend of increasing quartz and decreasing L, L_t, L_t + M ("maturation trend") is reversed approximately midsection, above the Moodies basaltic lava

Moodies depositional age and thus cannot have contributed. Whether there may have been other granites that are completely eroded is speculative.

Absence of metamorphic grains. A pervasive characteristic of Moodies sandstones, heavy mineral laminae and conglomerates is the absence of minerals, grains, and clasts, respectively, of metamorphic provenance. This absence virtually excludes the Ancient Gneiss Complex (AGC; Jackson et al. 1987; Kröner et al. 2014; van Schijndel et al. 2017), a complex group of 3.66–3.20 Ga predominantly granitoid and supracrustal metamorphic rocks in central Eswatini, from being a source terrane. An outlier of the AGC is preserved adjacent to the BGB, separated from it by the Phophonyane Shear Zone (Kröner et al. 1989; Kröner 2006; Schoene et al. 2008) and a poorly exposed fault. The oldest polydeformed banded and grey gneisses of the AGC at a roadcut near the Phophonyane waterfall locality northeast of Piggs Peak reach up to 3663 ± 0.5 Ma in age (Schoene et al. 2008; Kröner et al. 2014) and are thus significantly older than the most basal BGB strata of the Sandspruit/Theespruit formations (Kröner et al. 2016) which do not exceed ca. 3550 Ma in zircon crystallization age. If the AGC had been a major supplier of clastic sediment in Moodies time, it should have found its expression in proximal-to-distal facies distribution and wedge-shaped thickness pattern, conglomerate clast composition, sandstone petrography and a significant percentage of detrital zircons >3.58 Ga in age.

Zircon provenance. Age analysis of detrital zircons from sandstones and conglomerates of the Moodies Group is an active field of investigation (Zeh et al. 2013; Heubeck et al. 2013; Leisgang 2017; Drabon et al. 2017). Age spectra largely show a tripartite pattern (e.g., see Zeh et al. 2013) with peaks around 3530 Ma (corresponding to felsic magmatism at the base of the Onverwacht Group; Kröner et al. 2016), 3440 Ma (corresponding to felsic magmatism in the upper Hooggenoeg Formation, also variably referred to as the Noisy Formation, Noisy Complex or Buck Ridge Volcanic Complex; de Vries et al. 2010), and 3260–3220 Ma (corresponding to magmatism at Auber Villiers and Schoongezicht time; see Sanchez-Garrido et al. 2011 for a summary). The magnitude of these peaks is variable, but the oldest peak is usually the smallest. Many spectra also show ages around 3300 Ma (Mendon Formation ultramafic magmatism). Zircons >3570 Ma are very rare and may have been incorporated as xenocrysts into younger magmatic rocks. The very few pre-Kaapvaal-aged zircons as old as 3.8 Ga are all related to impact-related rocks (Lowe et al. 2014; Drabon et al. 2017). The youngest coherent cluster of concordant or nearly concordant zircons in detrital Moodies samples is never younger than 3215 ± 5 Ma; these zircon ages were interpreted by Heubeck et al. (2013) as representing crystallization during syn-Moodies volcanism. Such ages would also allow a plutonic contribution from the ca. 3.25-Ga TTGs along the northern margin but do not require it.

Shale geochemistry. Moodies shale geochemistry is consistent with derivation from a modelled source composed of about 60% tonalite, 12% komatiite, 18% felsic volcanic rock, and 10% granite (Hessler and Lowe 2006). Compared to reconstructed source composition from sandstone alone, this mixture represents a greater proportion of mafic and ultramafic lithologies. Hessler and Lowe (2006) argued convincingly that this contrast indicated preferential fractionation of the least resistant lithologies into the fine-grained clastic fraction by intensive Archaean weathering. It should be noted, however, that Moodies shales, with few exceptions, do not crop out well, and that Hessler and Lowe (2006) had to use shale samples near mines and from outcrop. Other samples cited in the literature (McLennan et al. 1983; Toulkeridis et al. 2015; Smits and Mezger 2017) were either taken from discarded drill core from mines or directly from underground mine exposures. Their stratigraphic position is doubtful and hydrothermal overprinting cannot be excluded.

Petrography south of the Inyoka Fault. Moodies petrography and provenance south of the Inyoka Fault has not been systematically studied. However, all sandstones appear to lack feldspar. Heubeck and Lowe (1999) defined these feldspar-free strata as the Angle Station Petrofacies. The strong petrographic contrast to Moodies sandstones north of the Inyoka Fault, where detrital feldspar reaches up to 22% in MdQ1 sandstones, may be explained as follows: (1) No feldspar-bearing rocks were available for erosion south of the Inyoka Fault and deposition occurred in separate basins; (2) detrital feldspar was lost due to intensive chemical weathering in the source region, during highly abrasive eolian transport or at the place of deposition; or (3) feldspar was removed south of the Inyoka Fault during hydrothermal alteration and/or due to a higher grade of metamorphism but preserved by selective very early silicification north of the Inyoka Fault. Further field, structural and petrographic studies are called for to evaluate these hypotheses.

6.4 Diagenesis, Alteration and Metamorphism

Many Moodies sandstones preserve delicate sedimentary structures. Nabhan et al. (2016a) documented widespread early-diagenetic sulfate concretions, now largely silicified, which poikilotopically overgrew uncompacted sand and predate compaction. Sandstones with microbial mats preserve abundantly elongate chert-filled cavities which were apparently filled prior to compaction. Homann et al. (2016) interpreted these features as having formed as fluid-filled "blisters" in sandy tidal flats, having been stabilized by early carbonate cement. Angular Moodies sandstone slabs of decimeter-extent are reworked within shallow channels, and angular broken chips of microbial mats fill intertidal channels. All these observations call for early stabilization of the sediment. Early silica cementation could have been aided by pore water acidification from decaying microbial mats under intensive chemical weathering or by devitrification reactions of interbedded thin tuffs. Preservation of the delicate microbial mats documents that carbonate and/or silica cementation preceded any compaction (Fig. 6.5); it may have occurred within days. Regionally, the best degree of textural preservation occurs within a belt ca. 1 km wide north of (and stratigraphically above) the Lomati River Sill in the central part of the BGB, suggesting a possible contribution by rising hydrothermal fluids above the cooling sill. Lapilli tuffs higher up in the section and elsewhere, however, also preserve near-spherical lapilli; they attest to early-diagenetic cementation reactions.

Preservation may have followed other, as yet poorly understood, pathways because some sandstones, in particular MdQ2, apparently free of microbial mats, volcanic contributions and distant from potential hydrothermal cells also show very low degrees of compaction. This suggests early cementation but pore spaces could alternatively have been kept open by pore space overpressure instead. Widespread early cementation is also suggested by a laterally extensive, stratiform breccia of sandstone slabs in sand matrix in the Masenjane Range (Fig. 6.15). Whether brecciation occurred due to seismic vibration, subsurface dissolution, or wave action (on "beach rock") is as yet unknown.

Most cement in Moodies sandstone and conglomerate consists of a mix of microcrystalline silica, carbonate and fine-grained sericite. This cement grades into a sericite-rich matrix, particularly in the basal Moodies Group; in compositionally more mature sandstones, clusters of intergrown fine-grained quartz and sericite of average grain size are interpreted best as altered metastable sand grains. Based in part on comparisons with weathered and altered Moodies

Fig. 6.15 Cross-sectional view of a stratiform sandstone-in-sandstone slab breccia, Masenjane Range, central Barberton Greenstone Belt. Stratigraphic top is down. This unit of about 2–4 m thickness can be followed over ca. 5 km along strike. It may represent an early-diagenetically, surficially cemented sandstone unit in a terrestrial environment which was subsequently broken up by subrosion, wave action or seismic shaking

conglomerate clasts, Heubeck and Lowe (1999) argued that this grain type likely resulted from the alteration of fine-grained felsic volcanic material by burial diagenesis (involving the conversion of volcanic smectite to illite) and hydrothermal alteration. Overall, Moodies sandstone diagenesis is complex and challenging (Fig. 6.16).

Nearly all Moodies strata dip subvertically because they have been deformed into tight synclines. Heubeck et al. (2013) argued that this deformation either occurred syndepositionally or shortly (within a few Ma) after deposition. Nevertheless, sandstones generally show virtually no strain, except at the tight hinges where shearing and faulting can be extensive. Some Moodies units may have escaped folding-related strain because they were lithified very early whereas other beds were incompletely lithified at the time of deformation and underwent a higher degree of deformation without showing solid-state strain. Hinge thickening is particularly widespread in fine-grained rocks, e.g., in siltstones and shales of the eastern Eureka and eastern Stolzburg Synclines, but can be observed in many other BGB units, including volcanics and cherts; it thus affected lithified rocks as well.

Alteration. Quartzose Moodies sandstones in the central BGB, and best exposed in the Masenjane Range just south of the Inyoka Fault, are not only altered and recrystallized due to the proximity to the Lomati River Sill but also by intrusion of a stockwork of peperitic, doleritic to basaltic subvolcanic rock (Hose 1990; Fugmann 2017). These lace through the Moodies strata over a stratigraphic thickness of ca. 800 m and at least 25 km strike length between Shiyalongubo Dam in the NE and Maid-of-the-Mists to the SW (Fig. 6.1). Thin section-, hand sample-, outcrop- and map-scale relationships show that subvolcanic magma

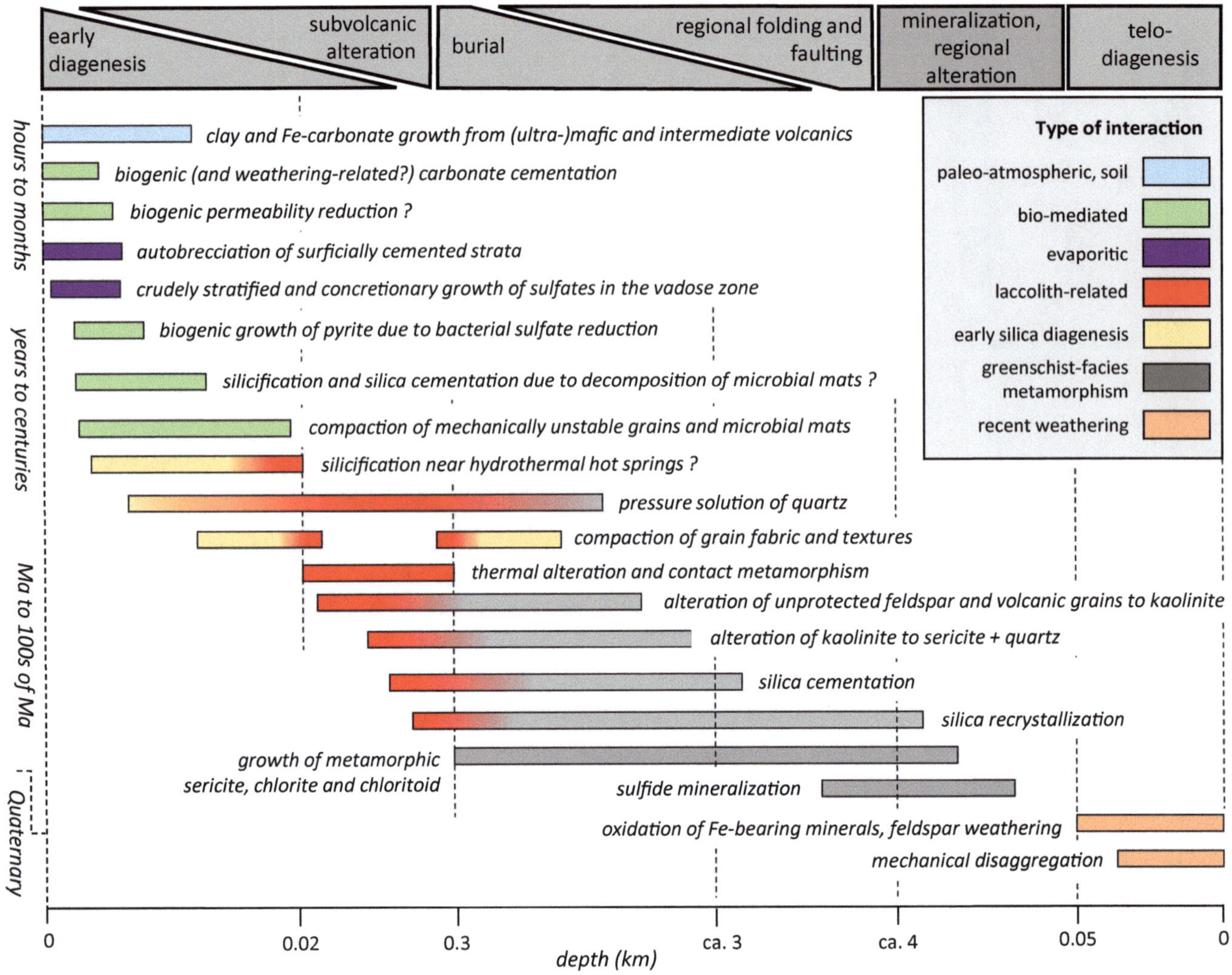

Fig. 6.16 Composite diagenetic sequence of Moodies Group sandstones

mingled intimately and extensively with unconsolidated Moodies sand and moderately cemented sandstones (Fugmann 2017; Fugmann et al. 2017; Fig. 6.17). In hand sample, trains of dispersed rounded quartz grains of clearly sedimentary origin are commonly incorporated into basaltic lava (Fig. 6.17e, f). In outcrop, basalt and dolerite dykes 1–100 m wide intrude Moodies sandstones either bedding-parallel or obliquely. Geologic mapping documents diffuse, multibranching dykes which interacted with readily deformable quartzose sandy Moodies sediments (Fig. 6.17). Locally abundant vesicles in fine-grained subvolcanic basalt suggest shallow intrusion depths. Dense networks of tapering, botryoidal, radiating or networked fractures and veins of mm- to cm-thickness suggest the involvement of steam. Commonly, outcrops of intermingled sandstone and lava are surrounded by halos several to tens of m thick in which silicified sandstone is stained intensely red or where staining extends along randomly-oriented joints. In such areas, depending on the degree of (sub)volcanic overprint, bedding and sedimentary structures are widely obscured and in places completely obliterated by recrystallization and random jointing (Fig. 6.17a, b).

Generally, many Moodies sandstones south of the Inyoka Fault show few primary sedimentary structures, appear thick-bedded or massive in outcrop, and show sutured grain contacts and polygonized monocrystalline-quartz grains in thin section.

Metamorphism and palaeotemperatures. Raman measurements on organic matter from the Moodies Group suggest peak metamorphic temperatures between 330 and 520 °C (Nabhan et al. 2017). The processes determining the maximum palaeotemperature and its variation in map pattern are not yet understood. They probably result from the varying superposition of burial metamorphism, local hydrothermal halos, and shearing along major fault zones.

Siltstones and thin metapelites of the Malolotsha Group (the Moodies-Group equivalent in Eswatini) contain abundant metamorphic chloritoid, whereas similar fine-grained

Fig. 6.17 Field aspects of subvolcanic stockwork intrusions into Moodies sands and semiconsolidated sandstones. All photographs from the Masenjane Range (Fig. 6.1): **a** rubble facies; **b** random-oriented quartz veins in silicified quartzites; **c** volcanic mafic dykes (red) in silicified quartzite; **d** radial dykes, possibly steam escape from pipe; **e** semi-digested, white, rounded sandstone "clasts" in red, weathered basaltic matrix; **f** individual sand grains (bottom) entrained in fine-grained basaltic lava (top). The subvolcanic stockwork extends between locations 1 and 2 (Fig. 6.1) along ca. 25 km strike length

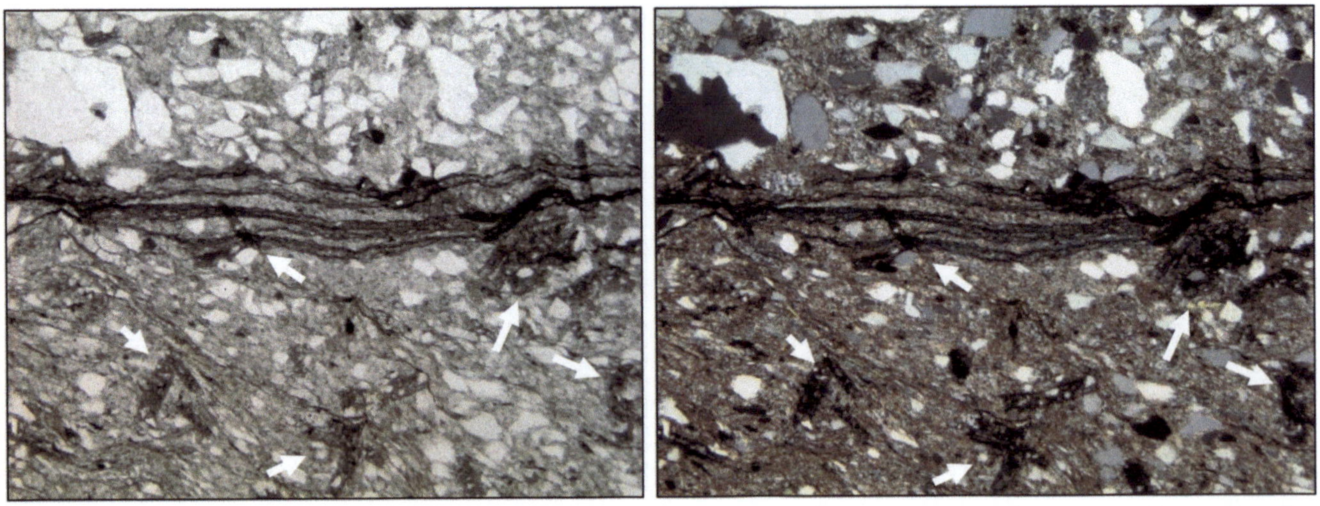

Fig. 6.18 Plane- (left) and cross-polarized (right) thin-section photomicrographs of Moodies sandstones of the Malolotsha Syncline (Eswatini) showing radial clusters of high-relief chloritoid (marked by white arrows) overgrowing poorly sorted, matrix-rich sandstone with shale laminations

Moodies strata north of the Inyoka Fault show newly grown sericite and chlorite only. The presence of chloritoid is rather non-specific. It indicates temperatures between 220 °C in Fe-rich pelites, which marks the chloritoid-in reaction in Fe-rich metapelites (the temperature is somewhat higher in "normal" metapelites), and 550 °C, which represents the chloritoid-out reaction (Ganguly 1969; Bucher and Frey 1994; Fig. 6.18). In general, the degree of recrystallization in Moodies/Malolotsha Group sandstones appears higher in the most southern outcrops than north of the Inyoka Fault.

Alteration is most intense near the shear zones, in particular where they splay near restraining bends. There, quartz grains are polygonized and fine-grained sericitic matrix is schistose. In outcrop, beds are permeated with chlorite-coated slickensides so that the rock takes on an overall yellow and grey appearance. A prime example of this process is seen in the open pits of the abandoned Svengali prospect in the eastern Saddleback Syncline.

6.5 Depositional Systems and Stratigraphic Architecture

The combination of lithology, diagnostic sedimentary structures and sand body geometry suggests that Moodies Group strata north of the Inyoka Fault include, fluvial, deltaic, tidal, shallow-marine, and possibly eolian settings (Eriksson 1978, 1979, 1980; Heubeck and Lowe 1994a, 1999; Simpson et al. 2012; Homann et al. 2015; Heubeck et al. 2016a; Nabhan et al. 2016a).

Alluvial facies, characterized by thick, commonly massive conglomerate in wide, shallow channels with erosional basal contacts is rare in the Moodies Group (but see Eriksson 1978; Kohler and Anhaeusser 2002). The best outcrops of such facies, indicating ephemeral subaerial transport of coarse clastic debris, are in unit MdQ3 in the Saddleback Syncline (Fig. 6.1). Thick alluvial conglomerates may also exist in the Three Sisters Syncline of the northeastern BGB, but geological mapping there is insufficient to allow a detailed assessment. A progressive inward-facing unconformity, overlain by a ca. 20 m-thick wedge-shaped conglomerate at Saddleback Pass and dated at 3219 ± 3 Ma (Heubeck et al. 2013), has been taken as the prime example of syndepositional uplift of the Moodies basin margins and erosion of Kaap Valley Tonalite cover strata (Heubeck and Lowe 1994a) (Fig. 6.19).

Fluvial facies is common. It shows thick- to thin-bedded, weakly to moderately erosive and in places gravelly sandstones with mudchips and abundant planar- and cross-stratification (Fig. 6.19). Evaporitic concretions, which presumably formed in sandy coastal plains (Nabhan et al. 2016a), are widespread. They exist throughout the stratigraphic column in nearly all structural belts of the Moodies Group in South Africa as well as in Eswatini.

T*idal facies* is widespread and represented by thick, medium-grained to gravelly sandstones (Eriksson 1977; Heubeck and Lowe 1994a; Eriksson et al. 2006; Heubeck 2009; Homann et al. 2015; Heubeck et al. 2016b) with bidirectional palaeocurrent patterns in cross-bedded sandstones, evidence of intermittent subaerial exposure, and rhythmic bedding. This facies hosts abundant and extensive microbial mats. Mat-bearing units reach >1100 m in stratigraphic thickness in the overturned limb of the Saddleback Syncline. The normal (Heubeck et al. 2016b) and overturned limbs of the Dycedale Syncline (Eriksson 1977; Eriksson et al. 2006; Noffke et al. 2006) show the transition from evaporitic-and-biomat-bearing fluvial and coastal strata to estuarine subtidal facies. Microbial-mat sandstones with

Fig. 6.19 Outcrop photographs of sedimentary structures common in the fluvial facies of the Moodies Group. **a** Single-clast-thick cobble trains in horizontally stratified sandstone; **b** isolated cobble clusters; **c** shale-coated current-ripple trains; **d** mudcracked shale coating on sandstone

their characteristic sedimentary structures also occur in the Masenjane Range and the The Heights Syncline.

Deltaic facies. Stutenbecker et al. (2019) documented a well-developed prograding delta complex in unit MdS1 of the Saddleback Syncline, characterized by a wedge-shaped geometry of a sandstone unit extending over ca. 7 km along strike, erosional trough-cross-bedded, coarse-grained chertarenites and several regressive surfaces represented by extensive shale-draped and mudcracked rippled sandstones and shale-clast conglomerates. In addition, parts of MdQ2 in the Saddleback Syncline and MdQ3 in the Eureka Syncline may represents stacked delta mouth bars; they are locally erosional into underlying and laterally equivalent horizontally stratified shoreline-facies sandstones. Prodelta facies are represented by monotonously laminated siltstones which exceed several 100 m in thickness in all Moodies tectonic blocks of the central BGB north of the Inyoka Fault; they are, however, generally poorly exposed. In the Eureka Syncline, slumped sandstone and "rafts" of soft-sediment-deformed sandstone beds, interbedded with thin-bedded, normally-graded T_{ab} sandy-silty turbidites, were evidently deposited in a prodelta setting (Fig. 6.20). Abundant mm-thick, distal, fine-grained turbidites in prodelta-facies dominate MdS1 in the Stolzburg Syncline and possibly also the Saddleback Syncline.

Eolian facies. Moodies sandstones south of the Inyoka Fault include at least one, possibly several, ca. 100 m thick quartzose sandstone bodies with large foresets that Simpson et al. (2012) interpreted as the world's oldest eolian dunes (Fig. 6.21). Field evidence, however, is ambiguous, as the occurrence of large foresets alone is not diagnostic. Thin sections of sandstones from the suspected units usually show highly compacted quartzarenites. They show poor rounding and lack the diagnostic small-scale inverse grading typical of eolian deposits.

The deepest-water and most distal deposition of Moodies sediments is represented by laminated shaly siltstone and interbedded jaspilitic chert (Fig. 6.20). Some of these occur

Fig. 6.20 Deep-water prodelta facies of unit MdS1 in the Clutha Creek section of Eureka Syncline. White arrows indicate stratigraphic-up direction: **a** normally graded siltstone-shale couplets, interpreted as T_{de} distal turbidites; **b** polished stream plate showing turbiditic section; unit in the center of outcrop between white dashed lines is a massive debris flow about 1.2 m thick; **c** slump-folded jaspilite unit encased in siltstones; **d** imbricated (top) and extended (middle) semi-lithified white chert (weathered and bleached jaspilite) in thin-bedded turbidites. Adjustments associated with compaction could have produced similar geometries; **e** "frozen" Rayleigh-Taylor instability (center) defined by density difference between siltstone (light-colored) and shale (dark) and amplified by mud diapir; **f** raft of laminated siltstone-shale couplets in thin-bedded turbidites. Both terminations of the raft are angular and at right angle to bedding. The raft is approximately 0.25 m thick by 2.5 m long

Fig. 6.21 Large-scale foresets in quartz-rich sandstones of the Moodies Group, Masenjane section, Makhonjwa Block. White triangles indicate stratigraphic-up direction. **a** Bedsets in preserved thickness of 1–3 m. **b** Close-up of thin-bedded foresets in coarse-grained sandstone. This sandstone lacks diagnostic eolian features at the macro-, meso- and microscale. It may represent a sandy channel fill of wide, braided-fluvial channels or shallow-marine, tide-related sand bars

basinwide in all four tectonic blocks north of the Inyoka Fault and serve as excellent correlation markers (Fig. 6.2). MdI1 and MdI2 are best exposed (Bontognali et al. 2013); MdI2 also occurs south of the Inyoka Fault. MdI3 and MdI4 are known only from the Saddleback Syncline. The jaspilites and shales clearly represent deposition below wavebase. Over- and underlying prodelta deposits suggests that water depths were at least several tens of m.

Stratigraphic architecture. It is unclear to what degree Moodies strata represent one or several coherent depositional system(s). North of the Inyoka Fault, units in the Stolzburg Syncline, Moodies Hills Block, Saddleback Syncline, Dycedale Syncline and Eureka Syncline can be reasonably well correlated (Fig. 6.2). The correlability of Moodies units MdbC, MdQ1, MdI1, MdS1, MdQ2, MdL2, and MdI2 north of the Inyoka Fault and the continuity of the Moodies basaltic lava and its topping jaspilite into northern Eswatini across strike (ca. 17 km) suggests that Moodies deposition initially occurred in a single basin although sediment sources were largely local. The difficulty of correlating stratigraphically higher units because of their variable lithologies and thicknesses, in contrast, indicate that the basin broke up following the outflow of the Moodies lavas and was tectonically segmented thereafter.

In the northeastern BGB, three major Moodies bodies north of the Inyoka Fault have not been stratigraphically correlated to Moodies units in the central BGB: (1) Moodies units in the narrow, elongate Lily Syncline are metamorphosed to albite-epidote-hornfels facies and considerably deformed (Anhaeusser 1963; Kohler and Anhaeusser 2002); in it, several hundred m of a basal conglomerate composed of chert, banded ironstone and quartzite clasts are overlain by impure shales and quartzites; (2) Moodies in the core of the Three Sisters Syncline includes thick conglomerates with a high proportion of quartzite clasts; and (3) little is known about the Big Buffalo Syncline (Kohler and Anhaeusser 2002). The stratigraphic knowledge south of the Inyoka Fault, in particular in Eswatini, is even poorer.

6.6 Age

The age of Moodies deposition is constrained by U–Pb zircon dates from volcanic beds near the top of the Schoongezicht Formation, by several thin dacitic air-fall tuffs interbedded with Moodies sandstones throughout the stratigraphic column, by a single cross-cutting dyke near the top of the stratigraphic sequence (Heubeck et al. 2013), and by a presumed feeder dyke to the Moodies Basaltic Lava (Fugmann 2017; Fig. 6.22). Stratigraphic ages from the base (3223.5 ± 0.3 Ma; Heubeck et al. 2013) and the top (3209 ± 14 Ma; unpublished sample 12–593 from MdS3 of the Eureka Syncline) of the Moodies Group are indistinguishable within analytical error, suggesting that Moodies strata were deposited within only a few million years. The age of a volcanic tuff immediately underlying a progressive angular unconformity in the middle Moodies Group at Saddleback Pass (3219 ± 3 Ma; Heubeck et al. 2013; Fig. 6.9) suggests that deformation by basin-shoulder uplift began at that time.

These depositional ages can be compared to those of nearby plutonic rocks. Moodies deposition north of the Inyoka Fault apparently began after crystallization of zircons in the adjacent plutons, including the Kaap Valley Tonalite

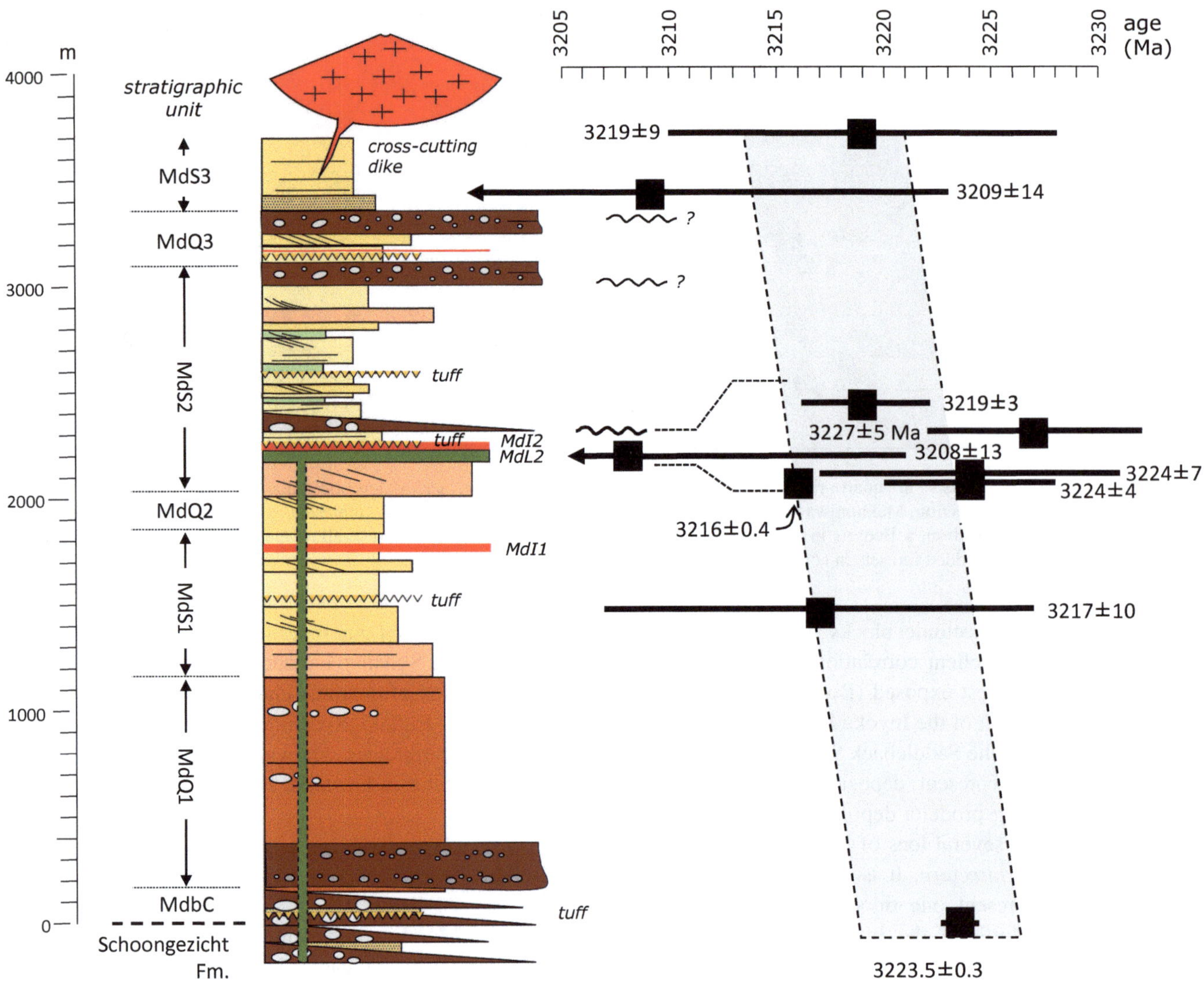

Fig. 6.22 Radiometric ages of dated units in the Moodies Group, including air-fall tuffs, water-reworked tuffites, and cross-cutting dikes. Moodies depositional ages at the base, within, and at the top of the 3.6 km thick sequence overlap within analytical error. For details of dating methods, see Heubeck et al. (2013). Moodies deposition probably started after 3223.5 ± 0.3 Ma. Strata began to be uplifted and eroded at mid-Moodies time, around 3219 ± 3 Ma. Deposition had ended by about 3209 ± 14 Ma, deformation had ended by 3219 ± 9 Ma. A plausible mean depositional rate, indicated by the dashed area, is about 1 mm/a. All of Moodies strata may represent only a few million years. Diagram modified from Heubeck et al. (2013)

and the Badplaas, Nelshoogte, and Stentor plutons at ca. 3229 Ma (best fit of 11 U–Pb zircon ages, compiled by Moyen et al. 2007; Moyen et al. 2019). The volcanic rocks of the Schoongezicht Formation, underlying the Moodies Group north of the Inyoka Fault and as young as 3225 ± 3 Ma (Byerly et al. 1996), are considered cogenetic with the KVT magma and contemporaneous with its crystallization (Kröner et al. 1991; van Kranendonk et al. 2009). Because all Moodies strata were clearly deformed by the KVT (Fig. 6.1), even the youngest strata (> 3209 ± 14 Ma; see above) must have been deposited prior to the KVT acting as a rigid body. This occurred at the latest when it cooled through the Ar–Ar isotherm (ca. 500 °C) at 3214 ± 4 Ma (Layer et al. 1992, 1996). The 3109 + 10/−8 Ma (Kamo and Davis 1994) and 3070 ± 16 Ma (Heubeck et al. 1993) ages of the Salisbury Kop pluton in the northeastern BGB, which crosscuts and thus postdates the Moodies fold train, do not provide tight constraints. Thus, KVT crystallization and cooling between ca. 3229 and 3214 ± 4 Ma, perhaps also uplift, appear to have been contemporaneous with Moodies deposition.

Moodies ages south of the Inyoka Fault generally resemble those from north of that fault: Onverwacht rocks in the southernmost Kromberg Syncline, folded in the same regional NE-SW trend as Moodies synclines, are truncated by the post-tectonic 3216 + 2/−1 Ma Dalmein Pluton (Kamo and Davis 1994); recently, this pluton has yielded a slightly younger age of 3203 ± 7 Ma (Lana et al. 2010). A felsic dyke in the The Heights Syncline, intrusive into basal Moodies strata, was dated at 3207 ± 2 Ma, providing a minimum age of sedimentation at the base of the Moodies Group (Heubeck and Lowe 1994a). A thin bed of water-reworked tuff (sample 15–103 of Heubeck; LA-ICP-MS; $n = 91$) from the Masenjane Range, about 1 km above the base of the Moodies, yielded a complex spread of zircon ages. Its youngest coherent cluster yielded a Concordia intercept age of 3208 ± 13 Ma ($n = 14$) and a similar, concordant age of 3206 ± 11 Ma ($n = 6$). Drabon et al. (2017) reported detrital zircon ages from four samples of Moodies sandstones south of the Inyoka Fault within South Africa (samples JAH-663-13, JAH-664-2, NAD-128, NAD-129). Except for sample JAH-664-2, which may have been hydrothermally altered, all samples show their youngest concordant zircon clusters near 3240 Ma. This is consistent with the age spectra of the two detrital Moodies samples from Eswatini: (1) Wiechert (2014; sample 13-342) dated a reworked metatuffite within thick quartzarenites from the Malolotsha Syncline at 3271 ± 6 Ma ($n = 16$); its youngest single grain was 3251 ± 9 Ma. (2) In Moodies basal conglomerates and sandstones of the Ntaba Mhlope Syncline west of Piggs Peak (Fig. 6.1), the majority of zircon grains is 3273 ± 20 Ma old, the youngest grain is 3232 ± 36 Ma (Leisgang 2017). Overall, with few exceptions, the youngest Moodies detrital zircons south of the Inyoka Fault appear to be ca. 20–30 Ma older than the youngest grains from populations north of the Inyoka Fault. This can be interpreted as reflecting an older depositional age (i.e., Moodies deposition south of the Inyoka Fault ended at ca. 3240 Ma) or as a lack of access to zircon-bearing units <3240 Ma in age. Because Moodies strata under- and overlying the Moodies lava and its overlying jaspilite MdI2 can be correlated across the Inyoka Fault into Eswatini and thus make a diachronous age of sedimentation unlikely, the latter interpretation appears to be correct. The postulate that basal fluvial-alluvial Moodies conglomerates in the south are correlative with submarine lobes of Fig Tree sandstones (Jackson et al. 1987; Eriksson et al. 1994) is thus incorrect on stratigraphic grounds. The hypothesis probably originated at least in part due to the erroneous assignment of thick chert-clast conglomerates at the South African—Eswatini border (Emlembe Mountain) to the Moodies Group rather than to the lower Fig Tree Group (Anhaeusser et al. 1981; Heubeck and Lowe 1994a; Harrington 2017; cp. Lowe et al. 2012). Likewise, there are currently no strong stratigraphic nor geochronologic arguments supporting the hypothesis that “Moodies Group-like sediments” developed at different times (de Wit 1991).

Several sills of mafic to intermediate (noritic-granodioritic) composition, up to 11 km long and up to 500 m thick, intruded the volcano-sedimentary sequence of the BGB during Moodies time: The Lomati River Sill intruded into the Schoongezicht Formation just north of the Inyoka Fault at 3211 ± 18 Ma (Fugmann et al. 2017); an unnamed and undated sill several km long of the same composition likely intruded into Moodies sandstones in the Masenjane Range at the same time; and the Sterkspruit Sill intruded at ca. 3228 Ma into Onverwacht rocks north of the Stolzburg Syncline (A. Hofmann, pers. comm.). Zircon ages of ca. 3180 ± 25 Ma from the Lomati River Sill are likely due to hydrothermal alteration. A sample from a mafic to intermediate peperitic stockwork which appears to represent feeder dykes to the Moodies lava yielded an age of 3208 ± 13 Ma (Fugmann 2017).

A synopsis of the available data (Fig. 6.22) suggests that Moodies deposition began around 3224 ± 1 Ma. Strata were intruded by sills around 3219 Ma; the Moodies lava erupted at about the same time. Strata were tilted, folded and uplifted by 3219 ± 3 Ma while sedimentation was still ongoing. By 3209 ± 14 Ma, even the stratigraphically highest Moodies strata had been tilted into the subvertical.

6.7 Structure and Tectonics; Relationship to Plutons

Moodies strata are preserved in a series of regional synclinal folds that extend SW-NE throughout the BGB and give the mountain range its dominant structural fabric. This structure is well known only in the central belt where it consists of large (5–15 km long, several km wide), tight to nearly isoclinal and northwestward-overturned synclines (Figs. 6.1 and 6.23). Overturned southeastern limbs are commonly much thicker than the upward-facing northwestern limbs. Fold plunges are moderate (ca. 35°) to overturned (ca. 100°) and commonly bidirectional; they are sketched in cross-section as “mushroom” and “canoe” folds (Reimer 1971; Heubeck and Lowe 1994b; Lowe et al. 1999) and were interpreted as being due to regional crossfolding (Herget 1966) or the diapiric rise of granitic plutons (e.g., Anhaeusser 1984).

Ongoing regional structural work by M. Schmitz (unpublished) shows that many fold axis plunges do not significantly shallow along the strike of the fold axial plane; thus, Moodies synclines appear to represent large sheath folds which implies that greenstone belt rocks had or still have a significant depth extent (Fig. 6.24).

The geometry of widespread anticline-syncline regional fold trains in the south-central part of the belt (Heubeck and Lowe 1994b; Lowe et al. 1999) may suggest a (late?) subhorizontal component of shortening above a relatively shallow subhorizontal detachment. Based in part on this geometry, Heubeck and Lowe (1994b) estimated a syn- and post-Moodies shortening of the cross section to 68% or less of its original length.

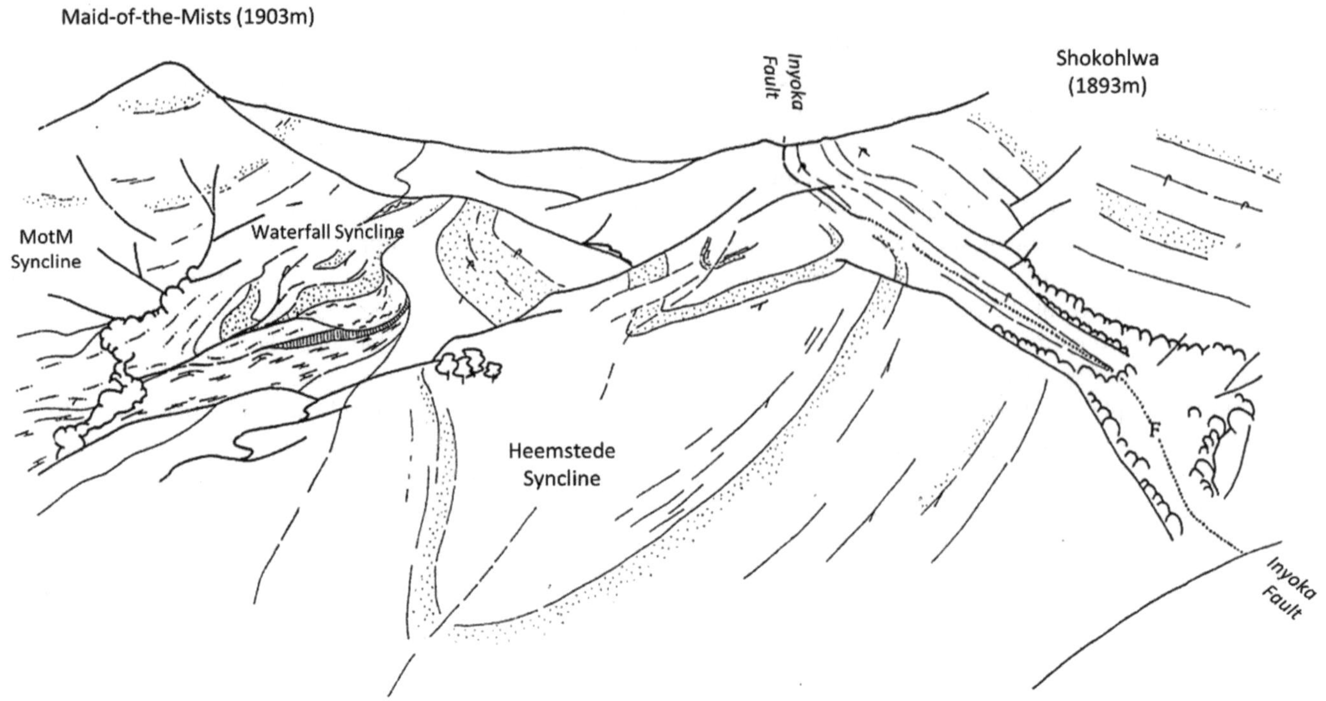

Fig. 6.23 Line drawing of fold style in the central BGB just south of the Inyoka Fault; view is westward along strike from a lookout location at lat -25.872193 °S, long 31.059536 °E. See Heubeck (2016) for a full discussion. Within the field of view, four synclines in the Moodies Group strata are juxtaposed, separated by subvertical faults. Fold axial plunge is between 67° and 30°, away from the viewer

Folds defined by Moodies strata partially wrap around the adjacent plutons (Kaap Valley Tonalite, Stentor Pluton, Nelshoogte Pluton), all of which crystallized in Moodies time. The greenstone-belt-wide fold train is only crosscut by the Salisbury Kop Pluton in the far northeastern BGB and the Dalmein Pluton in the southwest.

Ductile strain is very low to nearly absent in many units on the limbs of Moodies synclines, suggesting early lithification. Much postdepositional deformation associated with folding was likely accommodated along bedding planes, in ductilely deforming shales and siltstones, and in those units that were not silica-cemented early. In map view, these units show major hinge thickening and, where exposed, ptygmatic folding (Fig. 6.24). Ductile finite strain can be regionally quantified from Moodies conglomerate clasts (Anhaeusser 1963, 1969; Reimer 1967; Gay 1969; Lamb 1984a; Daneel 1987; Heubeck and Lowe 1994b; Fig. 6.25). Heubeck and Lowe (1994b) compiled all previous data and found that ductile strain mostly concentrated along the northern and southern greenstone belt margins and along the BGB-axial Inyoka Fault. Commonly, strained conglomerates are LS-tectonites: their clasts are moderately flattened perpendicular to the bedding plane and strongly elongated along subvertically plunging axes (Fig. 6.25). Among the clast types, black chert and shale show higher strain than feldspar porphyries whereas granitic and quartzitic clasts are least affected.

Building on this information, Heubeck and Lowe (1994b) and Lowe et al. (1999) derived a sequence of Moodies deformation events. They postulated initial syndepositional sagging, followed by substantial subhorizontal shortening and subsequent subvertical stretching, contemporaneous with refolding about the rising plutons. They interpreted these trends as representing an initially extensional setting of the greenstone belt in Schoongezicht and early Moodies time (ca 3230–3219 Ma; their D3 event) followed by syndepositional basin margin uplift and shortening in late Moodies time (D4; ca. 3219–3215 Ma).

6.8 Discussion

Moodies strata allow profound insights in three aspects of earth history: Weathering intensity and Archaean climate; tectonics and basin architecture; and depositional environments and the conditions of early life.

6.8.1 Weathering Intensity and Archaean Climate

Whether Moodies sediments can contribute to the discussion about the intensity of Archaean weathering depends in part on the degree to which such information has been preserved

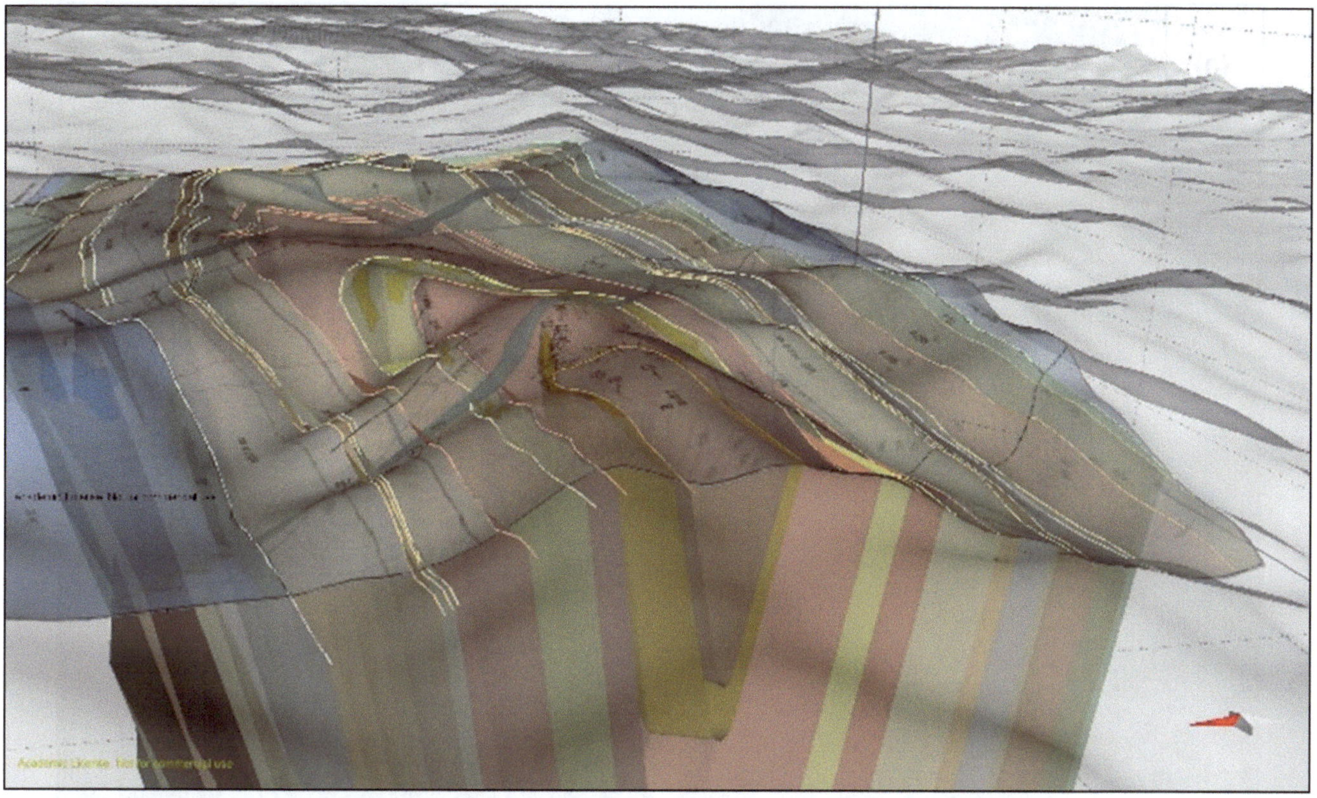

Fig. 6.24 Perspective view of the closure of the eastern Stolzburg Syncline (Fig. 6.1), looking to the SE; visualization by software 2-D Move. Fold axial plunges and bedding plane dips in the hinge region remain steep, usually ca. 50–80°, along the trace of the fold axis, suggesting that the concept of "canoe synclines" (Reimer 1971) is incorrect. Rather, Moodies synclines resemble large sheath folds; they may have (had) significant depth

or erased by transport and redeposition. As a case in point, Sleep and Hessler (2006) inferred that Archaean climate at Moodies time "was not especially hot" because Moodies conglomerates and sandstones contain a high percentage of thermally metastable chert grains and hardly any silt- and clay-size fragments, expected from the disintegration of chert. This conclusion can be criticized from several perspectives. Lowe (2007), in a comment to Sleep and Hessler (2006), disputed their conclusion by pointing out the syntectonic deposition of Moodies strata which would have overwhelmed its climatic information, invalidating the use of chert clasts as a palaeothermometer. Secondly, the absence of thick siltstones and shales alleged by Sleep and Hessler (2006) is not borne out by the thick prodelta deposits of the Eureka, Moodies Hills and Stolzburg Synclines. Thirdly, the noted absence of etch pits and quartz microkarst on grain surfaces (which should be expected under high temperatures) would rarely be preserved following greenschist-facies regional metamorphism, burial to 10–15 km, and polyphase alteration. An argument supporting clement Archaean temperatures, however, is the abundance of (likely photosynthesizing) biomats, few of which would have grown well at ambient temperatures >50 °C. Although a vigorous photosynthesizing benthic biosphere in the tidal zone would also have contributed to intensive chemical weathering, this intensity would have been unrelated to elevated temperatures.

Hessler and Lowe (2006) compared a quantitatively reconstructed Moodies source composition mix to the geochemistry of Moodies shales. They concluded that the shale contained a higher percentage of weathering products from chemically sensitive rock types and interpreted the dichotomy as a consequence of an aggressive weathering environment, probably caused by a combination of increased rainfall, higher temperatures, and/or higher atmospheric pCO_2 compared to that of today. Such conditions are also supported by the near-absence of unsilicified basalts or komatiites in conglomerate clasts and of feldspar grains in unsilicified sandstones. Similar effects could be generated by sedimentary recycling but much Moodies sediment, including at least the feldspar grains, juvenile volcanics and volcaniclastic grains are clearly first-cycle, derived from nearby outcrops of igneous rocks. In addition, the zircon population is mostly dominated by peaks at 3.23 and 3.42 Ga, suggesting that grains were abundantly derived from juvenile volcanic and volcaniclastic material of Fig Tree and upper Hooggenoeg Fm. (upper Onverwacht Group) strata; recycling appears minor. However, Hessler and Lowe (2006)

Fig. 6.25 Strain patterns in Moodies conglomerate. **a** Outcrop photograph of a loose block of strained Moodies basal conglomerate (a LS-tectonite showing dominant clast stretching, subordinate clast flattening) of Ntaba Mhlope Syncline, Eswatini, ca. 5 km west of Piggs Peak. Clast lithologies deform differentially, as already observed by Anhaeusser (1963) and Gay (1969), among others, in studies from the northern BGB margin. The in situ orientation of the stretching lineation is consistently subvertical all around the syncline which would be difficult to arrange if the clast fabric had been primary; **b** sample from the same syncline, sawed along axes of principal strain

also pointed out that Moodies conglomerates include a mean 6% of quartzose sandstone clasts, and the writer's own field observations indicate that these become more common to the northeast. In addition, the presence of at least one documented intraformational unconformity (at Saddleback Pass; Heubeck and Lowe 1994a) suggests that intraformational recycling and sediment reworking may have been significant, as do the thick tidal deposits with their common bidirectional paleocurrents. Appreciable feldspar content occurs only in early-silicified Moodies sandstones and is significantly depleted in plagioclase, suggesting a high Archaean weathering intensity. This translates into high values of the Chemical Index of Alteration (CIA) of 85 and 80 for Moodies shales and sandstones, respectively, compared to a baseline CIA of 45 for the source rock and modern values of weathering of around 60 (Hessler and Lowe 2006). Hessler et al. (2004) analysed the composition of Archaean siderite weathering crusts from a limited number of Moodies conglomerate clasts in drill cores and calculated a minimum pCO_2 pressure several times higher than at present-day temperatures, and increasing with increasing temperature. The calculated pCO_2 values, however, would likely be pedospheric values which may be significantly different from atmospheric values.

6.8.2 Tectonics and Basin Architecture

Assessment of the tectonic setting of the Moodies Group is hampered by the incomplete state of geological mapping, in particular in Eswatini, and the imperfect degree of stratigraphic correlation and, consequently, the partial knowledge of depositional architecture.

A considerable number of tectonic settings, ranging from passive margin (Eriksson 1980), foreland basin (Jackson et al. 1987; Hose 1990; Eriksson et al. 1994; Heubeck and Lowe 1999), subduction-related (de Wit 1991), strike-slip (de Wit 1991; Heubeck and Lowe 1994b), piggy-back (Lamb 1987) and hybrid settings (Eriksson et al. 2006) have been proposed for Moodies deposition. Kisters et al. (2003, 2010) merged actualistic and non-actualistic elements by postulating an initial extensional phase, followed by collapse and sagging; Lana et al. (2011) also favoured a strongly extensional basin. Although compressional and extensional vectors certainly applied, it may be inappropriate to force Moodies basin architecture into actualistic plate-tectonic settings because deposition may have occurred in environments largely controlled by vertical rather than horizontal tectonics (Bédard 2006; van Kranendonk et al. 2004; van Kranendonk 2011; Bédard 2019; van Kranendonk et al. in press). The contemporaneity of the solid-state pluton uplift with Moodies deposition, combined with its provenance and depositional architecture, supports deposition in one or several elongate subsiding troughs between the rising plutons and relates poorly to modern plate-tectonic settings (Fig. 6.26).

Such sedimentary basins would be created by mass withdrawal at depth between more-or-less-vertically rising, in map view approximately circular, tonalitic and granitic plutons. Sediment accommodation space would likely be elongate and sinuous, possibly branching or triangular in map view and drop- or crease-shaped in cross-section. Sedimentary basins would largely be fed from large radial point sources and be dominated by elongate, possibly single-strand dispersal systems (Fig. 6.26). At falling or low sea level, such basins would tend to be terrestrial ("overfilled basins"), closed, internally drained and possibly show stratigraphic architectures dominated by erosion and bypass. At times of rising and high sea level, basins would tend to become integrated, forming large, possibly tidal-dominated estuaries and experience rapid progradation of sedimentary systems along the axes of the basins. A balance between subsidence and fill would result in vertical stacking of facies tracts. In all cases, basin margins would be characterized by thinning strata and progressive angular unconformities, similar to the geometries known from modern salt mini-basins (Banham and Mountney 2013). The creation rate of accommodation space, rate of subsidence and deformation would be highly dynamic and, depending on the rates of uplift, subsidence and the intensity of weathering, possibly fast. At times of waning plutonic rise, horizontal (extensional and compressional) components of deformation could become dominant (Fig. 6.27).

At the current state of knowledge, Moodies strata indeed appear show properties predicted by such a model:

(1) the cuspate-lobate geometry of the BGB in map view shows an evident predominance of vertical tectonics although linear components of shortening and possibly extension are also apparent (Fig. 6.1);
(2) syndeformational deposition of the Moodies Group, established by radiometric dating, followed shortly after the crystallization of the KVT and other plutons (Fig. 6.22);
(3) sediment transport and dispersal systems in the Moodies Group are complex (in part unidirectional and in part bidirectional). Some units show strong progradational trends, others vertical stacking;
(4) mean depositional rates were high. A plausible mean value of about 1 km/Ma over 3.7 km stratigraphic thickness, equivalent to $\sim$1 m/1000a or $\sim$1 mm/a is comparable to net sedimentation rates in many Holocene passive-margin and orogenic settings such as the Holocene Po plain, Elbe estuary, or Mississippi delta (Heubeck et al. 2013);

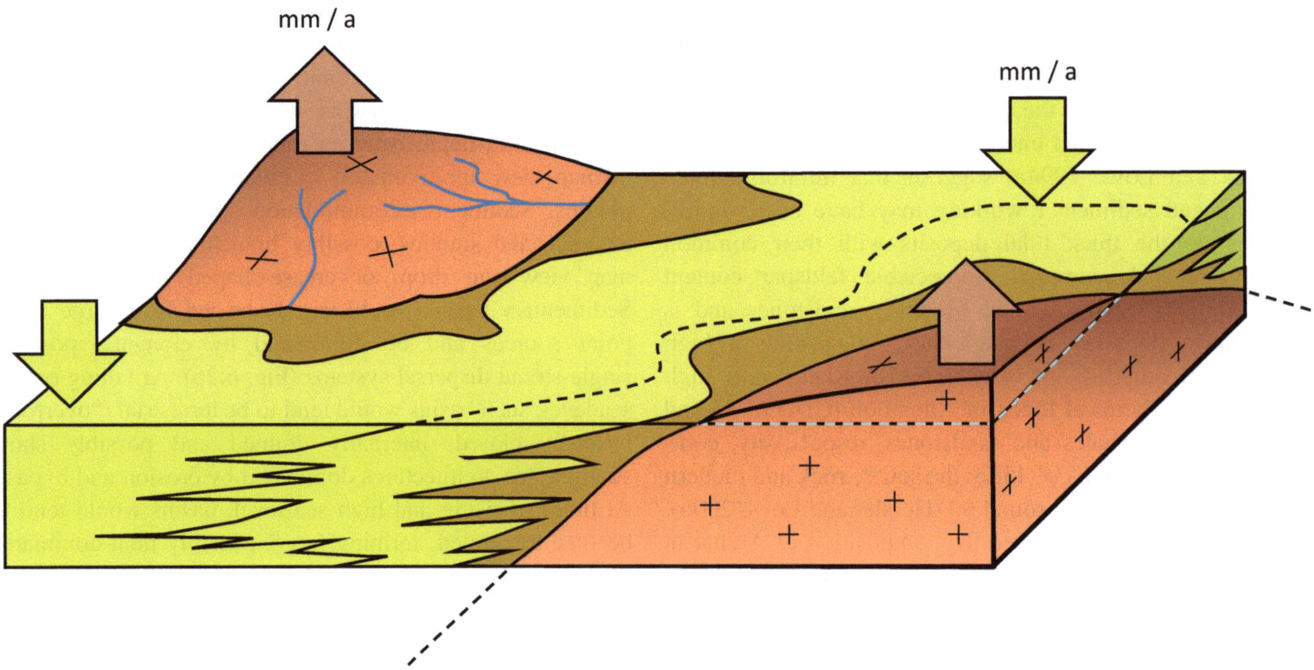

Fig. 6.26 Conceptual sketch of Moodies basin formation by sagging during contemporaneous rise, weathering and erosion of nearby plutons. Uplift and subsidence rates were roughly balanced at mm/a, equivalent to km/Ma

(5) The composition of Moodies clastic sediment at all scales (Hessler and Lowe 2006) suggests local derivation from nearby source areas (Figs. 6.13 and 6.14);
(6) The sheath-fold geometry of many Moodies synclines is indicative of major subvertical stretching (Figs. 6.24 and 6.25).

Historical geophysical data (summarized by de Beer et al. 1988; de Beer and Stettler 2009) suggest that the BGB is overall a shallow, not a deeply rooted orogen. Conventionally, this has been taken to support a fold-and-thrust-belt geometry (e.g., Heubeck and Lowe 1994b) but one could also argue that the former deep-reaching base may have been digested by younger plutons and is thus no longer apparent on geophysical data.

6.8.3 Depositional Environments and Conditions of Early Life

Moodies strata are the oldest sediments conducive for studying in detail the nature of terrestrial-to-marine transitions in a sedimentary context. Given the proper quantity and quality of samples and advanced methods of investigation, they allow to infer consequences of tidal ranges, radiation levels, degrees of desiccation, current strength, variations in sediment delivery rates, intensity of volcanism and processes of palaeosol formation as well as their influence on life across a range of adjacent depositional environments, rather than being limited to a single depositional setting. Few publications to date have focused on extracting these Archaean environmental parameters from the Moodies Group, aside from the detailed study on weathering intensity by Hessler and Lowe (2006). Eriksson and Simpson (2000) estimated Archaean lunar periodicity by measuring the variation in tidal-bundle thicknesses of a well-exposed subaqueous sand dune in the Eureka Syncline but other studies failed to replicate the postulated periodicity (Heubeck 2015). Several authors (Eriksson 1979; Heubeck et al. 2016a) reported that Moodies strata do not show much evidence of hyper- or even macrotidal conditions listed by Archer (2013). Homann et al. (2015, 2016) analyzed the micromorphology and depositional environment of the widespread microbial mats in the Saddleback Syncline and inferred that they possessed traits identical to those of tough and adaptable cyanobaterial mats in modern tidal settings. Nabhan et al. (2016b) documented evidence of Archaean microbial sulfate reduction in sulfate-rich fluvial palaeosols, and Homann et al. (2019) interpreted differences in C- and N-isotopes between microbial mats in bona-fide terrestrial facies and those in comparable subtidal sections to infer early metabolic adaptations of the biosphere to terrestrial settings. Simpson et al. (2012) identified Moodies eolian deposits in geographically and stratigraphically poorly resolved locations in the central BGB. They inferred that mechanical abrasion, rather than intensive chemical weathering, could have been responsible for part of the high mineralogical and textural maturity commonly observed in Archaean sandstones.

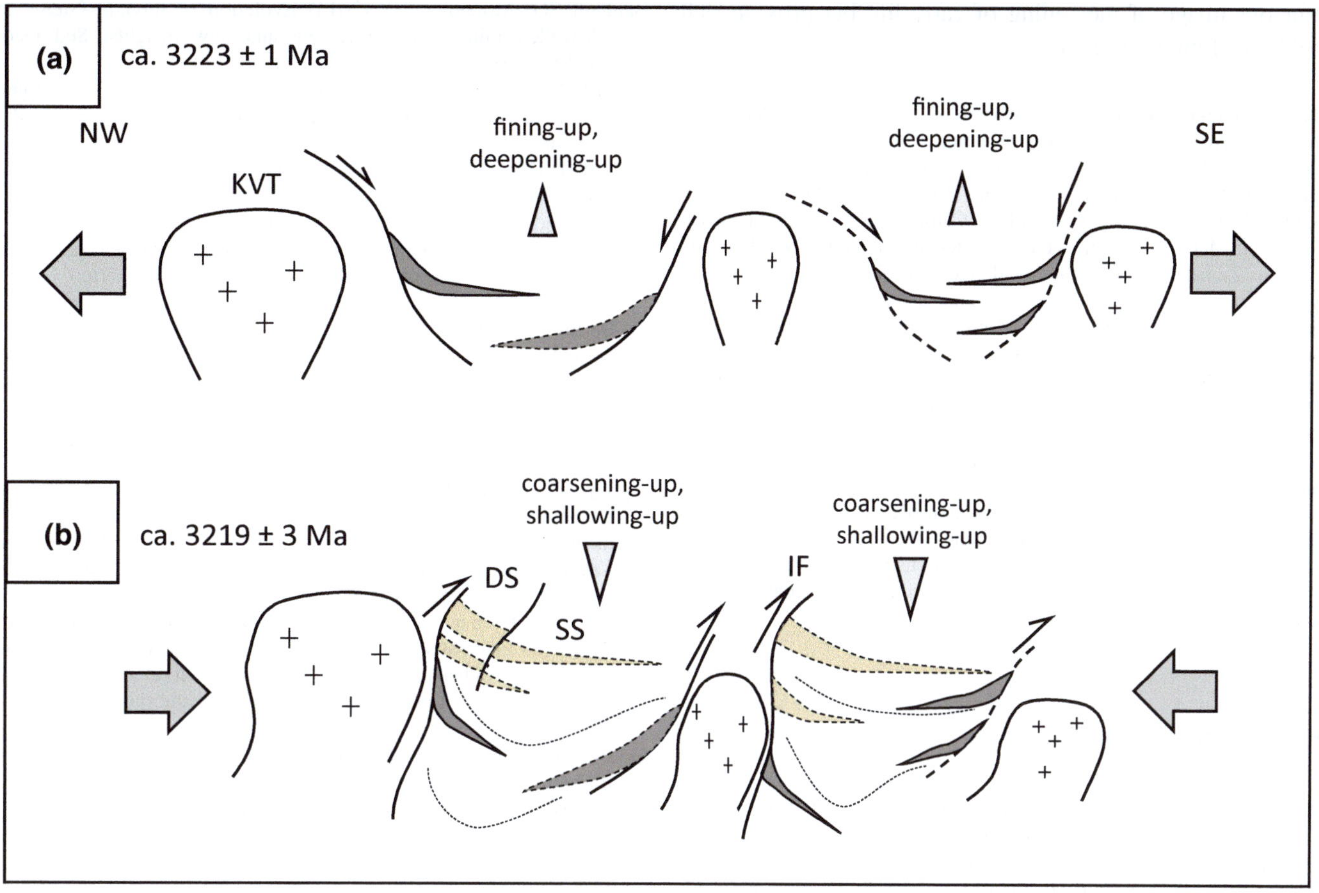

Fig. 6.27 Sketch (not to scale) emphasizing the horizontal component of Moodies basin formation. **a** Extensional or neutral setting dominated by regional extension and sagging between rising plutons. **b** Flip to regional compressional setting, converting former extensional faults into reverse faults, uplifting basin margins and steepening limbs of former "sag basins". *KVT* Kaap Valley Tonalite, *DS* Dycedale Syncline, *SS* Saddleback Syncline, *IF* Inyoka Fault

The abundance of microbial mats in the siliciclastic high-energy settings of unit MdQ1 in the Saddleback Syncline is, in some ways, surprising. Similar microbial mats are known only from four other locations in adjacent Moodies tectonic blocks, all within a few kilometers of the Lomati River Sill, and are not known from age- and facies-equivalent Pilbara units either. Facies and composition of the MdQ1 unit in the Saddleback Syncline differ little from those of the adjacent tectonic blocks; only its early diagenetic history, close to hydrothermal cells of the Lomati River Intrusive, sets this location apart. It appears that rapid burial in a subsiding basin and rapid reduction of permeability by several early-cementation pathways is necessary but perhaps still insufficient to ensure the preservation of microbial textures and materials. The restricted preservation of the microbial mats begs the question of whether the early biosphere had already widely colonized Archaean shorelines but its record obliterated everywhere except at those locations with rare and special circumstances (Schieber et al. 2007, Eriksson et al. 2013).

Although all of these studies made valuable inroads in understanding Archaean surface environments, we know as yet little or nothing about fundamental Archaean climatic parameters such as orbital parameters, the level, range and variability in temperature or precipitation, air density and composition, geochemistry of the oceans, and thus atmospheric and oceanic circulation and opacity. This affects numerous secondary parameters, such as the degree and variability of evaporation, the intensity of chemical weathering and the extent and depth of soil formation. Consequently, we know little about early diagenetic pathways and the chances of preservation of delicate features of the biosphere nor about many physical and biochemical characteristics of microbial mats, such as their oxygen production. Our estimates of continentality and topographic relief vary widely. The effects of meteorite impacts is an active but data-limited field of research. There is only one questionable data point on tidal range and lunar periodicity. We lack information about the ground effects and consequences of the high levels of solar UV radiation. All these inadequacies blur our image of the setting of early life but provide rich fields for future research.

Acknowledgements I thank my current and former students and postdocs Nora Bieker, Saskia Bläsing, Nadja Drabon, Jonathan Engelhardt, Paul Fugmann, Antonia Gamper, Marc Grund, Martin Homann, Jens Kirstein, Inga Köhler, Wigbert Lippold, Tim Luber, Sami Nabhan, Frank Ohnemüller, Caroline Rabethge, Sebastian Reimann, Matthias Schmitz, Laura Stutenbecker, Anne Wiechert and Andreas Zametzer for constant challenges and discussions in the field, lab and office. I thank the landowners and surface rights holders Sappi, Nico and Delia Oosthuizen from Mountainlands Nature Reserve, Barberton Mines, Taurus Estates, Nkomazi Wilderness, Swazi Timber, Malolotsha Nature Reserve and MTPA for access to their properties. Don Lowe and Gary Byerly took always time to critically test my findings and encourage new avenues of research. Duwies Duvenage from Barberton NTT Toyota provided mobility. Adriaan and Lili Nel, Sandy and Tony Ferrar, and Astrid Christianson are thanked for friendship and hospitality. Research was made possible by numerous grants by the DFG (He2418) and the EU COST Action "Life-ORIGINS" (TD1308). Detailed and constructive reviews by Carl Anhaeusser, Roger Buick and Don Lowe significantly improved the manuscript.

References

Agangi A, Hofmann A, Elburg MA (2019) A review of Palaeoarchaean felsic volcanism in the eastern Kaapvaal craton: linking plutonic and volcanic records. Geosci Front 9:667–688

Altermann W, Robinson T, Homann M, Lalonde S, Heubeck C (2017) First find of fossilized raindrop imprints in the Moodies Group (3.2 Ga), Barberton Greenstone Belt, South Africa. Geobull Geol Soc S Afr 60:46

Anhaeusser CR (1963) The geology of the Lily Syncline and portion of the Eureka Syncline between Sheba Siding and Louw's Creek Station, Barberton Mountain Land. Unpubl. MSc thesis, University of the Witwatersrand, Johannesburg

Anhaeusser CR (1969) The stratigraphy, structure, and gold mineralization of the Jamestown and Sheba Hills areas of the Barberton Mountain Land. Unpubl. PhD thesis, University of the Witwatersrand, Johannesburg

Anhaeusser CR (1972) The geology of the Jamestown Hills area of the Barberton Mountain Land, South Africa. Trans Geol Soc S Afr 75:225–263

Anhaeusser CR (1976) The geology of the Sheba Hills area of the Barberton Mountain Land, South Africa, with particular reference to the Eureka Syncline. Trans Geol Soc S Afr 79:253–280

Anhaeusser CR (1984) Structural elements of Archaean granite-greenstone terranes as exemplified by the Barberton Mountain Land, Southern Africa. In: Kröner A, Greiling R (eds) Precambrian Tectonics Illustrated. E. Schweizerbart'sche Verlagsbuchhandlung, Stuttgart, pp 57–78

Anhaeusser CR, Robb LJ, Viljoen MJ (1981) Provisional geological map of the Barberton greenstone belt and surrounding granitic terrane, eastern Transvaal and Swaziland, scale 1:250000. Geol Soc South Africa

Archer AW (2013) World's highest tides: Hypertidal coastal systems in North America, South America and Europe. Sed Geol 284–285:1–25

Banham SG, Mountney NP (2013) Evolution of fluvial systems in salt-walled mini-basins: a review and new insights. Sed Geol 296:142–166

Bédard JH (2006) A catalytic delamination-driven model for coupled genesis of Archaean crust and sub-continental lithospheric mantle. Geochim Cosmochim Acta 70:1188–1214

Bédard JH (2019) Stagnant lids and mantle overturns: Implications for Archaean tectonics, magma genesis, crustal growth, mantle evolution, and the start of plate tectonics. Geosci Front 9:19–49

Bläsing S (2015) Petrography and sedimentology of selected sandstones of the westernmost Dycedale Syncline, Barberton Greenstone Belt, South Africa. Unpubl. BSc thesis, Freie Universität Berlin, Germany

Bontognali TRR, Fischer WW, Föllmi K (2013) Siliciclastic associated banded iron formation from the 3.2 Ga Moodies Group, Barberton Greenstone Belt, South Africa. Precambrian Res. 226:116–124

Bucher K, Frey M (1994) Petrogenesis of metamorphic rocks, 6th ed. Springer, Berlin, 318p

Byerly GR, Lowe DR, Heubeck C (2019) Geologic evolution of the Barberton Greenstone Belt – a unique record of crustal development, surface processes, and early life 3.55 to 3.20 Ga. In: van Kranendonk MJ, Bennett VC, and Hoffmann JE (eds) Earth's Oldest Rocks – 2nd edn. Elsevier (Amsterdam), pp 569–613

Byerly GR, Kröner A, Lowe DR, Todt W, Walsh MM (1996) Prolonged magmatism and time constraints for sediment deposition in the early Archaean Barberton greenstone belt: evidence from the Upper Onverwacht and Fig Tree groups. Precambrian Res 78:125–138

De Beer JH, Stettler E (2009) Geophysics and the deep structure of the Barberton Belt: a review. In: 11th SAGA biennial technical meeting and exhibition, Swaziland, 16–18 Sept 2009

De Beer JH, Stettler E, du Plessis JG, Blume J (1988) The deep structure of the Barberton Greenstone Belt. S Afr J Geol 91:184–197

Daneel GJ (1987) The Structural Controls of Gold Mineralization in the Moodies Hills which surround the Agnes Gold Mine, Barberton Greenstone Belt. Unpubl. MSc Thesis, University of Natal, Durban

de Vries ST, Nijman W, de Boer PL (2010) Sedimentary geology of the Palaeoarchean Buck Ridge (South Africa) and Kittys Gap (Western Australia) volcano-sedimentary complexes. Precambrian Res 183:749–769

de Wit MJ (1991) Archaean greenstone belt tectonism and basin development: some insights from the Barberton and Pietersburg greenstone belts, Kaapvaal Craton, South Africa. J Afr Earth Sci 13:45–63

Dickinson WR, Beard LS, Brakenridge GR, Erjavec JL, Ferguson RC, Inman KF, Knepp RA, Lindberg FA, Ryberg PT (1983) Provenance of North American Phanerozoic sandstones in relation to tectonic setting. Geol Soc Am Bull 93:222–235

Drabon N, Lowe DR, Byerly GR, Harrington JA (2017) Detrital zircon geochronology of sandstones of the 3.6–3.2 Ga Barberton greenstone belt: No evidence for older continental crust. Geology 45:803–806

Engelhardt JF (2012) Constraints from high-resolution zircon geochronology on age and provenance of the Archean Moodies Group, Barberton Greenstone Belt, South Africa. Unpubl. MSc thesis, Freie Universität Berlin, Germany

Eriksson KA (1977) Tidal deposits from the Archean Moodies Group, Barberton Mountain Land, South Africa. Sed Geol 18:257–281

Eriksson KA (1978) Alluvial and destructive beach facies from the Archaean Moodies Group, Barberton Mountain Land, South Africa and Swaziland. In: Miall AD (ed), Fluvial Sedimentology, vol 5. Canadian Soc Petrol Geol Mem, pp 287–311

Eriksson KA (1979) Marginal marine depositional processes from the Archaean Moodies Group, Barberton Mountain Land, South Africa: Evidence and significance. Precambrian Res 8:153–182

Eriksson KA (1980) Transitional sedimentation styles in the Moodies and Fig Tree Groups, Barberton Mountain Land, South Africa: evidence favouring an Archean continental margin. Precambrian Res 12:141–160

Eriksson KA, Simpson EL (2000) Quantifying the oldest tidal record: The 3.2 Ga Moodies Group, Barberton Greenstone Belt, South Africa. Geology 28:831–834

Eriksson KA, Krapez B, Fralick PW (1994) Sedimentology of Archaean greenstone belts: signatures of tectonic evolution. Earth Sci Rev 37:1–88

Eriksson KA, Simpson EL, Mueller W (2006) An unusual fluvial to tidal transition in the meso-Archaean Moodies Group, South Africa: a response to high tidal range and active tectonics. Sed Geol 190:13–24

Eriksson PG, Banerjee S, Catuneanu O, Corcoran P, Eriksson KA, Hiatt EE, Laflamme M, Lenhardt N, Long DGF, Miall A, Mints MV, Pufahl PK, Sarkar S, Simpson EL, Williams GE (2013) Secular changes in sedimentation systems and sequence stratigraphy. Gondwana Res 24:468–489

Folk RL (1965) Petrology of sedimentary rocks. Hemphill Publishing, Austin, TX

Fugmann P (2017) Field geology, petrography, age and geochemistry of the Lomati River Complex: a unique late-phase igneous contribution to the interior of the Archean Barberton Greenstone Belt, South Africa. Unpubl. MSc thesis, Friedrich-Schiller-Universität Jena, Germany

Fugmann P, Heubeck C, Viereck L, Linnemann U, Hofmann M (2017) Field geology, petrography, age and geochemistry of the Lomati River Complex: a unique late-phase igneous contribution to the interior of the Archean Barberton Greenstone Belt, South Africa. In: Annual Meeting DGGV Bremen 2017, abstract volume, p 434

Gamper A, Heubeck C, Demske D, Hoehse M (2012) Composition and microfacies of Archean microbial mats (Moodies Group, ca. 3.22 Ga, South Africa). In: Noffke N, Chafetz H (eds) Microbial mats in siliciclastic depositional systems through time, vol 101. Society of Economic Paleontologists and Mineralogists (Society for Sedimentary Geology) Special publication, pp 65–74

Ganguly J (1969) Chloritoid stability and related parageneses: theory, experiments, and applications. Am J Sci 267:910–944

Gay NC (1969) Analysis of strain in Barberton Mountain Land, Eastern Transvaal, using deformed pebbles. J Geol 77:377–396

Grund M (2015) Stratigraphy and Structural Geology of the Westernmost Dycedale Syncline (Moodies Group, Barberton Greenstone Belt, South Africa). Unpubl. BSc thesis, Freie Universität Berlin, Germany

Hall AL (1918) The geology of the Barberton gold mining district. Geological Survey of South Africa Memoir, vol 9, 347p

Harrington JA (2017) Using sedimentology and provenance studies to determine depositional relationships between three structural belts of the ca. 3.22 Ga Fig Tree Group, Barberton Greenstone Belt, South Africa. Unpubl. MSc thesis, Stanford University, Palo Alto, USA

Herget R (1966) Beitrag zur Tektonik archaischer Sedimente im Barberton-Bergland (Transvaal-Südafrika). Geol. Rundschau (Int J Earth Sci) 55:768–775

Hessler AM, Lowe DR (2006) Weathering and sediment generation in the Archaean: An integrated study of the evolution of siliciclastic sedimentary rocks of the 3.2 Ga Moodies Group, Barberton Greenstone Belt, South Africa. Precambrian Res 151:185–210

Hessler AM, Lowe DR, Jones RL, Bird DK (2004) A lower limit for carbon dioxide levels 3.2 billion years ago. Nature 428:736–738

Heubeck C (2009) An early ecosystem of Archean tidal microbial mats (Moodies Group, South Africa, ca. 3.2 Ga). Geology 37:931–934

Heubeck C (2015) Reassessing evidence of Moon-Earth dynamics: no evidence of shorter lunar months from tidal bundles of the Moodies Group (Barberton Greenstone Belt, 3.22 Ga). In: Paper presented at the DGGV/DMG Annual Meeting, Berlin, Germany

Heubeck C (2016) Day 5 - Friday, Sept. 9: Sedimentation and deformation style in the central BGB. In: Heubeck, C (ed) Geology of the Barberton Greenstone Belt: processes on the early Earth. 35th International Geological Congress Field Trip Guide POST 7, 4–10 Sept 2016

Heubeck C, Lowe DR (1994a) Depositional and tectonic setting of the Archean Moodies Group, Barberton Greenstone Belt, South Africa. Precambrian Res 68:257–290

Heubeck C, Lowe DR (1994b) Late syndepositional deformation and detachment tectonics in the Barberton Greenstone Belt, South Africa. Tectonics 13:1514–1536

Heubeck C, Lowe DR (1999) Sedimentary petrology and provenance of the Archean Moodies Group, Barberton Greenstone Belt, South Africa. In: Lowe DR, Byerly GR (eds) Geologic Evolution of the Barberton Greenstone Belt, South Africa. Geol Soc America Spec Paper vol 329, pp 259–286

Heubeck C, Toulkeridis T, Wendt JI, Kröner A, Lowe DR (1993) Timing of deformation of the Archean Barberton greenstone belt, South Africa: constraints from zircon dating of the Salisbury Kop pluton. S Afr J Geol 96:1–8

Heubeck C, Engelhardt J, Byerly GR, Zeh A, Sell B, Luber T, Lowe DR (2013) Timing of deposition and deformation of the Moodies Group (Barberton Greenstone Belt, South Africa): very-high-resolution of Archaean surface processes. Precambrian Res 231:236–262

Heubeck C, Grosch E, Mason, P (2016a) Geology of the Barberton Greenstone Belt: Processes on the early Earth. In: 35th International geological congress field trip guidebook POST 7, 77p

Heubeck C, Bläsing S, Drabon N, Grund M, Homann M Nabhan S (2016b) Geological constraints on Archean (3.22 Ga) coastal-zone processes from the Dycedale Syncline, Barberton Greenstone Belt. S Afr J Geol 119:495–518

Homann M (2016) Depositional setting and metabolism of microbial mats in the Archean Moodies Group, Barberton Greenstone Belt, South Africa. Unpubl. doctoral thesis, Freie Universität Berlin, Germany

Homann M, Heubeck C, Airo A, Tice MM (2015) Morphological adaptations of 3.22 Ga-old microbial communities to Archean coastal habitats (Moodies Group, Barberton Greenstone Belt, South Africa). Precambrian Res 266:47–64

Homann M, Heubeck C, Bontognali TRR, Bouvier AS, Baumgartner LP, Airo A (2016) Evidence for cavity-dwelling microbial life in 3.22 Ga tidal deposits. Geology 44:51–54

Homann M, Sansjofre P, van Zuilen M, Heubeck C, Gong J, Killingsworth B, Foster I, Airo A, van Kranendonk MJ, Ader M, Lalonde SV (2019) Microbial life and biogeochemical cycling on land 3,220 million years ago. Nat Geosc 11:665–671

Hose LD (1990) The geology and stratigraphic evolution of the north-central part of the Barberton Greenstone Belt, South Africa. Unpubl. PhD thesis, Louisiana State University, Baton Rouge, USA

Hunter DR, Jones DH (1968) Swaziland 1: 25.000 geological series, sheet 1 Hhohho. Geol. Surv. Mines Dept., Swaziland

Hunter DR, Jones DH (1969) Swaziland 1: 25.000 geological series, sheet 2 Piggs Peak. Geol. Surv. Mines Dept., Swaziland

Jackson MPA, Eriksson KA, Harris CW (1987) Early Archean foredeep sedimentation related to crustal shortening: a reinterpretation of the Barberton Sequence, southern Africa. Tectonophysics 136:197–221

Kamo SL, Davis DW (1994) Reassessment of Archaean crustal development in the Barberton Mountain Land, South Africa. Tectonics 13:167–192

Kisters AFM, Stevens G, Dziggel A, Armstrong RA (2003) Extensional detachment faulting and core-complex formation in the southern Barberton granite-greenstone terrain, South Africa: evidence for a 3.2 Ga orogenic collapse. Precambrian Res 127: 355–378

Kisters AFM, Belcher RW, Pujol M, Dziggel A (2010) Continental growth and convergence-related arc plutonism in the Mesoarchaean: evidence from the Barberton granitoid-greenstone terrain, South Africa. Prec Res 178:15–26

Kohler EA (1994) The geology of the Archaean granitoid-greenstone terrain in the vicinity of three sisters, Barberton Greenstone Belt. Ph.D. Dissertation (unpublished), Johannesburg, University of the Witwatersrand, 309 p

Kohler E, Anhaeusser CR (2002) Geology and geodynamic setting of Archaean silicic metavolcaniclastic rocks of the Bien Venue Formation, Fig Tree Group, northeast Barberton greenstone belt, South Africa. Precambrian Res 116:199–235

Kröner A (2006) The Ancient Gneiss Complex of Swaziland and Environs: Record of Early Archaean Crustal Evolution in Southern Africa. In: Van Kranendonk M, Bennett V, Smithies H (eds) The world's oldest rocks, 1st edn. Elsevier, Amsterdam, pp 465–480

Kröner A, Compston W (1988) Ion microprobe ages of zircons from early Archean granite pebbles and greywacke, Barberton Greenstone Belt, Southern Africa. Precambrian Res 38:367–380

Kröner A, Compston W, Williams IS (1989) Growth of early Archaean crust in the Ancient Gneiss Complex of Swaziland as revealed by single-zircon dating. Tectonophysics 161:271–298

Kröner A, Byerly GR, Lowe DR (1991) Chronology of early Archaean granite-greenstone evolution in the Barberton Mountain Land, South Africa, based on precise dating by single zircon evaporation. Earth Planet Sci Lett 103:41–54

Kröner A, Hoffmann JE, Xie H, Münker C, Hegner E, Hofmann A, Liu D, Yang J (2014) Generation of early Archaean grey gneisses through melting of older crust in the eastern Kaapvaal craton, southern Africa. Precambrian Res 255:823–846

Kröner A, Anhaeusser CR, Hoffmann JE, Wong J, Geng H, Hegner E, Xie H, Yang J, Liu D (2016) Chronology of the oldest supracrustal sequences in the Palaeoarchaean Barberton Greenstone Belt, South Africa and Swaziland. Precambrian Res 279:123–143

Kröner A, Wong J, Xie H (2019) The oldest granite clast in the Moodies conglomerate, Barberton greenstone belt, South Africa, and its likely origin. S Afr J Geol 121:1–8

Lamb S (1984a) Geology of part of the Archean Barberton greenstone belt, Swaziland. Unpubl. PhD thesis, Cambridge University, UK

Lamb S (1984b) Structures on the eastern margin of the Archean Barberton greenstone belt, northwest Swaziland. In: Kröner A, Greiling R (eds) Precambrian Tectonics Illustrated. Schweizerbart Publishers, Stuttgart, pp 19–40

Lamb S (1987) Archean synsedimentary tectonic deformation—a comparison with the Quaternary. Geology 15:565–568

Lamb SH, Paris I (1988) Post-Onverwacht Group stratigraphy in the southeast part of the Archaean Barberton greenstone belt. J Afr Earth Sci 7:285–306

Lana C, Kisters A, Stevens G (2010) Exhumation of Mesoarchean TTG gneisses from the middle crust: insights from the Steynsdorp core complex, Barberton granitoid-greenstone terrain, South Africa. Geol Soc Am Bull 122:183–197

Lana C, Buick I, Stevens G, Roussow R, de Wet W (2011) 3230-3200 Ma post-orogenic extension and mid-crustal magmatism along the southeastern margin of the Barberton Greenstone Belt, South Africa. J Struct Geol 33:844–858

Layer PW, Kröner A, York D (1992) Pre-3000 Ma thermal history of the Archean Kaap Valley Pluton, South Africa. Geology 20:717–720

Layer PW, Kröner A, McWilliams M (1996) An Archaean geomagnetic reversal in the Kaap Valley Pluton, South Africa. Science 273:943–946

Leisgang I (2017) LA-ICP-MS zircon ages for clastic sediments of the Moodies Group, Barberton Greenstone Belt, South Africa and Swaziland. Unpubl. BSc thesis, Johannes-Gutenberg-University, Mainz, Germany

Lippold W (2010) Late-stage stratigraphic, petrographic and tectonic evolution of the upper Moodies Group, Saddleback Syncline, Barberton Greenstone Belt, South Africa. Unpubl. Mapping Report, Freie Universität Berlin, Germany

Lowe DR (2007) A comment on "Weathering of quartz as an Archean climatic indicator" by NH Sleep and AM Hessler [Earth Planet Sci Lett 241 (2006) 594–602]. Earth Planet Sci Lett 253:530–533

Lowe DR, Byerly GR, Heubeck C (1999) Structural divisions and development of the west-central part of the Barberton Greenstone Belt. In: Lowe DR, Byerly GR (eds) Geologic evolution of the Barberton Greenstone Belt, South Africa, Geol Soc America Spec Paper vol 329, pp 37–82

Lowe DR, Byerly GR, Heubeck C (2012) Geologic Map of the west-central Barberton Greenstone Belt, South Africa, scale 1:25,000. Geol Soc America Map and Chart Series No. 103

Lowe DR, Byerly GR, Kyte FT (2014) Recently discovered 3.42–3.23 Ga impact layers, Barberton Belt, South Africa: 3.8 Ga detrital zircons, Archean impact history, and tectonic implications. Geology 42:747–750

Luber T (2014) Archaean Geology of the east-central Stolzburg Syncline, Barberton Greenstone Belt, South Africa. Unpubl. MSc thesis, Freie Universität Berlin, Germany

McLennan SM, Taylor SR, Kröner A (1983) Geochemical evolution of Archean shales from South Africa. I. The Swaziland and Pongola Supergroups. Precambrian Res 22:93–124

Moyen JF, Stevens G, Kisters AF, Belcher RW (2007) TTG plutons of the Barberton Granitoid-Greenstone Terrain, South Africa. Dev Precambrian Geol 15:607–667

Moyen JF, Stevens G, Kisters AFM, Belcher RW, Lemirre B (2019) TTG plutons of the Barberton granitoid-greenstone terrain, South Africa. In: Van Kranendonk MJ, Bennett VC, and Hoffmann E (eds) Earth's oldest rocks, 2nd edn. Elsevier, Cambridge, MA, USA, pp 615–653

Nabhan S, Luber T, Scheffler F, Heubeck C (2016a) Climatic and geochemical implications of Archean pedogenic gypsum in the Moodies Group (~3.2 Ga), Barberton Greenstone Belt, South Africa. Prec Res 275:119–134

Nabhan S, Wiedenbeck M, Milke R, Heubeck C (2016b) Biogenic overgrowth on detrital pyrite in 3.2 Ga Archean paleosols. Geology 44:763–766

Nabhan S, Köhler I, Heubeck C (2017) Local and regional controls on the maturation state of carbonaceous matter in the Barberton Greenstone Belt. In: Annual Meeting DGGV Bremen 2017, abstract volume, p 449

Noffke N, Eriksson KA, Hazen RM, Simpson EL (2006) A new window into early Archaean life: microbial mats in Earth's oldest siliciclastic tidal deposits (3.2 Ga Moodies Group, South Africa). Geology 34:253–256

Reimer T (1967) Die Geologie der Stolzburg-Synklinale im Barberton Bergland (Transvaal-Südafrika). Unpubl. Doctoral dissertation, Goethe-Universität Frankfurt/Main, Germany

Reimer T (1971) Kanu-Falten - ein ungewöhnlicher Faltentyp aus geringmetamorphen präkambrischen Sedimenten. N Jb Geol Paläontol Monatshefte 1971:489–495

Reimer T, Condie KC, Schneider G, Georgi A (1985) Petrography and geochemistry of granitoid and metamorphite pebbles from the early Archaean Moodies Group, Barberton Mountainland/South Africa. Precambrian Res 29:383–404

Robb LJ, Barton JM, Kable EJD, Wallace RC (1986) Geology, geochemistry and isotopic characteristics of the Archaean Kaap Valley pluton, Barberton Mountain Land, South Africa. Precambrian Res 31:1–36

SACS (South African Committee on Stratigraphy) (1980) Stratigraphy of South Africa, Handbook 8, Part 1. Geol Surv South Africa, 690p

Sanchez-Garrido C, Stevens G, Armstrong RA, Moyen JF, Martin H, Doucelance R (2011) Diversity in Earth's early felsic crust: Paleoarchean peraluminous granites of the Barberton Greenstone Belt. Geology 39:963–966

Schieber J, Bose P, Eriksson PG, Altermann W, Catuneanu O (eds) (2007) Atlas of microbial mat features preserved within the Siliciclastic Rock Record. Elsevier Science, Amsterdam

Schoene B, de Wit M, Bowring SA (2008) Mesoarchean assembly and stabilization of the eastern Kaapvaal craton: a structural–thermochronological perspective. Tectonics 27:TC5010. https://doi.org/10.1029/2008tc002267

Simpson EL, Eriksson KA, Mueller W (2012) 3.2 Ga eolian deposits from the Moodies Group, Barberton Greenstone Belt, South Africa: implications for the origin of first-cycle quartz sandstones. Precambrian Res 214/215:185–191

Sleep NH, Hessler AM (2006) Weathering of quartz as an Archean climatic indicator. Earth Planet Sci Lett 241:594–602

Smits M, Mezger K (2017) Earth's early O_2 cycle suppressed by primitive continents. Nat Geosci 10:1–6

Stutenbecker L, Zeh A, Heubeck C (2019) The Lomati Delta Complex: A prograding tidal delta in the Archean Moodies Group, Barberton Greenstone Belt. S Afr J Geol 121

Tegtmeyer A, Kröner A (1987) U-Pb zircon ages bearing on the nature of early Archaean greenstone belt evolution, Barberton Mountainland, southern Africa. Precambrian Res 36:1–20

Toulkeridis T, Clauer N, Kröner A, Todt W (2015) A mineralogical, chemical and isotopic investigation of shales from the Barberton Greenstone Belt, South Africa, to constrain source materials and post-deposition evolution. S Afr J Geol 118:389–410

Urie JG (1970) Geological map series (1: 25,000) Sheet 3, Forbes Reef. Swaziland Geol Surv Mines Dept (Mbabane)

Urie JG (1971) Geological map series (1:25,000) Sheet 4, Motshane. Swaziland Geol Surv Mines Dept (Mbabane)

van Kranendonk MJ (2011) Cool greenstone drips and the role of partial convective overturn in Barberton greenstone belt evolution. J Afr Earth Sci 60:346–352

van Kranendonk MJ, Collins WJ, Hickman AH, Pawley MJ (2004) Critical tests of vertical vs. horizontal tectonic models for the Archaean East Pilbara granite greenstone terrane, Pilbara Craton, Western Australia. Precambrian Res 131:173–211

van Kranendonk M, Kröner A, Hegner E, Connelly J (2009) Age, lithology and structural evolution of the c. 3.53 Ga Theespruit Formation in the Tjakastad area, southwestern Barberton Greenstone Belt, South Africa, with implications for Archaean tectonics. Chem Geol 261:115–139

van Kranendonk MJ, Smithies RH, Champion DC (2019) Paleoarchean development of a continental nucleus: the East Pilbara Terrane of the Pilbara Craton, Western Australia. In: van Kranendonk MJ, Bennett VC, Hoffmann E (eds) Earth's oldest rocks, 2nd edn. Elsevier, Cambridge, MA, USA, pp 437–462

van Niekerk CB, Burger AJ (1978) The age of the Moodies conglomerate boulders. Geol Soc S Afr Spec Publ 4:99–106

van Schijndel V, Stevens G, Zeh A, Frei D, Lana C (2017) Zircon geochronology and Hf isotopes of the Dwalile Supracrustal Suite, Ancient Gneiss Complex, Swaziland: insights into the diversity of Palaeoarchaean source rocks, depositional and metamorphic ages. Precambrian Res 295:48–66

Visser DJL (ed, with reports by van Eeden OR, Joubert GK, Söhnge APG, van Zyl JS, Rossouw J, Taljaard JJ, Visser DJL) (1956) The geology of the Barberton area. Geol Survey South Africa Spec Publ 15

Wacey D, McLoughlin N, Green OR, Parnell J, Stoakes CA, Brasier MD (2006) The ~3.4 billion-year-old Strelley Pool Sandstone: a new window into early life on Earth. Int J Astrobiol 5:333–342

Wiechert A (2014) Contributions to the Geochronology of the northern and southern facies of the Archean Moodies Group (Barberton Greenstone Belt, South Africa and Swaziland). Unpubl. MSc thesis, Freie Universität Berlin, Germany

Zeh A, Gerdes A, Heubeck C (2013) U-Pb and Hf isotope data of detrital zircons from the Barberton Greenstone Belt: constraints on provenance and Archaean crustal evolution. J Geol Soc Lond 170:215–223

7 Mesoarchaean Gold Mineralisation in the Barberton Greenstone Belt: A Review

Andrea Agangi, Axel Hofmann, Benjamin Eickmann, and Johanna Marin-Carbonne

Abstract

The Barberton Greenstone Belt hosts abundant structurally controlled gold mineralisation of Mesoarchaean age. More than 300 gold occurrences have been reported, although most of the gold production so far (>350 tonnes Au) has come from a handful of deposits located along the northern margin of the greenstone belt. Most deposits are hosted by greenschist-facies metasedimentary and metamafic rocks, with the notable exception of the amphibolite-facies rocks at New Consort mine. Mineralisation is associated with quartz–carbonate veins that truncate major compressional structures at the greenstone belt scale. The age of mineralisation is loosely constrained at circa 3080–3030 Ma, based on U–Pb dating of hydrothermal rutile and titanite. In greenschist-facies deposits, the ore assemblage is dominated by pyrite and arsenopyrite, which contain up to thousands of ppm of 'invisible' gold, Ni–As–Sb sulphides and native gold. At New Consort mine, mineralisation includes massive replacement-style ore and vein-hosted or disseminated types. Both structural studies in the field and microstructural observation point to a multistage ore deposition process, which is reflected in the re-activation of brittle to ductile structures and the overprinting of sulphide assemblages. The presence of mass-independently fractionated S isotopes ($\Delta^{33}S$ = –0.6 to +1.0‰) in pyrite from Sheba and Fairview mines suggests that hydrothermal fluids mobilised S from volcanic and sedimentary rocks of the greenstone belt and places constraints on the origin of the Au itself.

Keywords

Archaean • Gold • Mineralisation • Multiple S isotopes • Barberton

7.1 Introduction

The Palaeoarchaean Barberton Greenstone Belt is host to some of the earliest known Au mineralisation (Anhaeusser 1976, 1986; de Ronde et al. 1991; Dziggel et al. 2010). The mineralisation is structurally controlled and occurs in strongly deformed greenschist- to amphibolite-facies metamorphic rocks. Deciphering mechanisms and timing of mineralisation in the Barberton Greenstone Belt and other Archaean terrains is important for understanding Archaean tectonic and hydrothermal processes (Anhaeusser 1986; Kolb et al. 2015; Sahoo and Venkatesh 2015; Hazarika et al. 2017; Mishra et al. 2017).

Mineralisation in the Barberton Greenstone Belt has significant economic relevance and, since the discovery of gold in this area in 1882, more than 350 tonnes of gold have been extracted (Anhaeusser 1976; Dirks et al. 2009; Pearton and Viljoen 2017). In plan view, the distribution of the main deposits reveals a heterogeneous distribution of gold mineralisation. Despite the fact that more than 300 deposits and prospects are known in the greenstone belt, the bulk of the gold production comes from a handful of mines, namely Fairview, Sheba, New Consort, Agnes and Princeton mines, all located in the northern portion of the greenstone belt near the contact between the Barberton Greenstone Belt and the Kaap Valley and Stentor plutons (Anhaeusser 1976; Ward 1995, 1999; Dirks et al. 2009) (Fig. 7.1). The fluids

A. Agangi (✉) · A. Hofmann
Department of Geology, University of Johannesburg, Johannesburg, South Africa
e-mail: aagangi@uj.ac.za

B. Eickmann
Isotopengeochemie, Universität Tübingen, Tübingen, Germany

J. Marin-Carbonne
Laboratoire Magmas et Volcans, UCA, IRD, CNRS, Université Lyon—UJM, Saint Étienne, France

Present Address:
J. Marin-Carbonne
Faculté des géosciences et de l'environnement, Institut des sciences de la Terre, University of Lausanne, Lausanne, Switzerland

A. Kröner and A. Hofmann (eds.), *The Archaean Geology of the Kaapvaal Craton, Southern Africa*, Regional Geology Reviews, https://doi.org/10.1007/978-3-319-78652-0_7

responsible for mineralisation are believed to have been low-salinity and H_2O-CO_2-rich (de Ronde et al. 1992) and, similar to other orogenic gold deposits, Au is interpreted to have been transported as complexes of reduced S such as $Au(HS)_2^-$ and AuHS (Pokrovski et al. 2014). Structural and microtextural evidence convincingly points at multiphase processes of mineralisation (Agangi et al. 2014; Munyai et al. 2011; Dziggel and Kisters 2019).

Some aspects of the genesis of this mineralisation are still controversial, and some of the open questions include the origin of H_2O-CO_2 mineralising fluids and, by inference, the origin of Au itself. Possible sources of fluids and Au include supracrustal lithologies (volcanic and sedimentary rocks), felsic intrusions, and the mantle. The mineralisation is believed to have occurred between circa 3080 and 3030 Ma, based on rutile and titanite U-Pb ages, but precise timing of mineral deposition is also uncertain, given the scarcity of available data (de Ronde et al. 1992; Dziggel et al. 2010; Dirks et al. 2013).

Several reviews have dealt with gold mineralisation in the Barberton Greenstone Belt and have mostly focussed on the meso- to mega-scale structural and metamorphic aspects of various mines (Anhaeusser 1986; Dziggel et al. 2007; Pearton and Viljoen 2017; Dziggel and Kisters 2019). Here, we review some of the main characteristics of gold deposits in the Barberton Greenstone Belt with an emphasis on geochemical and microstructural characteristics. We refer to the mentioned papers for more complete information on the control of structures and deformation on mineralisation. We distinguish mineralisation hosted in mostly greenschist-facies rocks from mineralisation hosted in amphibolite-facies rocks. We then evaluate different models of genetic mechanisms of mineralisation.

7.2 Geological Setting of the Barberton Greenstone Belt

The Palaeoarchaean Barberton Greenstone Belt is situated in the eastern part of the Kaapvaal craton, southern Africa (Fig. 7.1). Its ca. 3550–3220 Ma volcano-sedimentary succession, which forms the Barberton Supergroup (formerly Swaziland Supergroup), is preserved in a southwest-northeast-trending belt surrounded by granitoid rocks belonging to the trondhjemite-tonalite-granodiorite (TTG) and granite-granodiorite-syenogranite (GMS) series (Viljoen and Viljoen 1969; Lowe and Byerly 2007; de Wit et al. 2019). The supracrustal succession has been subdivided into three main lithostratigraphic units: the Onverwacht Group, the Fig Tree Group, and the Moodies Group, in ascending order (SACS 1980).

The Inyoka–Saddleback Fault System, a southwest to northeast-trending structure interpreted by some as a suture zone, separates a northern and a southern terrane of different age and geochemical characteristics (Kamo and Davis 1994; Kisters et al. 2003; Lowe and Byerly 2007). The Onverwacht Group consists mostly of komatiite, komatiitic basalt and basalt, with minor felsic volcanic rocks, and has been dated at circa 3550–3300 Ma in the southern terrane (Kröner et al. 1996, 2016). North of the Inyoka-Saddleback fault, the Onverwacht Group is composed of the Weltevreden Formation, which contains mafic-ultramafic volcanic rocks and numerous layered ultramafic complexes (Anhaeusser 2001; Stiegler et al. 2012). The Fig Tree Group is a largely marine, northwards deepening succession dominated by turbiditic greywackes, shales, banded iron formation and cherts (Lowe 1999; Hofmann 2005). The Moodies Group is a shallow marine to continental succession characterised by coarse-grained clastic sedimentary rocks, mainly sandstones and conglomerates, and only minor shale with a minimum depositional age of 3219 Ma (SACS 1980; Heubeck et al. 2013, 2016; Drabon et al. 2017). The metamorphic grade of the Barberton Greenstone Belt is generally low but increases towards the contacts with the surrounding gneiss domes (Dziggel et al. 2002, 2005; Diener et al. 2005). The structure of the greenstone belt is rather complex and is dominated by steeply plunging synforms separated either by thrust faults or narrow anticlines (Lowe et al. 2012).

The tectonic evolution of the Barberton Greenstone Belt is complex, and the multiple phases of deformation affecting these rocks have been described elsewhere (de Ronde and de Wit 1994; Kamo and Davis 1994; Lowe and Byerly 1999; de Ronde and Kamo 2000; Lana et al. 2010). According to de Ronde and de Wit (1994) and de Ronde and Kamo (2000), four main tectono-metamorphic events affected the greenstone belt. The first of these events (D1) remains to some extent enigmatic. D1 occurred between circa 3445 and 3416 Ma and was restricted to Onverwacht Group rocks in the southern part of the greenstone belt (de Ronde and de Wit 1994). D1 was coeval with the intrusion of TTGs along the southern margin of the greenstone belt and with an early phase of low-pressure amphibolite-facies metamorphism, with estimated peak conditions of ~550 °C and 4.5 kbar (Cutts et al. 2014).

The 3229–3227 Ma D2 event was responsible for the main regional strain and affected the entire greenstone belt (de Ronde and Kamo 2000; Schoene et al. 2008). Event D2 coincided with emplacement of TTG intrusions such as the 3227 Ma Kaap Valley Tonalite, the 3290–3230 Ma Badplaas pluton and the 3236 Ma Nelshoogte pluton (Kamo and Davis 1994; Kisters et al. 2010; Matsumura 2014). During this event, the rocks of the lowermost units of the Onverwacht Group, the Theespruit and Sandspruit Formations, experienced high-pressure amphibolite-facies metamorphism (Dziggel et al. 2002, 2005; Diener et al. 2005; Moyen et al. 2006). D2 marked the switch from sedimentation of the

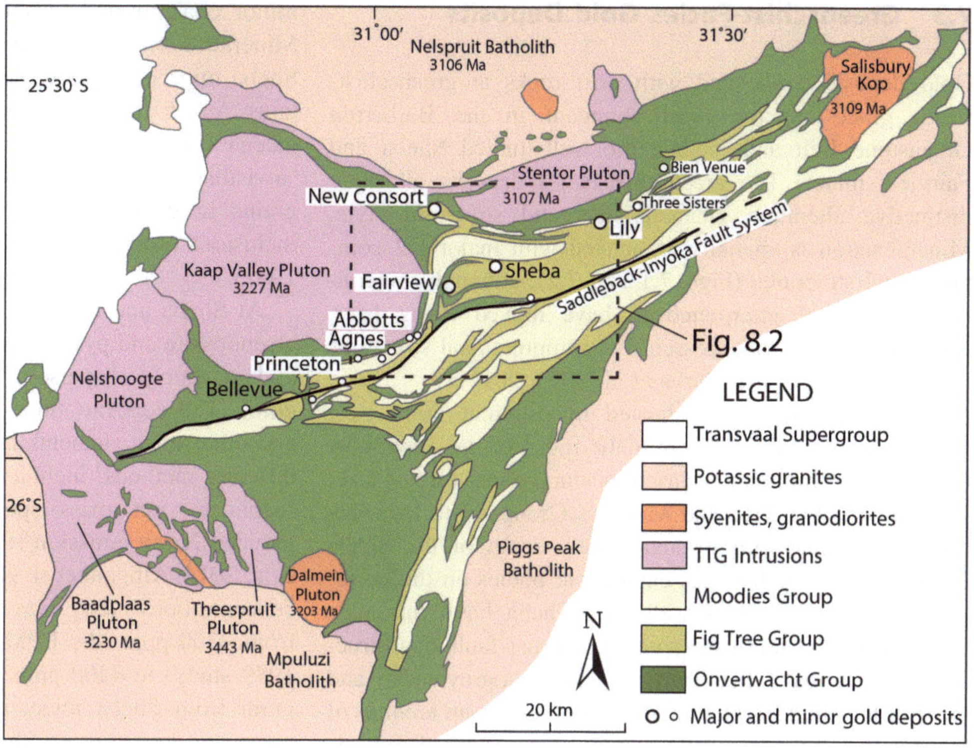

Fig. 7.1 Geological map of the Barberton Greenstone Belt and distribution of the main gold deposits (modified from de Ronde et al. 1992). Dashed line indicates position of Fig. 7.2

3260–3230 Ma Fig Tree Group in a relatively deep marine environment to the ~3225 Ma Moodies Group in a shallow marine to continental environment (de Ronde and de Wit 1994; Kamo and Davis 1994; de Ronde and Kamo 2000; Kisters et al. 2010; Heubeck et al. 2013). The structures associated with D2 are truncated by potassic intrusions such as the 3203 Ma Dalmein pluton in the southern part (Lana et al. 2010), which places constraints on the end of this phase of deformation. The tectonic circumstances that gave rise to the D2 event are controversial and directly mirror the controversy relating to the nature of Archaean tectonics and the origin of TTG magmas. Some authors interpreted D2 as the consequence of compression, subduction and accretion of the southern terrane along the Inyoka-Saddleback Fault during northwards-directed subduction (Moyen et al. 2006; Kisters et al. 2010). In this interpretation, the syntectonic granitoids to the north of the greenstone belt represent the roots of a volcanic arc. D2 was synchronous with, or immediately followed by, a period of syn-orogenic extension and solid-state doming that eventually resulted in the steepening of fabrics during the orogenic collapse of the belt (Kisters et al. 2003). Others interpret this deformation as a result of crustal overturn due to density inversions between buoyant (partially molten) TTG intrusions and cooler, denser greenstones (Anhaeusser 1984; Van Kranendonk et al. 2009). These authors argue that the relatively high-pressure conditions and the low temperature/pressure gradients can be achieved during sagging of the supracrustal succession.

The later, post-D2 events are less clear as geochronological constraints are scarce, and deformation gave rise to reactivation of earlier shear zones. The D3 tectonic event appears to have been related to NW-SE directed compression and orogen-parallel stretching accommodated by strike-slip shear zones (de Ronde and de Wit 1994). D3 structural elements thus parallel those of D2 (de Ronde and de Wit 1994). The timing of D3 has been constrained at <3126 Ma (de Ronde and Kamo 2000), although much looser time constraints (circa 3220–3080 Ma) were inferred for this event by other authors (de Ronde and de Wit 1994; Kamo and Davis 1994; Schoene et al. 2008). The D4 tectonic event was characterised by a switch to transtensional deformation (de Ronde and de Wit 1994). Geochronological constraints on this event are few and suggest that D4 occurred at around ~3080 Ma, based on U-Pb ages on hydrothermal rutile from Fairview mine (de Ronde et al. 1991; de Ronde and de Wit 1994).

Gold mineralisation is interpreted to be associated with D3 and/or D4 deformation stages. The age of ~3080 Ma (de Ronde et al. 1991), estimated on rutile in an altered, although not mineralised, sample offers a good indication of the timing of hydrothermal alteration at Fairview mine. However, much younger ages (3027 Ma) have been obtained from titanite associated with sulphide mineralisation at New Consort mine (Dziggel et al. 2010), and a maximum age of ~3015 Ma was estimated based on the zircon age of a felsic dyke interpreted to be syn- mineralisation at Fairview mine (Dirks et al. 2013).

7.3 Greenschist-Facies Gold Deposits

Gold deposits associated with host rocks at greenschist-facies grade are the most abundant in the Barberton Greenstone Belt and include the well-studied Sheba and Fairview mines. The greenschist-facies deposits all share distinctive alteration characteristics and structural style. Mineralisation is spatially associated with major D2 compressional structures (Figs. 7.1 and 7.2), although mineralising fluids are interpreted to have moved along later, extensional faults that crosscut D2 compressional structures (de Ronde et al. 1991; Dirks et al. 2009, 2013).

Gold mineralisation is hosted by different lithologies, ranging from metamafic–ultramafic rocks of the uppermost Onverwacht Group to metasedimentary rocks (greywacke, shale) of the Fig Tree and Moodies Groups. The Fairview and Sheba mines are located largely south of the Sheba Fault, although both mines exploit ore bodies on the northern side of the fault as well. The Sheba Fault has been interpreted as a north-west verging thrust fault that structurally superposed Fig Tree Group (meta)greywacke and shale of the Ulundi syncline onto Moodies Group arenites of the Eureka syncline (Anhaeusser 1976; Ward 1999; de Ronde and Kamo 2000) (Fig. 7.2). Minor metamafic rocks of the Onverwacht Group are preserved between the Ulundi and Eureka synclines as tight antiformal structures made of talc-carbonate schist of the Zwartkoppie Formation[1] (Schouwstra and De Villiers 1998; Schouwstra 1995).

Mineralisation occurs as auriferous quartz-carbonate ± rutile veins with associated semi-massive replacement sulphide bodies developing as part of wall-rock alteration (Schouwstra 1995). Alteration zones associated with mineralisation in mafic lithologies include, from distal to proximal, talc-carbonate, quartz-carbonate, fuchsite-quartz-carbonate ± sulphides, and fuchsite-quartz-sulphides ± carbonate (Schouwstra 1995). Sericite is present instead of fuchsite as the K-mica in the neighbouring greywacke and shale. Mineralised fractures in the Sheba–Fairview area extend for tens of metres, trending northeast to east-northeast, and are arranged in hundreds of metre-scale fracture zones (Dirks et al. 2009). In microtextural observations, veins directly associated with mineralisation appear as dilatant, mostly brittle structures, as demonstrated by euhedral prismatic quartz growing from the walls inwards (Agangi et al. 2014) (Fig. 7.3a).

The mineralisation occurs within or around quartz-carbonate veins that truncate the foliation (Fig. 7.3b, c). The mineralisation hosted in metapelite and greywacke at Fairview mine is dominated by arsenopyrite and pyrite, with minor chalcopyrite, gersdorffite, galena and Sb-sulphides. Mineralisation hosted in metamafic–ultramafic rocks at Sheba mine is typically dominated by pyrite and also contains As–Ni-sulphides, chalcopyrite, sphalerite and minor galena (Fig. 7.3d). Gold occurs in a variety of forms, as 'invisible' (or refractory) gold in sulphides, as free gold grains in quartz-carbonate veins, and as sulphide-hosted inclusions of native gold (Cabri et al. 1989; de Ronde et al. 1992).

At Sheba and Fairview mines, gold is mostly hosted in arsenopyrite and pyrite, both as micro-inclusions (Fig. 7.3e) and as finely dispersed, sub-microscopic Au (refractory gold). The concentrations and distribution of invisible Au and other trace elements in sulphides have been studied by different methods, including electron microprobe (EPMA), secondary ion mass spectrometry (SIMS) and proton-induced X-ray emission (PIXE) (Cabri et al. 1989; Agangi et al. 2014; Altigani et al. 2016). These studies found that Au concentrations are extremely heterogeneous and vary from <300 ppm (the EPMA detection limit in Cabri et al. 1989 study) to 4400 ppm Au even in a single arsenopyrite grain from Sheba mine, and up to 1020 ppm have been measured in pyrite at Sheba mine. In the Au versus As diagram (Fig. 7.4), pyrite compositions from mineralised samples at Sheba and Fairview mines plot both above and below the Au-saturation line of Reich et al. (2005). This is compatible with the presence of Au in different oxidation states and crystallographic positions, namely Au^{+} included in the crystal lattice and Au^{0} in gold inclusions. Mapping of trace element distributions at the micrometre scale has highlighted complex intragranular textures that indicate multiple phases of growth, resorption and recrystallisation of sulphide minerals (Fig. 7.5; Cabri et al. 1989; Agangi et al. 2014, 2015).

The largely refractory nature of gold (below the zone of surface oxidation) in these deposits requires the sulphide ore minerals to be oxidised during the process of extraction in order to liberate the gold. In the past, this was achieved by roasting, with consequent production of large amounts of SO_2 gas and volatile As oxides, a process that posed a considerable environmental hazard. Oxidation is now carried out through biogenic means using sulphide-oxidising bacteria, such as Thiobacillus ferrooxidans (Barberton Mines 2010).

7.4 Amphibolite-Facies Gold Deposits—The New Consort Mine

The New Consort mine (Figs. 7.1 and 7.2) represents an exception in the Barberton Greenstone Belt in being one of the few gold deposits hosted in amphibolite-facies rocks. It is located north of Sheba and Fairview mines and occurs in

[1]The Zwartkoppie Formation is not recognised by SACS and its use is informal.

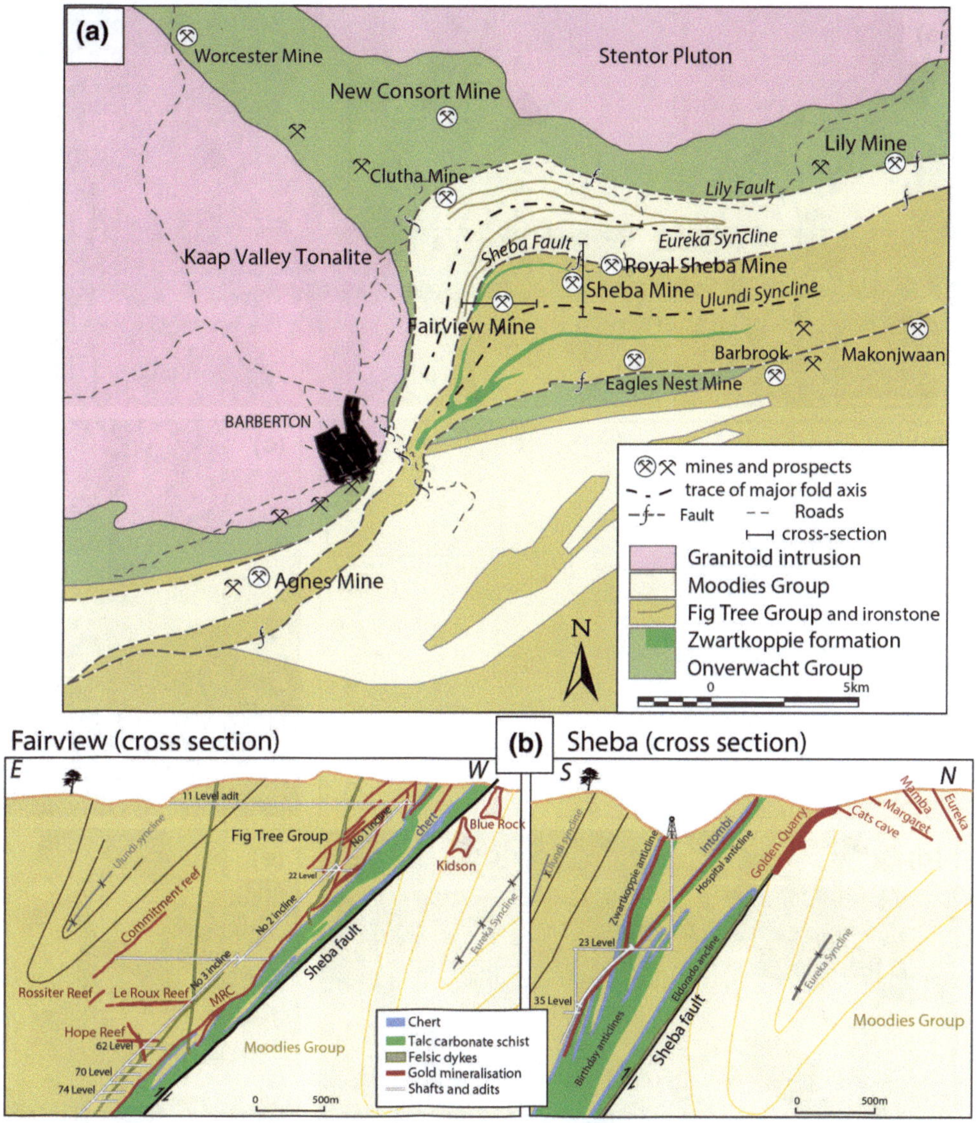

Fig. 7.2 **a** Map of the Eureka and Ulundi synclines and location of gold deposits (modified from Dirks et al. 2009), **b** cross sections of Fairview and Sheba mines (modified from Barberton gold mines 2014)

the hanging wall of the contact between greenstone lithologies and the Stentor Pluton. Mineralisation is hosted by upper greenschist- to upper amphibolite-facies rocks and is associated with the Consort Bar, a mylonitic shear zone separating (ultra)mafic rocks and chert of the Onverwacht Group from clastic sedimentary rocks of the Fig Tree Group (Otto et al. 2007; Munyai et al. 2011). Mineralisation is found in three bodies, the Seven Shaft Shoot, Isaura Shoot, and Prince Consort Shoot. The style of mineralisation in underground workings includes massive replacement-style ore and vein-hosted or disseminated mineralisation (Otto et al. 2007).

Two phases of mineralisation have been identified: the early phase in the footwall, characterised by löellingite-pyrrhotite ore assemblages and a high-temperature alteration calc-silicate assemblage composed of garnet, clinopyroxene, hornblende, K-feldspar, quartz, calcite and biotite (Dziggel et al. 2007; Otto et al. 2007; Dziggel and Kisters 2019). Pressure-temperature estimates for first-phase alteration assemblages in metamafic and intercalated metasedimentary rocks are circa 600–700 °C and 6–8 kbar (Otto et al. 2007).

The second, and main, phase of mineralisation has been attributed to oblique shear zones crosscutting the Consort Bar and was associated with the development of Cr-muscovite-K-feldspar-plagioclase-quartz or Cr-muscovite-tourmaline-plagioclase-rutile as typical alteration assemblages proximal to ore. This second mineralisation event occurred at temperatures ranging from 520 to 600 °C and pressures between 1 and 3 kbar; the ore assemblages vary from arsenopyrite-pyrrhotite to arsenopyrite-pyrrhotite-chalcopyrite-löellingite with increasing depth and temperature (Dziggel et al. 2006, 2010; Otto et al. 2007). Shearing was accompanied by emplacement of pegmatite that has been dated at

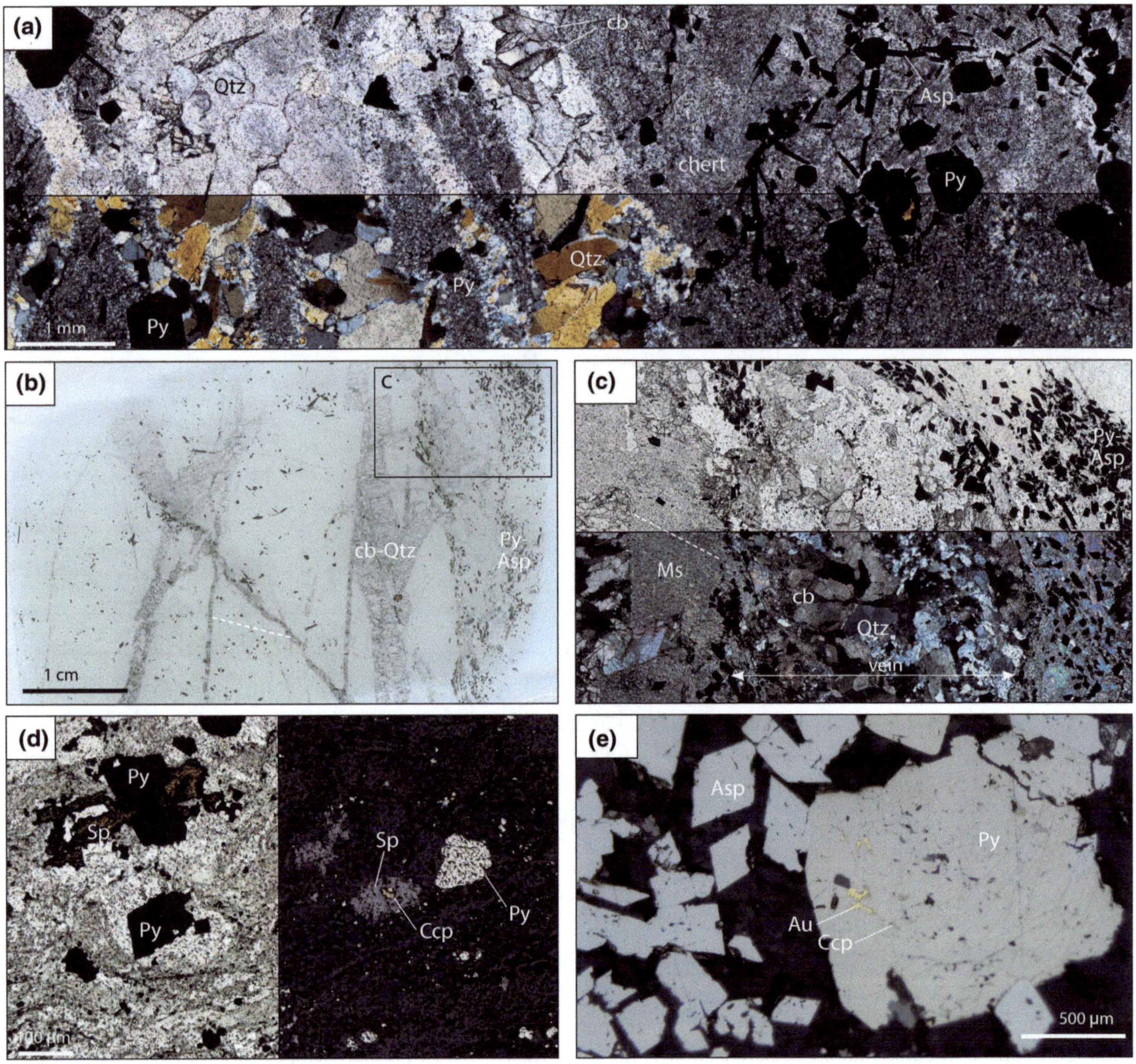

Fig. 7.3 Ore microtextures of samples from Sheba and Fairview mines. **a** Quartz-carbonate vein in chert, Fairview mine. Pyrite and arsenopyrite are present in the host rock. Note quartz crystals growing perpendicular to the vein walls, indicating dilatant behaviour of the cracks. **b**, **c** Carbonate-quartz veins in metagreywacke at Fairview mine. Arsenopyrite and pyrite are visible in both the vein and the alteration halo. Foliation is indicated by dashed lines. **d** Pyrite, sphalerite and chalcopyrite in metamafic schist at Sheba mine. **e** Pyrite-arsenopyrite ore with micro-inclusions of native gold and chalcopyrite, Fairview mine. All images transmitted polarised light, except for **a** and **c** (crossed polarisers in lower half), **d** (reflected light in right-hand half) and **e** (reflected light). Abbreviations: Asp arsenopyrite, cb carbonate, Ccp chalcopyrite, Ms K-mica, Py pyrite, Qtz quartz, Sp sphalerite

3040 ± 84 Ma (Rb–Sr whole-rock; Harris et al. 1995). Overall, textural relations and thermometric estimates indicate a clock-wise P-T path with a near-isobaric decompression to 520–600 °C and 1–3 kbar followed by near-isobaric cooling (Dziggel et al. 2006).

Munyai et al. (2011) and Dirks et al. (2013) focussed their observations on gold-associated fractures on surface exposure at, and in the vicinity of, New Consort mine. In stark contrast with what has been described earlier, they described gold mineralisation associated with low-temperature alteration assemblages and brittle deformation that developed during late-stage extensional tectonics. Thus, mineralisation at New Consort mine and surrounding areas is multiphase and may have formed at very different temperature conditions at different times (Dziggel and Kisters 2019).

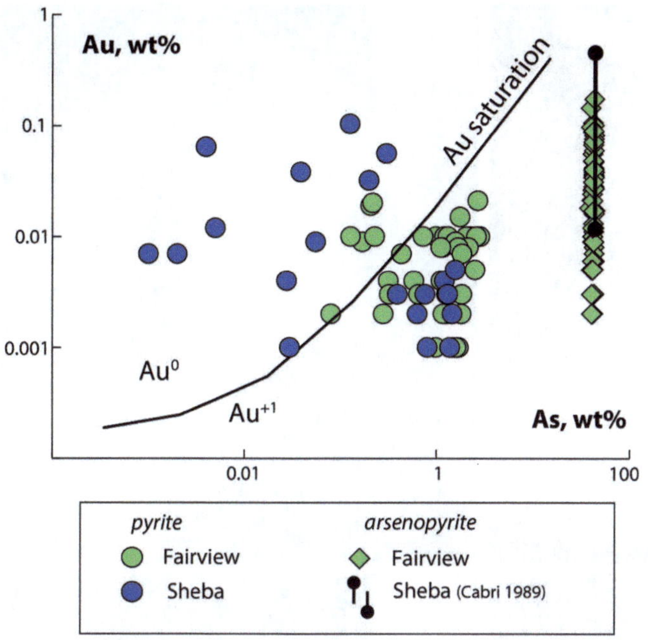

Fig. 7.4 Plot of Au versus As content of pyrite and arsenopyrite grains from Fairview and Sheba mines (data from Agangi et al. 2014 and Cabri et al. 1989). Gold saturation line in arsenian pyrite from Reich et al. (2005)

7.5 Fluid Inclusion and Mineral Stable Isotope Compositions

Fluid inclusion studies from the major deposits indicate low-salinity (NaCl eq = 5–6 wt%), H_2O-CO_2-rich fluids, and homogenisation temperatures in the T = 290 to 310 °C range (de Ronde et al. 1992). As-in-arsenopyrite geothermometry at Fairview mine indicates temperatures between <300 and 400 °C (Agangi et al. 2014). These temperature estimates are in broad agreement with the greenschist-facies alteration assemblages. Based on O, H and C isotope analyses of mineralisation-related quartz and carbonate, the ore fluid would have had narrow ranges of $\delta^{18}O$ (+4.7 to + 5.8‰), $\delta^{13}C$ (−4.5 to −2‰) and δD (−35 to −41‰), recalculated based on fluid-mineral equilibria at 300 °C (de Ronde et al. 1992). Hydrothermal sulphides have slightly positive $\delta^{34}S$ values (+1.2 to +3.9‰) for pyrite and arsenopyrite (Kakegawa and Ohmoto 1999). Overall, the fluid inclusion compositions and mineral stable isotope values indicate distinct homogeneity of mineralising fluids at the greenstone belt scale.

Despite the complexity of sulphide textures and trace element compositions, $\delta^{34}S$ values are confined to a narrow range of moderately positive values (de Ronde et al. 1992; Kakegawa and Ohmoto 1999) which limits the use of S isotopes as tracers of the fluid source. As a further complication, $\delta^{34}S$ values are affected by a complex range of processes during S transport and precipitation and associated redox reactions that can have overlapping effects. However, recent developments in S isotope studies have revealed that sulphate and sulphide minerals in the Archaean and early Palaeoproterozoic bear mass-independently fractionated S isotopes (MIF-S, indicated by the notation $\Delta^{33}S = \delta^{33}S - 0.515 \cdot \delta^{34}S$) (Farquhar et al. 2000; Ono et al. 2003; Johnston 2011). MIF-S signals are interpreted to be the footprint of photolytic reactions caused by UV irradiation of S gases (such as SO_2) in an O_2-free Archaean atmosphere, which would produce sulphide with $\Delta^{33}S > 0$ and sulphate with $\Delta^{33}S < 0$. Once produced, MIF-S is very robust and only slightly affected by most mass-dependent fractionation-inducing abiogenic and biogenic processes so that it can be used as a tracer of S processed in the atmosphere.

In situ multiple S isotope analyses of pyrite obtained by ion microprobe (SIMS) from samples of Fairview and Sheba mines revealed MIF-S with $\Delta^{33}S$ deviating significantly towards both positive and negative values ($\Delta^{33}S$ = −0.6 to +1.0‰) (Agangi et al. 2016). In the $\Delta^{33}S$ versus $\delta^{34}S$ plot, these values match the distribution peak of S compositions of pyrite and whole-rock analyses of the entire volcano-sedimentary succession and suggest derivation of S from the Barberton Supergroup (Fig. 7.6). Given the variability in $\Delta^{33}S$, S may have been scavenged from different rock types and then been transported by hydrothermal fluids responsible for Au mineralisation. More specifically, negative $\Delta^{33}S$ values have been measured in sulphides from volcanic-hosted massive sulphide deposits (VHMS) of the Bien Venue deposit of the Fig Tree Group and hydrothermally altered mafic and ultramafic rocks of the Komati Formation, Onverwacht Group (Montinaro et al. 2015). These rocks are interpreted to have derived their S isotopic signature from interaction with heated seawater and to have inherited the typically negative $\Delta^{33}S$ values of Archaean seawater sulphate (Bao et al. 2007; Montinaro et al. 2015). They are thus possible sources of some S (and Au) for the sulphides at Sheba and Fairview mines. Positive $\Delta^{33}S$ and $\delta^{34}S$ values were measured in pyrite from chert, conglomerate and dolomite from the Onverwacht and Fig Tree Groups, as well as from bulk shales from the Fig Tree Group (Grosch and McLoughlin 2013; Roerdink et al. 2013; Montinaro et al. 2015). In particular, some shale samples and pyrite from barite-free samples show a steep negative $\Delta^{33}S/\delta^{34}S$ slope that resembles the distribution of pyrite analyses at Sheba-Fairview. Irrespective of the origin of this negative trend, which may result from mixing of different S components or represent a primary photolytic signal (Roerdink et al. 2013), the similarity to the values observed for hydrothermal pyrite at Sheba-Fairview mines implies that similar rocks may be a good source of S in the Au deposits (Fig. 7.6b).

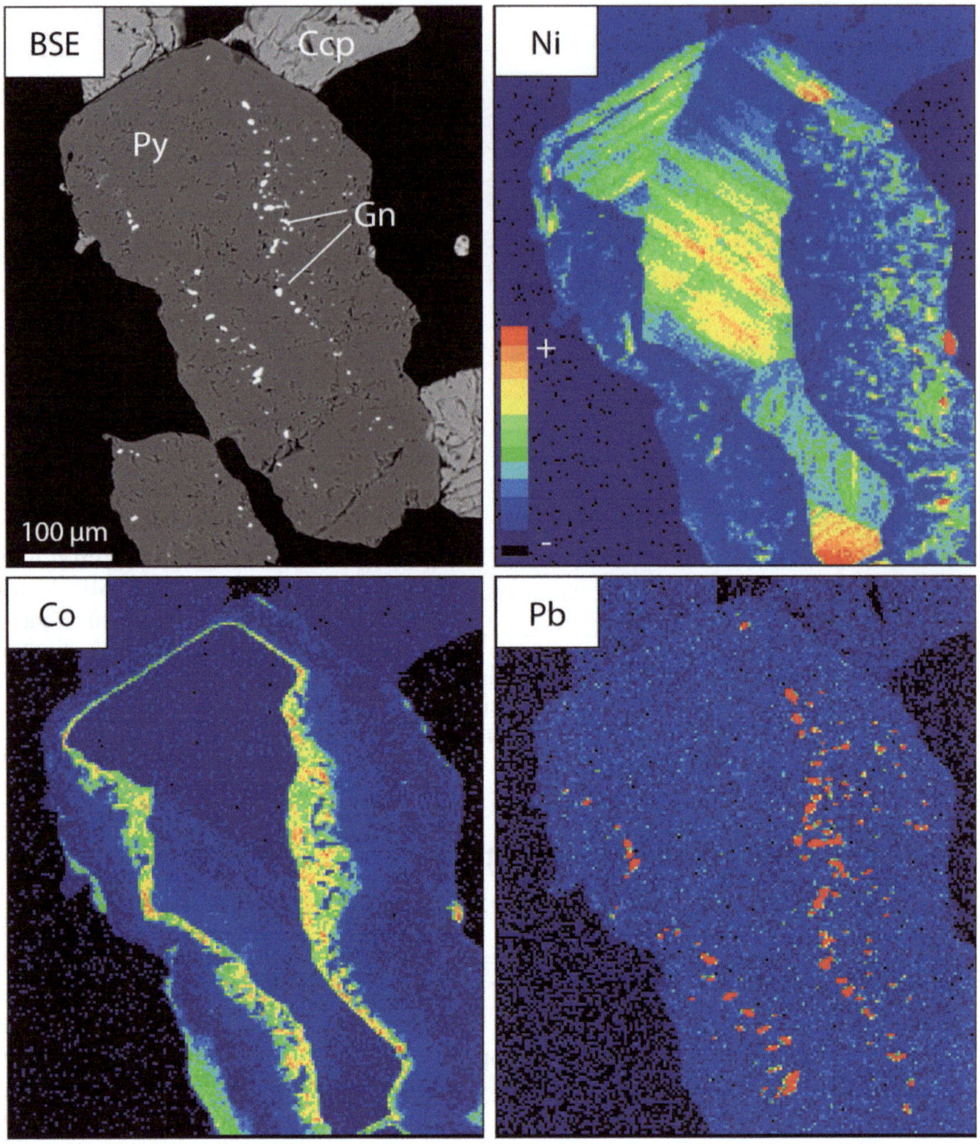

Fig. 7.5 BSE image and Ni-Co-Pb X-ray maps of pyrite from Sheba mine. Abbreviations: Ccp chalcopyrite, Gn galena, Py pyrite

7.6 Discussion and Conclusions

7.6.1 Sources of S, Fluids, and Gold: Internal or External to the Greenstone Belt?

The origin of mineralising fluids and Au in Archaean greenstone belts has proved to be elusive, and the various hypotheses proposed reflect the difficulty in the identification of such sources in structurally-controlled Au deposits, both in Precambrian and Phanerozoic ages (Hronsky et al. 2012; Gaboury 2013; Goldfarb and Groves 2015). Possible sources of sulphur in Archaean orogenic gold deposits include supracrustal rocks (Groves et al. 2003; Phillips and Powell 2009; Tomkins 2013), hydrothermal systems related to felsic magmatism (Salier et al. 2005; Doublier et al. 2014), the lower crust, and the mantle (Hronsky et al. 2012; Fu and Touret 2014).

The hypothesis of a magmatic origin for mineralising fluids has been proposed in several cases of Archaean structurally-controlled Au deposits, such as in Neoarchaean Au deposits of Western Australia, based on the presence of coeval magmatism (Wang et al. 1993; Doublier et al. 2014), trace element signature of accessory minerals (Bath et al. 2013), and Pb and noble gas isotope studies (Qiu and McNaughton 1999; Kendrick et al. 2011). A causal relationship between plutonism and Au mineralisation in the Barberton Greenstone Belt has been proposed mostly based on spatial distribution (Anhaeusser 1976, 1986). In the Barberton Greenstone Belt, mineralisation post-dates the main phase of potassic granite plutonism, which caused the emplacement of extensive batholiths, such as the

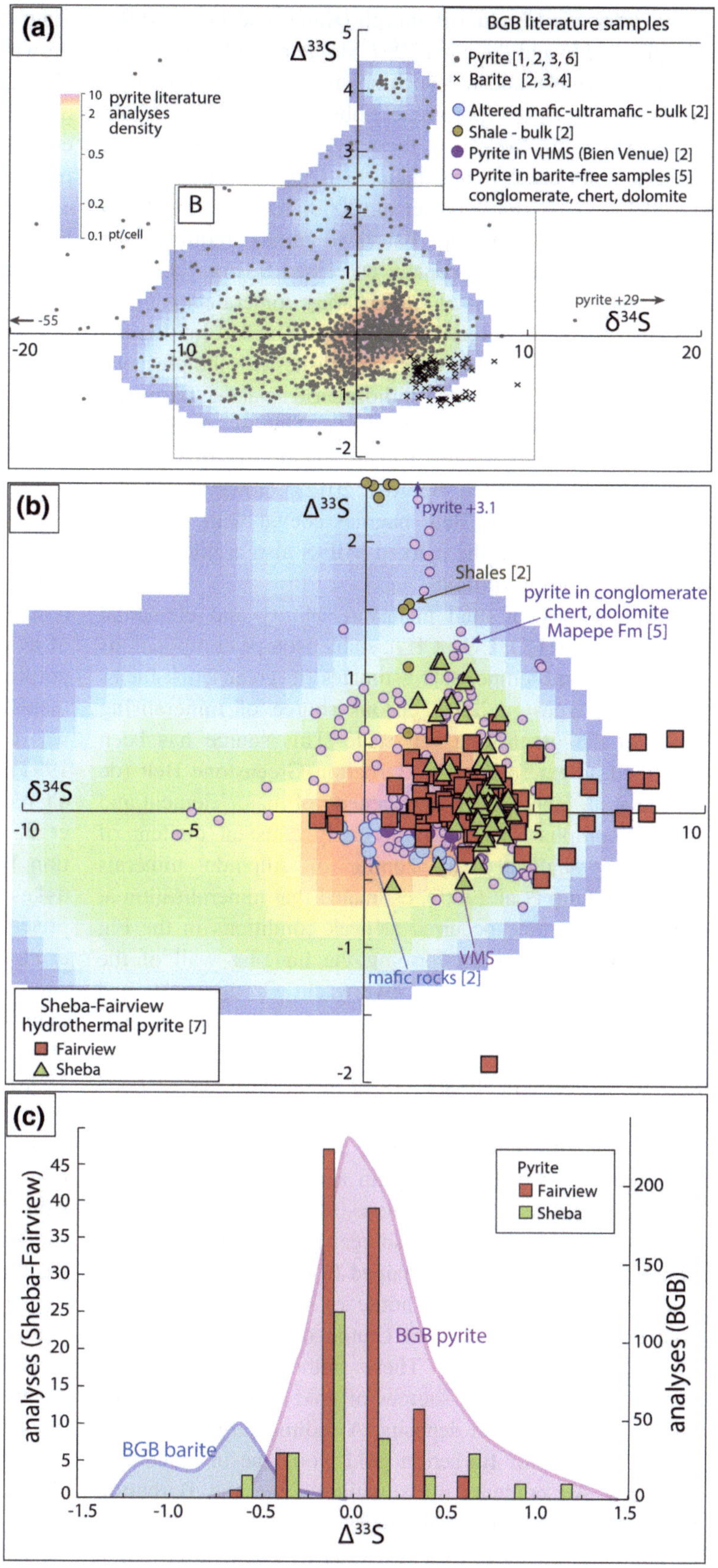

Fig. 7.6 Comparison of S isotope analyses of pyrite associated with gold mineralisation at Sheba and Fairview mines, and pyrite and barite from different units of the Barberton Greenstone Belt. **a** $\Delta^{33}S$ versus $\delta^{34}S$ plot of pyrite and barite from different units of the Barberton Greenstone Belt (n = 1680) and density distribution of pyrite data. **b** $\Delta^{33}S$ versus $\delta^{34}S$ plot of pyrite from Au-mineralised samples from Sheba and Fairview mines and some representative rocks of the Barberton Greenstone Belt. **c** Histogram of $\Delta^{33}S$ of pyrite from Sheba and Fairview mines compared with Barberton Greenstone Belt pyrite and barite. Data from: (1) Philippot et al. (2012), (2) Montinaro et al. (2015), (3) Roerdink et al. (2012), (4) Bao et al. (2007), (5) Roerdink et al. (2013), (6) Grosch and McLoughlin (2013), (7) Agangi et al. (2016)

3160–3090 Ma Mpuluzi Batholith (Kamo and Davis 1994; Murphy 2015) and the circa 3067 Ma Stentor pluton (Kamo and Davis 1994), and was accompanied by emplacement of small granitic dykes (porphyries) observed in different gold mines. However, the role of these dykes in the mineralising process is not clear. A magmatic source is expected to have 0 ± 0.2‰ $\Delta^{33}S$ (Farquhar et al. 2000) so that the MIF-S measured at Sheba and Fairview mines requires the involvement of atmospherically processed S (Fig. 7.6b).

A mantle derivation of CO_2-bearing mineralising fluids has been proposed in recent models that aimed at linking the presence of various types of gold deposits with the presence of "fertile' metasomatised lithospheric mantle (Hronsky et al. 2012; Fu and Touret 2014). This is particularly applicable in the presence of mafic mantle magmas coeval with mineralisation (de Boorder 2012). However, in the Barberton Greenstone Belt magmas coeval with mineralisation are felsic, and the marked MIF-S also argues against such a deep origin of fluids.

Homogeneity of fluid inclusion salinity and elemental compositions (H_2O, CO_2, CH_4), stable isotope compositions of minerals and temperature estimates (in greenschist-facies deposits) suggest a homogeneous source of mineralising fluids at the greenstone belt scale. This source has been identified as external to the Barberton Greenstone Belt (de Ronde et al. 1992). This homogeneity of major element and isotope compositions does not imply constant content of metals, as indicated by zoning of sulphide minerals (Fig. 7.5). Otto et al. (2007) estimated that mineralisation at New Consort mine occurred at peak conditions in the Fig Tree sedimentary rocks forming the hanging wall of the Consort Bar, whereas the Onverwacht Group rocks had already experienced metamorphic peak and were on their retrograde path. They concluded that, as a consequence, the host rocks are unlikely sources of fluids (also see Dziggel and Kisters 2019).

In apparent contrast, multiple S isotope analyses with non-zero MIF-S values imply an atmospheric source, and heterogeneous $\Delta^{33}S$ values suggest that a variety of rocks with different $\Delta^{33}S$ acted as S sources for mineralising fluids. Similar results have been obtained from multiple S isotope analyses of pyrite and pyrrhotite of various Neoarchaean gold deposits of the Yilgarn Craton of Western Australia (Selvaraja et al. 2017). These authors underlined the importance of shales as sources of gold in Archaean structurally controlled gold deposits. A sedimentary rather than igneous source for Barberton gold would be in agreement with low Au contents of komatiites from the Barberton Supergroup (Hofmann et al. 2017).

The microscale heterogeneity of $\Delta^{33}S$ values implies that the ore-forming fluid was isotopically heterogeneous, and variable MIF-S values were not completely homogenised during hydrothermal fluid flow. The isotopic heterogeneity is matched by trace element heterogeneity, which appears as growth zones, recrystallisation domains and veinlets, a characteristic also observed in several other orogenic gold deposits of Archaean and Palaeoproterozoic age (Morey et al. 2008; Fougerouse et al. 2016). This evidence is compatible with a pulsating fluid flow rather than a single large event and suggests that single fluid pulses had a very localised effect in terms of ore deposition. This is similar to what is observed in orogenic gold deposits and typically attributed to periodic fluid pressure build-up and earthquake-related release, also known as fault-valve mechanisms (Sibson et al. 1988).

7.6.2 Timing of Gold Mineralisation in the Barberton Greenstone Belt and Tectonic Context

The age of mineralisation in the Barberton Greenstone Belt is still rather poorly constrained, namely because of a dearth of available data (Fig. 7.7). At Fairview and Sheba mines, porphyry dykes crosscutting mineralisation give good estimates of minimum mineralisation ages. One of these dykes was dated at 3126 Ma (U-Pb on zircon; de Ronde et al. 1991). A rutile separate extracted from a pre-mineralisation, altered dyke at Fairview was dated at 3084 Ma (de Ronde et al. 1992), which may be a good estimate for mineralisation. However, recent U-Pb zircon dating of two mineralised dykes at Golden Quarry, near Sheba mine, has been interpreted to indicate mineralisation at circa 3015 Ma (Dirks et al. 2013). At New Consort mine, a 3027 ± 7 Ma U-Pb age on titanite (Dziggel et al. 2006) associated with main-stage mineralisation and a whole-rock Rb-Sr age of 3040 ± 84 Ma on a syn-mineralisation dyke (Harris et al. 1995) may indicate that mineralisation was caused by a protracted or episodic tectonic activity that lasted for several tens of million years. The multiphase nature of mineralisation at New Consort mine seems to support such a conclusion (Dziggel et al. 2007, 2010; Dziggel and Kisters 2019).

The tectonic regime in operation in the Kaapvaal craton at the time of Au mineralisation are inferred from a fragmentary record, and a complete picture has still to emerge. At the craton scale, mineralisation occurred shortly after the final consolidation of the Kaapvaal craton and at the beginning of an extension phase that led to bimodal volcanism at circa 3074 Ma (Dominion Group, Armstrong et al. 1991) in the central part of the craton and at 2980 Ma to the southeast (Nsuze Group, Hegner et al. 1984).

In the eastern part of the craton, felsic magmatism is indicated by emplacement of the Sinceni granite in Swaziland at circa 3070 Ma (Maphalala and Kröner 1993). A continent-wide felsic magmatic event at circa 3070 Ma is indicated by the abundance of detrital zircons of this age in

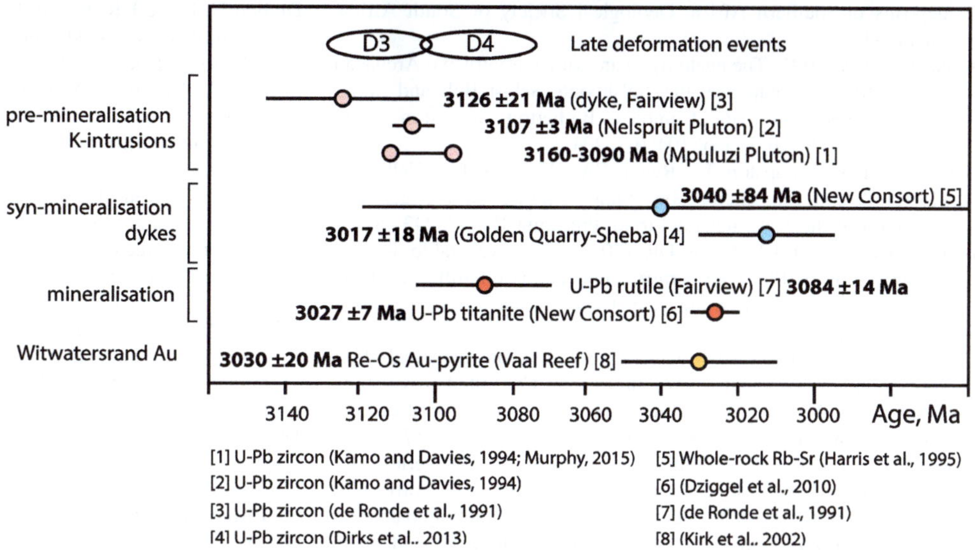

Fig. 7.7 Age constraints on Au mineralisation in the Barberton Greenstone Belt

the Witwatersrand and Pongola Supergroups (Kositcin and Krapez 2004; Wilson and Zeh 2019). At the same time in the north of the Kaapvaal craton, volcanic rocks and sedimentary rocks of the Murchison Greenstone Belt, such as the Weigel Formation and MacKop conglomerate, were deposited, possibly starting from circa 3090 Ma in what is interpreted as a convergent setting (Poujol et al. 1996). Following this magmatic event, volcanic rocks of the Rubbervale Formation in the Murchison Greenstone Belt and Rooiwater granitoids were emplaced at circa 2970 Ma. The emplacement of these volcanic rocks is taken as evidence of crust formation in an arc environment related to the accretion of the northern part of the Kaapvaal craton, the Pietersburg block, to the Witwatersrand block and collision at that time (Zeh et al. 2013).

Gold mineralisation in the Barberton Greenstone Belt is coeval with the tectono-magmatic event related to an extension that shortly followed the final phase of stabilisation of the central Kaapvaal craton. Greenstone gold mineralisation is frequently associated with the latest stage of stabilisation of cratons and can be related to the switch between compression-collision and extensions, when high-angle extensional faults allow deep fluids to flow across the crust (Goldfarb and Groves 2015; Groves et al. 2000). Intriguingly, the available ages for mineralisation partly overlap with Re–Os age determinations of gold and detrital pyrite grains from the Vaal Reef of the Witwatersrand basin (circa 3030 ± 20 Ma, Kirk et al. 2002), the largest gold deposit in the world (Frimmel this volume). Based on this age and an unradiogenic initial $^{187}Os/^{188}Os$ ratio of 0.109, Kirk et al. (2002) concluded that the Witwatersrand gold was eroded from a mantle-derived mafic-ultramafic source rock, although the specific origin of this detrital gold could not be identified. Therefore, gold mineralisation preserved in the Barberton Greenstone Belt may have formed as part of this major crust- and Au-forming event, of which it would represent but a remnant.

Acknowledgements This research was funded by SIEF (Science and Industry Endowment Fund), the NRF (National Research Foundation of South Africa), and the DST-NRF Centre of Excellence for Integrated Mineral and Energy Resource Analysis (CIMERA). JMC thanks the CNRS-INSU Programme National de Planétologie for their support. We acknowledge Chris Rippon (Barberton Mines (Pty) Limited) for providing sample material.

References

Agangi A, Hofmann A, Przybyłowicz W (2014) Trace element zoning of sulfides and quartz at Sheba and Fairview gold mines: clues to Mesoarchean mineralisation in the Barberton Greenstone Belt, South Africa. Ore Geol Rev 56:94–114

Agangi A, Przybyłowicz W, Hofmann A (2015) Trace element mapping of pyrite from Archean gold deposits—A comparison between PIXE and EPMA. Nucl Instrum Methods Phys Res, Sect B 348:302–306

Agangi A, Hofmann A, Eickmann B, Marin-Carbonne J, Reddy SM (2016) An atmospheric source of S in Mesoarchaean structurally-controlled gold mineralisation of the Barberton Greenstone Belt. Precambr Res 285:10–20

Altigani MAH, Merkle RKW, Dixon RD (2016) Geochemical identification of episodes of gold mineralisation in the Barberton Greenstone Belt, South Africa. Ore Geol Rev 75:186–205

Anhaeusser CR (1976) Archean metallogeny in southern Africa. Econ Geol 71:16–43

Anhaeusser CR (1984) Structural elements of Archaean granite-greenstone terranes as exemplified by the Barberton Mountain Land, Southern Africa. In: Kröner A, Greiling R (eds) Precambrian Tectonics Illustrated. E. Schweizerbart'sche Verlagsbuchhandlung, Stuttgart, pp 57–58

Anhaeusser CR (1986) Archaean gold mineralization in the Barberton Mountain Land. In: Anhaeusser CR, Maske S (eds) Mineral

deposits of southern Africa, Geological Society of South Africa, pp 113–154
Anhaeusser CR (2001) The anatomy of an extrusive-intrusive Archaean mafic-ultramafic sequence: the Nelshoogte Schist Belt and Stolzburg Layered Ultramafic Complex, Barberton Greenstone Belt, South Africa. S Afr J Geol 104:167–204
Armstrong RA, Compston W, Retief EA, Williams IS, Welke HJ (1991) Zircon ion microprobe studies bearing on the age and evolution of the Witwatersrand triad. Precambr Res 53:243–266
Bao H, Rumble D III, Lowe DR (2007) The five stable isotope compositions of Fig Tree barites: implications on sulfur cycle in ca. 3.2-Ga oceans. Geochimica et Cosmochimica Acta 71:4868–4879
Barberton Gold Mines (2010) Mineral resources and technical short form report. www.panafricanresources.com. Consulted Nov 2012
Barberton Gold Mines (2014) Mineral resources and mineral reserve report. www.panafricanresources.com. Consulted Feb 2017
Bath AB, Walshe JL, Cloutier J, Verrall M, Cleverley JS, Pownceby MI, Macrae CM, Wilson NC, Tunjic J, Nortje GS, Robinson P (2013) Biotite and apatite as tools for tracking pathways of oxidized fluids in the Archean East Repulse gold deposit, Australia. Econ Geol 108:667–690
Cabri LJ, Chryssoulis SL, de Villiers JPR, Laflamme JHG, Buseck PR (1989) The nature of "invisible" gold in arsenopyrite. The Canadian Mineralogist 27:353–362
Cutts KA, Stevens G, Hoffmann JE, Buick IS, Frei D, Münker C (2014) Paleo- to Mesoarchean polymetamorphism in the Barberton Granite-Greenstone Belt, South Africa: Constraints from U-Pb monazite and Lu-Hf garnet geochronology on the tectonic processes that shaped the belt. Geol Soc Am Bull 126:251–270
de Boorder H (2012) Spatial and temporal distribution of the orogenic gold deposits in the late Palaeozoic Variscides and Southern Tianshan: how orogenic are they? Ore Geol Rev 46:1–31
de Ronde CEJ, de Wit MJ (1994) Tectonic history of the Barberton greenstone belt, South Africa: 490 million years of Archean crustal evolution. Tectonics 13:983–1005
de Ronde CEJ, Kamo SL (2000) An Archaean arc-arc collisional event: a short-lived (ca 3 Myr) episode, Weltevreden area, Barberton greenstone belt, South Africa. J Afr Earth Sc 30:219–248
de Ronde CEJ, Kamo S, Davis DW, de Wit MJ, Spooner ETC (1991) Field, geochemical and U-Pb isotopic constraints from hypabyssal felsic intrusions within the Barberton greenstone belt, South Africa: implications for tectonics and the timing of gold mineralization. Precambr Res 49:261–280
de Ronde CEJ, Spooner ETC, de Wit MJ, Bray CJ (1992) Shear zone-related, Au quartz vein deposits in the Barberton greenstone belt, South Africa; field and petrographic characteristics, fluid properties, and light stable isotope geochemistry. Econ Geol 87:366–402
Diener JFA, Stevens G, Kisters AFM, Poujol M (2005) Metamorphism and exhumation of the basal parts of the Barberton greenstone belt, South Africa: constraining the rates of Mesoarchaean tectonism. Precambr Res 143:87–112
Dirks PHGM, Charlesworth EG, Munyai MR (2009) Cratonic extension and Archaean gold mineralisation in the Sheba-Fairview mine, Barberton greenstone belt, South Africa. S Afr J Geol 112:291–316
Dirks PHGM, Charlesworth EG, Munyai MR, Wormald R (2013) Stress analysis, post-orogenic extension and 3.01 Ga gold mineralisation in the Barberton Greenstone Belt, South Africa. Precambrian Res 226:157–184
Doublier MP, Thébaud N, Wingate MTD, Romano SS, Kirkland CL, Gessner K, Mole DR, Evans N (2014) Structure and timing of Neoarchean gold mineralization in the Southern Cross district (Yilgarn Craton, Western Australia) suggest leading role of late Low-Ca I-type granite intrusions. J Struct Geol 67, Part B: 205–221
Drabon N, Lowe DR, Byerly G, Harrington JA (2017) Detrital zircon geochronology of sandstones of the 3.6–3.2 Ga Barberton greenstone belt: no evidence for older continental crust. Geology 45:803–806
Dziggel A, Kisters AFM (2019) Chapter 26—Tectono-metamorphic controls on Archean gold mineralization in the Barberton Greenstone Belt, South Africa. In: Van Kranendonk RH, Bennet VC Hoffmann E (eds) Earth's oldest rocks 2nd Ed, Elsevier, pp 655–674
Dziggel A, Stevens G, Poujol M, Anhaeusser CR, Armstrong RA (2002) Metamorphism of the granite-greenstone terrane south of the Barberton greenstone belt, South Africa: an insight into the tectono-thermal evolution of the lower portions of the Onverwacht Group. Precambr Res 114:221–247
Dziggel A, Armstrong RA, Stevens G, Nasdala L (2005) Growth of zircon and titanite during metamorphism in the granitoid-gneiss terrane south of the Barberton greenstone belt, South Africa. Mineral Mag 69:1019–1036
Dziggel A, Knipfer S, Kisters AFM, Meyer FM (2006) P-T and structural evolution during exhumation of high-T, medium-P basement rocks in the Barberton Mountain Land, South Africa. J Metamorph Geol 24:535–551
Dziggel A, Otto A, Kisters AFM, Meyer FM (2007) Chapter 5.8 Tectono-metamorphic controls on Archean gold mineralization in the Barberton Greenstone Belt, South Africa: an example from the New Consort gold mine. In: Van Kranendonk, MJ, Smithies, RH & Bennet, VC (eds) Earth's oldest rocks, Elsevier, Amsterdam, pp 699–727
Dziggel A, Poujol M, Otto A, Kisters AFM, Trieloff M, Schwarz WH, Meyer FM (2010) New U-Pb and $^{40}Ar/^{39}Ar$ ages from the northern margin of the Barberton greenstone belt, South Africa: implications for the formation of Mesoarchaean gold deposits. Precambr Res 179:206–220
Farquhar J, Bao H, Thiemens M (2000) Atmospheric influence of Earth's earliest sulfur cycle. Science 289:756–758
Fougerouse D, Micklethwaite S, Ulrich S, Miller J, Godel B, Adams DT, McCuaig TC (2016) Evidence for two stages of mineralisation in west Africa's largest gold deposit: Obuasi, Ghana. Econ Geol 112:2–33
Frimmel HE, Groves DI, Kirk J, Ruiz J, Chesley J, Minter WEL (2005) The formation and preservation of the Witwatersrand goldfields, the world's largest gold province. Econ Geol 100th Anniversary Volume:769–698
Fu B, Touret JLR (2014) From granulite fluids to quartz-carbonate megashear zones: the gold rush. Geosci Front 5:747–758
Gaboury D (2013) Does gold in orogenic deposits come from pyrite in deeply buried carbon-rich sediments?: insight from volatiles in fluid inclusions. Geology 41:1207–1210
Goldfarb RJ, Groves DI (2015) Orogenic gold: common or evolving fluid and metal sources through time. Lithos 233:2–26
Grosch EG, McLoughlin N (2013) Paleoarchean sulfur cycle and biogeochemical surface conditions on the early Earth, Barberton, South Africa. Earth Planet Sci Lett 377–378:142–154
Groves DI, Goldfarb RJ, Knox-Robinson CM, Ojala J, Gardoll S, Yun GY, Holyland P (2000) Late-kinematic timing of orogenic gold deposits and significance for computer-based exploration techniques with emphasis on the Yilgarn Block, Western Australia. Ore Geol Rev 17:1–38
Groves DI, Goldfarb RJ, Robert F, Hart CJR (2003) Gold deposits in metamorphic belts: overview of current understanding, outstanding problems, future research, and exploration significance. Econ Geol 98:1–29
Harris PD, Robb LJ, Tomkinson MJ (1995) The nature and structural setting of rare-element pegmatites along the northern flank of the Barberton greenstone belt, South Africa. S Afr J Geol 98:82–94
Hazarika P, Mishra B, Pruseth KL (2017) Trace-element geochemistry of pyrite and arsenopyrite: ore genetic implications for late Archean orogenic gold deposits in southern India. Mineral Mag 81:661

Hegner E, Kröner A, Hofmann AW (1984) Age and isotope geochemistry of the Archaean Pongola and Usushwana suites in Swaziland, Southern Africa: a case for crustal contamination of mantle-derived magma. Earth Planet Sci Lett 70:267–279

Heubeck C, Engelhardt J, Byerly GR, Zeh A, Sell B, Luber T, Lowe DR (2013) Timing of deposition and deformation of the Moodies Group (Barberton Greenstone Belt, South Africa): very-high-resolution of Archaean surface processes. Precambr Res 231:236–262

Heubeck C, Bläsing S, Grund M, Drabon N, Homann M, Nabhan S (2016) Geological constraints on Archean (3.22 Ga) coastal-zone processes from the Dycedale Syncline, Barberton greenstone belt. S Afr J Geol 119:495

Hofmann A (2005) The geochemistry of sedimentary rocks from the Fig Tree Group, Barberton greenstone belt: implications for tectonic, hydrothermal and surface processes during mid-Archaean times. Precambr Res 143:23–49

Hofmann A, Pitcairn I, Wilson A (2017) Gold mobility during Palaeoarchaean submarine alteration. Earth Planet Sci Lett 462:47–54

Hronsky J, Groves D, Loucks R, Begg G (2012) A unified model for gold mineralisation in accretionary orogens and implications for regional-scale exploration targeting methods. Miner Deposita 47:339–358

Johnston DT (2011) Multiple sulfur isotopes and the evolution of Earth's surface sulfur cycle. Earth Sci Rev 106:161–183

Kakegawa T, Ohmoto H (1999) Sulfur isotope evidence for the origin of 3.4 to 3.1 Ga pyrite at the Princeton gold mine, Barberton greenstone belt, South Africa. Precambr Res 96:209–224

Kamo SL, Davis DW (1994) Reassessment of Archean crustal development in the Barberton Mountain Land, South Africa, based on U-Pb dating. Tectonics 13:167–192

Kendrick MA, Honda M, Walshe J, Petersen K (2011) Fluid sources and the role of abiogenic-CH_4 in Archean gold mineralization: Constraints from noble gases and halogens. Precambr Res 189(3–4):313–327

Kirk J, Ruiz J, Chesley J, Walshe J, England G (2002) A major Archean, gold- and crust-forming event in the Kaapvaal Craton, South Africa. Science 297:1856–1858

Kisters AFM, Stevens G, Dziggel A, Armstrong RA (2003) Extensional detachment faulting and core-complex formation in the southern Barberton granite-greenstone terrain, South Africa: evidence for a 3.2 Ga orogenic collapse. Precambr Res 127:355–378

Kisters AFM, Belcher RW, Poujol M, Dziggel A (2010) Continental growth and convergence-related arc plutonism in the Mesoarchaean: evidence from the Barberton granitoid-greenstone terrain, South Africa. Precambr Res 178:15–26

Kolb J, Dziggel A, Bagas L (2015) Hypozonal lode gold deposits: a genetic concept based on a review of the New Consort, Renco, Hutti, Hira Buddini, Navachab, Nevoria and the granites deposits. Precambr Res 262:20–44

Kositcin N, Krapez B (2004) Relationship between detrital zircon age-spectra and the tectonic evolution of the late Archaean Witwatersrand Basin, South Africa. Precambr Res 129:141–168

Kröner A, Hegner E, Wendt JI, Byerly GR (1996) The oldest part of the Barberton granitoid-greenstone terrain, South Africa: evidence for crust formation between 3.5 and 3.7 Ga. Precambr Res 78:105–124

Kröner A, Anhaeusser CR, Hoffmann JE, Wong J, Geng H, Hegner E, Xie H, Yang J, Liu D (2016) Chronology of the oldest supracrustal sequences in the Palaeoarchaean Barberton Greenstone Belt, South Africa and Swaziland. Precambr Res 279:123–143

Lana C, Tohver E, Cawood P (2010) Quantifying rates of dome-and-keel formation in the Barberton granitoid-greenstone belt, South Africa. Precambr Res 177:199–211

Lowe DR (1999) Geologic evolution of the Barberton Greenstone Belt and vicinity. Geological Society of America Special Papers 329:287–312

Lowe DR, Byerly GR (1999) Stratigraphy of the west-central part of the Barberton Greenstone Belt, South Africa. Geological Society of America Special Papers 329:1–36

Lowe DR, Byerly GR (2007) An overview of the geology of the Barberton Greenstone Belt and vicinity: implications for early crustal development. In: Van Kranendonk MJ, Smithies RH, Bennett VC (eds) Earth's oldest rocks. Elsevier, Amsterdam, pp 481–524

Lowe, DR, Byerly, GR, Heubeck, C (2012) Geologic Map of the west-central Barberton greenstone belt. In: South Africa, scale 1:25,000: Geological Society of America Map and Chart Series No. 103

Maphalala R, Kröner A (1993) Pb-Pb single zircon ages for the younger Archaean granitoids of Swaziland, southern Africa. In: Maphalala R, Mabuza M (eds) 16th Colloquium of African Geology. Swaziland Geological Survey and Mines 2: 201–206

Matsumura R (2014) The petrogenesis of the Nelshoogte pluton: the youngest and most compositionally variable TTG pluton in the Barberton Granite-Greenstone Terrain. Unpublished MSc thesis, Stellenbosch University, 108 p

Mishra B, Pruseth KL, Hazarika P, Chinnasamy SS (2017) Nature and source of the ore-forming fluids associated with orogenic gold deposits in the Dharwar Craton. Geosci Front 9:715–726

Montinaro A, Strauss H, Mason PRD, Roerdink D, Münker C, Schwarz-Schampera U, Arndt NT, Farquhar J, Beukes NJ, Gutzmer J, Peters M (2015) Paleoarchean sulfur cycling: multiple sulfur isotope constraints from the Barberton greenstone belt, South Africa. Precambr Res 267:311–322

Morey AA, Tomkins AG, Bierlein FP, Weinberg RF, Davidson GJ (2008) Bimodal distribution of gold in pyrite and arsenopyrite: examples from the Archean Boorara and Bardoc shear systems, Yilgarn Craton, Western Australia. Econ Geol 103:599–614

Moyen J-F, Stevens G, Kisters A (2006) Record of mid-Archaean subduction from metamorphism in the Barberton terrain, South Africa. Nature 442:559–562

Munyai MR, Dirks PHGM, Charlesworth EG (2011) Archaean gold mineralisation during post-orogenic extension in the New Consort gold mine, Barberton greenstone belt, South Africa. S Afr J Geol 114:121–144

Murphy RCL (2015) Stabilising a craton: the origin and emplacement of the 3.1 Ga Mpuluzi Batholith. Unpublished PhD thesis, Macquarie University, Sydney, Australia, 489p

Ono S, Eigenbrode JL, Pavlov AA, Kharecha P, Rumble D III, Kasting JF, Freeman KH (2003) New insights into Archean sulfur cycle from mass-independent sulfur isotope records from the Hamersley Basin, Australia. Earth Planet Sci Lett 213:15–30

Otto A, Dziggel A, Kisters A, Meyer F (2007) The New Consort Gold Mine, Barberton greenstone belt, South Africa: orogenic gold mineralization in a condensed metamorphic profile. Miner Deposita 42:715–735

Pearton T, Viljoen M (2017) Gold on the Kaapvaal Craton, outside the Witwatersrand Basin, South Africa. S Afr J Geol 120:101–132

Philippot P, van Zuilen M, Rollion-Bard C (2012) Variations in atmospheric sulphur chemistry on early Earth linked to volcanic activity. Nat Geosci 5:668–674

Phillips GN, Powell R (2009) Formation of gold deposits: review and evaluation of the continuum model. Earth Sci Rev 94:1–21

Pokrovski GS, Akinfiev NN, Borisova AY, Zotov AV, Kouzmanov K (2014) Gold speciation and transport in geological fluids: insights from experiments and physical-chemical modelling. Geol Soc Spec Pub 402:9–70

Poujol M, Robb LJ, Respaut J-P, Anhaeusser CR (1996) 3.07–2.97 Ga greenstone belt formation in the northeastern Kaapvaal Craton; implications for the origin of the Witwatersrand Basin. Econ Geol 91:1455–1461

Qiu Y, McNaughton NJ (1999) Source of Pb in orogenic lode-gold mineralisation: Pb isotope constraints from deep crustal rocks from the southwestern Archaean Yilgarn Craton, Australia. Mineralium Deposita 34(4):366–381

Reich M, Kesler SE, Utsunomiya S, Palenik CS, Chryssoulis SL, Ewing RC (2005) Solubility of gold in arsenian pyrite. Geochimica et Cosmochimica Acta 69:2781–2796

Roerdink DL, Mason PRD, Farquhar J, Reimer T (2012) Multiple sulfur isotopes in Paleoarchean barites identify an important role for microbial sulfate reduction in the early marine environment. Earth Planet Sci Lett 331–332:177–186

Roerdink DL, Mason PRD, Whitehouse MJ, Reimer T (2013) High-resolution quadruple sulfur isotope analyses of 3.2 Ga pyrite from the Barberton Greenstone Belt in South Africa reveal distinct environmental controls on sulfide isotopic arrays. Geochimica et Cosmochimica Acta 117:203–215

SACS (1980) Stratigraphy of South Africa Part 1: Lithostratigraphy of the Republic of South Africa, South West Africa/Namibia and the Republics of Bophuthatswana, Transkei and Venda. Geological Survey of South Africa, 690p

Sahoo PR, Venkatesh AS (2015) Constraints of mineralogical characterization of gold ore: implication for genesis, controls and evolution of gold from Kundarkocha gold deposit, eastern India. J Asian Earth Sci 97, Part A: 136–149

Salier BP, Groves DI, McNaughton NJ, Fletcher IR (2005) Geochronological and stable isotope evidence for widespread orogenic gold mineralization from a deep-seated fluid source at ca 2.65 Ga in the Laverton Gold Province, Western Australia. Econ Geol 100:1363–1388

Schoene B, de Wit MJ, Bowring SA (2008) Mesoarchean assembly and stabilization of the eastern Kaapvaal craton: a structural-thermochronological perspective. Tectonics 27:TC5010

Schouwstra RP (1995) Wall-rock alteration as a guide to gold-bearing fracture zones in the Zwartkoppie Section, Sheba gold mine, South Africa. S Afr J Geol 98:399–414

Schouwstra RP, De Villier J (1998) Gold mineralization and associated wallrock alteration in the Main Reef Complex at Sheba Mine, South Africa. Appl Earth Sci IMM Trans Sect B:B158–B170

Selvaraja V, Fiorentini ML, Jeon H, Savard DD, LaFlamme CK, Guagliardo P, Caruso S, Bui T-H (2017) Evidence of local sourcing of sulfur and gold in an Archaean sediment-hosted gold deposit. Ore Geol Rev 89:909–930

Sibson RH, Robert F, Poulsen KH (1988) High-angle reverse faults, fluid-pressure cycling, and mesothermal gold-quartz deposits. Geology 16:551–555

Stiegler MT, Cooper M, Byerly GR, Lowe DR (2012) Geochemistry and petrology of komatiites of the pioneer ultramafic complex of the 3.3 Ga Weltevreden formation, Barberton greenstone belt, South Africa. Precambr Res 212:1–12

Tomkins AG (2013) On the source of orogenic gold. Geology 41:1255–1256

Van Kranendonk MJ, Kröner A, Hegner E, Connelly J (2009) Age, lithology and structural evolution of the c. 3.53 Ga Theespruit Formation in the Tjakastad area, southwestern Barberton greenstone belt, South Africa, with implications for Archaean tectonics. Chem Geol 261:115–139

Viljoen MJ, Viljoen RP (1969) Introduction to the geology of the Barberton granite-greenstone terrain. Geological Society of South Africa Special Publication 2:9–28

Wang LG, McNaughton NJ, Groves DI (1993) An overview of the relationship between granitoid intrusions and gold mineralisation in the Archaean Murchison Province, Western Australia. Mineralium Deposita 28(6):482–494

Ward JHW (1995) Geology and metallogeny of the Barberton greenstone belt: a survey. J Afr Earth Sc 21:213–240

Ward JHW (1999) The metallogeny of the Barberton greenstone belt, South Africa and Swaziland. Geological Survey of South Africa Memoir 86, 108p

Wilson AH, Zeh A (2019) U-Pb and Hf isotopes of detrital zircons from the Pongola Supergroup: constraints on deposition ages, provenance and Archean evolution of the Kaapvaal craton. Precambrian Research 305 (Supplement C): 177–196

Wit De et al (2019) Paleoarchean bedrock lithologies across the Makhonjwa Mountains of South Africa and Swaziland linked to geochemical, magnetic and tectonic data reveal early plate tectonic genes flanking subduction margins. Geosci Front 9:603–665

Zeh A, Gerdes A, Heubeck C (2013) U-Pb and Hf isotope data of detrital zircons from the Barberton greenstone belt: constraints on provenance and Archaean crustal evolution. J Geol Soc 170:215–223

The Neoarchaean Limpopo Orogeny: Exhumation and Regional-Scale Gravitational Crustal Overturn Driven by a Granulite Diapir

8

Dirk D. van Reenen, C. Andre Smit, Alexei L. Perchuk, Jan M. Huizenga, Oleg G. Safonov, and Taras V. Gerya

Abstract

Integrated geological, geochronological and geophysical data from the Neoarchaean Limpopo Complex (LC) and adjacent granite-greenstone terrane of the Northern Kaapvaal Craton (NKVC) demonstrates the following three key features. (1) The LC-NKVC tectonic boundary defined by the Hout River Shear Zone (HRSZ) played a fundamental role in establishing high-grade metamorphic domains in the Southern Marginal Zone (SMZ) of the LC. These domains are characterized by their specific retrograde *P-T* evolution, (2) Near-isobaric southward thrusting of a hot allochthonous SMZ nappe with imbedded steeply plunging folds and steep shear zones against and over the NKVC along the HRSZ before ca. 2.68 Ga occurred as a consequence of steep exhumation within the Central Zone (CZ) at 2.72–2.62 Ga. (3) Initial exhumation of the CZ to mid-crustal levels before ca. 2.68 Ga was associated with isoclinal folding and melt-weakened domes and directed the southwards channelling of the hot nappe. Final exhumation of the CZ to the upper crust at 2.65–2.62 Ga was driven by granitic diapirism and the formation of shear-related, mega-scale closed structures. These features favour an intracrustal Neoarchaean Limpopo orogeny as the result of a granulite diapir triggered by mantle heat-fluid flow underneath the CZ at 2.72–2.62 Ga. The data is neither compatible with continent collisional models nor with an Archean accretionary orogen along the northern edge of the Kaapvaal Craton that is linked to the collision of the Pietersburg Block with the CZ.

Keywords

Limpopo orogeny • Granulite diapir • Granitic diapirism • Regional-scale gravitational crustal overturn • High-temperature retrograde hydration

D. D. van Reenen (✉) · C. A. Smit · A. L. Perchuk · J. M. Huizenga · O. G. Safonov
Department of Geology, University of Johannesburg, P.O. Box 524 Auckland Park, Johannesburg, 2006, South Africa
e-mail: dirkvr@uj.ac.za

A. L. Perchuk · O. G. Safonov · T. V. Gerya
Geological Faculty, Moscow State University, Leninskie Gory, Moscow, 119234, Russia

J. M. Huizenga
Economic Geology Research Institute (EGRU), College of Science and Engineering, James Cook University, Townsville, QLD 4811, Australia

A. L. Perchuk · O. G. Safonov
Institute of Experimental Mineralogy, Russian Academy of Sciences, Chernogolovka, 142432, Russia

T. V. Gerya
Department of Earth Sciences, Swiss Federal Institute of Technology, 8092 Zurich, Switzerland

8.1 Introduction

Developing a tectono-metamorphic model that explains the evolution of the Neoarchaean Limpopo granulite-facies terrane (the Limpopo Complex) located between the more ancient Kaapvaal and Zimbabwe granite-greenstone cratons of Southern Africa (Fig. 8.1) poses special challenges to researchers dealing with Archaean geology. The following principal features of the high-grade Limpopo Complex (LC) complicate this task. First, original stratigraphic relationships have been completely destroyed by intense deformation, high-grade metamorphism and associated magmatism (e.g. van Reenen et al. 1987; Roering et al. 1992b), and most of the upper crustal cap has been removed (e.g. Roering et al. 1992b). Second, there is no record of the prograde *P-T* path of the high-grade rocks, and no geological evidence for a geological process that might have been associated with the initial formation of the granulite-facies rocks at depth (e.g. van Reenen 1983; van Reenen et al. 1987, 1990; Perchuk et al. 2000a; Smit et al. 2011; Belyanin et al. 2014). Thirdly, the high-grade Central Zone (CZ) of the LC is a polymetamorphic terrane that was extensively overprinted in the Palaeoproterozoic by a younger mainly

A. Kröner and A. Hofmann (eds.), *The Archaean Geology of the Kaapvaal Craton, Southern Africa*, Regional Geology Reviews, https://doi.org/10.1007/978-3-319-78652-0_8

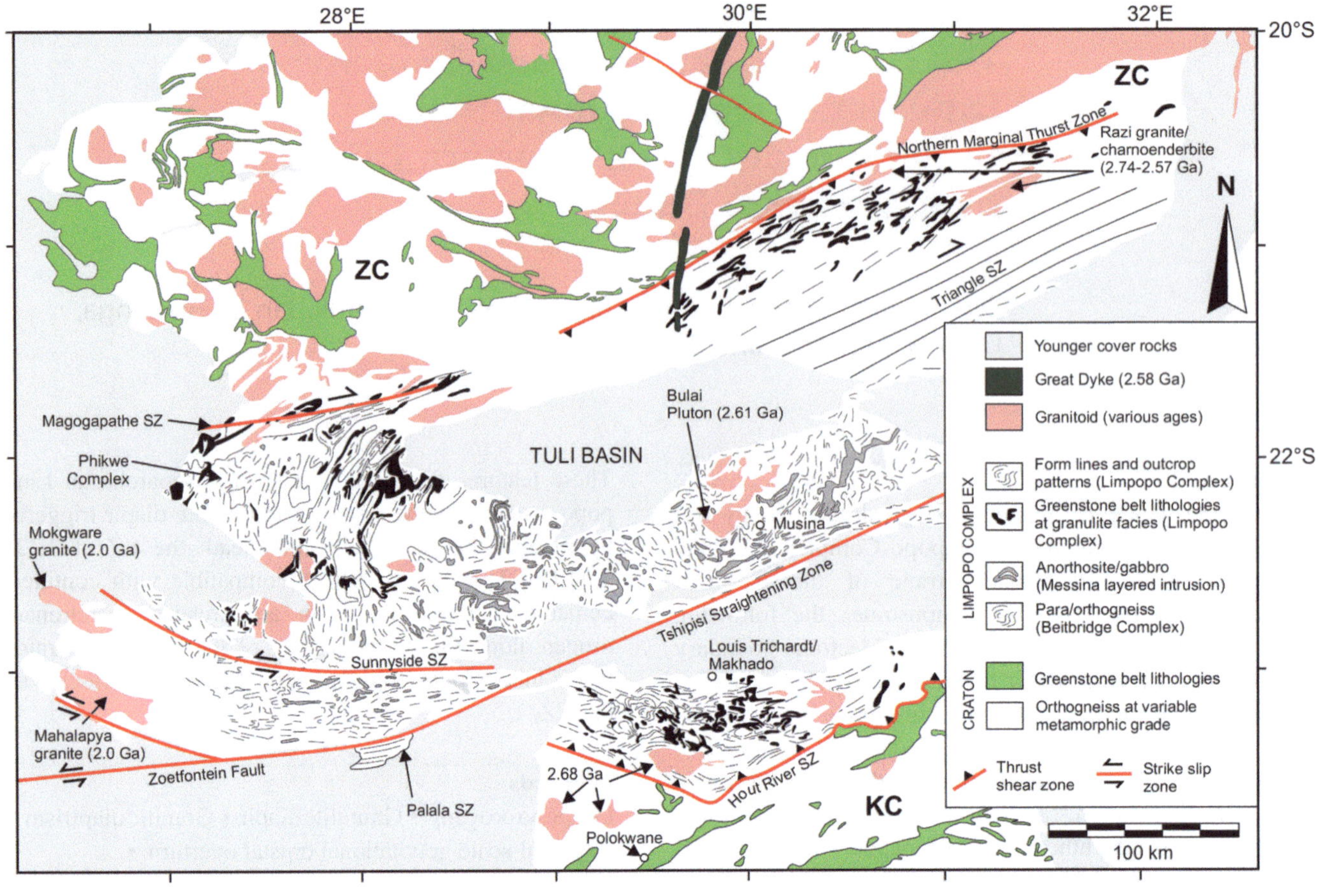

Fig. 8.1 Regional geological map of the Limpopo complex (after Smit et al. 2011) emphasizing the coherent structural style that characterizes the entire granulite-facies terrane, which is in strong contrast with the low-grade linear, greenstone belts on the adjacent cratons. Neoarchaean inward-dipping thrusts separate the granulite-facies terrane from the adjacent low-grade Kaapvaal (KC) and Zimbabwe (ZC) Cratons. The central zone is bounded in the south and the north by Neoarchaean dip-slip shear zones, Triangle SZ in the north and Tshipise Straightening Zone in the south, that were reactivated as strike-slip shear zones in the Palaeoproterozoic

thermal event associated with the formation of discrete shear zones (e.g. van Reenen et al. 2008; Mahan et al. 2011; Smit et al. 2011).

A credible orogenic building model for the LC not only depends on the geological controls imposed by this high-grade rock mass, but also by an understanding of the tectonic relationship with the juxtaposed granite-greenstone terranes, and especially the relationship of the Southern Marginal Zone (SMZ) with the Northern Kaapvaal Craton (NKVC) (e.g. McCourt and van Reenen 1992; Roering et al. 1992a, b; Perchuk et al. 1996, 2000b; van Reenen et al. 2011; Smit et al. 2014).

Kramers et al. (2011) appraised the diversity of published models that have been proposed for the evolution of the LC (Fig. 8.1) since the early 1980s including continent collisional models and gravitational redistribution models linked to the action of granulite diapirs. They concluded that only the model involving Archaean collision of the Kaapvaal and Zimbabwe Cratons (KVC and ZC, respectively) (e.g. van Reenen et al. 1987; De Wit et al. 1992c; Roering et al. 1992a; Treloar et al. 1992; Kramers et al. 2011; Smit et al. 2011) remains a viable working hypothesis. Kramers et al. (2011) further stated that gravitational redistribution in the context of crustal thickening (Gerya et al. 2000; Perchuk and Gerya 2011) is compatible with Neoarchaean collision and can explain important features at the contact of the LC with the granite-greenstone terrane of the NKVC (e.g. van Reenen et al. 2011; Smit et al. 2014).

However, if the LC was formed as the result of a collisional orogeny, a large volume of sediments (ca. 5 million km^3), should have accumulated within 2.72–2.65 Ga as result of erosion of the Limpopo Highland (Gerya et al. 2000; Perchuk and Gerya 2011; Dorland et al. 2004). These sediments have not been documented within the Precambrian successions on either the NKVC or the ZC. No Archaean age zircons were found within the Mesoproterozoic sedimentary formations on the adjacent NKVC, whereas ca. 2.0 Ga detrital zircons are very common in these rocks (Dorland et al. 2004). These observations imply that no voluminous mountains existed or were eroded above the Neoarchaean Limpopo Complex (Perchuk and Gerya 2011).

Smit et al. (2014) further proposed a link between crustal thickening and gravitational redistribution in the context of Neoarchaean collision by suggesting that southwards channelling of a hot SMZ nappe against and over the adjacent NKVC was controlled by a tectonic 'conveyor belt' mechanism. This concept, as it especially pertains to the SMZ and the relationship with the CZ and with the NKVC, is derived from studies of younger orogens such as the Himalayan–Tibetan and Variscan orogens (e.g. Jamieson et al. 2011 and references therein), and, in particular, the Bohemian Massif (Lexa et al. 2011; Schulmann et al. 2008). These studies suggest that thickened crust heated by radioactive decay would result in partial melting at mid to lower levels, producing melt-rich weakened migmatitic crust. This weakened crust is able to flow outwards onto the lower foreland areas through a process described as mid-crustal heterogeneous channel flow (e.g. Schulmann et al. 2008; Jamieson et al. 2011; Lexa et al. 2011).

Coupled petrological thermomechanical numerical modelling of Precambrian ultrahot orogenies (Perchuk et al. 2018) addresses the issue of the formation of granulite-facies rocks at the base of the crust, including granulites with ultrahigh-temperature (UHT) signatures (e.g. Belyanin et al. 2014; Rajesh et al. 2014). Numerical modelling thus provides a possible source for UHT granulites and a solution to a long-standing question on how granulite-facies terranes developed within moderately thickened crust (e.g. Ellis 1987; Thompson 1990), as has, for instance, been suggested for the LC (e.g. van Reenen et al. 1987; Roering et al. 1992a; Kramers et al. 2011; Smit et al. 2014).

8.1.1 Aims of This Study

The goal of this paper is to demonstrate, based on a large published geological, geochronological, and geophysical database (e.g. van Reenen et al. 1990, 2011, 2014; Kramers et al. 2006, 2011, 2014 and references therein) that exhumation of the LC as a geotectonic entity at 2.72–2.62 Ga was probably controlled by the action of a crustal-scale granulite diaper. This diaper was triggered by deep crustal fluid-heat flow that penetrated the upper crust (e.g. England and Thompson 1984; Pili et al. 1997; Perchuk and Gerya 2011) underneath the present CZ. We propose that this intracrustal process controlled exhumation of granulite-facies rocks that were formed at the base of the crust as the result of an earlier ultrahot orogeny. This caused a two-stage steep exhumation of the CZ at 2.72–2.62 Ga and regional-scale gravitational crustal overturn of complex morphology and dynamics that controlled the southwards channelling of a hot SMZ nappe against and over the adjacent NKVC before ca. 2.68 Ga (the conveyor belt tectonic mechanism of Smit et al. 2014). This gravity-driven tectonic model links the post-peak metamorphic exhumation of the SMZ, CZ and NMZ (Northern Marginal Zone) at 2.72–2.61 Ga and is adapted from gravitational redistribution models first considered by Gerya et al. (2000) and Perchuk and Gerya (2011). We thus propose that the formation and exhumation of the LC granulite-facies rocks should no longer be considered within the context of a collisional orogeny involving crustal thickening at ca. 2.72 Ga (e.g. van Reenen et al. 1987; De Wit et al. 1992c; Roering et al. 1992a; Treloar et al. 1992; Kramers et al. 2011; Smit et al. 2014) but rather a process driven by diapirism.

This paper takes recent publications by Vezinet et al. (2019) and Laurent et al. (2019) in consideration. These authors argue in favour of an Archean accretionary orogen at 2.95–2.75 Ga along the northern edge of the Kaapvaal Craton that involves collision of a proposed Pietersburg Block that includes the SMZ, with the CZ of the LC. Their model depends on a revised lithotectonic model for the SMZ of the LC in which the Nthabalala Shear Zone and not the Hout River Shear Zone (HRSZ) (Fig. 8.2) is taken as the tectonic boundary of the granulite-facies SMZ with the low-grade Pietersburg Block in the south (Laurent et al. 2019). We will demonstrate that the proposed revised lithotectonic model for the SMZ (Vezinet et al. 2018) is in complete disagreement with a large published geological, geochronological, and geophysical dataset showing (1) that the Nthabalala Shear Zone is totally developed within the granulite-facies domain of the SMZ, and (2) that the SMZ is separated from the NKVC in the south by the ca. 2.72 Ga HRSZ (Figs. 8.1, 8.2 and 8.3). We will show that there is no conflict between peak metamorphism dated at ca. 2.72 Ga in the SMZ and the time of a metamorphic event in the CZ that was recently dated at 2.64–2.62 Ga (Kröner et al. 2018). Finally, we will demonstrate that a proposed counterclockwise *P-T* path for the CZ based mainly on the interpretation of pseudosection modelling (Brandt et al. 2018) Ga and linked with burial and prograde heating at 2.64–2.62 Ga (Kröner et al. 2018), is not supported by published data. Integrated field, structural, metamorphic and geochronological data show that the CZ at this time was steeply exhumed towards the northeast (e.g. Smit et al. 2011).

8.1.2 Terminology

To avoid confusion, it is necessary that we explain the specific terminology used in this paper to describe important major geological features related to the evolution of the Limpopo Complex.

The term 'Limpopo Complex' was proposed by van Reenen and Du Toit (1977) to describe the crustal block referred to as the Limpopo Mobile Belt by Mason (1973). The rationale being that the rocks produced by the Limpopo orogeny covered a much bigger area than the linear zone (belt) between the (Palaeoproterozoic) Triangle and Palala Shear Zones as defined by Mason (1973). The Limpopo Belt of Mason (op cit) being the Central Zone of the Limpopo Complex.

The term 'orogeny' used in the context of the 'Limpopo Complex' has to be explained. Orogeny is used to describe the processes affecting continental lithosphere in convergent settings. In Phanerozoic orogens such processes are linked to horizontal shortening and significant topography, but for the Precambrian the term has been used in a broader sense to identify episodes of intense crustal deformation and tectono-magmatic reworking characterized by vertical tectonics and in places (e.g. Eastern Darwar Craton) flat topography (e.g. Chardon et al. 2009; Cagnard et al. 2011).

The term 'crustal-scale granulite diapir' (Perchuk and Gerya 2011) refers to the orogenic process triggered by mantle heat-fluid flow that in the context of the LC directed steep exhumation of the Central Zone of the LC and regional-scale gravitational crustal overturn of complex morphology and dynamics that controlled the southwards channelling of a hot SMZ nappe against and over the adjacent granite-greenstone craton before emplacement of the Matok pluton at ca. 2.68 Ga (Fig. 8.1) (Laurent et al. 2013, 2019; Smit et al. 2014).

The term 'melt-weakened domes' proposed by Jamieson et al. (2011) refers to large dome-like structural features cored by migmatites that were formed at deep crustal levels without evidence for significant segregation of the melt. In contrast, the term 'melt-cored domes' (e.g. Brown 2007; Rey et al. 2009) refers to the situation where larger volumes of melt are present and segregated, leading to emplacement of melt-cored domes in the upper crust. In this paper the term 'granitic diapirism' is equivalent to the term 'melt-cored domes'.

The term 'conveyor belt mechanism' was coined by Smit et al. (2014) to explain a tectonic mechanism linked to near-vertical tectonics in the CZ that controlled the southwards channelling of a hot SMZ nappe along the near-horizontal section of the Hout River Shear Zone (HRSZ), against and over the adjacent NKVC.

The term 'mega closed structures' refers to mega-scale structures in the CZ that are circular in outcrop pattern, and that were initially mapped as sheath folds (Roering et al. 1992a).

Finally, we use the term 'TSZ' in this paper to describe the Tshipise Straightening Zone, and not the Triangle Shear Zone, as is normally done (Fig. 8.1).

8.2 Regional Geological Setting of the Limpopo Complex of Southern Africa

The ca. 2.72 Ga LC of southern Africa (Fig. 8.1; van Reenen et al. 2011 and references therein) is a classic example of a Neoarchaean granulite-facies gneiss complex situated between two cratons. It provides a well-exposed crustal section from the low-grade granite-greenstone terrane of the NKVC in the south across the entire high-grade terrane of the LC and into the low-grade granite-greenstone terrane of the ZC in the north. The LC (Fig. 8.1) comprises three distinct high-grade zones namely NMZ, SMZ and CZ, which are separated from each other and from the juxtaposed cratons by complex inward-dipping Neoarchaean reverse-sense crustal-scale shear zones. Some elements of this Neoarchaean thrust sense system have been overprinted by strike-slip/transpressive activity during the Palaeoproterozoic (ca. 2.0 Ga) and are now represented by the Triangle and Palala Shear Zones (e.g. Kramers et al. 2011; Smit et al. 2011; van Reenen et al. 2011).

The major geological features of NMZ, CZ, SMZ, as well as those of the juxtaposed NKVC that collectively define the regional tectonic setting of the Limpopo Complex have been discussed in detail in numerous publications (e.g. van Reenen et al. 1990, 2011; Kröner et al. 1999; Kramers et al. 2006 and references therein). The following section presents a brief overview to explain important geological features of the LC high-grade terrane and adjacent cratons and serves as a general background for the main part of this paper.

8.2.1 The Northern Kaapvaal Craton (NKVC)

The NKVC comprises >3 Ga greenschist- to amphibolite-facies greenstone belts (Pietersburg, Giyani, Rhenosterkoppies) and associated granitoids located at the northern edge of the KVC (Fig. 8.1) (e.g. De Wit et al. 1992a; Vezinet et al. 2018; Laurent et al. 2019). It is separated from the ca. 2.72 Ga granulite-facies SMZ of the LC by the north-dipping reverse-sense Hout River Shear Zone (HRSZ) (Fig. 8.1).

Two post ca. 3.0 Ga tectono-metamorphic events are documented in the northern part of the NKVC: (1) the ca. 2.86 Ga Pietersburg greenstone belt-Lwaji (PGB-Lwaji) orogeny (Kramers et al. 2014) that affected much of the NKVC between the Giyani and Pietersburg greenstone belts (Figs. 8.1, 8.2 and 8.3), and (2) the ca. 2.72 Ga Limpopo orogeny that only affected a narrow zone in the footwall of the north-dipping HRSZ (e.g. McCourt and van Reenen 1992; Roering et al. 1992a, b; van Reenen et al. 2011, 2014; Kramers et al. 2014). Laurent et al. (2019) discuss a

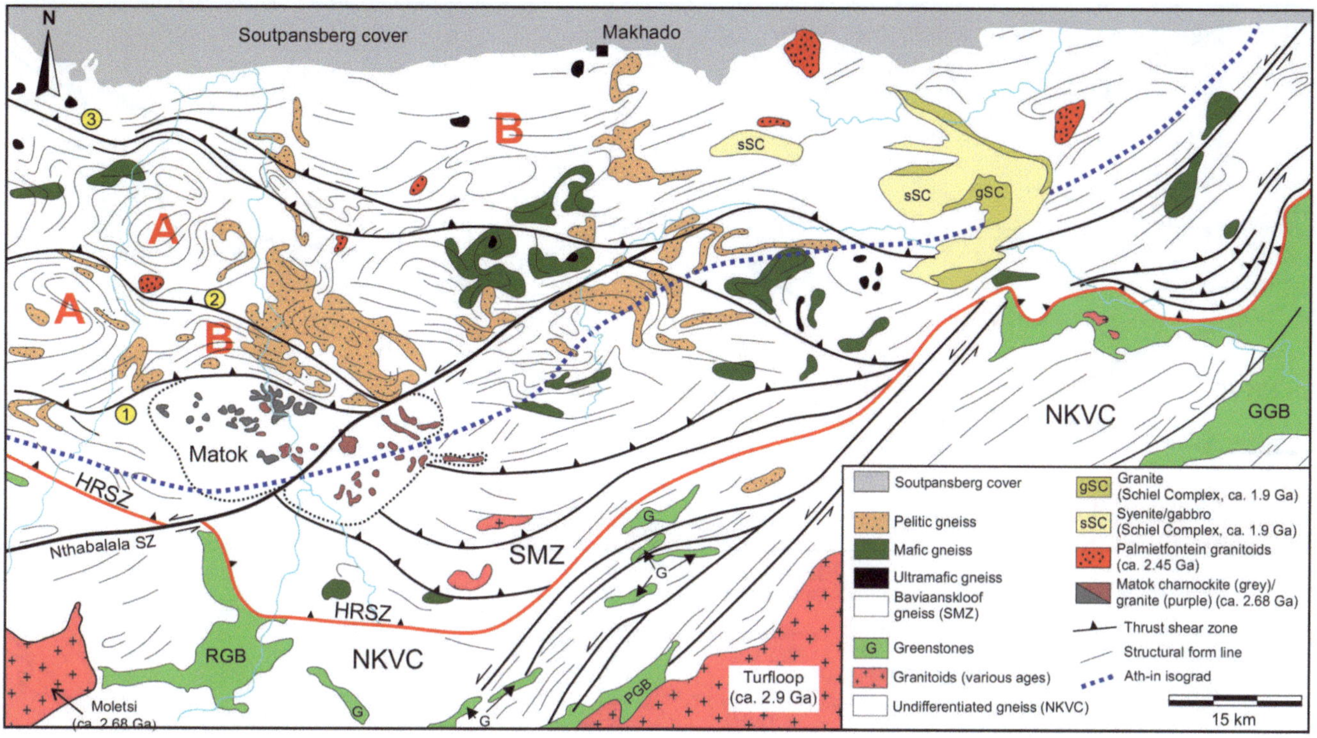

Fig. 8.2 Geological map of the SMZ (after van Reenen et al. 2011) showing: (1) North-dipping HRSZ that bounds the SMZ in the south, and presence of steeply dipping high-grade shear zones that internally subdivide the high-grade terrane into several large crustal blocks. (2) Ath-in isograd superimposed onto major folds that subdivides the SMZ into a northern granulite zone and a southern zone of retrograde hydrated granulite. (3) Crustal-scale blocks bounded by steeply dipping shear zones characterized by complex fold patterns depicting large migmatitic domes (**a**) and steeply plunging isoclinal folds (**b**). (4) Outcrop pattern of the Matok granitic complex comprising granitic (purple) and charnockitic (grey) rocks (compare with Fig. 8.3). (5) Large volumes of tonalitic orthogneisses (Baviaanskloof gneiss) with intercalated deformed slivers of Bandelierkop Formation lithologies comprising metapelitic, mafic and ultramafic gneisses characterize the SMZ. 1–3 (yellow circles) respectively refer to the Matok-, Petronella- and Annaskraal Shear Zones

geodynamic model for the Paleao- to Mesoarchaean evolution of the NKVC, which they also extends to the Neoarchaean involving the SMZ.

8.2.2 The Southern Marginal Zone (SMZ)

The SMZ of the LC (Figs. 8.2 and 8.3) occupies the hanging wall section of the north-dipping HRSZ that separates it from the NKVC and comprise intensely deformed granulite-facies gneisses that document evidence only for a high-temperature decompression-cooling metamorphic history (e.g. Roering et al. 1992b; van Reenen et al. 1990, 2011, 2014). The SMZ mainly comprises metasedimentary and mafic/ultramafic gneisses of the ca. 3.0 Ga Bandelierkop Formation structurally folded and intercalated with the volumetrically significant, enderbitic, Baviaanskloof Gneiss (Fig. 8.2) (van Reenen et al. 2011, 2014 and references therein). Granulite-facies metamorphism and deformation occurred between 2.72 and 2.68 Ga (e.g. Roering et al. 1992b; van Reenen et al. 2011, 2014), followed between ca. 2.68 Ga and ca. 2.64 Ga by emplacement of anatectic granitoids including diorite, granodiorite and monzogranite of the ca. 2.68 Ga Matok complex (e.g. Laurent et al. 2013; Figs. 8.2 and 8.3). Emplacement of the post-tectonic Palmietfontein granite (ca. 2.46 Ga) (Barton and van Reenen 1992) manifests the end of tectono-metamorphic activity related to the Neoarchaean Limpopo event.

The SMZ has been subdivided into a northern granulite domain separated from a southern domain of retrograde hydrated granulite by an anthophyllite-in (Ath-in) isograd (Figs. 8.2 and 8.3) (e.g. van Reenen 1986; van Reenen et al. 2014). The northern granulite domain comprises large crustal blocks characterized by complex fold deformation patterns separated from each other by steeply north-dipping high-grade shear zones (the Matok, Petronella and Annaskraal Shear Zones in Figs. 8.2 and 8.3) (Smit et al. 1992). The northern granulite domain is subdivided (Fig. 8.3) into a northern subdomain characterized by decompression-cooling pressure–temperature (*P-T*) paths and a southern

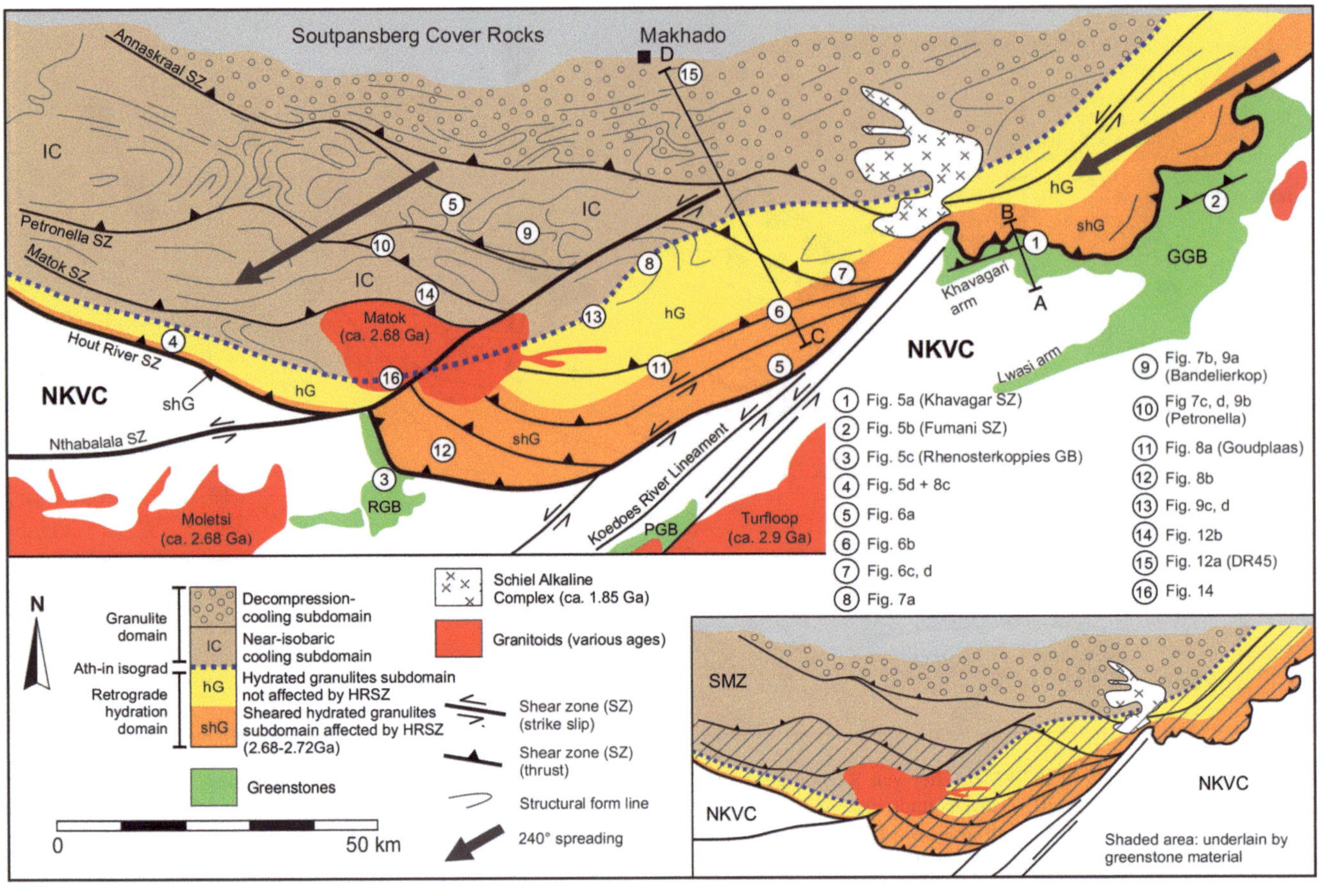

Fig. 8.3 SMZ metamorphic map (after van Reenen et al. 2014) showing (1) localities 1–16 referred to in the text; (2) two south–north-oriented traverses, A–B and C–D along which the SMZ crustal section (Fig. 8.4) has been constructed; (3) subdomain showing the flat section of the Hout River Shear Zone exposed in the deeply eroded Lowveld (orange); (4) subdomain of folded retrograde hydrated granulite exposed in the escarpment (yellow), (5) subdivision of the granulite domain exposed on the plateau (brown) into a northern subdomain characterized by decompression-cooling *P-T* paths, and a southern subdomain characterized by near-isobaric cooling *P-T* paths. The inset map shows the surface area (cross-hatched) of the Southern Marginal Zone that is underlain by greenstone belt material based on geophysical data (after De Beer and Stettler 1992)

subdomain characterized by near-isobaric cooling *P-T* paths (e.g. Perchuk et al. 2000a; Smit et al. 2001, 2014; van Reenen et al. 2011, 2014).

Additional important characteristics of the SMZ should be emphasized. First, syn-metamorphic granitoid diapirism is notably absent in the SMZ, whereas this phenomenon occurs extensively both in the CZ and the NMZ (see below). Secondly, the SMZ, like the NMZ, is dominated by magmatic rocks (i.e. tonalitic Baviaanskloof Gneiss), whereas meta-basic, meta-ultrabasic, and metasedimentary rocks of the Bandelierkop Formation (Fig. 8.2) are a minor constituent of the SMZ. This is in strong contrast with the CZ, which is characterized by a much larger volume of metasedimentary lithologies (see below). Third, metasedimentary rocks of both the Bandelierkop Formation (SMZ) (e.g. Rajesh et al. 2014) and the Beit Bridge Complex (CZ) (e.g. Kröner et al. 1999) record evidence for protolith ages greater than ca. 3.0 Ga.

8.2.3 The Central Zone (CZ)

The CZ (e.g. Hofmann et al. 1998; Kröner et al. 1999; Kramers et al. 2006; Smit et al. 2011, 2014 and references therein) forms the largest core portion of the LC (Fig. 8.1). It includes large volumes of metasedimentary granulite-facies rocks very different in character from those of either the SMZ or the NMZ. The ca. 3.3 Ga old Beit Bridge Complex is the most dominant unit in the CZ and comprises meta-quartzites, Mt-bearing quartzites (mineral abbreviations after Whitney and Evans 2010), marbles, and calc-silicate gneisses, mafic and ultramafic gneisses, metamorphosed iron formation and various Grt-Bt paragneisses interfolded with quartzofeldspathic orthogneisses. Precursors of the ca. 3.3 Ga Messina Suite and Sand River tonalite-trondhjemite orthogneisses intruded metasedimentary rocks of the Beit Bridge Complex. At 2.69–2.62 Ga this association was intruded by a number of tonalitic granitoids (precursors of the Verbaard and Alldays gneisses) and leucocratic quartzofeldspathic granitoids (precursors of the collectively termed Singelele-type granitoids). Singelele-type granitoids are considered to be the product of crustal anatexis (Bahnemann 1972; Fripp et al. 1979; Watkeys et al. 1983). The close association of the leucocratic granitoids with km-scale, steeply plunging closed structures mapped as sheath folds (e.g. Roering et al. 1992b; Perchuk et al. 2008a, b; Smit et al. 2011) support the concept of extensive granitic diapirism that accompanied final exhumation of the CZ. The slightly deformed Bulai pluton comprises a porphyritic phase dated at ca. 2.61 Ga (Millonig et al. 2008) and is an important time marker. It intruded high-grade metamorphic gneisses of the Beit Bridge Complex at the end of the Neoarchaean high-grade tectono-metamorphic event in the CZ. A second high-grade metamorphic event that is mainly expressed by a regional thermal overprint at ca. 2.0 Ga and associated with discrete linear ductile shear zones (e.g. van Reenen et al. 2008; Smit et al. 2011; Kramers and Mouri 2011) will not be further discussed since this paper only focuses on the Neoarchaean evolution of the LC.

8.2.4 The Northern Marginal Zone (NMZ)

The NMZ (Fig. 8.1) (e.g. Blenkinsop et al. 2004; Blenkinsop 2011 and references therein) consists mainly of charno-enderbites that diapirically intruded metasedimentary and meta-mafic rocks between 2.74 and 2.57 Ga, coeval with granulite-facies metamorphism. Similar to the proposed relationship of the SMZ with the NKVC, the NMZ exhibits lithologies that are geochemically similar to that of the ZC (Kreissig et al. 2000). Pervasive Neoarchaean horizontal shortening that occurred in association with the exhumation of the NMZ is expressed by steep to near-vertical gneissic banding striking in general east–northeast, parallel to the strike of the NMZ (Fig. 8.1) (e.g. Blenkinsop 2011). The highly radiogenic nature of the rocks of the NMZ, supplemented by heat from mantle melts, triggered heating and diapirism that culminated in the intrusion of distinctive porphyritic charnockites and granites (Blenkinsop 2011). This author proposed that crustal thickening during horizontal shortening and steep extrusion of the NMZ was limited by high geothermal gradients, in contrast with overthickening and gravitational collapse that characterizes younger orogens. The contact of the ZC with the NMZ (the North Marginal Thrust Zone) is a moderately south-dipping thrust along which NMZ granulites were thrust onto the ZC (Light 1982; Blenkinsop et al. 1995; Blenkinsop 2011). The NMZ was also affected by lower grade retrogression and transpressive deformation in localized shear zones at ca. 2.0 Ga, mainly in the southern part. Emplacement of the Great Dyke of Zimbabwe at 2.58 Ga (Armstrong and Wilson 2000) across the contact of the ZC with the NMZ (Fig. 8.1) signifies the end of all tectono-metamorphic activity related to the Neoarchaean Limpopo event in the NMZ.

8.2.5 The Zimbabwe Craton (ZC)

The south-dipping Neoarchaean North Marginal Thrust Zone (NMTZ) (Fig. 8.1) separates the NMZ from the ZC, which comprises low-grade greenstone belts and granitoids. The geometry of the south-dipping NMTZ is evidence for thrusting the NMZ against and over the ZC.

The rest of the paper focuses on a detailed discussion of published geological, geochronological, and geophysical data that forms the basis for the construction of a south–north crustal section (Fig. 8.4) along two linked traverses (sections A–B and C–D in Fig. 8.3) that connect the northern part of the Giyani greenstone belt in the south with the SMZ and CZ of the LC to the north (Fig. 8.1). The crustal section (Fig. 8.4) is the foundation for the analyses of the geological characteristics of the SMZ and its relationship with the NKVC in the south, and the CZ in the north. It also elucidates the arguments used to promote a gravity-driven diapiric model for the 2.72–2.62 Limpopo orogeny.

8.3 Northern Kaapvaal Craton—Southern Marginal Zone Relationship at ca. 2.72 Ga

Studies in the Giyani and Rhenosterkoppies greenstone belts located on the NKVC in the immediate footwall of the HRSZ (Figs. 8.1, 8.2 and 8.3) have proven to be critical to understanding the relationship between the NKVC and the SMZ, which is a major constraint on any geodynamic model proposed for the evolution of the LC at 2.72–2.62 Ga. The Giyani greenstone belt consists of a continuous central-northeastern part that in the west (Fig. 8.3) bifurcates into a northern (Khavagari) and a southern (Lwaji) arm separated by granitoid gneisses (McCourt and van Reenen 1992). The northern arm is dominated by ultramafic schists intercalated with minor mafic and metasedimentary rocks including BIF, ferruginous quartzite and metapelitic schist, but due to structural complexity, no stratigraphic sequence can be defined (McCourt and van Reenen 1992). In contrast, both the central section of the belt and the southern Lwaji arm, are characterized by metavolcanics of which mafic schists dominates (e.g. McCourt and van Reenen 1992; Kramers et al. 2014).

Kramers et al. (2014) recently showed that garnet-bearing mafic schists from the Lwaji arm (Fig. 8.3) document metamorphic ages of ca. 2.86 Ga that overlap with ages of 2.85–2.87 Ga for younger zircons in syntectonic granitic

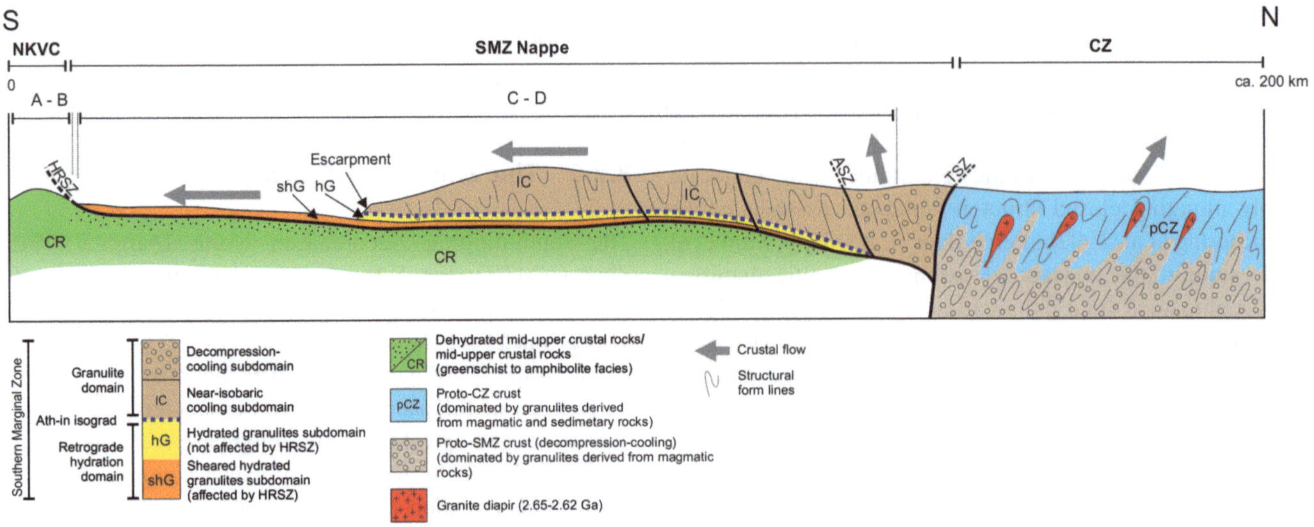

Fig. 8.4 Schematic south-north crustal section constructed along two linked traverses (A–B) and C–D (Fig. 8.3) respectively across the Khavagari arm of the Giyani greenstone belt (A–B) and through the central part of the SMZ (C–D). The section also includes the southern part of the CZ. ASZ: Annaskraal Shear Zone, HRSZ: Hout River Shear Zone; TSZ: Tshipise Straightening Zone. Decompression-cooling *P-T* paths (DR 45 in Fig. 8.11) in the decompression-cooling subdomain (brown coloured area with open circles) are in synchronicity with steeply plunging folds and steeply dipping shear zones that collectively document evidence for steep exhumation. In contrast, near-isobaric cooling *P-T* paths (B, C, and D in Fig. 8.11) in the near-isobaric cooling subdomain (brown coloured area) are not in synchronicity with the steep tectonic features, but documents metamorphic evidence that support spreading along the underlying sub-horizontal section of the HRSZ. See text for further discussion

intrusions within the Giyani and Pietersburg greenstone belts (Kröner et al. 2000; Jaguin et al. 2012; Laurent et al. 2013, 2019; Zeh et al. 2013; Kramers et al. 2014). Kramers et al. (2014) interpreted these ages to reflect a Pietersburg greenstone belt (PGB)-Lwaji orogeny that affected the Lwaji arm of the Giyani greenstone belt as well as the NKVC (Fig. 8.3) (also Vezinet et al. 2018; Laurent et al. 2019).

Ages of 2.85–2.87 Ga obtained from much of the NKVC are significantly older than peak metamorphic ages of 2.72–2.69 Ga obtained from the north-dipping Khavagari Shear Zone in the Khavagari arm (Kreissig et al. 2001; Kramers et al. 2014) of the Giyani greenstone belt (Fig. 8.3) and from the Rhenosterkoppies greenstone belt also located in the footwall of the HRSZ to the west of the Giyani greenstone belt (Figs. 8.2 and 8.3) (Miyano et al. 1992; Passeraub et al. 1999; Kramers et al. 2014). These younger ages have been interpreted as the timing of hot-iron metamorphism related to emplacement of SMZ granulites against and over the Giyani greenstone belt during the ca. 2.72 Ga Limpopo orogeny (McCourt and van Reenen 1992; Roering et al. 1992a; Perchuk et al. 1996, 2000b; Kramers et al. 2014) (see Sect. 8.4.3.3 and *P-T* path A in Fig. 8.11). Kramers et al. (2014) concluded that the ca. 120 Ma age difference between the ca. 2.86 Ga PGB-Lwaji orogeny and the ca. 2.72 Ga Limpopo orogeny 'precludes any causal link between these two events'.

8.4 Tectono-Metamorphic Characteristics of the SMZ and NKVC Linked to the North-Dipping HRSZ

8.4.1 Subdivision of the NKVC and Adjacent SMZ into Distinct Domains

The SMZ and juxtaposed granite-greenstone terrane of the NKVC (Fig. 8.3) comprise distinct structural and metamorphic domains that are relevant for the geodynamic models proposed for the evolution of the LC. The subdivision of the NKVC (Fig. 8.3) into a domain older than 2.85–2.86 Ga and a 2.72–2.69 Ga domain that occupies the footwall of the steep north-dipping crustal-scale HRSZ has already been discussed. The geological map (Figs. 8.2 and 8.3) (van Reenen et al. 2011, 2014; Smit et al. 2014) further shows that the SMZ is subdivided into a northern granulite domain separated from a southern domain of retrograde, hydrated, former granulite by a well-established Ath-in retrograde isograd. Both these domains can further be subdivided. The granulite domain with steeply plunging folds and steep dipping high-grade shear zones (Fig. 8.3) is subdivided into a northern subdomain characterized by decompression-cooling *P-T* paths preserved in metapelitic granulite, and a southern subdomain that abuts against the

Ath-in retrograde isograd characterized by near-isobaric cooling *P-T* paths, also preserved in metapelitic granulite.

The southern retrograde domain (Fig. 8.3) can also be subdivided into a northern and southern subdomain. The northern subdomain bordering the Ath-in retrograde isograd (yellow area in Fig. 8.3) is like the adjacent granulite domain characterized by steeply plunging folds and steep dipping shear zones but these structures deform retrograde, hydrated, former granulite. In contrast, the southern subdomain (orange area in Fig. 8.3) comprises flat-lying sheared retrograde hydrated former granulites that are considered to be closely associated with the underlying HRSZ as will be discussed. It is important to note that more than 60% of the SMZ surface area (inset map in Fig. 8.3) is underlain by rocks with a geophysical signature indistinguishable from the granite-greenstone material that characterizes the adjacent NKVC (De Beer and Stettler 1992).

Construction of a south–north crustal section (Figs. 8.3 and 8.4) explains the observed tectono-metamorphic characteristics of the SMZ (Fig. 8.3) and the relationship with the NKVC and CZ at 2.72–2.69 Ga as a direct consequence of the fundamental role that the HRSZ has played in the evolution of the SMZ. The crustal section (Fig. 8.4) is constructed along two linked south–north traverses (A–B and C–D) (Fig. 8.3). The short A–B traverse runs through the Khavagari arm to the steep north-dipping HRSZ demonstrating the interaction of the northern part of the Giyani greenstone belt with the SMZ at ca. 2.72 Ga.

The longer traverse C–D runs across the central part of the SMZ up to Makhado in the north (Fig. 8.3). The vertical topographical difference of ca. 600 m (Smit et al. 2014) between the plateau (Highveld) in the north (shown in brown in Fig. 8.3) and the Lowveld (shown in orange in Fig. 8.3) at the base of the escarpment (shown in yellow in Fig. 8.3) in the south allowed the reliable construction of this important crustal section (Fig. 8.4). Traverse C–D explains the regional distribution of metamorphic subdomains within the SMZ and their relationship with the underlying HRSZ (Fig. 8.3).

The construction and interpretation of the C–D crustal section is based on (1) *Field, structural and petrological data* (van Reenen 1983, 1986; van Reenen et al. 1990, 1995, 2008, 2011, 2014; van Reenen and Hollister 1988; Baker et al. 1992; De Wit et al. 1992c; McCourt and van Reenen 1992; Mahan et al. 2011; Miyano et al. 1992; Perchuk et al. 1996, 2000a, b, 2008a, b; Roering et al. 1992a, b, 1995; Van Schalkwyk and van Reenen 1992; Smit et al. 1992, 2011, 2014; Huizenga et al. 2014; Tsunogae and van Reenen 2014; Koizumi et al. 2014); (2) *Geochronological data* (Passeraub et al. 1999; Kreissig et al. 2001; Laurent et al. 2013; Belyanin et al. 2014, 2015; Kramers and Mouri 2011; Kramers and Zeh 2011; Kramers et al. 2014, *Oxygen isotope data* (Hoernes et al. 1995; Dubinina et al. 2015); (3) *Geophysical data* (De Beer and Stettler 1992 and references therein), and (4) *Numerical modelling data* (Gerya et al. 2000; Perchuk and Gerya 2011; Perchuk et al. 2018).

8.4.2 Steep North-Dipping HRSZ in Contact with the NKVC

8.4.2.1 Field and Structural Data

The HRSZ (Figs. 8.1, 8.2 and 8.3) in contact with the NKVC has a fault pattern that varies from steep southwest-verging thrust faults with steep dip-slip kinematics against both the Rhenosterkoppies (Fig. 8.5c) and Giyani (Fig. 8.5a, b) greenstone belts, to near-vertical southwest-verging sinistral strike-slip faults (Figs. 8.2 and 8.3). This typical frontal–lateral ramp geometry defines southwest (240°) movement of high-grade SMZ material against and over the NKVC (Fig. 8.3) (Smit et al. 1992, 2014).

Structural evidence for intense tectonic activity associated with steep north-over-south thrusting along the HRSZ against and over the Giyani greenstone belt is well documented by kinematic data acquired from underground workings at the Fumani Mine (locality 2 in Fig. 8.3) (Pretorius 1988; McCourt and van Reenen 1992) and the Klein Letaba and Frankie (Gan and van Reenen 1995, 1997) gold mines. Gold deposits in these gold mines occur in large structural shoots, respectively, associated with the steep north-dipping Fumani (locality 2 in Fig. 8.3) and Khavagari (locality 1 in Fig. 8.3) shear zones located in the immediate footwall of the HRSZ. The structural shoots hosting gold deposits plunge to the northeast, which is the same direction as the HRSZ.

Fault segments in the central part of the contact area with the NKVC (Figs. 8.2 and 8.3) display strike-slip kinematics characterized by left-lateral displacement (e.g. the Koedoes River Lineament, Figs. 8.2 and 8.3). They are particularly well exposed at the extreme western tip of the Khavagari arm of the Giyani greenstone belt near the village of Babango (see De Wit et al. 1992c). Based on data from surface exposures at the contact with the NKVC, Roering et al. (1992a, b) and Smit et al. (1992) initially interpreted the HRSZ as a steeply north-dipping thrust, even though geophysical data indicate that the shear zone flattens to the north (see Fig. 8.15).

8.4.2.2 Metamorphic Activity Associated with the Steep Section of the HRSZ

Data obtained from Grt-St-Ky (mineral abbreviations after Whitney and Evans 2010) bearing mica schists (Perchuk et al. 1996, 2000b) sampled from the Khavagari shear zone in the footwall of the HRSZ within the Khavagari limb of the Giyani greenstone belt (locality 1 in Fig. 8.3) provide direct evidence for a non-isobaric *P-T* loop comprising prograde

Fig. 8.5 Structural evidence highlighting the steep attitude of the HRSZ at the contact with the NKVC. All localities are shown in Fig. 8.3. **a** Khavagari Hills in the footwall of the HRSZ southwest of the town of Giyani. View to the west. Note the steeply north-plunging thrust duplexes associated with the steep north-dipping section of the HRSZ. **b** Outcrops of the steep north-dipping Fumani Shear Zone exposed in the spillway of the Hudson Ntsanwisi dam (locality 2 in Fig. 8.3) in the northeastern part of the Giyani greenstone belt. The sheared rocks comprise quartz-sericite schists and mafic schists. View is to the west. **c** Rhenosterkoppies greenstone belt (locality 3 in Fig. 8.3) as viewed towards the west. Near-horizontal structures in the central part of the belt (left side of photograph) have been rotated to dip steeply north below the steep north-dipping section of the HRSZ. **d** Steeply north-plunging minor closed folds developed within retrograde hydrated metapelitic gneiss in the Hout River (locality 4 in Fig. 8.3)

and retrograde segments (the 'hot iron effect', see *P-T* path A in Fig. 8.11). The loop shows that prograde and peak metamorphic conditions, reached at 2.72–2.69 Ga within the footwall greenschist, are coeval with attainment of peak metamorphism at ca. 2.72 Ga documented by hot overriding SMZ granulite in the hanging wall (Kreissig et al. 2001; Kramers et al. 2014; Rajesh et al. 2014; Taylor et al. 2014; Nicoli et al. 2015). Data obtained from the central and northern portions of the Rhenosterkoppies greenstone belt (Figs. 8.3 and 8.5c) document a similar scenario in which slightly higher peak *P-T* conditions (ca. 7 kbar and 700 °C) resulted from the thrusting of hot granulites over relatively cold rocks at ca. 2.72 Ga (Miyano et al. 1992; Passeraub et al. 1999; Kramers et al. 2014). The age of 2.7 Ga provides a constraint on deformation along the HRSZ that is compatible with that obtained by Kreissig et al. (2001) from the Khavagari section of the Giyani greenstone belt.

A minimum age for the juxtaposition of the hot SMZ nappe with the granite-greenstone terrane of the NKVC along the steeply north-dipping HRSZ is also accurately constrained by emplacement at ca. 2.68 Ga (Laurent et al. 2013) of geochemically similar and mainly undeformed Matok- and Moletsi granitic plutons, respectively, on either side of the HRSZ (Figs. 8.2 and 8.3). This data provides unequivocal evidence that the SMZ was already juxtaposed at the position of the HRSZ against the NKVC before ca. 2.68 Ga.

In conclusion, published structural, metamorphic, and geochronological data supported by geophysical data (see Fig. 8.15) confirm the status of the HRSZ as a crustal-scale tectonic feature that bounds the SMZ in the south. These data contradict the revised lithotectonic model proposed for the SMZ in which Vezinet et al. (2018) and Laurent et al. (2019) claims that that the Nthabalala Shear Zone (Fig. 8.2), and not the HRSZ, is the tectonic boundary of the SMZ with what these authors term the low-grade Pietersburg block. With reference to the published geological of the SMZ (Fig. 8.2) it is clear that the Nthabalala Shear Zone is located entirely within the granulite domain of the SMZ, 5 km north of the Ath-in retrograde isograd. Large folds comprising granulite-facies metapelitic gneisses can thus be followed uninterruptedly across this oblique sinistral shear zone with little left-lateral displacement.

8.4.3 Flat-Lying Section of the HRSZ

8.4.3.1 Field and Structural Data

The long south–north traverse (C–D, Fig. 8.3) reveals field and structural data (Figs. 8.5, 8.6, 8.7 and 8.8) demonstrating important features of the SMZ that are critical to understand the concept of a hot SMZ granulite nappe with imbedded steeply plunging folds and steep dipping high-grade shear zones that was thrust southwards along a near-isobaric (ca. 6 kbar) crustal plane defined by the near-horizontal section of the HRSZ (Fig. 8.4) before ca. 2.68 Ga (van Reenen et al. 2011; Smit et al. 2014).

First, it shows that the attitude of the HRSZ (see Fig. 8.4) changes over a short distance from a steep north-dipping dip-slip fault (Fig. 8.5a–d) at the direct contact with the NKVC in the south, through a moderately north-dipping thrust still close to the contact with the NKVC (Fig. 8.6a, locality 5 in Fig. 8.3) into a sub-horizontal thrust (Fig. 8.6b–d, localities 6 and 7 in Fig. 8.3) that underlies the entire Lowveld up to the base of the escarpment in the north (orange area in Fig. 8.3). Flat-lying and intensely sheared and retrograde gneisses that have been shown to represent the retrograde hydrated equivalents of former metapelitic-, mafic-, ultramafic granulite of the Bandelierkop Formation and Opx-bearing tonalitic gneisses (Fig. 8.6b–d) (van Reenen 1986) are considered to be closely linked to the underlying sub-horizontal section of the HRSZ (Fig. 8.3) (Smit et al. 2014). This conclusion follows from the observation that outcrops of flat-lying sheared gneisses occupy large areas in the Lowveld (orange area in Fig. 8.3) up to the base of the escarpment (yellow area in Fig. 8.3) in the north.

Second, traverse C–D (Fig. 8.3) exposes the steeply dipping gneissic foliation (Fig. 8.7a) associated with folded retrograde hydrated granulites that outcrop in the escarpment towards the north (yellow area in Figs. 8.3 and 8.4).

Third, the traverse continuous across the retrograde Ath-in isograd located at the top of the escarpment (Figs. 8.2 and 8.3) and into the overlying northern granulite domain (brown area in Figs. 8.3 and 8.4) of the Highveld. The granulite domain (Fig. 8.3) is also characterized by a steeply dipping gneissic foliation (Fig. 8.7b) associated with steeply plunging folds and steeply dipping high-grade shear zones (Matok-, Petronella, Annaskraal shear zones) (Fig. 8.7c, d).

Gravel roads that connect the Lowveld in the south at an elevation of ca. 660 m (orange area in Figs. 8.3 and 8.4) with the Highveld in the north at an elevation of 1240 m (brown area in Figs. 8.3 and 8.4) demonstrate that the flat-lying shear fabric of gneisses (Fig. 8.6b–d) that outcrop in the deeply eroded Lowveld (Figs. 8.3 and 8.4) is superimposed onto and thus younger, than the steep gneissic fabric of folds and shear zones (Fig. 8.8a–c) exposed in road cuttings through the escarpment in the north (coloured yellow in Figs. 8.3 and 8.4). This observation strongly suggests that the sub-horizontal section of the HRSZ (Fig. 8.4) projects northwards underneath the granulite domain exposed on the plateau (brown area in Fig. 8.3), with its steep tectonic features that include the Matok, Petronella, and Annaskraal Shear Zones (Figs. 8.3, 8.4 and 8.7c).

Our overall conclusion is that the granulite domain of the SMZ underlain by retrograde hydrated granulite both with embedded steeply plunging folds and steep high-grade shear zones (Matok-, Petronella-, Annaskraal-) that respectively occupy the Highveld and the escarpment (Fig. 8.3) represents an allochthonous granulite nappe that is riding on the younger underlying sub-horizontal section of the HRSZ (Fig. 8.4). This proposal is supported by *P*-*T* and geophysical data (Figs. 8.11 and 8.15) as will be further discussed.

8.4.3.2 Metamorphic Evolution of the Retrograde Hydrated Domain

The Ath-in retrograde isograd overprints major folds located south of the Nthabalala Shear Zone (Figs. 8.2 and 8.3) confirming its status as a true metamorphic feature that reflects the continuous transition from the granulite domain

Fig. 8.6 Structural evidence of the HRSZ changing its attitude from a steep north-dipping shear zone in direct contact with the NKVC into a near-horizontal shear zone that underlies much of the SMZ in the east (Figs. 8.3 and 8.4). Localities are shown in Fig. 8.3. **a** Shallow north-dipping sheared retrograde hydrated tonalitic gneiss exposed in a gravel road cutting immediately north of the HRSZ. View to the east (locality 5 in Fig. 8.3). **b** Flat-lying outcrops of sheared and retrograde hydrated tonalitic gneiss exposed at locality 6 (Fig. 8.3). View is towards the east. **c** Flat-lying outcrops of sheared and hydrated metapelitic, mafic, ultramafic gneisses and BIF of the Bandelierkop Formation exposed in Shiphophi Hill on farm Hartebeesfontein (locality 7 in Fig. 8.3). View is towards the west. **d** Sheared retrograde hydrated outcrops of tonalitic gneiss exposed in a dry river bed between localities 6 and 7 (Fig. 8.3). The thrust dips at 6° to the north. View is towards the west. This locality and localities 6 and 7 are considered type localities for the sub-horizontal section of the HRSZ

in the north to the retrograde domain in the south (van Reenen 1986; van Reenen et al. 2011, 2014). There is no genetic relationship with the Nthabalala Shear Zone (Fig. 8.2) as is demonstrated by the fact that major folds comprising Bandelierkop Formation granulite-facies gneisses continue across the oblique-slip Nthabalala Shear Zone with little left-lateral displacement. Published geological maps thus contradict the claims by Vezinet et al. (2018) and Laurent et al. (2019) that the sinistral oblique-slip Nthabalala Shear Zone (Fig. 8.2) is a tectonic boundary that separates the granulite-facies SMZ from their low-grade Pietersburg Block to the south.

Fig. 8.7 Steeply plunging folds and steeply dipping high-grade shear zones exposed in the folded retrograde hydrated and granulite domains. **a** Steeply dipping migmatitic metapelitic gneisses that outcrop at the position of the Ath-in isograd (locality 8 in Fig. 8.3). Note undeformed leucocratic trondhjemite dykes that intrude the migmatitic gneiss. **b** Steeply dipping pristine foliated migmatitic metapelitic granulite at Bandelierkop quarry (locality 9 in Fig. 8.3). **c** Steeply dipping granulite-facies shear fabric developed within migmatitic metapelitic granulite in the km-wide Petronella Shear Zone (locality 10 in Fig. 8.3). This steep fabric characterizes all shear zones developed within the granulite-facies domain of the SMZ. View is towards the west. **d** Same locality as (**a**), but with view parallel to the steep plunging lineation showing evidence of heterogeneous deformation of metapelitic granulite in which the early folded gneissic fabric is transposed and partly destroyed by the Petronella shear zone

Fig. 8.8 Flat-lying shear zones superimposed onto the steep gneissic fabric of folds exposed near the contact of the zone of hydration with the underlying flat-lying HRSZ (Fig. 8.3). **a** Steeply dipping foliation developed in retrograde hydrated migmatitic tonalitic gneiss cut by near-horizontal shear fabric at the classic Goudplaats locality exposed in a river bed (locality 11 in Fig. 8.3). View to the north. **b** Instructive example in which the early steeply dipping shear fabric and steeply dipping foliation (bottom right side) preserved in retrograde hydrated tonalitic gneiss is cut by the flat-lying shear fabric in the Sand River on farm Breipaal (locality 12 in Fig. 8.3). View is to the east. **c** Steeply dipping foliation of retrograde hydrated tonalitic gneiss (bottom) cut by flat-lying shear fabric (Hout River, locality 4 in Fig. 8.3). **d** Aerial photograph taken with a drone of a steeply dipping high-grade shear zone developed in retrograde hydrated tonalitic and mafic gneiss exposed in the Sand River on farm Breipaal. Heterogeneous deformation reflected by large boudins of folded tonalitic and mafic gneiss wrapped by steep anastomosing shear zones is also reflected on a small scale in Fig. 8.7d

As previously mentioned, the Ath-in retrograde isograd (Figs. 8.2 and 8.3) is a true metamorphic feature that subdivides the SMZ into a northern granulite domain characterized by pristine metapelitic-, mafic- and ultramafic granulite gneisses of the Bandelierkop Formation and large volumes of Opx-bearing tonalitic gneiss (Baviaanskloof gneiss), and a southern domain of retrograde hydrated equivalents of these former granulites. This fact has been thoroughly documented in numerous publications based on field, petrographic, petrological, and O-isotopic studies (e.g. van Reenen 1986; van Reenen et al. 1990, 2011, 2014; Van Schalkwyk 1991; Baker et al. 1992; Van Schalkwyk and van Reenen 1992; Hoernes et al. 1995; Koizumi et al. 2014). Unfortunately, this information is ignored by Vezinet et al. (2018) and Laurent et al. (2019).

Metapelitic granulites at the position of the Ath-in retrograde isograd (Figs. 8.2 and 8.3) are characterized by coexisting Opx + Ath + Qz + Bt + Pl + Qz in Fe-rich (Crd-free) varieties (Fig. 8.9c), and additionally by Ged + Ky after Crd in Mg-rich (Crd-bearing) varieties (Fig. 8.9d) (van Reenen 1986). Retrograde hydrated equivalents of former metapelitic granulites that outcrop south of the Ath-in retrograde isograd are characterized by the complete replacement of Opx + Crd by fine-grained intergrowths of Ged + Ky (Fig. 8.9e). Two major hydration reactions in Opx + Crd + Qz-bearing metapelitic granulites resulted in the production of anthophyllite and gedrite in the same thin sections (Fig. 8.9d) (e.g. van Reenen 1986; Huizenga et al. 2014):

$$3.5\ \mathrm{Opx} + \mathrm{Qz} + \mathrm{H_2O} \rightarrow \mathrm{Ath} \quad (1a)$$

$$5\ \mathrm{Crd} + 2\ \mathrm{H_2O} \rightarrow 2\ \mathrm{Ged} + 6\ \mathrm{Ky} + 7\ \mathrm{Qz} \quad (1b)$$

Reaction (1a), which occurs in both Fe-rich Crd-free (Fig. 8.9c) and in Mg-rich Crd-bearing (Fig. 8.9d) granulites is univariant in the subsystem $MgO–FeO–SiO_2–H_2O$. This feature is evidenced by the fact that Opx − Ath − Qz − bearing metapelitic granulites accurately define the position of the Ath-in isograd in the field (van Reenen 1986; van Reenen et al. 2011, 2014; Huizenga et al. 2014). In contrast, reaction (1b) (Fig. 8.9d) is divariant in the subsystem $MgO–FeO–Al_2O_3–SiO_2–H_2O$ and occurs over a much larger area in the field as is demonstrated by partially hydrated Crd-bearing gneisses that straddles the position of the Ath-in isograd (van Reenen 1986; van Reenen et al. 2011, 2014; Huizenga et al. 2014). The significance of this observation will be further discussed.

Mafic granulite in the zone of hydrated granulite often still contains relict orthopyroxene that is replaced by hornblende (e.g. van Reenen et al. 2011), while tonalitic Baviaanskloof gneiss in the hydrated zone is characterized by the complete absence of orthopyroxene. Ultramafic rocks in the zone of hydration show evidence for extensive interaction of the peak metamorphic assemblage Opx + Ol + Sp + Ca-Amp with $H_2O–CO_2$ fluids (e.g. Van Schalkwyk 1991; Van Schalkwyk and van Reenen 1992; van Reenen et al. 2011, 2014).

Petrological, fluid inclusion and stable oxygen isotope data show that the fluid responsible for establishing the Ath-in isograd in quartz-bearing metapelites at ca. 6 kbar and 600–630 °C comprised 70–90 mol% CO_2-rich (Fig. 8.10) and 30–10 mol% H_2O and brine fluids (van Reenen 1986; van Reenen and Hollister 1988; Baker et al. 1992; Hoernes and van Reenen 1992; Hoernes et al. 1995; Huizenga et al. 2014; Koizumi et al. 2014; van Reenen et al. 2011, 2014; Koizumi et al. 2014; Smit et al. 2014; Tsunogae and van Reenen 2014). Direct evidence for the passage of an externally derived CO_2-rich fluid through granulites is clearly reflected in granoblastic ultramafic rocks (Van Schalkwyk and van Reenen 1992; van Reenen et al. 2014). Here, the peak assemblage Ol + Hbl-1 + Opx-1 + Spl is partially replaced by the retrograde assemblage Ol + Hbl-2 + Opx-2 + Mgs + Dol + Chl + Tlc at ca. 600 °C and 6 kbar.

Temperature control on the regional hydration of granulite-facies rocks by low H_2O activity fluids (Fig. 8.10) is provided by the observation that at temperatures above ca. 650 °C, fluids moved through the metapelitic granulite rocks without interacting with orthopyroxene. However, when the temperature dropped to 600–630 °C, the infiltrating fluid reacted with orthopyroxene to establish the Ath-in isograd (Fig. 8.3) in metapelitic granulite (Fig. 8.9c, d) (e.g. van Reenen 1986; Baker et al. 1992; Huizenga et al. 2014; Koizumi et al. 2014; Safonov et al. 2014; van Reenen et al. 2014). That water-bearing fluids must have been present at temperatures above ca. 630 °C is evidenced by fluid-inclusion data (van Reenen and Hollister 1988; Huizenga et al. 2014), and by the replacement of cordierite coexisting with orthopyroxene and quartz by gedrite plus kyanite whereas orthopyroxene remains completely stable (Fig. 8.9f) (see also van Reenen et al. 2014, Fig. 8.6a, b).

Additional evidence that hot granulite interacted with cooler infiltrating water-bearing fluids at T ca 620 °C is documented by oxygen isotope data obtained from pristine granulites and their retrograde rehydrated equivalents (Hoernes et al. 1995), and from pristine metapelitic granulite and associated intrusive leucocratic granitoids at different localities that include the Bandelierkop (locality 9 in Fig. 8.3) (Dubinina et al. 2015; Safonov et al. 2018a, b).

As mentioned previously, more than 60% of the SMZ surface area based on their geophysical signature (De Beer and Stettler 1992) is underlain by lower grade greenstone material (inset map in Fig. 8.3). This observation provides a credible source for infiltrating fluids that interacted with the overlying hot granulites to establish the Ath-in retrograde isograd and associated zone of retrograde hydrated granulite. In this scenario the HRSZ acted as a conduit for fluid

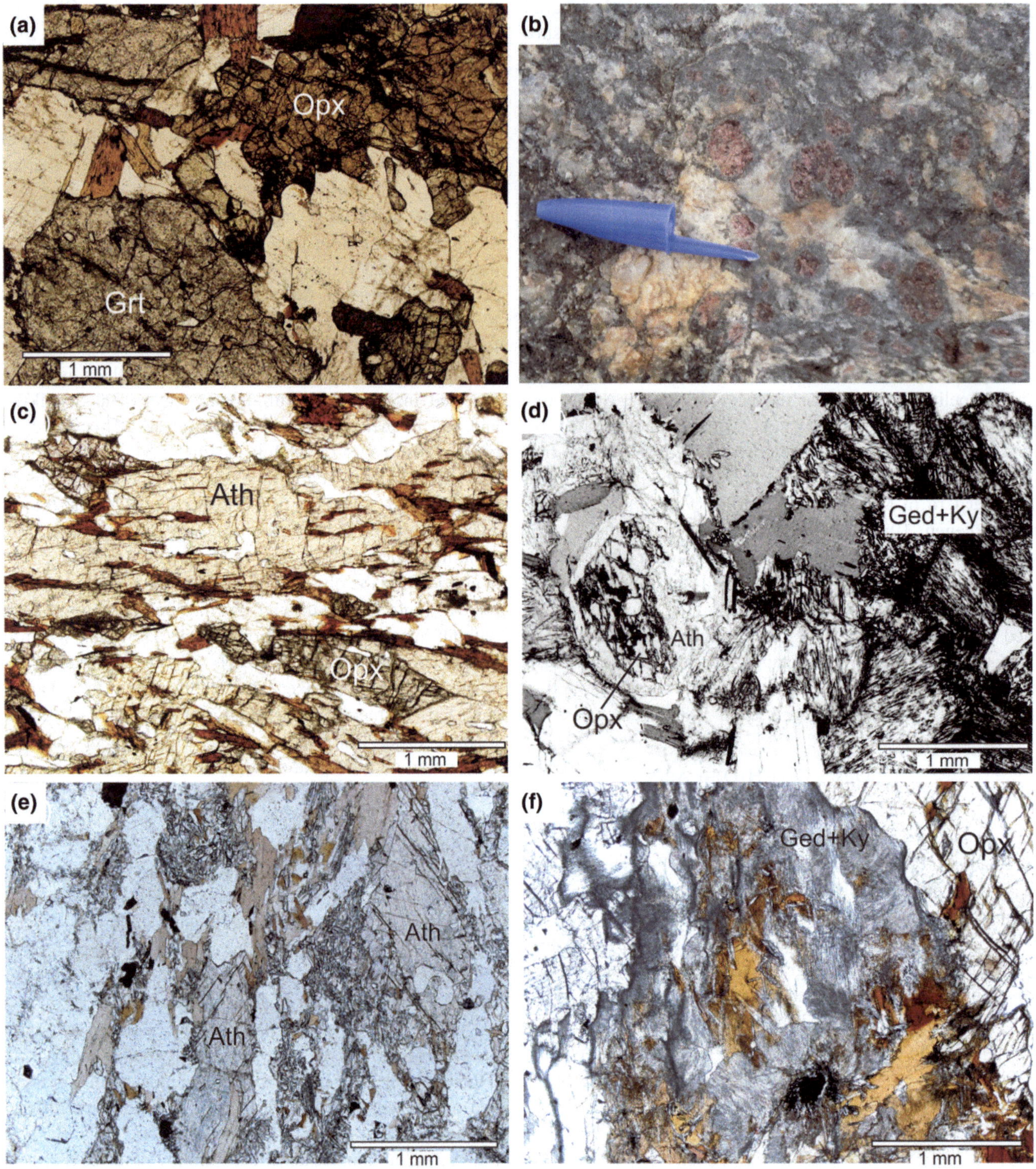

Fig. 8.9 Metamorphic evolution of metapelitic granulite illustrated by photomicrographs (plane-polarized light) (a, c, d, e) and an outcrop photograph (b). **a** Fe-rich (Crd-free) pristine granoblastic Grt-Opx-Bt-Pl-Qz-bearing granulite. **b** Outcrop of pristine granoblastic metapelitic granulite comprising Grt + Crd + Opx at the Petronella locality (locality 7 in Fig. 8.3). Note dark blue cordierite rimming garnet porphyroblasts reflecting evidence for the decompression reaction 2 Grt + 3 Qz → Crd + 2 Opx. **c** Fe-rich Opx-Ath-Qz-bearing equivalent of (a) at the position of the Ath-in isograd, reflecting evidence for the isograd-producing reaction (1a) (see text). Note equilibrium textural relationships of anthophyllite and orthopyroxene. **d** Mg-rich (Crd-bearing) metapelitic granulite at the Ath-in isograd. Orthopyroxene is rimmed by anthophyllite, whereas cordierite is extensively replaced by a fine-grained intergrowth of gedrite and kyanite, reflecting evidence for reaction (1b) (see text). **e** Retrograde hydrated equivalent of (b) and (d) south of the Ath-in isograd. Note large aligned grains of anthophyllite (Ath) and fine-grained intergrowth of gedrite and kyanite completely replacing cordierite in a matrix of quartz and plagioclase. **f** Localized cordierite hydration at Petronella locality (locality 10 in Fig. 8.3) in the granulite domain. Cordierite is replaced by fine-grained intergrowth of gedrite + kyanite, and by coarser grained biotite + kyanite, while orthopyroxene remains unaltered. See text for discussion

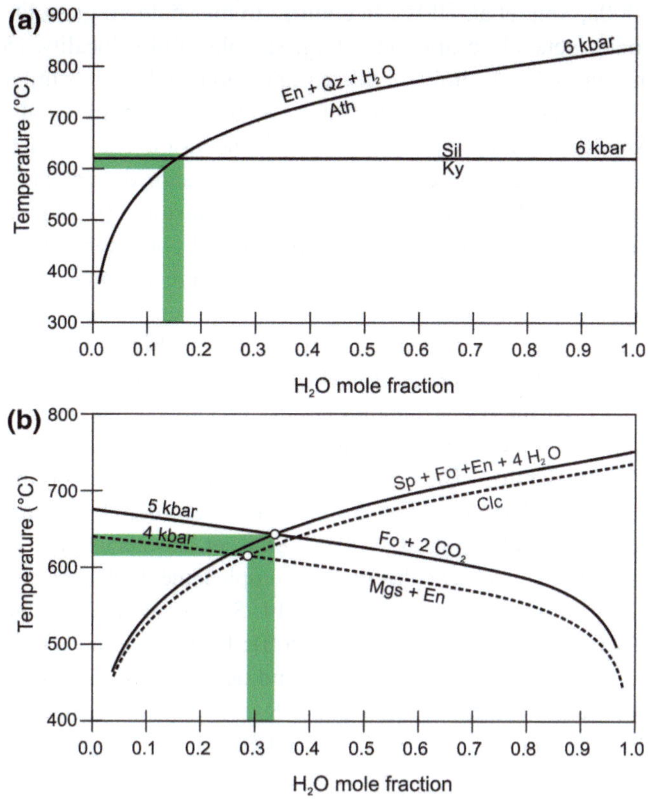

Fig. 8.10 **a** Composition of the hydrating fluid constrained by *P-T* conditions of ca. 6 kbar, 600–630 °C calculated for the retrograde Ath-in isograd assuming Mg-end members. **b** Composition of externally derived fluid phase established from partially hydrated and carbonated ultramafic assemblages. All reaction curves are calculated using the dataset of Holland and Powell (1998). Mineral compositions are given by Van Schalkwyk and van Reenen (1992) (sample DR191). Activities for the end members were calculated using the program AX developed by T. Holland. See text for discussion. Note that the H_2O activity is recalculated to H_2O mole fraction using the H_2O–CO_2 activity-composition model from Aranovich and Newton (1999) (modified after van Reenen et al. 2014)

circulation through the overlying granulite-facies crust at 2.68–2.62 Ga and is manifest in the petrography of the ca. 2.68 Ga Matok granitoid body (Fig. 8.3) (Bohlender et al. 1992; Laurent et al. 2013) and by published age data on fluid events that affected the SMZ up to ca 2.62 Ga (Kreissig et al. 2001; Belyanin et al. 2015).

Evidence that the entire SMZ consisted of granulite-facies rocks prior to ca. 2.68 Ga and before retrograde hydration is demonstrated by partially retrograde hydrated mafic, ultramafic granulite and BIF of the Bandelierkop Formation associated with completely retrograde hydrated metapelitic granulite that outcrop in the Lowveld at locality 7 (Fig. 8.3). De Wit et al. (1992c) also described outcrops of partly retrograde hydrated ultramafic granulite that outcrop at the extreme western tip of the Giyani greenstone belt near Babango (Fig. 8.3) (De Wit et al. 1992c). Partially hydrated ultramafic granulites also outcrop at the extreme western tip of the Rhenosterkoppies greenstone belt (Fig. 8.3).

Finally, it should be noted that the regional retrograde rehydration event that affected large areas of the SMZ in the hanging wall of the sub-horizontal HRSZ occurred after emplacement of Matok pluton at ca 2.68 Ga and reflects the static infiltration of hot water-bearing fluids that reacted with the overlying cooling granulites.

8.4.3.3 Metamorphic Evolution of the Granulite Domain North of the Retrograde Ath-in Isograd

Traverse C–D (Fig. 8.3) continues northwards into the Highveld that exposes steeply foliated pristine granulites (Fig. 8.7b, c) that are subdivided into two distinct metamorphic subdomains located south and north of Annaskraal Shear Zone (Figs. 8.3 and 8.4). Metapelitic assemblages characterized by Grt + Opx + Crd + Bt + Pl + Qz ± Sil in rocks from the southern subdomain document clear evidence for near-isobaric cooling (IC) *P-T* paths superimposed onto earlier decompression-cooling *P-T* paths (DR19 in Fig. 8.11). In contrast, samples with similar mineralogical compositions, but lacking sillimanite and collected from the northern subdomain document only evidence for a two-stage decompression-cooling *P-T* history (DR45 in Fig. 8.11). High-grade gneisses in both subdomains are devoid of kinematic indicators being characterized by macroscopic and microscopic granoblastic textures (Figs. 8.9a, b and 8.12a, b) while only a few rodding structures and small-scale fold axes were identified in the field (Du Toit et al. 1983; Roering et al. 1992b; Smit et al. 1992, 2001, 2011). Steeply north-dipping high-grade shear zones characterized by straight gneisses (Fig. 8.7c) (Smit and van Reenen 1997; Smit et al. 2001) that bound the complexly folded crustal blocks (Figs. 8.2 and 8.3) locally preserve kinematic indicators related to thrust sense shear deformation and record evidence for near-isobaric cooling *P-T* paths similar to that of unsheared granulites (Smit et al. 1992, 2001; Smit and van Reenen 1997). The granoblastic textures of these high-grade gneisses are interpreted to reflect evidence for extensive recrystallization and annealing during initial exhumation to the mid-crustal level as is suggested by metamorphic data (Fig. 8.11) (van Reenen 1983; van Reenen et al. 1987, 1990, 2011, 2014; Roering et al. 1992b; Smit et al. 2001, 2011, 2014).

P-T paths (Fig. 8.11) for rocks from both zones are derived from common Grt + Opx + Crd + Qz ± Sil bearing mineral assemblages, which document evidence for two major net transfer reactions (Fig. 8.12a, b) (Perchuk et al. 1996, 2000a; Smit et al. 2001; van Reenen et al. 2011, 2014):

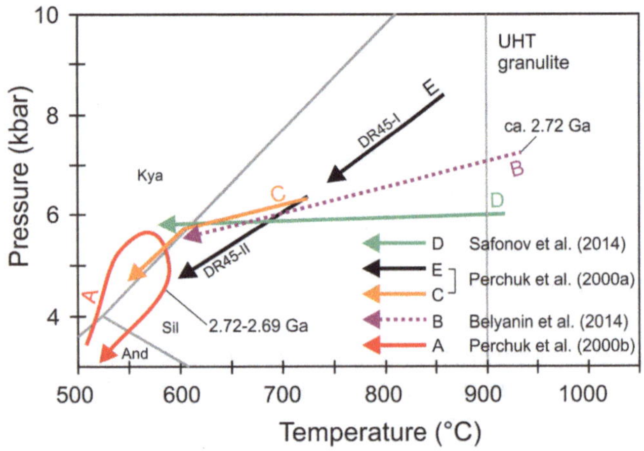

Fig. 8.11 *P-T* paths (A–E) compiled from a large published database (e.g. Perchuk et al. 2000a, b; Smit et al. 2001; Belyanin et al. 2014; Safonov et al. 2014) that highlights important metamorphic features of the SMZ and the juxtaposed Khavagari arm of the Giyani greenstone belt located in the footwall of the steeply north-dipping HRSZ (locality 1 in Fig. 8.3). A: *P-T* loop constructed from Grt-Ky-bearing schists exposed in the Khavagari Shear Zone in the Khavagari arm of the Giyani greenstone belt (locality 1 in Fig. 8.3). B: Near-isobaric cooling *P-T* path that characterizes the southern granulite domain (Figs. 8.3 and 8.4) constructed from MgO–Al_2O_3-rich zones in sample DR19 (Belyanin et al. 2014). C: Near-isobaric cooling *P-T* path that characterizes the southern granulite domain constructed from the common mineral assemblage Grt + Opx + Crd + Qz in sample DR19 (Perchuk et al. 2000a; Smit et al. 2001). This assemblage characterizes metapelitic granulite throughout the granulite subdomain of the SMZ. D: Evidence for ultrahigh-temperature near-isobaric cooling from >900 °C calculated from partially assimilated metapelitic granulite that interacted with intruding ultrahigh-temperature Grt-bearing trondhjemites at the Petronella locality (locality 10 in Fig. 8.3). E: Multi-stage decompression-cooling *P-T* path that characterizes the northern granulite subdomain (DR45) (Figs. 8.3 and 8.4)

$$2\ \mathrm{Grt} - 1 + 3\ \mathrm{Qz} \rightarrow \mathrm{Crd} + 2\ \mathrm{Opx} \qquad (2a)$$

$$3\ \mathrm{Crd} \rightarrow 2\ \mathrm{Grt} - 2 + 4\ \mathrm{Sil} + 5\ \mathrm{Qz} \qquad (2b)$$

Garnet-consuming reaction (2a) (Fig. 8.12a) documents a decompression-cooling history, whereas reaction (2b) (Fig. 8.12b) that is superimposed onto the products of reaction (2a) represents new garnet growth with sillimanite at the expense of cordierite during near-isobaric cooling. For all rocks, *P-T* calculations were done using the same pressure-sensitive reactions (2a) and (2b) combined with temperature-sensitive Fe–Mg cation exchange reactions (Perchuk et al. 2000a; Smit et al. 2001; see Perchuk 2011 for a detailed explanation).

A large number of samples collected throughout the southern granulite subdomain that include straight gneisses sampled from shear zones, show near-isobaric *P-T* paths. These *P-T* paths typically start at a pressure of ca. 6 kbar (ca. 20 km depth) and a temperature of ca. 700 °C, and end at 5.0–5.5 kbar and 570–600 °C (Fig. 8.11) (Perchuk et al. 2000a; Smit et al. 2001). In contrast to the southern granulite zone, metapelitic granulites (e.g. sample DR45, locality 15 in Fig. 8.3) of similar bulk composition and with similar mineral assemblages (but lacking sillimanite) north of the Annaskraal Shear Zone show no evidence for a near-isobaric cooling stage (Perchuk et al. 2000a) and document a two-stage decompression-cooling *P-T* history (DR45 in Fig. 8.11). The rocks record maximum *P-T* conditions of 6.7 kbar and 850 °C, respectively. They were exhumed and emplaced at the mid-crustal level at a pressure of ca. 6 kbar and a temperature of ca. 720 °C (DR45 in Fig. 8.11) (e.g. Perchuk et al. 2000a; Smit et al. 2001; van Reenen et al. 2011). At these *P-T* conditions, the rocks were cooled and annealed before being finally exhumed to 4.5 kbar (ca. 15 km depth). The gap in the decompression-cooling *P-T* path (DR45 in Fig. 8.11) is not recorded by garnet zoning (Perchuk et al. 2000a; Smit et al. 2001) and is interpreted to reflect the time of emplacement of anatectic melts (van Reenen et al. 2011, 2014; Smit et al. 2014) dated at 2.64–2.68 Ga (Kreissig et al. 2001; Belyanin et al. 2014).

The contrasting near-isobaric and decompression-cooling *P-T* paths (compare DR19 and DR45 in Fig. 8.11) suggest that the entire SMZ, with its embedded steeply plunging folds and steeply dipping high-grade shear zones was initially uplifted to the mid-crustal level (ca. 20 km depth), being accompanied by emplacement of leucocratic trondhjemitic magmas at 2.68–2.64 Ga (e.g. Kreissig et al. 2001; Belyanin et al. 2014; Safonov et al. 2014, 2018b; van Reenen et al. 2014), where both granulites and magmas were concomitantly cooled and annealed (Safonov et al. 2014). Consequently, these granoblastic rocks are devoid of mineral lineations (Du Toit et al. 1983; Roering et al. 1992b; Smit et al. 1992).

The presence of near-isobaric cooling *P-T* paths (Fig. 8.11) documented by granulites and their sheared equivalents south of the Annaskraal Shear Zone is in support of structural (Fig. 8.4) and geophysical data (see Fig. 8.15a, b) that a hot SMZ nappe with imbedded steeply plunging folds and steeply dipping shear zones was channelled southwards for a distance of ca. 30 km (inset map in Fig. 8.3) against and onto the adjacent NKVC along a ca. 6 kbar near-isobaric (20 km depth) crustal plane. This plane is represented by the flat-lying HRSZ (Figs. 8.4 and 8.6b, c) and is underlain by lower grade greenstone material (inset map in Fig. 8.3) (De Beer and Stettler 1992; Smit et al. 2014; van Reenen et al. 2014). Near-isobaric cooling *P-T* paths (Fig. 8.11) demonstrate that these rocks had to cool rapidly. The only mechanism to facilitate such cooling of hot rock masses at mid-crustal level and far from the exposed contact with the juxtaposed granite-greenstone craton (Figs. 8.3 and 8.4) is the pervasive and channelled infiltration of large volumes of relatively cold fluids derived from devolatilization of greenstone belt material that underlies much of the

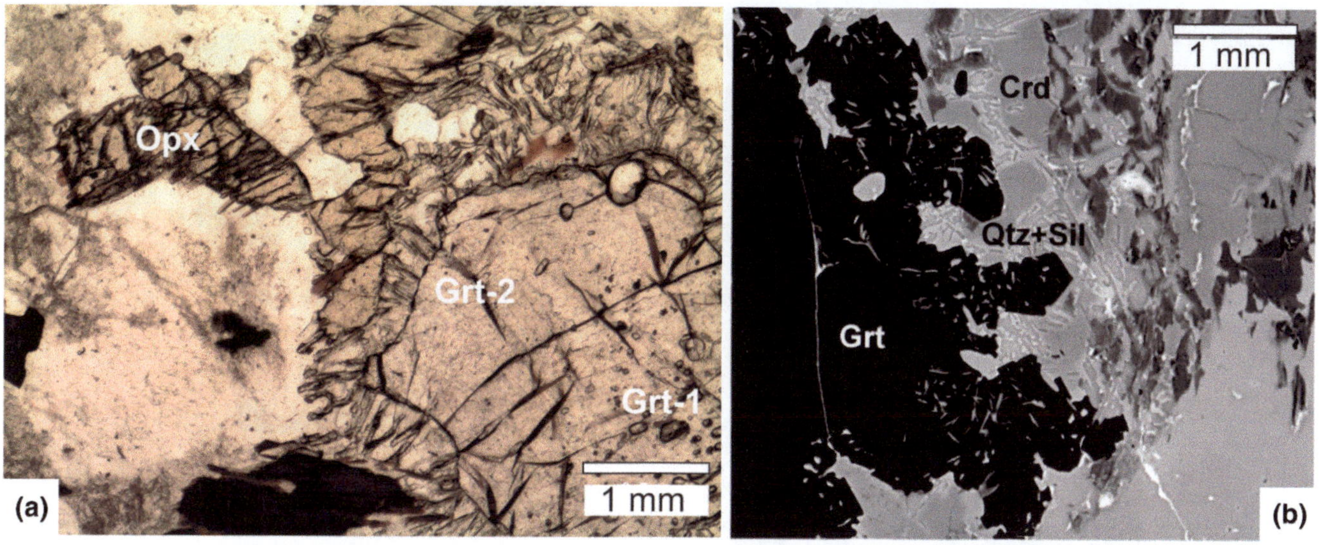

Fig. 8.12 **a** Photomicrograph of reaction (2a): 2 Grt-1 + 3 Qz → Crd + 2 Opx. This reaction documents evidence for a two-stage decompression-cooling *P-T* path (DR45 in Fig. 8.11). The early stage is documented by the poikiloblastic core of Grt-1 in equilibrium with coarse-grained Opx-1 and Crd-1 in the matrix of the rock. The second stage is reflected by the inclusion-free Grt-2 rim in contact with symplectitic intergrowths of Opx-2 and Crd-2 (e.g. Smit et al. 2001). **b** Backscatter electron image of reaction (2b): 3 Crd → 2 Grt-2 + 4 Sil + 5 Qz preserved in sample DV101 (locality 14 in Fig. 8.3). This reaction demonstrates evidence for near-isobaric cooling (e.g. Smit et al. 2001). See text for discussion

SMZ (inset map in Figs. 8.3 and 8.4) (De Beer and Stettler 1992; van Reenen et al. 2011, 2014; Smit et al. 2014).

8.4.3.4 Fluid-Induced Melting and Generation of Granitic Magma at UHT Conditions in the SMZ

High-grade gneisses exposed throughout the SMZ (Figs. 8.2 and 8.3) are migmatitic. This is expressed in outcrops by the presence of small volumes (<10%) of leucocratic Pl + Qz ± perthite/antiperthite anatectic material enhancing the gneissic and shear fabric (e.g. Figs. 8.7a–c and 8.8a) (Du Toit et al. 1983; van Reenen et al. 2014). Migmatization that affected high-grade gneisses is interpreted to have occurred at the deep crustal level during peak metamorphism and deformation (e.g. van Reenen et al. 2011, 2014), resulting in melt-weakened crust from which small amounts of melt were not removed (e.g. Jamieson et al. 2011; Smit et al. 2014).

In addition to the fabric sub-parallel anatectic material, large volumes of mainly discordant and weakly to non-foliated granodioritic-trondhjemitic veins and bodies (Du Toit et al. 1983) also intrude and partially assimilated (Safonov et al. 2014, 2018a) the deformed and metamorphosed granulite at ca. 2.67–2.64 Ga (Kreissig et al. 2001; Belyanin et al. 2014; van Reenen et al. 2011, 2014). These large undeformed bodies (e.g. Fig. 8.13a) reflect a major pulse of anatexis linked to fluid-induced decompression melting that post-date peak metamorphism and deformation at ca. 2.72 Ga and occur at different localities in the SMZ (Dubinina et al. 2015; Safonov et al. 2018b). Safonov et al. (2014, 2018a) studied the large (ca. 30 × 60 m) and almost undeformed granodiorite-trondhjemite body dated at ca. 2.67 Ga that intrude sheared migmatitic granulite (Fig. 8.13a) on the Farm Petronella (locality 10 in Fig. 8.3). The partially assimilated migmatitic gneisses preserve evidence for the near-isobaric cooling of hot magma from >900 to ca. 600 °C (at a pressure of 5.5–6.5 kbar) (D in Fig. 8.11), that also affected the host metapelitic granulite. CO_2 fluid inclusions in quartz and garnet in trondhjemites show that the magma transported CO_2 fluids with densities corresponding to the late stages of magma cooling at 5.5–6.5 kbar and 600–650 °C. CO_2 fluids coexisted with low H_2O activity brine fluids (Safonov et al. 2014). Carbon isotope compositions of graphite and fluid inclusions from different garnet-sillimanite-bearing leucocratic tonalites, trondhjemites and granites associated with orthopyroxene-bearing granulite metapelites (including Bandelierkop, locality 9 in Fig. 8.3) in the SMZ, also indicate that fluids associated with the granitoid magmas of the SMZ originated from a source unrelated to the host metapelites (Safonov et al. 2018b).

Evidence for deep crustal embrittlement (Fig. 8.13b) at the Petronella locality documented by brittle deformation in which granulite-facies breccia are cemented by Opx-bearing leucocratic material (Roering et al. 1995) confirms the action of fluids/melts at near-lithostatic pressures during granulite-facies metamorphism in the SMZ (Roering et al. 1995).

Fig. 8.13 **a** Undeformed leucocratic trondjhemite (ca. 2.67 Ga) with partially assimilated migmatitic metapelite granulite (dated at ca. 2.72 Ga) at Petronella (locality 10 in Fig. 8.3). **b** Exposure of metapelitic granulite that contains two distinct foliation trends at high angles to each other. The leucocratic vein separating the two blocks of foliated granulite is composed of orthopyroxene. The relationship is interpreted as evidence for deep crustal embrittlement under granulite-facies conditions within the Petronella Shear Zone, whereby fractured blocks of granulite gneiss have been cemented together along the orthopyroxene vein (Roering et al. 1995); see text for discussion

Roering et al. (1995) showed that the presence of such fluids/melts along thrust planes resulted in a lubrication effect due to reduction of brittle/plastic strength of rocks subjected to fluid/melt percolation at near-lithostatic pressures. It would have enhanced (and effectively permitted) near-horizontal channelling of a hot granulite nappe for more than 35 km over relatively cold greenstone material by reducing friction along the thrust plane (Roering et al. 1995; Smit et al. 2014). Devolatilization of underlying greenschist material persisted as long as southwards channelling of the hot nappe continued (Koons et al. 1998; Yardley 2009; Smit et al. 2014).

8.4.3.5 Timing of Major Events that Affected the SMZ and Juxtaposed NKVC

Geochronological data obtained by different researchers provide the following important constraints on the timing of major events that affected the SMZ and the juxtaposed NKVC.

(1) Different authors used zircon SHRIMP age data obtained from metamorphic charnockite (Retief et al. 1990) and from metapelitic granulite at different localities in the granulite zone south of the Annaskraal Shear Zone (Rajesh et al. 2014; Taylor et al. 2014; Nicoli et al. 2015) to accurately constrain the timing of the peak metamorphic and deformational event in the SMZ to ca. 2.72 Ga.

(2) Thermal and dynamic interaction of SMZ hanging wall granulite with relatively cold greenstone material in the footwall of the steeply north-dipping HRSZ at the Khavagari arm of the Giyani greenstone belt at 2.72–2.69 Ga accurately dates the age of the HRSZ and the time of emplacement of the SMZ against and over the adjacent NKVC (Figs. 8.3, 8.4 and 8.11) (e.g. Perchuk et al. 2000b; Kreissig et al. 2001; Kramers et al. 2014).

(3) Growth of monazite associated with the common decompression metapelitic assemblage Grt + Crd + Opx + Qz from the Bandelierkop locality at 2.69 Ga accurately dates the early post-peak stage of exhumation (Fig. 8.11) (Kreissig et al. 2001).

(4) Emplacement at ca. 2.67 Ga of the granodioritic-tonalitic body (Fig. 8.13a) into already deformed and sheared migmatitic granulite associated with the steeply dipping Petronella Shear Zone provides a minimum age for peak metamorphism and deformation (van Reenen et al. 2011, 2014; Belyanin et al. 2014; Safonov et al. 2014).

(5) Intrusions of geochemically similar granitoid plutons into the SMZ (Matok) (Figs. 8.2, 8.3 and 8.13) and in the adjacent NKVC (Moletsi) north and south of the HRSZ at ca. 2.68 Ga (Figs. 8.2, 8.3 and 8.14) (Laurent et al. 2013), provide unequivocal evidence that the SMZ nappe was already juxtaposed at ca. 2.72 Ga against and over the NKVC.

Fig. 8.14 Matok granite intrudes a steeply dipping high-grade shear zone developed in sheared Opx-bearing metapelitic granulite on Farm Tarentaaldraai (locality 16 in Fig. 8.3). This high-grade shear zone was subsequently reactivated to produce the oblique-slip upper crustal Nthabalala mylonitic shear zone (e.g. van Reenen et al. 2011) that displaces the Matok pluton (Figs. 8.2 and 8.3). This is shown by superimposition of the younger shear fabric onto the intrusive granite (this figure), by left-lateral displacement of the Matok pluton, and by similar sense displacement of large folds in the area east of the Matok pluton (Fig. 8.2) (van Reenen et al. 2011)

(6) Regional infiltration of aqueous-carbonic fluids into the overlying hot granulite nappe that established the Ath-in retrograde isograd and associated zones of retrograde hydrated granulite (Fig. 8.3) also affected the Matok pluton and occurred at 2.68–2.62 Ga (Kreissig et al. 2001; Laurent et al. 2013).

(7) Emplacement of the post-kinematic Palmietfontein granite into migmatitic granulite at ca. 2.45 Ga indicates the end of major tectono-metamorphic events in the SMZ (Barton and van Reenen 1992).

(8) Rajesh et al. (2014) and Nicoli et al. (2015) reported ages older than 3.0 Ga (as old as 3.44 Ga, Rajesh et al. 2014) for detrital zircon from metapelites of the Bandelierkop Formation, whereas Retief et al. (1990) reported similar ages from zircon in metamorphic charnockite. These ages are consistent with ages reported for greenstone successions on the adjacent NKVC (Zeh et al. 2013).

8.4.4 Geophysical Constraints on the Nature of the SMZ and the Relationship with the NKVC and the CZ

Aeromagnetic-, vibro seismic-, gravity-, and geoelectrical data obtained from the LC and adjacent NKVC (Figs. 8.1, 8.3 and 8.15a, b) (Stuart and Zengeni 1987; Kleywegt et al. 1987; Kleywegt 1988; De Beer and Stettler 1988, 1992; Stettler et al. 1989; Roering et al. 1990; De Beer et al. 1991; Durrheim et al. 1992) provide important information regarding the complex geological relationship of a hot allochthonous SMZ granulite nappe with both the CZ in the north and the NKVC in the south.

Detailed geophysical data (Fig. 8.15a, b) collected across the southern boundary (HRSZ) of the SMZ (Kleywegt et al. 1987; De Beer and Stettler 1988, 1992; De Beer et al. 1991; Klemperer 1992) showed that SMZ granulites were thrust southwards over the low-grade NKVC along the entire lateral length of the HRSZ (Figs. 8.2 and 8.3), confirming the allochthonous nature of the SMZ nappe based on structural and *P-T-t* studies (e.g. Miyano et al. 1992; Roering et al. 1992a; Perchuk et al. 1996, 2000b; Kreissig et al. 2001; Kramers et al. 2014). Furthermore, geoelectrical data (inset map in Fig. 8.3; De Beer and Stettler 1992) show that more than 60% of the surface area of the SMZ is underlain at depth by a material with physical characteristics identical to those of greenstone material on the NKVC. In addition, geoelectrical, seismic, gravity and aeromagnetic data from the NKVC, LC, and ZC allowed modelling the crust with three distinct units (Fig. 8.15a) (e.g. De Beer and Stettler 1992). These units comprise (1) the two granite-greenstone cratons (NKVC and ZC) with similar lithologies and gravity signatures, (2) the SMZ and NMZ both consisting mainly of meta-igneous rocks with identical gravity signatures, and (3) the CZ dominated by metasedimentary rocks with gravity properties that differ significantly from that of the two marginal zones, but that are identical to that of the two cratons (De Beer et al. 1992).

The following section is a detailed discussion of the two-stage steep exhumation of the CZ at 2.72–2.62 Ga that is interpreted to have directed southwards channelling of a hot allochthonous SMZ nappe by means of a conveyor belt mechanism as previously discussed by Smit et al. (2014).

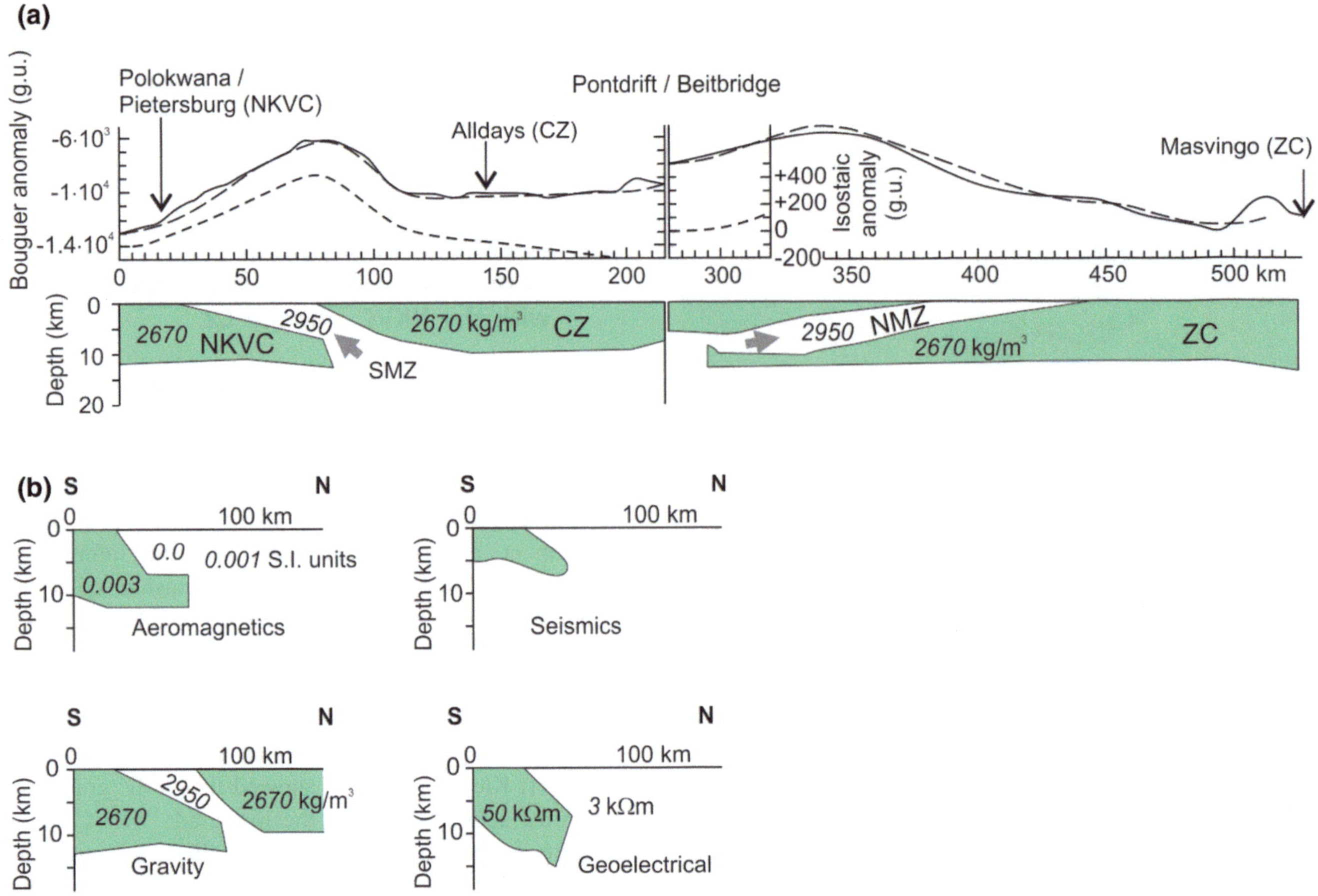

Fig. 8.15 Geophysical data obtained from the Limpopo Complex and adjacent KVC and ZC cratons (modified after De Beer and Stettler 1992). **a** Bouguer gravity models for the Limpopo complex and adjacent granite-greenstone cratons (Fig. 8.1). Densities are in kg/m^3 and Bouguer (solid line) and isostatic (broken line) anomalies are in gravity units (1 g.u. = 10^{-6} m/s^2 = 0.1 mgal). **b** Comparison of the geoelectrical, gravity, aeromagnetic, and seismic reflection data across the SMZ of the LC. Density and susceptibility are expressed in kg/m^3 and S.I. units, respectively, the resistivity is expressed in kΩ·m. See text for discussion

8.5 Two-Stage Exhumation History of the CZ at 2.72–2.61 Ga

The two-stage exhumation of the CZ at ca. 2.72–2.62 Ga is discussed by considering the most relevant published regional structural, magmatic, metamorphic, and geochronological features of the CZ (e.g. Roering et al. 1992a, b; Perchuk et al. 2008a, b; van Reenen et al. 2008; Kramers and Mouri 2011; Kramers and Zeh 2011; Kramers et al. 2011; Smit et al. 2011; Smit et al. 2014; Kröner et al. 2018).

Decompression-cooling *P-T-t* paths constructed for metapelitic gneisses (of which sample TOV13 in Fig. 8.17 is a prime example) that are associated with distinct regional structural features (isoclinal folds and closed structures) of the CZ confirmed the two-stage steep *P-T* evolution of this high-grade gneiss terrane in the Neoarchaean (e.g. Smit et al. 2011). Steeply plunging isoclinal folds and melt-weakened domes were initially emplaced at the mid-crustal level before ca. 2.65 Ga. This was followed by emplacement at the upper crustal level of steeply southwest-plunging mega closed structures at 2.65–2.62 Ga (Figs. 8.17, 8.18, 8.19, 8.23, 8.24a).

P-T-t paths for these distinct major structural features are derived from Grt − Crd − Sil − Bt − Kfs − Qz ± Opx-bearing metapelites of comparable bulk chemical compositions and based on the concept of local chemical equilibrium (e.g. Perchuk et al. 2008a, b; Perchuk and van Reenen 2008; Smit et al. 2011). This approach also recognizes the important role of heterogeneous deformation, which resulted in the formation and preservation of discrete Neoarchaean (and Palaeoproterozoic) high-grade events on scales that vary from the regional scale (Fig. 8.19) to outcrop scale (Fig. 8.20a–c) and also the thin section scale in the CZ (Perchuk and van Reenen 2008; Perchuk et al. 2008a; Mahan et al. 2011; Smit et al. 2011). *P-T-t* paths (Fig. 8.17) linked to

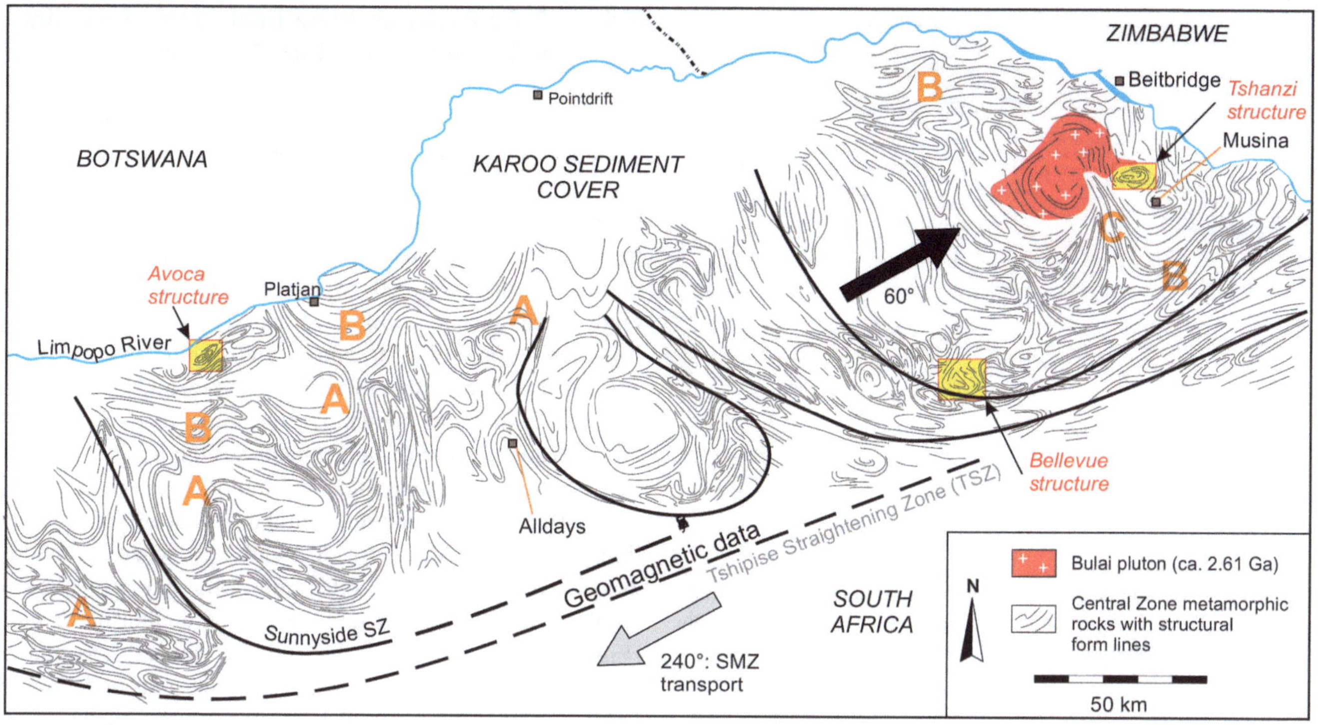

Fig. 8.16 Structural map of the central zone (CZ) based on aerial photographic interpretation (after Smit et al. 2011). Structural form lines (thin black lines) highlight large melt-weakened migmatitic domes (A) and regional-scale isoclinal folds developed in granoblastic gneisses that collectively reflect the oldest deformational features of the CZ (B). These early structural features are overprinted by large north–south trending fold structures referred to as cross folds (C), and by large closed structures mapped as sheath folds (Roering et al. 1992b), of which the Avoca, Tshanzi, and Bellevue structures (highlighted in yellow) have been studied in detail. Note the presence of a large number of closed structures developed in the vicinity of the Bellevue structure. The southwest–northeast-trending Tshipise Straightening Zone (here termed TSZ) bounds and truncates the CZ in the south. Also shown is the position of the post-tectonic ca. 2.61 Ga Bulai granitic pluton as well as the northeast direction of exhumation of CZ rocks. The SMZ was channelled in the exact opposite direction (Roering et al. 1992b). North–south trending shear zones related to the TSZ (thick black lines) subdivide the CZ into large crustal blocks

discrete fabric-forming events associated with the formation of regional isoclinal folds and with mega-scale closed structures (Fig. 8.16) provided, for the first time, unequivocal evidence for the high-grade polymetamorphic evolution of the CZ. This complex evolutionary history is reflected by a multi-stage high-grade tectono-metamorphic event at ca. 2.72–2.62 Ga in the Neoarchaean (Fig. 8.17, TOV13-1 and TOV13-2) followed by a high-grade and mainly thermal overprint associated with discrete (metres wide) shear zones in the Palaeoproterozoic at ca. 2.0 Ga (Perchuk and van Reenen 2008; van Reenen et al. 2008; Kramers and Mouri 2011; Kramers and Zeh 2011; Kramers et al. 2011; Mahan et al. 2011; Smit et al. 2011). Kröner et al. (2018) recently published detailed U-Pb age data obtained from deformed high-grade gneisses preserved as a large enclave in the ca. 2.61 Ga Bulai pluton that we interpret to provide direct evidence for the timing of the second stage of exhumation of the CZ (Fig. 8.17, TOV13-2) in the Neoarchaean at ca. 2.62 Ga. This new metamorphic data finally put to rest the conclusion of a single Palaeoproterozoic Limpopo orogeny at ca. 2.0 Ga as has been argued for by different researchers (e.g. Zeh and Klemd 2008 and references therein). However, the proposal by Brandt et al. (2018) that the Neoarchaean high-grade event at 2.64–2.62 Ga (Kröner et al. 2018) is linked to a counterclockwise *P-T* path is not in accordance with our structural-metamorphic and magmatic data as will be further discussed in Sect. 8.5.2.

Evidence for the steep two-stage exhumation of CZ granulites in the interval 2.72–2.61 Ga is demonstrated by linking specific stages of two-stage *P-T-t* paths (TOV13-I and TOV13-II in Fig. 8.17) to the successive evolution of major fold- and shear-deformational events in the CZ (Figs. 8.18, 8.19 and 8.22). Granoblastic metapelitic granulite associated with regionally developed steeply plunging isoclinal folds and dome-like structures cored by migmatitic gneisses (Fig. 8.16) are associated with decompression-cooling *P-T* paths (TOV13-1 in Fig. 8.17) that accompanied the initial exhumation of the CZ at the mid-crustal level

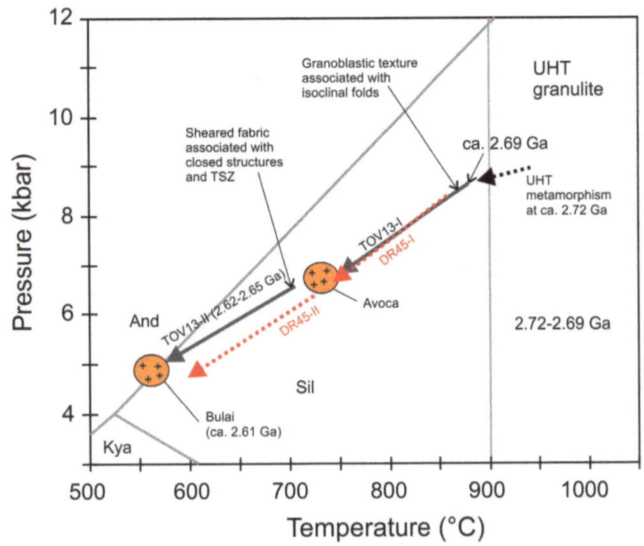

Fig. 8.17 *P-T-t* path constructed for metapelitic sample TOV13 in the rim of the Tshanzi closed structure (Fig. 8.16) is an example of the two-stage decompression-cooling (DC) metamorphic evolution that characterizes the CZ (Perchuk et al. 2008b; Smit et al. 2011). The first stage of the *P-T-t* path (TOV13-I) is constructed from a Grt-1 + Crd-1 + Sil-1 + Bt-1 + Qtz assemblage preserved in a relict granoblastic portion of sample TOV13. TOV13-I demonstrates the initial exhumation of early isoclinal folds and dome-like structures at the mid-crustal level before ca. 2.65 (emplacement of Avoca closed structure). The second stage of the *P-T-t* path (TOV13-II) is constructed from a Grt-2 + Crd-2 + Sil-2 + Bt-2 + Qtz-bearing assemblage preserved in the superimposed shear fabric developed in sample TOV13 (Smit et al. 2011). TOV13-II reflects the subsequent decompression-cooling *P-T* evolution of sheared gneisses associated with formation of closed structures (and with the TSZ) at 2.65–2.62 Ga, prior to emplacement of the Bulai pluton at ca. 2.61 Ga. Note the close similarity of *P-T* paths constructed for metapelitic granulite from the granulite subdomain north of the Annaskraal Shear Zone in the SMZ (DR45 in Fig. 8.11), and *P-T* paths constructed for CZ metapelitic granulite (TOV13 in Fig. 8.17). TSZ: Tshipise Straightening Zone. See text for discussion

before ca. 2.65 Ga. In turn, shear deformation is reflected by large closed structures cored by granitoids and previously mapped as steeply southwest-plunging mega sheath folds (Figs. 8.16, 8.19 and 8.22) (Roering et al. 1992b). These structures record evidence for granitic diapirism that accompanied the steep NE-directed exhumation of the CZ at 2.65–2.62 Ga (Perchuk et al. 2008a; Smit et al. 2011). *P-T* evidence presented for the multi-stage decompression-cooling evolution of the CZ (Fig. 8.17) is discussed with reference to the prime example of sample TOV13 collected from the rim of the Tshanzi closed structure (Fig. 8.16). The reader might also consult publications that discuss similar *P-T* data collected from other localities in the CZ that are up to 150 km apart (e.g. Perchuk et al. 2008a; Huizenga et al. 2011; Smit et al. 2011).

8.5.1 *P-T-t* Paths and the Evolution of Isoclinal Folds and Dome-like Structures

Small-scale to kilometre-scale isoclinal folds and closely associated dome-like structures cored by migmatitic gneisses occur throughout the CZ (Fig. 8.16). These folds deform a steep gneissic/migmatitc foliation (e.g. Smit et al. 2011) that trends in general in an east–west direction sub-parallel to the regional structural trend of the CZ (Figs. 8.1 and 8.16). High-grade gneisses within complexly deformed isoclinal folds and dome-like structures are granoblastic in hand specimen and under the microscope and characterized by the absence of mineral stretching lineations (e.g. Smit et al. 2011). The only kinematic elements visible in such gneisses in the field are quartz rodding and minor fold axes (e.g. Smit et al. 2011). This situation is similar to that of the granulite domain of the SMZ (Figs. 8.2 and 8.3) where granulites associated with isoclinal folds and dome-like structures of similar scale are also characterized by granoblastic textures (Figs. 8.9a, b and 8.12a, b) and the general absence of kinematic indicators. These features allow areas underlain by isoclinal folds and dome-like structures to be easily distinguished in the field from areas underlain by large closed structures with abundant kinematic indicators (compare Figs. 8.16, 8.20a–c and 8.21).

Isoclinal folds are developed in all rocks comprising the Beit Bridge Complex, the Messina Layered Complex, the Sand River Gneiss, and the Alldays/Verbaard gneiss. They provide metamorphic evidence for the early stage of the two-stage exhumation history of the CZ. Peak metamorphism has been reached in the core of the orogen at ca. 2.72 Ga (Fig. 8.17). This was followed by the early decompression-cooling stage, which is reflected by a decompression-cooling *P-T* path constructed for granoblastic metapelitic granulite (TOV 13-I in Fig. 8.17). Decompression-cooling ended concomitant with the onset of granitic diapirism at 2.65–2.63 Ga (Fig. 8.17) and the formation of major closed structures typified by a high-grade shear fabric (Figs. 8.20c and 8.21) (e.g. Smit et al. 2011). *P-T* conditions during the early exhumation stage of the decompression-cooling history calculated from the common metapelitic assemblage Grt + Crd + Sil + Qz + Kfs (Fig. 8.17) range from ca. 6.3 kbar and ca. 830 °C to 6.5 kbar and 700 °C. Timing of peak metamorphism in the CZ has not been accurately established, due to the almost pervasive superimposed thermal event that affected the CZ at ca. 2.01 Ga (van Reenen et al. 2008; Kramers and Mouri 2011; Smit et al. 2011) (see also discussion by Kröner et al. 2018). However, Kröner et al. (1998, 1999) provided accurate age data obtained from two samples of quartzofeldspathic granitoids collected from road cuttings southwest of Musina (TR115 and TR130, Fig. 8.2; Kröner et al. 1999) that constrain the minimum age of peak metamorphism and deformation in the CZ. The first sample (TR115) described as a well layered, fine-grained, biotite-rich quartzofeldspathic gneiss with granoblastic texture, gave a SHRIMP age of 2717 ± 3 Ma considered by Kröner et al. (1999) to reflect the formation age of the gneiss protolith. The second sample (TR130) considered by these authors to represent anatectic Singelele-type gneiss, gave a SHRIMP age of 2681 ± 8 Ma. Similar rocks also considered as anatectic granitoids have at numerous localities in the CZ been shown to intrude already deformed and metamorphosed supracrustal rocks. Singelele-type granitoids, in general, show little or no evidence of deformation after emplacement (e.g. Fig. 8.18), comparable to similar field relationships described in the SMZ (e.g. Fig. 8.13a). These results, and similar results obtained from similar anatectic granitoids in the SMZ, constrain the minimum time of peak metamorphism and deformation that affected both domains at ca. 2.68 Ga. In fact, the time of peak metamorphism in the CZ based on the age of sample TR115 (2717 ± 3 Ma) is the same as the time of peak metamorphism in the SMZ dated at ca. 2.72 Ga. The proposed linked evolution of these two geotectonic domains is thus strongly supported by geochronological data obtained from intrusive granitoids.

High-grade gneisses associated with steeply plunging isoclinal folds seldomly show evidence of intense deformation on the outcrop scale (Fig. 8.20a) but are always granoblastic in hand specimen and under the microscope (e.g. van Reenen et al. 2008; Mahan et al. 2011; Smit et al. 2011). This suggests, as the case is in the SMZ, that hot granulites were exhumed before ca. 2.65 Ga and emplaced at the mid-crustal level (ca. 20 km depth) (DR45-I, II in Fig. 8.11, and TOV13-I, II in Fig. 8.17), where the rocks were cooled and annealed before being further exhumed along decompression-cooling *P-T* paths.

8.5.2 *P-T-t* Paths and the Evolution of Mega Closed Structures

Mega-scale steeply southwest-plunging folds (Figs. 8.19 and 8.22) that are characteristically circular or eye-shaped in the plan have been recognized throughout the CZ (Fig. 8.16) (e.g. Roering et al. 1992b). Closed structures deforming early isoclinal folds (Figs. 8.19 and 8.20a–c) were initially mapped as sheath folds (Roering et al. 1992b) due to the fact that such well-defined structural features are characterized by identical steeply northeast-verging kinematic indicators (Figs. 8.19 and 8.22). High-grade gneisses located in the

Fig. 8.18 Undeformed Singelele-type leucogranite (dated at 2627 ± 2.5 Ma) that intrudes migmatitic metapelitic granulite directly north of the Bulai Pluton (Fig. 8.15) (Boshoff et al. 2006; Smit et al. 2011). This body has the same age as the tonalitic L-tectonite (2626.8 ± 5.4 Ma) that intrudes the core of the Avoca circular structure (Fig. 8.23a, c), located more than 150 km to the west. Note also the similarity with undeformed leucocratic trondjhemite that intruded the SMZ (Fig. 8.13a)

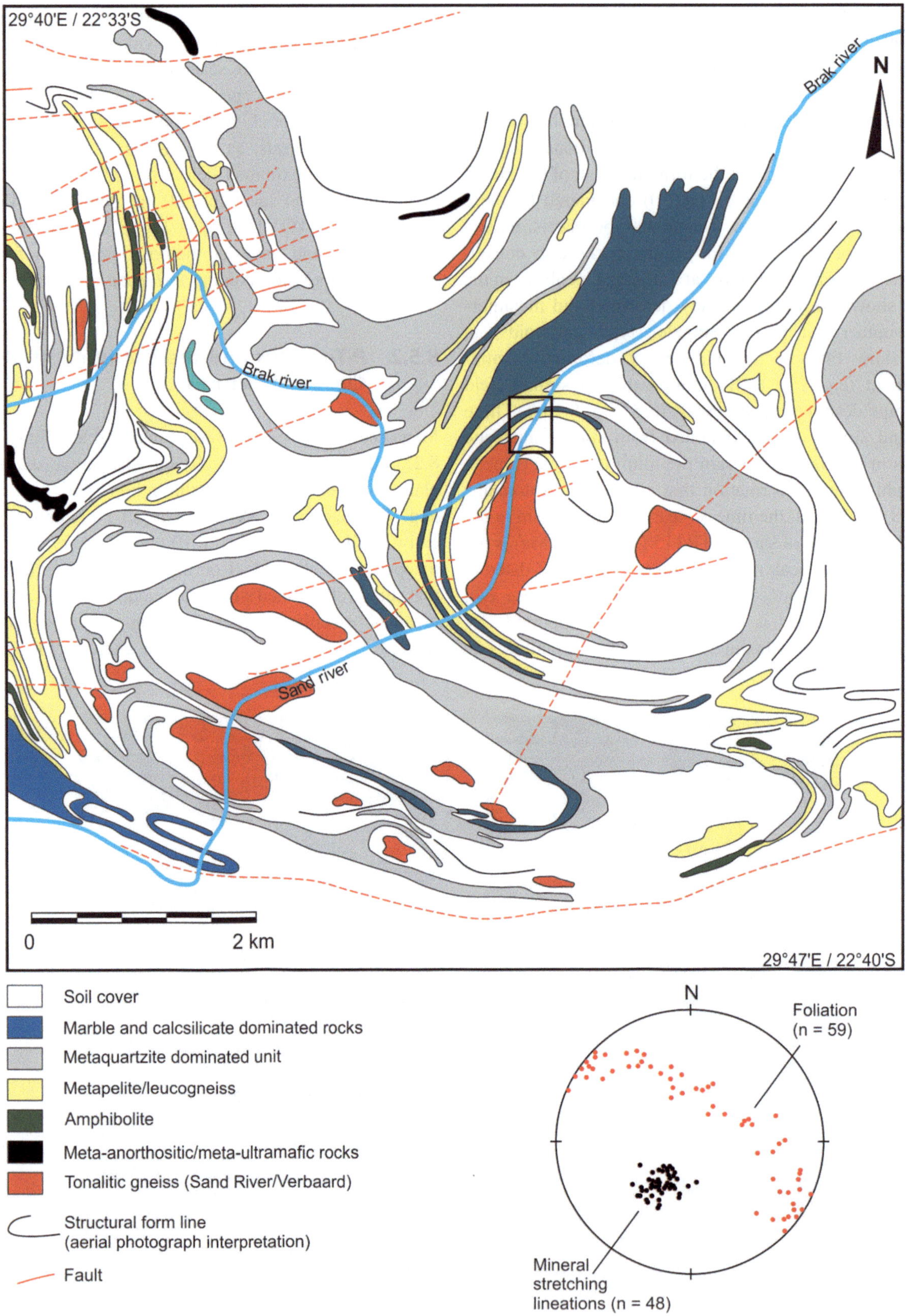

Fig. 8.19 Geological map of the steep southwest-plunging (inset: lower hemisphere stereonet projection) Bellevue closed structure developed on Farm Bellevue southwest of Musina in the CZ (Fig. 8.16) (after Roering et al. 1992b). The map clearly shows transposition of regional-scale isoclinal folds into parallelism with the shear fabric in the rim of the closed structure (Fig. 8.20a–c). Also shown is the location of a well-exposed traverse (rectangular box) through the rim of the structure along the Brak River (Fig. 8.19)

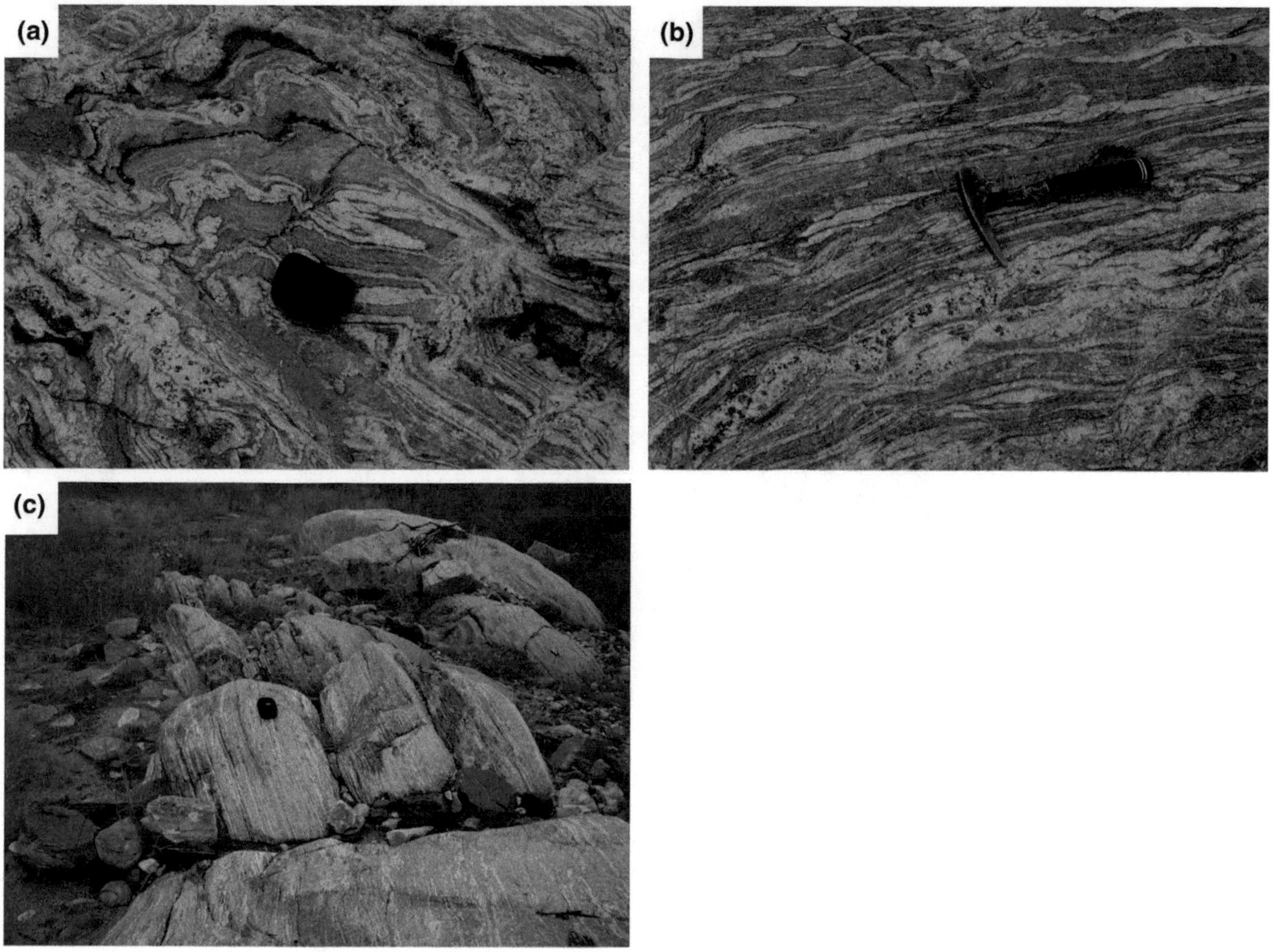

Fig. 8.20 South-north traverse across the rim of the Bellevue closed structure (Fig. 8.19) exposes continuous outcrops of grey tonalitic gneiss. The traverse starts in folded migmatitic gneiss associated with isoclinal folding (**a**) outside the closed structure, and demonstrates the progressive transposition (**b**) of the earlier folds into parallelism with the intense shear fabric developed in the rim (**c**) of the closed structure. This instructive example demonstrates the principle of heterogeneous deformation that controlled the formation and preservation of distinct granulite assemblages respectively, associated with isoclinal folds (**a**) and with closed structures (**c**) and of the TSZ. Gneisses associated with isoclinal folds (**a**) are granoblastic in thin sections in contrast with highly sheared gneisses (**b, c**) associated with closed structures. The principal of heterogeneous deformation allowed the construction of the two-stage D-*P*-*T*-*t* evolution of the CZ (e.g. TOV13-I, II in Fig. 8.17) metapelitic granulite in the rim of Tshanzi closed structure where an early granoblastic fabric is also overprinted by the younger shear fabric (Perchuk et al. 2008b; Mahan et al. 2011; Smit et al. 2011)

outer rims of the closed structures are high-temperature sheared rocks (straight gneisses) characterized by kinematic elements (Fig. 8.21) and differ significantly from annealed granoblastic rocks associated with isoclinal folds and dome-like structures that are largely devoid of kinematic elements. All *P*-*T* paths constructed for closed structures document nearly identical maximum *P*-*T* conditions of ca. 6 kbar and ca. 700 °C for the onset of shear deformation and related granitic diapirism and minimum *P*-*T* conditions of ca. 4.5 kbar and ca. 600 °C (Fig. 8.17) during emplacement into the upper crust. The latter being constrained to before intrusion of the Bulai pluton at ca. 2.61 Ga (Millonig et al. 2008; Smit et al. 2011). Sheared metapelitic gneisses associated with major circular structures retain no *P*-*T* memory of the early exhumation history (TOV13-I in Fig. 8.17), which is defined by the development of isoclinal folds and dome-like structures (Perchuk et al. 2008a, b; Smit et al. 2011). This is due to reworking of the early granoblastic fabric during the superimposed high-grade shear-deformational event (Fig. 8.20a–c).

The direct link between the onset of granitic diapirism and the formation of major closed structures is illustrated by the Avoca and Tshansi structures (Figs. 8.16 and 8.22) (Boshoff et al. 2006; Perchuk et al. 2008b; van Reenen et al. 2008; Smit et al. 2011). The steeply southwest-plunging oval-shaped closed Avoca structure (Fig. 8.22) located

Fig. 8.21 Steeply southwest-dipping planar shear fabric of migmatitic gneiss defining the rim of the Bellevue closed structure (Fig. 8.19). See text for discussion

northwest of the town of Alldays (Fig. 8.16) is cored by a ca. 2.63 Ga trondhjemitic L-tectonite (Fig. 8.23a) that intrudes (Fig. 8.23c) ca. 2.65 Ga penetrative foliated and lineated Singelele-type gneiss in the rim (Fig. 8.23b) (e.g. Boshoff et al. 2006). This relationship provides a precise link between granitic diapirism and the onset of the shear deformation at 2.65–2.62 Ga that controlled the final steep exhumation of the CZ during the Neoarchaean Limpopo orogeny (TOV13-II in Fig. 8.17) (Perchuk et al. 2008a; van Reenen et al. 2008; Smit et al. 2011). Superimposition of shear deformation onto previously folded lithologies is revealed by a south–north traverse that starts in strongly folded Alldays gneiss exposed directly south of the closed structure (Fig. 8.22), continues across intensely sheared and lineated Singelele-type granite in the rim of the circular structure (Fig. 8.23b) and ends in non-foliated but penetratively lineated trondhjemite in the core of the Avoca closed structure (Fig. 8.23a) (e.g. Boshoff et al. 2006; van Reenen et al. 2008; Smit et al. 2011). The origin and mechanism of emplacement of melt-cored diapirs in the CZ have been numerically modelled (Perchuk et al. 2008b), explaining the relationship between the onset of granitic diapirism and the development of closed structures associated with intense shear deformation.

The Tshanzi closed structure (Fig. 8.16) demonstrates the critical role of heterogeneous strain in the construction of two-stage *P-T-t* paths from distinct reaction textures preserved in a single thin section (TOV 13-I, II in Fig. 8.17) of a metapelitic granulite (Perchuk et al. 2008a; Mahan et al. 2011; Smit et al. 2011). The role of widespread granitic diapirism reflected by mega closed structures during the near-vertical exhumation of the CZ is furthermore documented by emplacement of large volumes of Singelele-type granitoid after 2.69 Ga.

The only examples of regional-scale Neoarchaean lineaments associated with shear deformation in the CZ are represented by the Triangle Shear Zone and Tshipise Straightening Zone (here identified by the acronym TSZ) that respectively bound the CZ in the north and in the south (Fig. 8.1). The 35 km wide southwest–northeast striking and steeply southeast-dipping high-grade TSZ (Fig. 8.1) separates the CZ from the SMZ. Sheared gneisses associated with this major shear zone are characterized by a steep to moderate southwest-plunging lineation and shear sense indicators (Horrocks 1983; Smit et al. 2011) that document reverse-sinistral displacement of the CZ rocks to the northeast (Smit et al. 2011), parallel to the direction of movement indicated by the plunge of mega-scale closed structures in the CZ. Decompression-cooling *P-T* paths (Perchuk et al. 2008a; Smit et al. 2011) derived from highly sheared metapelitic gneiss records *P-T* data that are almost identical to decompression-cooling *P-T* paths constructed for mega-scale closed structures in the CZ (TOV13-II in Fig. 8.17). Sheared rocks from the TSZ also retain no memory of the early stage of exhumation (TOV13-I in Fig. 8.17) recorded by their granoblastic equivalents associated with isoclinal folds. The similar direction of exhumation and *P-T-t* evolution of the TSZ and of closed

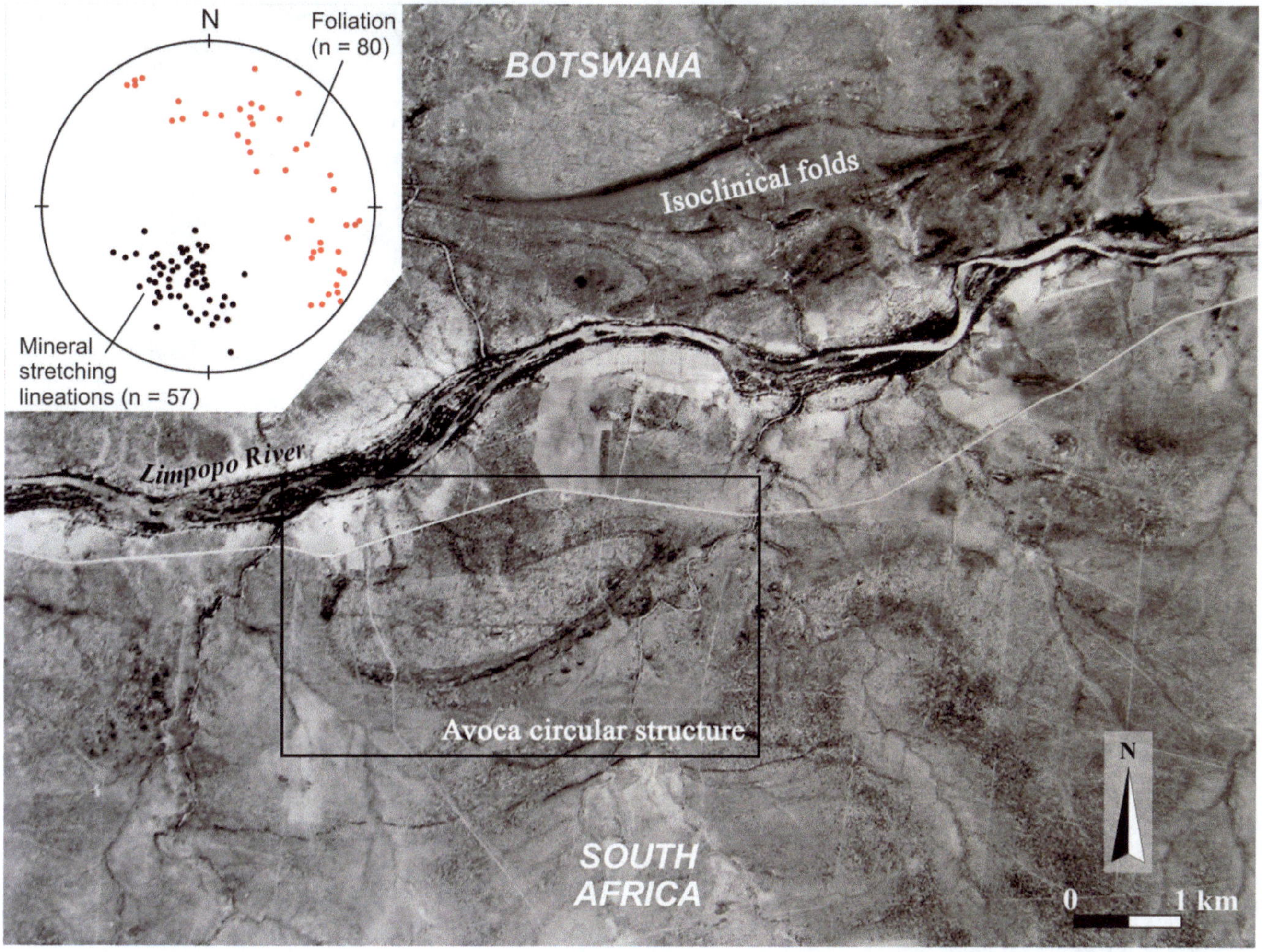

Fig. 8.22 Aerial photograph of the ca. 2.5 km in length southwest–northeast-directed oval-shaped closed Avoca structure located west of Alldays (Fig. 8.16) (modified after Boshoff et al. 2006). This sheared closed structure is closely associated with isoclinal folds of which the regional-scale isoclinal fold developed in Alldays tonalitic gneiss across the Limpopo River in Botswana is an example. The rim of the moderately southwest-plunging structure (inset structural stereogram) comprises penetrative lineated and foliated Singelele-type leucogranite dated at ca. 2.65 Ga. The core comprises unfoliated but intensely lineated (L-tectonite) leucotonalite dated at ca. 2.63 Ga that intrudes Singelele-type leucogranite (Fig. 8.23) (Boshoff et al. 2006; Smit et al. 2011). See text for discussion

structures within the CZ is strong evidence for the steep northeast-directed exhumation of the entire CZ at 2.65–2.62 Ga (e.g. Perchuk et al. 2008b; Smit et al. 2011).

Successful linking *P-T* data (TOV13-I, II in Fig. 8.17) obtained from high-grade metapelitic gneisses associated with distinct regional structural elements in the CZ (respectively, isoclinal folds and mega closed shear-related structures) (Figs. 8.16, 8.17 and 8.21) with time (*D-P-T-t* paths) is supported by field data. These data show that isoclinal folds are deformed by younger mega-scale closed shear-related structures (Figs. 8.16, 8.19 and 8.22). These younger structural features document kinematic, metamorphic and geochronological (Figs. 8.17, 8.19, 8.21 and 8.22) evidence for the final steep northeast-directed exhumation of the CZ at 2.65–2.62 Ga. Integrated geological data thus provides unequivocal evidence that argues against the proposal by Brandt et al. (2018) for a counterclockwise *P-T* path associated with the burial of rocks and prograde heating at exactly the same time and from the same general area.

It can thus be concluded that during initial exhumation proto-CZ and proto-SMZ material were both characterized by isoclinal folds and large dome-like structures (compare Figs. 8.2 and 8.16) that formed at ca. 2.72 Ga under peak metamorphic conditions in the lower crust. This melt-weakened crust (e.g. Jamieson et al. 2011) reflected by migmatitic gneisses in both the CZ (e.g. Figs. 8.20a–c) and the SMZ (Figs. 8.7a–c and 8.8a, b) documents evidence for steep exhumation of granulites at post-peak metamorphic

Fig. 8.23 **a** Humpback outcrop of ca. 2.63 Ga Avoca L-tectonite developed in tonalite that comprises the core of the oval-shaped closed structure. A penetrative lineation defined by quartz rodding and mineral elongation lineations is demonstrated in sections parallel to the plunge (inset photo). **b** Sheared and lineated Avoca granitic gneiss (ca. 2.65 Ga) in rim of the oval structure. **c.** Xenolith of ca. 2.65 Ga Avoca granitic gneiss enclosed by the ca. 2.63 Ga L-tectonic in the core of the structure. This important diapiric structural feature accurately constrains the timing of regional shear deformation at 2.65–2.62 Ga associated with final steep NE-directed exhumation of the CZ (Fig. 8.16). See text for further discussion

conditions (DR45-I in Fig. 8.11, TOV 13-I in Fig. 8.17) to the mid-crustal level (6 kbar, ca. 20 km level) before ca. 2.65 Ga (formation of mega closed structures). At this crustal level, the hot rocks cooled and annealed, resulting in widespread development of granoblastic textures that characterize much of the CZ and SMZ. The SMZ with its imbedded steeply plunging folds and steep shear zones (Fig. 8.4) was then extruded onto and over the NKVC while the CZ (and northern part of the SMZ north of Annaskraal Shear Zone) still resided at the mid-crustal level. At this level granitic diapirism accompanied by the formation of major closed structures (Figs. 8.19 and 8.22) controlled the final steep northeast-directed emplacement of the CZ at the 15 km (ca. 4.5 kbar) crustal level at 2.65–2.61 Ga (Fig. 8.17) (Smit et al. 2011). At this time the hot SMZ nappe interacted with infiltrating water-bearing fluids that resulted in widespread retrograde hydration (Figs. 8.3 and 8.9c–e).

The CZ was finally exhumed to the upper crustal level in the Palaeoproterozoic at ca. 2.01 Ga as the result of a superimposed high-grade shear event linked to discrete (metres-wide) fabric-parallel shear zones associated with widespread thermal overprint (e.g. Perchuk et al. 2008a; Perchuk and van Reenen 2008; van Reenen et al. 2008; Kramers and Zeh 2011; Mahan et al. 2011; Smit et al. 2011).

8.6 Linked Exhumation Histories of the SMZ and CZ at 2.72–2.62 Ga

Distinct *P-T-t* paths (Figs. 8.11 and 8.17) deduced for metapelitic granulite with similar principal mineral assemblages (Grt + Opx/Sil + Crd + Bt + Kfs + Qz) from the SMZ and CZ using a single methodology (e.g. Perchuk et al. 2000a; Smit et al. 2001, 2011; Perchuk and van Reenen 2008; Perchuk 2011) support a linked post-peak metamorphic evolution of the SMZ and CZ at ca 2.72–2.62 Ga that is supported by geophysical data (Fig. 8.15). Two-stage decompression-cooling *P-T* paths constructed for metapelitic gneisses from the CZ and the SMZ north of the Annaskraal Shear Zone (Figs. 8.1 and 8.2) are characterized by distinct gaps (Figs. 8.11 and 8.17). These gaps are in each case located at a pressure of 6-6.5 kbar (Figs. 8.11 and 8.17) and separate two distinct stages of decompression-cooling. In the case of the CZ, the first stage of decompression-cooling at 2.69–2.65 Ga is documented by emplacement of regional-scale isoclinal folds and dome-like structures cored by migmatitic gneisses at the mid-crustal level (Fig. 8.17), followed by cooling and annealing. Exactly the same situation applies to the SMZ north of Annaskraal Shear Zone (Figs. 8.3 and 8.11). The second decompression-cooling stage of steep exhumation of the CZ at ca. 2.65–2.62 Ga (Fig. 8.17) is documented by granitic diapirism that was accompanied by the formation of mega-scale closed structures associated with intensely sheared high-grade gneisses and the establishment of the TSZ. Sheared granulite (Figs. 8.20a–c and 8.21) associated with major closed structures and with the TSZ (Figs. 8.1 and 8.16) has no *P-T* memory of the early decompression-cooling stage due to intense shear deformation and high-temperature reworking of granoblastic gneisses (e.g. Fig. 8.20a–c) that resulted in intensely sheared high-grade gneisses (e.g. Figs. 8.20c and 8.21) (e.g. Perchuk et al. 2008a, b; Perchuk and van Reenen 2008; van Reenen et al. 2008; Mahan et al. 2011; Smit et al. 2011).

In contrast to the situation described for the CZ and for the SMZ north of the Annaskraal Shear Zone (Figs. 8.3 and 8.4), metapelitic granulites that outcrop south of this shear zone in the SMZ document evidence for near-isobaric cooling that commenced at exactly the same mid-crustal level (Figs. 8.11 and 8.16). In keeping with the rocks of the CZ, however, these granulites have little memory of an earlier decompression-cooling stage that accompanied emplacement at the mid-crustal level. Belyanin et al. (2014) and Safonov et al. (2014, 2018a, b) have shown that near-isobaric cooling of the SMZ nappe probably started at ultrahigh-temperatures above 900 °C and at 6–7 kbar pressure (Fig. 8.11).

Evidence for the near-isobaric cooling *P-T* path is supported by field, structural (Fig. 8.4) and geophysical data (Fig. 8.15) showing that the hot nappe was southwards transported over a distance of ca. 35 km along a near-isobaric (ca. 6 kbar) crustal plane (Fig. 8.11) underlain by relative cool granite-greenstone material (inset map in Fig. 8.3) against and over the adjacent NKVC at ca. 2.72 Ga. The near-isobaric crustal plane is defined by the flat section of the HRSZ (Fig. 8.4). The timing is constrained by metamorphism related to the hot-iron effect in the footwall of the HRSZ at 2.72–2.69 Ga (Kreissig et al. 2001; Kramers et al. 2014) and by the time of emplacement of the Matok pluton at ca. 2.68 Ga (Laurent et al. 2013).

The observation that a hot granulite nappe with its imbedded steeply plunging folds and steeply dipping shear zones is allochthonous with respect to the underlying near-horizontal shear fabric associated with the flat section of the HRSZ (Figs. 8.4 and 8.8a–c) has important constraints on the concept of a conveyor belt mechanism linked to steep tectonics in the CZ. Firstly, it demonstrates that the granulite-facies nappe could not have been formed and been exhumed in situ at mid-crustal level above the flat-lying HRSZ that is underlain by cooler greenstone belt material (inset map in Fig. 8.3). Second, the conveyor belt mechanism requires that the SMZ nappe be transported from a position in the core of the orogen below the present CZ in the north. The implication that proto CZ lower crustal material with a large metasedimentary component was underlain by proto-SMZ crustal material dominated by magmatic granulites (Fig. 8.24a) is supported by geophysical data (Fig. 8.15).

The evolution of the hot nappe after emplacement at 2.68 Ga is expressed by widespread retrograde hydration that established the Ath-in isograd and the regional zone of retrograde hydrated granulite (Figs. 8.3 and 8.4). The evolution of the CZ after emplacement at the mid-crustal level at 2.65–2.62 Ga (Kröner et al. 2018) was controlled by granitic diapirism and the formation of major closed structures associated with intense shear deformation.

The time difference between emplacement of the mainly undeformed Bulai pluton at ca. 2.61 Ga in the CZ (Millonig et al. 2008), and of the mainly undeformed Matok pluton at ca. 2.68 Ga in the SMZ (Laurent et al. 2013) is interpreted to be the direct result of the linked evolution of these two high-grade terranes. Whereas emplacement of the Bulai pluton signifies the end of steep granitic diapirism at ca. 2.61 Ga in the CZ, emplacement of the Matok pluton, at ca. 2.68 Ga occurred after southwards horizontal channelling of a hot SMZ nappe from underneath the CZ, prior to granitic diapirism. Granitic diapirism in the CZ at 2.65–2.63 Ga was thus complemented by a static high-temperature widespread

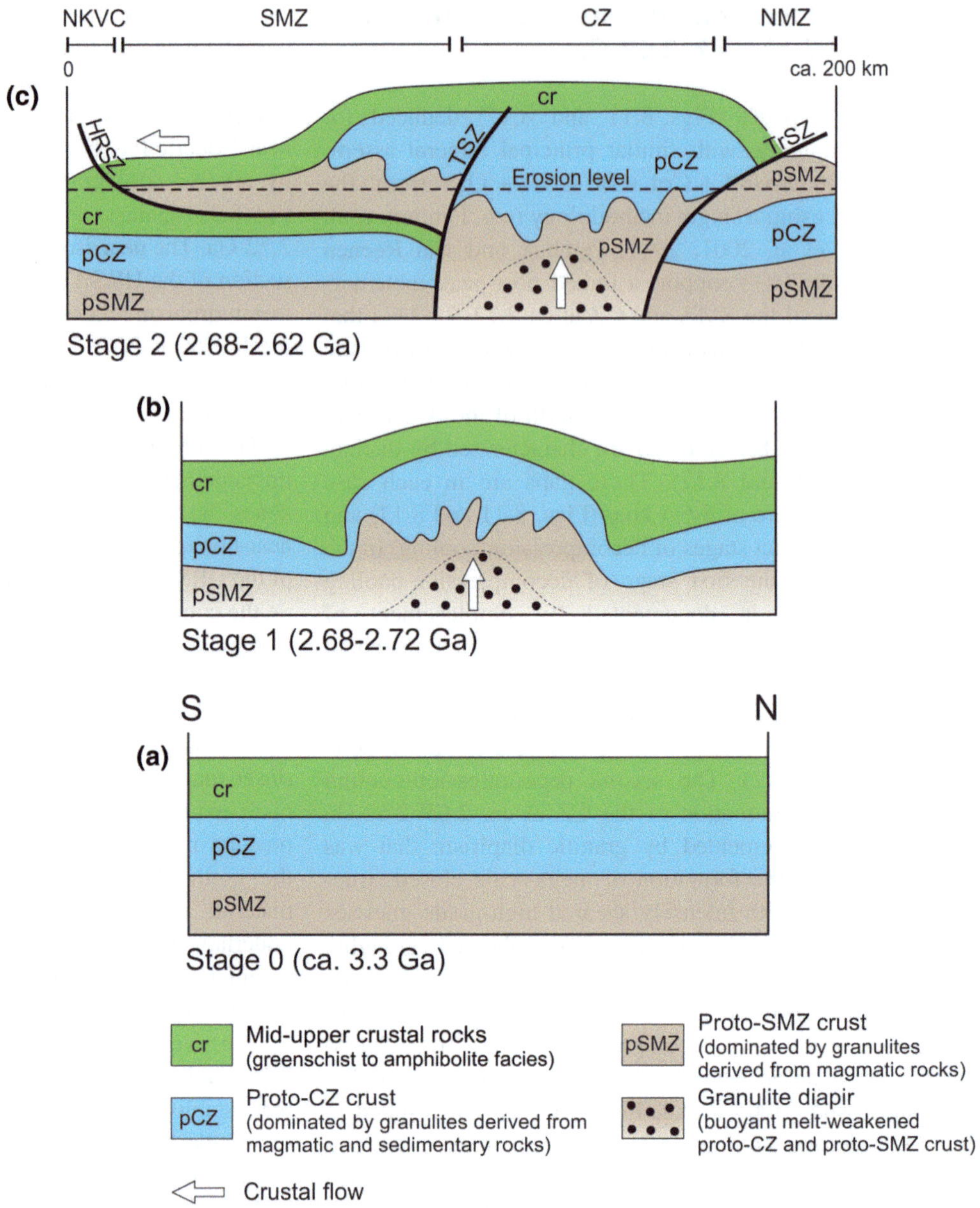

Fig. 8.24 Cartoon showing a crustal-scale granulite diapir penetrating into the upper crust at 2.72–2.62 Ga. **a** Stage 0: rhythmic layered successions reflecting a hypothetical granite-greenstone crustal section formed at ca. 3.3. Ga. **b** Stage 1: penetrating granulite diapir triggered steep exhumation of proto-SMZ and proto-CZ granulite-facies rocks at the mid-crustal section at 2.72–2.68 Ga (Fig. 8.17). **c** Stage 2: regional-scale gravitational crustal overturn of complex morphologies and dynamics that controlled the southwards channelling of a hot SMZ nappe along a near-isobaric crustal section (HRSZ) against and over the adjacent NKVC before 2.68 Ga (Fig. 8.4). Final steep exhumation of the CZ to the upper crustal level at 2.65–2.62 Ga was controlled by granitic diapirism associated with the formation of closed structures before emplacement of the Bulai pluton at ca. 2.61 Ga. HRSZ: Hout River Shear Zone; TSZ: Thipise Straightening Zone; TrSZ: Triangle Shear Zone. See text for further discussion

retrograde hydration event that established the Ath-in isograd in the SMZ after ca. 2.68 Ga. The two high-grade terranes were therefore subjected to different but temporally related processes during the final stages of the same orogenic event.

An important conclusion that follows from this discussion is that the SMZ conveyor belt mechanism (Fig. 8.4) was not driven by granitic diapirism in the CZ as proposed by Smit et al. (2014). Granitic diapirism only controlled the final emplacement of the CZ i.e., after ca. 2.65 Ga (Fig. 8.17), while tectonism in. the SMZ already stopped before emplacement of Matok pluton at ca. 2.68 Ga. Diapirism in the CZ and regional rehydration of the SMZ occurred at the same time.

In the section to follow we examine a proposed link between the conveyor belt mechanism (Fig. 8.4) and the action of a crustal-scale granulite diapir triggered by mantle heat-fluid flow that penetrated the upper crust underneath the present CZ (Fig. 8.24) at or before ca. 2.72 Ga.

8.7 Conveyor Belt Tectonic Mechanism and the Link with a Crustal-Scale Granulite Diapir

The numerical gravitational redistribution model for the LC proposed by Gerya et al. (2000) discusses important data that supports the concept of a conveyer belt mechanism (Fig. 8.4) that controlled emplacement of a hot SMZ nappe against and over the adjacent NKVC before ca. 2.68 Ga. The model of Gerya et al. (2000) explains the buoyant exhumation of a granulite complex separated from two granite-greenstone cratons by inward-dipping crustal-scale shear zones up to 10 km wide and several hundred kilometres long via gravitational redistribution linked to the action of a granulite diaper within metastable hot and soft early Precambrian crust that was subjected to HT and UHT metamorphism. This scenario, where two stiff cratons are separated by a weak pre-existing zone developed within relatively thin (ca. 30 km thick) crust, produces an asymmetrical exhumation pattern (Perchuk and Gerya 2011) that accurately describes an analogue to thrusting the near-horizontal SMZ granulite-facies terrane (granulite nappe) against and onto the adjacent NKVC. It explains important aspects of the SMZ and its relationship with the NKVC (see Kramers et al. 2011). This asymmetrical numerical model (Perchuk and Gerya 2011) strongly supports the proposed conveyor belt mechanism as it pertains to the SMZ and contact with the NKVC (Fig. 8.4) but does not explain the linked evolution of the magmatic dominated SMZ and the CZ with large volumes of metasedimentary rocks as is suggested by the conveyor belt mechanism (Fig. 8.4).

A revision of the Gerya et al. (2000) gravitational redistribution model by Perchuk and Gerya (2011) discusses a different scenario in which relatively thick (ca. 40 km thick) sedimentary-dominated crust is overlain by dense greenstone material as part of a single crustal plate affected by mantle-derived fluid-heat flow (crustal-scale granulite diapir). In this scenario, near-vertical exhumation dominates and hot granulite-facies rocks are transported from a depth of ca. 40 km to mid- and upper crustal levels. Due to the absence of pre-existing weak zones, lateral channelling of hot granulite (representing the SMZ granulite nappe) is subordinate (Perchuk and Gerya 2011). The model illustrates the gravitational redistribution in multi-layered crust in which a crustal-scale granulite diapir penetrates into the upper crust.

We propose that the Perchuk and Gerya (2011) model offers a credible explanation for the exhumation of the CZ, with its large metasedimentary component and widespread granitic diapirism (Roering et al. 1992b; Perchuk et al. 2008a). It also offers an explanation for the Venetia Klippen, (Zeh et al. 2004; Perchuk and Gerya 2011) where sagduction enclosed typical low-grade Archaean greenschist-facies rocks within granulite-facies CZ lithologies. However, the Perchuk and Gerya (2011) numerical model does not explain the southward spreading of a hot SMZ nappe dominated by magmatic-derived granulites from underneath the CZ against and over the adjacent NKVC along a near-isobaric (ca. 6 kbar) crustal section, and the 'hot-iron' relationship with the NKVC (Fig. 8.11).

8.8 Gravity-Driven Geodynamic Model for Exhumation of the LC

We address the issue of the joint evolution of the SMZ (and NMZ) and CZ by considering a combination of the Gerya et al. (2000) and Perchuk and Gerya (2011) numerical gravitational models that link the action of a crustal-scale granulite diaper triggered by heat-fluid flow that penetrated the crust underneath the CZ at or before ca. 2.72 Ga.

8.8.1 Pre-Limpopo Complex Crustal Architecture

The assumption that the Mesoarchaean Kaapvaal Craton was underlain by granulite-facies rocks is supported by different data. First, the tilted crustal section through the Mesoarchaean Basement Complex in the core of the Vredefort Dome exposes the transition from upper amphibolite to granulite-facies rocks at 3130–3080 Ma (e.g. Gibson 2019, this volume). Second, Schmitz and Bowring (2003) suggested that much of the central KVC outside the Vredefort Dome is underlain by ca. 2.72 Ga ultrahigh-temperature granulite-facies rocks including metasedimentary varieties that were exhumed as xenoliths in kimberlites. Thirdly, the hypothetical crustal section that includes a significant volume of metasedimentary rocks at depth is supported by the results of thermomechanical numerical modelling of a Precambrian ultra-hot accretionary orogen (Perchuk et al. 2018) that suggests a link between the formation of the extensive Paleoarchaean granite-greenstone terrane of the central Kaapvaal craton and ultrahigh-temperature granulites at the base of the crust.

The pre-Limpopo Complex crustal architecture shown in Fig. 8.24a is, therefore, adapted from the crustal section of an ultrahot orogeny modelled by Perchuk et al. (2018). Our hypothetical ca. 35 km thick crustal section comprises (1) an igneous-dominated lower granulite-facies unit (termed proto-SMZ crustal material) unit at the base of the crust, followed by (2) a crustal unit still at granulite grade that includes large volumes of metasedimentary rocks (termed proto CZ crustal material), and (3) upper crustal section of greenschist-amphibolite-facies upper crustal material forming a granitoid-greenstone terrane. We propose that the igneous-dominated lower granulite-facies unit represents proto-SMZ crustal rocks, which on the surface today is represented by tonalitic orthogneisses (Baviaanskloof Gneiss) with minor volumes of metapelitic, mafic and ultramafic granulites (Bandelierkop Complex) (Fig. 8.2). The overlying granulite-facies metasedimentary component is interpreted to represent proto-CZ crustal material, while part of the mid- to upper crustal material represents the granite-greenstone terrane presently exposed at surface on both the Kaapvaal and Zimbabwe Cratons.

The hypothetical crustal section (Fig. 8.24a) suggests that a pre-LC Archaean granite-greenstone terrane that included both the KVC and ZC (De Wit et al. 1992b) was underlain at depth by granulite-facies rocks that were exhumed at ca. 2.72 Ga and today are exposed at the surface as the LC.

8.8.2 Gravity-Driven Geodynamic Model of the Limpopo Complex

We speculate that the hypothetical pre-Limpopo crustal architecture (Fig. 8.24a) was disrupted at or before ca. 2.72 Ga by mantle fluid-heat flow that triggered a buoyant crustal-scale granulite diapir that penetrated the pre-Limpopo Complex crustal architecture underneath the present CZ.

8.8.2.1 Early Exhumation Stage Prior to ca 2.68 Ga

Fluid-heat flow enhanced by crustal radiogenic decay (e.g. Andreoli et al. 2011) and the release of hydrous fluid from metamorphic minerals at or before ca. 2.72 Ga resulted in partial melting of rocks at lower crustal levels with formation of melt-weakened migmatitic crust with reduced viscosity (e.g. Jamieson et al. 2011; Lexa et al. 2011). The resultant buoyant crustal-scale granulite diapir comprising hot lower (proto-SMZ) crustal material overlain by cooler proto-CZ crustal material (Fig. 8.24b) systematically disrupted the pre-existing stratigraphy and penetrated the mid- to upper crust (e.g. Perchuk and Gerya 2011) (Fig. 8.24a). The ascending melt-weakened migmatitic crust initially resulted in the formation of melt-weakened gneiss domes and associated steeply plunging regional isoclinal folds that currently characterize both the SMZ and CZ at the surface (compare Figs. 8.2 and 8.16). They eventually cooled at a depth of ca. 20 km (ca. 6 kbar) (Figs. 8.17 and 8.24b) beneath a stiffer mid- to upper crust before ca. 2.68 Ga (constrained by emplacement of the Matok pluton into the extruded the SMZ nappe). At this crucial crustal level, hot granulites cooled and were annealed as demonstrated by regionally developed granoblastic textures.

Metamorphic temperatures conditions of at least 900 °C at a pressure of 9 kbar (Tsunogae et al. 2004; Tsunogae and van Reenen 2011; Belyanin et al. 2014) were reached at the base of the crustal section (Fig. 8.24). This resulted in the production of large volumes of leuco-granitoid that intruded CZ migmatitic crust at 2.69–2.62 Ga (Singelele-type) (e.g. Kröner et al. 1999; Smit et al. 2011). This process further affected density, crustal strength, and tectonic style of the melt-weakened migmatitc crust (e.g. Jamieson et al. 2011; Lexa et al. 2011; Sawyer et al. 2011).

8.8.2.2 Second Exhumation Stage at 2.65–2.62 Ga Driven by Granitic Diapirism

The early exhumation stage (TOV 13-I in Fig. 8.17) was followed by final exhumation of the CZ to the upper crustal level, a process driven by granitic diapirism and associated with the formation of mega-scale closed structures dated at 2.65–2.62 Ga (Figs. 8.16, 8.19 and 8.22) and intensely sheared high-grade gneisses. Closed structures cored by granitoids are developed throughout the entire CZ in South Africa (Fig. 8.16) and are all characterized by similar features. These include evidence for intense shear deformation (Figs. 8.20c and 8.21), constant steep southwest-plunging fold axes that support exhumation towards the northeast (Figs. 8.19 and 8.22), and the fact that sheared metapelitic gneisses have no memory of the early decompression-cooling stagthe e of exhumation but only record *P-T-t* evidence for decompression-cooling after emplacement at the mid-crustal level (TOV13-II in Fig. 8.17) (e.g. Smit et al. 2011). The timing of a major event at ca. 2.65–2.62 Ga was recently confirmed by a detailed study of highly deformed high-grade gneisses from an enclave within the ca. 2.61 Ga Bulai pluton (Kröner et al. 2018).

8.8.2.3 Regional-Scale Gravitational Crustal Overturn of Complex Morphology and Tectonics Before ca. 2.68 Ga

Granulites comprising the proto-SMZ cooled after emplacement at the mid-crustal level (Fig. 8.11) underneath cooler and stiffer CZ material (Fig. 8.24b) and cover rocks before ca. 2.68 Ga. We proposed that rheological differences at this crustal level between hot underlying proto-SMZ rocks and cooler and stiffer overlying proto-CZ rocks facilitated regional-scale gravitational crustal overturn of complex dynamics and morphology (Figs. 8.4 and 8.24b) that initiated and controlled the southward channelling of a hot and denser granulite nappe with imbedded steep structures (the SMZ underlain by the sub-horizontal HRSZ) from underneath CZ material.

Near-horizontal channelling of a hot SMZ nappe at the mid-crustal level is supported by geophysical (Fig. 8.15a) and structural data (Fig. 8.6b–d) and constrained by *P-T-t* data (Fig. 8.11). It explains the preserved (imbedded) steeply plunging folds and steeply dipping shear zones that currently characterize the SMZ granulite nappe. These steeply dipping imbedded structures, that are allochthonous with respect to the younger near-horizontal underlying HRSZ (Figs. 8.6 and 8.8), preserve evidence for the initial near-vertical stage of exhumation of the SMZ from underneath the CZ (Figs. 8.11 and 8.24b). The timing of near-horizontal channelling of the hot SMZ nappe is accurately constrained by different lines of evidence. First, thermal and dynamic interaction at 2.72–2.69 Ga of overriding hot granulite with underlying relatively cool greenschist along the contact of the steeply north-dipping HRSZ with the adjacent NKVC (e.g. Figs. 8.5a and 8.6a). Secondly, emplacement of the mainly undeformed Matok granite into already deformed and metamorphosed SMZ granulites at ca. 2.68 Ga (Fig. 8.3) (Laurent et al. 2013).

Jamieson et al. (2011) used the analogue 'of pressing on an egg sandwich' to explain a similar process that operated in the Moldanubian region of the Bohemian Massif in the Variscan Orogen of central Europe. In this example, melt-weakened migmatitic crust flowed outwards onto the lower foreland area, a process described as mid-crustal heterogeneous channel flow (Schulman et al. 2008). In the channel flow model, the tongue of melt-weakened crust is often bounded above and below by ductile shear zones and often preserves its imbedded structures such as shear zones and their initial orientation within the nappe (Jameson et al. 2011).

8.8.2.4 Regional Retrograde Hydration of the Hot SMZ Nappe at ca. 2.68–2.62 Ga

Emplacement of hot granulite for a distance of at least 35 km over cool underthrust greenschists (inset map in Figs. 8.3, 8.4 and 8.15) of the NKVC resulted in devolatilization of the underlying greenschist and the persistent production of large volumes of fluids at 2.72–2.68 Ga as long as the conveyor mechanism operated (Fig. 8.4) (Koons and Craw 1991; Koons et al. 1998; Yardley 2009). Fluids thus derived (e.g. van Reenen et al. 2014) infiltrated and interacted with cooling SMZ granulites in the overriding nappe, establishing the retrograde Ath-in isograd and the zone of retrograde hydrated granulite that bounds the SMZ in the south (yellow and orange areas in Fig. 8.3).

Retrograde hydration occurred at 2.68–2.62 Ga after emplacement of the nappe and during the same general time interval as granitic diapirism that controlled final emplacement of the CZ after ca. 2.65 Ga. Different processes operating at different times in the different zones of the LC are therefore linked to different stages of the same orogeny that controlled the Neoarchaean evolution of this granulite-facies terrane.

8.9 Conclusions

Published geological, geochronological and geophysical data show that the complexly deformed and intensely metamorphosed LC is separated from lower grade large continuous linear greenstone belts that outcrop on the adjacent KC and ZC cratons (Fig. 8.1) by inward-dipping crustal-scale Northern Marginal Thrust Zone and the HRSZ (Fig. 8.1). These inward-dipping dip-slip crustal-scale shear zones controlled thrusting high-grade LC rocks onto adjacent lower grade granite-greenstone lithologies during the Limpopo orogeny. In the case of the SMZ, the timing of this event is accurately constrained at 2.72–2.69 Ga. The entire LC (Fig. 8.1) is furthermore characterized by steeply plunging folds and steeply dipping high-grade shear zones linked to decompression-cooling *P-T* paths (e.g. Blenkinsop 2011; Smit et al. 2011; van Reenen et al. 2011) that attest to initial regional steep exhumation of the LC before 2.65 Ga. A hot granulite nappe with imbedded steep tectonic features comprising the SMZ south of Annaskraal Shear Zone was channelled southwards against and over the NKVC at the mid-crustal level before ca. 2.68 Ga along the sub-horizontal HRSZ. Evidence for low-angle tectonic features in the LC are restricted to the deeply eroded central and eastern parts of the SMZ that expose sheared flat-lying retrograde hydrated former granulites that are closely associated with the flat-lying section of the HRSZ (orange area in Figs. 8.3 and 8.4). Metamorphic, structural and geochronological data provide credible evidence only for a linked post-peak metamorphic decompression-cooling metamorphic evolution directed by steep tectonic features that affected both the CZ and SMZ at 2.72–2.62 Ga. On the other hand, both the NKVC and the ZC (Figs. 8.1 and 8.2) are typical low- to medium grade granite greenstone terrains with no evidence for shallow north-verging structures that might have been associated with crustal thickening related to continent–continent collision at ca. 2.72 Ga (Kramers et al. 2014). In fact, only the most northern edge of the NKVC in the immediate footwall of the HRSZ documents evidence for interaction with the SMZ during the ca. 2.72 Ga Limpopo orogeny (McCourt and van Reenen 1992; Roering et al. 1992a; Kramers et al. 2014). Finally, the most significant feature that characterizes the entire LC is the coherent and dramatic difference between the style of deformation and intensity of metamorphism compared with the adjacent lower grade granite-greenstone cratons (Fig. 8.1).

There is no credible published data that argues against the proposed linked tectono-metamorphic evolution of the SMZ and CZ at 2.72–262 Ga. Evidence for the timing of peak metamorphism in the SMZ at ca. 2.72 Ga is in the first place constrained by emplacement of the undeformed Matok granitoid pluton (Laurent et al. 2013) and Petronella anatectic tonalite (Belyanin et al. 2014) at ca. 2.68 Ga into already deformed and metamorphosed granulite gneisses. Metamorphic age data obtained directly from granulite gneisses precisely date the peak metamorphic event in the SMZ at ca. 2.72 Ga (Rajesh et al. 2014; Taylor et al. 2014; Nicoli et al. 2015). In the case of the CZ, Kröner et al. (1999, 2018) dated anatectic Singelele-type granite that intrudes already deformed and metamorphosed high-grade supracrustal gneisses at ca 2.68 Ga. Furthermore, the same authors provide age data for a granoblastic quartzofeldspathic gneiss dated at ca. 2.72 Ga. Finally, the results of a detailed geochronological study of a large supracrustal enclave in the ca. 2.61 Ga Bulai pluton that predominantly records evidence for a high-grade metamorphic event at 2.64–2.62 Ga (Kröner et al. 2018) is in good agreement with our published *P-T-t* data for the final steep northeast-directed exhumation of the CZ associated with formation of mega closed structures (e.g. Smit et al. 2011). This contradicts linking the Neoarchaean evolution of the CZ to a counterclockwise *P-T* loop associated with burial and prograde metamorphism (Brandt et al. 2018). We thus propose that published data on the SMZ and CZ strongly support the working hypothesis of an intracrustal Limpopo orogeny at 2.72–2.62 Ga that is linked to the action of a granulite diapir.

This proposed gravity-driven intracrustal orogenic model is demonstrated with reference to a hypothetical crustal section (Fig. 8.24a) that comprises magmatic dominated lower crustal (proto-SMZ) granulites overlain by a crust that includes a large volume of metasedimentary granulites (proto-CZ rocks) and capped by mid-upper crustal greenschist-amphibolite grade granite-greenstone lithologies. The proposed crustal section is supported by published data on the nature of the Mesoarchaean Kaapvaal crust (e.g. Gibson 2019, this volume), and is not in conflict with published geophysical data on the LC (e.g. De Beer and Stettler 1992). The model is also supported by the results of thermomechanical numerical modelling of a Precambrian ultra-hot accretionary orogen that explains the formation of an extensive granite-greenstone terrane linked to the formation of ultrahigh-temperature granulite at the base of the crust (Perchuk et al. 2018).

The hypothetical Archaean crustal section (Fig. 8.24a) was disrupted at ca. 2.72 Ga by a buoyant crustal-scale granulite diapir that penetrated the upper crust underneath the present CZ. This diaper, triggered by mantle fluid-heat flow, resulted in emplacement of melt-weakened migmatitic crust characterized by large dome-like structures and regional isoclinal folds at the mid-crustal level before ca. 2.68 Ga. Regional-scale gravitational crustal overturn of complex morphology and dynamics at this crucial mid-crustal level directed southwards near-horizontal channelling of a hot SMZ nappe from underneath the CZ and along the flat-lying HRSZ against and over the adjacent NKVC for a distance of ca. 35 km before ca. 2.68 Ga (time of emplacement of the Matok pluton). The subsequent exhumation of the CZ at 2.65–2.61 Ga to the upper crustal level was controlled by extensive granitic diapirism associated with the formation of steep SW-plunging mega closed structures characterized by intensely sheared gneisses.

After southward thrusting, the hot SMZ nappe experienced static high-temperature retrograde hydration at 2.68–2.62 Ga related to infiltration of aqueous-carbonic fluids along the underlying flat-lying HRSZ and derived from devolatilization of lower grade greenstone material that underlies more than 60% of the SMZ (inset map in Fig. 8.3). The observation that high-grade SMZ lithologies and typical granite-greenstone lithologies of the NKVC share similar geochemical signatures (Kreissig et al. 2000) is a direct consequence of the proposed orogenic model that links the exhumation history of the SMZ to a pre-existing granite-greenstone-type crust.

Although this paper focuses on evidence for a linked evolution of the SMZ and CZ at 2.72–2.61 Ga, and the relationship of the SMZ with the NKVC, the proposed crustal-scale granulite diapir can be easily expanded (Fig. 8.24) to include exhumation of the NMZ, that is shown to also have been driven by extensive granitic diapirism in the interval 2.74–2.57 Ga (Blenkinsop 2011).

The proposed gravity-driven Neoarchaean Limpopo orogeny reflected by exhumation and regional-scale gravitational crustal overturn of complex morphology and dynamics might also be applicable to the evolution of other Archaean granulite-facies terranes located within granite-greenstone cratons. Data presented and discussed in this paper argues against the concept of a continental collision at ca. 2.72 Ga associated with burial and crustal thickening.

Acknowledgements This study was supported by grants from the National Science Foundation (NRF) of South Africa (GUN: 81040 to DvR), the University of Johannesburg (DvR), and James Cook University (JMH). The authors would like to thank Steve McCourt and Leonid Aranovich for their time and effort to produce highly informed objective and critical reviews of the manuscript. Steve McCourt is especially thanked for being willing to review a completely restructured version of the paper.

References

Andreoli MAC, Brandl G, Kramers JD, Mouri H (2011) Intracrustal radioactivity as an important heat source for Neoarchean metamorphism in the Central Zone of the Limpopo complex. In: van Reenen DD, Kramers JD, McCourt S, Perchuk LL (eds) Origin and evolution of Precambrian high grade gneiss terranes, with special emphasis on the Limpopo complex of Southern Africa, vol 207. Geological Society of America memoirs, pp 143–161

Aranovich LY, Newton RC (1999) Experimental determination of CO2–H2O activity-composition relations at 600–1000 °C and 6–14 kbar by reversed decarbonation and dehydration reactions. Am Mineral 84:1319–1332

Armstrong RA, Wilson AH (2000) SHRIMP U-Pb study of zircons from the layered sequence of the Great Dyke, Zimbabwe, and a granitoid anatectic dyke. Earth Planet Sci Lett 180:1–12

Bahnemann KP (1972) A review of the structure, the stratigraphy and the metamorphism of the basement rocks of the Messina District, Northern Transvaal. Dissertation, University of Pretoria

Baker J, van Reenen DD, Van Schalkwyk JF, Newton RC (1992) Constraints on the composition of fluids involved in retrograde anthophyllite formation in the Limpopo Belt, South Africa. Precambr Res 55:327–336

Barton JM Jr, van Reenen DD (1992) When was the Limpopo Orogeny? Precambr Res 55:7–16

Belyanin G, van Reenen DD, Safonov OG (2014) Response to comments on "ultrahigh-temperature metamorphism from an unusual corundum-orthopyroxene intergrowth bearing Al-Mg granulite from the southern Marginal Zone, Limpopo complex, South Africa (Belyanin et al. 2012)" by Nicoli et al. (2014). Contrib Mineral Petrol 167:1–5

Belyanin GA, Kramers JD, Vorster C, Knoper MD (2015) The timing of successive fluid events in the Southern Marginal Zone of the Limpopo complex, South Africa: constraints from ^{40}Ar-^{39}Ar geochronology. Precambr Res 254:169–193

Blenkinsop TG, Mkweli, S, Rollinson H, Fedo CM, Paya BK, Kamber BS, Kramers JD, Berger M (1995) The North Limpopo Thrust Zone (NLTZ): the northern boundary of the Limpopo belt in Zimbabwe and Botswana. In: Extended abstracts, Centennial Geocongress 95, geological society of South Africa, pp 174–177

Blenkinsop TG, Kröner A, Chiwara V (2004) Single stage, late Archaean exhumation of granulites in the Northern Marginal Zone, Limpopo Belt, Zimbabwe, and relevance to gold mineralization at Renco Mine. S Afr J Geol 107:377–396

Blenkinsop TG (2011) Archean magmatic granulites, diapirism, and Proterozoic reworking in the Northern Marginal Zone of the Limpopo Belt. In: van Reenen DD, Kramers JD, McCourt S, Perchuk LL (eds) Origin and evolution of precambrian high grade gneiss terranes, with special emphasis on the Limpopo complex of Southern Africa, vol 207. Geological society of America memoirs, pp 245–267

Bohlender F, van Reenen DD, Barton JM Jr (1992) Evidence for metamorphic and igneous charnockites in the Southern Marginal Zone of the Limpopo Belt. Precambr Res 55:429–449

Boshoff R, van Reenen DD, Smit CA, Perchuk LL, Kramers JD, Armstrong R (2006) Geologic history of the central zone of the Limpopo complex: the west Alldays area. J Geol 114:699–716

Brown M (2007) Crustal melting and melt extraction, ascent and emplacement in orogens: mechanisms and consequences. J Geol Soc Lond 164:709–730

Brandt S, Klemd R, Li Q, Kröner A, Brandl G, Fischer A, Bobek P, Zhou T (2018) Pressure-temperature evolution during two granulite-facies metamorphic events (2.62 and 2.02 Ga) in rocks from the central zone of the Limpopo Belt. South Africa Precambr Res 310:471–506

Cagnard F, Barbey P, Gapais D (2011) Transition between "Archaean-type" and "modern-type" tectonics: insights from the Finnish Lapland Granulite Belt. Precambr Res 187:127–142

Chardon D, Gapais D, Cagnard F (2009) Flow and ultra-hot orogens: a view from the Precambrian, clues for the Phanerozoic. Tectonophysics 477:105–118

De Beer JH, Stettler EH (1988) Geophysical characteristics of the southern African Continental Crust. J Petrol 1:163–184

De Beer JH, Stettler EH (1992) The deep structure of the Limpopo Belt from geophysical studies. Precambr Res 55:173–186

De Beer JH, Le Roux CL, Hanstein T, Stack KM (1991) DC resistivity and LOTEM model for the deep structure of the northern edge of the Kaapvaal craton, South Africa. Phys Earth Planet Inter 66:51–61

De Wit MJ, Jones MG, Buchanan DL (1992a) The geology and tectonic evolution of the Pietersburg greenstone belt, South Africa. Precambr Res 55:123–153

De Wit MJ, van Reenen DD, Roering C (1992b) Geologic observations across a tectono-metamorphic boundary in the Babangu area, Giyani (Sutherland) greenstone belt, South Africa. Precambr Res 55:111–122

De Wit MJ, Roering C, Hart RJ, Armstrong RA, De Ronde CEJ, Green RWE, Tredoux M, Peberdy E, Hart RA (1992c) Formation of an Archean continent. Nature 357:553–562

Dorland HC, Beukes NJ, Gutzmer J, Evans DAD, Armstrong RA (2004) Trends in detrital zircon provenance from Neoarchaean Paleoproterozoic sedimentary successions in the Kaapvaal Craton. Geoscience Africa 2004, Abstract Volume, University of the Witwatersrand, Johannesburg, South Africa, pp 176–177

Du Toit MC, van Reenen DD, Roering C (1983) Some aspects of the geology, structure and metamorphism of the Southern Marginal Zone of the Limpopo metamorphic complex. Spec Publ Geol Soc S Afr 8:89–102

Dubinina EO, Aranovich LYA, van Reenen DD, Avdeenco AS, Varlamov DA, Kurdyukov EB, Shaposhnikov VV (2015) Oxygen isotope systematics of the high-grade rocks of the Southern Marginal Zone of the Limpopo complex. Precambr Res 256:48–61

Durrheim RJ, Barker WH, Green RWE (1992) Seismic studies in the Limpopo belt. Precambr Res 55:187–200

Ellis DJ (1987) Origin and evolution of granulites in normal and in thickened crust. Geology 15:167–170

England PC, Thompson AB (1984) Pressure-temperature-time paths of regional metamorphism. I. Heat transfer during the evolution of regions of thickened continental crust. J Petrol 25:894–928

Fripp REP, Lilly PA, Barton JM Jr (1979) The structure and origin of the Singelele Gneiss, Limpopo mobile belt, South Africa. Trans Geol Soc S Afr 82:161–167

Gerya TV, Perchuk LL, van Reenen DD, Smit CA (2000) Two-dimensional numerical modeling of pressure-temperature-time paths for the exhumation of some granulite facies terranes in the Precambrian. J Geodyn 30:17–35

Gan SF, van Reenen DD (1995) Geology of gold deposits in the Southern Marginal Zone of the Limpopo Belt and the adjacent Sutherland Greenstone Belt, South Africa: Franke mine. S Afr J Geol 98:263–275

Gan SF, van Reenen DD (1997) Geology of gold deposits in the Southern Marginal Zone of the Limpopo Belt and adjacent Sutherland Greenstone Belt, South Africa: Klein Letaba. S Afr J Geol 100:73–83

Gibson RL (2019) The Mesoarchaean Basement Complex of the Vredefort Dome—A Mid-Crustal Section Through the Central Kaapvaal Craton Exposed by Impact. In: Kröner A, Hofmann A (eds) The Archaean Geology of the Kaapvaal Craton, Southern Africa. Regional geology reviews. Springer International Publishing AG

Hoernes S, van Reenen DD (1992) The oxygen isotopic composition of granulites and retrogressed granulites from the Limpopo Belt as a monitor of fluid-rock interaction. Precambr Res 55:353–364

Hoernes S, Lichtenstein U, van Reenen DD, Mokgatla KP (1995) Whole rock/mineral O-isotope fractionations as a tool to model fluid-rock interaction. Precambr Res 55:353–364

Hofmann A, Kröner A, Brandl G (1998) Field relationships of mid-to late Archaean high-grade gneisses of igneous and sedimentary parentage in the Sand River, Central Zone of the Limpopo Belt, South Africa. S Afr J Geol 101:185–200

Holland TJB, Powell R (1998) An internally consistent thermodynamic data set for phases of petrological interest. J Metamorph Geol 16:309–343

Horrocks PC (1983) The Precambrian geology of an area between Messina and Tshipise, Limpopo mobile belt. In: Van Biljon WJ, Legg JH (eds) The Limpopo Mobile Belt, vol 8. Special publication of the geological society of South Africa, pp 81–88

Huizenga JM, Perchuk LL, van Reenen DD, Varlamov DA, Smit CA, Gerya TV (2011) Granite emplacement and the retrograde P-T-fluid evolution of Neoarchean granulites from the CZ of the Limpopo complex. Geol Soc Am Mem 207:125–142

Huizenga JM, van Reenen DD, Touret JLR (2014) Fluid-rock interaction in retrograde granulites of the Southern Marginal Zone, Limpopo high-grade terrain, South Africa. Geosci Front 5:673–682

Jaguin J, Gapais D, Poujol M, Boulvais P, Moyen J-F (2012) The Murchison greenstone belt (South Africa): a general tectonic framework. S Afr J Geol 115:65–76

Jamieson RA, Unsworth MJ, Harris NB, Rosenberg CL, Schulmann K (2011) Crustal melting and the flow of mountains. Elements 7:253–260

Klemperer SL (1992) Introduction: deep crustal probing. Precambr Res 55:169–172

Kleywegt RJ (1988) Gravity signature of the Limpopo-Kaapvaal fossil plate boundary in southern Africa-discussion. Tectonophysics 145:349–350

Kleywegt RJ, De Beer JH, Stettler EH, Brandl G, Duvenhage AWA, Day RW (1987) The structure of the Giyani greenstone belt, as derived from geophysical studies. S Afr J Geol 90:282–295

Koizumi T, Tsunogae T, van Reenen DD (2014) Fluid evolution of partially retrogressed pelitic granulite from the Southern Marginal Zone of the Neoarchean Limpopo complex, South Africa: Evidence from phase equilibrium modelling. Precambr Res 253:146–156

Koons PO, Craw D (1991) Evolution of fluid driving forces and composition within collisional orogens. Geophys Res Lett 280: 935–938

Koons PO, Craw D, Cox S, Upton P, Templeton A, Chamberlain CP (1998) Fluid flow during active oblique convergence: a Southern Alps model from mechanical and geochemical observations. Geology 26:159–162

Kramers JD, Mouri H (2011) The geochronology of the Limpopo complex: a controversy solved. In: van Reenen DD, Kramers JD, McCourt S, Perchuk LL (eds) Origin and evolution of Precambrian high grade gneiss terranes, with special emphasis on the Limpopo complex of Southern Africa, vol 207. Geological Society of America Memoirs, pp 85–106

Kramers JD, Zeh A (2011) A review of Sm-Nd and Lu-Hf isotope studies in the Limpopo complex and adjacent cratonic areas, and their bearing on models of crustal evolution and tectonism. In: van Reenen DD, Kramers JD, McCourt S, Perchuk LL (eds) Origin and evolution of Precambrian high grade gneiss Terranes, with special emphasis on the Limpopo complex of Southern Africa, vol 207. Geological Society of America memoirs, pp 163–188

Kramers JD, Henzen M, Steidle L (2014) Greenstone belts at the northernmost edge of the Kaapvaal Craton: timing of tectonic events and a possible crustal fluid source. Precambr Res 253:96–113

Kramers JD, McCourt S, van Reenen DD (2006) The Limpopo Belt. In: Johnson MR, Anhaeusser CR, Thomas RJ (eds) The geology of South Africa. Geological Society of South Africa, Council for Geoscience, Johannesburg, Pretoria, pp 209–236

Kramers JD, McCourt S, Roering C, Smit CA, van Reenen DD (2011) Tectonic models proposed for the Limpopo complex: mutual compatibilities and constraints. In: van Reenen DD, Kramers JD, McCourt S, Perchuk LL (eds) Origin and evolution of Precambrian high grade gneiss Terranes, with special emphasis on the Limpopo complex of Southern Africa, vol 207. Geological Society of America memoirs, pp 311–324

Kreissig K, Nägler TF, Kramers JD, van Reenen DD, Smit CA (2000) An isotopic and geochemical study of the northern Kaapvaal craton and the Southern marginal zone of the Limpopo Belt: are they juxtaposed terranes? Lithos 50:1–25

Kreissig K, Holzer L, Frei IM, Villa JD, Kramers JD, Kröner A, Smit CA, van Reenen DD (2001) Geochronology of the Hout River Shear Zone and the metamorphism in the Southern Marginal Zone of the Limpopo belt, Southern Africa. Precambr Res 109:145–173

Kröner A, Jaeckel P, Brandl G (2000) Single zircon ages for felsic to intermediate rocks from the Pietersburg and Giyani greenstone belts and bordering granitoid orthogneisses, northern Kaapvaal Craton, South Africa. J Afr Earth Sci 30:773–793

Kröner A, Jaeckel P, Hofmann A, Nemchin AA, Brandl G (1998) Field relationships and age of supracrustal Beit Bridge complex and associated granitoid gneisses in the central zone of the Limpopo Belt, South Africa. S Afr J Geol 101:201–213

Kröner A, Jaeckel P, Brandl G, Nemchin AA, Pidgeon RT (1999) Single zircon ages for granitoids gneisses in the central zone of the Limpopo belt, southern Africa and geodynamic significance. Precambr Res 93:299–337

Kröner A, Brandl G, Brandt S, Klemd R, Xie H (2018) Geochronological evidence for three granulite-facies tectono-metamorphic events in the central zone of the Limpopo Belt, South Africa. Precambr Res 310:320–347

Laurent O, Paquette JL, Doucelance R, Martin H, Moyen JF (2013) Crustal growth and evolution in the northern Kaapvaal craton inferred by LA-ICP-MS dating of zircons from Meso- and Neoarchaean granitoids. Precambr Res 230:209–226

Laurent O, Zeh A, Brandl G, Vezinet A, Wilson A (2019) Granitoids and Greenstone Belts of the Pietersburg Block—Witnesses of an Archaean Accretionary Orogeny Along the Northern Edge of the Kaapvaal Craton. In: Kröner A, Hofmann A (eds) The Archaean Kaapvaal Craton, Southern Africa. Regional Geology Reviews. Springer International Publishing AG

Lexa O, Schulmann K, Janoušek V, Štípská P, Guy A, Racek M (2011) Heat sources and trigger mechanisms of exhumation of HP granulites in Variscan orogenic root. J Metamorph Geol 29:79–102

Light MPR (1982) The Limpopo Belt, Southern Africa: a result of continental collision. Tectonics 1:325–342

Mahan KH, Smit CA, Williams ML, Dumond G, van Reenen DD (2011) Scales of heterogeneous strain and polymetamorphism in high-grade terranes: insight from the Athabasca Granulite Terrane, western Canada and the Limpopo complex, southern Africa. In: van Reenen DD, Kramers JD, McCourt S, Perchuk LL (eds) Origin and evolution of Precambrian high-grade gneiss terranes, with special emphasis on the Limpopo complex of Southern Africa, vol 207. Geological Society of America memoirs, pp 269–287

Mason R (1973) The Limpopo mobile belt: southern Africa. Philos Trans R Soc Lond Series A 273:463–485

McCourt S, van Reenen DD (1992) Structural geology and tectonic setting of the Sutherland greenstone belt, Kaapvaal Craton, South Africa. Precambr Res 55:93–110

Millonig L, Zeh A, Gerdes R, Klemd R (2008) Neoarchaean high-grade metamorphism in the central zone of the Limpopo Belt (South Africa): Combined petrological and geochronological evidence from the Bulai pluton. Lithos 103:333–351

Miyano T, Ogata H, van Reenen DD, Van Schalkwyk JF, Arawaka Y (1992) Peak metamorphic conditions of sapphirine-bearing rocks in the Rhenosterkoppies greenstone belt, Northern Kaapvaal Craton, South Africa. In: Clover JE, Ho SE (eds) The Archean terranes, processes and metallogeny, vol 27. Geology Department (Key Centre), the University of Western Australia, pp 73–87

Nicoli G, Stevens G, Moyen JF, Frei D (2015) Rapid evolution from sediment to anatectic granulite in an Archean continental collision zone: the example of the Bandelierkop Formation metapelites, South Marginal Zone, Limpopo Belt, South Africa. J Metamorph Geol 33:177–202

Passeraub M, Wuest T, Kreissig K, Smit CA, Kramers JD (1999) Structure, metamorphism, and geochronology of the Rhenosterkoppies greenstone belt, South Africa. S Afr J Geol 102:323–334
Perchuk AL, Safonov OG, Smit CA, van Reenen DD, Zakharov VS, Gerya TV (2018) Precambrian ultra-hot orogenic factory: making and reworking of continental crust. Tectonophysics 746:572–586
Perchuk LL (2011) Local mineral equilibria and P-T paths: fundamental principles and their application for high-grade metamorphic terrains. In: van Reenen DD, Kramers JD, McCourt S, Perchuk LL (eds) Origin and evolution of Precambrian high-grade gneiss terranes, with special emphasis on the Limpopo complex of Southern Africa, vol 207. Geological Society of America memoirs, pp 61–84
Perchuk LL, van Reenen DD (2008) Reply to "comments on "P–T record of two high-grade metamorphic events in the central zone of the Limpopo complex, South Africa" by A Zeh and R Klemd. Lithos 106:403–410
Perchuk LL, Gerya TV (2011) Formation and evolution of Precambrian granulite terranes: a gravitational redistribution model. In: van Reenen DD, Kramers JD, McCourt S, Perchuk LL (eds) Origin and evolution of Precambrian high grade gneiss terranes, with special emphasis on the Limpopo complex of Southern Africa, vol 207. Geological Society of America memoirs, pp 289–310
Perchuk LL, Gerya TV, van Reenen DD, Safanov OG, Smit CA (1996) The Limpopo metamorphic complex, South Africa: 2 decompression/cooling regimes of granulites and adjacent rocks of the Kaapvaal Craton. Petrology 4:571–599
Perchuk LL, Gerya TV, van Reenen DD, Krotov AV, Safonov OG, Smit CA, Shur MYu (2000a) Comparative petrology and metamorphic evolution of the Limpopo (South Africa) and Lapland (Fennoscandia) high-grade terrains. Mineral Petrol 69:69–107
Perchuk LL, Gerya TV, van Reenen DD, Smit CA, Krotov AV (2000b) P-T paths and tectonic evolution of shear zones separating high-grade terrains from cratons, examples from Kola Peninsula (Russia) and Limpopo region (South Africa). Mineral Petrol 69:109–142
Perchuk LL, van Reenen DD, Smit CA, Boshoff R, Belyanin GA, Yapaskurt VO (2008a) Role of granite intrusions for the formation of ring structures in granulite complexes: examples from the Limpopo Belt, South Africa. Petrology 16:652–678
Perchuk LL, van Reenen DD, Varlamov DA, van Kal SM, Boshoff R (2008b) P-T record of two high-grade metamorphic events in the central zone of the Limpopo complex, South Africa. Lithos 103:70–105
Pili E, Shepard SMF, Lardeaux JM, Martelat JF, Nicollete C (1997) Fluid flow versus scale of shear zones in the lower continental crust and the granulite paradox. Geology 25:15–18
Pretorius JM (1988) BIF-hosted gold mineralization at the Fumani Mine, Sutherland greenstone belt, South Africa. S Afr J Geol 91:429–438
Rajesh HM, Santosh M, Wan Y, Liu SJ, Belyanin GA (2014) Ultrahigh temperature granulites and magnesium charnockites: evidence for Neoarchean accretion along the northern margin of the Kaapvaal Craton. Precambr Res 246:150–159
Retief EA, Compston W, Armstrong RA, Williams IS (1990) Characteristics and preliminary U-Pb ages of zircons from Limpopo Belt lithologies. In: Barton JM Jr (ed) Extended abstract volume, the Limpopo Belt: a field workshop on granulites and deep crustal tectonites, pp 100–101
Rey PF, Teyssier C, Whitney DL (2009) Extension rates, crustal melting, and core complex dynamics. Geology 37:391–394
Roering C, De Beer JH, van Reenen DD, Wolhuter LE, Baton JM Jr, Smit CA, McCourt S, Du Plessis A, Stettler EH, Brandl G, Pretorius SJ, Van Schalkwyk JF, Geerthsen K (1990) A geotransect across the Limpopo belt. In: Barton JM Jr (ed) Extended abstract volume, the Limpopo Belt: a field workshop on granulites and deep crustal tectonites, pp 105–106
Roering C, van Reenen DD, De Wit MJ, Smit CA, De Beer JH, Van Schalkwyk JF (1992a) Structural geological and metamorphic significance of the Kaapvaal Craton/Limpopo Belt contact. Precambr Res 55:69–80
Roering C, van Reenen DD, Smit CA, Barton JM Jr, De Beer JH, De Wit MJ, Stettler EH, Van Schalkwyk JF, Stevens G, Pretorius S (1992b) Tectonic model for the evolution of the Limpopo Belt. Precambr Res 55:539–552
Roering C, van Reenen DD, Smit CA, Du Toit R (1995) Deep crustal embrittlement and fluid flow during granulite metamorphism in the Limpopo Belt, South Africa. J Geol 103:673–686
Safonov OG, Tatarinova DS, van Reenen DD, Golunova MA, Yapaskurt VO (2014) Fluid-assisted interaction of peraluminous metapelites with tonalitic magma in the Southern Marginal Zone of the Limpopo complex. Precambr Res 253:114–145
Safonov OG, Tatarinova DS, Japaskurt WO, van Reenen DD, Gognova, MA, Varlamov DA, Smit CA (2018a) In source melting of metapelite from the Petronella shear zone, Limpopo complex South Africa: constraints from petrological studies and experiments. J Petrol. https://doi:10.1093/petrology/egy052
Safonov OG, Vadim N, Reutsky VN, Varlamov DA, Yapaskurt VA, Golunova MA, Shcherbakov VD, van Reenen DD, Smit CA, Butvina VG (2018b) Composition and source of fluids in high-temperature graphite-bearing granitoids associated with granulites: examples from the Southern Marginal Zone, Limpopo complex, South Africa. Gondwana Res 60:129–152
Sawyer EW, Cesare B, Brown M (2011) When the continental crust melts. Elements 7:229–234
Schmitz MD, Bowring SA (2003) Utrahigh-temperature metamorphism in the lower crust during Neoarchean Ventersdorp rifting and magmatism, Kaapvaal Craton, southern Africa. GSA Bull 115:533–548
Schulmann K, Lexa O, Štípská P, Racek M, Tajčmanová L, Konopásek J, Edel JB, Pechler A, Lehmann J (2008) Vertical extrusion and horizontal channel flow of orogenic lower crust: key exhumation mechanisms in large hot orogens? J Metamorph Geol 26:273–297
Smit CA, van Reenen DD (1997) Deep crustal shear zones, high-grade tectonites and associated alteration in the Limpopo belt, South Africa: implication for deep crustal processes. J Geol 105:37–57
Smit CA, Roering C, van Reenen DD (1992) The structural framework of the southern margin of the Limpopo Belt, South Africa. Precambr Res 55:51–67
Smit CA, van Reenen DD, Roering C (2014) Role of fluids in the exhumation of the Southern Marginal Zone of the Limpopo complex, South Africa. Precambr Res 253:81–95
Smit CA, van Reenen DD, Gerya TV, Perchuk LL (2001) P-T conditions of decompression of the Limpopo high-grade terrain: record from shear zones. J Metamorph Geol 19:249–268
Smit CA, van Reenen DD, Roering C, Boshoff R, Perchuk LL (2011) Neoarchean to Paleoproterozoic evolution of the polymetamorphic central zone of the Limpopo complex. In: van Reenen DD, Kramers JD, McCourt S, Perchuk LL (eds) Origin and evolution of Precambrian high grade gneiss terranes, with special emphasis on the Limpopo complex of Southern Africa, vol 207. Geological Society of America memoirs, pp 213–244
Stettler EH, De Beer JH, Blom MP (1989) Crustal domains in the northern Kaapvaal Craton as defined by magnetic lineaments. Precambr Res 45:263–276
Stuart GW, Zengeni T (1987) Seismic crustal structure of the Limpopo mobile belt, Zimbabwe. Tectonophysics 144:323–335
Taylor J, Nicole G, Stevens G, Fry ED, Moyne JF (2014) A process that controls leucosome composition in metasedimentary granulites: perspective from the Southern Marginal Zone, Limpopo complex, South Africa. J Metamorph Geol 32:713–742

Thompson AB (1990) Heat, fluids and melting in the granulite facies. In: Vielzeuf D, Vidal PH (eds) Granulites and crustal evolution. Kluwer Academic Publishers, pp 37–57

Treloar P, Coward MP, Harris NBW (1992) Himalayan-Tibetan analogies for the evolution of the Zimbabwe Craton and the Limpopo Belt. Precambr Res 55:571–587

Tsunogae T, van Reenen DD (2011) High-pressure and high-temperature metamorphism of Precambrian high-grade terranes: case study of the Limpopo Complex. In: van Reenen DD, Kramers JD, McCourt S, Perchuk LL (eds) Origin and evolution of Precambrian high grade gneiss terranes, with special emphasis on the Limpopo complex of Southern Africa, vol 207. Geological Society of America memoirs, pp 107–1214

Tsunogae T, van Reenen DD (2014) High- to ultrahigh-temperature metasomatism related to brine infiltration in the Neoarchaean Limpopo complex. Petrology and phase equilibrium modeling. Precambr Res 253:157–170

Tsunogae T, Miyano T, van Reenen DD, Smit CA (2004) Ultrahigh-temperature metamorphism of the southern marginal zone of the Archean Limpopo Belt, South Africa. J Miner Petrol Sci 99:213–224

van Reenen DD (1983) Cordierite+garnet+hyperthene+biotite-bearing assemblages as a function of changing metamorphic conditions in the Southern Marginal Zone of the Limpopo metamorphic complex, South Africa. In: Van Biljon WJ, Leg JH (eds) The Limpopo Belt, vol 8. Special publication of the Geological Society of South Africa, pp 143–167

van Reenen DD (1986) Hydration of cordierite and hypersthenes and description of the retrograde orthoamphibole isograd in the Limpopo Belt, South Africa. Am Mineral 71:900–915

van Reenen DD, Du Toit MC (1977) Mineral reactions and the timing of metamorphic events in the Limpopo Metamorphic complex south of the Soutpansberg. Bull Bots Geol Surv 12:107–128

van Reenen DD, Hollister LS (1988) Fluid inclusions in hydrated granulite facies rocks, Southern Marginal Zone of the Limpopo Belt, South Africa. Geochim Cosmochim Acta 52:1057–1064

van Reenen DD, McCourt S, Smit CA (1995) Are the Southern and Northern Marginal Zones of the Limpopo Belt related to a single continental collisional event? S Afr J Geol 98:489–504

van Reenen DD, Barton JM Jr, Roering C, Smit CA, Van Schalkwyk JF (1987) Deep crustal response to continental collision: the Limpopo Belt of South Africa. Geology 15:11–14

van Reenen DD, Roering C, Brandl G, Smit CA, Barton JM Jr (1990) The granulite facies rocks of the Limpopo Belt, southern Africa. In: Vielzeuf D, Vidal PH (eds) Granulites and crustal evolution. Kluwer Academic Publishers, pp 257–289

van Reenen DD, Boshoff R, Smit CA, Perchuk LL, Kramers JD, McCourt SM, Armstrong RA (2008) Geochronological problems related to polymetamorphism in the Limpopo Complex, South Africa. Gondwana Res 14:644–662

van Reenen DD, Smit CA, Perchuk LL, Roering C, Boshoff R (2011) Thrust exhumation of the Neoarchean ultrahigh-temperature Southern Marginal Zone, Limpopo Complex: Convergence of decompression-cooling paths in the hanging wall and prograde P-T paths in the footwall. In: van Reenen DD, Kramers JD, McCourt S, Perchuk LL (eds) Origin and evolution of precambrian high grade gneiss terranes, with special emphasis on the Limpopo Complex of Southern Africa, vol 207. Geological Society of America memoirs, pp 189–212

van Reenen DD, Huizenga JM, Smit CA, Roering C (2014) High-temperature fluid-rock interaction: instructive examples from the Southern Marginal Zone of the Limpopo complex. Precambr Res 253:63–80

Van Schalkwyk JF (1991) Metamorphism of ultramafic rocks during the Limpopo orogeny: evidence from the timing and significance of CO_2-rich fluids. Rand Afrikaans University

Van Schalkwyk JF, van Reenen DD (1992) High temperature hydration of ultramafic granulites from the Southern Marginal Zone of the Limpopo Belt by infiltration of CO_2-rich fluid. Precambr Res 55:337–352

Vezinet A, Moyen JF, Stevens G, Gautier N, Laurent O, Couzinié S, Frei D (2018) A record of 0.5 Ga of evolution of the continental crust along the northern edge of the Kaapvaal Craton, South Africa: consequences for the understanding of Archean geodynamic processes. Precambr Res 305:310–326

Watkeys MK, Light MPR, Broderick RJ (1983) A retrospective view of the central zone of the Limpopo Belt, Zimbabwe. In: Van Biljon WJ, Leg JH (eds) The Limpopo Belt, Special publication of the Geological Society of South Africa, vol 8, pp 65–80

Whitney DL, Evans BW (2010) Abbreviations for names of rock-forming minerals. Am Mineral 95:185–187

Yardley BWD (2009) The role of water in crustal evolution. J Geol Soc Lond 166:585–600

Zeh A, Klemd R, Buhlmann S, Barton JM (2004) Prograde-retrograde P-T evolution of granulites of the Beit Bridge Complex (Limpopo Belt, South Africa): Constraints from quantitative phase diagrams and geotectonic implications. J Metamorph Geol 22:79–95

Zeh A, Klemd R (2008) Comments on "P-T record of two high-grade metamorphic events in the central zone of the Limpopo complex, South Africa" by LL Perchuk, DD van Reenen, DA Varlamov, SM van Kal, Tabatabaeimanesh. R Boshoff Lithos 106:403–410

Zeh A, Jaguin J, Poujol M, Boulvais P, Block S, Paquette JL (2013) Juvenile crust formation in the Northeastern Kaapvaal Craton at 2.97 Ga, implications for Archean terrane accretion, and the source of the Petersburg gold. Precambr Res 233:2–43

9 The Pongola Supergroup: Mesoarchaean Deposition Following Kaapvaal Craton Stabilization

Casey Luskin, Allan Wilson, Digby Gold, and Axel Hofmann

Abstract

The Pongola Supergroup, deposited between 2.99 and 2.87 Ga on the southeast margin of the Kaapvaal Craton, is one of the least tectonically disturbed Mesoarchaean supracrustal sequences on Earth. Two groups comprise the Pongola Supergroup: the lower Nsuze Group, dominated by volcanic rocks ranging from basalt to basalt-andesite through intermediate to rhyolite compositions, and the upper Mozaan Group, dominated by sedimentary units, which may correlate with portions of the Witwatersrand Supergroup. Pongola Supergroup rocks are largely pristine, frequently preserving original sedimentary and volcanic structures and primary compositions. The depositional environment was an epicontinental sea subjected to sea-level oscillations, resulting in predominantly marginal marine siliciclastic deposits. BIFs, stromatolitic carbonates, glacial diamictites, Mn-rich shales and palaeosols elucidate Mesoarchaean surface processes and suggest shallow-marine oxygen oases under an anoxic atmosphere. The tectonic setting involved a marginal basin undergoing progressive thermal subsidence after magmatism. Palaeomagnetic results are inconsistent but suggest that the Kaapvaal Craton underwent significant wandering. The Pongola Supergroup exhibits two dominant structural domains, including north–northwest-trending rifts and thrust-related structures in the Northern and Central domains, and east–west-trending folds and thrust faults in the south. Economically, the Pongola Supergroup is insignificant apart from small gold deposits, and subeconomic deposits of iron, manganese and uranium.

Keywords

Pongola supergroup • Archaean • Mesoarchaean crust • Kaapvaal craton • Archaean volcanism • Archaean palaeomagnetism • Pongola structure • Mesoarchaean surface processes

9.1 Introduction

The Pongola Supergroup is a Mesoarchaean sequence of volcanic and sedimentary rocks, outcropping within a north–south-trending belt, representing a remnant basin about 100 km wide by 270 km long, situated on the southeast margin of the Kaapvaal Craton in South Africa and Swaziland (now renamed to Eswatini) (Matthews 1990; Dixon 2004; Gold 2006; Hicks and Hofmann 2012; Wilson et al. 2013; Wilson and Zeh 2018). Deposited between 2990 and 2870 Ma (Mukasa et al. 2013; Gumsley et al. 2013, 2015), with a thickness up to, and in places potentially exceeding 11,000 m, and an inferred lateral extent of 32,500 km^2 (Button et al. 1981), it represents one of Earth's oldest Archaean intracontinental supracrustal sequences (Wilson et al. 2013; Wilson and Hofmann 2013). While Earth's contemporaneous crust as preserved in greenstone belts was easily recycled and was characterized by extensive submarine komatiitic and tholeiitic extrusions, the Pongola Supergroup provides some of the earliest evidence that the Kaapvaal Craton had stabilised, as it was accumulating mature clastic sediments in slowly subsiding, subaerial or shallow-water tidal environments, interspersed by episodes of intracontinental volcanism (Hunter et al. 1986; Wilson and Jackson 1988; de Wit et al. 1992; Wilson and Grant 2006; Wilson et al. 2013; Mukasa et al. 2013).

C. Luskin (✉) · A. Hofmann
Department of Geology, University of Johannesburg, Auckland Park, 2006, South Africa
e-mail: caseyl@uj.ac.za

A. Wilson
School of Geosciences, University of the Witwatersrand, Johannesburg, Wits 2050, South Africa

D. Gold
CopperLeaf Consulting, PO Box 519 Hilton, 3245, South Africa

A. Kröner and A. Hofmann (eds.), *The Archaean Geology of the Kaapvaal Craton, Southern Africa*, Regional Geology Reviews, https://doi.org/10.1007/978-3-319-78652-0_9

Pongola Supergroup rocks nonconformably overlie a basement of ≥ 3.1 Ga Archaean granitoid gneisses, greenstone belts and post-tectonic granites, including the Ancient Gneiss Complex of Swaziland, the Nondweni greenstone belt and the Mpuluzi batholith (Hofmann et al. 2019). Immediately below the Pongola Supergroup is a locally developed 2–10 m thick palaeosol (Matthews and Scharrer 1968; Grandstaff et al. 1986; Kimberley and Grandstaff 1986).

Two groups comprise the Pongola Supergroup. The lower Nsuze Group (~2.99–2.96 Ga; Gumsley et al. 2015) has a thickness of ~4600 m in the Pongola Supergroup's type area, the Hartland area of the main Pongola Basin (Cole 1994; Gold 2006), and is characterized by subaerial lavas of bimodally distributed basaltic to rhyolitic chemistries, interbedded with sedimentary units. Nsuze Group lavas may represent the oldest continental flood basalts on Earth (Gumsley et al. 2015). Some authors (Beukes and Cairncross 1991; Cole 1994; Gumsley et al. 2015) have correlated Nsuze Group volcanic rocks with the Dominion Group, a volcano-sedimentary sequence which underlies the Witwatersrand Supergroup; however, as the Dominion Group has been dated at 3074 ± 6 Ma (Armstrong et al. 1991), correlation relates rather to similarity in processes of intracratonic basin evolution in different parts of the Kaapvaal Craton. The younger Mozaan Group (~2.95–2.87 Ga) has a type area thickness of ~4800 m (Beukes and Cairncross 1991; Gold 2006), is characterized primarily by sedimentary units, including sandstones, shales and conglomerates, as well as two thin volcanic horizons, and may bear the world's oldest glacial sequence (Young et al. 1998). The Mozaan Group has been stratigraphically correlated with the West Rand and Central Rand groups of the Witwatersrand Supergroup (Beukes and Cairncross 1991). The contact between the Nsuze and Mozaan groups is largely conformable; an angular unconformity with a palaeosol dividing the two groups is present in the south (SACS 1980; Nhleko 2003; Delvigne et al. 2016).

Given its age, the Pongola Supergroup is remarkably well preserved, consisting largely of undeformed rocks subjected to lower greenschist-facies metamorphism (Armstrong et al. 1986; Wilson and Grant 2006). Original volcanic and sedimentary structures are frequently preserved. Its widespread and diverse volcanic and sedimentary units resemble modern continental successions and may have been among the first of their kind on Earth (Wilson et al. 2013). Nsuze Group volcanic rocks were fed by magmas associated with the Usushwana Complex, a northwest-orientated system of ~2.99–2.96 Ga mafic layered intrusions, about 100 km long by 6.5 km wide, straddling the South Africa–Swaziland border near Piet Retief (Burke et al. 1985; Anhaeusser 2006; Gumsley et al. 2015). Mozaan Group lavas are contemporaneous with intrusion of the Hlagothi Complex, a ~2.87 Ga complex of mafic-ultramafic sills intruding the southeast Kaapvaal Craton (Groenewald 1984; Gumsley et al. 2013, 2015).

The Pongola Supergroup is partitioned into three structural domains, a Northern, Central and Southern Domain, with the Southern Domain separated from the former by a basement high, the Melmoth Basement Arch (Matthews 1990; Gold 2006; Wilson et al. 2013). The Northern Domain forms the locus of deposition within the main basin of the Pongola Supergroup (Gold and von Veh 1995) with the Mozaan Group structurally separated into the Hartland sub-basin and the Piensrand and Tobolsk structural basins (Wilson et al. 2013). The Southern Domain within the Nkandla area comprises thinner successions, which have been subjected to stronger deformation than the Northern and Central Domains. Although Wilson et al. (2013) suggested that correlation of strata across the basement high has proven difficult, researchers such as Cole (1994) and Hicks et al. (2011) tentatively correlated formations within the Southern Domain with units in the north, based on lithostratigraphic similarities. Thickness estimates of Pongola strata in different domains vary greatly in the literature. Most authors, however, have proposed that the Pongola Supergroup is thickest in the Hartland type area in the Northern Domain of the main Pongola Basin, and isopach lines were drawn parallel to the Melmoth Basement Arch (Fig. 9.1; Matthews 1990; Cole 1994; Wilson et al. 2013). The Nsuze Group is thickest in the Magudu area, and thick accumulations are also found in the Southern Domain, south of the Melmoth Basement Arch, in the Nkandla sub-basin and the Buffalo River Inlier (Cole 1994).

It is thought that Pongola Supergroup rocks were originally deposited in a single basin, perhaps as large as 50,000–90,000 km^2 (Gold 2006; Wilson and Hofmann 2013). There may have been different depocentres, however, with the Hartland sub-basin in the Northern Domain, and the Nkandla sub-basin, equated with the Southern Domain (Wilson et al. 2013). Much original Pongola Supergroup strata have been eroded or disrupted by igneous intrusions such that, today, only about 30% is exposed by aerial extent, typically as inliers surrounded by Phanerozoic strata, including the Natal Group and the Karoo Supergroup (Wilson et al. 2013). A notably pristine and extensive section through the Pongola Supergroup is found at the White Mfolozi Inlier, which bears large portions of the Nsuze Group (Fig. 9.2b) and part of the Mozaan Group along a surface exposure at the White Mfolozi River Gorge, making it one of the best-preserved Mesoarchaean supracrustal sections.

The depositional environment of the Pongola Supergroup is thought to have varied from fluvial to shallow-marine and outer shelf, controlled by relative sea-level changes in an

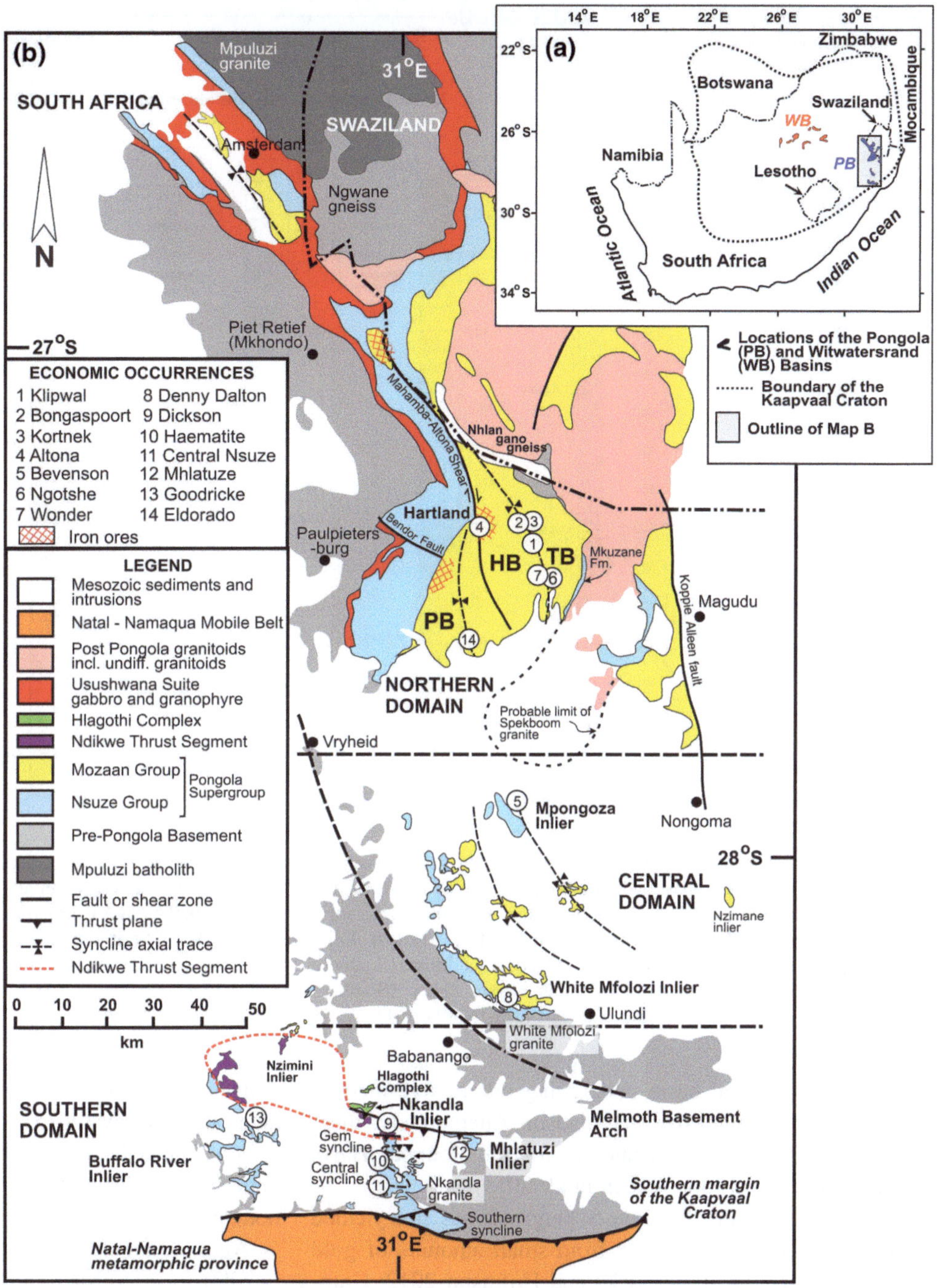

Fig. 9.1 Geological map of the Pongola Supergroup modified after Wilson and Zeh (2018), Gold (2006), and Matthews (1990). **a** Location of the main Pongola Basin (PB) and approximately coeval Witwatersrand Basin (WB) within the Kaapvaal Craton of southern Africa; **b** General geology of the Pongola Supergroup in South Africa and Swaziland. The main Pongola Basin is subdivided into the Northern, Central, and Southern domains; the latter two are separated by the northwest-trending arcuate Melmoth Basement Arch. The Southern Domain is also referred to as the Nkandla sub-basin. Probable limit of Spekboom granite is after Matthews (1990). The occurrences and inliers of the Pongola Supergroup rocks are named according to the area in which they occur. In the Hartland area, the structural trends and various basins (*PB = Piensrand structural basin; TB = Tobolsk structural basin; HB = Hartland sub-basin*; HB comprises both PB and TB) are from Gold (2006). Numbers 1–14 indicate localities of gold occurrences

epicontinental sea (Watchorn and Armstrong 1980; Beukes and Cairncross 1991; Gold 1993; Alexander et al. 2008; Noffke et al. 2008; Siahi et al. 2016). There is debate, however, about the tectonic setting. Burke et al. (1985), Matthews (1990), and Olsson et al. (2010) proposed a continental extensional rift to passive margin basin, whereas Wilson et al. (2013) and Mukasa et al. (2013) suggested formation in a continental back-arc extensional basin. While not favouring a particular tectonic model, Klausen et al. (2010) noted that a continental back-arc extensional basin behind a subduction zone located along the Kaapvaal Craton's northern margin is consistent with the sub-alkaline, wide-spectrum chemistry of Nsuze Group lavas (also pointing out that non-subduction-related settings could produce these chemistries). However, they also observed that the Pongola Supergroup's distance from the proposed subduction zone, and the high angle of the subduction zone to the network of 2.95 Ga southeast-trending dykes that fed Nsuze Group volcanics, could suggest a failed volcanic rift. Nhleko (2003) and Gumsley et al. (2015) cautiously

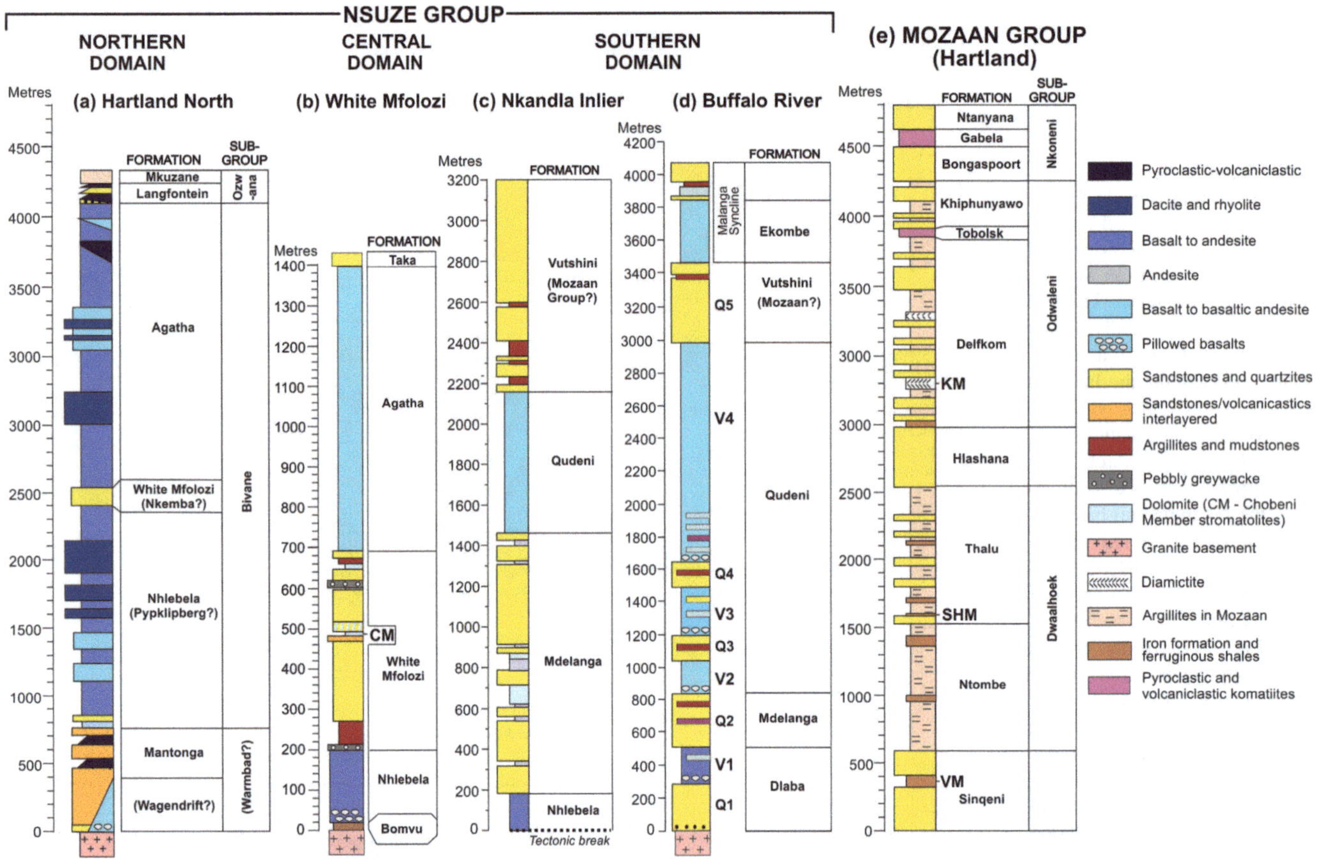

Fig. 9.2 Stratigraphy of the Nsuze and Mozaan groups of the Pongola Supergroup, and rock types and stratigraphic subdivisions of the Pongola Supergroup. **a** Nsuze Group in the Hartland area north (Northern Domain); **b** Nsuze Group at the White Mfolozi Inlier (Central Domain); **c** Nsuze Group at the Nkandla Inlier (Southern Domain); **d** Nsuze Group at the Buffalo River Inlier (Southern Domain); **e** Mozaan Group in the Hartland area (Northern Domain). Stratigraphic sections are adapted and modified after Wilson and Zeh (2018), as well as Wilson et al. (2013), Gold (2006), Cole (1994), Gold (1993), Beukes and Cairncross (1991), Armstrong et al. (1982), and SACS (1980). Stratigraphic revisions proposed by Wilson et al. (2013) shown in parentheses (see also Sect. 9.3.3). *CM = Chobeni Member; VM = Vlakhoek Member; SHM = Scotts Hill BIF Member; KM = Klipwal Member Diamictite*

proposed a marginal basin undergoing progressive thermal subsidence after extensive magmatism. Alternatively, there may be no modern-day analogue for the Pongola Basin (Armstrong et al. 1986; Wilson et al. 2013).

Economically, the Pongola Supergroup has not yielded many mineral resources apart from small amounts of gold (Gold 2006; Hicks 2009; Hicks and Hofmann 2012; Chinnasamy et al. 2015). The Mozaan Group was explored in the early twentieth century, but no major economic deposits were found. Gold (2006) and Beukes and Cairncross (1991) suggested that presently buried Mozaan Group rocks may bear gold, especially among stratigraphically higher units.

9.2 Geochronology

The depositional age of the Pongola Supergroup is broadly constrained to 3.0–2.8 Ga (Table 9.1). The Mpuluzi batholith, which underlies the Pongola Supergroup in the Amsterdam area of the Northern Domain (Fig. 9.1), was dated by Murphy (2015) to ~3.12 Ga. The White Mfolozi granite, dated by Reinhardt et al. (2015) at 3.25–3.23 Ga, underlies the Nsuze Group in the Central Domain, where the nonconformable contact is locally well-exposed.

Ages obtained for the Nsuze Group generally range from ~2990 to 2970 Ma. The Usushwana Complex, a layered ultramafic, mafic and felsic intrusive suite (Hammerbeck 1982; Anhaeusser 2006), which has been genetically linked to Nsuze volcanic rocks (Klausen et al. 2010), was dated by Gumsley et al. (2015) at 2978 ± 2 Ma and 2989 ± 1 Ma.

The age of the Mozaan Group is less well-constrained. The Mooihoek pluton, an undeformed Mswati Suite granite intrusion into the Mozaan Group, dates to 2824 ± 6 Ma (Maphalala and Kröner 1993), meaning that most Mozaan deformation occurred prior to that time. However, deformation within the 2700–2730 Ma Spekboom granite (Reimold et al. 1993), consistent with D_3 (Gold 1993), indicates a separate tectonic event. A 2837 ± 5 Ma quartz

Table 9.1 Ages relevant to the Pongola Supergroup and related intrusions and igneous complexes

Age (Ma)	Unit	Method	Reference
2837 ± 4.7	Post-Mozaan quartz porphyry in Swaziland	SHRIMP U-Pb zircon	Gutzmer et al. (1999)
2866 ± 2	Hlagothi Complex sill intruding Mozaan Group	U-Pb baddeleyite	Gumsley et al. (2013)
2869 ± 5	Thole Suite dolerite sill intruding Mozaan Group	ID-TIMS U-Pb baddeleyite	Gumsley et al. (2015)
2954 ± 9	Mozaan Group andesite (Tobolsk) in Swaziland	U-Pb zircon	Mukasa et al. (2013)
2965.9 ± 0.7	NW-trending dolerite dyke from Badplaas Dyke Swarm	U-Pb baddeleyite	Olsson et al. (2010)
2968 ± 6	Nsuze Group rhyolite from Hartland area	U-Pb zircon	Mukasa et al. (2013)
2977 ± 5	Rhyolite from mid-Agatha Formation in Swaziland	SHRIMP U-Pb zircon	Nhleko (2003)
2978 ± 2	Piet Retief Suite of Usushwana Complex	ID-TIMS U-Pb baddeleyite	Gumsley et al. (2015)
2978 ± 2	Nsuze Group lava	Single grain zircon Pb evaporation	Walraven and Pape (1994)
2980 ± 10	Nsuze Group andesite in Hartland area	U-Pb zircon	Mukasa et al. (2013)
2985 ± 1	Nsuze Group rhyodacite from Swaziland	U-Pb zircon	Hegner et al. (1994)
2989 ± 1	Piet Retief Suite of Usushwana Complex	ID-TIMS U-Pb baddeleyite	Gumsley et al. (2015)
3165-3090 (~3120)	Mpuluzi batholith	U-Pb zircon	Murphy (2015)
3230 ± 80	Pre-Nsuze granite	Rb-Sr isochron	Burger and Coertze (1973)
3234 ± 5 –254 ± 7	White Mfolozi granite	U-Pb zircon	Reinhardt et al. (2015)
3254 ± 8	Nkandla granite	U-Pb zircon	Hicks et al. (2015)

porphyry sill intruding upper Mozaan Group sedimentary strata (Gutzmer et al. 1999), which was deformed along with other Mozaan strata, provides an even earlier minimum age for the Mozaan Group. Mafic dykes and sills associated with the Hlagothi Complex and intruding into Mozaan Group strata, dated to 2866 ± 2 Ma, likely fed upper Mozaan Group lavas of the Gabela and Tobolsk formations (Gumsley et al. 2013, 2015). However, Mukasa et al. (2013) dated zircons extracted from an andesite of the Tobolsk Formation, near the top of the Mozaan Group to a much older age of 2954 ± 9 Ma. These zircons likely represent inherited grains.

9.3 Stratigraphy

The evolution of Pongola Supergroup stratigraphic terminology is recounted by SACS (1980), which notes that the term ‘Pongola’ was first used by Humphrey (1913) and later by Krige in the 1930s. Krige identified two major Pongola units, the ‘Lower’ and ‘Upper’ Pongola Series. An ‘Insuzi Series’ was first described by Hatch (1910) as composed of quartzites, grits and volcanic rocks; the Upper Pongola was named the ‘Mozaan Series’ in 1940 by the South African Geological Survey (SACS 1980). Matthews (1967) similarly split Pongola strata into the lower ‘Insuzi Formation’ and the upper ‘Mozaan Formation.’ SACS (1980) adopted the terms ‘Nsuze Group’ (changing the spelling from ‘Insuzi’ to ‘Nsuze’) and ‘Mozaan Group,’ noting that their thicknesses, lithostratigraphy and exposure can vary. Supergroup designation was first proposed by von Brunn (1974).

Today, it is generally agreed that the Pongola Supergroup consists of two groups, the Nsuze and Mozaan groups. While both contain volcanic and sedimentary formations, the lower Nsuze Group is dominated by volcanic sequences and the upper Mozaan Group is dominated by sedimentary rocks. The Mozaan Group’s stratigraphy is relatively well-constrained, based upon BIF, diamictite and volcanic marker horizons; however, stratigraphic discrepancies remain within the Nsuze Group. A variety of stratigraphic schemes has been applied to these groups, particularly the Nsuze Group, and thicknesses of units show extensive lateral variation (Fig. 9.2), frequently resulting in conflicting nomenclature when the literature describes formations within these groups (see Sect. 9.3.3 and reviews by Beukes and Cairncross 1991; Cole 1994; Gold 2006; Wilson et al. 2013). For example, SACS (1980) described the Nsuze Group at the White Mfolozi Inlier as beginning with the sedimentary Bomvu Formation (quartzite), followed by the Nhlebela Formation (lavas), followed by shales of the Thembeni Formation, followed by the sedimentary (and carbonate-bearing) Chobeni Formation, followed by another thick lava unit, the Bivane Formation, and capped by the Taka Formation, a sedimentary unit. This precise scheme is rarely used, however, as current literature typically calls the Bivane Formation the ‘Agatha Formation,’ and the name

'White Mfolozi Formation' is often—though not always (Bolhar et al. 2015; Siahi et al. 2016)—used in place of the Chobeni Formation (Cole 1994; Gold 2006; Wilson et al. 2013).

9.3.1 Stratigraphy of the Nsuze Group

The predominantly volcanic ~4400 m thick Nsuze Group has three subgroups and seven formations in the Hartland type area of the Northern Domain. It contains lava flows ranging from basalt to rhyolite in composition. Generally, the Nsuze Group is thought to have been deposited in a subsiding basin that commenced with braided fluvial sand deposition (Mantonga Formation), followed by shallow-water and subaerially erupted lavas (Nhlebela Formation), transitioning into a near-shore, shallow tide-dominated marine environment (White Mfolozi Formation), followed by more subaerial lava eruptions (Agatha Formation), ending with a shallow-marine shelf environment (Ozwana Subgroup; Cole 1994; Wilson and Grant 2006; Wilson et al. 2013).

9.3.1.1 Mantonga Formation

The ~800 m thick Mantonga Formation (SACS 1980) at the base of the Nsuze Group is a primarily sedimentary unit with quartz arenites and volcaniclastic rocks. Deposited in a distal braided fluvial to near-littoral environment, Mantonga Formation sedimentary rocks may be derived from the erosion of older sedimentary rocks such as the Moodies Group of the Barberton Greenstone Belt (Matthews 1967; Wilson and Zeh 2018).

9.3.1.2 Bivane Subgroup

The Bivane Subgroup contains the Pongola Supergroup's most-reputed lava units. The lower Nhlebela Formation is ~1600 m thick, bearing basaltic, basaltic andesite and andesitic amygdaloidal lavas, interbedded with sandstone. Pillows are present at the base of this unit, from which the sequence transitions upwards into subaerial lavas. The upper Agatha Formation is ~1600 m thick with subaerial basaltic, andesitic, dacitic and rhyolitic lavas as well as pyroclastic components (Wilson and Grant 2006; Wilson et al. 2013). These volcanic rocks, including their well-preserved textures, are detailed in Sect. 9.4. Gumsley et al. (2015) proposed at least two magmatic episodes: one at ~2.99–2.98 Ga associated with Nhlebela Formation volcanics, and another at ~2.97–2.96 Ga associated with Agatha Formation lavas. These lava units are separated by a sedimentary unit, the ~200 to 500 m thick shallow-marine and tide-dominated White Mfolozi Formation (Cole 1994; Siahi et al. 2016), which contains cross-bedded sandstones, quartz- and greywackes, carbonates, shales and breccias. The carbonates, known as the Chobeni Member, are dolomite-bearing stromatolites (Mason and von Brunn 1977; Beukes and Lowe 1989; Siahi et al. 2016). A siliciclastic unit with mudstones, sandstones, as well as phyllite, possibly correlating with the White Mfolozi Formation, outcrops in the Mahlangatsha-Ntungulu area of Swaziland (Cole 1994; Nhleko 2003).

9.3.1.3 Ozwana Subgroup

The primarily sedimentary Ozwana Subgroup, ranging from 200 to 2800 m in thickness, is present at the top of the Nsuze Group in a sharp but conformable contact with underlying Agatha Formation lavas (Gold 2006). Its lower Langfontein Formation is composed of sandstone, conglomerates, siltstone, as well as pyroclastic rocks (Gold 2006). At the base of the Langfontein Formation is the Roodewal Member, bearing volcaniclastics and a lahar-breccia (Gold 1993; Nhleko 2003). The uppermost Mkuzane Formation is primarily siltstone. According to Wilson and Grant (2006), the distal tidal and shallow-marine shelf sedimentary strata of the Ozwana Subgroup in the Central and Southern domains formed through basinal subsidence following Agatha Formation eruptions.

9.3.2 Stratigraphy of the Mozaan Group

The ~4800 m thick Mozaan Group (Fig. 9.2e) disconformably overlies the Nsuze Group and is subdivided into three subgroups and ten formations. This predominantly sedimentary sequence primarily includes shallow marine and fluvial sedimentary rocks such as sandstones, quartz arenites, conglomerates, diamictites and shales, as well as two volcanic units (Beukes and Cairncross 1991; Gold 2006). Although unit thicknesses vary greatly between the Hartland type area (discussed below), and other exposures in the Central and Northern Domains, correlation of units is possible throughout the main Pongola Basin. Nearly the entire Mozaan Group also crops out in the Kubuta-Mooihoek area of southern Swaziland in an open synclinal structure with a northwest–southeast-trending fold axis (Nhleko 2003; Wilson et al. 2013).

9.3.2.1 Sinqeni Formation

The base of the Mozaan Group is defined by the ~650 m thick Sinqeni Formation, composed of two major fluvial to shallow-marine sandstone units, the lower Dipka and upper Kwaaiman members, separated by an 80 m thick unit of shale and iron formation known as the Vlakhoek Member (Beukes and Cairncross 1991; Gold 1993, 2006; Hicks and Hofmann 2012). Within the White Mfolozi Inlier, the basal portion of the Sinqeni Formation is defined by an auriferous and uraniferous conglomeratic succession of fluvial origin known as the Denny Dalton Member. Although discontinuous, similar basal conglomerates have been identified in the Hartland region (Hicks 2009), Gold (2006) included these units within the Dipka Member in this area. The Dipka Member is about 400 m thick and contains upward-fining quartz arenite, interbedded with subordinate conglomerate lenses and arkoses. A sharp and locally sheared contact separates the Dipka Member from the overlying Vlakhoek Member that consists of shale and banded iron formation (BIF), known as the Ijzermijn BIF (Hicks 2009; Albut et al. 2018). Its overlying Kwaaiman Member sandstone unit is a 170 m thick quartz arenite representing an intertidal unit (Beukes and Cairncross 1991; Hicks and Hofmann 2012).

9.3.2.2 Dwaalhoek Subgroup

This ~2000 m thick sequence of sandy-to-muddy marine shelf deposits begins with the Ntombe Formation, a ~1300 m thick transgressive sequence of ferruginous shales, siltstones and sandstones (Beukes and Cairncross 1991). This is overlain by the ~700 m thick Thalu Formation, a highstand deposit composed of several mud-silt-sandstone cycles, as well as the Scotts Hill BIF Member, a widespread marker horizon which likely correlates with the Contorted Bed BIF of the West Rand Group of the Witwatersrand Supergroup (Smith 2017).

9.3.2.3 Hlashana Formation

Above the Dwaalhoek Subgroup is the Hlashana Formation, a prominent ~400 m thick unit composed of cross-bedded sandstones deposited in a current-dominated sandy shelf environment (Beukes and Cairncross 1991).

9.3.2.4 Odwaleni Subgroup

The base of the ~1550 m thick Odwaleni Subgroup is defined by the ~1000 m thick Delfkom Formation, a succession of sandstone and mudstone with subordinate diamictite. The Delfkom Formation is overlain by the ~50 m thick Tobolsk Formation, a volcanic unit bearing amygdaloidal flood basalts, pillow basalts as well as tuffs. The ~500 m thick Khiphunyawo Formation, a transgressive marine sequence comprised of quartz arenite, quartzwacke, mudstone and diamictite, caps the Odwaleni Subgroup (Beukes and Cairncross 1991; Strik et al. 2007).

9.3.2.5 Nkoneni Subgroup

The top of the Mozaan Group consists of three formations that constitute the ~600 m thick Nkoneni Subgroup. Its lowermost unit is the 250 m thick Bongaspoort Formation, composed of coarsening-upwards metasandstones and quartzwackes deposited on a sandy fluvial braid plain (Beukes and Cairncross 1991). The central 100–800 m thick and predominantly volcanic Gabela Formation is characterized by pyroclastic rocks, including andesitic and dacitic tuffs as well as fragmental komatiite, interbedded among sandstones and shales (Wilson and Zeh 2018). The uppermost 150 m thick Ntanyana Formation is composed of both coarse-grained sandstone and magnetic shale of fluvial and marine origin (Gold 2006).

9.3.3 Stratigraphic Issues

This review takes a deliberately conservative approach to Pongola Supergroup stratigraphic nomenclature, though acknowledging that disagreement exists and new proposals (e.g. Wilson et al. 2013) may shift the field's terminology in the future. While the bulk of this review adopts the most commonly used stratigraphic divisions for the Pongola Supergroup, it is worth noting some important outstanding questions.

9.3.3.1 Status of the 'Wagendrift Formation'

Following Armstrong (1980) and Armstrong et al. (1982), Wilson et al. (2013) proposed that the basaltic Wagendrift Formation in the Hartland/Piet Retief area forms the base of the Nsuze Group and that it, along with the overlying sedimentary Mantonga Formation, together compose the Warmbad Subgroup, which underlies the Bivane Subgroup. A counter-view is that it is difficult to determine this unit's exact position in the field as contacts are either obscured by intrusion of the Usushwana Complex or by Phanerozoic cover.

9.3.3.2 'Nhlebela Formation' Versus 'Pypklipberg Formation'

Following Cole and Beukes (1995), Wilson et al. (2013) proposed that in the Hartland/Piet Retief area, the Nhlebela Formation should be renamed the Pypklipberg Formation. They described these formations as having different lithologies and compositions (the Nhlebela Formation lacks rhyolites, which are present in the proposed Pypklipberg Formation), reflecting different petrogenesis and depositional environments (the Pypklipberg Formation is entirely subaerial, unlike the Nhlebela Formation). A counter-view might state that while the lithologies of the Nhlebela and proposed Pypklipberg formations are clearly distinct, these units seem to occupy the same stratigraphic position as the lower major lava package of the Nsuze Group and may represent lateral variation within a chronostratigraphic unit.

9.3.3.3 'White Mfolozi Formation'/'Chobeni Formation' Versus 'Nkemba Formation'

As noted, there is no overall agreement on the naming of the sedimentary package between the two major lava sections of the Nsuze Group, as either the 'Chobeni Formation' (Siahi et al. 2016; Eickmann et al. 2018) or the 'White Mfolozi Formation' (which contains the carbonate Chobeni Member; Gold 2006; Wilson et al. 2013). Wilson et al. (2013) proposed that this formation should be called the Nkemba Formation in the Hartland/Piet Retief area, due to lithological differences that reflect different depositional environments from its presumed stratigraphic equivalent at the White Mfolozi Inlier. Specifically, they noted that in the Hartland/Piet Retief area this unit is almost entirely volcaniclastic or composed of terrigenous sediments and lacks carbonates while being interlayered with thin dacitic and andesitic lava units, whereas at the White Mfolozi Inlier this unit lacks lava flows and is rich in carbonates. An alternative possibility is that these represent the same chronostratigraphic unit with significant lateral facies changes.

9.3.3.4 Pongola Terminology in the Southern Domain

It is widely agreed that the stratigraphy of the Pongola Supergroup in the Southern Domain is complex, making correlations with northern Pongola Supergroup rocks difficult. In particular, the Nsuze Group in the Nkandla Inlier of the Southern Domain uses a different stratigraphic nomenclature (Fig. 9.2c; Wilson et al. 2013; Gold 2006) and is characterized by a greater proportion of sedimentary and pyroclastic deposits than in the Northern Domain. Nhlebela Formation basalts are likely present, and the overlying sedimentary Mdelanga Formation, which bears a carbonate unit similar to the Chobeni Member (Groenewald 1984), may correlate with the White Mfolozi Formation to the north (Hicks et al. 2011; Wilson et al. 2013). Above this is the Qudeni Formation, bearing basaltic to andesitic lavas that may correlate with the Agatha Formation. The overlying Vutshini Formation may either correlate with the Langfontein Formation or represent younger Mozaan Group strata. Whereas this review adopts the Southern Domain nomenclature of Wilson et al. (2013), this terminology should be taken tentatively due to the need for further work to elucidate stratigraphic relationships. For further details on the stratigraphy of the Southern Domain and related nomenclature, see Wilson et al. (2013) and Dixon (2004).

9.4 Overview of the Volcanic Rocks

Volcanic rocks are an important component of the Pongola Supergroup and occur throughout the exposed area. The style of volcanism and the compositional ranges are unlike other volcanic successions of this age, which were mainly constituents of greenstone belts and dominated by submarine basalt and komatiite. The Nsuze Group exhibits great lateral variation in the thickness of the volcanic pile as well as in the morphological character of individual flows. The volcanic pile is thickest ($\sim$4200 m) in the type Hartland area of the Northern Domain (Fig. 9.1). From the northern (Hartland) to the southern (Nkandla) part of the province ($\sim$300 km), there is a progressively greater proportion of interbedded clastic sediments as well as different volcanic rock types. Thus, while the northern type area exhibits a compositionally wider range of subaerial volcanic rocks, ranging from basalt through more intermediate compositions to rhyolite, the southern areas (such as the White Mfolozi Inlier) contain more massive lava and pillow basaltic flows. Intermediate and felsic rock types are predominantly pyroclastic in the northern area, but lava flows of these compositional types also occur. Spectacular pahoehoe lava flows occur in all volcanic units of the Pongola Supergroup and are particularly well-developed in the White Mfolozi Inlier. The dominantly sedimentary Mozaan Group contains two volcanic units (Tobolsk and Gabela formations; Gold 2006) near the top, which contain evidence of ultramafic lavas that may be linked to a late-stage period of ultramafic magmatism.

9.4.1 The Hartland (Type) Area

The type area of the Nsuze Group's volcanic rocks lies in the Hartland area, east of Piet Retief (now renamed to Mkhondo) (Armstrong et al 1982; Cole and Beukes 1995; Gold 2006; Linström 1987a; Nhleko 2003; Wilson et al. 2013), where their greatest thickness (>4000 m) and diversity of rock compositions are found. Here, a continuous suite from basalt to rhyolite occurs (Fig. 9.2a). Combined, the intermediate rock types (basaltic andesite and andesite) constitute 52–54% of all rock types, whereas rhyolite and basalt constitute only about 20 and 15%, respectively. Dacite is the least abundant rock type, however, leading to an overall compositionally bimodal distribution (Armstrong 1980; Wilson et al. 2013). The stratigraphically lowest volcano-sedimentary Mantonga Formation contains compositionally intermediate lava flows and volcaniclastic rocks that are interbedded with sandstones. A laterally separate volcanic section, northwest of this type area, contains pillowed lavas close to its base as well as basalt and is not typical of the Mantonga Formation. It has limited exposure and is truncated by the Usushwana intrusion and may be ascribed to an independent Wagendrift Formation (Fig. 9.2a; Wilson et al. 2013).

The main succession of lava flows in this area is contained within the Nhlebela Formation (Pypklipberg Formation according to Wilson et al. 2013; Fig. 9.2a) and Agatha Formation of the Bivane Subgroup, described in detail by Armstrong et al. (1986) and Wilson et al. (2013). Both lava sequences are made up of flow cycles, ranging from basalt through to rhyolite but also intermittently interbedded by flows of various compositions. Typical textures and

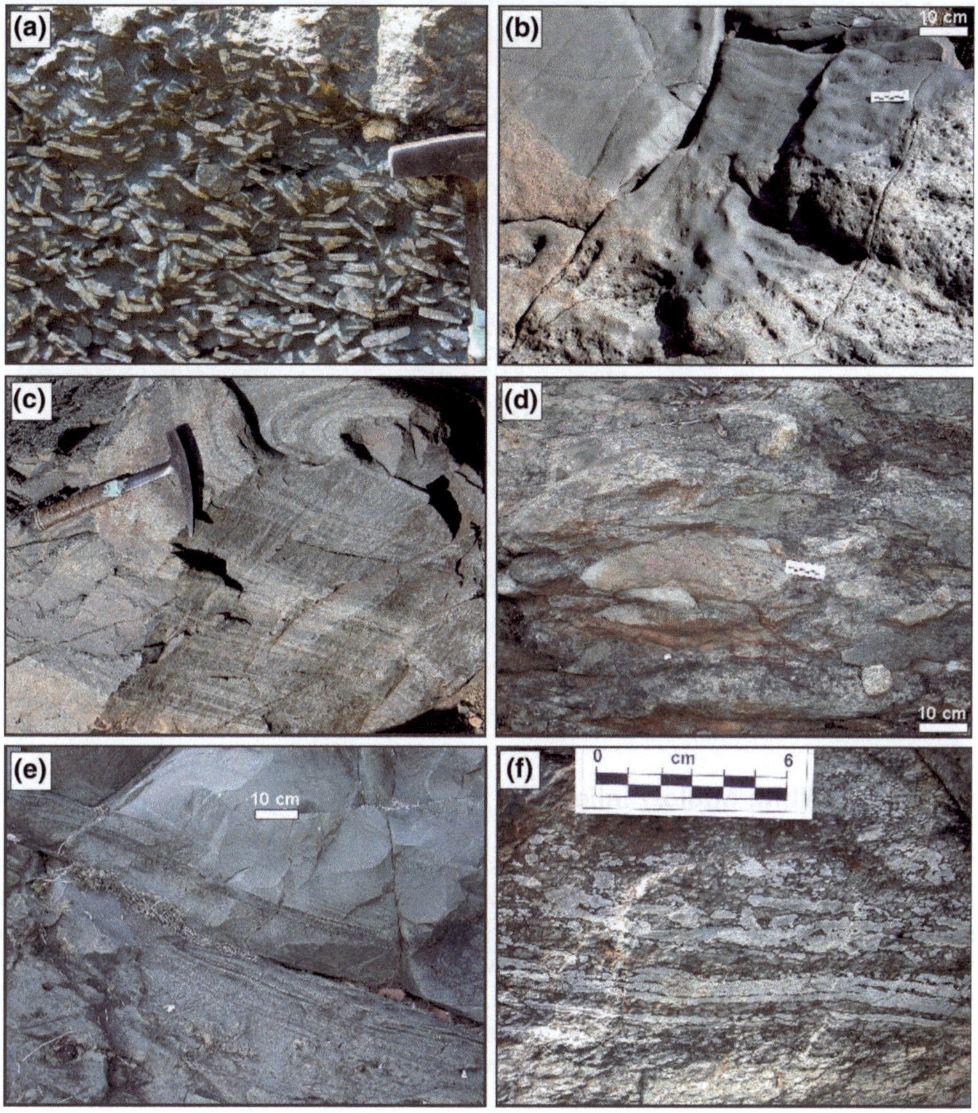

Fig. 9.3 Field photographs of volcanic textures in the Hartland area from Wilson et al (2013). **a** Porphyritic andesite lava flow; **b** Contact of porphyritic andesite and aphyric basaltic andesite flow units; **c** Rhyolite unit with flow banding in the Nhlebela Formation; **d** Agglomerate in the Nhlebela Formation; **e** Andesitic tuff with laminations and vesicles; **f** Immiscible segregations of dacitic and basaltic magmas within the same composite flow. Photographs reproduced by permission of the South African Journal of Geology; Wilson et al. (2013)

structures are shown in Fig. 9.3. Towards the north, some porphyritic andesites are characterized by exceptionally large (1–3 cm) plagioclase phenocrysts (Fig. 9.3a), absent in southern localities.

Agglomerates and bomb tuffs occur in close association with banded rhyolite in the Nhlebela Formation. Many seemingly massive andesites are finely banded and are indicative of ash deposits. Towards the top of the Agatha Formation in the Hartland area is a distinct volcaniclastic unit (andesite and dacite) that is up to 300 m thick and extends for 16 km.

9.4.2 The Mpongoza Inlier

The Mpongoza Inlier (11 × 3 km in areal extent) is the most northerly occurrence of volcanic rocks in the Central Domain and exposes nearly 3 km of stratigraphy, dipping westward at 20–45° (Preston 1987; Wilson et al. 2013). The inlier is dominated by felsic pyroclastic rocks. Basalts comprise only 15% of the volcanic rocks. The sequence is made up of two cycles (lower cycle 1500 m thick; upper cycle 1200 m thick), both commencing with massive or amygdaloidal basalt flows and progressing into a variety of pyroclastic rock types that include block and ash deposits and felsic ignimbrites. Cross-bedded ash layers are indicative of surge deposits resulting from Plinian-type eruptions. The two volcanic cycles are separated by the sedimentary Mpongoza Formation (correlated with the White Mfolozi Formation to the south), composed mainly of sandstone with minor interbedded siltstone. The basalt commonly contains chert fragments derived from an unconsolidated substrate. Typical micro-textures and structures are shown in Fig. 9.4.

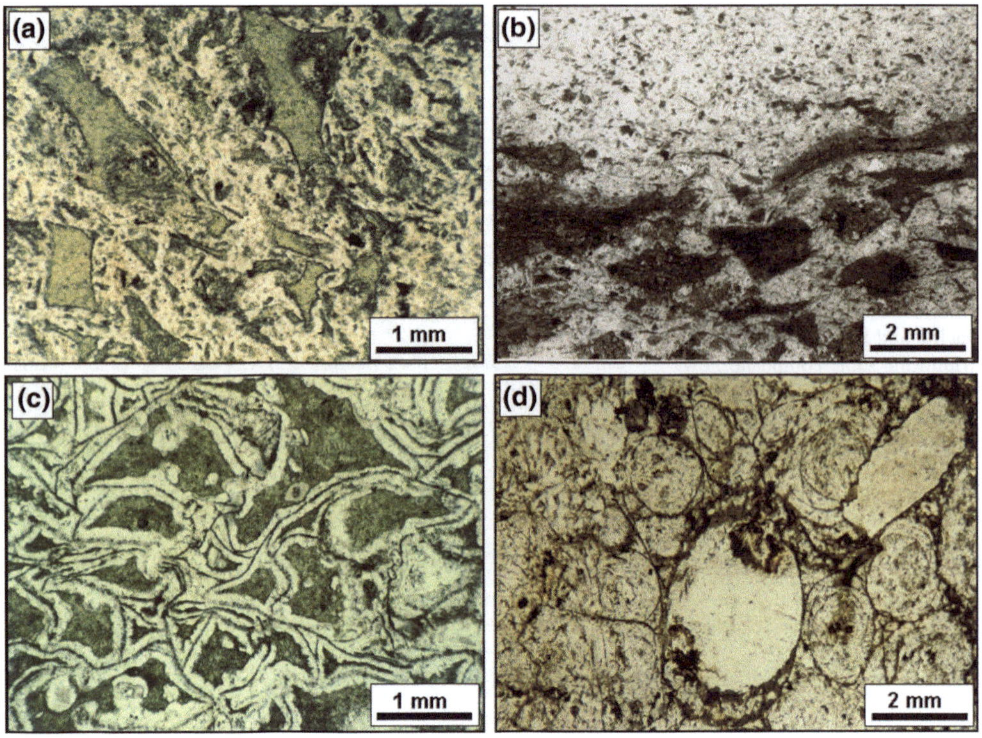

Fig. 9.4 Photomicrographs under plane light showing primary volcanic and devitrification textures of volcanic glass in the Mpongoza Inlier from Wilson et al (2013) and Preston (1987). **a** Vitric crystal-tuff showing angular and lunate-shaped glass shards; **b** Tuff overlying andesite containing angular juvenile glass shards; **c** Axiolitic texture resulting from devitrification of an ash-tuff; **d** Conchoidal perlitic cracks developed during contraction and combined with hydration of metastable glass. Photographs reproduced by permission of the South African Journal of Geology; Wilson et al. (2013)

9.4.3 The White Mfolozi Inlier

The continuous exposure of volcanic rocks in the White Mfolozi River gorge is one of the best-preserved successions in the Archaean record. These volcanic rocks comprise a wide range of textural types including pillow lavas, pahoehoe flows and mafic pyroclastic rocks. Unlike volcanic rocks to the north (Hartland area and Mpongoza Inlier), this succession lacks felsic rocks and is made up of andesite, basaltic andesite and minor basalt, within upper and lower volcanic successions (Fig. 9.2b).

The lower volcanic succession, the Nhlebela Formation (200 m thick), is mainly basaltic andesite, commencing with well-developed pillows overlying the sedimentary Bomvu Formation (correlated with the Mantonga Formation in the north). The lowermost pillow unit grades upwards into more bulbous and ovoid structured units, which are highly fragmented and most likely formed in shallow water. Peperite textures further indicate hydro-magmatic interaction and mingling of lava and wet sediments. The upper 40 m erupted subaerially forming pahoehoe flow units with amygdaloidal lower contacts and chilled ropey upper surfaces. The top of the lava sequence has a sharp contact against shales of the overlying White Mfolozi Formation.

The upper volcanic Agatha Formation ($\sim$700 m thick) overlies coarse-grained sandstone of the White Mfolozi Formation. The lowest 150 m of the succession comprises hydroclastic tuff and fragment lava-breccia deposits. Overlying is a succession, 165 m thick, made up of 14 lava units exhibiting a range of lava textures and structures, including ropey surfaces, vesicular tops, lava toes and flow terminations (Fig. 9.5b–f). Pipe vesicles occur within both the basal and upper chill zones of each flow, as well as up to 20 cm large gas blisters below the upper chill zone. Dykes (10–25 cm wide) cut through the more fragmented deposits, which could be invasive lava flows as well as younger feeders to overlying units.

The upper section (320 m) of the Agatha Formation comprises pahoehoe basalt flows (2–10 m thick) that can be traced laterally (on the north side) for several hundred metres. Termination of flow units is commonly seen as sharply curved lava toes overlain by thicker flows with well-developed columnar jointing. In this section there is an upward change to more silica-rich and lower magnesian lava compositions (Wilson and Grant 2006).

9.4.4 The Southern Domain

The Southern Domain comprises a series of volcano-sedimentary sequences which do not have stratigraphic equivalents to the north. The most easterly is the Mhlatuze Inlier of which little is known but which appears broadly similar to other Nsuze Group occurrences in the south. The most extensive Southern Domain inliers are the easterly Nkandla Inlier and the Buffalo River Inlier to the west. In the Nkandla Inlier three major synclines incorporate the entire sections of the Nsuze and Mozaan groups, but

Fig. 9.5 Volcanic textures of the White Mfolozi Inlier. **a** Hydroclastic volcanic breccia in the Nhlebela Formation. All within the Agatha Formation: **b** Section through a basalt flow unit, showing a cracked ropy surface and upward accumulation of vesicles (amygdales) below that chill zone; **c** Large ropy surface in another pahoehoe lava flow lobe; **d** Flow structures inside basalt flow, together with lava breakouts (centre-top); **e** Lava toe below an overlying columnar jointed flow lobe; **f** Cross section through several pahoehoe flow lobes, where concentric zonation and lack of radiating cooling joints distinguish these from pillows. Photographs b and d from Wilson et al. (2013); reproduced by permission of the South African Journal of Geology; Wilson et al. (2013)

tentative correlations suggest that the volcanic sequence only equates with the upper Agatha Formation (here called the Qudeni Formation; Fig. 9.2c). Equivalent volcanic successions in the southern syncline are currently unknown. Dixon (2004) mapped the Buffalo River Inlier in detail and identified four fault blocks with folded successions of the Nsuze and Mozaan groups. Within each fault block are four volcano-sedimentary successions of lava flows separated by sandstone units, giving a total thickness of nearly 4000 m (Fig. 9.2d). The volcanic sequence (Fig. 9.6) consists of pillows of basaltic andesite (Fig. 9.6), overlain by flows that exhibit an upward compositional progression to andesite and in some cases dacite near the top.

The Ndikwe Thrust Segment (Fig. 9.1) occurs in the northern part of the Southern Domain and is part of an allochthonous nappe structure comprising sedimentary and volcanic rocks. The southern part of the thrust sequence was originally mapped by Groenewald (1984) who used the term Ndikwe Formation as supposedly being part of the Nsuze Group. The northern part of the nappe was recognised by Matthews (1990) and Dixon (2004), who originally named it the Nsuze Nappe. Geochemical evidence shows the entire section to be part of the same structure, and detrital zircons from this sequence with concordant ages as young as 2888 $\pm$ 6 Ma show that it is part of the Mozaan Group (Wilson and Zeh 2018) and not the Nsuze Group.

9.4.5 Volcanic Rocks of the Mozaan Group

Two volcanic formations in the Mozaan Group are exposed in the Tobolsk syncline, in the Hartland area (Fig. 9.1), and

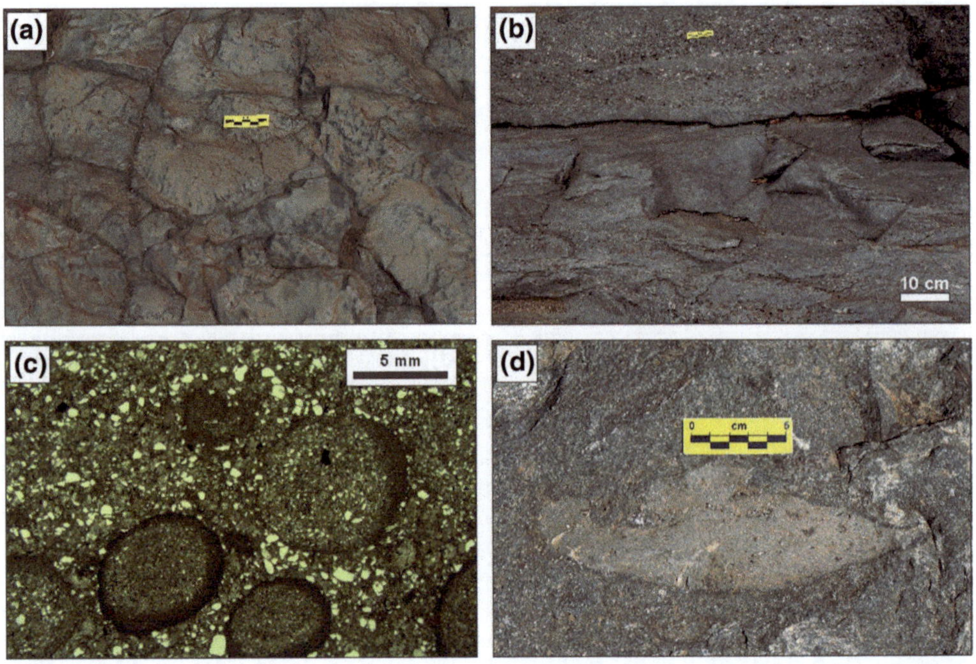

Fig. 9.6 Volcanic and volcaniclastic textures in the Southern Domain, from Wilson et al. (2013). In the Buffalo River Inlier: **a** Shallow-water pahoehoe flow lobes with radiating amygdales in the top of a 5 m thick basaltic andesite lava unit; **b** Massive interior of a basaltic andesite flow lobe in the Ndikwe Thrust segment; **c** Armoured lapilli of andesitic ash together with fine sand grains in a coarse ash-sandy matrix; **d** Irregularly shaped bombs of high-Mg lava in a matrix of both volcanic ash and clastic sand grains. Photographs reproduced by permission of the South African Journal of Geology; Wilson et al. (2013)

in southern Swaziland (Nhleko 2003), known as the lower Tobolsk Formation and upper Gabela Formation. Both exposed formations are only 40–120 m thick, but drilling has revealed these combined volcanic sequences to be at least 800 m thick in places (Wilson et al. 2013), comprising both lava flows and fragmental pyroclastics of contrasting compositions. As discussed above, the Ndikwe Thrust Segment, which contains abundant volcanic rocks, is part of the Mozaan Group volcanic succession. Well-developed pillows occur in the basalts of the Tobolsk Formation, but pyroclastic rocks commonly associated with ultramafic fragments indicate associated high-Mg picritic or komatiitic volcanism. Ultramafic ash particles are commonly intermixed with sand grains in cross-bedded deposits. Groenewald (1984) considered the Hlagothi Complex (Fig. 9.1) to have a komatiitic parental magma composition. The precise age of 2866 ± 2 Ma (Gumsley et al. 2013) for this complex confirms it to be of Mozaan Group age and thereby establishing a further link with the ultramafic volcanic component.

9.5 Intrusions into the Pongola Supergroup

The Usushwana Complex (~2.99–2.96 Ga; Gumsley et al. 2015) is one of the oldest known layered intrusions that developed within stable continental crust. It was initially named by Hunter (1950) and identified as a northwest-orientated complex of mafic/ultramafic (the Piet Retief Suite) and felsic (the Hlelo Suite) layered components. Being approximately 100 km long and about 6.5 km wide (Burke et al. 1985; Anhaeusser 2006), it straddles the South Africa–Swaziland border in the Piet Retief (Mkhondo) area. The complex appears to have intruded along northwest-striking basement faults of the Mahamba-Altona shear zone (Fig. 9.1; Matthews 1990; Gold and von Veh 1995), and along the base of the Pongola Supergroup (Burke et al. 1985). Hunter (1970) suggested that the intrusions were emplaced during fault reactivation, subsequent to deposition of the Pongola Supergroup. However, Gumsley et al. (2015) precisely dated igneous baddeleyite from the Piet Retief Suite gabbro at 2.99–2.98 Ga (Table 9.1) and thereby showed that the Usushwana Complex represents a magmatic feeder chamber for Nsuze Group volcanics, also indicating that its felsic Hlelo Suite component was derived from partial melting of Pongola host rocks.

The Hlagothi Complex is a 2866 ± 2 Ma series of layered mafic-ultramafic sills of gabbro, pyroxenite and peridotite intruding into the Pongola Supergroup in the Nkandla sub-basin (Groenewald 1984; Gumsley et al. 2013, 2015). A related event is the Thole Complex, a suite of ultramafic sills that intrude into the Pongola Supergroup in the Amsterdam area and the main Pongola Basin, also outcropping in southwest Swaziland (Gumsley et al. 2013, 2015). A pyroxenite assigned to the Thole Complex by Gumsley et al. (2015) was dated by Hegner et al. (1984) at 2871 ± 30 Ma, suggesting that the Thole and Hlagothi complexes are coeval. The Amsterdam Formation is a ~2.8 Ga felsic suite of northwest-striking dykes in the Amsterdam area, consisting of dacite, rhyolite and tuff (Hammerbeck 1982; Anhaeusser 2006; Gumsley et al. 2015). Gumsley et al. (2013; 2015) proposed that a widespread ~2.87 Ga magmatic event across the south-eastern Kaapvaal Craton—possibly a large igneous province—produced the Hlagothi Complex, which fed Mozaan Group volcanics and also generated the Thole Complex, the Amsterdam Formation and portions of the granophyres previously ascribed to the Hlelo Suite of the Usushwana Complex.

Multiple ~2.7–2.9 Ga anorogenic granite plutons intrude the eastern side of the Pongola Supergroup in southern Swaziland and northern Mpumalanga Province in the Hartland area (Fig. 9.1; Matthews 1985, 1990; Mukasa et al. 2013), disrupting, metamorphosing and deforming Pongola Supergroup strata in those areas (see Sect. 9.8.7; Wilson et al. 2013; Hofmann et al. 2015). These largely post-Pongola granitoid bodies include the Kwetta/Mtombe granite (~2.67–2.72 Ga; Reimold et al. 1993; Maphalala and Kröner 1993; Mukasa et al. 2013), Spekboom granite (~2.70–2.73 Ga; Reimold et al. 1993), Hlatikulu granite (~2.72–2.73 Ga; Maphalala and Kröner 1993; Schoene and Bowring 2010), Mswati granitoid suite (~2.72–2.84 Ga; Maphalala and Kröner 1993; Mukasa et al. 2013; Dlamini et al. 2017), Godlwayo granite (~2.86 Ga; Reimold et al. 1993) and the Nhlangano gneiss (~2.92–2.99 Ga; Maphalala and Kröner 1993; Mukasa et al. 2013; Hofmann et al. 2015). The geodynamic setting of these intrusions remains to be established.

9.6 Palaeomagnetism

Published palaeomagnetic studies of the Pongola Supergroup and related Mesoarchaean intrusions have produced relatively sparse data and inconsistent results (Fig. 9.7; Table 9.2), but individually suggest that the Kaapvaal Craton drifted over an active mantle at this time (Strik et al. 2007; Lubnina et al. 2010; Gumsley et al. 2013). As reviewed below, six main palaeomagnetic investigations have been conducted since 2003 on the Pongola Supergroup and related rocks.

Strik et al. (2007) collected more than 90 palaeomagnetic samples from 13 Pongola Supergroup sites and found both medium temperature (MT) and high temperature (HT) magnetic remanence components. The MT component post-dated structural tilt, which they attributed to a massive regional magnetic overprint at around 180 Ma, during the breakup of Gondwana. Differing HT components were found in the Nhlebela Formation basalts as compared to Agatha Formation basalts, yielding virtual geomagnetic poles (VGPs) in broad agreement with Layer et al.'s (1988, 1989) palaeopoles for the roughly coeval Usushwana Complex (Gumsley et al. 2015) but different enough to suggest possible later remagnetization or craton drift. They concluded that the palaeolatitude of the Pongola Supergroup was near 30°, and the Ventersdorp Supergroup's palaeolatitude was near 55°, suggesting that the Kaapvaal Craton drifted at least 3000 km within ~200 Ma. However, their study lacked successful field tests to confirm whether the palaeopoles were derived from primary magnetic components, and they concluded that the Kaapvaal Craton's Meso-Neoarchaean apparent polar wander path is 'poorly defined.'

Lubnina et al. (2010) obtained a palaeopole for Agatha Formation basalts near a VGP they obtained for ~2.95 Ga, southeast-trending Badplaas Dyke Swarm feeders of Nsuze Group lavas. A baked contact test indicated this component to be coeval with the dykes, just as a conglomerate test implied the directions were primary. Their Agatha Formation VGP is significantly different from that of Strik et al. (2007), but likewise they inferred significant Meso-Neoarchaean drifting of the Kaapvaal Craton, concluding that 'paleolatitude changes significantly, moving from near the equator at 2.95 Ga, southward to >60° S at 2.78 Ga and again northward to about 40° S at 2.65 Ga.' Maré and Fourie (2012) obtained three poorly constrained VGPs for 2.95 Ga dykes of the Badplaas Dyke Swarm. One of these lies very close to Layer et al.'s (1988, 1989) Usushwana Complex palaeopole, thought to be consistent with the results of Strik et al. (2007), though it was significantly different from the VGP obtained by Lubnina et al. (2010).

Two studies by Nhleko (2003) and Lechekoane (2016) investigated the Klipwal Member diamictite of the Mozaan Group's Delfkom Formation (Fig. 9.2e) and obtained discrepant VGPs which were unlike any other known palaeopoles for the Pongola Supergroup (Fig. 9.7; Table 9.2). The results of Nhleko (2003) correspond to a palaeolatitude of 47.7°, which he suggested 'represents polar glaciation in the Archaean similar to today where glacial events are confined in higher latitudes.' Nhleko's (2003) palaeolatitude is similar to the 52.6° palaeolatitude obtained by Lechekoane (2016) but distinct from a palaeolatitude of 10.0° obtained by Strik (2004) and Strik et al. (2007) for diamictite of the younger Khiphunyawo Formation, which was likely remagnetized.

Gumsley et al. (2013) sampled the Hlagothi Complex, directly adjacent to Nsuze Group volcanic rocks, and sedimentary strata in the Nkandla sub-basin. Their palaeomagnetic study recovered five components, one of which was interpreted as primary. This primary palaeopole (Fig. 9.7; Table 9.2) was similar to much younger palaeopoles associated with the Bushveld Complex and Vredefort impact

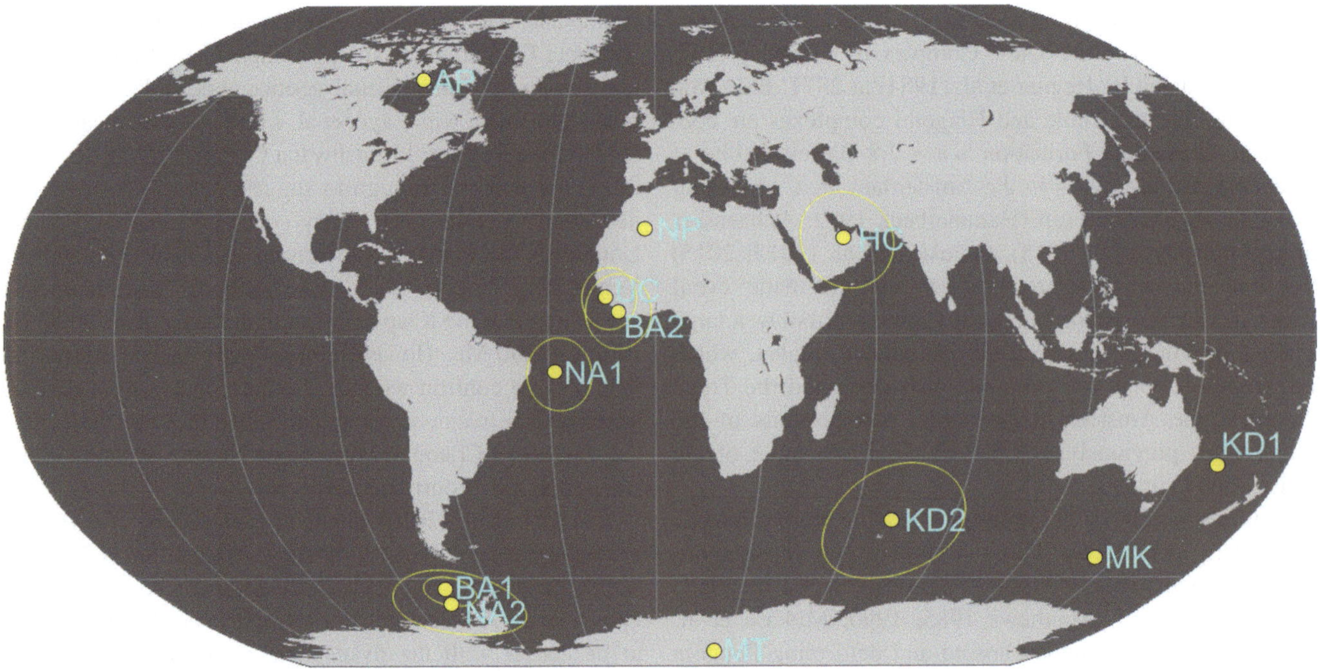

Fig. 9.7 Palaeopoles of Pongola Supergroup rocks and related intrusive complexes. Label abbreviations according to Table 9.2

Table 9.2 VGPs and palaeolatitudes obtained by eight different palaeomagnetic studies of Pongola Supergroup rocks and related intrusive complexes

Unit (Map Label)	VGP latitude°	VGP longitude°	D_p°	D_m°(or A95 in °)	Palaeolatitude°	Age of magnetization (Ma)	Reference
Usushwana Complex (UC)	9.2	347		(7.6)	33.3	2990–2860	Layer (1986); Layer et al. (1988, 1989)
Nsuze Group Nhlebela Formation Volcanics (NP)	25.9	358.4			27.4	2984 ± 2.6	Strik et al. (2007)
Nsuze Group Agatha Formation Basalts (NA1)	−9.4	333		(8.9)	32.2	2977 ± 5	Strik et al. (2007)
NW–SE-Trending Badplaas Dolerite Dykes (BA1)	−63.6	285.4	2.3	4	18.5	~2950	Lubnina et al. (2010)
Nsuze Group Agatha Formation Basalts (NA2)	−67	285.6	5.3	9.2	20	~2950	Lubnina et al. (2010)
NW–SE-Trending Badplaas Dolerite Dykes (BA2)	6.0	349.6		9.2	37.8	2984 ± 2.6	Maré and Fourie (2012)
All Pongola (AP)	63.1	278.2			33.2	~2900	Strik (2004)
Mozaan Group Khiphunyawo Formation Diamictite (MK)	−55.3	145.2			10	~2900	Strik et al. (2007)
Mozaan Group Klipwal Member Diamictite (KD1)	−32.3	162			47.7	~2900	Nhleko (2003)
Mozaan Group Klipwal Member Diamictite (KD2)	45.7	252.8		(14.5)	52.5	~2900	Lechekoane (2016)
Mozaan Group Tobolsk Formation Lava (MT)	−82.7	27.9			34.8	~2900	Strik (2004)
Hlagothi Complex (HC)	23.4	53.4	8.2	11.8	13.2	2866 ± 2	Gumsley et al. (2013)

VGP virtual geomagnetic pole, D_p semi-axis of ellipse of confidence along great circle from site to pole, D_m semi-axis of ellipse of confidence perpendicular to great circle defining D_p, *A95* radius of 95% confidence circle around pole

—similarities regarded as coincidental as these events are too geographically distant to influence remanence carried by the Hlagothi Complex. Because their primary palaeopole differed from other Archaean poles on the Kaapvaal Craton, Gumsley et al. (2013) argued 'against a single relatively stationary mantle plume' feeding intrusion events, concluding that the Kaapvaal Craton was drifting over an active mantle, experiencing many smaller plumes during the Mesoarchaean.

9.7 Aspects of Mesoarchaean Surface Processes

The vast majority of Pongola sedimentary rocks are siliciclastic, comprising sandstones and shales, but chemical and biogenic deposits also occur. Sedimentary strata have provided a wealth of information on Mesoarchaean surface processes, specifically BIFs, stromatolitic carbonates, glacial diamictites, palaeosols and siliciclastic rocks deposited in an intertidal setting. Some of the best-preserved- and most easily accessible sedimentary rocks of the Pongola Supergroup are exposed in the White Mfolozi Inlier, where much recent research has been undertaken as summarised below.

9.7.1 Stromatolitic Carbonates

Several stratigraphic intervals of mixed carbonate-siliciclastic facies occur in the White Mfolozi Formation of the Nsuze Group (Figs. 9.2b and 9.8a). In this unit, the combination of both symmetric and asymmetric ripple marks, abundant mud drapes as well as flaser, lenticular and wavy bedding and other evidence for bidirectional currents, are consistent with deposition in a tide-dominated shallow-marine environment (Mason and von Brunn 1977; Beukes and Lowe 1989; Siahi et al. 2016).

Several stratigraphic intervals in the White Mfolozi Formation host carbonates, which are otherwise rare in 3.0 Ga strata. Facies consist of mixed siliciclastic and carbonate rocks and include sub-tidal cross-bedded sandstone and oncolitic dolostone, intertidal wave-rippled oodolarenite (Fig. 9.8c) and supratidal wrinkly laminated dolarenite-dololutite (Fig. 9.8e). Sedimentary facies are partially arranged in a shallowing-upward pattern, reflecting cyclic variations of Mesoarchaean sea level (Fig. 9.8b). Microbialites occur in a variety of sedimentary sub-environments and include laminated as well as structureless, lenticular-domal types (Fig. 9.8d; Siahi et al. 2016). Differences in microbialite morphology, macrofabric and size, suggest potentially distinct assemblages of microorganisms responding to different physicochemical and environmental conditions. Despite their great age, Pongola Supergroup carbonates are compositionally and texturally similar to much younger carbonate successions. Abundant stratiform stromatolite facies with characteristic assemblages of structures are indicative of intermittent exposure on widespread intertidal flats. Coarse-grained clastic subtidal facies, the presence of ooids (Fig. 9.8f; Siahi et al. 2017), evidence for storm deposits and abundant clastic detritus in carbonate sections, suggest deposition along a tide-influenced clastic-carbonate shoreline rather than a restricted lagoonal setting (Siahi et al. 2016). Trace element and stable (C, O) and radiogenic (Sr) isotope geochemistry of carbonate rocks supports deposition in a shallow-water epicontinental basin with restricted but variable exchange with the open ocean (Bolhar et al. 2015; Siahi et al. 2018). Multiple sulphur and Fe isotope data of early diagenetic pyrites indicate microbial sulphate reduction of sulphur species cycled through an anoxic atmosphere and record partial Fe oxidation in oxygenated shallow oceans (Eickmann et al. 2018). This suggests a reduced Mesoarchaean atmosphere, while free oxygen accumulated in near-shore marine environments due to the evolution of oxygenic photosynthesis at or before 3 Ga.

9.7.2 Palaeosols

Several sites exist where the Nsuze Group overlies a granite peneplain onto which a palaeosol has developed (Grandstaff et al. 1986; Matthews and Sharrer 1968). Higher in the succession, the Denny Dalton palaeosol occurs directly below the unconformity that separates the Mozaan Group from the Nsuze Group in the White Mfolozi Inlier (Fig. 9.9; Nhleko 2003; Crowe et al. 2013; Delvigne et al. 2016). The Denny Dalton palaeosol is developed on amygdaloidal basaltic andesite of the Nsuze Group, which was tectonically tilted $\sim 10°$ and eroded prior to palaeosol formation and then unconformably overlain by the Mozaan Group (Hicks and Hofmann 2012).

In drill core from the Denny Dalton Mine area, Delvigne et al. (2016) indicated that the uppermost part of the altered volcanic rock to about 190 cm below the palaeosurface consists mainly of sericite with minor chlorite and rutile (Sericite-Dominated Zone, SDZ; Fig. 9.10). Immediately below the SDZ is a Chlorite-Dominated Zone (CDZ) with quartz, epidote and rutile. Similar downward zonation from an SDZ into a CDZ is known from other Precambrian palaeosols, developed on mafic volcanic protoliths, where the two distinct zones are inherited from pedogenic processes. Whereas kaolinite and/or smectite are expected during andesite weathering, post-weathering processes likely

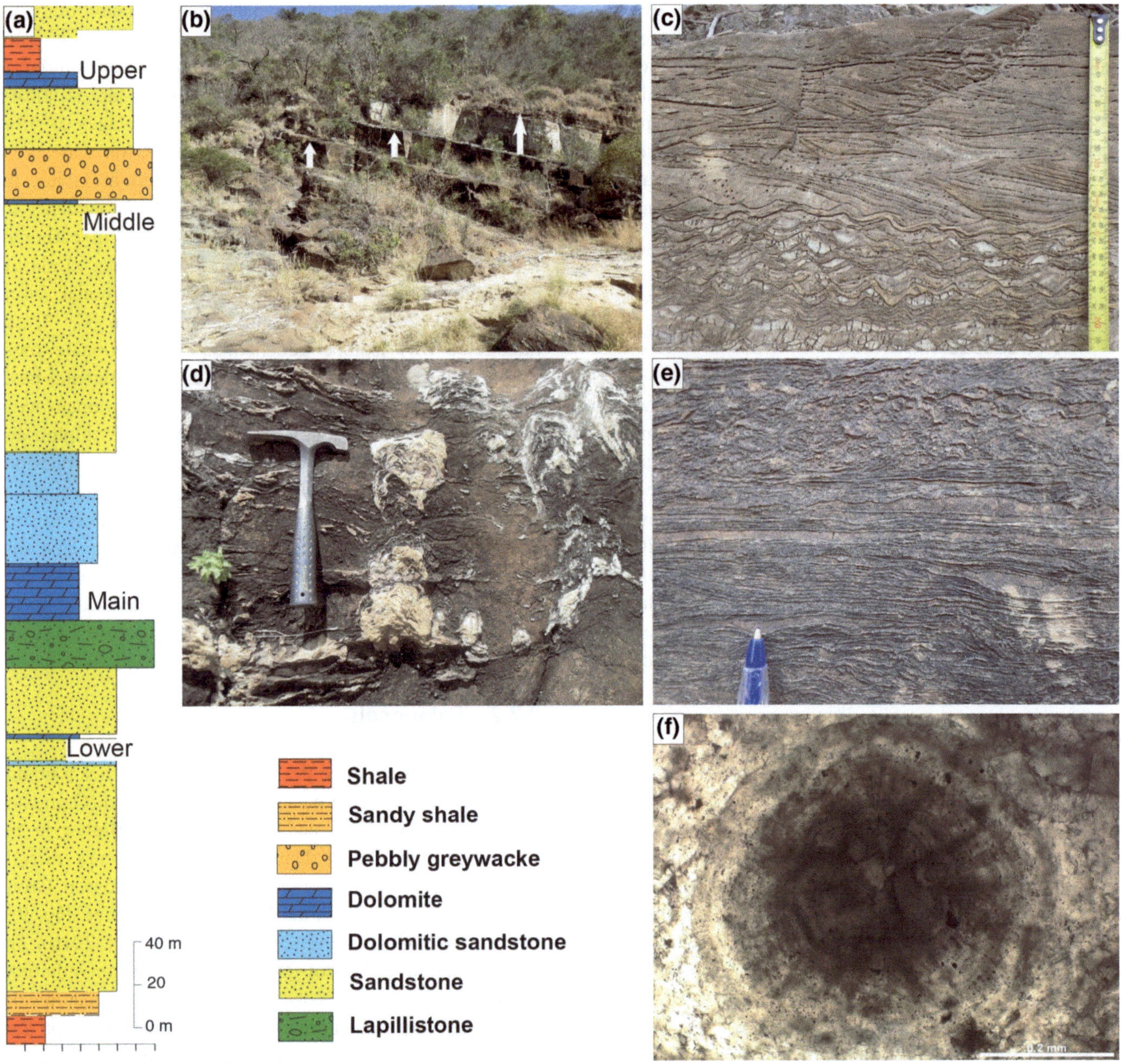

Fig. 9.8 **a** Stratigraphic column of the White Mfolozi Formation, with dolostone units labelled following Siahi et al. (2016); **b** Shallowing-upward sedimentary cycles (white arrows) near the base of the main carbonate section; **c** Wave-ripple cross-lamination and trough cross-bedding of subtidal dolarenite facies; **d** Columnar stromatolites in subtidal channel; **e** Supratidal stratiform stromatolite facies. Note disrupted laminae and intraclasts at top of photo; **f** Photomicrograph of partially recrystallized radial-concentric ooid

explain the association of sericite and chlorite. Sericitization is attributed to metasomatism of the upper part of the soil profile through the percolation of K-rich waters during burial diagenesis or low-temperature metamorphism. Greenalite and/or Fe(II)-rich smectite are considered as pedogenic precursor minerals that transformed into chlorite plus quartz during diagenesis and greenschist-facies metamorphism (Delvigne et al. 2016).

Fractionation of Cr isotopes and fluctuations in U concentrations in the palaeosol horizon was observed by Crowe et al. (2013) and indicate the mobilisation of redox sensitive elements through oxidative weathering, implying appreciable levels of atmospheric oxygen during soil formation at levels of 3×10^{-4} PAL. In contrast, Albut et al. (2018) attributed the Cr isotope fractionation as a product of modern oxidative weathering.

Fig. 9.9 Angular unconformity between the Nsuze and Mozaan groups

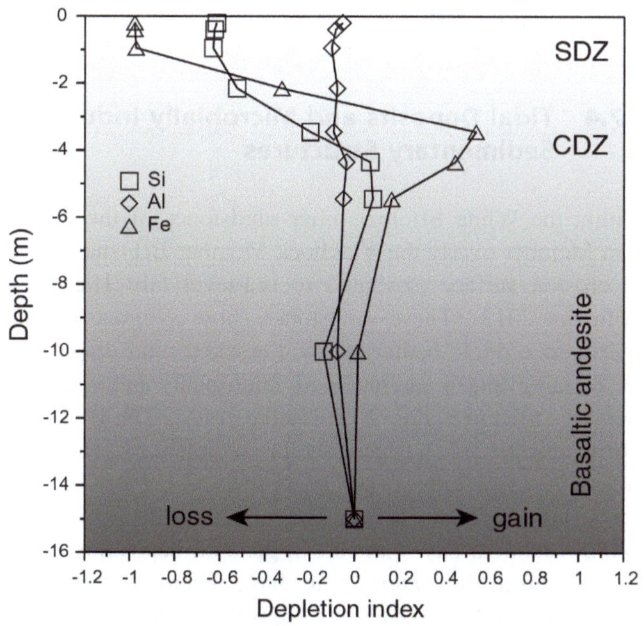

Fig. 9.10 Depletion index (τ_j) of major elements (j = Si, Fe, and Al) relative to Ti in the Denny Dalton palaeosol. Approximate position of parent basaltic andesite and CDZ and SDZ are indicated. Modified after Delvigne et al. (2016)

9.7.3 Banded Iron Formation

Within the Sinqeni Formation (Mozaan Group), a succession of shale and BIF (Vlakhoek Member; Fig. 9.2e) separating Dipka Member quartz arenites from sandstones of the Kwaaiman Member, has been the focus of several studies (von Brunn and Hobday 1976; Alexander et al. 2008; Hicks and Hofmann 2012; Planavsky et al. 2014; Bolhar et al. 2015; Ossa et al. 2016). Within the White Mfolozi Inlier, the shale-BIF succession is represented, from base to top, by a 12 m thick succession of dark grey shale, BIF and a dark to light green ferruginous shale (Hicks and Hofmann 2012). This succession is overlain by intertidal to shallow subtidal sandstone of the Kwaaiman Member. The lower shale is characterized by planar, laminated mudstone which is interbedded with silt-rich layers that locally exhibit ripple-lamination. The BIF consists of mesobands enriched in iron oxide, Mn-carbonate and chert. The contact from shale to BIF indicates an overall deepening of the environment from a marginal marine to a shallow outer shelf environment within a likely epicontinental sea starved of siliciclastic input during transgression (Hicks and Hofmann 2012; Planavsky et al. 2014). The upper ferruginous shale

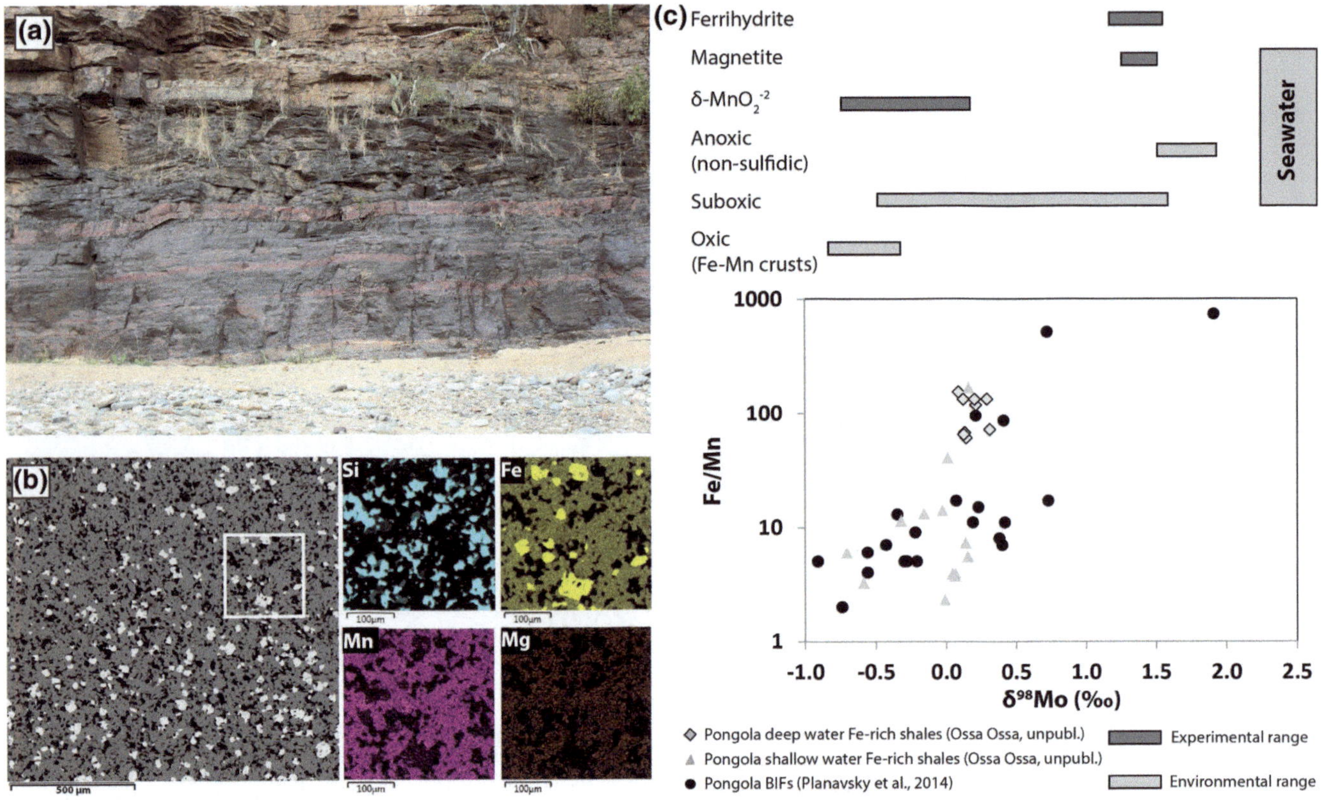

Fig. 9.11 **a** Field photograph of BIF (height of section is ~7 m); **b** petrographic data (backscattered electron image and element maps) of magnetite-bearing Fe–Mn-carbonate in a chert matrix; **c** Mo isotope behaviour during Fe and Mn oxide formation and δ^{98}Mo (‰) versus Fe/Mn in Sinqeni BIF and shale (refer to Planavsky et al. 2014 for details)

shows planar lamination and some silt- and carbonate-rich layers. BIF and shale samples both show a positive correlation between Fe/Mn ratios and Mo isotope values (Fig. 9.11c; Planavsky et al. 2014). Large fractionations of Mo isotopes were related to sorption of Mo onto Mn oxyhydroxides, which would require substantial concentrations of free oxygen in the shallow-marine water, likely derived from oxygenic photosynthesis (Planavsky et al. 2014).

Ossa et al. (2016) noted strong enrichment in manganese concentrations in shales and BIF of the Mozaan Group in the White Mfolozi Inlier, representing the oldest known examples of primary sedimentary Mn deposition. This Mn enrichment was related to oxidation of dissolved Mn(II) by free oxygen in a shallow-marine oxygen oasis. Although marine hydrothermal systems are thought to have been the main sources of Mn and Fe during the Archaean, a contribution from continent-derived solutes cannot be discounted for sediments that formed within an intracontinental setting such as the Pongola basin. Indeed, rare earth element (REE) patterns and Nd isotope ratios of the BIF suggested that continental derivation of Nd made a significant contribution to seawater (Alexander et al. 2008). On the basis of REE systematics, Delvigne et al. (2012) argued for a minor (~10%) freshwater component in the depositional basin precipitating the BIF.

9.7.4 Tidal Deposits and Microbially Induced Sedimentary Structures

Within the White Mfolozi Inlier sandstones of the Kwaaiman Member overly the Vlakhoek Member BIF/shale along an erosion surface attributed to sea-level fall (Hicks and Hofmann 2012). These sandstones show symmetrical and interference ripple bedforms, mud-cracked mud drapes, flaser bedding and a multitude of microbially induced sedimentary structures (Fig. 9.12; Noffke et al. 2008; Hicks and Hofmann 2012). A decrease in sedimentary structures indicative of periodic exposure upsection suggests deposition in an intertidal environment gradually deepening to a shallow subtidal environment. In general, deposition of the Sinqeni Formation appears to have been controlled by (eustatic?) sea-level changes. Sandstones illustrate deposition in a tide-influenced, shallow-marine environment under a micro- to mesotidal range (von Brunn and Hobday 1976; Beukes and Cairncross 1991; Noffke et al. 2008; Hicks and

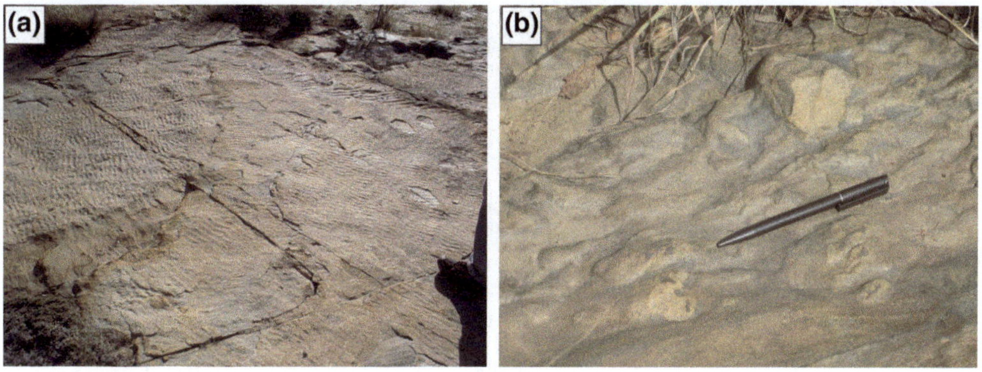

Fig. 9.12 **a** Pavement of wave-ripple bedforms. Multidirectional ripple marks were described by Noffke et al. (2008) from this outcrop; **b** Upper bedding plane of sandstone with wrinkle structure pointing to microbial mat cover of an erosional remnant

Hofmann 2012). The Ntombe Formation overlies the Sinqeni Formation and is made up predominantly of ferruginous and non-ferruginous shales with minor siltstones and sandstones (Gold 2006). Sedimentary structures and microbial mat fabrics record a variation between shallow-marine, storm-dominated shelf and tide-influenced environments (Noffke et al. 2003).

9.7.5 Glacial Diamictites

There are four diamictite units (including the Klipwal Member—which is the most prominent, Fig. 9.2e) found within the Delfkom Formation which display a highly varied suite of clasts, some of which are striated and faceted (von Brunn and Gold 1993). Young et al. (1998) presented chemical evidence for a glacial origin of the diamictites, further supported by the presence of dropstones, which they argued represent the world's oldest glaciation. Stabilisation of the Kaapvaal Craton at ~3.1 Ga may have contributed to the reduction of atmospheric CO_2 and could thereby have provided suitable conditions for the development of continental glaciers (Young et al. 1998).

9.8 Structure and Metamorphism

Two dominant structural domains are recognised across present-day outcrops of the Pongola Supergroup. The area encompassing the Northern and Central domains (including the Hartland sub-basin) is typified by both north- and northwest-orientated structural trends—namely thrusts, folds and strike-slip shear zones (Gold and von Veh 1995; Gold 2006). In contrast, the much smaller Southern Domain (Nkandla sub-basin) is characterized by east-west-trending folds and associated zones of thrust faults (Groenewald 1984; Matthews 1990; Thomas et al. 1994; Dixon 2004).

The northwest-striking Mahamba shear zone (Fig. 9.1; Hatfield 1990; Matthews 1990) developed during extension and probably defined the eastern boundary of a half-graben during the extrusion of Nsuze Group volcanics (Matthews 1990) as they significantly decrease in thickness eastward. Abrupt thickening and coarsening of the basal quartzites east of the Mahamba shear zone indicate that the fault was active during Mozaan times. The ~40 km sinistral strike separation of the basement and Nsuze-Mozaan group contacts (Fig. 9.1; Hunter and Wilson 1988) can be attributed to the downthrow of the northeastern side. Dykes and sills intruded syntectonically along this discontinuity (Hunter and Wilson 1988).

9.8.1 Amsterdam Area

Matthews (1990) described a well-developed, 20 km-wide graben structure in the Amsterdam area, some 80 km northwest of the Hartland sub-basin. The graben hosts some 4.7 km of both Nsuze Group and Mozaan Group strata and links up southwards with the Mahamba-Altona structure, which extends southeastwards into the Hartland sub-basin (Humphrey and Krige 1931; Matthews 1990; Gold and von Veh 1995).

9.8.2 Northern Domain

The large-scale structural morphology of the Northern Domain is dominated by two major peri-synclinal folds, the Piensrand and Tobolsk structural basins, separated by an anticlinal saddle, the Bethu anticline, which are the result of interference folding of northwest-trending F_1 and east-northeast-trending F_2 folds (Fig. 9.13; Watchorn 1978; Matthews 1990; Gold 1993; Gold and von Veh 1995). The intensity of the latter phase of folding is considered to

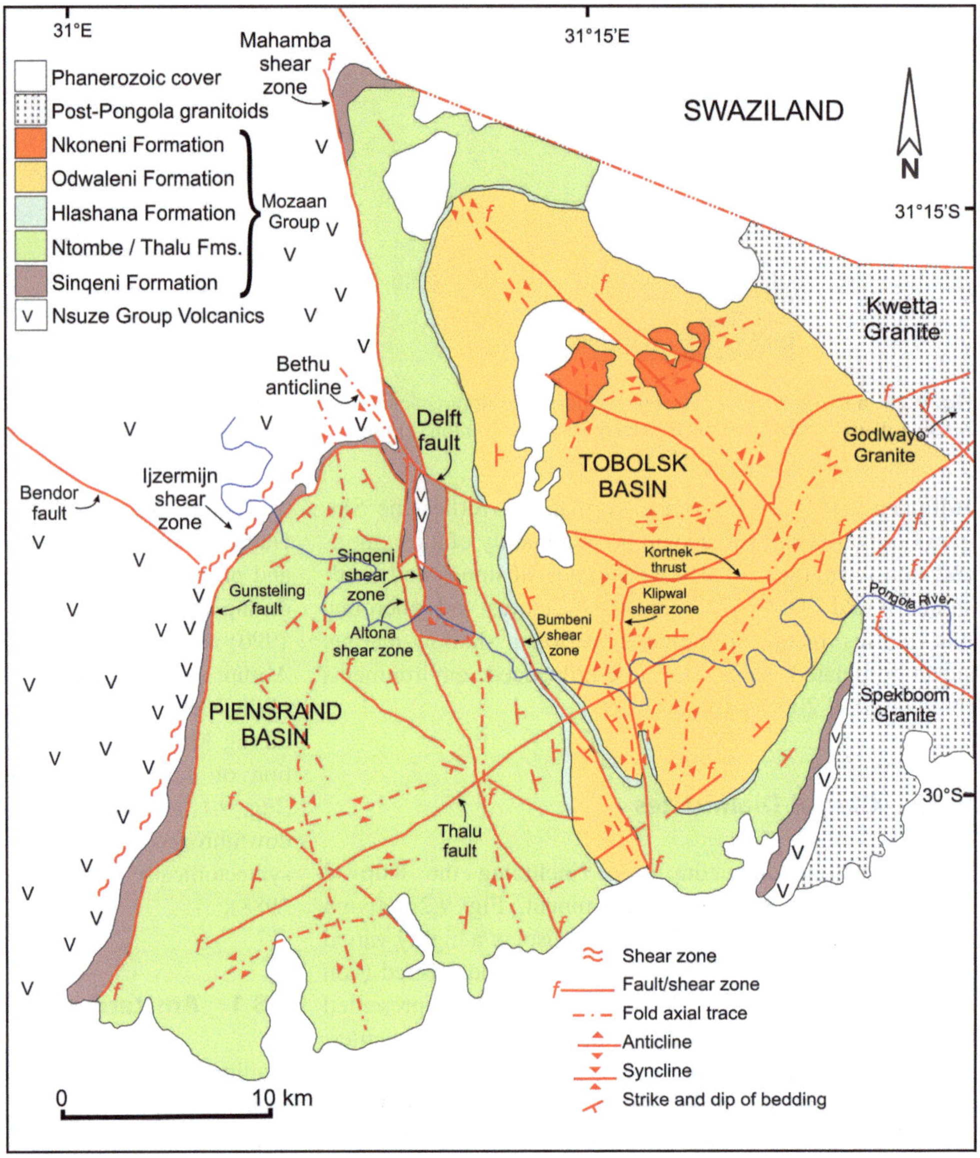

Fig. 9.13 Geological map of the Hartland sub-basin in the Northern Domain showing major structures. The basin is dominated by two northwesterly aligned, doubly-plunging synforms (Tobolsk and Piensrand structural basins), which are separated by the sheared Bethu anticline. Modified after Gold and von Veh (1995)

decrease away from the intrusive contact of the extensive Spekboom granite (Matthews 1985). The northwest-striking Bendor fault (Figs. 9.1 and 9.13; Matthews and Scharrer 1968; Armstrong 1980) appears to be a pre-Mozaan fault because it offsets the basal Nsuze Group contact in a sinistral sense by about 5 km but does not affect the basal Mozaan Group contact.

Four separate phases of deformation have been identified in the Hartland sub-basin which deform the Pongola Supergroup, as described below.

9.8.2.1 Early North–Northwest-Directed Thrusting (D_1)

An early phase of top-to-north–northwest-directed thrusting and reverse-faulting-generated shear zones have been further deformed by major D_2 folding and faulting. The Ijzermijn shear zone, exposed along the contact of the Nsuze and Mozaan groups in the Pongola Valley (Fig. 9.13), is one such major bedding-parallel thrust that can be followed throughout the Northern Domain. Consequently, the thick Sinqeni Formation was thrust northwards over the upper Nsuze Group units.

Within the core of the Bethu anticline is the Altona shear zone, which appears to be the southeasterly extension of the Mahamba shear zone mapped by Matthews (1990) and Hatfield (1990) to the east of Piet Retief (Mkhondo). The Mahamba shear zone was interpreted by Hatfield (1990) as a transpressional structure. In the Swaziland area, sinistral transpression seems to be partitioned into an easterly dipping thrust and an adjacent sinistral shear zone. Hatfield (1990) postulated that the two structures were generated during the same D_1 deformational event. The ca. 15-km long Mahamba shear zone is associated with about 40 km of sinistral displacement, accompanying large-scale drag and attenuation of the Pongola Supergroup strata, particularly the Nsuze Group (Hunter and Wilson 1988).

The contact between the Nsuze Group and the Sinqeni Formation in the Central Domain is an angular unconformity of some 10–15° (see Sect. 9.7.2; Matthews 1967; Fig. 9.9). In the Northern Domain, the contact between the upper Nsuze Group (the Ozwana Subgroup) and the lowermost Mozaan Group (Sinqeni Formation) is sheared and displays deformation within a wide (~100 m) zone of ductile-brittle deformation within the upper Nsuze Group's volcano-sedimentary package (Gold 1993).

The Klipwal shear zone (in the Tobolsk structural basin; Figs. 9.13 and 9.14), is a north–south-trending D_1 structure that cuts progressively upwards through the overturned forelimb of a shallow fault-propagation fold with a north-plunging fold axis. A sinistral strike-slip separation of ~3.3 km with a reverse oblique component is indicated by the displacement of the Klipwal Member diamictite (Gold and von Veh 1995). The irregular trace of the shear zone is attributed to oblique lateral ramping. The Kortnek thrust forms the frontal ramp of the same structure (Fig. 9.13).

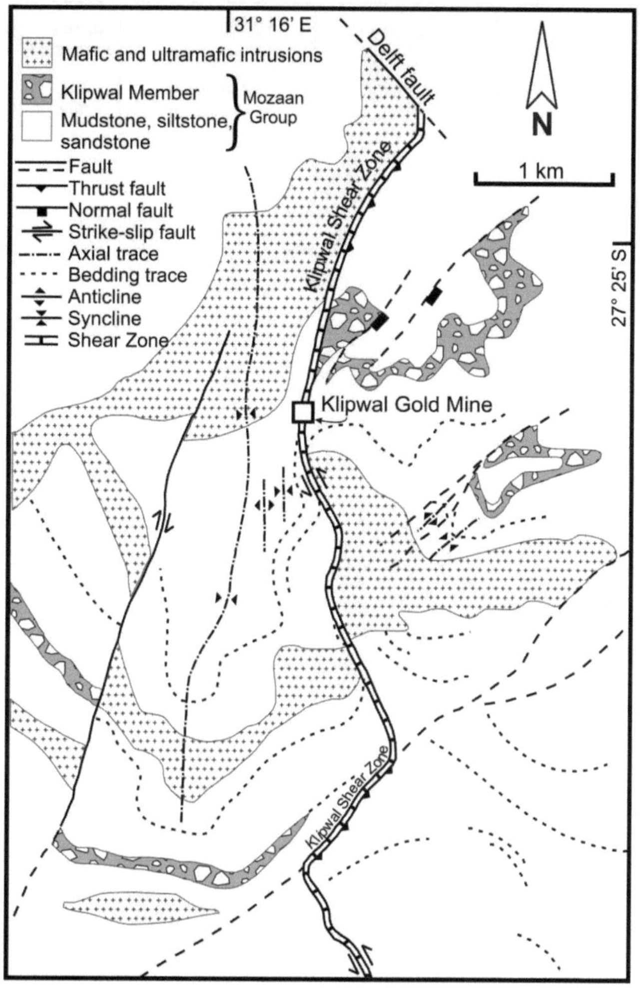

Fig. 9.14 The Klipwal shear zone. Modified after Chinnasamy et al. 2015; Gold 1993; Gold and von Veh 1995). The irregular trace of this sinistral transpressional shear zone is attributed to oblique lateral ramping (Gold 1993; Gold and von Veh 1995; Chinnasamy et al. 2015)

9.8.2.2 Early Extensional Tectonics (D_2)

The Northern Domain was subjected to southeast-directed extensional tectonic activity associated with the extrusion of upper Mozaan Group volcanic rocks and the emplacement of the ~2.8–2.9 Ga Thole Complex as well as the Amsterdam Formation and other coeval mafic dykes and sills including the Hlagothi Complex to the south (Hunter 1970; Hammerbeck 1982; Hunter and Wilson 1988; Gumsley et al. 2013, 2015). The age relationship of this event with respect to the north–northwest-directed D_1 thrusting is uncertain (Gold and von Veh 1995).

The Gunsteling fault (Figs. 9.13 and 9.15; Humphrey and Krige 1931; Gold and von Veh 1995) is a D_2 normal fault running along the upper contact of Sinqeni Formation quartzites in the Northern Domain. During D_1 compression, the Sinqeni Formation acted as a thick, rigid unit that was thrust northwards, whereas during D_2 extension it acted as a rigid unit which resulted in displacement along the hanging wall of the massive sandstone. The Gunsteling fault is folded around the Bethu anticline (Fig. 9.15). A large-scale roll-over anticline with an approximately north–northeast-trending axis in the hanging wall suggests a listric geometry at depth.

9.8.2.3 Northeast-Southwest Compression (D_3)

Major upright folds with northwesterly axial traces deformed the early faults and refolded the east–northeast folds to produce a dome-and-basin interference pattern. The latter folds open-up southwards, and the axial traces diverge with the more westerly folds, curving into a north–northeast trend that is sub-parallel with the western margin of the Piensrand structural basin (Fig. 9.13).

The Mahamba-Altona shear zone system (referred to as the Keerom fault in Humphrey and Krige (1931), Altona shear zone in Carter (1964), Mahamba fault belt in Matthews (1990) and Hatfield (1990); Figs. 9.13 and 9.15), lies in the hinge zone of the Bethu anticline. The tectonic history of the Mahamba-Altona shear zone is complex and long-lived and may have affected deposition of the Pongola Supergroup as stratigraphic units are thinner northeast of the structure. The Nhlangano gneiss dome was emplaced into this region during northwest–southeast extension at 2.72 Ga (Hofmann et al. 2015), probably during the D_3 event.

9.8.2.4 Late Brittle Extension (D_4)

The area is transected by clusters of northerly striking faults and fractures. These discontinuities are often associated with

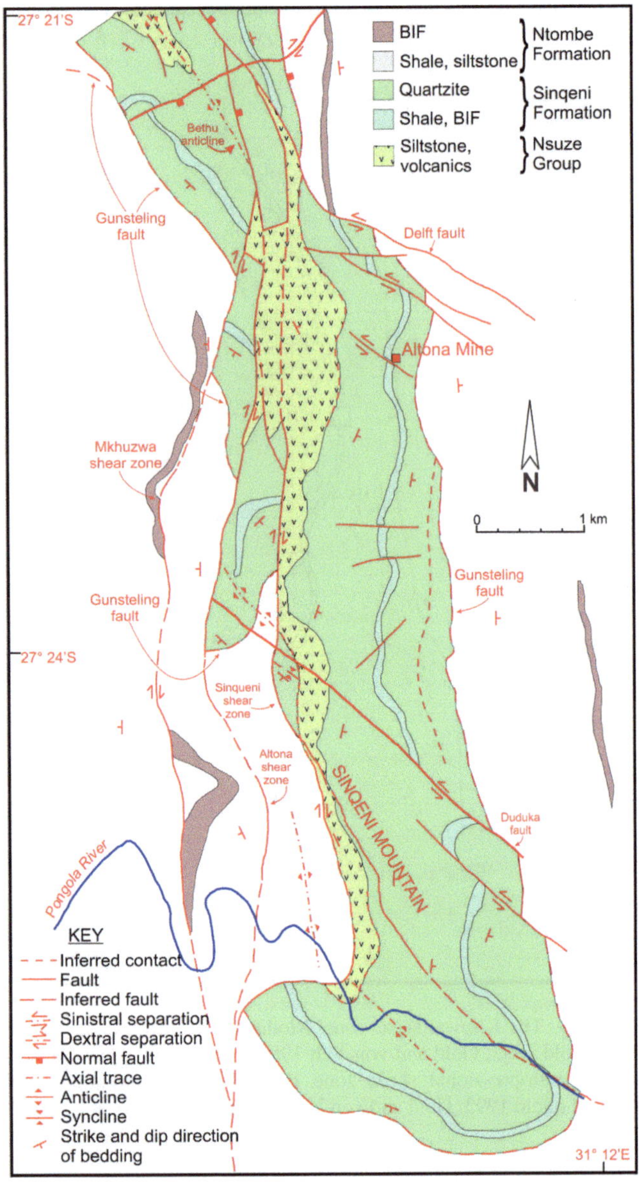

Fig. 9.15 Geological map of the area around Sinqeni Mountain showing the Altona-Sinqeni-Mkhuzwa shear zone system, the Bethu anticline and the eastern extension of the Gunsteling fault. Modified after Gold and von Veh (1995)

a fracture cleavage and may be intruded by dolerite dykes or filled with vein quartz. Faults with a northwesterly alignment within the hinge zone and flanks of the Bethu anticline repeatedly offset strata in a dextral sense (Fig. 9.15). Normal faults with a north–south strike are common near the eastern margin of the basin (Linström 1987b). Among these, the Koppie Alleen fault has an eastward normal displacement of ~1000 m with a dextral component (Fig. 9.1; Humphrey and Krige 1931; Gold and von Veh 1995). The relative and absolute ages of these fault sets are uncertain; however, they displace the late Archaean Spekboom, Kwetta and Godlwayo granitoids, and at least some of these postdate deposition of the Phanerozoic Karoo Supergroup. These cross-cutting relationships and their overall passive margin-parallel trends are consistent with extension associated with the breakup of Gondwana.

9.8.3 Swaziland

Although northwest-trending structures are most commonly developed in Swaziland (Fig. 9.16), the most intense deformation is displayed by northerly striking dextral shear zones. The most prominent is the north–northeast-striking Sibowe shear zone, which truncates the Mozaan Group strata in the Kubuta area (Hunter 1953, 1991; Hepworth 1973; Geological Survey Swaziland 1980; Hunter and Gold 1993). It is a broad mylonite zone with lineations and fold axes typically plunging towards the southeast (Hepworth 1973). Large amplitude folds with axes also plunging towards the southeast deform the Mozaan Group to the east. The Mozaan Group sandstone east of the Sibowe shear zone is preserved within a broad synclinorium, the northeastern limb of which is truncated by the northwest-trending Siyalo shear zone. The Siyalo shear zone contains sinistral movement indicators and sub-horizontal lineations. The synclinorium has a sub-vertical northeastern limb, which Hunter (1963) attributed to the forceful emplacement of a granite plug on the northeastern flank.

Several phases of deformation have been recognised in the Mahlangatsha area (Roering 1968). A large northeast-trending synclinal structure resulted from two coaxial phases of folding. Later northeast-trending cross-folding produced several subsidiary basins (Roering 1968). A final phase of deformation involved westward overthrusting. The contact between the Nsuze and Mozaan groups appears to be tectonic because Mozaan Group beds are truncated at the contact (Roering 1968; Hunter 1991).

The regional strike of the Pongola Supergroup sequence swings southeast in the southernmost part of the Mahlangatsha area where it is truncated by the Mahamba shear zone. Whereas a substantial thickness of Nsuze Group strata is preserved to the west of the Mahlangatsha plateau, it is not found on the eastern side of the regional syncline, where Mozaan Group sandstone is intruded by post-Pongola granite.

9.8.4 Central Domain

Structural studies of inliers of the Pongola Supergroup in the central area (including White Mfolozi) are limited. Matthews (1990) reported that in this area there is a series of rotated/tilted fault blocks which display a general northwest–southeast structural trend. Unmapped northerly and

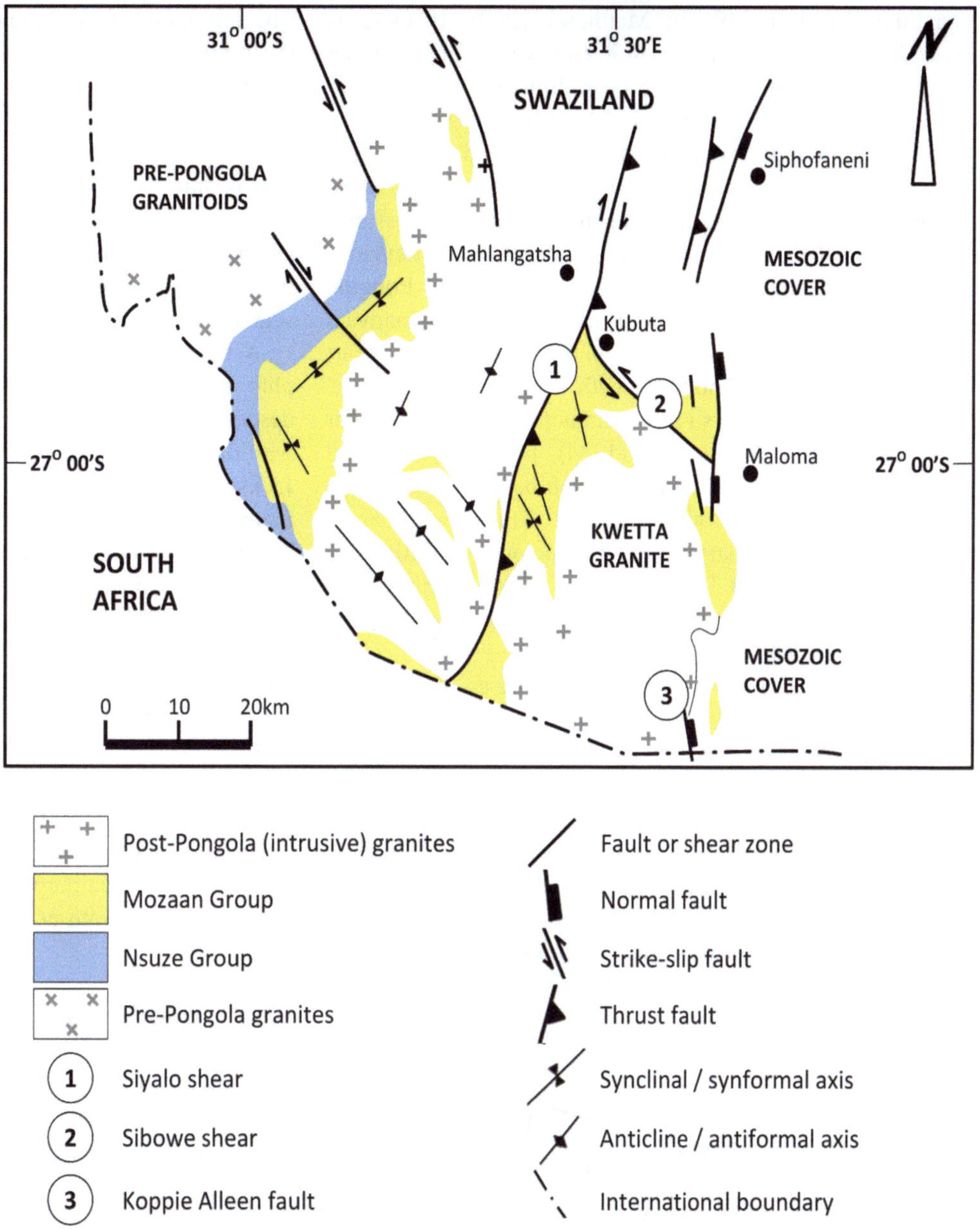

Fig. 9.16 Map showing the distribution of the Pongola Supergroup in Swaziland and the main structural features (after Hunter and Gold 1993; Mabuza and von Veh 1996)

northwesterly- orientated shear zones are present in the White Mflozi Inlier. The Mpongoza Inlier is elongated in a northwest direction and displays an anomalous westerly dip direction (Preston 1987). Within the Nsuze Group at the White Mfolozi Inlier, Hicks (2009) reported two phases of deformation which pre-date the Mozaan Group—both of which involved extensional faulting. The third phase of deformation described by Hicks (2009) resulted in the development of bedding-parallel thrusting, probably equated with the D_2 compression observed in the Northern Domain. Subsequent events resulted in a series of extensional structures.

9.8.5 Southern Domain

The style and extent of deformation within the Southern Domain differs significantly from that of the Northern and Central domains. Groenewald (1984) and Matthews (1990) described several phases of extension and compression, all of which seem aligned approximately north–south. This deformation resulted in the development of two east-trending synclinal structures which incorporate Nsuze Group strata. A well-developed tectonic mélange represents a north-verging thrust that separates the anomalous stratigraphy of the Southern Syncline from that of the Central

Syncline (Linström 1987b; Matthews 1990). If there is indeed northward thrusting, then it may even be possible that older Nsuze Group rocks have been thrust northwards onto younger Mozaan Group rocks. Matthews (1990) claimed that 'the tectonic development of the southern domain of the Pongola Supergroup occurred long before the evolution of the mobile belt to the south' and that 'several lines of evidence suggest that the initial tectono-depositional development and subsequent tectonic deformation of the Pongola Supergroup along the southern margin of the Kaapvaal Craton took place when this region was situated along part of a transform plate boundary with a major ocean basin.' Unfortunately, Matthews (1990) did not postulate an age for this deformation. Groenewald (1984), however, attributed the deformation of the Nkandla sub-basin to the close proximity of the ~1000 Ma Natal Thrust-Fault Front (Fig. 9.1). This is supported by isotopic resetting of Archaean minerals at 950 Ma up to 25 km north of the Kaapvaal Craton margin along the Natal Thrust Front (Thomas et al. 1994).

Groenewald (1984) described three phases of deformation in the Southern Domain. The first (D_1) is defined by a crenulation cleavage. The second (D_2) is the intense deformation of the basal Nsuze Group rocks adjacent to the Natal Thrust Front and from where the metamorphic grade diminishes northwards. The geometry of associated folds is consistent with northwards compression that may be synchronous with the ~1000 Ma Namaqua-Natal orogen. The third phase (D_3) developed brittle, conjugate strike-slip fault sets with dextral northwest–southeast and sinistral faults. Late (D_4) brittle extensional faults are also observed to displace the Karoo Supergroup.

9.8.6 Metamorphism

Greenschist-facies metamorphism was reported from Pongola Supergroup rocks in the Northern Domain by Armstrong (1980), Gold (1993), Matthews (1990) and Cole and Beukes (1995) and from the Central Domain by Dix (1984), Thomas et al. (1997) and Ossa et al. (2016). This metamorphism is seen in the recrystallization of quartz-rich rocks and in the formation of chlorite and chloritoid in shales. Low-grade metamorphism was also reported in the Southern Domain, most likely resulting from tectonic loading of the crust by northward-thrust Tugela nappe sheets related to the Namaqua-Natal orogeny (Groenewald 1984; Matthews 1990; Jacobs and Thomas 1994; Dixon 2004).

Elworthy et al. (2000) obtained a Rb-Sr biotite age of 2614 ± 74 Ma for granitoids exposed more than 50 km north of the Natal Thrust Front, including those of the White Mfolozi Inlier, whereas granitoids located within 25 km of the thrust front gave Rb-Sr biotite ages of ~933–990 Ma. The authors regarded these older ages as cooling ages, below the biotite closure temperature of 300 °C, following a late Archaean regional metamorphic episode. The younger Rb-Sr biotite ages were ascribed to thermal overprinting during the Namaqua-Natal orogeny. In addition to regional metamorphism, the Pongola Supergroup has also been locally contact-metamorphosed by post-Pongola granites, but usually only to a relatively low grade. Adjacent to the Spekboom granite, however, there are well-developed andalusite schists of the Mkuzane Formation (Fig. 9.17).

A much higher amphibolite- to granulite-facies metamorphic overprint of Pongola Supergroup strata has been described from several localities in southern Swaziland and was dated at ~2.72 Ga (Condie et al 1996; Wilson and Jackson 1988; Taylor et al. 2010; Horváth et al. 2014; Hofmann et al. 2015; Dlamini et al. 2017). For example, the Nhlangano dome in southwest Swaziland is a migmatitic core complex, wrapped by high-grade Pongola Supergroup strata exhumed during crustal extension affecting the eastern part of the Kaapvaal Craton. This resulted in the formation of north–south-orientated extensional normal faults and the intrusion of broadly syn- to post-tectonic (post-Pongola) granite plutons at ~2.7 Ga (Meyer et al. 1994; see sect. 9.8.7), possibly related to Neoarchaean failed rifting of the Grunehogna Craton (Hofmann et al. 2015). Phase equilibria modelling of Pongola Supergroup metapelites and ironstones from the Nhlangano dome indicate metamorphic peak conditions around 6–8 kbar and 850 °C (Horváth et al. 2014).

9.8.7 Geodynamic Synthesis

Matthews (1990) proposed that the Pongola Supergroup accumulated in a rift environment. The Nsuze Group was seen as an initial phase of subsidence and filling of a rift as a

Fig. 9.17 Andalusite schist of the Mkuzane Formation (Nsuze Group), exposed adjacent to the Spekboom granite in the Ithala Game Reserve. Note the steeply dipping crenulation cleavage

result of uplift and crustal extension caused by lithospheric heating. This assumption was based on the recognition of characteristics similar to those of more recent rift deposits including large variations in lateral thickness of sedimentary facies, syndepositional basement faulting, irregular basement topography, immature basement derived deposits, faulted (at least locally) margins and bimodal volcanic rocks. The accumulation of ~4800 m of sedimentary rocks with only a minor component of volcanic rocks during Mozaan Group deposition was attributed by Burke et al. (1985) and Matthews (1990) to a thermal subsidence phase. A subsequent resurgence of crustal extension and thermal activity in the northern parts of the basin is evidenced by several volcanic events. A late Mozaan Group episode of volcanism is represented by the Tobolsk and Gabela Formations. Following these events, the basin was subjected to several phases of deformation.

The first (D_1) involved compression from the southeast (e.g. the Kortnek and Klipwal shear zones). The second phase of deformation (D_2) involved the development of extensional faults with displacement towards the southeast (e.g. the Gunsteling fault) and to the northeast (e.g. reactivation along the Klipwal shear zone). The third (D_3), and most intensely developed event, resulted in the formation of northwest-orientated folds, north-trending dextral and northwest-striking sinistral shear zones.

The tectonic stresses that produced the dome-and-basin interference pattern in the main Pongola Basin were considered by Matthews (1990) to have a largely local origin related to the forceful emplacement of the adjacent post-Pongola granitoid plutons, i.e. the Spekboom, Godlwayo and Kwetta granites. Gold (1993) suggested that field evidence offers limited support for this mechanism as the contacts are generally discordant, the contact aureoles are narrow and the deformational imprint is regionally persistent. These post-Pongola granites (Spekboom, Kwetta and Godlwayo) appear to be relatively unaffected by phases D_1–D_3; however small-scale conjugate strike-slip structures are observed in the Spekboom granite (this may, however, not be significant as these granites may resist deformation more than surrounding supracrustal rocks of the Pongola Supergroup). This indicates that reactivation of these basement dislocations was long-lived—initiated prior to deposition of the Pongola Supergroup and still active following the emplacement of the post-Pongola granites.

Matthews (1990) envisaged the Mahamba rift fault as extending southwards along a line coinciding roughly with the present-day position of the Mahamba-Altona shear zones. Reactivation of this basement fault during the northwest–southeast compression was probably responsible for the nucleation of the Bethu anticline. The fault then propagated upwards into the hinge zone, forming the Mahamba-Altona shear zone system. The latest event (D_4) was the development of northerly aligned normal faults attributed to extension related to the breakup of Gondwana.

9.9 Economic Geology

The Pongola Supergroup contains small gold deposits. Non-economic deposits of iron, uranium and manganese are also found sporadically throughout the Pongola Basin.

9.9.1 Gold

Gold is the only mineral of economic significance that has been mined from the Pongola Supergroup (Gold 2006). Gold occurrences are widespread throughout the Nsuze Group in the Southern Domain and from the Mozaan Group in the Northern Domain (Fig. 9.1). Three different styles of gold mineralization are found within the Pongola Supergroup: placer gold associated with conglomerates, shear zone hosted gold associated with compressional deformation structures and non-stratiform iron formation deposits.

9.9.1.1 Placer Gold

Placer gold occurrences include the Dickson Reef, Speedwell North, South Reefs, the Central Nsuze Deposit and the Mozaan Contact Reef at Denny Dalton (Hammerbeck 1976; Saager et al. 1986; Ward and Wilson 1998). The largest of these deposits was the Denny Dalton Mine of the White Mfolozi Inlier. Mining was commissioned in 1893, ultimately producing ~100 kg from the Mozaan Contact Reef (MCR), a lenticular, poorly sorted, clast-supported polymictic conglomerate with detrital pyrite (Fig. 9.18a), yielding grades of 4 g/t (Saager et al. 1986; Hicks 2009; Hofmann et al. 2009; Hicks and Hofmann 2012). MCR gold occurs as rare inclusions inside rounded massive pyrite grains (Hicks 2009) and as particles associated with gersdorffite aggregates within the matrix (Hicks and Hofmann 2012). According to Hicks and Hofmann (2012), the former suggests a placer model of gold deposition where pyrites were derived from erosion of shales or syn-sedimentary pyrites, whereas the latter indicates remobilization and precipitation of gold within the conglomerate.

9.9.1.2 Lode Gold

The largest extraction of gold from the Pongola Supergroup has come from lode deposits within the Klipwal-Kortnek thrust system (Figs. 9.13, 9.14 and 9.18b). Prior to 2003, 5.7 tonnes of gold were produced from the Klipwal Gold Mine (Chinnasamy et al. 2015). Records are not accessible post-2003, but 60–80 kg of gold was produced per month during the early 2000s. There are several mineralized reefs within the system, all associated with D_1 deformation.

Fig. 9.18 **a** Detrital pyrite in a conglomerate from Denny Dalton Mine; **b** Native gold (black arrow) within a carbonate-quartz vein from the Klipwal Gold Mine; **c** Pyrite veins formed in iron formation of the Vlakhoek Member from the Altona Mine

9.9.1.3 Iron Formation-Hosted Gold Deposits

Gold was mined from Altona from the Vlakhoek Member of the Sinqeni Formation, where gold is associated with pyrite that is disseminated within the iron formation (Fig. 9.18c). The Altona mine was abandoned due to structural complexities and the fractured nature of the iron formation in the hanging wall (Gold 1993).

9.9.2 Iron

Ferruginous mudstones and BIFs are widespread throughout the Mozaan Group, and many within the Northern Domain have been explored for iron ore. According to Astrup et al. (1998), Fe-rich lithologies are located in three principal areas (Fig. 9.1): (1) north of the Pivaan River on the Ijzermijn Farm within the Vlakhoek Member of the Sinqeni Formation (35–40% Fe content), (2) the Scotts Hill BIF Member (28–35% Fe content) and (3) the Redcliff Formation north–northwest of Piet Retief (31.8% Fe content).

9.9.3 Uranium

Placer uranium mineralization of subeconomic grade has been described mainly from the Mozaan Group with the highest grade reported from the Mozaan Contact Reef at the Denny Dalton Mine. The principal U-bearing phase is found in leucoxene and brannerite, with brannerite being non-detrital and only existing as overgrowths on the edges of leucoxene grains (Saager et al. 1986; Hicks and Hofmann 2012).

9.9.4 Manganese

Shales and BIFs of the Mozaan Group are locally enriched in Mn with primary sedimentary MnO contents as high as 8.5 wt% (Horváth et al. 2014; Ossa et al. 2016). High primary Mn contents could make the Mozaan Group a potential Mn ore producer where it has been affected by prolonged secondary enrichment processes. Limited exploration indicates high-grade ores up to 47% Mn content in Mozaan Group shales within the Vryheid-Piet Retief area and at Bellevue, northern KwaZulu-Natal (De Villiers 1960).

9.10 Conclusions

1. The Pongola Supergroup is a remarkably pristine Mesoarchaean sequence of volcanic and sedimentary rocks that provides some of the earliest evidence that the Kaapvaal Craton had stabilised.
2. Volcanic rocks of the Pongola Supergroup bear diverse chemistries ranging from basalt to basalt-andesite through intermediate to rhyolite compositions, and

volcanic textures indicate shallow-water or subaerial extrusion, distinguishing them from then-contemporary greenstone belt volcanic rocks which were dominated by submarine basalts and komatiites.
3. Carbonate and siliciclastic units frequently bear original sedimentary structures indicating deposition in a tide-dominated shallow-marine environment. Stromatolites, BIF and palaeosol chemistries provide important constraints on oxygen levels and the presence of oxygen oases in the Mesoarchaean.
4. Four phases of deformation have been identified within the Pongola Supergroup's Northern Domain, including early north–northwest-directed thrusting (D_1), southeast-directed extensional tectonic activity (D_2) associated with Mozaan Group volcanism, northeast–southwest compression (D_3) at 2.72 Ga, and northerly aligned normal faults attributed to extension related to the break-up of Gondwana (D_4).

Acknowledgements Research on the Pongola Supergroup was funded by the National Research Foundation (NRF) of South Africa and the DST-NRF Centre of Excellence for Integrated Mineral and Energy Resource Analysis (CIMERA). CL thanks M. de Kock for input and feedback. AH thanks Acclaim Exploration NL, AngloGold-Ashanti, J. Hancox, and N. Hicks for access to drill core samples. Most of all we thank the reviewers—M. Klausen, N. Hicks, and J. Reinhardt—for their many helpful comments and improvements.

References

Albut G, Babechuk MG, Kleinhanns IC et al (2018) Modern rather than Mesoarchaean oxidative weathering responsible for the heavy stable Cr isotopic signatures of the 2.95 Ga old Ijzermijn iron formation (South Africa). Geochim Cosmochim Acta 228:157–189

Alexander BW, Bau M, Andersson P, Dulski P (2008) Continentally-derived solutes in shallow Archean seawater: rare earth element and Nd isotope evidence in iron formation from the 2.9 Ga Pongola Supergroup. S Afr Geochim Cosmochim Acta 72:378–394

Anhaeusser CR (2006) Ultramafic and mafic intrusions of the Kaapvaal Craton. In: Johnson MR, Anhaeusser CR, Thomas RJ (eds) The geology of South Africa. Geological Society of South Africa/ Council for Geoscience, Johannesburg/Pretoria, pp 95–134

Armstrong NV (1980) The geology and geochemistry of the Nsuze Group in northern Natal and the southeastern Transvaal. Unpublished PhD thesis, University of Natal (Pietermaritzburg)

Armstrong NV, Hunter DR, Wilson AH (1982) Stratigraphy and petrology of the Archaean Nsuze Group, northern Natal and southeastern Transvaal. Precambrian Res 19:75–107

Armstrong NV, Wilson AH, Hunter DR (1986) The Nsuze Group, Pongola sequence, South Africa: geochemical evidence for Archaean volcanism in a continental setting. Precambrian Res 34:175–203

Armstrong RA, Compston W, Retief EA et al (1991) Zircon ion microprobe studies bearing on the age and evolution of the Witwatersrand triad. Precambrian Res 53:243–266

Astrup J, Hammerbeck ECI, van den Berg H (1998) Iron. In: Wilson MGC, Anhaeusser CR (eds) The mineral resources of South Africa, Handbook 16, 6th edn. Council for Geoscience, Pretoria, pp 402–416

Beukes NJ, Cairncross B (1991) A lithostratigraphic-sedimentological reference profile for the late Archaean Mozaan Group, Pongola sequence: application to sequence stratigraphy and correlation with the Witwatersrand Supergroup. S. African J. Geology 94:44–69

Beukes NJ, Lowe DR (1989) Environmental control on diverse stromatolite morphologies in the 3000 Myr Pongola Supergroup, South Africa. Sedimentology 36:383–397

Bolhar R, Hofmann A, Siahi M et al (2015) A trace element and Pb isotopic investigation into the provenance and deposition of stromatolitic carbonates, ironstones and associated shales of the ~3.0 Ga Pongola Supergroup, Kaapvaal Craton. Geochimica et Cosmochimica Acta 158:57–78

Burger AJ, Coertze FJ (1973) Radiometric age measurements on rocks from southern Africa to the end of 1971. Bull Geological Surv S Afr 58:1–46

Burke K, Kidd WSF, Kusky TM (1985) The Pongola structure of southeastern Africa: the world's oldest preserved rift? J Geodyn 2:35–49

Button A, Pretorius DA, Jansen H et al (1981) The cratonic environment. In: Hunter DR (ed) Precambrian of the Southern Hemisphere, Developments in Precambrian geology 2. Elsevier, Amsterdam, pp 501–639

Carter NG (1964) The geology of an area between the Pongola and Mozaan rivers, SE Transvaal. Unpublished MSc thesis, University of Natal (Durban)

Chinnasamy SS, Uken R, Reinhardt J et al (2015) Pressure, temperature, and timing of mineralization of the sedimentary rock-hosted orogenic gold deposit at Klipwal, southeastern Kaapvaal Craton, South Africa. Miner Deposita 50:739–766

Cole EG (1994) Lithostratigraphy and depositional environment of the Archaean Nsuze Group, Pongola Supergroup. Unpublished MSc thesis, Rand Afrikaans University, Johannesburg

Cole EG, Beukes NJ (1995) Lithostratigraphy of the Archaean Nsuze Group of the Pongola Supergroup in South Africa and Swaziland. In: Barton JM Jr, Copperthwaite YE (eds) Centennial Geocongress 1995 extended abstracts, vol II. Geological Society of South Africa, Johannesburg, pp 817–819

Condie KC, Kröner A, Milisenda CC (1996) Geochemistry and geochronology of the Mkhondo suite, Swaziland: evidence for passive-margin deposition and granulite facies metamorphism in the late Archean of southern Africa. J Afr Earth Sci 21:483–506

Crowe SA, Døssing LN, Beukes NJ et al (2013) Atmospheric oxygenation three billion years ago. Nature 501:535–538

Delvigne C, Cardinal D, Hofmann A et al (2012) Stratigraphic changes of Ge/Si, REE + Y and silicon isotopes as insights into the deposition of a Mesoarchaean banded iron formation. Earth Planet Sci Lett 355–356:109–118

Delvigne C, Opfergelt S, Cardinal D et al (2016) Desilication in Archean weathering processes traced by silicon isotopes and Ge/Si ratios. Chem Geol 420:139–147

De Villiers J (1960) The manganese deposits of the Union of South Africa. In: Geological Survey of South Africa 2:263

De Wit MJ, Roering C, Hart RJ et al (1992) Formation of an Archaean continent. Nature 357:553–562

Dix OR (1984) Early Proterozoic braided stream, shelf and tidal deposition in the Pongola Sequence, Zululand. S. African J. Geol 87:1–10

Dixon JGP (2004) Archaean geology of the Buffalo River Gorge, KwaZulu-Natal. Unpublished PhD thesis, University of Natal (Durban)

Dlamini N, Hofmann A, Belyanin G et al (2017) Supracrustal gneisses in southern Swaziland: a basalt-sandstone assemblage of the upper Mozaan Group deformed in the Neoarchaean. S Afr J Geol 120:477–498

Eickmann B, Hofmann A, Wille M et al (2018) Isotopic evidence for oxygenated Mesoarchaean shallow oceans. Nat Geosci 11:133–138
Elworthy T, Eglington BM, Armstrong RA et al (2000) Rb-Sr isotope constraints on the timing of late to post-Archaean tectonometamorphism affecting the southeastern Kaapvaal Craton. J Afr Earth Sci 30:641–650
Geological Survey of Swaziland (1980) 1:50000 Geological Series, 3rd edn. Sheets 2631 CA, CD and 2731 AA, AB. Crown copyright. Government of Swaziland, Mbabane
Gold DJC (1993) The geological evolution of a part of the Pongola Basin, southeastern Kaapvaal Craton. Unpublished PhD thesis, University of Natal (Pietermaritzburg)
Gold DJC, von Veh MW (1995) Tectonic evolution of the late Archaean Pongola-Mozaan basin, South Africa. J Afr Earth Sci 21:203–212
Gold DJC (2006) The Pongola Supergroup. In: Johnson MR, Anhaeusser CR, Thomas RJ (eds) The geology of South Africa. Geological Society of South Africa/Council for Geoscience, Johannesburg/Pretoria, pp 135–147
Grandstaff DE, Edelman MJ, Foster RW et al (1986) Chemistry and mineralogy of Precambrian paleosols at the base of the Dominion and Pongola groups (Transvaal, South Africa). Precambrian Res. 32:97–131
Groenewald PB (1984) The lithostratigraphy and petrogenesis of the Nsuze Group, northwest of Nkandla, Natal. Unpublished MSc thesis, University of Natal (Pietermaritzburg)
Gumsley AP, de Kock MO, Rajesh HM et al (2013) The Hlagothi Complex: the identification of fragments from a Mesoarchaean large igneous province on the Kaapvaal Craton. Lithos 174:333–348
Gumsley A, Olsson J, Söderlund U et al (2015) Precise U-Pb baddeleyite age dating of the Usushwana Complex, southern Africa—Implications for the Mesoarchaean magmatic and sedimentological evolution of the Pongola Supergroup, Kaapvaal Craton. Precambrian Res 267:174–185
Gutzmer J, Nhleko N, Beukes NJ et al (1999) Geochemistry and ion microprobe (SHRIMP) age of a quartz porphyry sill in the Mozaan Group of the Pongola Supergroup: implications for the Pongola and Witwatersrand supergroups. S Afr J Geol 102:139–146
Hammerbeck ECI (1976) Gold outside the Witwatersrand triad. In: Coetzee CB (ed) Mineral resources of the Republic of South Africa. Department of Mines and the Geological Survey 7:75–92
Hammerbeck E (1982) The geology of the Usushwana Complex and associated formations—southeastern Transvaal. Geol Surv S Afr Mem 70:1–119
Hatch FH (1910) Report on the mines and mineral resources of Natal (other than coal). R. Clay and Sons Ltd, London
Hatfield RA (1990) The geology and structure of the Pongola sequence east of Piet Retief, southeastern Transvaal. Unpublished MSc thesis, University of Natal (Pietermaritzburg)
Hegner E, Kröner A, Hofmann AW (1984) Age and isotope geochemistry of the Archaean Pongola and Usushwana suites in Swaziland, southern Africa: a case for crustal contamination of mantle-derived magma. Earth Planet Sci Lett 70:267–279
Hegner E, Kröner A, Hunt P (1994) A precise U-Pb zircon age for the Archaean Pongola Supergroup volcanics in Swaziland. J Afr Earth Sci 18:339–341
Hepworth JV (1973) A photogeological investigation of the Ancient Gneiss complex and the Mozaan of Swaziland. Institute of Geological Sciences, Overseas Division, Swaziland. Report 20
Hicks N (2009) A combined sedimentological-mineralogical study of sediment-hosted gold and uranium mineralization at Denny Dalton, Pongola Supergroup, South Africa. Unpublished MSc thesis, University of KwaZulu-Natal (Durban)
Hicks N, Dunlevey JN, Liu KW (2011) A new stromatolite occurrence in the Nsuze Group, Pongola Supergroup of northern Kwazulu-Natal, South Africa. S Afr J Geol 114:195–200
Hicks N, Hofmann A (2012) Stratigraphy and provenance of the auriferous-uraniferous, fluvial to shallow-marine Sinqeni Formation, Mozaan Group, northern KwaZulu-Natal, South Africa. S Afr J Geol 115:327–344
Hicks N, Elburg M, Anderson T (2015) U-Pb and Hf isotope constraints for emplacement of the Nkandla Granite, southeastern Kaapvaal Craton, South Africa. S Afr J Geol 118:119–128
Hofmann A, Bekker A, Rouxel O et al (2009) Multiple sulphur and iron isotope composition of detrital pyrite in Archaean sedimentary rocks: a new tool for provenance analysis. Earth Planet Sci Lett 286:436–445
Hofmann A, Kröner A, Xie H et al (2015) The Nhlangano gneiss dome in south-west Swaziland—A record of crustal destabilization of the eastern Kaapvaal craton in the Neoarchaean. Precambrian Res 258:109–132
Horváth P, Reinhardt J, Hofmann A (2014) High-grade metamorphism of ironstones in the Mesoarchaean of southwest Swaziland. Miner Petrol 108:589–605
Humphrey WA (1913) The geology of a portion of northern Natal between Vryheid and the Pongola River. Annu Rep Geol Surv S Afr 1912:99–123
Humphrey WA, Krige LA (1931) The geology of the country south of Piet Retief: explanation of Sheet 68 (Piet Retief). Geological Survey of South Africa, Pretoria
Hunter DR (1950) Annu Rep Geological Surv Dep., Swaziland
Hunter DR (1953) Annu. Rep. Geological Surv. Dep., Swaziland
Hunter DR (1963) The Mozaan Series in Swaziland, 5-16. Geological Surv. Mines Dep., Swaziland, p 3
Hunter DR (1970) The Usushwana Complex in Swaziland. Spec. Publication Geological Society of S. Afr. 1:645–660
Hunter DR, Wilson AH, Versfeld JA et al (1986) A continuous record of tectonic evolution from 3.5 Ga to 2.6 Ga in Swaziland and northern Natal. In: de Wit MJ, Ashwal LD (eds) Workshop on tectonic evolution of greenstone belts. Houston, Lunar and Planetary Institute, pp 127–131
Hunter DR, Wilson AH (1988) A continuous record of Archaean evolution from 3.5 Ga to 2.6 Ga in Swaziland and northern Natal. S Afr J Geol 91:57–74
Hunter DR (1991) Excursion guide for the Pongola sequence: conference on Precambrian sedimentary basins of southern Africa. Geological Society of South Africa, Johannesburg
Hunter DR, Gold DJC (1993) A re-examination of the Pongola sequence in Swaziland and northern Natal. In: Maphalala R, Mabuza M (eds) 16th Colloquium of African geology 14–16th September, 1993, Mbabane, Swaziland, abstracts vol. I. Jubilee Printing, Matsapha, pp 159–160
Jacobs J, Thomas RJ (1994) Oblique collision at about 1.1 Ga along the southern margin of the Kaapvaal continent, south-east Africa. Geol Rundsch 83:322–333
Kimberley MW, Grandstaff DE (1986) Profiles of elemental concentrations in Precambrian paleosols on basaltic and granitic parent materials. Precambrian Res. 32:133–154
Klausen MB, Söderlund U, Olsson JR et al (2010) Petrological discrimination among Precambrian dyke swarms: eastern Kaapvaal Craton (South Africa). Precambrian Res. 183:501–522
Layer P (1986) Archean paleomagnetism of southern Africa. Unpublished PhD thesis, Stanford University, Palo Alto, USA
Layer PW, Kröner A, McWilliams M et al (1988) Regional magnetic overprinting of Witwatersrand Supergroup sediments, South Africa. J Geophys Res 93:2191–2200
Layer PW, Kröner A, McWilliams M et al (1989) Elements of the Archean thermal history and apparent polar wander of the eastern Kaapvaal Craton, Swaziland, from single grain dating and paleomagnetism. Earth Planet Sci Lett 93:23–34
Lechekoane T (2016) Paleomagnetism and geochemistry of the Mesoarchean Klipwal Diamictite, Mozaan Group, Pongola

Supergroup, South Africa. Unpublished MSc thesis, University of Johannesburg, South Africa
Linström W (1987a) The geology of the Dundee area: Explanation of Sheet 2830. Geological Survey of South Africa, Pretoria
Linström W (1987b) Die geologie van die gebied Vryheid. Toeligting van blad 2730. Department of Minerals and Energy, Pretoria
Lubnina N, Ernst R, Klausen M et al (2010) Paleomagnetic study of Neoarchean-Paleoproterozoic dykes in the Kaapvaal Craton. Precambrian Res 183:523–552
Mabuza MH, von Veh MW (1996) A note on the structure of the late Archaean Pongola sequence in the Sigwe Hills area, southeastern Swaziland evidence from a road cutting. S Afr J Geol 99:97–101
Maphalala RM, Kröner A (1993) Pb-Pb single zircon ages for the younger Archaean granitoids of Swaziland, southern Africa. In: Maphalala R, Mabuza M (eds) 16th Colloquium of African geology 14–16th September, 1993, Mbabane, Swaziland, extended abstracts vol. II. Jubilee Printing, Matsapha, pp 201–206
Maré LP, Fourie CJS (2012) New geochemical and palaeomagnetic results from Neoarchean dyke swarms in the Badplaas-Barberton area, South Africa. S Afr J Geol 115:145–170
Mason TR, von Brunn V (1977) 3-Gyr-old stromatolites from South Africa. Nature 266:47–49
Matthews PE (1967) The pre-Karoo formations of the White Umfolozi Inlier, northern Natal. Trans Geol Soc S Afr 70:39–63
Matthews PE, Scharrer RH (1968) A graded unconformity at the base of the early Precambrian Pongola system. Tran Geol Soc S Afr 71:257–272
Matthews PE (1985) Archaean post-Pongola granites in the southeastern Kaapvaal Craton. S Afr J Sci 81:479–484
Matthews PE (1990) A plate tectonic model for the late Archaean Pongola Supergroup in southeastern Africa. In: Sychanthavong SPH (ed) Crustal evolution and orogeny. Oxford Publishing, New Delhi, pp 41–73
Meyer FM, Robb LJ, Reimold WU et al (1994) Contrasting low- and high-Ca granites in the Archaean Barberton Mountain Land, southern Africa. Lithos 32:63–76
Mukasa SB, Wilson AH, Young KR (2013) Geochronological constraints on the magmatic and tectonic development of the Pongola Supergroup (central region), South Africa. Precambr Res 224:268–286
Murphy RCL (2015) Stabilising a craton: the origin and emplacement of the 3.1 Ga Mpuluzi batholith. Unpublished PhD thesis, Macquarie University, Sydney, Australia
Nhleko N (2003) The Pongola Supergroup in Swaziland. Unpublished PhD thesis, Rand Afrikaans University, Johannesburg, South Africa
Noffke N, Hazen R, Nhleko N (2003) Earth's earliest microbial mats in a siliciclastic marine environment (2.9 Ga Mozaan Group, South Africa). Geology 31:673–676
Noffke N, Beukes N, Bower D et al (2008) An actualistic perspective into Archean worlds –(cyano-)bacterially induced sedimentary structures in the siliciclastic Nhlazatse Section, 2.9 Ga Pongola Supergroup, South Africa. Geobiology 6:5–20
Olsson JR, Söderlund U, Klausen MB et al (2010) U-Pb baddeleyite ages linking major Archean dyke swarms to volcanic-rift forming events in the Kaapvaal craton (South Africa), and a precise age for the Bushveld Complex. Precambr Res 183:490–500
Ossa FO, Hofmann A, Vidal O et al (2016) Unusual manganese enrichment in the Mesoarchean Mozaan Group, Pongola Supergroup, South Africa. Precambr Res 281:414–433
Preston VA (1987) The geology and geochemistry of the Mpongoza Inlier, northern Natal. Unpublished MSc thesis, University of Natal (Pietermaritzburg)
Planavsky N, Asael D, Hofmann A et al (2014) Evidence for oxygenic photosynthesis half a billion years before the Great Oxidation Event. Nat Geosci 7:283–286
Reimold WU, Meyer FM, Walraven F, et al. (1993) Geochemistry and chronology of pre- and post- Pongola granitoids from northeastern Natal. In: Maphalala R, Mabuza M (eds) 16th Colloquium of African geology 14–16th September, 1993, Mbabane, Swaziland, extended abstracts vol. II. Jubilee Printing, Matsapha, pp 294–296
Reinhardt J, Elburg MA, Anderson T (2015) Zircon U-Pb age data and Hf isotopic signature of Kaapvaal basement granitoids from the Archaean White Mfolozi Inlier, northern Kwazulu-Natal. S Afr J Geol 118:473–488
Roering C (1968) Non-orogenic granites and the age of the Precambrian Pongola sequence. Economic Geology Res. Unit, University of the Witwatersrand, Information Circular 46
Saager R, Stupp HD, Utter T et al (1986) Geological and mineralogical notes on placer occurrences in some conglomerates of the Pongola Sequence. In: Anhaeusser CR, Maske S (eds) Mineral deposits of southern Africa. Geological Society of South Africa, Johannesburg, pp 473–487
SACS (South African Committee For Stratigraphy) (1980) Stratigraphy of South Africa part 1: lithostratigraphy of the Republic of South Africa, South West Africa/Namibia, and the Republics of Bophuthatswana, Transkei and Venda. Geological Survey of South Africa, Johannesburg
Schoene B, Bowring SA (2010) Rates and mechanisms of Mesoarchean magmatic arc construction, eastern Kaapvaal craton, Swaziland. Geol. Soc. Am Bull 122:408–429
Siahi M, Hofmann A, Hegner E et al (2016) Sedimentology and facies analysis of Mesoarchaean carbonate rocks of the Pongola Supergroup, South Africa. Precambr Res 278:244–264
Siahi M, Hofmann A, Master S, Mueller CW, Gerdes A (2017) Carbonate ooids of the Mesoarchaean Pongola Supergroup, South Africa. Geobiology 15:750–766
Siahi M, Hofmann A, Master S et al (2018) Trace element and stable (C, O) and radiogenic (Sr) isotope geochemistry of stromatolitic carbonate rocks of the Mesoarchaean Pongola Supergroup: implications for seawater composition. Chem Geol 476:389–406
Smith A (2017) The iron formations of southern Africa. In: Siegesmund S, Basei MAS, Oyhantçabal P, Oriolo S (eds) Geology of southwest Gondwana. Springer, Cham, pp 469–491
Strik GM (2004) Palaeomagnetism of late Archaean flood basalt terrains: implications for early Earth geodynamics and geomagnetism. Unpublished PhD thesis, University of Utrecht, The Netherlands
Strik G, de Wit MJ, Langereis CG (2007) Palaeomagnetism of the Neoarchaean Pongola and Ventersdorp Supergroups and an appraisal of the 3.0–1.9 Ga apparent polar wander path of the Kaapvaal Craton, southern Africa. Precambr Res 153:96–115
Taylor J, Stevens G, Armstrong R et al (2010) Granulite facies anatexis in the Ancient Gneiss Complex, Swaziland, at 2.73 Ga: mid-crustal metamorphic evidence for mantle heating of the Kaapvaal craton during Ventersdorp magmatism. Precambr Res 177: 88–102
Thomas RJ, Agenbacht ALD, Cornell DH et al (1994) The Kibaran of southern Africa: tectonic evolution and metallogeny. Ore Geol Rev 9:131–160
Thomas RJ, Gold DJC, Verbeek JA et al (1997) Geology of the Archaean Nzimane Inlier, Zululand. S Afr J Geol 100:123–136
Von Brunn V (1974) Tidalites of the Pongola Supergroup (early Precambrian) in the Swart-Mfolozi area, northern Natal. Bull. Precambr Res, vol 15. Unit, Univ. Cape Town , pp 107–122

Von Brunn V, Hobday DK (1976) Early Precambrian tidal sedimentation in the Pongola Supergroup of South Africa. J Sedim Petrol 46:670–679
Von Brunn V, Gold DJC (1993) Diamictite in the Archean Pongola Sequence of southern Africa. J Afr Earth Sci 16:367–374
Walraven F, Pape J (1994) Pb-Pb whole-rock ages for the Pongola Supergroup and the Usushwana Complex, South Africa. J Afr Earth Sci 18:297–308
Ward JHW, Wilson MGC (1998) Gold outside the Witwatersrand basin. In: Wilson MGC, Anhaeusser CR (eds) The mineral resources of South Africa. Council for Geosciences, Pretoria, pp 350–386
Watchorn MB (1978) Sedimentology of the Mozaan Group in the southeastern Transvaal and northern Natal. Unpublished MSc thesis, University of Natal (Pietermaritzburg)
Watchorn MB, Armstrong NV (1980) Contemporaneous sedimentation and volcanism at the base of the early Precambrian Nsuze Group, South Africa. Trans Geol Soc S Afr 83:231–238
Wilson AC, Jackson MPA (1988) Mantled gneiss domes in southern Swaziland and the concept of 'stable' Pongola cratonic cover. S Afr J Geol 91:404–414
Wilson AH, Grant CE (2006) Physical volcanology and compositions of the basaltic lava in the Archean Nsuze Group, White Mflozi Inlier, South Africa. In: Reimold WU, Gibson RL (eds) Processes on the early Earth: special paper, vol 405. Geological Society of America, pp 255–289
Wilson A, Groenewald B, Palmer C (2013) Volcanic and volcaniclastic rocks of the Mesoarchaean Pongola Supergroup in South Africa and Swaziland: distribution, physical characteristics, stratigraphy and correlations. S Afr J Geol 116:119–168
Wilson A, Hofmann A (2013) June 2013 LIP of the month: the Meso-Archaean Pongola Supergroup in South Africa and Swaziland—The oldest known large igneous province on stable continental crust? Large Igneous Provinces Commission. http://www.largeigneousprovinces.org/13jun. Accessed 10 June 2018
Wilson AH, Zeh A (2018) U-Pb and Hf isotopes of detrital zircons from the Pongola Supergroup: constraints on deposition ages, provenance and Archean evolution of the Kaapvaal Craton. Precambr Res 305:177–196
Young GM, Brunn V, Gold DJC et al (1998) Earth's oldest reported glaciation: physical and chemical evidence from the Archean Mozaan Group (~2.9 Ga) of South Africa. J Geol 106:523–538

The Witwatersrand Basin and Its Gold Deposits

10

Hartwig E. Frimmel

Abstract

The Mesoarchaean Witwatersrand Basin represents the largest known gold anomaly and has produced more gold than any other ore province in the world. Deposition of the predominantly siliciclastic basin fill began shortly after 2985 Ma in a shallow marine environment, maybe along a passive margin to an old continent to the north (west), and under generally cool climatic conditions (West Rand Group). Tectonic inversion following the extrusion of 2914 Ma andesitic lava led to a shift towards continental sedimentation. Following a basin-wide hiatus at around 2900 Ma, the largely arenitic Central Rand Group was deposited into a retroarc foreland basin that regressed as it was filled between <2902 and >2780 Ma in response to crustal accretion along the western and northern margins of the basin. This stage was accompanied by the largest known concentration of gold in Earth's history, initially by microbial fixation on emerging colonies of probably cyanobacteria in near-coastal environments, subsequently by mechanical reworking of the gold-rich microbial mats to form conglomerate-hosted placer deposits. The source of the huge amount of gold in the Witwatersrand is thought to be the entire greenstone-dominated Archaean cratonic surface, which was subjected to intensive chemical weathering permitting large-scale leaching of gold by contemporaneous surface waters. Syn-depositional tectonism that extended from the hinterland into the foreland basin led to repeated further physical reworking of gold-rich sediments to form more placer deposits higher up in the stratigraphy, even above the Witwatersrand Supergroup. The Witwatersrand Basin fill was subjected to a series of alteration events, ranging from burial and regional low-grade metamorphism to heating in the course of the emplacement of the 2054 Ma Bushveld Igneous Complex and catastrophic shattering during the 2023 Ma Vredefort impact, which enabled renewed fluid flow long after primary rock porosity had been obliterated. This caused some short-range mobilisation of ore components, including gold, but without changing the overall sedimentological and stratigraphic control on ore grade.

Keywords

Witwatersrand • Gold • Mesoarchaean • Kaapvaal Craton

10.1 Introduction

The Witwatersrand Basin near the centre of the Kaapvaal Craton (Fig. 10.1) is a domain of many superlatives and holds several world records. Foremost it has become famous for representing the by far largest known gold anomaly in the world. The Witwatersrand goldfields (Fig. 10.2) have produced, with some 53,000 t Au, about one-third of all the gold that has ever been mined in history, and estimated remaining resources account for close to 30% of global known resources (Frimmel 2014), though much of this is beyond the reach of conventional and currently economic mining methods. For many decades, they were the world's leading producer of gold, although the more easily accessible ores have largely been exploited, and annual production has followed a steady downwards trend since 1970. Apart from the exceptional endowment in gold, the Witwatersrand ores also represent one of the world's major uranium resources. Within some of the gold-bearing conglomerates, a total resource of >200 kt U_3O_8 is still present, but most of the uranium, where it has not been extracted, is now

H. E. Frimmel (✉)
Department of Geodynamics and Geomaterials Research, Bavarian Georesources Centre, Institute of Geography and Geology, University of Würzburg, Am Hubland, 97074 Würzburg, Germany
e-mail: hartwig.frimmel@uni-wuerzburg.de

H. E. Frimmel
Department of Geological Sciences, University of Cape Town, Rondebosch, 7701, South Africa

A. Kröner and A. Hofmann (eds.), *The Archaean Geology of the Kaapvaal Craton, Southern Africa*, Regional Geology Reviews, https://doi.org/10.1007/978-3-319-78652-0_10

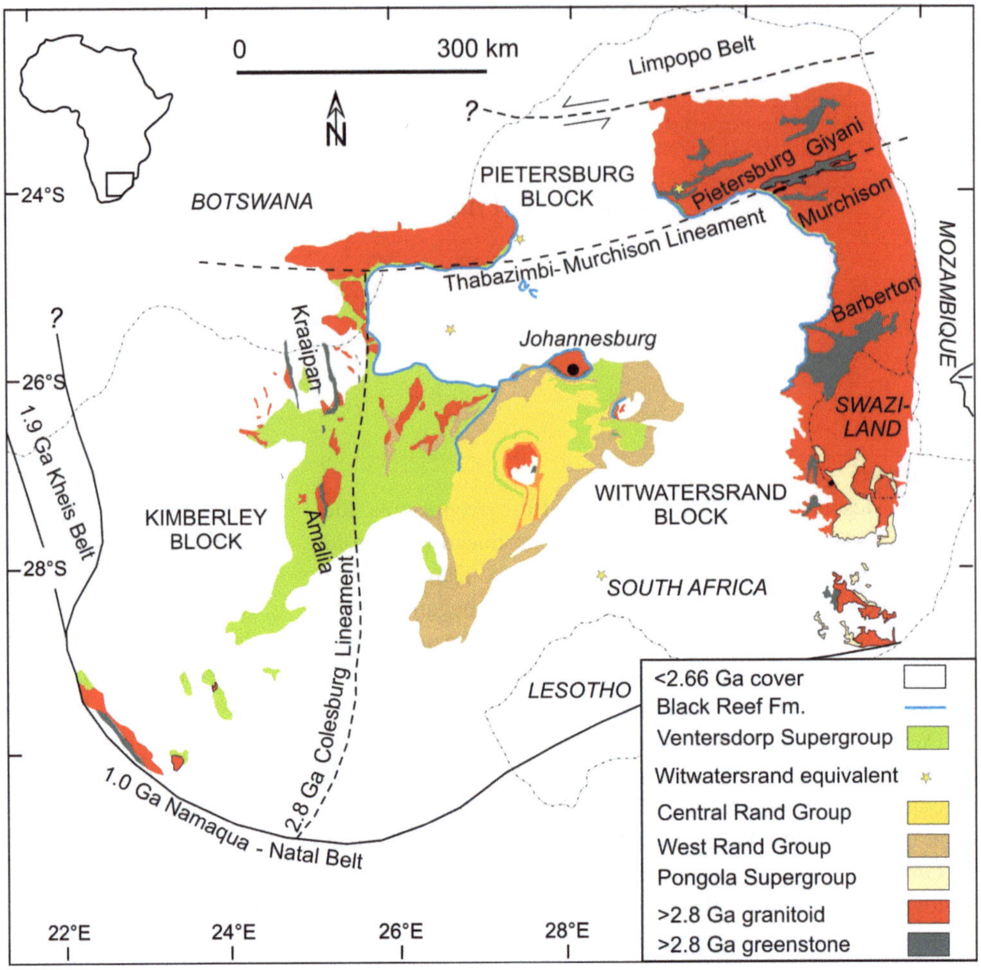

Fig. 10.1 Surface and sub-surface distribution of the main Archaean stratigraphic units of the Kaapvaal Craton, showing the known extent of the Witwatersrand Basin, the fill of which comprises the West Rand and Central Rand groups; blue line delineates trace of the Black Reef Quartzite Formation at the base of the Neoarchaean to early Palaeoproterozoic Transvaal Basin; from Frimmel and Hennigh (2015); stars indicate positions of Witwatersrand-equivalent rocks having been intersected in drill core

concentrated in voluminous tailings dams, totalling close to 1 Mt U_3O_8. This huge amount of uranium constitutes, with close to 10%, the world's third largest uranium resource in terms of deposit type (Frimmel 2010) and poses one of the largest mining-induced environmental health risks.

Thanks to more than a century of mining and exploration, there is hardly any other area in the world whose underground geology has been three-dimensionally documented to the same extent as in the Witwatersrand. From a technical point of view, some of the leading tools in the exploration industry have been developed there. As early as 1889 diamond drilling led to the discovery of sub-surface extensions of gold-bearing conglomerates. The discovery of the Carletonville and Welkom goldfields in the 1930s and in 1946, respectively, was based on the pioneering use of magnetometer and gravimeter. With close to 4 km below surface, today some of the mines in the Witwatersrand are the deepest in the world.

Although rocks of the Witwatersrand Supergroup are hardly exposed on surface, they can be regarded as the best-preserved and most intensely investigated Mesoarchaean sediment succession thanks to not only the intensity of mining and exploration but also a unique geological history that prevented total erosion or tectonic recycling of these ancient deposits. Thus the Witwatersrand Basin fill has become the best available record of Mesoarchaean environmental conditions and serves as a reference for the reconstruction of ancient secular changes in the composition of the atmosphere, hydrosphere and biosphere.

Last but not least, the Witwatersrand Basin hosts in its centre the world's largest and oldest known impact structure, now evident in the form of the 2023 ± 2 Ma (Kamo et al. 1996) Vredefort Dome (Fig. 10.3). Although the exceptional metal endowment of the Witwatersrand strata cannot be related to this impact because it predates this catastrophic event by some 800–900 million years, the impact had some major consequences for the auriferous sediments. The resulting impact melt sheet probably helped in preserving the underlying, much older sediments from subsequent erosion. The impact shattered the surrounding crustal rocks, especially those of the Witwatersrand Basin, whose areal extent corresponds closely to the diameter of the original astrobleme, estimated to have been some 300 km (Therriault et al. 1997). This enabled the widespread percolation of fluids and

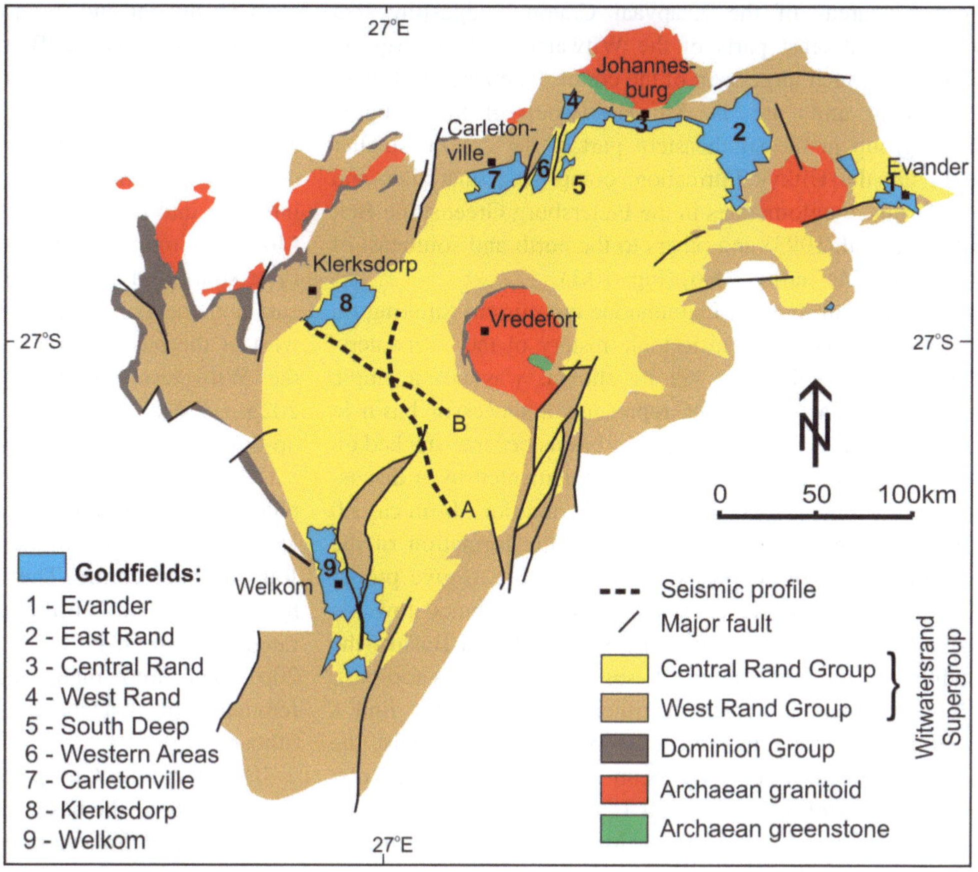

Fig. 10.2 Simplified surface and sub-surface geological map of the Witwatersrand Basin, also showing the distribution of Archaean granitoid domes, the location of the goldfields, major faults, and position of seismic profiles shown in Fig. 10.6; from Frimmel and Minter (2002)

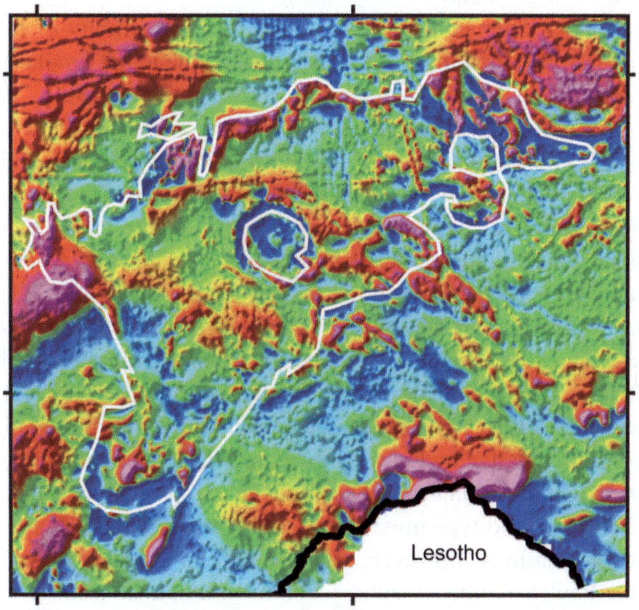

Fig. 10.3 Aeromagnetic map of the Witwatersrand Basin (extracted from map constructed by H. Jelsma based on data provided by Anglo American Corp. Ltd. for the Kaapvaal Craton Project—see South African Journal of Geology, vol. 107, p. 2, 2004); high magnetic signatures reflect ferruginous (magnetic) shales in the West Rand Group. Note the circular magnetic structure defined by the Vredefort Dome in the centre

partial remobilisation of ore components, including gold, which, in turn, led to textural and geochemical features that confused many of those who have attempted to establish the genesis of this gold province.

Almost every imaginable genetic model has been proposed to explain the Witwatersrand gold deposits, ranging from syn-sedimentary placer to epigenetic magmatic or metamorphic origins. More than 50 years ago the question of the Witwatersrand gold genesis was coined 'the most debated issue in economic geology' (Davidson 1965), and looking at the huge amount of literature on this topic since then, it appears as if little has changed. In spite of well over 1000 publications on the geology of the Witwatersrand, a number of fundamental questions remain unanswered. Nevertheless, tremendous advances could be made in our understanding of the geological evolution of the Witwatersrand strata, and more recently especially on their bearing on our knowledge of the evolution of the atmosphere and biosphere, both of which can be directly linked to the formation of the gold deposits.

For obvious economic reasons the focus has been on the Witwatersrand Basin as known from surface and sub-surface observations, stretching for about 350 km in a northeasterly and about 200 km in a northwesterly direction (Fig. 10.1). Stratigraphically equivalent units have been recognised also in other areas of the Kaapvaal Craton, suggesting that originally at least parts of the Witwatersrand Supergroup covered a much larger part of the craton. The largest of these equivalent units are the Mozaan Group of the Pongola Supergroup in the southeastern part of the craton, smaller ones are the Uitkyk Formation, comprising quartzites and auriferous conglomerates in the Pietersburg Greenstone Belt (de Wit et al. 1993) and others to the north and southeast of the basin (McCarthy 2006) (Fig. 10.1).

In this contribution, I summarise our current thinking on the sedimentological and tectonic history of the 'Witwatersrand Basin' which, as will be shown, was not a single coherent basin but, in fact, represents a successor basin in which the fills of different types of basins became stacked on top of each other. This will be complemented by a discussion of the changes in environmental conditions from ca. 3.0 to 2.6 Ga and their consequences on the formation of the uniquely rich gold deposits. It is impossible to give proper credit to every single worker who has contributed towards advancing our knowledge of the Witwatersrand Basin—the number of publications and theses on this topic is too vast. Review papers in which the interested reader can find a wealth of further references were published repeatedly, the more recent ones being by Robb and Meyer (1995), Robb and Robb (1998), Phillips and Law (2000), Frimmel et al. (2005), Hayward et al. (2005), McCarthy (2006), Frimmel (2014) and Tucker et al. (2016).

10.2 Litho- and Chronostratigraphy

The stratigraphic sub-division of the Witwatersrand Basin fill, in its essential parts, goes back to Mellor (1917) whose broad division into 'Lower Witwatersrand System' with the 'Hospital Hill, Government Reef' and 'Jeppestown Series', and 'Upper Witwatersrand System' with the 'Main-Bird Series' and the 'Kimberley-Elsburg Series' has withstood the test of time, albeit with slightly modified names of units and stratigraphic ranks. Mellor's two systems were renamed by SACS (1980) to West Rand Group and Central Rand Group, respectively, and his various series are now ranked as subgroups (see Fig. 10.4). The latter have been further sub-divided into formations based on basin-wide disconformities. Stratigraphic correlation between the various goldfields at the subgroup level poses no problem but is highly problematic at formation level. McCarthy (2006) provided an attempt of a unified lithostratigraphic scheme for the Witwatersrand Supergroup, and this is adopted here (Fig. 10.4).

The pre-Witwatersrand basement comprises Palaeo- to Mesoarchaean granitoid-greenstone terranes (Fig. 10.1) that are described in greater detail elsewhere in this book. The most prominent and intensely studied amongst them is the Barberton Greenstone Belt, which is well exposed farther east of today's extent of the Witwatersrand Basin fill. Its maximum and minimum ages are constrained by the 3660 ± 4 Ma Ancient Gneiss Complex in the southern part of the belt and the 3106 ± 3 Ma Nelspruit Batholith that is intrusive into the 3226 to 3209 Ma foreland deposits of the Moodies Group (Heubeck 2019). Others are the Murchison, Pietersburg and Giyani greenstone belts towards the northeast and the Amalia and Kraaipan greenstone belts to the west of the basin. Remnants of the immediate basement of the Witwatersrand Basin are today exposed, thanks to 2023 ± 2 Ma (Kamo et al. 1996) impact-induced up-doming, as a variety of high-grade metamorphic ca. 3.1 Ga gneisses and migmatites and a low-grade metamorphic mafic to ultramafic volcanic unit within the Vredefort Dome, as well as in several other domes, notably the Johannesburg Dome. The latter consists of tonalitic, trondhjemitic to granodioritic gneisses (TTG suite) with ages ranging from 3340 to 3114 Ma (Poujol and Anhaeusser 2001) and minor mafic and ultramafic rocks interpreted as remnants of Palaeoarchaean upper mantle or oceanic lithosphere.

The Dominion Group constitutes the earliest supracrustal sequence above the basement rocks. Although not part of the Witwatersrand Supergroup, it reflects first intracontinental basin formation after crustal amalgamation at around 3.1 Ga. The bimodal nature of magmatism in the up to 2250 m-thick volcanic-dominated sequence and a tholeiitic affinity of the mafic rocks therein have been used as evidence of emplacement due to lithospheric thinning (Marsh et al. 1989). Its maximum age is constrained by the youngest age obtained on pre-Dominion basement, i.e. 3086 ± 3 Ma (Robb et al. 1992), and the best available age for the volcanism is 3074 ± 6 Ma (Armstrong et al. 1991). Granite and coeval hornblende gabbro, intersected in drill core through a basement horst on the flank of the Witwatersrand Basin, yielded a U-Pb zircon age of 3062 ± 5 Ma and geochemical/isotopic signatures typical of a volcanic arc setting (Frimmel et al. 2009). They might be related to the Dominion Group volcanic rocks. At its base the group contains a thin basal siliciclastic unit that includes pyritic quartz-pebble conglomerate beds that are of interest in so far as they represent the oldest Witwatersrand-type mineralisation (Dominion Reef). The Au tenor therein is, however, much lower (on average 0.6 g/t), but they constitute a locally economic uranium concentration with a reported resource of 256 Mt @ 0.4 kg/t U_3O_8 (RMG 2015). Rocks of the Dominion Group crop out, or have been intersected in drill core, to the west of the Witwatersrand Basin (Fig. 10.2). Palaeocurrent data consistently indicate a sediment source to the north or northeast (Frimmel and Minter 2002).

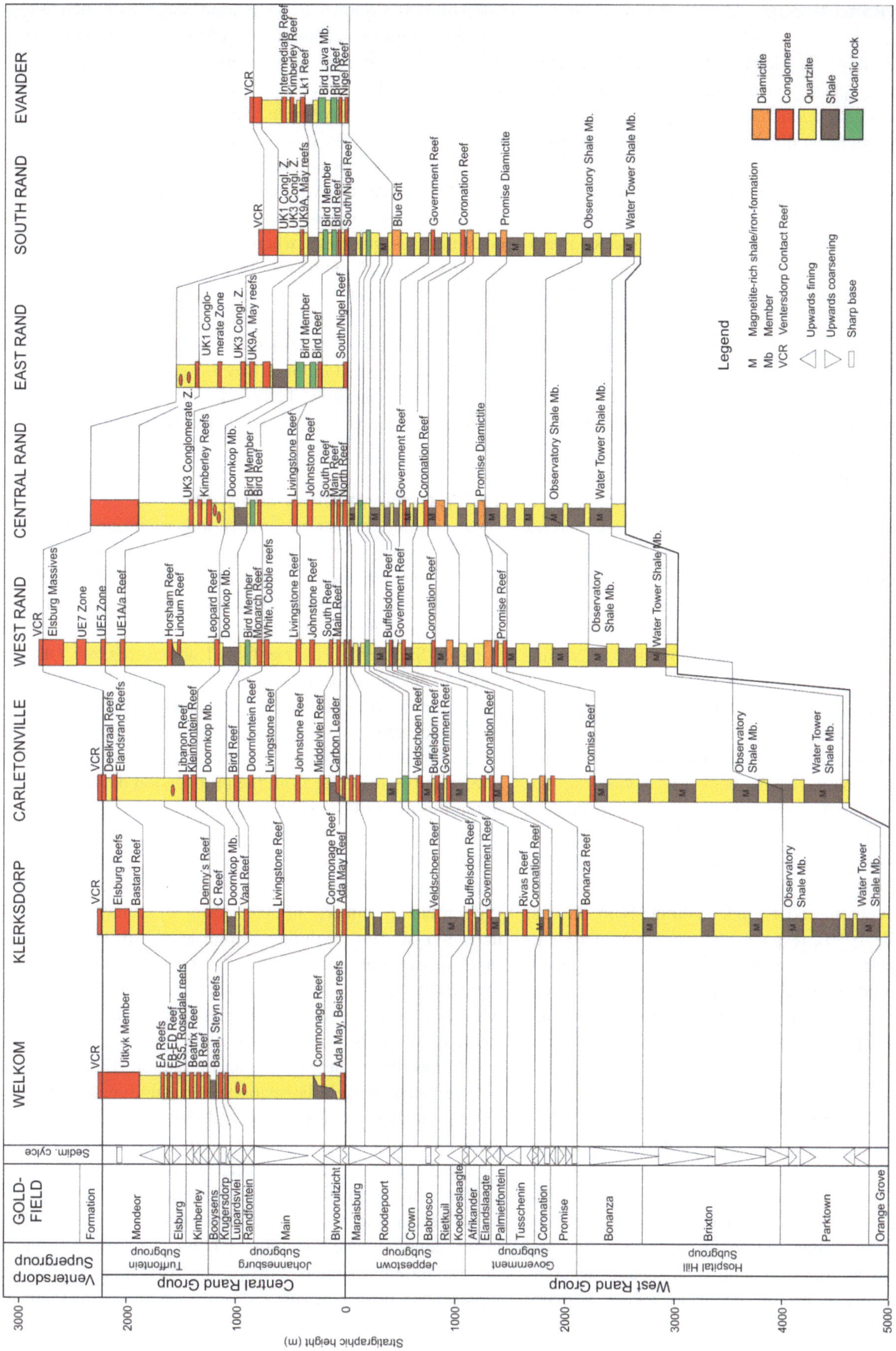

Fig. 10.4 Stratigraphy of the Witwatersrand Supergroup and attempted correlation between the various goldfields. Modified after McCarthy (2006)

10.2.1 West Rand Group

The West Rand Group's lithology is dominated by roughly equal proportions of quartzite and shale. Conglomerate beds are subordinate, largely restricted to the upper parts of the group (Fig. 10.4) and generally distal. Only in one area, west of Klerksdorp, are they proximal in character, and this is the only place where they were mined. Many of the quartzite beds have been interpreted as shallow marine shelf sand accumulations (Eriksson et al. 1981; Winter and Brink 1991). The shale beds are marine, generally rich in iron, with some of them representing proper iron-formation. Three iron-formation marker beds are laterally persistent: (i) in the Water Tower Shale Member, (ii) the Contorted Bed in the Observatory Shale Member, and (iii) the Silverfield iron-formation in the Coronation Formation (Fig. 10.4). A fourth iron-formation bed has been identified in drill core east of Johannesburg within the Promise Formation (Smith et al. 2013). The Promise and Silverfield iron-formations are of particular interest as they overlie diamictite beds that can be correlated with similar marker beds in the Mozaan Group of the Pongola Supergroup (Fig. 10.1) and have become famous as the oldest known glacial deposits in the world (Young et al. 1998). A third diamicite unit, referred to as the Blue Grit Member, at the top of the Afrikander Formation in the South Rand (Fig. 10.4) lacks a corresponding overlying iron-formation, which may be due to erosion prior to deposition of the Jeppestown Subgroup. The contact between the two is a major erosional unconformity.

Noteworthy is the abundance of magnetite as principal Fe-mineral in the marine shale beds as opposed to the dominance of pyrite in the coarser grained, fluvial to shallow marine conglomerates and quartzite beds. The Fe-rich shales have played a major role in sub-surface exploration as magnetic markers (Fig. 10.3) and made possible the development of the Welkom and Evander goldfields in the late 1940s and middle 1950s, respectively, when new aeromagnetic and gravimetric survey technology led to the discovery of the first significant blind ore district. The West Rand Group contains only a single volcanic unit, a basaltic andesite that constitutes the Crown Formation near the top of the group.

10.2.1.1 Hospital Hill Subgroup

The subgroup starts with a thin basal conglomerate and fluvial sandstone (for clarity, the sedimentary protolith names are used here instead of their low-grade metamorphic equivalents) that, over short distance, grade into shallow marine sub-tidal sandstone, fining upwards into silt- and mudstone as well as iron-formation of the Parktown Formation. This sequence is considered to reflect a major transgression of the craton from the southwest. The Parktown Formation is dominated by marine shale with two major iron-formation intercalations, the older one being massive to laminated magnetite-siderite/ankerite-chert bands, the younger one haematite-magnetite rhythmite grading into magnetite-siderite rhythmite. The iron-formation and magnetite mudstone beds are part of upwards coarsening sequences with sharp transgressive upper contacts (Smith et al. 2013). The overlying Brixton Formation contains more sub-tidal sandstone and less mudstone in several upwards coarsening sequences, whereas the Bonanza Formation comprises fluvial sandstone. In the more proximal depositional environments in the west, the latter has few intercalated polymictic, pyritic conglomerate beds (Fig. 10.4). Four cycles of shoreline progradation leading to fluvial deposits can be distinguished in these two formations, each separated by rapid transgressions. Palaeocurrent directions indicate a sediment source area to the north and northwest (McCarthy 2006).

10.2.1.2 Government Subgroup

Great lithological variety marks this subgroup, reflecting repeated cycles of rapid transgression and regression. The base of the subgroup is defined by a laterally extensive diamictite of the Promise Formation (Fig. 10.4). It contains striated clasts and is, together with its correlatives in the Pongola Supergroup as mentioned above, regarded as the earliest glacial deposit recorded so far on Earth. The diamictite is overlain by littoral sandstone and shallow marine ferruginous shale, locally iron-formation, followed by an upwards coarsening sequence into sandstone that is interpreted as resulting from the propagation of a braid-plain at the top of the Promise Formation.

The overlying Coronation Formation consists of shale with a second, intercalated glaciogenic diamictite unit (Kensington Member) that is overlain by iron-formation (Silverfield Member) in the form of finely laminated magnetite-silicate rhythmite (Smith et al. 2013). It reflects rapid marine transgression, only to be followed by a rapid regression that is evident from the disconformable contact with the overlying fluvial conglomerate of the Coronation Reef, followed by fluvial braid-plain quartz wacke, all of which constitutes the Tusschenin Formation (Fig. 10.4).

The Palmietfontein Formation, with a dominance of more or less magnetic marine shale and marine sandstone, indicates renewed marine transgression that was followed again by rapid regression and emergence, evident from the disconformably overlying fluvial conglomerate of the Government Reef at the base of the Elandslaagte Formation. The latter conglomerate is followed by sandstone from a braid-plain environment. Finally, the Afrikander Formation represents a further marine transgression and is made up of mainly marine mud- and sandstone. In the eastern part, it contains a further diamictite bed, the Blue Grit Member. A major hiatus defines the boundary between the Government and Jeppestown subgroups as much of the Afrikander Formation has been eroded prior to Jeppestown deposition.

10.2.1.3 Jeppestown Subgroup

A fluvial conglomerate, the Buffelsdoorn Reef, rests above the basin-wide disconformity at the base of the Koedoeslaagte Formation at the bottom of the subgroup. It is overlain by an upwards fining sequence of littoral sandstone, leading to marine magnetite mudstone of the Rietkuil Formation (Fig. 10.4). The record of the next phase of shoreline progradation is largely absent as the Rietkuil Formation is erosionally truncated by a further disconformity, on top of which follows a conglomerate of the Veldskoen Reef and feldspathic sandstone of the Babrosco Formation (Fig. 10.4). The latter is overlain, across the entire known extent of the West Rand Group, by up to 250 m-thick former lava flows of basaltic andesite of the Crown Formation, followed by sandstone, fining upwards into marine ferruginous, partly magnetic mudstone of the Roodepoort Formation. Much of the Jeppestown Subgroup reflects marine conditions with the basin deepening towards the east. The Maraisburg Formation at the top of the subgroup marks, however, a major regression with propagation of braid-plains from the northwest, leaving behind thick sandstone and subordinate conglomerate deposits (Fig. 10.4).

10.2.1.4 Geochronology and Tectonic Evolution

The maximum age of West Rand Group sedimentation is set by the youngest concordant U-Pb detrital zircon age of 2985 $\pm$ 14 Ma from the basal Orange Grove Formation (Kositcin and Krapez 2004). The Rietkuil Formation must be younger than 2931 $\pm$ 8 Ma (Kositcin and Krapez 2004) but older than 2914 $\pm$ 8 Ma, the age of Crown Formation volcanism (Armstrong et al. 1991). This leaves a maximum time span of some 71 million years for the deposition of as much as 4200 m of sediment (in the Klerksdorp goldfield), that is all of the group up to the Crown Formation, corresponding to a sedimentation rate of 59 m/myr. Considering the existence of several disconformities in the succession, the true maximum integrated sedimentation rate should be even higher. It reflects a combination of initial drowning of the post-Dominion cratonic land surface, followed by repeated cycles of regression and transgression, which at least at Government Subgroup times were largely eustatic. Deepening of the basin towards the south and southeast is deduced from lateral distribution of proximal and distal sedimentary facies (Fig. 10.4). The extent of the West Rand Basin remains unknown but was, in all likelihood, much greater than suggested by the known outcrop and sub-crop of West Rand Group strata. The Mozaan Group of the Pongola Supergroup farther east has been correlated with the West Rand Group (Beukes and Cairncross 1991), thus extending the size of the original basin considerably to the southeast. The tectonic nature of this basin has been a matter of debate. Thermal subsidence may have been responsible for the development of a passive margin to the south(east) of a Mesoarchaean land mass north(west) of today's Witwatersrand, as suggested by uniform palaeocurrent directions from the north, northeast and northwest (Frimmel and Minter 2002) and an upwards decrease in complexity of detrital zircon age spectra (Kositcin and Krapez 2004). The general deepening of the basin, though repeatedly interrupted by eustatic sea level changes, became reversed after extrusion of the Crown Formation lavas. Shoreline propagation and eventual fluvial and braid-plain deposition recorded by the Maraisburg Formation portends the onset of the predominantly continental sedimentation of the Central Rand Group. This development is best explained by a foreland basin superimposed on the older passive margin deposits.

10.2.2 Central Rand Group

The Central Rand Group is unified by the dominance of fluvial to fluvio-deltaic sandstone and conglomerate as opposed to the marine shale-dominated West Rand Group. Principal depositional environments were alluvial braid-plains, especially in the lower Central Rand Group (Els 1998), and alluvial fans, especially towards the top of the group (Kingsley 1987). A further difference to the West Rand Group lies in the sandstone mineralogy. Feldspar is conspicuously absent in the Central Rand Group (Law et al. 1990). The only shale unit of basin-wide extent within the Central Rand Group (Booysens Formation) led Mellor (1917) to a sub-division of the group into two units, now referred to as the Johannesburg and Tuffontein subgroups. Two diamictite units are present within this group, but they differ from those in the West Rand Group by being non-glaciogenic debris flow deposits.

In contrast to the West Rand Group, many of the conglomerate beds in the Central Rand Group are exceptionally rich in gold, with ca. 93% of all the Witwatersrand gold produced so far coming from the Central Rand Group (H.E. Frimmel, unpubl. data). This holds true especially for the proximal areas along the basin edge, where a number of palaeoriver entry points could be mapped out. Most of the conglomerate beds are due to uplift triggered by intermittent tectonic events in the hinterland, followed by periods of tectonic stability and thus low-energy degradation of the conglomerate beds.

10.2.2.1 Johannesburg Subgroup

This subgroup is mainly an aggradational sequence of arenitic fluvial braid-plain deposits. Occasional erosional episodes resulted in widespread peneplanation and thus disconformities that are overlain by conglomeratic placer deposits. Local concentration of gold to ore grade was achieved both by high-energy rivers within incised valleys as well as in thin gravel lags by aeolian transport on exposed surfaces on the interfluves. The latter were also sites of thin layers of carbonaceous material (so-called 'carbon seams'), which represent former microbial mats now consisting of kerogen (Mossman et al. 2008). Several such degradation cycles are distinguished as reflected by the various formations, from bottom to top, the Blyvooruitzicht, Main, Randfontein, Luipaardsvlei and Krugersdorp formations. Each of these typically starts with a basal, in many places auriferous, conglomerate (Fig. 10.4), in one case of (non-glaciogenic) diamictite where argillaceous sediment of the exposed Jeppestown Subgroup became reworked as debris flows into channels cutting through the Main Reef (Martin et al. 1989). From an economic point of view, two degradation episodes stand out, an older one that led to the Main Reefs at the base of the Main Formation (Carbon Leader, Main, South, North, Nigel reefs) and a younger one at the base of the Krugersdorp Formation, especially in the west and northwest of the basin, leaving behind the Basal and Steyn reefs in the Welkom goldfield and the Vaal Reef in the Klerksdorp goldfield (Fig. 10.4). These reefs have constituted the by far richest ore bodies of the Witwatersrand.

Eruption of basaltic lava (Bird Member) during deposition of the Krugersdorp Formation is the only magmatic event recorded in the Central Rand Group and was limited to the northeastern portion of the basin (Fig. 10.4). This was followed by a basin-wide transgression that eventually led to deposition of the only laterally extensive marine shale succession (Booysens Formation) in the Central Rand Group. While fluvial environments clearly dominated in the Central Rand Basin, the Booysens Formation attests to occasional marine incursion in places whereas in others evidence of sub-aerial exposure exists in the form of abundant ventifacts (Minter 1999). Many of the economically important conglomerate bodies take the form of thin laterally extensive sheets, which may be interpreted as result of marine reworking of initially fluvial deposits.

10.2.2.2 Turffontein Subgroup

Deposition of this uppermost subgroup within the Witwatersrand Supergroup set in after a further hiatus at the end of Booysens Formation sedimentation. Progradation of braid-plains led to a coarse-grained sequence with several conglomerate beds, unified as Kimberley Formation. Repeated erosional degradation of these conglomerates left behind deep erosion channels in places and several economic placers (Kimberley Reefs). Further erosional reworking of the older stratigraphy is indicated by the overlying Elsburg Formation which locally contains several sub-economic to economic placers (Fig. 10.4). Towards the end of basin evolution, alluvial fans prograded as the hinterland encroached, leaving behind massive conglomerate and boulder beds (Mondeor Formation) with local placer formation where coarse-grained sediment became reworked at the fan-heads.

10.2.2.3 Geochronology and Tectonic Evolution

Palaeocurrent directions are from all around the current known surface and sub-surface extent of the Central Rand Group (Minter and Loen 1991). Several former river entry points into the basin have been identified thanks to underground exploration for gold for more than a century, because they tend to be sites of maximum gold grades in the fluvial to fluvio-deltaic conglomerates. The positions of the goldfields on the eastern, northern and northwestern to southwestern margin of the area over which the Central Rand Group is known (Fig. 10.2) delineate the former basin margin. This is supported by the spatial distribution of sedimentary facies within a given stratigraphic level, specifically grain size and sorting trends downslope (Minter 1978), all of which point to a deepening of the Central Rand Basin towards the centre of today's known domain occupied by Central Rand Group rocks, which approximates the original size and extent of the depositional basin.

Most of the degradation cycles mentioned above can be explained by syndepositional tectonism in the hinterland, which resulted in a number of largely low-angle unconformities within the Central Rand Group. Major such unconformities exist at the bottom of the Krugersdorp, Booysens, Kimberley and Elsburg formations, with increasing intensity of tilting towards the top of the succession. In the Welkom goldfield, the Elsburg Reefs truncate most of the older strata as they had been up-turned already prior to and during Elsburg deposition (Fig. 10.5). This provides evidence of syndepositional uplift of the hinterland to the west, which involved also progressive unroofing of Neoarchaen granites. The latter is indicated by detrital zircon age spectra (Kositcin and Krapez 2004) as well as changes in sediment chemistry (Nwaila et al. 2017).

The maximum age of sedimentation is given by the age of the Crown Formation volcanic rocks, that is 2914 ± 8 Ma (Armstrong et al. 1991), in the underlying West Rand Group. Further age constraints are from U-Pb detrital zircon ages (Kositcin and Krapez 2004): The bottom of the Central Rand Group is younger than 2902 ± 13 Ma, the Krugersdorp Formation younger than 2872 ± 6 Ma, and the upper Elsburg Formation younger than 2840 ± 3 Ma. Lower age limits of the group are given by a U-Pb age on authigenic xenotime of 2780 ± 3 Ma (Kositcin et al. 2003) and by a U-Pb baddeleyite age from the overlying Klipriviersberg Group, lower Ventersdorp Supergroup, that is

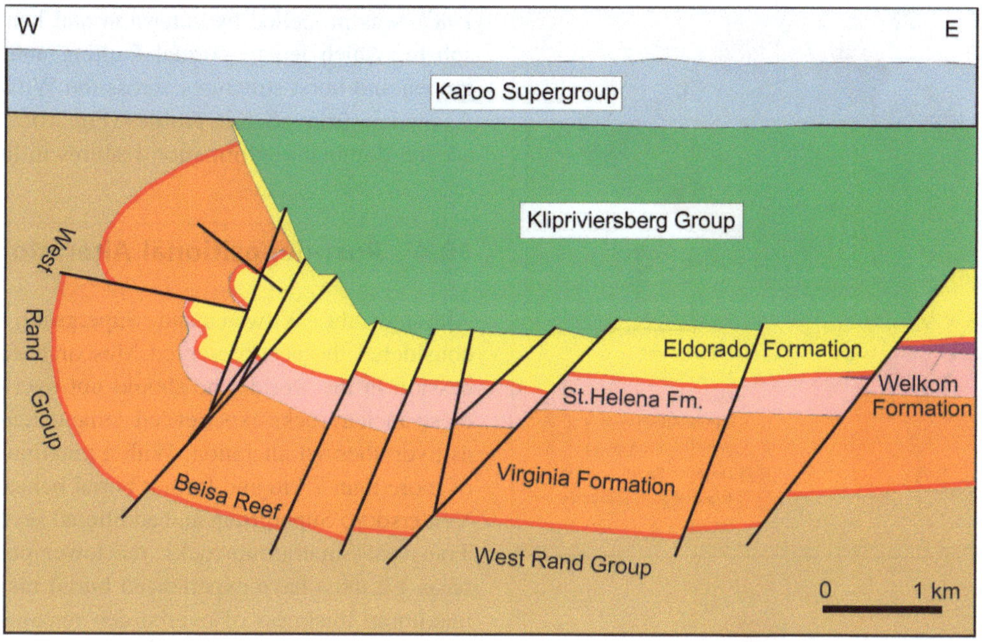

Fig. 10.5 Schematic cross-section through the western margin of the southern part of the Welkom goldfield, showing the up-turning of older Witwatersrand strata prior to and during deposition of the upper Eldorado Formation sediments; note that the Virginia Formation corresponds to the lower Johannesburg Subgroup, St. Helena to the middle Johannesburg Subgroup, i.e. Randfontein to Krugersdorp formations, Spes Bona and Aandenk to Kimberley Formation and Eldorda to the Elsburg and Mondeor formations in Fig. 10.4; from Frimmel and Minter (2002)

2785 ± 1 Ma (Gumsley et al. 2018). Thus, evolution of the Central Rand Basin probably lasted from ca. 2900 to ca. 2785 Ma, which overlaps in time with the Limpopo orogeny, accretion of the Murchison granitoid-greenstone belt to the north and the Kraaipan-Amalia granitoid-greenstone belt to the west (see reviews by Frimmel 2005 and Zeh et al. 2013).

A retroarc foreland setting has been suggested by most workers (e.g. Catuneanu 2001) for several reasons: (i) The numerous low-angle unconformities within the group and syn-sedimentary folding of the older Central Rand Group strata towards the western margin of the basin; (ii) the overall continental composition of most of the sediments and geochemical affinity to arc-derived material (Nwaila et al. 2017); (iii) the large range in palaeocurrent directions from all around the basin (Minter and Loen 1991); (iv) geochemical changes in the sediments, suggesting progressive unroofing of granites in the course of sedimentation (Nwaila et al. 2017); and (v) an overall increase in the complexity of detrital zircon age spectra upsection (Kositcin and Krapez 2004). Taking the above age constraints leaves a time span of some 120 million years, which corresponds to an integrated sedimentation rate of ca. 23 m/myr for the up to 2800 m-thick Central Rand Group. The true sedimentation rate must have been, however, considerably higher, bearing in mind the many hiatuses (of unknown duration) within the group.

10.3 Structure

The structural make-up of the Witwatersrand Basin is, in spite of the scarcity of surface outcrops, comparatively well understood, thanks to many decades of daily underground mapping and surveying by numerous mine geologists (summarised by Dankert and Hein 2010) as well as, more recently, several regional deep seismic (Tinker et al. 2002) and more local 3-D seismic reflection surveys (Manzi et al. 2013). The multitude of data reveals a complex tectonic history with both compressional and extensional forces having shaped the basin to different extents at various times. The seismic data confirmed the unconformity between the West Rand and Central Rand groups as well as the unconformable relation between Witwatersrand and Ventersdorp supergroups, both of which represent periods of tilting and large-scale peneplanation of the old land surfaces.

The above mentioned syn-depositional tectonism in the hinterland during Central Rand Group times, intensifying from Johannesburg to Turffontein Subgroup times, had a profound influence on the structure and sedimentation patterns of the Witwatersrand Basin fill. Within an overall compressional regime, regional uplift of basement domes in the hinterland, development of basin edge faults and uplifted blocks (Fig. 10.6a) became the controlling factors for the source of conglomerates and the location of fluvial channels.

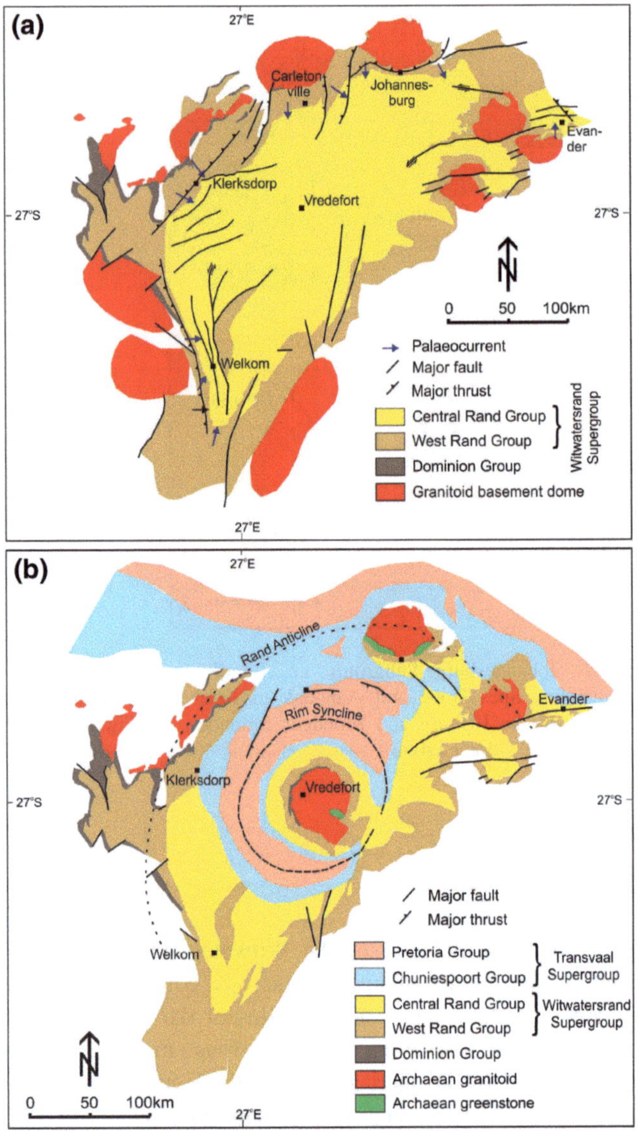

Fig. 10.6 **a** Syn-depositional structures during Central Rand Group sedimentation, showing entry points of major rivers (palaeocurrent directions); **b** Vredefort impact-related structures in the Transvaal and Witwatersrand basins; Ventersdorp and Karoo Supergroup rocks are omitted to better reveal the underlying strata. Modified after McCarthy (2006)

This compressional phase led to first-order folds, thrusts, and reverse faults, especially along the western basin margin, whereas strike-slip movements dominated on the eastern side of the basin (Fig. 10.6a).

After the Central Rand foreland basin had eventually been filled and become subjected to erosion and peneplanation at ca. 2.785 Ga, a major change in the stress field of the central Kaapvaal Craton set in with the outpouring of massive flood basalt and later more evolved volcanic rocks of the Ventersdorp Supergroup. This probably mantle plume-induced episode of volcanism (Crow and Condie 1987) was propelled by extension and heating of the lithosphere, which led to normal faulting and development of graben-and-horst structures across the Witwatersrand Basin. As evident from seismic profiles (Fig. 10.7), these structures are the dominant deformation features in the basin.

10.4 Post-depositional Alteration

Although the Witwatesrand Supergroup has often been considered the best-preserved Mesoarchaean sediment succession in the world, one should not overlook the fact that these ancient rocks experienced some metamorphism as well as hydrothermal alteration. With a maximum total thickness of more than 7 km and further burial beneath some 3 km of Ventersdorp Supergroup and additional several kilometres of Transvaal Supergroup rocks, the lower parts of the former basin fill must have experienced burial metamorphism. The maximum thickness of overburden remains unknown. Integrating the maximum thicknesses reported for the various units of the Transvaal Supergroup yields ca. 16 km but the thickest portions of each unit were never stacked on top of each other in the same area. Furthermore, major erosion on top of the Witwatersrand and the Ventersdorp supergroups would have removed large portions of older stratigraphy prior to the deposition of younger strata. Further burial beneath the Mesozoic Karoo Supergroup rocks probably did not add to the overall maximum metamorphic grade because it occurred after parts of the Palaeoproterozoic succession had already been removed by erosion. Although we do not know the exact maximum depth to which the Witwatersrand Supergroup rocks had been buried, phase relations indicate lower greenschist-facies conditions (Wallmach and Meyer 1990; Frimmel 1994). Tectonic shortening in the course of thrusting of the Limpopo Belt's Southern Marginal Zone onto the Kaapvaal Craton between 2690 and 2670 Ma affected at least the northern parts of the Witwatersrand Basin. The emplacement of the world's largest layered intrusive complex, the 2054 Ma Bushveld Complex, to the immediate north of the Witwatersrand Basin must have had at least some thermal effect on the latter's fill. Of probably even greater consequence was the 2023 Ma Vredefort impact that left behind a 90 km-wide domal structure in the middle of the known area of Central Rand Group rocks. The original diameter of the impact structure has been modelled from geophysical data as 250–280 km, which corresponds closely to the size of the Witwatersrand Basin (Henkel and Reimold 1998). An impact-induced basin-wide network of fractures and low-angle thrusts, now partially filled with pseudotachylyte, created a secondary permeability long after primary permeability of the sedimentary rocks had been eliminated by diagenesis and metamorphism. This made it

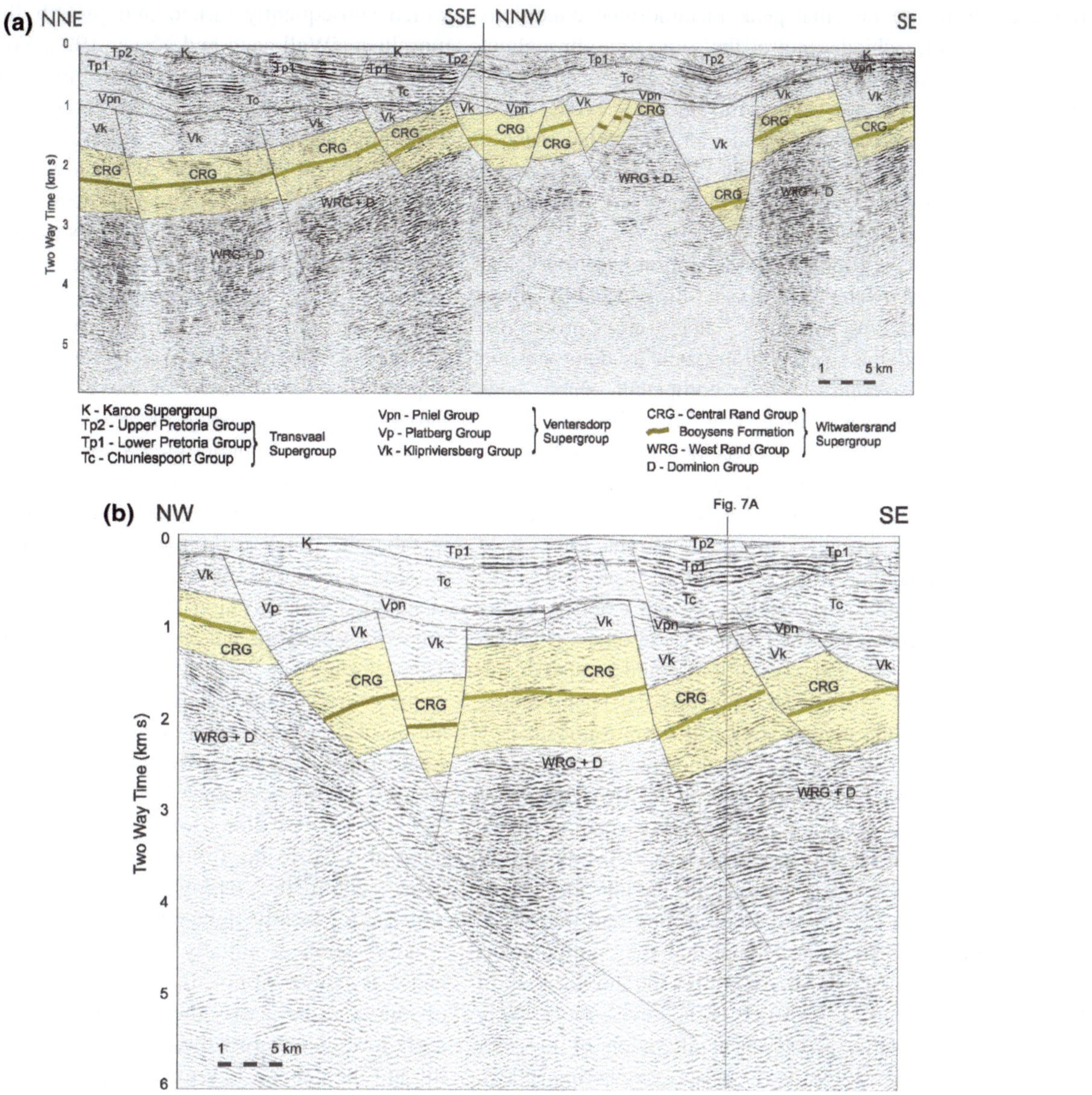

Fig. 10.7 Stacked and migrated reflection seismic profiles through parts of the Witwatersrand Basin with stratigraphic interpretation from Tinker et al. (2002); highlighted are the Central Rand Group and the Booysens Formation shale in the middle of that group; for lines of sections see Fig. 10.2

possible for fluids to circulate throughout the former basin fill—at least for a short time—and cause hydrothermal alteration along the fractures (Frimmel et al. 1999). Although the impact cannot have anything to do with the exceptional gold endowment of the Witwatersrand rocks (the mineralisation took place some 900 million years before the impact occurred), it played a decisive role in the preservation of these Archaean sedimentary rocks. The infolding of the Witwatersrand strata in a concentric synclinorium around the Vredefort Dome (Fig. 10.6b) helped to preserve them from later erosion, thus providing an explanation for the apparently fortuitous position of the world's largest known impact structure in the centre of the area of preserved Central Rand Basin fill (McCarthy 2006).

The net result of the above events was generally low-grade metamorphism, reaching even higher grade in the lower parts of the basin fill, which are now exposed up-turned to almost vertical position along the collar of the impact-induced Vredefort Dome (Gibson and Wallmach 1995). The generally low metamorphic grade, quantified by thermodynamic modelling of observed phase relationships (Wallmach and Meyer 1990; Frimmel 1994; Phillips and Law 1994) as 350 ± 50 °C at ca. 3 kbar, seems fairly uniform across the former basin. This should not distract, however, from the fact that peak metamorphic conditions were most likely achieved not at the same time throughout the basin but at different times, depending on position (Frimmel 1997). Various tectono-metamorphic events along the margins of the Kaapvaal Craton after the emplacement of the Bushveld Igneous Complex had, by and large, only very little influence on the Witwatersrand Basin fill and are not further considered here.

Evidence of hydrothermal alteration is prevalent throughout the Witwatersrand Basin fill, especially in the more competent rock types that are prone to fracture-controlled fluid percolation, such as quartzite and conglomerate. In contrast, less competent shale beds remained barely affected. In general, the observed metamorphic mineral assemblages, especially the abundance of pyrophyllite, indicate largely acidic fluid conditions. This has led some workers to argue for basin-wide H^+-metasomatism that also supposedly caused the introduction of gold into the Witwatersrand strata (Barnicoat et al. 1997). As opposed to fluid infiltration from a source external to the basin as suggested by the latter authors and others (Phillips and Powell 2011), the acidic conditions could be also due to diagenetic flow of groundwater towards the basin centre, following a hydraulic gradient that had been established by uplift along the basin margin. With increasing burial, maturation of organic matter and release of H^+ by dehydration reactions could have lowered the pH, thus leading to the dissolution of detrital feldspars. Such a development of groundwater and formation water could explain the lack of detrital feldspar in the Central Rand Group and the observation, based on geochemical data, that the arenitic rocks remained largely isochemical, except for the mobilisation of potassium (Sutton et al. 1990). It is, however, more likely that, at least in parts, geochemical alteration of the predominantly arenitic rocks is a palaeo-weathering effect. A wealth of geochemical analyses of arenitic units, across the Central Rand and subordinately from the West Rand Group, document progressive removal of Ca and Na (related to plagioclase weathering), followed by removal of K (due to K-feldspar weathering) from the average composition of Archaean upper continental crust (Frimmel 2005). The Chemical Index of Alteration (CIA), as defined by Nesbitt and Young (1982), of the various sedimentary units varies greatly with stratigraphic depth and reaches particularly high values in the Central Rand Group (Frimmel 2005). Peaks in CIA are distributed not randomly but systematically in the footwall of old erosion surfaces. This strongly points at intense chemical weathering on the old land surfaces that were exposed in Central Rand Group times with alteration of detrital feldspar to kaolinite (and similar clay minerals), which subsequently turned into pyrophyllite during metamorphism (Wallmach and Meyer 1990; Frimmel 1994).

Chemical weathering over tens of metres into the footwall below conglomerate beds (palaeoerosion surfaces) notwithstanding, there is also evidence of more or less bedding-parallel fluid flow along low-angle thrust planes within some of the conglomerate beds (Jolley et al. 1999). A prime example is the post-Witwatersrand Ventersdorp Contact Reef, an up to several metres thick fluvial conglomerate with minor quartzite intercalations at the base of the Ventersdorp Supergroup. It forms the richest ore body outside of the Witwatersrand Supergroup. Capping by impermeable flood basalt of the 2.785 Ga Klipriviersberg Group enabled focusing of post-depositional fluids that emanated from the underlying Witwatersrand strata into this conglomerate unit. This caused initially considerable potassic alteration along the conglomerate and into the overlying metabasalt as well as the underlying quartzitic footwall over a vertical distance of a few metres, followed by more channelled chloritisation along the conglomerate and the immediate hanging and footwall (Gartz and Frimmel 1999). Thus, throughout the Witwatersrand, alteration phenomena in the coarser grained sedimentary units comprise a combination of large-scale, more or less stratiform, deep chemical weathering below palaeoerosion surfaces and, locally, highly focused post-depositional potassic alteration (sericitisation) and chloritisation along some of the conglomerate beds. The latter also led to the formation of quartz veins (and calcite veins within the overlying Klipriviersberg metabasalt), which are locally mineralised, thus testifying to some remobilisation of initially conglomerate-hosted ore components. It should be noted, however, that the overall amount of quartz veins across the Witwatersrand strata is surprisingly low, considering the sheer amount of SiO_2 in the basin fill and the general low metamorphic grade.

Witwatersrand shales show little evidence of hydrothermal alteration, except in the immediate vicinity of faults, dykes, veins, and along contacts with adjacent more permeable arenitic rocks. They retained mass-independent sulphur isotope signals (uniform negative $\Delta^{33}S$ and largely negative $\delta^{34}S$ ratios) that are indicative of Mesoarchaean atmospheric sulphur speciation (Farquhar et al. 2007)—signals that would not survive significant post-depositional hydrothermal overprint (Hofmann et al. 2009; Guy et al. 2010). Consequently, the composition of marine Witwatersrand shales can be used as proxy of provenance and intensity of chemical weathering in the source areas. Geochemical data of such shales reveal high CIA values and thus provide evidence of intense chemical weathering in their source areas (Nwaila et al. 2017).

10.5 Sediment Provenance

Most of the Witwatersrand conglomerates have a pebble assemblage that is strongly dominated by quartz, much of which is derived from granite and pegmatite (55%) and from greenstone-hosted vein quartz and chert (45%), as indicated by oxygen isotope data (Vennemann et al. 1992, 1995). Volumetrically minor clast types are quartz porphyry (3%) and metamorphic clasts (2%). Exceptions are the Steyn Reef in the Welkom goldfield with up to 30% porphyry clasts and the Kimberley Reef in the Evander goldfield with as much as 50% chert clasts (Tweedie 1986). A mix of (ultra)mafic and felsic source rocks is indicated by the heavy mineral populations, which include detrital chromite and PGE-minerals from former and detrital zircon, uraninite, columbite, molybenite and cassiterite from latter sources, respectively (Feather and Koen 1975). Such a lithological mix in the hinterland is also indicated by the geochemistry of the arenitic fraction, which reflects variable degrees of chemical weathering of mafic to felsic source rocks (Frimmel 2005). The same applies to shale beds in the West and Central Rand groups (Nwaila et al. 2017). Several elements, typically concentrated in detrital phases, can be used as proxies for the relative contribution of certain source rock types to the sediment mix. Examples are Zr for felsic sources and Cr and Sc for mafic sources. Furthermore, a negative correlation between Ni and total organic carbon (TOC) attests to the Ni budget in the pelitic sediments to be principally controlled by the proportion of mafic/ultramafic rocks in the hinterland (Nwaila et al. 2017).

As mafic and ultramafic rocks are more magnesian, pelitic sediments from such source rocks have a higher X_{Mg} (= Mg/(Mg + Fe) ratio. Thus, using the ratios of elements typical of felsic and mafic/ultramafic rocks, respectively, the relative proportion of principal source rock types for different stratigraphic levels across the Witwatersrand Supergroup have been calculated (Nwaila et al. 2017), according to which the lowermost West Rand Group sediments were derived from granitic sources, with a roughly equal proportion of TTG and granitic source in the upper Hospital Hill Group, again an almost exclusively granitic source for the lower Government Subgroup, changing to predominantly TTG and minor granitic sources for the upper part of that subgroup. The Jeppestown Subgroup reflects a major change in sediment provenance towards a mix of TTG and mafic source. Across the Johannesburg to lower Turffontein subgroups, mafic rocks played a particularly dominant role as source, whereas towards the top of the latter group, TTGs and granitic sources became dominant again. Significant deviations from this general trend can be observed within a given stratigraphic unit between different goldfields. For example, the provenance of the Booysens Formation shale in the Welkom goldfield was felsic whereas farther north(east) it was predominantly mafic. At a given locality, the proportion of mafic to felsic detrital sediment input can change substantially within a single formation, e.g. Zr/Ni increasing from < 0.2 (mafic) to > 0.8 (felsic) from the bottom to the top of the Booysens Formation shale (Nwaila et al. 2017). Detrital zircon ages in the Witwatersrand metasedimentary rocks mirror the age distribution of the surrounding Archaean granitoid-greenstone belts (Frimmel et al. 2005; Koglin et al. 2010b), thus strongly confirming that these belts acted as principal source domains for the Witwatersrand sediments.

10.6 Mode of Mineralisation

Most of the gold occurs, together with other heavy minerals of undoubtedly allogenic nature, in fluvial conglomerates, pebbly arenite and pebble lag surfaces associated with trough cross-bedded quartz arenite (Fig. 10.8). The orebodies are typically bounded by a basal degradation surface, usually an angular unconformity, and an upper planar bedding surface that is due to either aeolian deflation (Minter 1999) or shoreline encroachment (Buck and Minter 1985). Thicker (a few metres) units represent multichannel sequences with the heavy minerals, including gold, being concentrated along the basal degradation surface of each graded bed. Elsewhere, a few centimetres-thin pebble lags define degradation surfaces that can contain very high gold grades (Frimmel et al. 2005). Furthermore, elevated heavy placer mineral contents are found at the base of clast-supported oligomictic to loosely packed conglomerates, rarely in immature debris flow deposits (Minter et al. 1988). The contacts of the ore bodies are always sharp with regard to ore grade. Whereas the Au grade within the conglomerates can attain several grams to tens of grams per tonne, it drops by a few orders of magnitude across the boundaries, be it an unconformity or bedding plane, against the arenitic footwall and hanging wall.

In addition, a considerable proportion of gold is hosted by millimetre- to centimetre-thick kerogen layers, locally referred to as 'carbon seams' (Fig. 10.9), which in some instances, such as the Carbon Leader Reef, hosted as much as 70% of the entire gold (Hallbauer and Joughin 1973). Such gold-rich kerogen mats occur essentially in the lower Central Rand Group, good examples being the mentioned Carbon Leader, the Basal Reef or the Vaal Reef.

For decades it has been repeatedly observed that, under the microscope, much of the gold occurs in late paragenetic textural positions (Ramdohr 1958), together with

Fig. 10.8 **a** Gold concentrated along crossbeds in pebbly quartz arenite at the base of the Basal Reef, Free State Geduld mine, Welkom goldfield; **b** Attenuation radioscopic X-ray image of gold particles along the crossbeds; **c** 3-D detail of the middle of radiograph shown in (**b**); width of photo is 4 mm; from Hölzing et al. (2015)

hydrothermal bitumen, metamorphic chlorite (Gartz and Frimmel 1999), pyrophyllite (Barnicoat et al. 1997), or as inclusions in euhedral, secondary pyrite (Frimmel 2005). This type of gold is, without doubt, the product of precipitation from hydrothermal/metamorphic fluids, which has led several researchers to suggest an overall epigenetic, hydrothermal origin of the gold (e.g. Phillips and Law 2000; Phillips and Powell 2011). It should be noted, however, that the above late paragenetic position of the gold does not provide any indication as to the distance of hydrothermal gold transport, whether the gold was introduced into the conglomerates from an external source far away or whether it represents locally remobilised placer gold within the conglomerates. Several observations speak for the latter case. Some of the gold occurs in textural positions and morphological forms that leave little doubt as to their detrital nature. These are mechanically rounded, spheroidal to toroidal particles that are associated with other heavy minerals as placer concentrates, a prime example of which is illustrated in Fig. 10.8. With typical diameters of some 100–150 μm they can be regarded as micronuggets. Larger, proper nuggets are notably absent.

In spite of the exceptional endowment, visible gold is rare in the Witwatersrand. It is found either as placer concentrate (Fig. 10.8) or as secondary, hydrothermal gold in quartz veins, micro-shears and fractures. Gold in the latter, cross-cutting features is typically confined to domains in the immediate vicinity of gold-rich conglomeratic reefs. It thus reflects mobilisation of gold from the conglomerate beds into whatever cross-cutting, fluid-filled feature, whereas nowhere is an example known that would illustrate dispersion of gold from a cross-cutting feeder channel into the siliciclastic wall rocks. Although various epigenetic-hydrothermal models, analogous to orogenic gold models, have been suggested for decades by some workers (e.g. Phillips and Powell 2011) as prospective guide for future exploration, not a single cross-cutting ore body has been mined or discovered over the century-long mining history of the Witwatersrand.

Besides gold, U-minerals, pyrite and hydrocarbons are important constituents of the ore whose genesis has been a matter of debate as much as that of the gold itself. Their mode of occurrence has direct bearing on our understanding of Witwatersrand gold genesis. The U-minerals can be locally of economic significance, with some 28 mines having been producers of uranium in the past, mainly from the Bird Reefs in the West Rand and Klerksdorp goldfields (Tucker et al. 2016). Uraninite is the by far most abundant U-mineral in the gold fields and occurs mainly as semi-rounded, millimetre-sized grains whose chemical composition speaks for an evolved granitic/pegmatitic provenance but against a hydrothermal origin (Frimmel et al. 2014). The latter study also revealed systematic differences in the trace element distribution within uraninite populations from different former river systems (Basal and Steyn placers, Welkom goldfield), all of which leaves little doubt on the detrital nature of these grains. Minute, in places sub-microscopic U-minerals, predominantly brannerite, subordinately secondary uraninite, are doubtlessly the product of hydrothermal mobilisation of some of the uranium by post-depositional fluids (Depiné et al. 2013; Fuchs et al. 2016). The variety of pyrite types is even wider: pre-, syn- and post-depositional types have been distinguished, initially merely on morphological grounds (Hallbauer 1986), later on also based on sulphur isotopic evidence (England et al. 2002; Hofmann et al. 2009) and trace element data (Koglin et al. 2010a; Guy et al. 2010; Agangi et al. 2015). These studies document the presence of (i) rounded, mechanically abraded, evidently detrital pyrite derived from hydrothermal veins, volcanic-hosted massive sulphide deposits, sulfidised Fe-oxides or black shales; (ii) concentrically laminated pyrite rich in trace elements, including Au, that formed within the wet sediment and/or during early diagenesis, and (iii) epigenetic-hydrothermal pyrite in the form of euhedral cubes or overgrowths around the previous two types. Detrital pyrite is the by far most abundant type. Locally, some reefs, such as the Ventersdorp Contact Reef, seemingly contain more euhedral pyrite, but careful examination by etching revealed that many of the euhedral pyrite grains represent hydrothermal overgrowths around detrital pyrite cores (Gartz and Frimmel 1999). Repeated mechanical reworking of ore constituents is beautifully illustrated by a mechanically rounded polymictic pyritic pebble with inclusions of rounded uraninite, rounded pyrite, rounded pyrrhotite and rounded chromite all of which are bound together by secondary interstitial pyrite (Fig. 10 in Tucker et al. 2016).

Hydrocarbons occur in two principal forms, the above-mentioned stratiform 'carbon seams' on top of erosion surfaces and millimetre- to centimetre-sized nodules (so-called 'flyspeck carbon'). Although some workers (e.g. Gray et al. 1998; Drennan and Robb 2006; Fuchs et al. 2016) argued for both of these being products of migrating oils (thus representing pyrobitumen), field relations and isotopic data argue against such an interpretation. Whereas the nodules are without doubt pyrobitumen and related to post-depositional fluid flow, as they also occur within quartz veins, the 'carbon seams' are interpreted as former microbial mats as originally suggested by Hallbauer (1975). This is supported by a wealth of very delicate sedimentary structures defined by the 'carbon seams', and field evidence of intraformational sedimentary reworking of such seams (Mossman et al. 2008; Frimmel 2014). In addition, C isotope data (Spangenberg and Frimmel 2001) reveal a lack of fractionation across the range of *n*-alkanes, which argues against any long-range mobilisation of bitumen (Frimmel and Hennigh 2015). The texture of the 'carbon seams' has been affected, however, by post-depositional strain, and original microbial microstructures were lost.

An intimate spatial association of uraninite and hydrocarbons, evident in many reefs, can be explained by three independent processes, that is (i) microbial growth on pebble beds rich in heavy detrital minerals, including uraninite, (ii) polymerisation and cross-linking of liquid hydrocarbons on the surface of detrital uraninite grains (Schidlowski 1981), and (iii) remobilisation of U by carbonic fluids giving rise to secondary U-minerals intergrown with pyrobitumen (Fuchs et al. 2016). A positive correlation exists between Au and U as expected for elements principally housed in co-existing detrital minerals, but in most areas this correlation is relatively weak because of short-range dispersion of both elements by post-depositional fluids (Frimmel and Minter 2002). Similarly, a positive correlation exists between Au and Zr in some instances. A larger data set, however, reveals many conglomerate samples rich in Zr and poor in Au (as expected for heavy mineral concentrates that lack detrital gold) but, with a single exception, no sample that is rich in Au and poor in Zr, as would be expected for hydrothermal systems in which Zr is typically not very mobile (Frimmel et al. 2005).

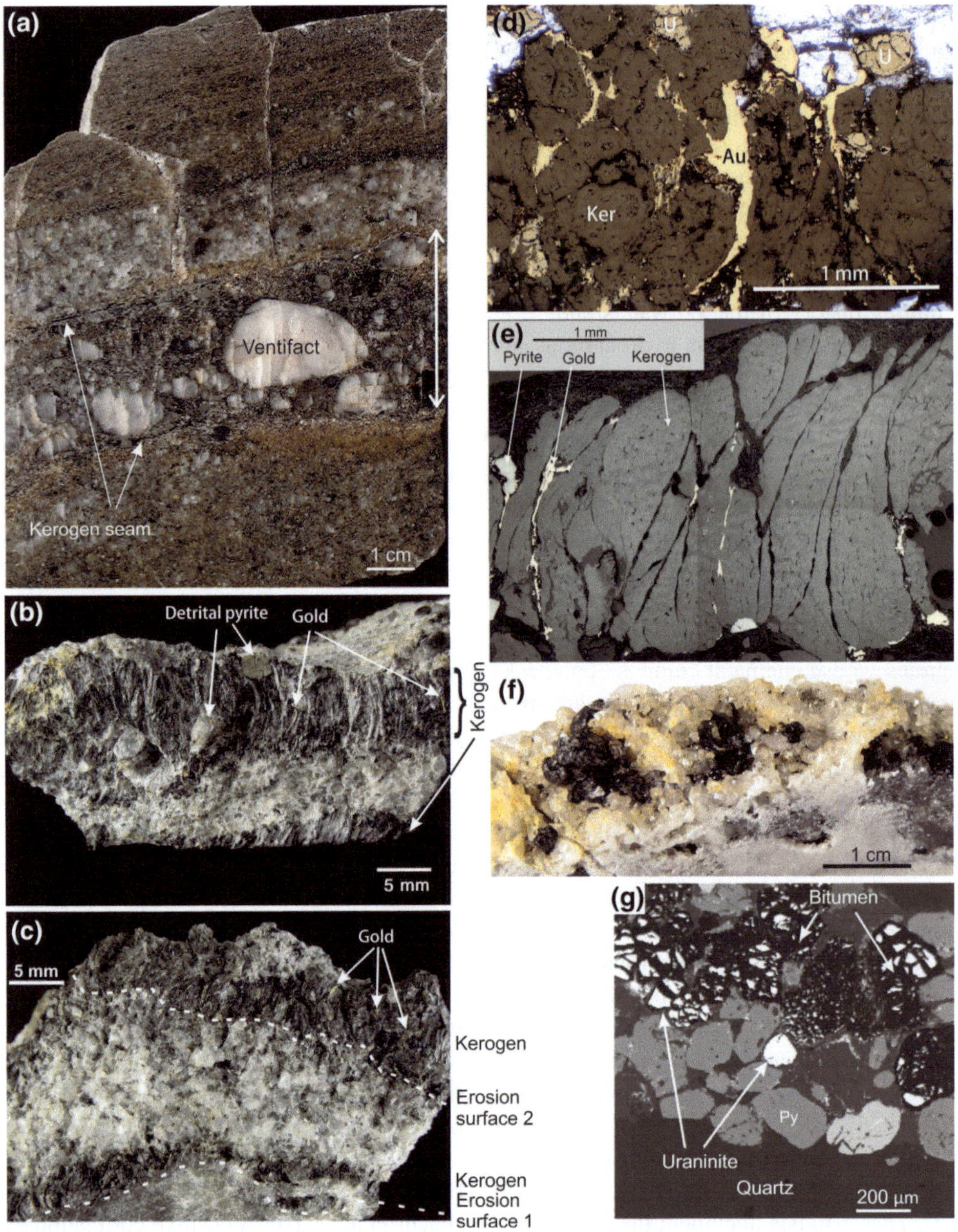

Fig. 10.9 Mode of hydrocarbon occurrence in Witwatersrand rocks: **a** Thin kerogen seams on scour surfaces at bottom of pebble lags, Steyn Reef, President Steyn mine, Welkom goldfield; pebble lag represents an aeolian deflation surface (vertical double arrow) as indicated by ventifact; **b**,**c** Gold-rich kerogen seams with interstitial heavy minerals (pyrite) at the base of the Vaal Reef, Stilfontein mine, Klerksdorp goldfield; **d** microphotograph under combined transmitted and reflected light of kerogen seam (Ker) rich in platy gold (Au) and intercalated detrital uraninite (U) from base of the B-Reef, Tshepong mine, Welkom goldfield; **e** reflected light microphotograph of columnar kerogen with gold platelets between columns from base of Carbon Leader, West Driefontein mine, Carletonville goldfield (from Frimmel and Hennigh 2015); **f** pyrobitumen globules (black) on drusy quartz vein, Ventersdorp Contact Reef, Tau Lekoa mine, Klerksdorp goldfield; **g** back-scatter electron image of pyrobitumen (black) enclosing broken-up detrital uraninite fragments (white) in quartz-pebble conglomerate, note that some uraninite (rounded grain in the middle) is not enclosed by pyrobitumen, Vaal Reef, Great Noligwa mine, Klerksdorp goldfield; from Depiné et al. (2013)

10.7 Towards a Holistic Metallogenetic Model

A purely epigenetic, hydrothermal model for the Witwatersrand gold can be discarded for many reasons, including the presence of evidently detrital micronuggets (Fig. 10.8), the strong sedimentological control on ore grade and the lack of cross-cutting ore bodies. A purely syngenetic placer model defies, however, the observation of much of the gold occurring in late paragenetic textural positions. This apparent contradiction has been overcome by the currently favoured 'modified placer model', according to which placer mineralisation was followed by mobilisation of ore components, including gold, by post-depositional fluids (Robb and Meyer 1991; Frimmel et al. 1993). It should be emphasised, however, that the mobilisation of gold took place, in most instances, over only very short distances of a few millimetres to maybe metres, largely confined to within the original host conglomerate beds, and is thus of little economic significance. While consensus on this genetic model seems to be reached, the ultimate question as to the source of all the implied detrital gold in the Witwatersrand Basin has remained unresolved and posed, in fact, the possibly biggest challenge to a (modified) placer model.

Numerous discrete sources have been suggested in the past, foremost greenstone-hosted auriferous vein deposits in the hinterland. Apart from a gold chemistry that is at odds with a hydrothermal source (Witwatersrand gold contains orders of magnitude more Os than any other hydrothermally derived gold analysed so far), simple mass balance excludes such a source (Frimmel 2014). Other suggestions, such as hydrothermally altered granites in the hinterland or syn-depositional exhalites along the basin margin, had to be discarded when the granite alteration was found to be younger than the gold mineralisation in the Witwatersrand and due to a lack of any geological evidence of the suggested exhalites along the basin margin, respectively. No kind of discrete gold deposit in the Mesoarchaean hinterland can explain the huge amounts of postulated detrital gold in the Central Rand Basin. This leaves only the mobilisation of background concentrations of Au from the hinterland or underlying crust as alternative. Such large-scale leaching could have been achieved by regional metamorphic devolatilisation, as suggested by Phillips and Powell (2011), but this would imply an overall epigenetic gold mineralization, which has been discarded as likely genetic model (see above). Large et al. (2013) proposed pyritic shales farther down in the Witwatersrand stratigraphy as important source of gold (and pyrite) in the Central Rand Group conglomerates, but problems in both mass balance and timing of sediment deposition, mineralisation and fluid flow render this hypothesis unlikely. The only remaining option is syn-depositional leaching of background Au by surface waters.

Reconstruction of the likely environmental conditions during the Mesoarchaean, combined with thermodynamic modelling, led Frimmel (2014) and, independently, Heinrich (2015) to suggest that gold solubility in river waters at that time was orders of magnitude greater than today. This, combined with a period of intense chemical weathering under an aggressive acid atmosphere, enabled craton-wide leaching of gold from the granitoid-greenstone terrains in the hinterland and a tremendous flux of Au, largely in dissolved form, off the land into the ocean. Thus the entire Mesoarchaean land surface becomes a potential gold source. Only a small portion of all this gold was needed to be trapped to form the first major gold deposits. And the most effective trap was constituted by early microbial mats in coastal wetlands, shallow marine to littoral positions, and on mudflats along rivers. The geological evidence of the oldest, very rich ore bodies in the Witwatersrand (e.g. Carbon Leader) attests to the efficiency of such microbial mats in fixing gold from the surrounding river water or seawater. This could have been achieved either by reduction on the organic surface as suggested by Heinrich (2015) or by oxidation in an overall O_2-deficient atmosphere, triggered by the first O_2-producing photosynthesisers, most likely cyanobacteria, as suggested by Frimmel (2014). Whatever the redox-reaction might have been, microbial fixation of large amounts of gold is likely and, considering the extremely low preservation potential of such microbial mats, provides an ideal source for the detrital gold in the Witwatersrand conglomerates (Fig. 10.10). It also explains the lack of proper larger nuggets because of the generally very small size of the delicate gold filaments and platelets that drape kerogen columns in the 'carbon seams'.

All in all, the exceptional gold endowment of the Witwatersrand is now seen as the product of the fortuitous interplay of changes in atmospheric conditions, chemical weathering rate, abundance of reduced S-species in the meteoric environment, and the evolution of life forms, specifically the emergence of photosynthesising microbes at around 2.9 Ga. It is the product of a unique window of opportunity, never to be repeated again, when the conditions were ideal for continent-wide leaching and large-scale trapping of gold in and around coastal environments. Mechanical reworking of the gold-bearing microbial mats led to initially very rich placer deposits (Fig. 10.10).

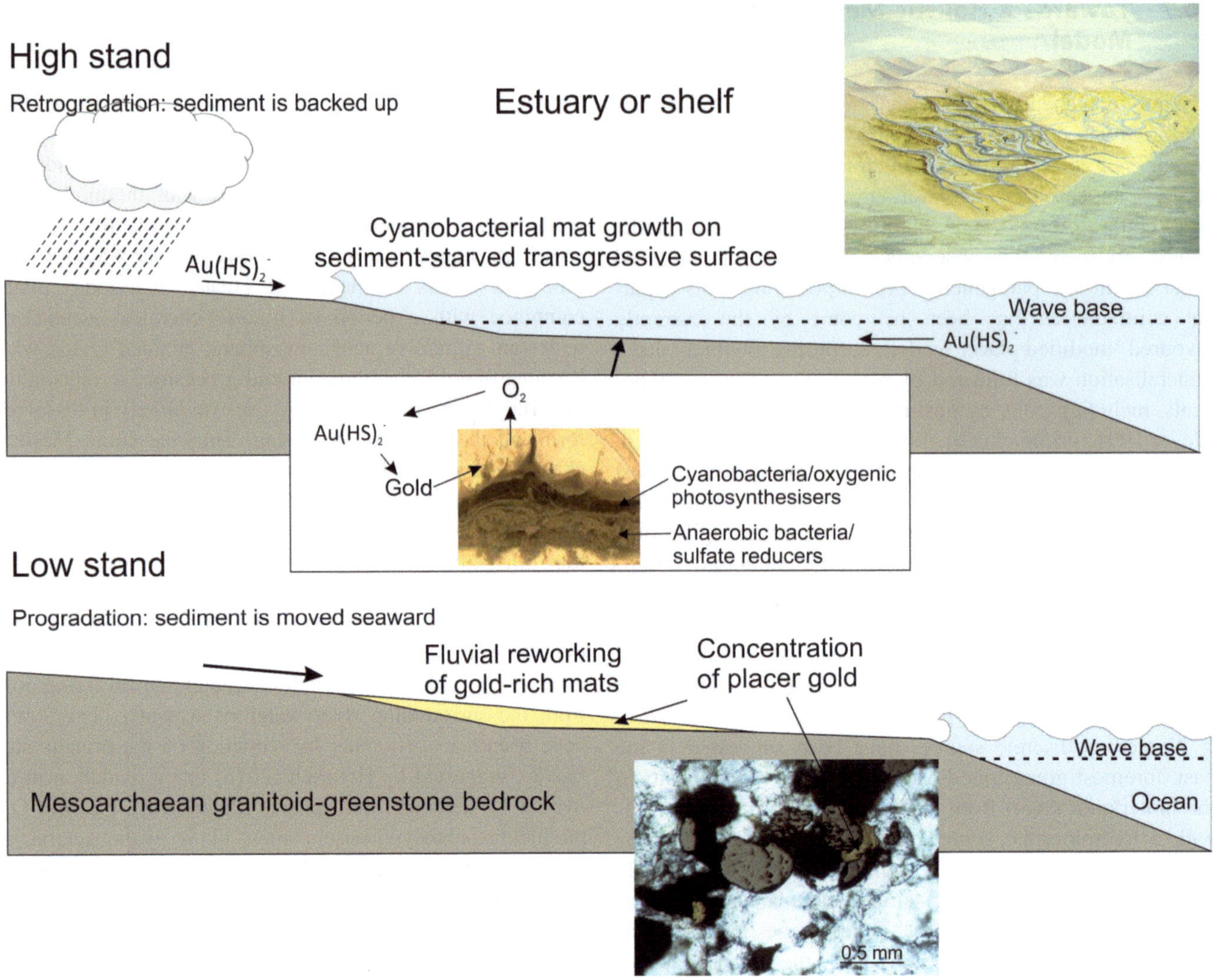

Fig. 10.10 Schematic cartoon illustrating the two principal gold concentration mechanisms in the Mesoarchaean Central Rand Basin: (i) microbial gold fixation from $Au(HS)_2$—bearing meteoric waters and shallow seawater and (ii) sedimentary reworking of that gold by rivers

The potential to form such rich placers diminished over time when less and less of the microbial mats and early placers became available for erosion (Frimmel 2018).

The above processes of gold accumulation were surely not restricted to the Kaapvaal Craton but should have operated on all Mesoarchaean cratons and their margins. In most areas, the corresponding sediments became reworked during subsequent erosion or tectonic, orogenic, magmatic or metamorphic overprints. The uniqueness of the Witwatersrand is rooted in its exceptional preservation, which is owed to the incidental combination of three geological factors: (i) the Witwatersrand sediments became covered to a large extent by a thick pile of extremely competent, erosion-resistant basalt (Ventersdorp Supergroup), (ii) the subsequent deformation into bowl-shape by the Vredefort impact and further protection by a likely veneer of hardened impact melt, and (iii) its position in the middle of one of the oldest and most buoyant cratons where it largely escaped later orogenic overprints.

Acknowledgements Numerous mining companies and their mine geologists are thanked for providing access to underground workings and samples over a > 25 years research period. W.E.L. Minter is thanked for passing on to me valuable collections of rock samples from old mines that are not accessible anymore. Parts of the thoughts expressed in this paper are based on research funded by the South African National Research Foundation (NRF) and the Deutsche Forschungsgemeinschaft (DFG grant FR2183/3).

References

Agangi A, Hofmann A, Rollion-Bard C, Marin-Carbonne J, Cavalazzi B, Large R, Meffre S (2015) Gold accumulation in the Archaean Witwatersrand Basin, South Africa—Evidence from concentrically laminated pyrite. Earth-Sci Rev 140:27–53

Armstrong RA, Compston W, Retief EA, William LS, Welke HJ (1991) Zircon ion microprobe studies bearing on the age and evolution of the Witwatersrand triad. Precambr Res 53:243–266

Barnicoat AC, Henderson IHC, Knipe RJ, Yardley BWD, Napier RW, Fox NPC, Kenyon AK, Muntingh DJ, Strydom D, Winkler KS, Lawrence SR, Cornford C (1997) Hydrothermal gold mineralization in the Witwatersrand basin. Nature 386:820–824

Beukes NJ, Cairncross B (1991) A lithostratigraphic-sedimentological reference profile for the late Mozaan Group, Pongola Sequence: application to sequence stratigraphy and correlation with the Witwatersrand Supergroup. S Afr J Geol 94:44–69

Buck SG, Minter WEL (1985) Placer formation by fluvial degradation of an alluvial fan sequence: the Proterozoic Carbon Leader placer, Witwatersrand Supergroup, South Africa. J Geol Soc London 142:757–764

Catuneanu O (2001) Flexural partitioning of the late Archaean Witwatersrand foreland system. Sed Geol 141–142:95–112

Crow C, Condie KC (1987) Geochemistry and origin of late Archaean volcanic rocks from the Rhenosterhoek Formation, Dominion Group, South Africa. Precambr Res 37:217–229

Dankert BT, Hein KAA (2010) Evaluating the structural character and tectonic history of the Witwatersrand Basin. Precambr Res 177: 1–22

Davidson CF (1965) The mode and origin of banket orebodies. Institute of Mining and Metallurgy, London, Transcripts 74:319–338

de Wit MJ, Armstrong RA, Kamo SL, Erlank AJ (1993) Gold bearing sediments in the Pietersberg greenstone belt: age equivalents of the Witwatersrand Supergroup sediments. Econ Geol 88:1242–1252

Depiné M, Frimmel HE, Emsbo P, Koenig AE, Kern M (2013) Trace element distribution in uraninite from Mesoarchaean Witwatersrand conglomerates (South Africa) supports placer model and magmatogenic source. Miner Deposita 48:423–435

Drennan GR, Robb LJ (2006) The nature of hydrocarbons and related fluids in the Witwatersrand Basin, South Africa: Their role in metal redistribution. Geol Soc Am Spec Pap 405:353–385

Els BG (1998) The question of alluvial fans in the auriferous Archaean and Proterozoic successions of South Africa. S Afr J Geol 101:17–25

England GL, Rasmussen B, Krapez B, Groves DI (2002) Palaeoenvironmental significance of rounded pyrite in siliciclastic sequences of the Late Archaean Witwatersrand Basin: oxygen-deficient atmosphere or hydrothermal evolution. Sedimentology 49:1122–1156

Eriksson KA, Turner BR, Vos RG (1981) Evidence of tidal processes from the lower part of the Witwatersrand Supergroup, South Africa. Sed Geol 29:309–325

Farquhar J, Peters M, Johnston DT, Strauss H, Masterson A, Wiechert U, Kaufman AJ (2007) Isotopic evidence for Mesoarchaean anoxia and changing atmospheric sulphur chemistry. Nature 449:706–709

Feather C, Koen GM (1975) The mineralogy of the Witwatersrand reefs. Minerals Science and Engineering 7:189–224

Frimmel HE (1994) Metamorphism of Witwatersrand gold. Explor Min Geol 3:357–370

Frimmel HE (1997) Chlorite thermometry in the Witwatersrand basin: constraints on the Paleoproterozoic geotherm in the Kaapvaal Craton, South Africa. J Geol 105:601–615

Frimmel HE (2005) Archaean atmospheric evolution: evidence from the Witwatersrand gold fields, South Africa. Earth Sci Rev 70:1–46

Frimmel HE (2010) Verfügbarkeit von natürlich vorkommendem Uran. Unpublished Report, Office of Technology Assessment at the German Bundestag (TAB), Berlin, p 123

Frimmel HE (2014) A giant Mesoarchean crustal gold-enrichment episode: possible causes and consequences for exploration In: Kelley K, Golden HC (eds) Building exploration capability for the 21st Century. Society of Economic Geologists, Special Publication 18, pp 209–234

Frimmel HE (2018) Episodic concentration of gold to ore grade through Earth's history. Earth Sci Rev 180:148–158

Frimmel HE, Le Roex AP, Knight J, Minter WEL (1993) A case study of the postdepositional alteration of the Witwatersrand Basal reef gold placer. Econ Geol 88:249–265

Frimmel HE, Hallbauer DK, Gartz VH (1999) Gold mobilizing fluids in the Witwatersrand Basin: composition and possible sources. Mineral Petrol 66:55–81

Frimmel HE, Minter WEL (2002) Recent developments concerning the geological history and genesis of the Witwatersrand gold deposits, South Africa. In: Goldfarb RJ, Nielsen RL (eds) Integrated methods for discovery: global exploration in the twenty-first century. Society of Economic Geologists, Littleton, Special Publication 9, pp 17–45

Frimmel HE, Groves DI, Kirk J, Ruiz J, Chesley J, Minter WEL (2005) The formation and preservation of the Witwatersrand goldfields, the largest gold province in the world. In: Hedenquist JW, Thompson JFH, Goldfarb RJ, Richards JP (eds) Economic Geology One Hundredth Anniversary Volume. Society of Economic Geologists, Littleton, Colorado, pp 769–797

Frimmel HE, Zeh A, Lehrmann B, Hallbauer DK, Frank W (2009) Geochemical and geochronological constraints on the nature of the immediate basement beneath the Mesoarchaean auriferous Witwatersrand Basin, South Africa. J Petrol 50:2187–2220

Frimmel HE, Schedel S, Brätz H (2014) Uraninite chemistry as forensic tool for provenance analysis. Appl Geochem 48:104–121

Frimmel HE, Hennigh Q (2015) First whiffs of atmospheric oxygen triggered onset of crustal gold cycle. Miner Deposita 50:5–23

Fuchs S, Williams-Jones AE, Jackson SE, Przybylowicz WJ (2016) Metal distribution in pyrobitumen of the Carbon Leader Reef, Witwatersrand Supergroup, South Africa: Evidence for liquid hydrocarbon ore fluids. Chem Geol 426:45–59

Gartz VH, Frimmel HE (1999) Complex metasomatism of an Archean placer in the Witwatersrand basin, South Africa: The Ventersdorp Contact reef - a hydrothermal aquifer? Econ Geol 94:689–706

Gibson RL, Wallmach T (1995) Low pressure-high temperature metamorphism in the Vredefort Dome, South Africa—Anticlockwise pressure-temperature path followed by rapid decompression. Geol J 30:121–135

Gray GJ, Lawrence SR, Kenyon K, Cornford C (1998) Nature and origin of carbon in the Archean Witwatersrand Basin, South Africa. J Geol Soc, London 155:39–59

Gumsley A, Stamsnijder J, Larsson E, Söderlund U, Naeraa T, de Kock MO, Ernst R (2018) The 2789-2782 Ma Klipriviersberg large igneous province: implications for the chronostratigraphy of the Ventersdorp Supergroup and the timing of Witwatersrand gold deposition. Geocongress 2018, 18-20 July 2018, Johannesburg, Geological Society of South Africa, Abstract Book, p 133

Guy BM, Beukes NJ, Gutzmer J (2010) Paleoenvironmental controls on the texture and chemical composition of pyrite from non-conglomeratic sedimentary rocks of the Mesoarchean Witwatersrand Supergroup, South Africa. S Afr J Geol 113:195–228

Hallbauer DK, Joughin NC (1973) The size distribution and morphology of gold particles in the Witwatersrand reefs and their crushed products. J S Afr Inst Min Metall 1973:395–405

Hallbauer DK (1975) The plant origin of Witwatersrand carbon. Miner Sci Eng 7:111–131

Hallbauer DK (1986) The mineralogy and geochemistry of Witwatersrand pyrite, gold, uranium, and carbonaceous matter. In: Anhaeusser CR, Maske S (eds) Mineral deposits of Southern Africa. Geological Society of South Africa, Johannesburg, pp 731–752

Hayward CL, Reimold WU, Gibson RL, Robb LJ (2005) Gold mineralization within the Witwatersrand basin, South Africa: evidence for a modified placer origin, and the role of the Vredefort impact event. In: MacDonald I, Boyce AJ, Butler IB, Herrington RJ, Polya DA (eds) Mineral deposits and earth evolution. Geological Society, London, Special Publication 248, pp 31–58

Heinrich CA (2015) Witwatersrand gold deposits formed by volcanic rain, anoxic rivers and Archaean life. Nat Geosci 8:206–209

Henkel H, Reimold WU (1998) Integrated geophysical modelling of a giant, complex impact structure: anatomy of the Vredefort Structure, South Africa. Tectonophysics 287:1–20

Heubeck C (2019) The Moodies Group—a High-Resolution Archive of Archean Surface Processes and Basin-Forming Mechanisms. In: Kröner A, Hofmann A (eds) The Archaean Geology of the Kaapvaal Craton, Southern Africa. Springer, Heidelberg, Chap. 6, pp 203–239

Hölzing A, Frimmel HE, Voland V, Dremel K, Zabler S, Minter WEL (2015) The cover of Mineralium Deposita's anniversary volume uncovered. In: André-Mayer A-S, Cathelineau M, Muchez P, Pirard E, Sindern S (eds) Mineral resources in a sustainable world, Proceedings of 13th Biennial SGA Meeting, 24–27 August 2015. Université de Lorraine, Nancy, vol. 4, pp 1407–1410

Hofmann A, Bekker A, Rouxel O, Rumble D, Master S (2009) Multiple sulphur and iron isotope composition of detrital pyrite in Archaean sedimentary rocks: A new tool for provenance analysis. Earth Planet Sci Lett 286:436–445

Jolley SJ, Henderson HC, Barnicoat AC, Fox NPC (1999) Thrust-fracture network and hydrothermal gold mineralization: Witwatersrand Basin, South Africa. In: McCaffrey KJW, Lonergan L, Wilkinson JJ (eds) Fractures, Fluid Flow and Mineralization. Geological Society, London, Special Publication 155: 153–165

Kamo SL, Reimold WU, Krogh TE, Colliston WP (1996) A 2.023 Ga age for the Vredefort impact event and a first report of shock metamorphosed zircons in pseudotachylitic breccias and granophyre. Earth Planet Sci Lett 144:369–387

Kingsley CS (1987) Facies changes from fluvial conglomerate to braided sandstone of the early Proterozoic Eldorado Formation, Welkom Goldfieldd, South Africa. SEPM Spec Publ 39:359–370

Koglin N, Frimmel HE, Minter WEL, Brätz H (2010a) Trace-element characteristics of different pyrite types in Mesoarchaean to Palaeoproterozoic placer deposits. Miner Deposita 45:259–280

Koglin N, Zeh A, Frimmel HE, Gerdes A (2010b) New constraints on the auriferous Witwatersrand sediment provenance from combined detrital zircon U-Pb and Lu-Hf isotope data for the Eldorado Reef (Central Rand Group, South Africa). Precambr Res 183:817–824

Kositcin N, McNaughton NJ, Griffin BJ, Fletcher IR, Groves DI, Rasmussen B (2003) Textural and geochemical discrimination between xenotime of different origin in the Archaean Witwatersrand Basin, South Africa. Geochim Cosmochim Acta 67:709–731

Kositcin N, Krapez B (2004) SHRIMP U-Pb detrital zircon geochronology of the Late Archaean Witwatersrand Basin of South Africa: relation between zircon provenance age spectra and basin evolution. Precambr Res 129:141–168

Large RR, Meffre S, Burnett R, Guy B, Bull S, Gilbert S, Goemann K, Danyushevsky L (2013) Evidence for an intrabasinal source and multiple concentration processes in the formation of the Carbon Leader Reef, Witwatersrand Supergroup, South Africa. Econ Geol 108:1215–1241

Law JDM, Bailey AC, Cadle AB, Phillips GN, Stanistreet IG (1990) Reconstructive approach to the classification of Witwatersrand 'quartzites'. S Afr J Geol 93:83–92

Manzi MSD, Hein KAA, King N, Durrheim RJ (2013) Neoarchaean tectonic history of the Witwatersrand Basin and Ventersdorp Supergroup: new constraints from high-resolution 3D seismic reflection data. Tectonophysics 590:94–105

Marsh JS, Bowen MP, Rogers NW, Bowen TB (1989) Volcanic rocks of the Witwatersrand Triad, South Africa. II: Petrogenesis of mafic and felsic rocks of the Dominion Group. Precambr Res 44:39–65

Martin DM, Stanistreet IG, Camden-Smith PM (1989) The interaction between tectonics and mudflow deposits within the Main Conglomerate formation in the 2.8–2.7 Ga Witwatersrand Basin. Precambr Res 44:19–38

McCarthy TS (2006) The Witwatersrand Supergroup. In: Johnson MR, Anhaeusser CR, Thomas RJ (eds) The geology of South Africa. Geological Society of South Africa, Johannesburg, pp 155–186

Mellor ET (1917) The geology of the Witwatersrand. explanation to the geological map. Spec Publi Geol Survey S Afr 3:1–46

Minter WEL (1978) A sedimentological synthesis of placer gold, uranium and pyrite concentrations in Proterozoic Witwatersrand sediments. In: Miall AD (ed) Fluvial sedimentology. Canadian Society of Petroleum Geologists, pp 801–829

Minter WEL (1999) Irrefutable detrital origin of Witwatersrand gold and evidence of eolian signatures. Econ Geol 94:665–670

Minter WEL, Feather CE, Glathaar CW (1988) Sedimentological and mineralogical aspects of the newly discovered Witwatersrand placer deposit that reflect Proterozoic weathering, Welkom gold field, South Africa. Econ Geol 83:481–491

Minter WEL, Loen JS (1991) Palaeocurrent dispersal patterns of Witwatersrand gold placers. S Afr J Geol 94:70–85

Mossman DJ, Minter WEL, Dutkiewicz A, Hallbauer DK, George SC, Hennigh Q, Reimer TO, Horscroft FD (2008) The indigenous origin of Witwatersrand "carbon". Precambr Res 164:173–186

Nesbitt HW, Young GM (1982) Early Proterozoic climates and plate motions inferred from major element chemistry of lutites. Nature 299:715–717

Nwaila G, Frimmel HE, Minter WEL (2017) Provenance and geochemical variations in shales of the Mesoarchaean Witwatersrand Supergroup. J Geol 125:399–422

Phillips GN, Law JDM (1994) Metamorphism of the Witwatersrand gold fields: a review. Ore Geol Rev 9:1–31

Phillips GN, Law JDM (2000) Witwatersrand gold fields: geology, genesis and exploration. SEG Reviews 13:439–500

Phillips GN, Powell R (2011) Origin of Witwatersrand gold: a metamorphic devolatilisation-hydrothermal replacement model. Appl Earth Sci (Trans Inst Min Metall B) 120:112–129

Poujol M, Anhaeusser CR (2001) The Johannesburg Dome, South Africa: new single zircon U-Pb isotopic evidence for early Archaean granite-greenstone development within the central Kaapvaal Craton. Precambr Res 108:139–158

Poujol M, Kiefer R, Robb LJ, Annhaeusser CR, Armstrong RA (2005) New U-Pb data on zircons from the Amalia greenstone belt, Southern Africa: insights into the Neoarchaean evolution of the Kaapvaal Craton. S Afr J Geol 108:317–332

Ramdohr P (1958) New observations on the ores of the Witwatersrand in South Africa and their genetic significance. Trans Geol Soc South Africa 61:1–50

RMG (2015) Raw Materials Data Base. © Intierra Raw Materials Group, Stockholm

Robb LJ, Meyer FM (1991) A contribution to recent debate concerning epigenetic versus syngenetic mineralization processes in the Witwatersrand Basin. Econ Geol 86:396–401

Robb LJ, Davis D, Kamo SL, Meyer FM (1992) Ages of altered granites adjoining the Witwatersrand Basin with implications for the origin of gold and uranium. Nature 357:677–680

Robb LJ, Meyer FM (1995) The Witwatersrand Basin, South Africa: geological framework and mineralization processes. Ore Geol Rev 10:67–94

Robb LJ, Robb VM (1998) Gold in the Witwatersrand Basin. In: Wilson MGC, Anhaeusser CR (eds) The mineral resources of South Africa. Council for Geoscience, Pretoria, pp 294–349

SACS, South African Committee for Stratigraphy (1980) Part 1: Lithostratigraphy of the Republic of South Africa, South West Africa/Namibia and the Republics of Bophuthatswana, Transkei and Venda. Department of Mineral and Energy Affairs, Geological Survey, Pretoria

Schidlowski M (1981) Uraniferous constituents of the Witwatersrand conglomerates: ore-microscopic observations and implications for Witwatersrand metallogeny. US Geol Surv Prof Pap 1161: N1–N29

Smith AJB, Beukes NJ, Gutzmer J (2013) The composition and depositional environments of Mesoarchean iron formations of the West Rand Group of the Witwatersrand Supergroup, South Africa. Econ Geol 108:111–134

Spangenberg J, Frimmel HE (2001) Basin-internal derivation of hydrocarbons in the Witwatersrand Basin, South Africa: Evidence from bulk and molecular $\delta^{13}C$ data. Chem Geol 173:339–355

Sutton SJ, Ritger SD, Maynard JB (1990) Stratigraphic control of chemistry and mineralogy in metamorphosed Witwatersrand quartzites. J Geol 98:329–341

Therriault A, Grieve R, Reimold WU (1997) Original size of the Vredefort impact structure: implications for the geological evolution of the Witwatersrand Basin. Meteorit Planet Sci 32:71–77

Tinker J, de Wit M, Grotzinger J (2002) Seismic stratigraphic constraints on Neoarchean-Paleoproterozoic evolution of the western margin of the Kaapvaal Craton, South Africa. S Afr J Geol 105:107–134

Tucker RF, Viljoen RP, Viljoen MJ (2016) A review of the Witwatersrand Basin—The world's greatest goldfield. Episodes 39:105–133

Tweedie EB (1986) The Evander goldfield. In: Anhaeusser CR, Maske S (eds) Mineral deposits of Southern Africa. Geological Society of South Africa, Johannesburg, pp 705–730

Vennemann TW, Kesler SE, O'Neil JR (1992) Stable isotope compositions of quartz pebbles and their fluid inclusions as tracers of sediment provenance: Implications for gold- and uranium-bearing quartz pebble conglomerates. Geology 20:837–840

Vennemann TW, Kesler SE, Frederickson GC, Minter WEL, Heine RR (1995) Oxygen isotope sedimentology of gold- and uranium-bearing Witwatersrand and Huronian Supergroup quartz-pebble conglomerates. Econ Geol 91:322–342

Wallmach T, Meyer FM (1990) A petrogenetic grid for metamorphosed aluminous Witwatersrand shales. S Afr J Geol 93:93–102

Winter HdlR, Brink MC (1991) Chronostratigraphic subdivision of the Witwatersrand Basin based on Western Transvaal composite columns. S Afr J Geol 94:191–203

Young GM, von Brunn V, Gold DJC, Minter WEL (1998) Earth's oldest reported glaciation: physical and chemical evidence from the Archean Mozaan Group (−2.9 Ga) of South Africa. J Geol 106:523–538

Zeh A, Jaguin J, Poujol M, Boulvais P, Block S, Paquette J-L (2013) Juvenile crust formation in the northeastern Kaapvaal Craton at 2.97 Ga—Implications for Archean terrane accretion, and the source of the Pietersburg gold. Precambr Res 233:20–43

11 Neoarchaean to Early Palaeoproterozoic Within-Plate Volcanism of the Kaapvaal Craton: Comparing the Ventersdorp Supergroup and the Ongeluk and Hekpoort Formations (Transvaal Supergroup)

Fabien Humbert, Michiel de Kock, Nils Lenhardt, and Wladyslaw Altermann

Abstract

The Archaean-Palaeoproterozoic Kaapvaal craton (South Africa) records several voluminous intra-plate volcanic events. Here we present a petrologic and geochemical characterisation of the Neoarchaean Ventersdorp Supergroup and the early Palaeoproterozoic Ongeluk and Hekpoort formations. These Large Igneous Provinces (LIP) represent episodes of widespread continental flood volcanism and localized rift-related volcanism punctuating an interval of more than 500 million years. Our study is based on a data compilation from literature and new analyses. In addition, the palaeoenvironmental conditions of these events are evaluated. The Ventersdorp Supergroup is the most extensive volcano-sedimentary succession preserved on the craton and represents several discrete volcanic events with a complex geochemical history. They seem to represent a progressive differentiation series from a single parental magma with variable degrees of crustal assimilation, strongly inconsistent with their extrusion order. All the samples in our compilation show arc-like signatures, mainly represented by negative Nb–Ta anomalies, which we relate to metasomatism of the subcontinental lithospheric mantle by past subduction (s). The Hekpoort and Ongeluk formations exhibit relatively high Th contents and high Th/Nb ratios, which sharply contrasts those of Archaean units. The data suggest that an important subduction event marked the Archaean-Proterozoic boundary in the Kaapvaal craton.

F. Humbert (✉) · M. de Kock
PPM Group, Geology Department, University of Johannesburg, PO Box 524 Auckland Park, Johannesburg, 2006, South Africa
e-mail: fhumbert@uj.ac.za

N. Lenhardt · W. Altermann
Department of Geology, University of Pretoria, Pretoria, 0002, South Africa

Keywords

Kaapvaal craton • Large igneous province
Ventersdorp supergroup • Hekpoort formation
Ongeluk formation • Geochemistry

11.1 Introduction

The Kaapvaal craton (Fig. 11.1) is one of the best-preserved Archaean cratons, and exhibits a remarkable record of Palaeoarchaean to Palaeoproterozoic sedimentation that is punctuated by several episodes of voluminous magmatic activity (e.g. Johnson et al. 2006).

The predominantly granite-greenstone basement of the Kaapvaal craton formed between 3.6 and 2.9 Ga (e.g. Poujol et al. 2003 and references therein). By *ca.* 3.1 Ga, the nucleus of the craton had formed by magmatic accretion and amalgamation of several smaller crustal blocks (Lowe and Byerly 2007; Bumby et al. 2012). The craton serves as the basement to a number of successive volcano-sedimentary basin remnants of Archaean to Mesoproterozoic age. Among these, the three largest and best preserved are represented by the *ca.* 3.0–2.78 Ga Pongola-Witwatersrand Supergroup (e.g. De Wit et al. 1992; Robb and Meyer 1995; Zeh et al. 2009; Bumby et al. 2012), the *ca.* 2780–2709 Ma Ventersdorp Supergroup (Armstrong et al. 1991; de Kock et al. 2012; Cornell et al. 2017) and the *ca.* 2.66–2.05 Ga Transvaal Supergroup (Altermann and Nelson 1998; Buick et al. 2001; Mapeo et al. 2006), respectively.

In this contribution, we compare the petrology, geochemistry and palaeoenvironmental setting of volcanic units of the Kaapvaal craton that straddle the Neoarchaean-Palaeoproterozoic boundary, i.e. the Neoarchaean Ventersdorp Supergroup (Fig. 11.1) and Palaeoproterozoic volcano-sedimentary successions of the Ongeluk and Hekpoort formations (Fig. 11.2). They represent voluminous volcanic events on the craton and can be classified as large igneous provinces or LIPs. LIPs are defined as bodies of

A. Kröner and A. Hofmann (eds.), *The Archaean Geology of the Kaapvaal Craton, Southern Africa*, Regional Geology Reviews, https://doi.org/10.1007/978-3-319-78652-0_11

Fig. 11.1 Distribution of the Ventersdorp Supergroup (including correlative units) and coeval dyke swarms on the Kaapvaal craton. The location of the craton is specified in the bottom right corner. Gp. = Group, Fm. = Formation

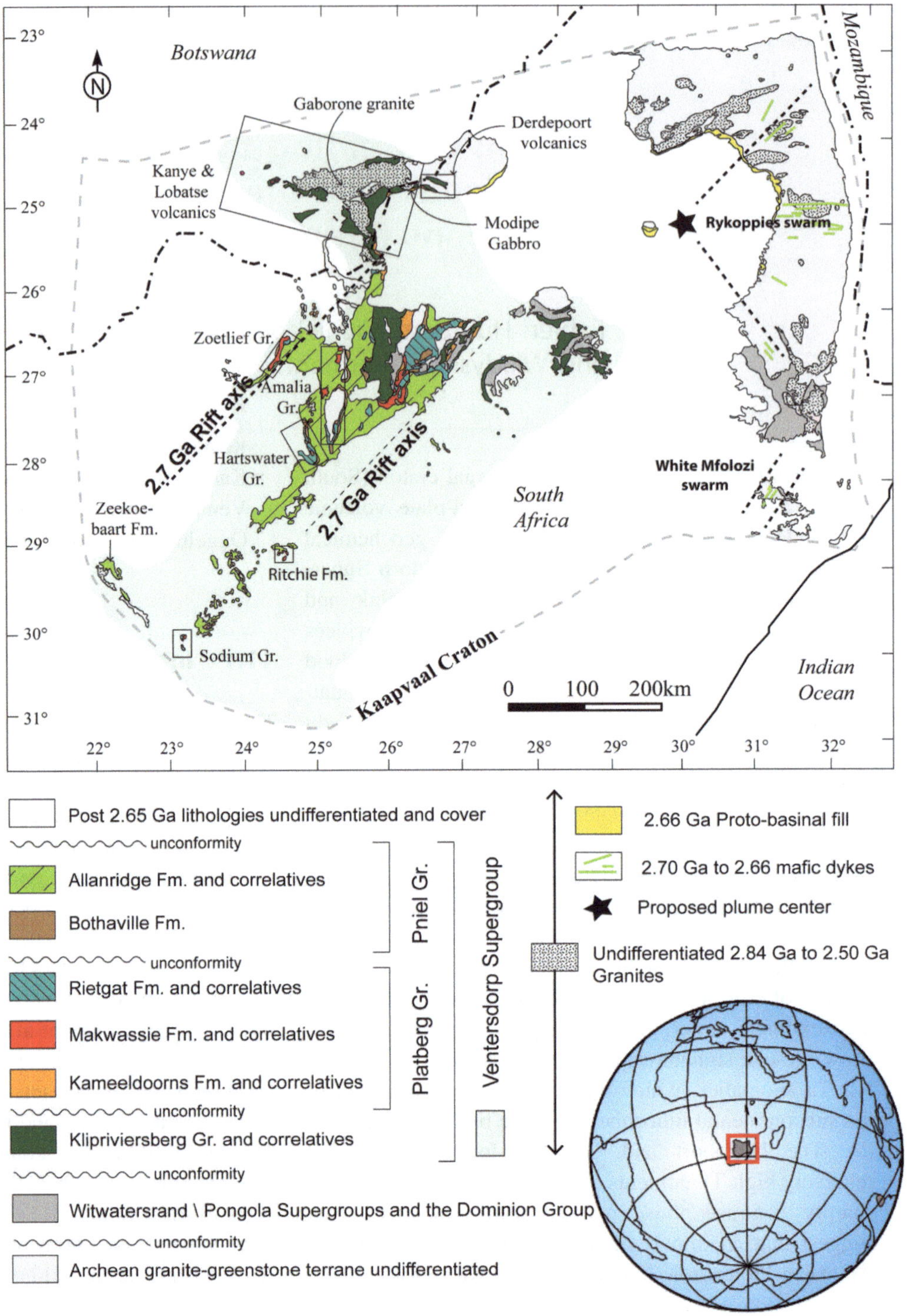

mostly mafic extrusive and related intrusive rocks with volumes greater than 100,000 km^3 that were emplaced over a relatively short period of time or as several discrete pulses (Coffin and Eldholm 1994; Ernst 2014). Other Neoarchaean-Palaeoproterozoic magmatic events recorded on the Kaapvaal craton are either related feeder-systems (e.g. 2.70–2.66 Ga Rykoppies Dyke Swarm as documented by Olsson et al. 2016; and the 2.66 Ga White Umfolozi Dyke Swarm identified by Gumsley et al. 2016), or less voluminous events, such as the extrusion of the Pre-Transvaal 2.66 Ga lava units (i.e. Buffelsfontein Group and Abel Erasmus Member, Nelson et al. 1999; Barton et al. 1995; Dorland 2004), and the syn-Transvaal volcanic rocks of the Bushy Bend Lava Member (Eriksson et al. 1994) and the Machadodorp Volcanic Member (Lenhardt et al. 2012; Fig. 11.2).

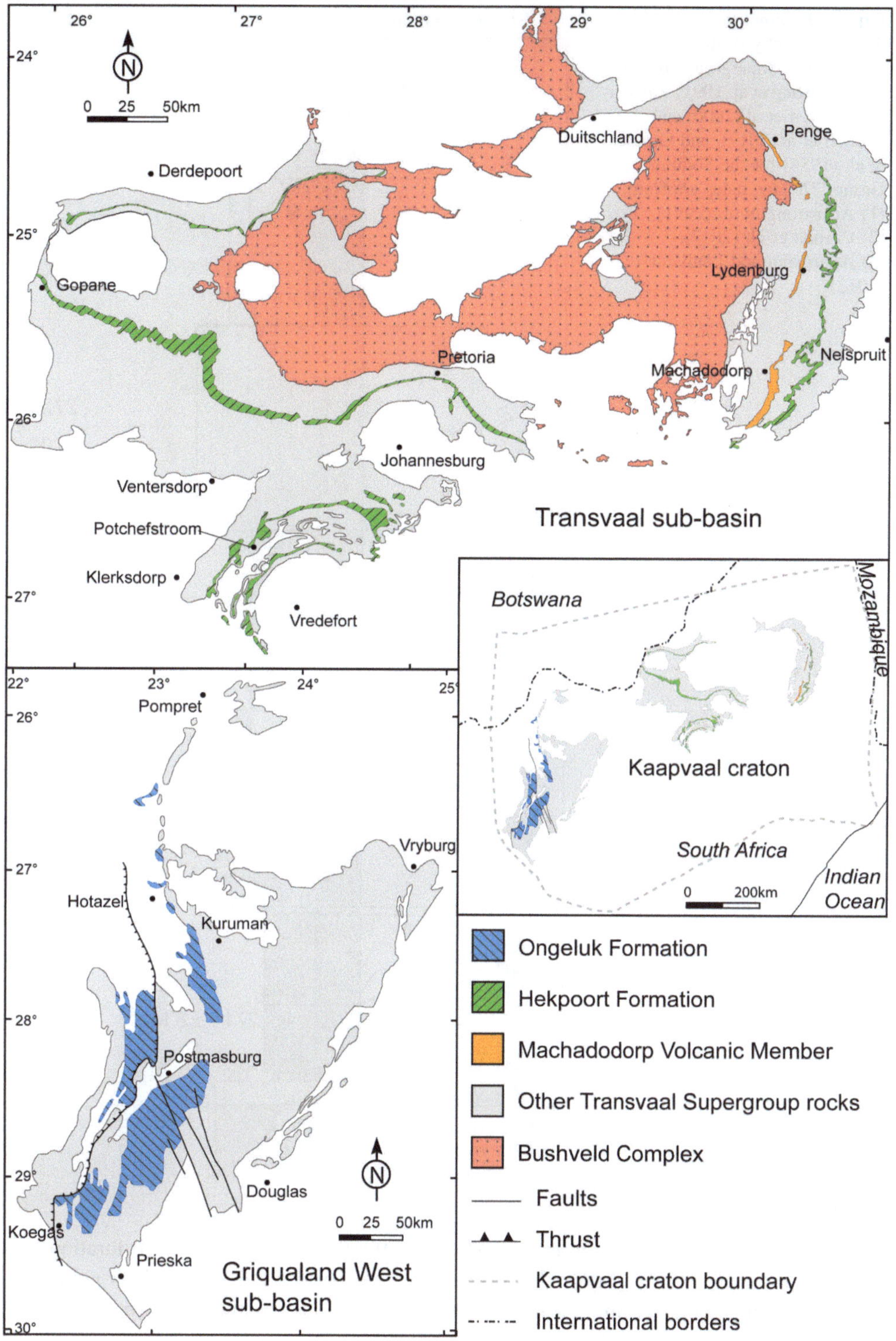

Fig. 11.2 Distribution of the outcrops of the Bushveld Complex and the Transvaal Supergroup showing the Hekpoort Formation and Machadodorp Volcanic Member in the Transvaal sub-basin and the Ongeluk Formation in the Griqualand West sub-basin (compiled from 1:250,000 South African geological maps)

11.2 Geological Setting

11.2.1 The Ventersdorp Supergroup

With a preserved thickness of 5100 m, the Neoarchaean Ventersdorp Supergroup (Van der Westhuizen et al. 1991, 2006) represents the largest, thickest and most dominantly volcanic succession preserved on the craton (Fig. 11.1). It represents multiple volcanic pulses forming a succession of continental flood basalts, andesitic to felsic volcanic rocks, and associated sedimentary rocks (Fig. 11.3) that extend across an area greater than 300,000 km^2 (e.g. Van der Westhuizen et al. 1991). The Ventersdorp Supergroup is

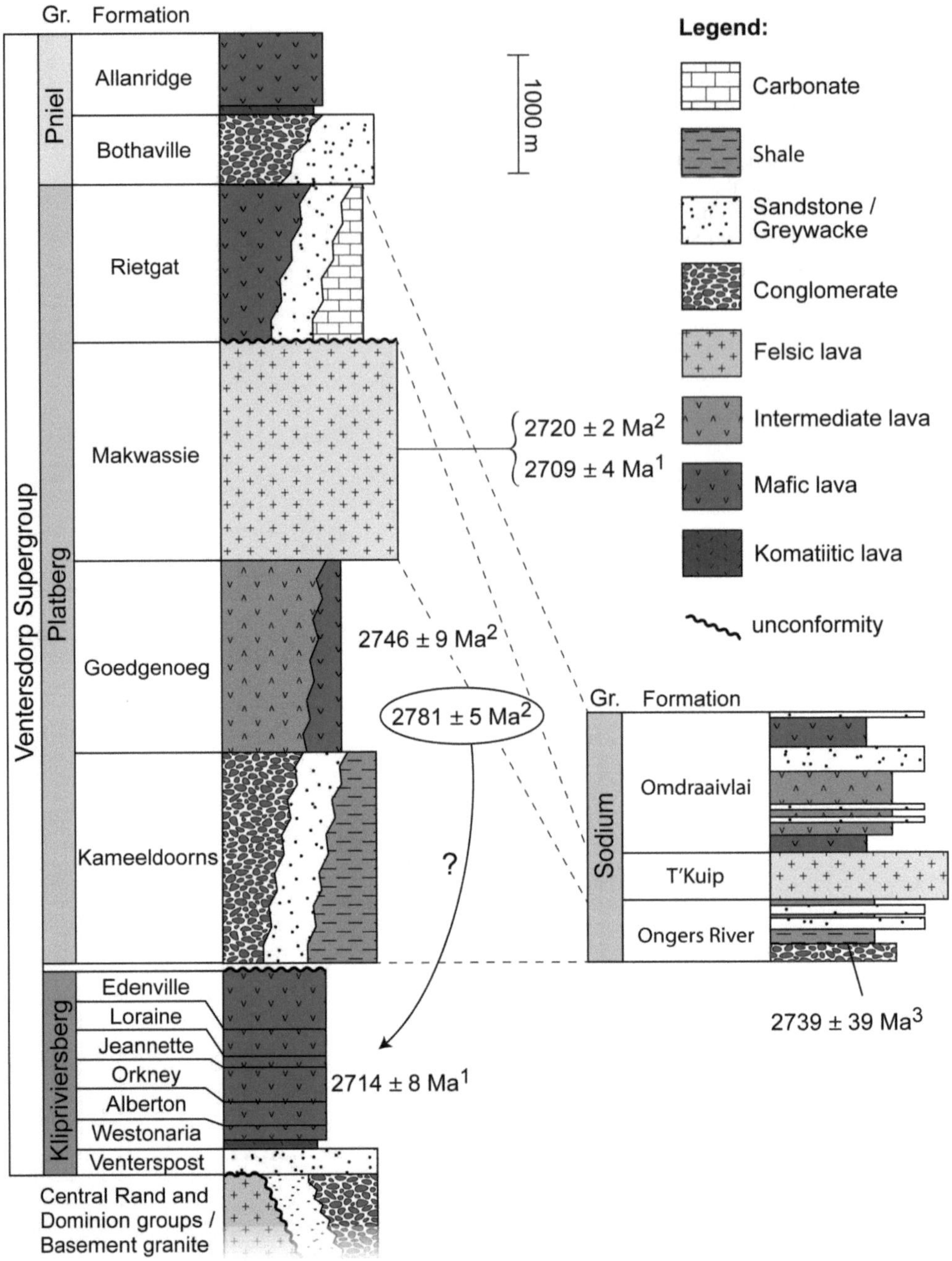

Fig. 11.3 Simplified lithostratigraphy of the Ventersdorp Supergroup (based on Armstrong et al. 1991; Van der Westhuizen et al. 2006; and Cornell et al. 2017), and Grobler et al. (1989) for the Sodium Group. The age dates are from: (1) Armstrong et al. (1991), (2) Cornell et al. (2017), (3) Altermann and Lenhardt (2012)

subdivided into three stratigraphic units (Winter 1976; Van der Westhuizen et al. 2006). Exposures of the basal Klipriviersberg Group is limited to the central parts of the craton, whereas the overlying Platberg and Pniel groups (Fig. 11.3; Table 11.1) are best exposed along the western part of the craton (Van der Westhuizen et al. 1991). Subdivision of these groups and correlation of the units amongst individual outcrop areas is largely based on lithostratigraphy with only few isotopic ages being available.

Two U–Pb zircon ages by Armstrong et al. (1991) of 2714 ± 8 Ma for the Klipriviersberg Group and 2709 ± 4 Ma for the Platberg Group from the central part, and the short duration these ages imply for extrusion, have guided opinions around the Ventersdorp Supergroup development for many years. However, U–Pb zircon dates of 2720 ± 2 Ma and 2746 ± 9 Ma for the Makwassie and Goedgenoeg formations supersedes the 2709 Ma age of the Platberg Group, and is in agreement with ages from equivalents from the western parts of the craton (Cornell et al. 2017). These ages are generally older than the Klipriviersberg Group's 2714 Ma age determination (Table 11.1). A Pb–Pb evaporation zircon date of 2714 ± 3 Ma from the Zoetlief Group (Walraven et al. 1991); a U–Pb zircon date of 2729 ± 3 Ma from the Amalia Group (Poujol et al. 2005);

Table 11.1 Lithotratigraphy of the Ventersdorp Supergroup in the central part of the Kaapvaal craton (after Armstrong et al. 1991), its correlatives along the western part of the craton (after SACS 1980), and the Sodium Group (after Grobler et al. 1989)

Witwatersrand block		Main lithologies	Age (Ma)	Mg#	εNd (t)	Kimberley block equivalents	Main lithologies	Age (Ma)	Sodium Group formations	Main lithologies	Age (Ma)	Mg#
Groups	Formations											
Pniel	Allanridge	Basalt		43.1	−2.7[7]	Allanridge	Basalt	<2720 ± 4[7]				
						Zeekoeboard Fm.						
	Bothaville	Quartzite; conglomerate; tuffaceous shales	–	–	–	Bothaville	Quartzite; conglomerate; tuffaceous shales					
Platberg	Rietgat	Basaltic and porphyritic andesite; various sedimentary rocks	–	49.1	−3.3[7]	Hartswater Group (Phokwane Fm.)	Andesite; tuffaceous shale; limestone	–	Omdraaivlai	Mafic to dacite lavas; volcanoclas-tic sediments	–	41.3
	Makwassie	Quartz-porphyry	2709 ± 4[1] 2720 ± 2[2]	46.8			Porphyritic dacite	2724 ± 6[3]	T'Kuip	Quartz-porphyry	–	
						Zoetlief Group (Kareefontein Fm.) Amalia Group (Paardefontein Fm.) Ritchie Fm.	Quartz-porphyry	2714 ± 3[2] 2729.1 ± 3[5]				
	Goedgenoeg	Basaltic and porphyritic andesite	2746 ± 9[2] (2781 ± 5[2]?)	51.3		–	–		Ongers River	Mafic lava, arkosic sandstones; tuffaceous shale	2739 ± 39[6]	
	Kameeldoorns	Quartzite, tuffaceous shale, conglomerate	–	–	–	Hartswater Group (Mohle Fm.)	Quartzite; tuffaceous shale; conglomerate	2735 ± 3[3]				
Klipriviersberg	Edenville	Porphyritic and amygdaloidal basalt	–	62.2	−0.6[7]							
	Loraine				−0.5[8]							
	Jeannette	Amygdaloidal basalt		45.7	−0.4[7] 0.5[8]							
	Orkney		2714 ± 8[1]									
	Alberton	Porphyritic basalt	–									
	Westonaria	Basaltic; komatiitic lavas	–	68.2	–							
	Venterspost	Quartzite; conglomerate	–	–	–							

Dating are from: (1) Armstrong et al. (1991), (2) Cornell et al. (2017), (3) De Kock et al. (2012), (4) Walraven et al. (1991), (5) Poujol et al. (2005), (6) Altermann and Lenhardt (2012). Nd isotopes data are from: (7) Nelson et al. (1992) and (8) Marsh et al. (1992). Average Mg# from: Crow and Condie (1988), Grobler et al. (1989), Nelson et al. (1992), Marsh et al. (1992) and Altermann and Lenhardt (2012) two U–Pb zircon dates of 2735 ± 3 Ma and 2724 ± 6 Ma from the Hartswater Group (de Kock et al. 2012); and a 2739 ± 39 Ma U–Pb zircon date for the T'Kuip Formation (Altermann and Lenhardt 2012) all suggest the Klipriviersberg Group to be older than 2714 Ma. Wingate (1998) highlighted possible lead loss affecting the 2714 Ma date of Armstrong et al. (1991) and regarded it as a minimum constraint for the group. de Kock et al. (2012) suggested that the Klipriviersberg Group is perhaps better correlated with 2.78 Ga volcanic and magmatic units in the northwest of the craton, such as the 2782 ± 5 Ma Derdepoort volcanics (Wingate 1998) or the 2784 ± 1 Ma Modipe Gabbro (Denyszyn et al. 2013)—Fig. 11.1. Other possible correlative units are the 2785 ± 5 Ma to 2781 ± 2 Ma Kanye Formation north of Mafikeng in South Africa (Grobler and Walraven 1993; Moore et al. 1993), and the Lobatse Group volcanic rocks in Botswana (Walraven et al. 1996)—see Fig. 11.1. It is interesting to note that 2781 ± 5 Ma volcanism was also identified at Kimberley (previously regarded as Goedgenoeg Formation, Cornell et al. 2017). The Ventersdorp Supergroup thus likely presents two pulses of continental flood basalt eruption at 2.78 and at 2.70 Ga separated by an intervening period of rift-related volcanism and sedimentation (Ernst and Buchan 2001; de Kock et al. 2012; Ernst 2014).

During the development of the Ventersdorp Supergroup the eastern and northern margins of the Kaapvaal craton witnessed plutonic activity with the emplacement of numerous granite plutons (Eglington and Armstrong 2004). Exposed in the basement of the eastern Kaapvaal craton are two prominent Neoarchaean dyke swarms. These are the radiating 2.70–2.66 Ga Rykoppies Dyke Swarm (Olsson et al. 2016), and the northeast trending 2.66 Ga White Umfolozi Dyke Swarm (Gumsley et al. 2016). The Rykoppies Dyke Swarm, in particular, may represent magmatic feeders to the 2.70 Ga Allanridge Formation lavas (Olsson et al. 2016).

11.2.1.1 The Klipriviersberg Group

The 2000-m-thick Klipriviersberg Group consists of predominantly basaltic lava with minor Mg-rich lavas near its base (Table 11.1; McIver et al. 1982) and rare sedimentary intercalations (Van der Westhuizen et al. 2006). Based on characteristic features such as colour or the existence of amygdales, it is subdivided into six formations (i.e. the Westonaria, Alberton, Orkney, Jeanette, Loraine and Edenville formations; Fig. 11.3; Winter 1976; SACS 1980). The aerial extent of the Klipriviersberg Group is limited to the central Kaapvaal craton with some isolated occurrences towards the northwest of Klerksdorp (Van der Westhuizen et al. 1991). Some 2.78 Ga lava units, previously assigned to the Platberg Group, intersected by drill core at Kimberley and dated by Cornell et al. (2017) likely belong to the Klipriviersberg Group. It has been suggested that the Klipriviersberg lavas erupted from a NE–SW trending feeder dyke swarm (McCarthy et al. 1990), but very little is known about these dykes.

11.2.1.2 The Platberg Group

The Platberg Group overlies the Klipriviersberg Group with a prominent angular unconformity (Van der Westhuizen et al. 1991) and in its type locality of the central Kaapvaal craton it consists of clastic and chemical-sedimentary rocks (Kameeldoorns Formation) with mafic (Goedgenoeg and Rietgat formations) and felsic (Makwassie Formation) volcanic rocks deposited during graben development (Table 11.1; Fig. 11.3; Winter 1976). Outcrops of the Platberg Group are also found between Kimberley, Ventersdorp, Mafikeng and Vryburg on the western part of Kaapvaal craton (Keyser 1998). This includes extensive outcrops mapped near Taung (Liebenberg 1977) and along the Orange River between Douglas and Hopetown (Potgieter 1973). Felsic porphyry, overlain by andesitic lava, crops out in the vicinity of Ritchie (Potgieter and Lock 1978). Farther south rocks of the Sodium Group form another isolated outcrop of the Ventersdorp Supergroup (Grobler et al. 1989; Altermann and Lenhardt 2012). These occurrences are correlated lithostratigraphically by SACS (1980), but chronostratigraphic correlation suggests their deposition in isolated grabens that developed diachronously between 2.74 and 2.72 Ga (de Kock et al. 2012; Cornell et al. 2017).

11.2.1.3 The Pniel Group

In contrast to the wedge-like geometries and abrupt facies changes that characterize the graben fills of the Platberg Group, the strata of the Bothaville and Allanridge formations (i.e. the Pniel Group) have blanket-like geometries, unconformably covering the Platberg Group, and can be lithostratigraphically correlated over large distances (Table 11.1; Visser et al. 1976). The arenaceous Bothaville Formation records a transgressive sedimentary package that developed along a N–S trending shoreline on the western part of the craton (Visser et al. 1976; Van der Westhuizen et al. 1991; de Kock et al. 2009, 2012). The 750-m-thick Allanridge Formation consists predominantly of amygdaloidal and porphyritic basaltic andesites (Van der Westhuizen et al. 2006). Basaltic komatiites are reported from in the lower part of the Allanridge Formation (Van der Westhuizen et al. 2006). These are interpreted as either the result of magma mixing between a komatiitic basalt and less mafic magma (Keyser 1998), or they are a consequence of liquid immiscibility in the source magma (Cawthorn et al. 1979). The Allanridge Formation, while younger than the underlying 2720 Ma Makwassie Formation, remains undated. It has, however, been linked geochemically and palaeomagnetically to the 2.7–2.65 Ga radiating Rykoppies Dyke Swarm of the eastern Kaapvaal craton (Klausen et al. 2010; Lubnina et al. 2010; Olsson et al. 2010, 2011, 2016). The general NE striking structure of the Ventersdorp Supergroup outcrops (Stanistreet and McCarthy 1991) trends towards the radiating centre of the Rykoppies swarm and is suggestive of a triple-junction rift system (Fig. 11.1). A geochemically distinct 2.66 Ga swarm of NW trending phenocrystic dykes (White Mfolozi Dyke Swarm) was identified in the extreme south-east of the craton (Gumsley et al. 2016). Although the Allanridge Formation may be partly coeval with the Rykoppies and White Mfolozi swarms, a direct connection between the swarms and the Ventersdorp volcanics is not apparent. Alternatively, the Rykoppies Dyke Swarm may be related to the deposition of the ~2.66 Ga Wolkberg Group (Barton et al. 1995), which occurs as discrete sedimentary basins exposed along the base of the Transvaal Supergroup. The swarms are also partly coeval with volcanism at the base of the eastern Transvaal Supergroup, correlative to the 2.64 Ga Vryburg lavas of the western Transvaal Supergroup in the Griqualand West sub-basin (Walraven et al. 1994; Altermann 1996).

Towards the western margin of the craton, foliated lava of the Zeekoebaart Formation (Altermann and Hälbich 1991) is regarded as a correlative of the Allanridge Formation (SACS 1980). The Zeekoebaard Formation has yielded a single detrital zircon age population of 2720 ± 4 Ma (Cornell et al. 2018).

11.2.2 The Transvaal Supergroup

The Transvaal Supergroup rests on the Ventersdorp Supergroup with an angular unconformity (Fig. 11.4) that represents a minimum of a few tens of millions of years (Armstrong et al. 1991; Walraven and Martini 1995; Altermann 1996). The Transvaal Supergroup is preserved in two areas known as the Transvaal and Griqualand West sub-basins. Isolated outcrops of the Transvaal Supergroup in Botswana in the so-called Kanye sub-basin represent outliers of the Transvaal sub-basin stratigraphy (Mapeo et al. 2006). The lower, chemical-sedimentary succession of the Transvaal Supergroup hosts the best-preserved carbonate ramp and platform developed early in Earth's history (Beukes 1987; Altermann and Nelson 1998; Sumner and Grotzinger 2004; Sumner and Beukes 2006). This succession spans the Archaean-Proterozoic boundary and is correlated between the Transvaal and Griqualand West sub-basins (Sumner and Beukes 2006). A minor volcanic unit is described near the base of the succession in the Griqualand West sub-basin (i.e. the Vryburg lava, see above). Volcanic units are more significant constituents of the upper Transvaal Supergroup (i.e. Pretoria and Postmasburg groups; Fig. 11.4). In the Griqualand West sub-basin, the lower Transvaal chemical-sedimentary succession culminates in increasingly upward shallowing, iron-rich, chemical to siliciclastic-tuffaceous deposits of the Koegas Group. Glacial diamictite of the Makganyene Formation, a Palaeoproterozoic Snowball Earth event (Evans et al. 1997), discordantly to concordantly caps the Koegas Group. The Makganyene Formation is conformably to unconformably overlain by mafic lava of the Ongeluk Formation (Visser 1971, 1999; Altermann and Hälbich 1991; Moore et al. 2001; Melezhik et al. 2013; Altermann et al. 2018).

Traditionally the Pretoria Group of the Transvaal sub-basin was correlated with the Postmasburg Group of Griqualand West sub-basin based mainly on the lithostratigraphic correlation of the Hekpoort and the Ongeluk formations between the two sub-basins (Beukes et al. 2002; Eriksson et al. 2006). The correlation is supported by a 2.22 Ga Rb–Sr date of the Ongeluk Formation and 2.32–2.22 Ga detrital zircon age populations from the Hekpoort Formation (Cornell et al. 1996; Schröder et al. 2016). However, Moore et al. (2001) suggested that the two successions cannot be correlated and that the Pretoria Group is younger than the Postmasburg Group. The publication of an age of 2426 ± 3 Ma for the Ongeluk Formation by Gumsley et al. (2017) supports the view of Moore et al. (2001). However, the relationship between the Pretoria and Postmasburg groups is not fully resolved yet as a 2.25–2.22 Ga detrital zircon age population was found in the Makganyene Formation immediately below the Ongeluk Formation (Beukes et al. 2013).

In the Transvaal sub-basin, a glacial-fluvial conglomerate of the Boshoek Formation is overlain by the Hekpoort Formation basalts (Fig. 11.4; Pretoria Group; Eriksson and Reczko 1995). A correlative unit of the Hekpoort Formation, known as the Tsatsu Formation occurs within the smaller Kanye sub-basin of Botswana. The age of the Hekpoort Formation is constrained by a whole-rock Rb–Sr isochron of 2224 ± 21 Ma (Walraven and Martini 1995) and a whole-rock Pb–Pb isochron of 2236 ± 38 Ma (Cornell et al. 1996). The 2.22 Ga age is further supported by 2225 ± 3 Ma and 2247 ± 10 Ma detrital zircon ages (Dorland 2004; Schröder et al. 2016). Another mafic volcanic unit, the Machadodorp Volcanic Member occurs stratigraphically higher in the Pretoria Group, but its occurrence is limited to the eastern parts of the Transvaal sub-basin (Fig. 11.2). The minimum age of the Machadodorp Volcanic Member is ca. 2.2 Ga which correspond to the dating of the Mashishing dyke swarm which cross cut its lower part (Wabo et al. 2019). In the centre of the Transvaal sub-basin, the Pretoria Group has been intruded and locally contact metamorphosed by plutonic rocks of the Bushveld Complex. In most areas, however, the Pretoria Group has only experienced lower greenschist facies regional metamorphism.

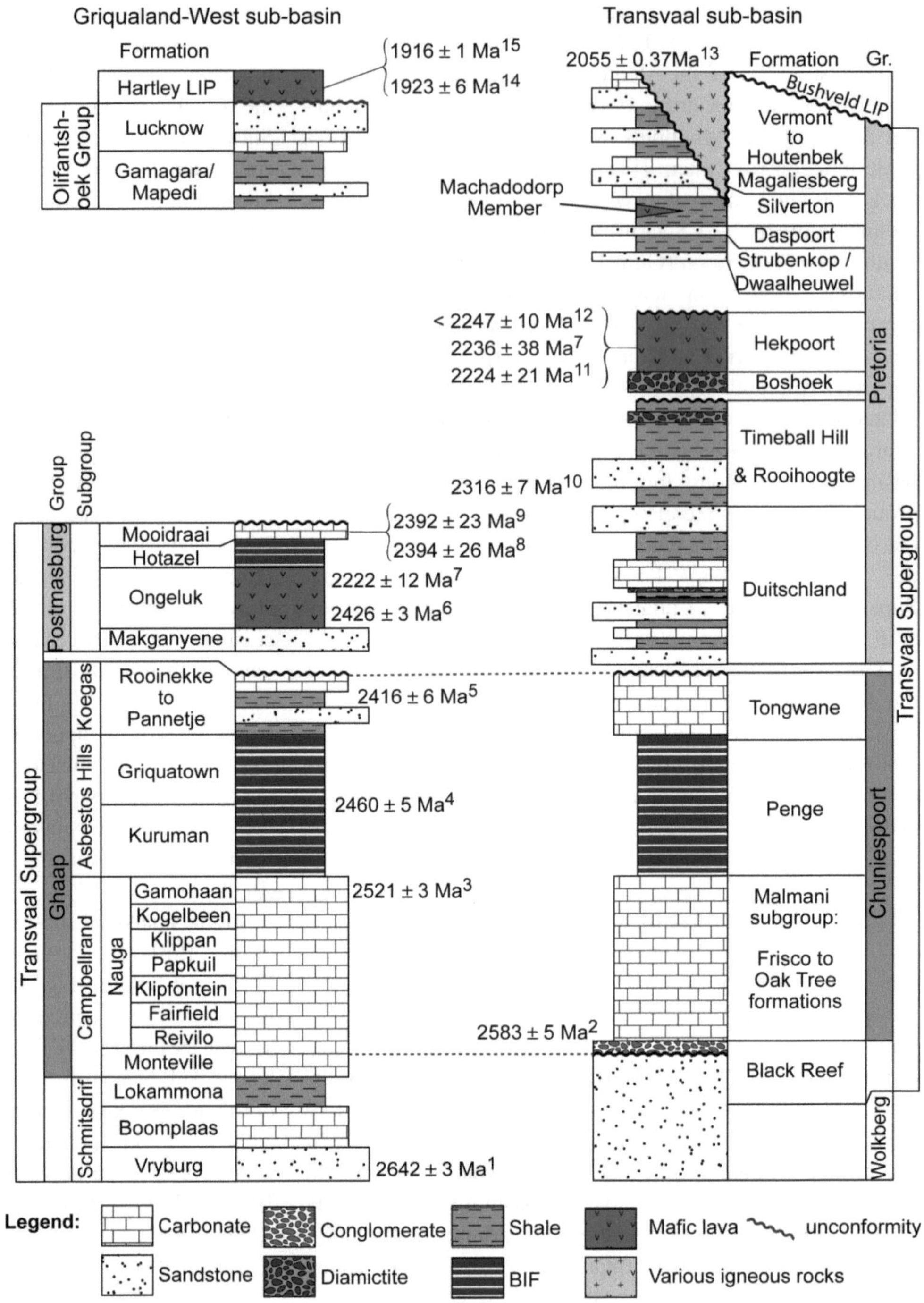

Fig. 11.4 Simplified lithostratigraphy of the Transvaal Supergroup (based on Beukes et al. 2002; Dorland 2004; Guo et al. 2009). The age dates are from: (1) Walraven and Martini (1995), (2) Martin et al. (1998), (3) Sumner and Bowring (1996), (4) Pickard (2003), (5) Gutzmer and Beukes (1998), (6) Gumsley et al. (2017), (7) Cornell et al. (1996), (8) Bau et al. (1999), (9) Fairey et al. (2013), (10) Hannah et al. (2004), (11) Burger and Coertze (1973), (12) Schröder et al. (2016), (13) Zeh et al. (2015), (14) Alebouyeh Semami et al. (2016), (15) Cornell et al. (2016). The Kanye sub-basin stratigraphy is not shown

11.2.2.1 The Ongeluk Formation

The Ongeluk Formation is characterized by an up to 900 m-thick succession of massive basaltic andesites, pillow lavas and hyaloclastites indicating terrestrial and subaqueous extrusion of the volcanic rocks along the western margin of the Kaapvaal craton (Fig. 11.2; Grobler and Botha 1976; Cornell et al. 1996; Altermann and Hälbich 1990; Gutzmer et al. 2003). The Ongeluk Formation extruded onto the glacial Makganyene Formation (Altermann and Hälbich 1991), and is overlain by banded iron formations, manganese-formations and 2.39 Ga dolostones of the Voëlwater Subgroup of the Transvaal Supergroup (Fig. 11.4; Beukes 1983; Bau et al. 1999; Fairey et al. 2013).

Dark-green, massive basaltic-andesitic lava flows are the dominant rock type, but amygdaloidal and pillow lavas occur as well. Basal portions of flows are generally free of amygdales. The first lava flow developed on the Magkanyene diamictite, however, occasionally display up to 20 cm-long pipe amygdales at its base, which suggests a wet substrate (Cornell et al. 1996). Pahoehoe lavas are present in the first flows as well (Altermann et al. 2018). Flow breccias are often found near the tops of the flows (Cornell et al. 1996). According to Gutzmer et al. (2003), the basaltic andesites of the Ongeluk Formation are typically microcrystalline and dark grey to greenish-grey in colour with well-preserved primary igneous textures. Pillow lavas occur in laterally restricted outcrops, towards the top of the Ongeluk Formation and are characterized by very dark-green rims, cooling cracks and rounded three-dimensional forms, with either vesicular or massive cores (Cornell et al. 1996). Associated hyaloclastites are non-graded and consist of fragments of massive lava in a matrix of angular glass shards with altered rims. Tuffaceous, neptunian sand dykes occur between the pillows. Alteration is characterized by brown palagonite and epidote (or piedmontite) along boundaries, whereas the matrix consists largely of zeolites and chlorite (Cornell et al. 1996).

11.2.2.2 The Hekpoort Formation

The Hekpoort Formation marks a major volcanic event in the Pretoria Group of the Transvaal Supergroup (Figs. 11.2 and 11.4). The Hekpoort Formation is studied mainly in isolated outcrops or in borehole cores. Consequently, large parts of the formation remain not described. This formation exhibits gentle dips of no more than $\sim 25°$ and is characterized by a complex interplay of subaerial lava flows, pyroclastic deposits and their reworked counterparts, representing deposition in a terrestrial palaeoenvironment (Button 1973; Eriksson and Reczko 1995; Oberholzer 1995; Humbert et al. 2018a). Although, so far, there has been no regional study to establish the proportions of lava flows and volcaniclastic detritus in the Hekpoort Formation, Sharpe et al. (1983) suggested that lava flows predominate. The thickness of the Hekpoort Formation decreases from 1100 m in the south of the Transvaal sub-basin to 800 m in the western Transvaal sub-basin, and to the northeast less than 50 m of thickness are found (Button 1973; Lenhardt et al. 2012). The total thickness, however, may be overestimated, especially in the southern part as the true inclination of boreholes is mostly unknown and possible thrust duplication was not taken into account. The southern boundary of the Hekpoort Formation is close to the central rim of the Vredefort impact crater, thus duplication of some part of the Hekpoort Formation by thrust faulting can be expected. The Hekpoort Formation is best studied in the south-western part of the Transvaal sub-basin, between Skeerpoort and Hartebeeshoek, where it reaches thicknesses of up to 600 m. Oberholzer and Eriksson (2000) described it as follows: stratigraphically, a lower unit of ~ 100 m thickness and comprising of lava flows is succeeded by lenticular volcaniclastic rocks, with a maximum thickness of 50–100 m and extending along strike for ~ 5000 m. A 400–450 m-thick succession follows, composed of lava flows in the southwest, and lenticular volcaniclastic rocks in the northeast, laterally extending for at least 8 km to the southwest. The volcanic succession is overlain by a 9-m-thick tuffaceous mudrock, followed by 50–100 m thick lava, with a laterally restricted, 100–150-m thick lens of volcaniclastic rocks, ranging from lapilli tuffs and pyroclastic breccia to coarse-grained conglomerates and lahars. At the scale of the whole Hekpoort Formation, lava flows with a thickness up to 60 m appear massive to amygdaloidal, and can locally develop several metres thick flow top breccias. Amygdales are typically developed in the upper part of the flows and are often coalesced and filled by quartz or lined by chlorite and zeolites. The relatively fine-grained lavas contain microphenocrysts of feldspar and chloritized pyroxene set in a groundmass of chlorite and feldspar laths. Secondary phases are characterized by chlorite, K-feldspar, quartz, actinolite, epidote and pyrite (Humbert et al. 2017). The alteration is more pervasive in amygdaloidal lavas than it is in massive lavas (Coetzee 2001). A complete description of the breccias and primary and secondary volcaniclastic rocks of the Hekpoort Formation is available in Lenhardt et al. (2012) and Humbert et al. (2018a).

11.3 Review of the Geochemistry of Major Volcanic Units of the Ventersdorp and Transvaal Supergroups

11.3.1 The Ventersdorp Supergroup

11.3.1.1 Introduction

The geochemical data of the Ventersdorp Supergroup presented in this contribution (Figs. 11.5, 11.6, 11.7, 11.8) are all derived from published sources, i.e. Crow and Condie (1988), Grobler et al. (1989), Nelson et al. (1992), Marsh et al. (1992) and Altermann and Lenhardt (2012). They include samples from the Klipriviersberg Group (Westonaria, Alberton, Orkney, Jeannette, Loraine, and Edenville formations), the Platberg Group (Goedgenoeg, Makwassie and Rietgat formations), and the Pniel Group (Allanridge Formation)—see Table 11.1. The Sodium Group, including the volcanics from the Ongers River, T'Kuip and Omdraaivlai formations, is considered as an outlier of the Platberg Group; however, as Altermann and Lenhardt (2012) provided detailed geochemistry of this group we have decided to treat the Sodium Group separately in all diagrams presented here.

Due to alkali and silicon mobility during metamorphism and/or metasomatic alteration (which is common in Precambrian rocks) the Zr/TiO_2 versus Nb/Y plot (Fig. 11.5; Pearce 1996) is preferred to the TAS diagram (Cox et al. 1979). Similarly, we selected the magnesium number as a differentiation index instead of bulk MgO (Fig. 11.6). Although silicification will inevitably also affect

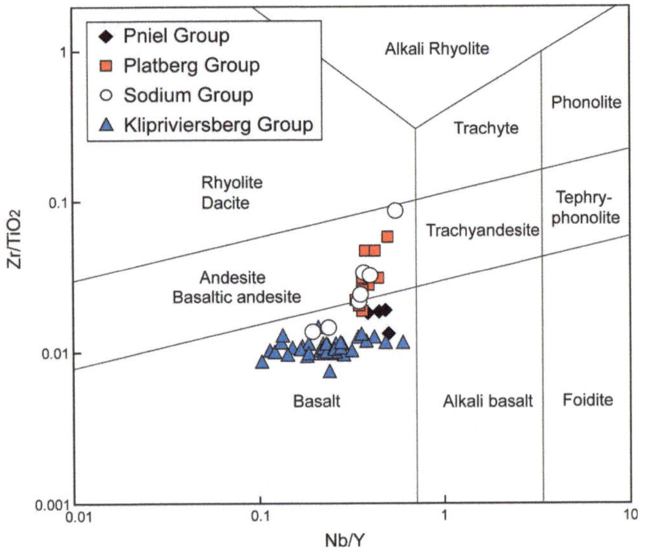

Fig. 11.5 Classification diagram of Winchester and Floyd (1977), modified by Pearce (1996) of the Ventersdorp Supergroup lavas. See text for references of the data

concentrations of other elements, ratios will not be influenced to the same extent. The magnesium number is defined by Mg# = molar [MgO/(MgO + FeO*)] and it decreases with increasing fractionation. The increase of Zr with Mg# (Fig. 11.6) for the samples of the Klipriviersberg and Platberg groups supports the idea that Mg# can be used as a proxy for differentiation.

The element mobility is clear in the Ventersdorp Supergroup lavas. The scatter of both Na_2O (especially below Mg# = 60) and K_2O strongly suggest that these elements have been affected by alteration (Fig. 11.6). Samples of Platberg and Sodium groups document a large variation in SiO_2 content possibly related to a high degree of silicification. Fluid-mobile element compositions (i.e. Si, Na, Ca, and large-ion lithophile elements) are considered with caution when interpreted here.

11.3.1.2 Volcanic Lithologies

The lavas of the Ventersdorp Supergroup are characterized by a wide range of volcanic lithologies, evolving from basalt up to the boundary between andesites and dacites. The most mafic rocks are of the Klipriviersberg Group that cluster in the range of basalt. This statement is in agreement with all previous studies of this formation (e.g. Marsh et al. 1992; Van der Westhuizen et al. 2006). This uniform geochemistry is surprising as the Mg# is relatively variable in this group (Table 11.1; Fig. 11.6). Mg# is higher than 65 for the Westonaria and Edenville formations, between 65 and 51 for the Loraine Formation and below 51 for the Alberton, Orkney and Jeannette formations.

Samples of the Allanridge Formation plot in the basalt range (Fig. 11.5) although they have been described as basaltic andesites in the past (Van der Westhuizen et al. 2006 and reference therein). This may be due to the limitations of the Zr/TiO_2 versus Nb/Y plot. Indeed, the 90% probability contour of the basaltic-andesite range (see Pearce 1996) overlap the range of basalt in the vicinity of where the samples are displayed. Moreover, the MgO content of the Allanridge Formation samples is not high (between 3 and 4.5%; Mg# $\approx$ 43), and is more in agreement with basaltic andesites. In addition to basaltic andesites, a komatiitic basalt is described from the base of the formation (Van der Westhuizen et al. 2006).

The Sodium and Platberg groups present a clear increase in the Zr/TiO_2 content, linked to high values of Zr (up to 540 ppm). This ratio is relatively constant for samples of the Klipriviersberg and Pniel groups. Makwassie Formation samples show the highest Zr/TiO_2 content paired with moderate Mg# between 51 and 42. The two other formations (*i.e.* Goedgenoeg and Rietgat with Mg# > 45) are basaltic andesites to dacites (Van der Westhuizen et al. 2006) and exhibit similar Zr/TiO_2 contents. The Sodium Group samples are characterized by intermediate compositions and plot mainly in the same area as the Platberg Group. However, unlike the Platberg Group, no clear variation is observed in Zr/TiO_2 versus Nb/Y (Fig. 11.5) or Mg# (Fig. 11.6) among the formations of the Sodium Group.

11.3.1.3 Magmatic Evolution and Differentiation by Crystal Fractionation

Chondrite-normalized (after Palme and O'Neill 2004) rare earth element (REE) diagrams are plotted for the Pniel, Platberg, Klipriviersberg and Sodium groups in Fig. 11.7a–d. One can observe that (1) in each group, all samples are showing comparable patterns (with the exception of some subtle divergence in the Sodium Group) and (2), all groups are showing different patterns, mainly linked to the varying enrichment in REE. Two types of patterns can be defined: the first type is represented by the Platberg, Sodium and Pniel groups which show strong fractionation of light rare earth elements or LREE ($[La]_N > 100$) and high La/Lu values (*ca.* 9.5–11.5; Table 11.2). These samples further present a moderate decreasing fractionation between the LREE (from 2.80 to 3.40) and the heavy rare earth element (HREE) (from 1.65 to 2.00). The samples from the felsic Makwassie Formation are slightly more enriched in REE compared to the Rietgat and Goedgenoeg formations. The second type only encompasses Klipriviersberg Group samples, characterized by moderate fractionation of LREE ($[La]_N < 100$), moderate enrichment of La relative to Lu (*ca.* 4), and a similar fractionation of HREE relative to LREE as

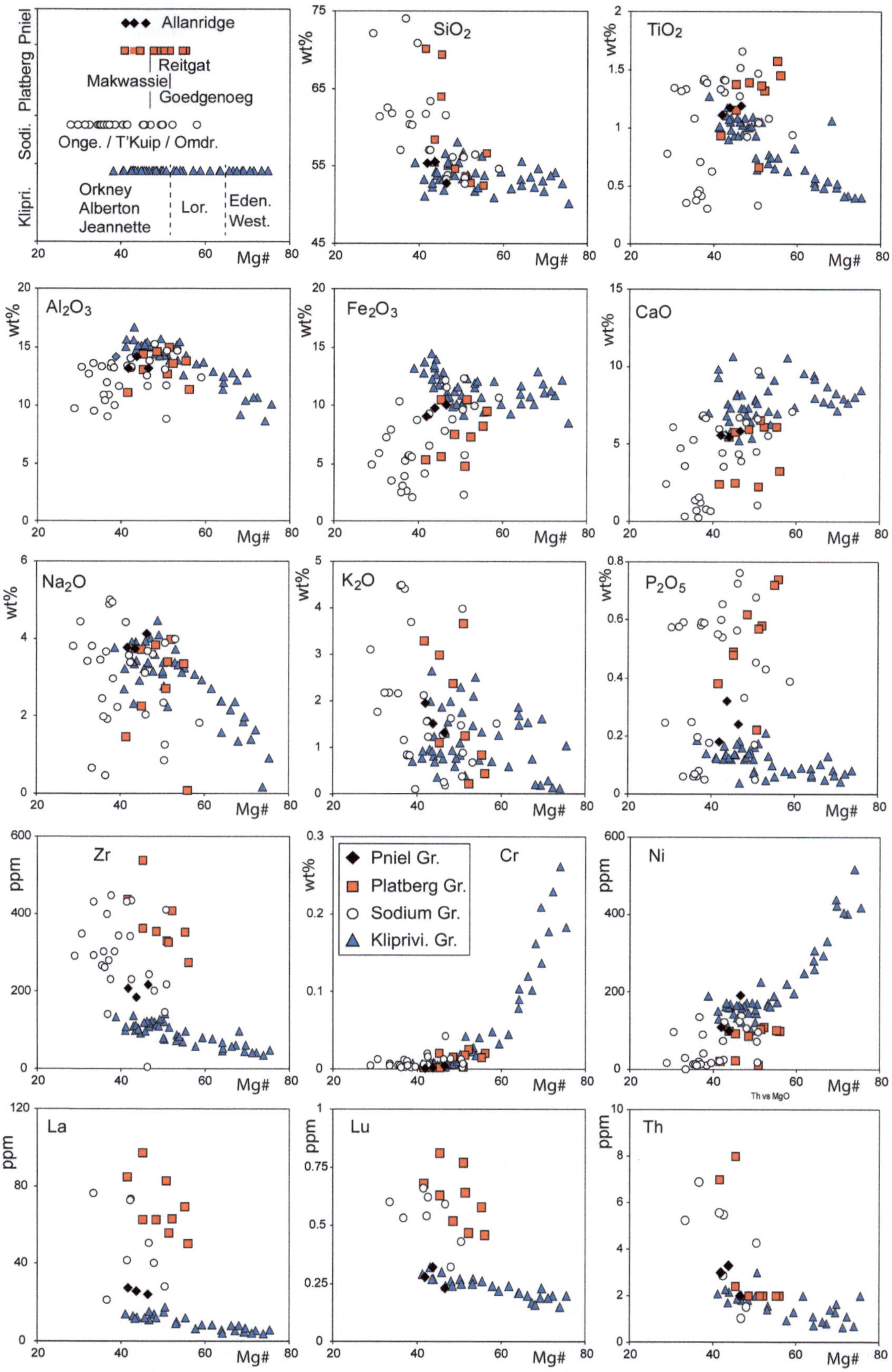

Fig. 11.6 Variation of major oxides and some traces elements of the Ventersdorp Supergroup lavas versus the Mg# (same samples as in Fig. 11.5). Onge. = Ongers River; Omdr. = Omdraaivlai; Lor. = Loraine; Eden. = Edenville; West. = Westonaria

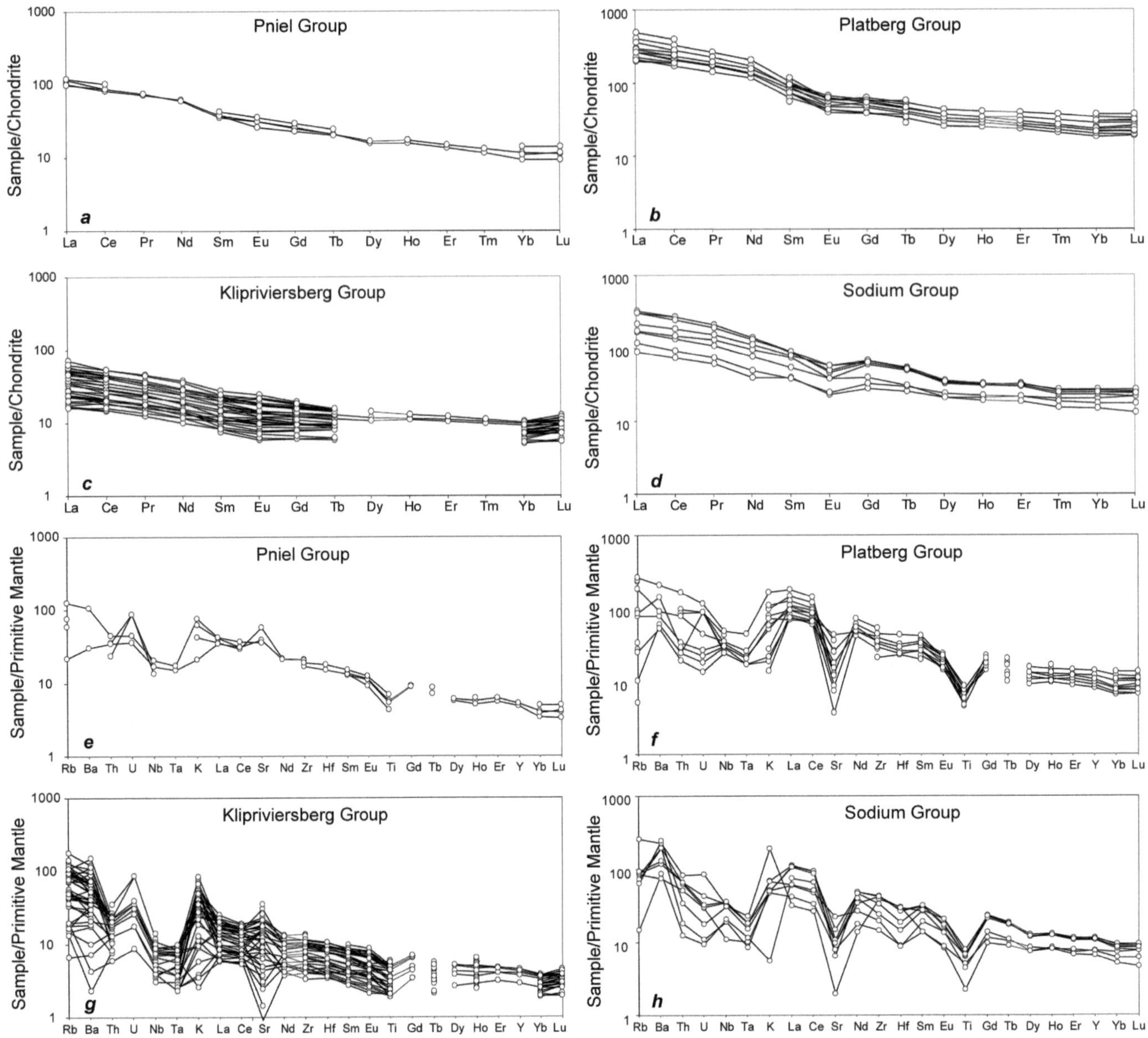

Fig. 11.7 REE (**a–d**) and multi-element diagrams (**e–f**) of the same samples as in Figs. 11.5 and 11.6. The samples were respectively normalized to chondrites (REE diagrams) and to primitive mantle (multi-element diagrams) after Palme and O'Neill (2004)

the rest of the Ventersdorp Supergroup. Yet, the overall fractionation of HREE relative to LREE, here represented by $[La/Sm]_N/[Tb/La]_N$ (Table 11.2), is relatively similar throughout the Ventersdorp Supergroup.

Klipriviersberg Group

Due to element mobility the fractionating mineral assemblage of the Ventersdorp Supergroup lavas is difficult to constrain. However, some trends are observed in the variation diagrams, notably for the Klipriviersberg Group. If fluid-mobile elements (Si, Na, K, potentially Ca) are omitted, the remaining elements present positive linear trends with Mg# and Al_2O_3, Zr, La, Lu and Th. Fe_2O_3, Cr and Ni display linear trends with clear inflection points associated with slope changes. TiO_2 and P_2O_5, display clear offsets around Mg# = 50 for most formations. This value of Mg# = 50 corresponds to the boundary between the Edenville, Loraine and Westonaria formations with the Orkney, Alberton and Jeannette formations. The clear linear trends confirm that all these formations are probably comagmatic and evolved largely by crystal fractionation. However, the presence of less differentiated lavas (Edenville and Loraine

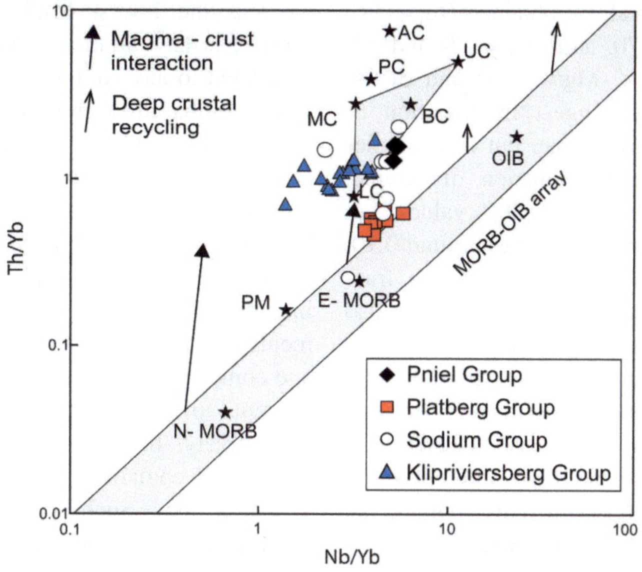

Fig. 11.8 Th/Yb–Nb/Yb ratio of Pearce (2008) of the same samples as in Figs. 11.5, 11.6, 11.7. LC = Lower Crust, MC = Middle Crust, UC = Upper Crust, BC = Bulk Crust, PC = Phanerozoic Crust, AC = Archaean Crust

Table 11.2 Feature of the shape of REE pattern of Fig. 11.7a–d; values concerning Hekpoort and Ongeluk formations of Fig. 11.11a, b are added as comparison

	[La]N range	[La/Lu]N	[La/Sm]N	[Tb/Lu]N	[La/Sm]N/ [Tb/Lu]N
Pniel Group	98–118	9.50 ± 1.28	2.82 ± 0.26	1.90 ± 0.25	1.48
Platberg Group	195–493	11.34 ± 1.94	3.42 ± 0.48	1.66 ± 0.24	2.07
Sodium Group	89–317	9.86 ± 3.73	3.01 ± 0.51	1.97 ± 0.39	1.53
Klipriviersberg Group	16–72	4.06 ± 1.26	2.29 ± 0.30	1.27 ± 0.27	1.81
Hekpoort Formation	72–117	7.64 ± 0.87	4.25 ± 0.22	1.26 ± 0.09	3.38
Ongeluk Formation	64–101	7.07 ± 1.83	3.85 ± 0.75	1.00 ± 0.41	3.66

formations) above more differentiated lavas (Edenville, Loraine and Westonaria formations) hint at a more complex evolution possibly involving magma replenishment of the magmatic chamber or evolution in different magma chambers.

We estimated the extent of fractional crystallization for the Klipriviersberg Group by using the Rayleigh Equation: $C_L/C_0 = F^{(Di-1)}$ with C_0, the initial concentration of an element in the parental or less differentiated melt, C_L, the concentration in the residual liquid, F the fraction of melt remaining and D_i the partition coefficient of an element i. For highly incompatible elements D can be set to zero, and the equation reduces to $F = C_0/C_L$. We calculated F based on Th. Th is preferred over Cs, Rb and Ba which are nominally more incompatible but also quite mobile. Using the Th value of the sample with the highest Mg# (i.e. ≈ 75) for C_O and an average of the four samples with the lowest Mg# (i.e. between 54 and 50) for C_L, we obtained $F = 0.44$ and thus, $1\text{–}F = 56\%$ of crystal fractionation for the whole Klipriviersberg Group.

The linear trends with inflection points show that the crystal fractionation changed during the differentiation. The decreases of Cr and Ni with decreasing Mg# (Fig. 11.6) respectively suggest fractionation of pyroxene, olivine, Cr-spinel, or all three phases. The variation of Cr and Ni displays an inflection of slope around Mg# = 50. Near constant concentrations below Mg# = 50 suggest that pyroxene, olivine or both fractionated from the parental magma until its Mg# decreased to 50. A second break of slope is present around Mg# = 62 in the variation of Cr.

By recalculating *F* for each interval, we obtained $D_{Ni} = 5.6$, for the interval $50 \leq Mg\# \leq 70$, and $D_{Cr} = 4.9$ and 12, respectably for the interval $62 \leq Mg\# \leq 70$ and $50 \leq Mg\# \leq 62$. These data mainly suggest fractionation of pyroxenes, possibly together with olivine, but in a smaller relative amount. Indeed, in the fractionation of olivine, orthopyroxene and clinopyroxene, the D_{Ni} values are respectively around 14, 5 and 7 and D_{Cr} values around 0.70, 10 and 34 (Rollinson 1993). Below Mg# = 62, the fractionation of clinopyroxene became prevalent, which can also explain the decrease of CaO, although that may also be attributed to the mobility of Ca. At Mg# ≈ 50, the fractionation of pyroxenes and olivine seems to have strongly decreased. This point is marked by an increase in Fe_2O_3 with decreasing Mg#. At this threshold the Al_2O_3 content variation is subdued with decreasing Mg#, which may point to the fractionation of plagioclase and would explain both the decrease of CaO and the increase of Fe_2O_3. The case of plagioclase remains complex, because most of its constituents are mobile (i.e. Si, Na, Ba, Sr, Eu^{2+}). Most differentiated Klipriviersberg Group samples display slightly negative Eu-anomalies (Fig. 11.7c, g). The remainder of samples shows either slightly positive Eu-negative or no anomalies. Variation in Eu, however, can result from this element's mobility. Finally, the increasing content of Dy with decreasing Mg# (not shown) argues against amphibole fractionation (Davidson et al. 2007) at least until Mg# = 48 as no Dy data are available for samples with lower Mg#. Similarly, the increase of TiO_2 with decreasing Mg# shows that titanomagnetite did not fractionate from the magma.

In conclusion, if we approximate all the samples of the Klipriviersberg Group to be originated from the same magmatic chamber, our interpretations match the conclusions of Marsh et al. (1992), for the Loraine and Edenville formations: '*the lavas are consistent with fractional crystallization processes dominated by orthopyroxene, with clinopyroxene and plagioclase playing a role in the late stages*'. These 'late stages' can be extended to the Orkney, Alberton and Jeannette formations. Note that, according Marsh et al. (1992) phase equilibria suggest that fractionation occurred at moderate pressure.

Platberg and Sodium Groups

The samples from the Platberg Group present a large degree of scattering in some element contents, but particularly for fluid-mobile elements (Fig. 11.6). The scattering is present for much of the Sodium Group, regardless of the source of the data (i.e. Grobler et al. 1989; or Altermann and Lenhardt 2012). It is a real possibility that the Platberg and Sodium groups are moderately altered. The Sodium Group contains samples showing a wide range of Mg#, down to Mg# = 29, whereas the lowest Mg# value for the Platberg Group samples is 42. Variation of Mg# in the Sodium Group cannot be related to any stratigraphic position.

The global linear trend for the Platberg Group samples in several plots confirms their comagmatic origin. The Rietgat and Goedgenoeg formations of the Platberg Group have higher Mg# ($45 < Mg\# \leq 56$) than the felsic Makwassie Formation ($42 \leq Mg\# \leq 51$). In the Platberg Group samples, the feldspar-porphyritic Makwassie Formation also includes the richest samples in SiO_2 and incompatible elements (i.e. Zr, La, Lu, Th) and the lowest content of Fe_2O_3 and compatible elements (i.e. Cr, Ni). The average Cr and Ni contents of the Makwassie Formation are 14 and 19, respectively. In comparison, the Goedgenoeg and Rietgat formations contain on average 155 and 213 ppm Cr, and 100 ppm Ni. Such chemical difference between the Goedgenoeg and Rietgat formations and Makwassie Formation suggest a large degree of mineral fractionation separate the Makwassie Formation from the Goedgenoeg and Rietgat formations. By contrast to the differentiation index, Mg#, does not evolve drastically and some samples between the three formations even present the same Mg#. Thus, it seems that the source magma of the felsic Makwassie Formation lava evolved separately from the two other formations.

We did not calculate the extent of fractional crystallization *F* for the Platberg Group given the large variation in Th between the Makwassie and other formations. We also did not calculate *F* for the Sodium Group samples due to their scattering. For both groups, the general decrease of Al_2O_3 and CaO below Mg# = 50 (Fig. 11.6) coupled with clear negative Sr- and Eu-anomalies (Fig. 11.7e–f) suggest the fractionation of plagioclase from their source magma. A decrease of P_2O_5 in Platberg Group samples suggests the fractionation of apatite. As detailed in the previous paragraph to explain the decrease in Fe_2O_3, olivine and pyroxene likely fractionated from the source magma of the Makwassie Formation. A decrease in Fe_2O_3 can also be related to fractionation of titanomagnetite, which would explain the decrease of TiO_2. Finally, as with the Klipriviersberg Group, the increasing content of Dy with decreasing Mg# (not shown) argues against amphibole fractionation.

Pniel Group

Due to a limited database very little can be said of the Allanridge Formation. It is, however, interesting to observe that the samples of this group are quite different from the other groups. Although the Pniel Group samples tend to cluster with the most differentiated samples from the Klipriviersberg and Platberg groups (e.g. TiO_2, Al_2O_3, Cr Vs. Mg#; Fig. 11.6), specific elemental content is quite different for similar Mg# (e.g. Fe_2O_3, P_2O_5, Zr, La Vs. Mg#). This might point to a very different magma evolution or origin for the Allanridge Formation. Samples from the Sodium Group provide the best match to samples of the Pniel Group.

11.3.1.4 Magma Evolution

Basaltic lavas of the Klipriviersberg Group exhibit a well-developed geochemical stratigraphy characterized by the initial eruption of primitive lavas of the Westonaria Formation, followed by differentiated lavas of the Alberton and Orkney formations. The units became geochemically more primitive upwards through the Loraine and Edenville formations (Table 11.1). This may indicate cyclicity in petrogenic processes, which can also be observed in the Platberg Group, where the felsic Makwassie Formation is surrounded by the intermediate Goedgenoeg and Rietgat formations. Moreover, in each group, all the formations share common features like their trace elements patterns (Fig. 11.7). Cyclicity might also be present in the Sodium Group. According to Grobler et al. (1989) and Altermann and Lenhardt (2012), the lavas of the T'Kuip Formation are more differentiated than those of the surrounding Ongers River and Omdraaivlei formations. The significant scatter present in the dataset could mask more prominent cyclicity (Fig. 11.6).

Crow and Condie (1988) proposed a model of magma evolution to explain the cyclicity in the Klipriviersberg Group, involving batch melting and olivine fractional crystallization, applied on a garnet lherzolite source. In this model all Klipriviersberg lavas descend from the same magmatic source that split into two magmas that evolve separately. The first magma led to the origin of the Westonaria Formation, while all other volcanic formations result from the second magma. The decrease of differentiation from Alberton, Orkney and Jeannette formations to Loraine and Edenville formations by a continuous increasing batch melting in the source from 10 to 20% is thus explained. We disagree with that model because the trace element patterns clearly show that all the formations forming the Klipriviersberg Group result from the same degree of melting (Fig. 11.7c, g). The conclusion by Marsh et al. (1992) that the '*Alberton and Orkney lavas are indicative of magma mixing […] the mixing end-members are evolved Loraine/Edenville and evolved Westonaria magmas*' appears more likely. Note that, according to the data presented in this contribution (Figs. 11.6 and 11.7d, g), the simplest model to explain all compositional variation in Klipriviersberg Group would be through fractional crystallization in a single magma chamber, with sporadic magma replenishment. But, for Marsh et al. (1992), unlike the Loraine/Edenville lavas which can be explained by fractional crystallization alone, the composition of the Alberton and Orkney lavas results from magma mixing based on the unusual inter-element correlations (see Marsh et al. 1992).

Crow and Condie (1988) proposed that the Pniel Group is genetically linked to the Klipriviersberg Group which is quite unlikely inasmuch as (1) the Platberg Group is present in between these two groups and (2) a significant difference in age also probably separate them (Fig. 11.3; Table 11.1). It is yet true that Pniel Group samples partly resemble the most differentiated Klipriviersberg Group samples (Fig. 11.6). The two groups also share similar Th/Yb values (Fig. 11.8). But, for several elements, the Pniel Group is comparable with some samples of the Platberg and Sodium groups (Fig. 11.6). These three groups also share similar Nb/Yb values (and Th/Yb values with some Sodium Group samples). The correspondence between the trace element patterns of the three groups demonstrates that they issued from the same degree of partial melting. Furthermore, the Platberg and Pniel groups present similar epsilon Nd data (see next paragraphs), which differ from those of the Klipriviersberg Group. The difference of Zr, La and Lu content between the Pniel Group and Platberg/Sodium groups (Fig. 11.6) can also be explained by different degrees of fractionation, linked to different evolutionary paths from a shared source magma, or by source heterogeneity.

Despite of similarities, a direct comparison between Platberg and Sodium groups remains ambiguous. Did magmas responsible for volcanic rocks of these groups differentiate in the same magma chamber? Crow and Condie (1988) concluded that it was not the case, whereas Grobler et al. (1989) correlated the two groups. Indeed, the large spatial separation between present-day outcrops (Fig. 11.1) and their occurrence in isolated grabens that were possibly active at different times, suggest that they did not share a magma chamber. The large scatter of geochemical data of the Sodium Group samples (Figs. 11.5, 11.6 and 11.8) clouds any interpretation. Nevertheless, the difference between the two groups in some incompatible element content (i.e. Zr, La, Lu) for similar Mg#, coupled with the fact that Sodium Group encompasses significantly stronger differentiated samples than the Platberg Group, is in favour of origins from different magmatic evolutionary paths.

Multi-element diagrams, normalized to primitive mantle values (Palme and O'Neill 2004), are presented for the same samples in Fig. 11.7e, f. Beyond the difference of overall trace elements content, the patterns of the four different groups are relatively similar. Samples exhibit clearly negative anomalies in some high field strength elements (HFSE) like Nb, Ta and Ti. Such anomalies can point towards metasomatism of the mantle source via crustal input (Pearce 2008). Such 'subduction-related' (e.g. Condie and Crow 1990; Marsh et al. 1992) or 'arc-like' (e.g. Kamber 2015) signatures are quite common in Archaean and Palaeoproterozoic mafic rocks. Different authors have already suggested the implication of subduction processes in the subcontinental lithospheric mantle (SCLM) in their models of magma origin for the Ventersdorp Supergroup volcanic rocks. For example: an '*enriched within-plate source, perhaps the mantle lithosphere, to which a subduction zone component was later added*' (Crow and Condie 1988), or '*interaction of asthenospheric mantle plumes with subcontinental lithospheric mantle sources which have been modified by subduction processes*' (Nelson et al. 1992).

Nevertheless, the majority of Precambrian mafic igneous rocks across the Kaapvaal craton show the same characteristics (e.g. Klausen et al. 2010; Humbert and Elburg 2017; Humbert et al. 2018a) whereas none of them presents any obvious geodynamic evidence of being related to a subduction. For instance, the Ventersdorp Supergroup is rather linked to continental scale rifting process, i.e. the break-up of the Vaalbara supercontinent (de Kock et al. 2009, 2012). Most authors agree that the Ventersdorp Supergroup represents continental flood basalt volcanism (e.g. Marsh et al. 1992; Nelson et al. 1992). Furthermore, it seems unlikely that widespread volcanic activity (Fig. 11.1) on the exhumed Archaean craton can be subduction-related. The subduction-related views stem from a formerly suggested link of the Ventersdorp Supergroup basin formation to the Limpopo orogeny in the north, which as discussed by Altermann and Lenhardt (2012) seems unrealistic and has been refuted by isotopic ages for the Ventersdorp and the Limpopo orogeny (Kramers et al. 2006; Kramers and Mouri 2011).

Crustal input can also be linked to crustal contamination or to deep recycling of crustal material. In a Th/Yb–Nb/Yb diagram (Pearce 2008) samples plots above the mantle array, near or overlapping with the field of continental crust. This suggests crustal contamination as the possible source of the HFSE anomalies. In addition, most of the samples have moderate Nb/Yb values, which is inadequate for deep crustal recycling. Epsilon Nd can be used to highlight crustal contamination. Defined as $\varepsilon Nd(t) = 10^4 \times [(^{143}Nd/^{144}Nd)^{sample} - (^{143}Nd/^{144}Nd)^{CHUR}]/(^{143}Nd/^{144}Nd)^{CHUR}$ calculated at the age of the sample, with CHUR for Chondritic Uniform Reservoir. Typically mantle samples show a $^{143}Nd/^{144}Nd$ ratio higher than CHUR ($\varepsilon Nd > 0$), whereas samples from the continental crust display lower $^{143}Nd/^{144}Nd$ compared to CHUR ($\varepsilon Nd < 0$). The Platberg and Pniel groups show clear negative values εNd_{2709Ma} −3.3 and −2.7, respectively (Table 11.1). The Klipriviersberg Group has average εNd_{2714Ma} values close to 0 (Table 11.1), with $\varepsilon Nd_{2714Ma} = 0.5$ (Marsh et al. 1992) or −0.4 (Nelson et al. 1992) for the Alberton, Orkney and Jeannette formations, and $\varepsilon Nd_{2714Ma} = -0.5$ (Marsh et al. 1992) or −0.6 (Nelson et al. 1992) for the Loraine and Edenville formations. The Klipriviersberg Group does not show a depleted mantle signature, which should be $\varepsilon Nd_{2714Ma} \approx 2.2$ based on the current depleted mantle Nd values (DePaolo 1988), nor a depleted mantle source of komatiites with εNd_{2714Ma} up to 4 (Marsh et al. 1992). Note that if we attribute to the Klipriviersberg Group the same age of the Derderpoort basalts (see previous sections), i.e. 2780 Ma instead of 2714 Ma the εNd values increase only minimally. The previous values become $\varepsilon Nd_{2780Ma} = 1.1$ and −0.2 (Alberton, Orkney and Jeannette formations) and $\varepsilon Nd_{2780Ma} = 0.0$ and −0.3 (Loraine and Edenville formations).

In conclusion, crustal contamination can be inferred for the Pniel and Platberg groups, but not for the Klipriviersberg Group. Many workers believed that the geochemical features of the volcanic Ventersdorp rocks, and in a general way, of all volcanic rocks older than 1.8 Ga, arise from crustal contamination of komatiitic magmas (Sun et al 1989; Condie and Crow 1990). But, based on modelling, Marsh et al. (1992) demonstrated that a komatiite cannot be the parental source formations in a contamination process for the origin of the Loraine and Edenville lavas, neither for their εNd, nor for their geochemical features, whatever the degrees of differentiation may be.

The negative HFSE anomalies and high Th/Nb in the Klipriviersberg Group samples are therefore certainly related to metasomatism of the SCLM from past subduction, i.e. the SCLM was already metasomatically enriched before the Ventersdorp Supergroup volcanism. It is unlikely that crustal contamination has influenced the composition of the Klipriviersberg Group lavas to any significant extent. This explains to some degree the elevation of Th/Nb of the Platberg and Pniel groups (and also possibly the Sodium Group) above mantle array (Fig. 11.8). It is possible that metasomatism of the subcontinental lithospheric mantle also contributed to the 'arc-like' signature of the Platberg, Sodium and Pniel groups.

11.3.2 The Ongeluk and Hekpoort Formations of the Transvaal Supergroup

11.3.2.1 Volcanic Lithology

The geochemical data from the Ongeluk and Hekpoort formations discussed herein are from new and previously published data: Engelbrecht (1986), Oberholzer (1995) and Coetzee (2001) for the Hekpoort Formation, and Schütte (1992) and Cornell et al. (1996) for the Ongeluk Formation. Data from these publications were added to more recent data for both formations (on samples devoid in amygdales and veins) from Humbert et al. (2018a). All previous works highlighted greenschist facies metamorphism for both the Hekpoort and Ongeluk formations. In addition, metasomatism and mobility of some elements has occurred in the Hekpoort Formation. Fluid-mobile element data are interpreted with caution.

In the Zr/TiO_2 versus Nb/Y diagram (Pearce 1996), all samples generally are of basaltic composition, but near the boundary with basaltic andesites. In previous studies the Hekpoort Formation displayed variation of Nb/Y values (Fig. 11.9b) not seen in the dataset of Humbert et al. (2018a) (Fig. 11.9a). The latter show a fair amount of differentiation ($50 \leq Mg\# \leq 69$; Fig. 11.10). Similar differences can also be observed between the data of Engelbrecht (1986), Schütte (1992), Cornell et al. (1996), and Humbert et al. (2018a) for the Ongeluk Formation (Fig. 11.9c). We assign such variations to technical limits (i.e. trace element data only acquired by optical or atomic emission spectrometry) or difficulties during the measurements of these data. For this reason, much of data is omitted in further discussion, except that of Schütte (1992), Cornell et al. (1996) and Humbert et al. (2018a).

The classification of these rocks as basalt is in disagreement with all previous studies on Hekpoort Formation samples, which typically classified the rocks as basaltic–andesites or andesites (e.g. Button 1973; Sharpe et al. 1983; Engelbrecht 1986; Oberholzer 1995; Reczko et al. 1995; Oberholzer and Eriksson 2000). Cornell et al. (1996) also classified the Ongeluk Formation lavas as andesites. Although one Hekpoort Formation sample plots in the field of basaltic andesites (Fig. 11.9a) the remainder are clearly basalts. The difference is mainly due to the fact that we are using the Zr/TiO_2 vs. Nb/Y diagram with the boundaries between magmatic lithologies defined by Pearce (1996), instead of the original ones by Winchester and Floyd (1977) where the area of basalt is less prominent (see Fig. 11.5). Based on SiO_2 contents alone, with an average of 54.52 wt% ± 1.09 and 54.24 wt% ± 2.80 (Fig. 11.10), these samples are too differentiated to be basalts. However, fairly invariable SiO_2 content versus the evolution of Mg# strongly suggests that the samples of the two formations were affected by silicification (already suggested by Reczko et al. 1995 and Humbert et al 2018a, b). This assumption is supported by the fact that Humbert et al. (2018a), observe only secondary quartz in the Hekpoort and Ongeluk formations lavas. Silicification can be expected based on the age of the units and surrounding siliciclastics rocks. Because of the high MgO content of the Hekpoort and Ongeluk samples, between 5.0 and 11.3 wt% (mean: 7.0 ± 1.7 wt%; $50 < Mg\# < 69$) and 5.2 and 8.7 wt% (mean: 6.3 ± 0.9 wt%; $50 < Mg\# < 56$) respectively, it seems most appropriate to classify the Hekpoort and Ongeluk formations as basaltic. Note that the relatively high silica content of the two formations can also be primary if their lavas are anorogenic boninites, also called boninitic-noritic (BN) rocks. Anorogenic boninites are common during the 2.5–2.1 Ga period and have a $SiO_2 > 52$ wt% (Srivastava and Ernst 2013; Ernst 2014). However, the fairly invariable SiO_2 versus Mg# still suggest that the samples were at least partly silicified (Humbert et al. 2018a, b).

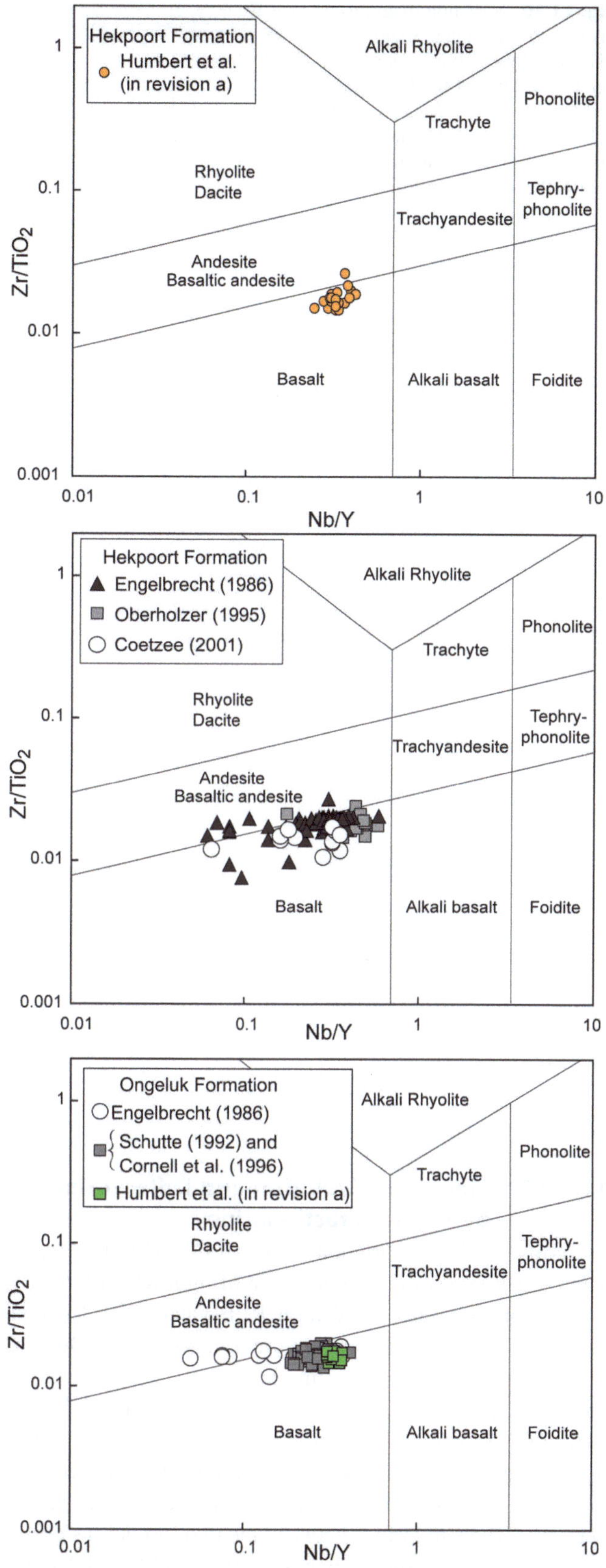

Fig. 11.9 Classification diagram the Hekpoort (**a**, **b**) and Ongeluk samples (**c**) from literature. Only data obtained with Instrumental Neutron Activation Analysis or ICP mass spectrometry from Schütte (1992) are represented here

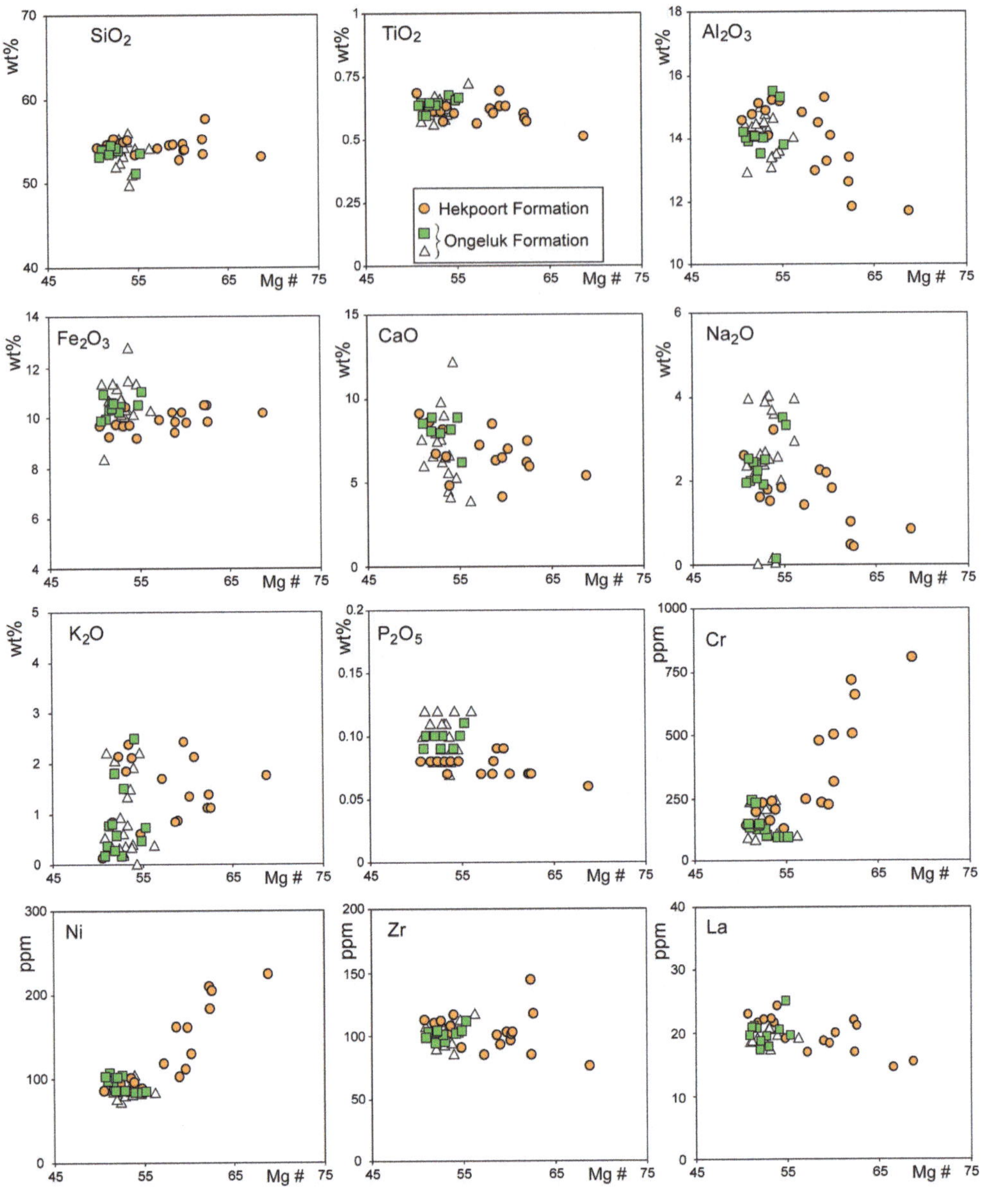

Fig. 11.10 Variation diagrams displaying the content major oxides and some traces elements versus the Mg# of the Hekpoort and Ongeluk lavas. Data are from Schütte (1992) and Cornell et al. (1996) (triangles), and Humbert et al. (2018a) (squares and circles)

11.3.2.2 Magmatic Evolution and Differentiation by Crystal Fractionation

In terms of REE variation (Table 11.2; Fig. 11.11), all samples show a strong fractionation of light REE of 70–100 times chondrite values and high REE values (10–13 times chondrite values). Samples of the Hekpoort and Ongeluk formations further display remarkably similar trace element profiles with just a slight scattering of the HREE for the Ongeluk Formation. The similarity between the two formations is again observed when comparing the multi-element diagrams (Fig. 11.11c, d); the two trends match almost perfectly. The similarities between the two formations can be also observed in other diagrams (Figs. 11.9, 11.10 and 11.12). In conclusion, as already suggested by Cornell et al. (1996) and Humbert et al. (2018a) the Ongeluk samples resemble the most differentiated Hekpoort samples.

As with the Klipriviersberg Group samples, we estimated the extent of fractional crystallization for the Hekpoort Formation by using the Rayleigh Equation, using the Th value of the sample with the highest Mg# (69) for C_O and an average of the samples with the lowest Mg# (i.e. between 53 and 50) for C_L, we obtained $F = 0.58$ and thus 42% of crystal fractionation for the Hekpoort Formation. F is 0.97 for the Ongeluk Formation samples due to their similar Mg#. The decrease of Cr and Ni with decreasing Mg# (Fig. 11.9) respectively suggest fractionation of either pyroxenes, olivine, Cr-spinel, or all three of these phases. A break in slope is actually observed in the two plots at Mg# = 55, notably in

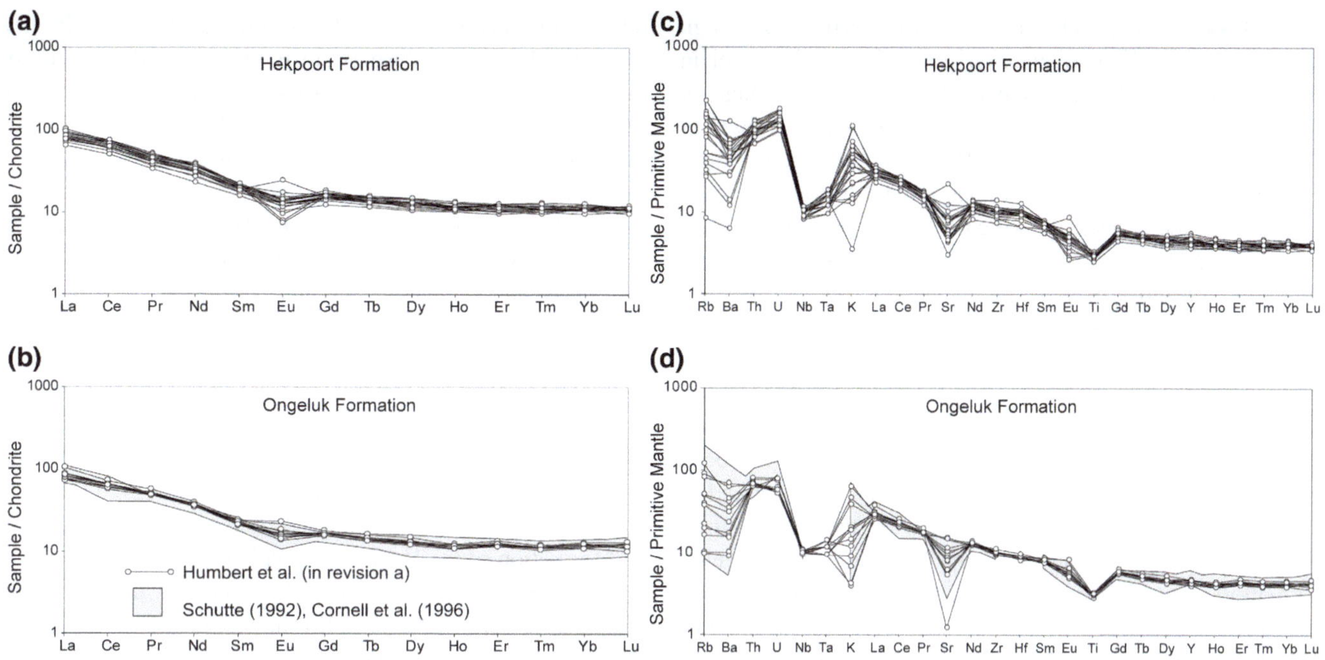

Fig. 11.11 REE (**a**, **b**) and multi-element diagrams (**c**, **d**) of the same samples as in Fig. 11.10. The samples were respectively normalized to chondrites (REE diagrams) and to primitive mantle (multi-element diagrams) after Palme and O'Neill (2004)

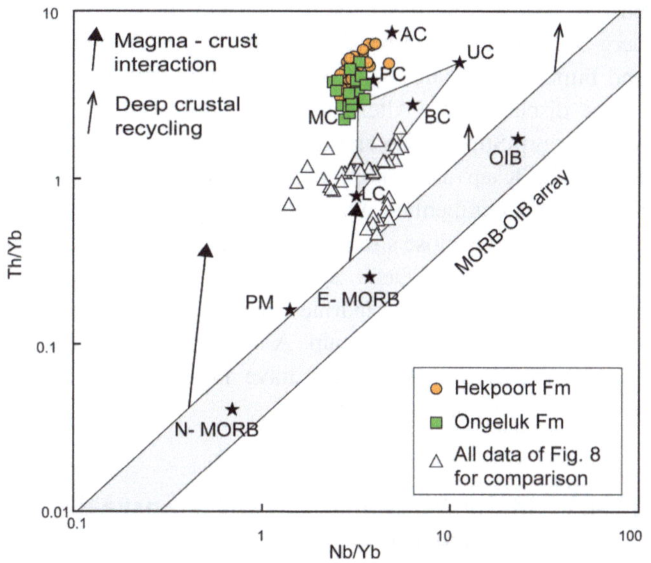

Fig. 11.12 Th/Yb–Nb/Yb ratio of Pearce (2008) of the same samples as in Figs. 11.10 and 11.11 (all Ongeluk samples are plotted together for simplification) added to the samples from Fig. 11.8. PM = Primordial Mantle, LC = Lower Crust, MC = Middle Crust, UC = Upper Crust, BC = Bulk Crust, PC = Phanerozoic Crust, AC = Archaean Crust

the case of Ni versus Mg#, suggesting that pyroxenes, olive, or both fractionated from the parental magma until its Mg# decreased to 55. We calculated $D_{Ni} = 3.5$ and $D_{Cr} = 4.9$ in the Hekpoort Formation lavas which suggest mostly fractionation of pyroxenes and chromite, possibly together with minor olivine (see Sect. 11.3.1.3.1). The increasing content of Dy with decreasing Mg# (not shown) argues against amphibole fractionation. Similarly, the increase of TiO_2 with decreasing Mg# shows that (titano-)magnetite did not fractionate from the magma. Finally, the increase of Al until Mg# = 55, followed by a slight decrease, indicates that plagioclase started fractionating at a more advanced stage of differentiation. Negative Eu-anomalies in most of the samples with Mg# $\leq$ 55 (Fig. 11.11) for both the Hekpoort and Ongeluk formations, coupled with the Na decrease in the Ongeluk Formation (despite being fluid-mobile), support the fractionation of plagioclase.

11.3.2.3 Magma Evolution

Because the ratios of incompatible elements do not change much with crystal fractionation, the correspondence of REE patterns between the Hekpoort and Ongeluk formations strongly suggests similar magma sources and subsequent evolution. Cornell et al. (1996) suggested a lithospheric mantle source whereas Reczko et al. (1995) favoured an asthenospheric source. The asthenospheric influence on the source of the Hekpoort and Ongeluk formations is likely, but only as a heat source. In the Th/Yb–Nb/Yb plot (Fig. 11.12), all samples of the Hekpoort and Ongeluk formations cluster in a similar area marked by a high Th/Yb ratio (significantly higher than those of all Ventersdorp Supergroup lavas) and not characteristic of an asthenosphere source. The Th content in the Hekpoort and Ongeluk samples explains such ratios (i.e. Th = 8.36 ppm ± 1.46 and Th = 6.8 ppm ± 1.97 for the Hekpoort and Ongeluk formations respectively). A high degree of crustal contamination can theoretically explain these values, but the positive εNd values of the Ongeluk Formation ($\varepsilon Nd_{2426Ma} = 0.9$ on average based on values of four samples; Cornell et al. 1996), argue against this. Notwithstanding, these εNd values remain too low to stem from a depleted mantle, for which $\varepsilon Nd_{2426Ma} = 2.7$ would be expected (DePaolo 1988). The Th/Yb of the samples is quite different with those of depleted mantle (Fig. 11.12). No εNd data exist for the Hekpoort Formation.

The Hekpoort and Ongeluk samples exhibit trace element patterns similar to those of the Ventersdorp Supergroup lavas, i.e. an 'arc-like' geochemical signature. As discussed before, it is very improbable that this anomaly is produced by magmatic differentiation. More likely it can be linked to the magma source. Moreover, as with the Ventersdorp Supergroup lavas, the lavas of the Hekpoort and Ongeluk formations are the products of continental flood basalt magmatism. No evidence for subduction during the time of their extrusion has been described (e.g. Eriksson et al. 2006). A simple 'subduction-related' origin of this lava is thus unlikely.

In summary the Ventersdorp Supergroup volcanics and lavas of the Hekpoort and Ongeluk formations have a shared SCLM source, which was likely to have been metasomatised by past subduction. All Ventersdorp, Hekpoort and Ongeluk samples present an 'arc-like' signature, the higher Th/Yb ratios of all Hekpoort and Ongeluk samples imply that an episode, or multiple episodes of metasomatism of the SCLM occurred close to the Archaean-Proterozoic transition.

11.4 Palaeoenvironmental Conditions During Volcanism of the Ventersdorp and Transvaal Supergroups

11.4.1 The Ventersdorp Supergroup

So far, sparse information on the palaeoenvironment during Ventersdorp Supergroup deposition is available and more studies are needed on the different sub-basins of this volcano-sedimentary sequence, its lithofacies distribution, and geochronology in order to correlate different basins and outliers. Nevertheless, a relatively generalized palaeoenvironmental reconstruction can be established.

According to Clendenin et al. (1988a, b) and Burke et al (1985), large fractures developed in the crust during Ventersdorp Supergroup times in response to collision between the Kaapvaal and Zimbabwe cratons along the Limpopo Metamorphic Complex, *ca.* 2.65 and 2.0. Ga, Kramers et al. 2006; Kramers and Mouri 2011). This supposedly lead to the formation of graben-like basins on the Kaapvaal craton, hundreds of kilometres south of the mobile belt, and the release of extensive lava flows through fissures. In this scenario, the Ventersdorp Basin is viewed as the second of three successor basins including the Dominion-Witwatersrand and Transvaal basins. The location of depocentres, morphology and magmatism within these basins are assumed to reflect different angles and velocities of subduction beneath the Kaapvaal craton. However, despite many studies, it is still difficult to precisely constrain the geodynamic history of the Limpopo Metamorphic Complex (Nicoli et al. 2015 and references therein), but it is now well accepted that the Limpopo Metamorphic Complex is significantly younger than the Ventersdorp Supergroup volcano-sedimentary succession (e.g. Kramers et al. 2006), and therefore the collision scenario can be largely discarded as discussed by Altermann and Lenhardt (2012).

An alternative explanation for the graben development and formation of the Ventersdorp volcano-sedimentary basin is a mantle plume (Hatton 1995), causing the Kaapvaal craton to undergo intense uplift and rupturing at the end of Witwatersrand Supergroup sedimentation (Eriksson et al. 2001). According to this hypothesis, crustal extension caused by mantle plume-related thermal elevation led to the development of a set of graben and half-graben depositional basins, forming the basis for the ~2-km thick Klipriviersberg Supergroup flood basalts and the wedge-shaped clastic and bimodal volcanic rocks of the Platberg Group.

As discussed by Altermann and Lenhardt (2012), the Ventersdorp Supergroup and its outliers represent the largest LIP of the Kaapvaal craton. These authors emphasised that the volcano-sedimentary Ventersdorp Supergroup sequence of rocks displays close similarity to graben fills of a large rift system such as the Dead Sea and the East African Rift system, although clearly marine-oceanic realms are missing in the Ventersdorp Supergroup. A scenario of a failed continental rift system that might have lasted for 70 million years or more was proposed.

11.4.2 Major Volcanisms in the Transvaal Supergroup

The Hekpoort and the Ongeluk formations of the Transvaal and Griqualand West sub-basins, respectively, are seen as major continental flood basalt events on the Kaapvaal craton that occurred mainly under subaerial conditions in the east and under at least partly subaquatic (probably lacustrine) conditions at times, particularly along the western margin of the present-day Kaapvaal craton. With this in mind, it should be remembered that the two formations are separated by 200 myr and thus their palaeoenvironment cannot be directly related.

The extension of the Transvaal sub-basin into southern Botswana indicates no change in the subaerial nature of the

Hekpoort Formation volcanics towards the northwest (Key 1983). Nevertheless, the occurrence of intercalated sedimentary rocks such as cross-bedded sandstones and laminated to cross-laminated mudstones (e.g. Button 1973) that probably marked intermittent hiatuses in volcanism, suggests the presence of open water bodies such as localized ponds and lakes (Button 1973; Humbert et al. 2018a, b) where these sediments could have been deposited (Lenhardt et al. 2012).

The predominance of lavas over pyroclastic rocks favours an origin from fissure eruptions (White et al. 2009) typical for the source vents of continental flood basalts (Bryan and Ferrari 2013). Presumably, the basaltic character of the Hekpoort Formation lavas and the concomitant low volatile content were responsible for the apparent lack of pyroclastic rocks on a regional scale (Oberholzer and Eriksson 2000). The slightly calc-alkaline basaltic-andesitic Hekpoort Formation magmas reported by Oberholzer (1995) in the south-western part of the Transvaal sub-basin may have promoted a more explosive style of volcanism that allowed for the formation of varying proportions of lava flows and pyroclastic rocks. This may have led to frequent changes from effusive to explosive activity along a fissure vent or the localized formation of point vents such as smaller cone structures and stratovolcanoes (Bryan and Ferrari 2013). The occurrence of lahar deposits in the south-western part of the Transvaal sub-basin implies an elevated relief or a hilly morphology that would make the formation of muddy volcanic debris-flows possible, and is therefore in support of the idea of the existence of localized stratovolcanoes.

11.5 Discussion and Conclusions

Conform with the geochemical data presented in this manuscript, we can conclude that all Ventersdorp Supergroup lavas are different from volcanic rocks of the Ongeluk and Hekpoort formations. The geochemical data presented in this work show obvious differences for the Zr/TiO_2–Nb/Y and Th/Yb–Nb/Yb plots (Figs. 11.5, 11.8, 11.9 and 11.12) or for the trace element patterns (Figs. 11.7 and 11.11). Such heterogeneity cannot only be explained by a varying differentiation in a magma chamber, as the Ventersdorp Supergroup includes both, stronger and less differentiated volcanic rocks than those of the Ongeluk and Hekpoort formations. It rather highlights either heterogeneities of evolution of the magma source (by different degrees of fractionation, crustal contamination, and recharging), or magma descent from different chambers. The Ventersdorp Supergroup exhibits a polyphased volcanic history, unlike the Hekpoort and Ongeluk formations, which are uniform and which share remarkably similar geochemical signatures. This is not surprising when the time and duration of extrusion and the volume of magmas are considered. The 2.78–2.70 Ga Ventersdorp Supergroup is several kilometres thick and extruded over a period of at least 70 million years (but likely much longer) in a graben setting, whereas the maximally 1 km thick lavas of the 2.46 Ga Ongeluk and the 2.23 Ga Hekpoort formations must have extruded over a much shorter period of time and with a significantly lower volume.

It may seem superfluous to conclude that mafic units separated by more than 300 Ma (Ventersdorp Supergroup and Ongeluk Formation) or 500 myr (Ventersdorp Supergroup and Hekpoort Formation) are geochemically different. But, on the other hand 200 Ma separate the Ongeluk Formation from the Hekpoort Formation, and they are geochemically very similar (Figs. 11.9, 11.10, 11.11 and 11.12). One can still consider this equivalence as an argument in favour of the 'uniform trace element pattern' hypothesis cited earlier. However, Humbert and Elburg (2017) and Humbert et al. (2018a) show that this geochemical equivalence between two mafic units remains a unique case in the whole igneous history of the Kaapvaal craton.

With the exception of the volcanic formations belonging to the Platberg/Sodium and the Pniel groups, the εNd(t) of all other lava formations indicates a mantle source that is either close to chondritic (i.e. $\varepsilon Nd \approx 0$; Table 11.1) or enriched compared to the depleted mantle (i.e. depleted mantle > εNd > 0). A magma derived from the depleted upper mantle (i.e. with εNd(t) > + 2.2; +2.7 and + 3 around $t = 2.72$; 2.43 and 2.25 Ga, respectively; DePaolo 1988), contaminated by continental crust (with εNd(t) of ca. −5 to −15) can provide an εNd(t) close to 0. Yet, by their Th/Yb and Nb/Yb ratios (Figs. 11.8 and 11.12) all the lavas discussed in the contribution were more likely generated from enriched sources than depleted ones. By their high enrichment in most incompatible elements, the REE and trace element patterns of the all lavas (Figs. 11.7 and 11.11) also suggest that they were generated from enriched sources. All the things considered, a negligible effect of crustal contamination is recorded by the lavas of Klipriviersberg Group and lower than those of the Platberg/Sodium and the Pniel groups.

The Ventersdorp Supergroup, Ongeluk and Hekpoort formations also exhibit geochemical signatures quite similar to those of modern 'arcs', mainly in the form of pronounced negative anomalies in Nb, Ta and Ti. Yet, based on field evidence and large spatial extend, these volcanic events can be interpreted as within-plate magmatism. The magmatic events of the Ventersdorp Supergroup, Ongeluk Formation and Hekpoort Formation cannot be linked to any marked tectonic events and thus may represent failed rifting events grounded in mantle dynamics (e.g. Eglington and Armstrong 2004). We conclude that this 'arc-like' signature highlights a metasomatism of the SCLM.

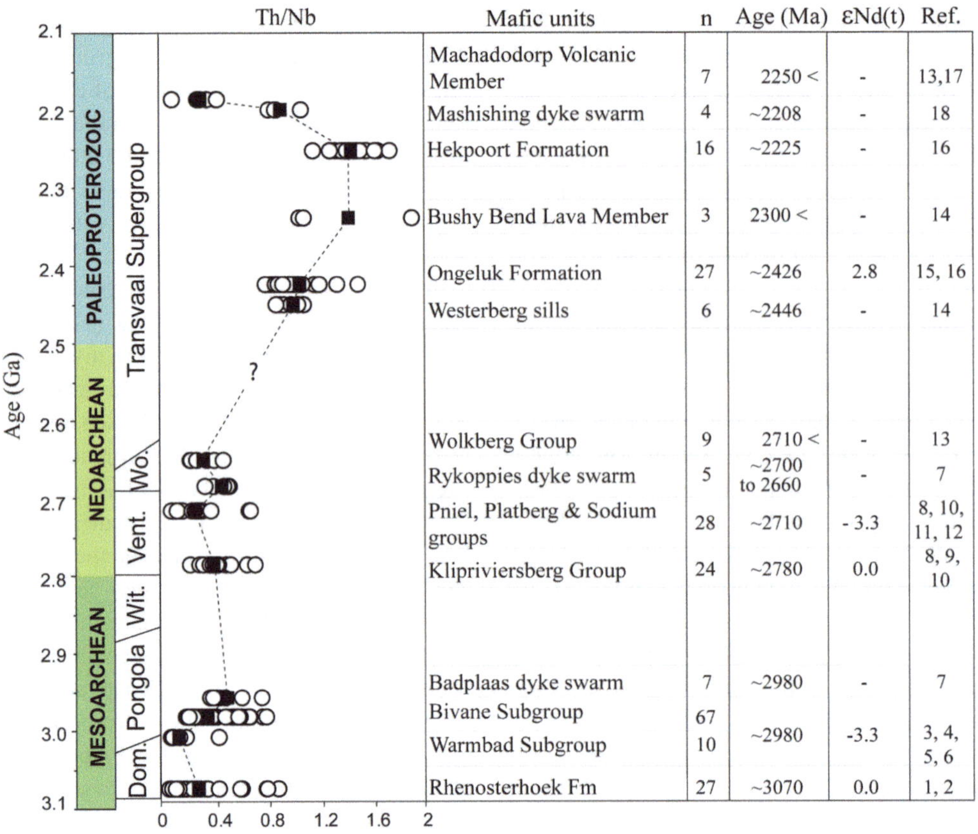

Fig. 11.13 Th/Nb ratio of Neoarchaean to Palaeoproterozoic mafic units in the Kaapvaal craton. n = number of samples plotted. Dom. = Dominion Group; Pon. = Pongola Supergroup; Wit. = Witwatersrand Supergroup, Vent. = Ventersdorp Supergroup. The open circles represent the data and the black squares the average value (calculated for units with more than 2 samples). The εNd(t) values were calculated for the age in the age column. References are (1) Crow and Condie (1987), (2) Marsh et al. (1989), (3) Armstrong et al. (1986), (4) Bowen et al. (1986), (5) Grant (2003), (6) Mukasa et al. (2013), (7) Klausen et al. (2010), (8) Crow and Condie (1988), (9) Marsh et al. (1992), (10) Nelson et al. (1992), (11) Grobler et al. (1989), (12) Altermann and Lenhardt (2012), (13) Crow and Condie (1990), (14) own data, (15) Schütte (1992), (16) Humbert et al. (2018a), (17) Wabo et al. (2019)

In the Th/Yb versus Nb/Yb plots (Figs. 11.8 and 11.12) the only common point between Ventersdorp Supergroup lavas and lavas of the Hekpoort and Ongeluk formations lavas is that some samples from the Klipriviersberg Group share similar Nb/Yb ratios with Hekpoort and Ongeluk samples. Overall, the Ventersdorp Supergroup lavas present significantly lower Th/Yb ratios. The difference between the Ventersdorp Supergroup and the two Transvaal formations is clear when the Th/Nb ratio of most of the mafic units in the craton during a period from the ~3.07 Ga Dominion Group to the ~2.2 Ga Machadodorp Volcanic Member lavas is compares (Fig. 11.13). All Archaean mafic units of the Kaapvaal craton show low Th/Nb ratios, whereas mafic units in the first part of the Palaeoproterozoic, which include the Hekpoort and Ongeluk formations, present relatively higher Th/Nb. To make the offset of Th/Nb around the Archaaean—Proterozoic boundary clearer, this ratio has respectively a mean of 0.31 ± 0.18 with $0.05 < Th/Nb < 0.88$ in the Archaean (based on 164 samples) and a mean of 1.21 ± 0.26 with $0.80 < Th/Nb < 1.90$ in the Palaeoproterozoic from the 2.46 Ga Westerberg sill to the Mashishing dyke swarm (based on 76 samples).

The high Th/Yb (Fig. 11.12), Th/Nb (Fig. 11.13) and moderate Nb/Yb (Fig. 11.12) ratios of both the Hekpoort and Ongeluk formations suggest preferential enrichment in Th, which is in disagreement with plume-related magmatism and more in favour of an inherited signature from a metasomatic enrichment of the SCLM as a result of an older subduction of enriched, likely crustal, material. In other words, the geochemical data for both the Hekpoort and Ongeluk formations probably show that an important subduction event marked the Archaean-Proterozoic boundary in the Kaapvaal craton. Note that Zirakparvar et al. (2014) already proposed that a subduction occurred around 2.7–2.6 Ga through their study of Bushveld Complex.

Yet, this episode of fertilization of the SCLM can only explain the 'arc-like' signature in the first part of the Palaeoproterozoic, whereas this signature is also found in all Ventersdorp Supergroup lavas, even if these lavas present lower Th/Nb ratios. This signature is actually found in all

Archaean and Palaeoproterozoic mafic units in the Kaapvaal craton (Humbert and Elburg 2017; Humbert et al. 2018a). The Archaean 'arc-like' signatures in the Kaapvaal craton may thus be related to one or several and pre-Ventersdorp episode(s) of fertilization of the SCLM, e.g. before the Dominion Group, possibly during a cratonisation events prior to 2.9 Ga, by a process able to carry fluid to a mantle at that period: no matter whether subduction or sagduction.

In conclusion, the data presented here seem to indicate that the SCLM below the Kaapvaal craton was metasomatically enriched since its origin. It changed close to the Archaean—Proterozoic transition, and remained stable for at least 200 Ma.

Acknowledgements We thank the two reviewers of this contribution, Richard Ernst and David R. Nelson, whose comments and suggestions have improved the manuscript considerably. Funding was provided by grants from the South African Department of Science and Technology and the National Research Foundation (DST-NRF)-funded Centre of Excellence for Integrated Mineral and Energy Resource Analysis (CIMERA). Funding to Wladyslaw Altermann for fieldwork by the team of the University of Pretoria (UP) was from the NRF incentive funding for rated researchers and by the Kumba-Exxaro Chair at UP.

References

Alebouyeh Semami F, de Kock MO, Söderlund U, Gumsley A, Da Silva R, Beukes N, Armstrong R (2016) New U–Pb geochronology and paleomagnetic constraints on the late Paleoproterozoic Hartley magmatic event: evidence for a large igneous province in the Kaapvaal craton during Kalahari assembly, South Africa. GFF 138:164–182

Altermann W (1996) Discussion of zircon Pb-evaporation age determinations of the oak tree formation Chuniespoort Group Transvaal Sequence: implications for the Transvaal-Griqualand West basin correlations. S Afr J Geol 99:337–338

Altermann W, Hälbich IW (1990) Thrusting folding and stratigraphy of the Ghaap Group along the south-western margin of the Kaapvaal craton. S Afr J Geol 93:553–566

Altermann W, Hälbich IW (1991) Structural history of the south-western corner of the Kaapvaal craton and the adjacent Namaqua realm: new observations and reappraisal. Precambrian Res 52:133–166

Altermann W, Lenhardt N (2012) The volcano-sedimentary succession of the Archean sodium group ventersdorp supergroup, South Africa: volcanology sedimentology and geochemistry. Precambrian Res 214–215:60–81

Altermann W, Nelson DR (1998) Sedimentation rates basin analysis and regional correlations of three Neoarchaean and Palaeoproterozoic sub-basins of the Kaapvaal craton as implied by precise SHRIMP U–Pb zircon ages from volcanic sediments. Sed Geol 120:225–256

Altermann W, van Kranendonk MJ, Humbert F, Cornell DH (2018) The end of the Early Proterozoic glaciation and the Ongeluk—Hekpoort Large Igneous Province of the Kaapvaal Craton, South Africa. 27th Colloquium of African Geology Abstract Volume. Aveiro, Portugal, July (2018)

Armstrong NV, Wilson AH, hunter DR (1986) The Nsuze Group Pongola Sequence, South Africa geochemical evidence for Archean volcanism in a continental setting. Precambrian Res 34:175–203

Armstrong RA, Compston W, Retief EA, Williams IS, Welke HJ (1991) Zircon ion microprobe studies bearing on the age and evolution of the Witwatersrand triad. Precambrian Res 53:243–266

Barton JMJ, Blignaut E, Salnikova EB, Kotov AB (1995) The stratigraphical position of the Buffelsfontein Group based on field relationships and chemical and geochronological data. S Afr J Geol 98:386–392

Bau M, Romer RL, Lüders V, Beukes NJ (1999) Pb, O and C isotopes in silicified Mooidraai dolomite (Transvaal Supergroup, South Africa): implications for the composition of Paleoproterozoic seawater and 'dating' the increase of oxygen in the Precambrian atmosphere. Earth Planet Sci Lett 174:43–57

Beukes NJ (1983) Palaeoenvironmental setting of iron formations in the depositional basin of the Transvaal Supergroup, South Africa. In: Trendall AF, Morris RC (eds) Iron formations: facts and problems, Elsevier, Amsterdam, pp 131–209

Beukes NJ (1987) Facies relations depositional environments and diagenesis in a major early Proterozoic stromatolitic carbonate platform to basinal sequence Campbellrand Subgroup, Transvaal Supergroup, Southern Africa. Sediment Geol 54:1–46

Beukes NJ, Dorland H, Gutzmer J, Nedachi M, Ohmoto H (2002) Tropical laterites life on land and the history of atmospheric oxygen in the Paleoproterozoic. Geology 30:491–494

Beukes NJ, Frei D, Vorster C (2013) Detrital zircon age constraints of the Makganyene Snowball Earth Event (South Africa) at 225–222 Ga: the youngest of four Huronian glaciations? In: Geological society of America abstracts with programs, vol 7, Denver, Colorado, USA, p 340

Bowen TB, Marsh JS, Bowen MP, Eales HV (1986) Volcanic rocks of the Witwatersrand triad, South Africa. I: description classification and geochemical stratigraphy. Precambrian Res 31:297–324

Bryan SE, Ferrari L (2013) Large igneous provinces and silicic large igneous provinces: progress in our understanding over the last 25 years. Geol Soc Am Bull 125:1053–1078

Buick IS, Maas R, Gibson R (2001) Precise U–Pb titanite age constraints on the emplacement of the Bushveld Complex, South Africa. J Geol Soc London 158:3–6

Bumby AJ, Eriksson PG, Catuneanu O, Nelson DR, Rigby MJ (2012) Meso-Archaean and Palaeo-Proterozoic sedimentary sequence stratigraphy of the Kaapvaal Craton. Mar Petrol Geol 33:92–116. https://doi.org/10.1016/jmarpetgeo201109010

Burger AJ, Coertze FJ (1973) Radiometric age determinations on rocks from southern Africa up to the end of 1971. Geol Surv S Afr Bull 58:1–46

Burke K, Kidd WSF, Kusky TM (1985) The Pongola structure of southeastern Africa: the world's oldest preserved rift? J Geodyn 2:35–49

Button A (1973) A regional study of the stratigraphy and development of the Transvaal Basin in the eastern and north-eastern Transvaal. Ph.D. thesis, University of Witwatersrand, Johannesburg

Cawthorn RG, Mciver JR, McCarthy TS, Wyatt BA, Ferguson J, Barnes SJ (1979) Possible liquid immiscibility textures in high-magnesia basalts from the Ventersdorp Supergroup. S Afr J Geol 87:105–113

Clendenin CW, Charlesworth EG, Maske S (1988) An early Proterozoic three-stage rift system Kaapvaal craton, South Africa. Tectonophysics 156:275–291

Coetzee LL (2001) Genetic stratigraphy of the Paleoproterozoic Pretoria Group in the Western Transvaal. M.Sc. thesis, Rand Afrikaans University

Coffin MF, Eldholm O (1994) Large igneous provinces: crustal structure dimensions and external consequences. Rev Geophys 32:1–36

Condie KC, Crow C (1990) Early Precambrian within-plate basalts from the Kaapvaal craton in southern Africa: a case for crustally contaminated komatiites. J Geol 98:100–107

Cornell DH, Schütte SS, Eglington BL (1996) The Ongeluk basaltic andesite formation in Griqualand West, South Africa: submarine alteration in a 2222 Ma Proterozoic sea. Precambrian Res 79:101–123

Cornell DH, Zack T, Anderson T, Corfu F, Frei D, Van Schjindel V (2016) Th–U–Pb zircon geochronology of the Palaeoproterozoic Hartley Formation porphyry by six methods with age uncertainty approaching 1 Ma. S Afr J Geol 119:473–494

Cornell DH, Meintjies PG, Van der Westhuizen WA, Frei D (2017) Microbeam U–Pb Zircon dating of the Makwassie and Goedgenoeg formations in the Ventersdorp Supergroup of South Africa. S Afr J Geol 120:525–540. https://doi.org/10.25131/gssajg.120.4.525

Cornell DH, Minaar H, Frei D, Kristoffersen M (2018) Precise microbeam dating defines three Archaean granitoid suites at the southwestern margin of the Kaapvaal craton. Precambrian Res 304:21–38

Cox KG, Bell JD, Pankhurst RJ (1979) The Interpretation of Igneous Rocks. George Allen & Unwin, London

Crow C, Condie KC (1987) Geochemistry and origin of late Archean volcanics from the Rhenosterhock Formation Dominion Group, South Africa. Precambrian Res 37:217–229

Crow C, Condie KC (1988) Geochemistry and origin of the late Archean volcanics from the Ventersdorp Supergroup, South Africa. Precambrian Res 42:19–37

Davidson J, Turner S, Handley H, Macpherson C, Dosseto A (2007) Amphibole "sponge" in arc crust? Geology 35:787–790

de Kock MO, Evans DAD, Beukes NJ (2009) Validating the existence of Vaalbara in the late Neoarchean. Precambrian Res 174:145–154

de Kock MO, Beukes NJ, Armstrong RA (2012) New SHRIMP U–Pb zircon ages from the Hartswater Group South Africa: implications for correlations of the Neoarchean Ventersdorp Supergroup on the Kaapvaal craton and with the Fortescue Group on the Pilbara craton. Precambrian Res 204–205:66–74

De Wit MJ, Roering C, Hart RJ, Armstrong RA, De Ronde CEJ, Green RWE, Tredoux M, Peberdy E, Hart RA (1992) Formation of an Archaean continent. Nature 357:553–562

Denyszyn SW, Feinberg JM, Renne PR, Scott GR (2013) Revisiting the age and paleomagnetism of the Modipe Gabbro of South Africa. Precambrian Res 238:176–185

DePaolo DJ (1988) Neodymium Isotope Geochemistry. An Introduction. In: Wyllie PJ (ed) Minerals and Rocks Series, 20. Springer, p 187

Dorland HC (2004) Provenance ages and timing of sedimentation of selected Neoarchean and Paleoproterozoic successions on the Kaapvaal craton. Unpublished Ph.D. thesis, Rand Afrikaans University Johannesburg

Eglington B, Armstrong R (2004) The Kaapvaal Craton and adjacent orogens southern Africa: a geochronological database and overview of the geological development of the craton. S Afr J Geol 107:13–32

Engelbrecht JP (1986) The Bushveld Complex and its floor rocks in the area of Nietverdiend Western Transvaal (Translated from the Afrikaans: Die Bosveld Kompleks en sy vloergesteentes in die omgewing van Nietverdiend Wes-Transvaal). Ph.D. thesis, University of Pretoria

Eriksson PG, Reczko BFF (1995) The sedimentary and tectonic setting of the Transvaal Supergroup floor rocks to the Bushveld Complex: tectonic setting derived from basin-fill geometry inferred volcanic processes and sedimentation systems. J Afr Earth Sci 21:487–504

Eriksson PG, Engelbrecht JP, Res M, Harmer RE (1994) The Bushy Bend lavas a new volcanic member of the Pretoria Group Transvaal Sequence South African. J Geol 97:1–7

Eriksson PG, Martins-Neto MA, Nelson DR, Aspler LB, Chiarenzelli JR, Catuneanu O, Sarkar S, Altermann W, Rautenbach CJ de W (2001) An introduction to Precambrian basins: their characteristics and genesis. Sed Geol 141–142:1–35

Eriksson PG, Altermann W, Hartzer FJ (2006) The Transvaal Supergroup and its precursors. In: Johnson MR, Anhaeusser CR, Thomas RJ (eds) The Geology of South Africa. Geological Society of South Africa and the Council for Geoscience, Pretoria, pp 237–260

Ernst RE (2014) Large igneous provinces. Cambridge University Press, Cambridge, United Kingdom. https://doi.org/10.1017/CBO9781139025300

Ernst RE, Buchan KL (2001) Large mafic magmatic events through time and links to mantle-plume heads. In: Ernst RE, Buchan KL (eds) Mantle plumes: their identification through time, Boulder, CO: Geol Soc Am Special Paper 352:483–575

Evans DA, Beukes NJ, Kirschvink JL (1997) Low-latitude glaciation in the Palaeoproterozoic era. Nature 386:262–266

Fairey B, Tsikos H, Corfu F, Polteau S (2013) U–Pb systematics in carbonates of the Postmasburg Group Transvaal Supergroup South Africa: primary versus metasomatic controls. Precambrian Res 231:194–205

Grant CE (2003) The physical volcanology and geochemistry of the Nsuze Group Pongola Supergroup of northern KwaZulu-Natal and southeastern Mpumalanga. M.Sc. thesis, University of Pretoria

Grobler NJ, Botha BJV (1976) Pillow lavas and hyaloclastites in the Ongeluk andesite formation in a road-cut west of Griquatown, South Africa. Trans Geol Soc SA 79:53–57

Grobler DF, Walraven F (1993) Geochronology of Gabarone Granite Complex extensions in the area north of Mafikeng, South Africa. Chem Geol 105:319–337

Grobler NJ, Van der Westhuizen WA, Tordiffe EAW (1989) The Sodium Group South Africa; reference section for late Archaean-early Proterozoic cratonic cover sequences. Aust J Earth Sci 36:41–64

Gumsley AP, Rådman J, Söderlund U, Klausen M (2016) U–Pb baddeleyite geochronology and geochemistry of the White Mfolozi Dyke Swarm: unravelling the complexities of 270–266 Ga dyke swarms across the eastern Kaapvaal Craton, South Africa. GFF 138:115–132

Gumsley AP, Chamberlain KR, Bleeker W, Söderlund U, de Kock MO, Larsson ER, Bekker A (2017) Timing and tempo of the Great Oxidation Event. Proc Natl Acad Sci USA 114:1811–1816

Guo Q, Strauss H, Schröder S, Gutzmer J, Wing BA, Baker M, Bekker A, Jin Q, Kim S, Farquhar J (2009) Reconstructing Earth's surface oxidation across the Archean- Proterozoic transition. Geology 37:399–402

Gutzmer J, Beukes NJ (1998) High-grade manganese ores in the Kalahari manganese field: characterization and dating of the ore forming events. Unpublished Report, Rand Afrikaans University Johannesburg

Gutzmer J, Banks DA, Lüders V, Hoefs J, Beukes NJ, von Bezing KL (2003) Ancient sub-seafloor alteration of basaltic andesites of the Ongeluk Formation South Africa: implications for the chemistry of Paleoproterozoic seawater. Chem Geol 201:37–53

Hannah JL, Bekker A, Stein HJ, Markey RJ, Holland HD (2004) Primitive Os and 2316 Ma age for marine shale: implications for Paleoproterozoic glacial events and the rise of atmospheric oxygen. Earth Planet Sc Lett 225:43–52

Hatton CJ (1995) Mantle plume origin for the Bushveld and Ventersdorp magmatic provinces. J Afr Earth Sci 21:571–577

HR Winter (1976) A lithostratigraphic classification of the ventersdorp succession. Trans Geol Soc SA 79:31–48

Humbert F, Elburg M (2017) Large Igneous Provinces (LIP) and the Subcontinental Lithosphere of the Kaapvaal Craton (South Africa) from the Mesoarchean to the Present. Goldschmidt Abstracts 2017, 1744

Humbert F, Sonnette L, de Kock MO, Robion P, Horng CS, Cousture A, Wabo H (2017) Palaeomagnetism of the early Palaeoproterozoic volcanic Hekpoort Formation (Transvaal Supergroup) of the Kaapvaal craton, South Africa. Geophys J Int 209:842–865. https://doi.org/101093/gji/ggx055

Humbert F, de Kock MO, Altermann W, Elburg MA, Lenhardt N, Masango S (2018a) Petrology physical volcanology and geochemistry of a Palaeoproterozoic large igneous province: the Hekpoort Formation in the southern Transvaal sub-basin (Kaapvaal Craton). Precambrian Res 315:232–256. https://doi.org/10.1016/j.precamres.2018.07.022

Humbert F, Elburg M, Ossa Ossa F, de Kock MO, Robion P (2018b) Variolites of the Paleoproterozoic Hekpoort Formation (Transvaal sub-basin, Kaapvaal craton): multistage undercooling textures? Lithos 316–317:48–65. https://doi.org/10.1016/j.lithos.2018.07.005

Johnson MR, Anhaeusser CR, Thomas RJ (2006) The Geology of South Africa. Council for Geoscience, Pretoria, 691p

Kamber BS (2015) The evolving nature of terrestrial crust from the Hadean through the Archaean into the Proterozoic. Precambrian Res 258:48–82

Key RM (1983) The geology of the area around Gaborone and Lobatse Kweneng Kgatleng Southern and South East Districts. District Memoir, Geological Survey Botswana 5:230

Keyser N (1998) The Geology and geochemistry of the Ventersdorp Supergroup between Vryburg Ottosdal and Mafikeng. Geol Surv SA:9–33

Klausen MB, Söderlund U, Olsson JR, Ernst RE, Armoogam M, Mkhize SW, Petzer G (2010) Petrological discrimination among Precambrian dyke swarms: Eastern Kaapvaal craton (South Africa). Precambrian Res 183:501–522

Kramers JD, Mouri H (2011) The geochronology of the Limpopo Complex: a controversy solved. In: Van Reenen DD, Kramers JD, McCourt S, Perchuk LL (eds) Origin and Evolution of Precambrian High-Grade Gneiss Terranes with Special Emphasis on the Limpopo Complex of South Africa: Geological Society of America Memoir 207, pp 85–106

Kramers JD, McCourt S, van Reenen DD (2006) The Limpopo Belt. In: Johnson MR Anhaeusser CR Thomas RJ (eds) The geology of South Africa Geol Soc SA and the Council for Geoscience Pretoria, pp 209–236

Lenhardt N, Eriksson PG, Catuneanu O, Bumby AJ (2012) Nature of and controls on volcanism in the ca 232-206 Ga Pretoria Group Transvaal Supergroup Kaapvaal craton, South Africa. Precambrian Res 214–215:106–123

Liebenberg JJ (1977) *The geology of the area 2724D (Andalusia)* from the Afrikaans: Die geologie van die gebied 2724D (Andalusia). M. Sc. thesis, University of the Orange Free State

Lowe DR, Byerly GR (2007) An overview of the geology of the Barberton Greenstone Belt and vicinity: implications for early crustal development. In: Van Kranendonk MJ, Smithies RH Bennett VC (eds) Earth's Oldest Rocks Developments in Precambrian Geology, 15. Elsevier, Amsterdam, pp 481–526

Lubnina N, Ernst RE, Klausen M, Söderlund U (2010) Paleomagnetic study of Neoarchean-Paleoproterozoic dykes in the Kaapvaal craton. Precambrian Res 183:523–552

Mapeo RBM, Armstrong RA, Kampunzu AB, Modisi MP, Ramokate LV, Modie BNJ (2006) A ca. 200 Ma hiatus between the Lower and Upper Transvaal Groups of southern Africa: SHRIMP U–Pb detrital zircon evidence from the Segwagwa Group Botswana: implications for Palaeoproterozoic glaciations. Earth Planet Sci Lett 244:113–132

Marsh JS, Bowen MP, Rogers NW, Bowen TB (1989) Volcanic rocks of the Witwatersrand Triad, South Africa II: Petrogenesis of mafic and felsic rocks of the Dominion Group. Precambrian Res 44:39–65

Marsh JS, Bowen MP, Rogers NW, Bowen TB (1992) Petrogenesis of late Archean flood-type basic lavas from the Klipriviersberg Group Ventersdorp Supergroup, South Africa. J Petrol 33:817–847

Martin DMB, Clendenin CW, Krapez B, McNaughton NJ (1998) Tectonic and geochronological constraints on Late Archaean and Palaeoproterozoic stratigraphic correlation within and between the Kaapvaal and Pilbara cratons. J Geol Soc London 155:311–322

McCarthy TS, McCallum K, Myers RE, Linton P (1990) Stress states along the northern margin of the Witwatersrand Basin during Klipriviersberg Group volcanism. S Afr J Geol 93:245–260

McIver JR, Cawthorn RG, Wyatt BA (1982) The Ventersdorp Supergroup—the youngest komatiitic sequence in South Africa. In: Arndt NT, Nisbet EG (Eds) Komatiites. George Allen and Unwin, London, pp 81–90

Melezhik VA, Young GM, Eriksson PG, Altermann W, Kump LR, Lepland A (2013) Huronian-age glaciations Chapter 72. In: Melezhik VA, Prave AR, Fallick AE, Kump LR, Strauss H, Lepland A, Hanski EJ (eds) Reading the Archives of Earth's Oxygenation, Vol 3; Frontiers in Earth Sciences, Vol 8. Springer, pp 1059–1109

Moore M, Davis DW, Robb LJ, Jackson MC, Grobler DF (1993) Archean rapakivi granite-anorthosite-rhyolite complex in the Witwatersrand basin hinterland southern Africa. Geology 21:1031–1034

Moore JM, Tsikos H, Polteau S (2001) Deconstructing the Transvaal Supergroup South Africa: implications for Palaeoproterozoic palaeoclimate models. J Afr Eart Sci 33:437–444

Mukasa SB, Wilson AH, Young KR (2013) Geochronological constraints on the magmatic and tectonic development of the Pongola Supergroup (Central Region), South Africa. Precambrian Res 224:268–286

Nelson DR, Trendall AF, de Laeter JR, Grobler NJ, Fletcher IR (1992) A comparative study of the geochemical and isotopic systematics of late Archean flood basalts from the Pilbara and Kaapvaal cratons. Precambrian Res 54:231–256

Nelson DR, Trendall AF, Altermann W (1999) Chronological correlations between the Pilbara and Kaapvaal cratons. Precambrian Res 9:165–189

Nicoli G, Stevens G, Moyen JF, Frei D (2015) Rapid evolution from sediment to anatectic granulite in an Archean continental collision zone: the example of the Bandelierkop Formation metapelites South Marginal Zone Limpopo Belt, South Africa. J Metamorph Geol 33:177–202

Oberholzer JD (1995) *The geology of the pyroclastic rocks of the Hekpoort Formation Transvaal succession* Translated from the Afrikaans: Die geologie van die piroklastiese gesteentes in die Hekpoort Formasie Transvaal Opeenvolging. M.Sc. thesis, University of Pretoria

Oberholzer JD, Eriksson PG (2000) Subaerial volcanism in the Palaeoproterozoic Hekpoort Formation (Transvaal Supergroup) Kaapvaal craton. Precambrian Res 101:193–210

Olsson JR, Söderlund U, Klausen MB, Ernst RE (2010) U–Pb baddeleyite ages linking major Archean dyke swarms to volcanic-rift forming events in the Kaapvaal craton (South Africa) and a precice age for the Bushveld Complex. Precambrian Res 183:490–500

Olsson JR, Söderlund U, Hamilton MA, Klausen MB, Helffrich GR (2011) A late Archean radiating dyke swarm as possible clue to the origin of the Bushveld Complex. Nat Geosci 4:865–869

Olsson JR, Klausen MB, Hamilton MA, März N, Söderlund U, Roberts RJ (2016) Baddeleyite U–Pb ages and geochemistry of the 1875–1835 Ma Black Hills Dyke Swarm across north-eastern South Africa: Part of a trans-Kalahari Craton back-arc setting? GFF 138:183–202. https://doi.org/10.1080/11035897201511 03781

Palme H, O'Neill HStC (2004) Cosmochemical estimates of mantle composition In: Holland HD, Turrekian KK (eds) Treatise on Geochemistry. Elsevier, Amsterdam, The Netherlands 2:1–38

Pearce JA (1996) A user's guide to basalt discrimination diagrams. In: DA Wyman (eds) Trace Element Geochemistry of Volcanic Rocks: applications for Massive Sulphide Exploration. Geological Association of Canada Short Course Notes 12:79–133

Pearce JA (2008) Geochemical fingerprinting of oceanic basalts with applications to ophiolite classification and the search for Archean oceanic crust. Lithos 100:14–48

Pickard AL (2003) SHRIMP U–Pb zircon ages for the Palaeoproterozoic Kuruman Iron Formation Northern Cape Province South Africa: evidence for simultaneous BIF deposition on Kaapvaal and Pilbara Cratons. Precambrian Res 125:275–315

Potgieter CD (1973) *The geology of an area south of Douglas Northern Cape* Translated from the Afrikaans: Die geologie van 'n gebied suid van Douglas Noord-Kaapland. Department of Geology University of the Orange Free State Bloemfontein, p 235

Potgieter GJA, Lock BE (1978) Correlation and lithology of the Ritchie Quartz Porphyry Formation along the Riet River near Kimberley. Trans Geol Soc S Afr 81:41–46

Poujol M, Robb LJ, Anhaeusser CR, Gericke B (2003) A review of the geochronological constraints on the evolution of the Kaapvaal craton, South Africa. Precambrian Res 127:181–213

Poujol M, Kiefer R, Robb LJ, Anhaeusser CR, Armstrong RA (2005) New U–Pb data on zircons from the Amalia greenstone belt Southern Africa: insights into the Neoarchean evolution of the Kaapvaal craton. S Afr J Geol 108:317–332

Reczko BFF, Oberholzer JD, Res M, Eriksson PG, Schreiber UM (1995) A revaluation of the volcanism of the Palaeoproterozoic Pretoria Group (Kaapvaal craton) and a hypothesis on basin development. J Afr Eart Sci 21:505–519

Robb LJ, Meyer FM (1995) The Witwatersrand Basin, South Africa: geological framework and mineralization processes. Ore Geol Rev 10:67–94

Rollinson HR (1993) Using geochemical data: Evaluation presentation interpretation Longman Scientific and Technical, p 352

SACS: South African Committee for Stratigraphy (1980) Stratigraphy of South Africa Part 1: Lithostratigraphy of the Republic of South Africa South West Africa/Namibia and the Republics of Bophuthatswana Transkei and Venda (Compiled by Kent LE). Handbook 8 Geol Surv SA

Schröder S, Beukes NJ, Armstrong RA (2016) Detrital zircon constraints on the tectonostratigraphy of the Paleoproterozoic Pretoria Group, South Africa. Precambrian Res 278:362–393

Schütte SS (1992) Ongeluk volcanism in relation to the Kalahari manganese deposits. Unpublished Ph.D. thesis, University of Natal

Sharpe MR, Brits R, Engelbrecht JP (1983) Rare earth and trace element evidence pertaining to the petrogenesis of 23 Ga old continental andesites and other volcanic rocks from the Transvaal Sequence, South Africa. University of Pretoria Institute for Geological Research on the Bushveld Complex Research Report 40, p 63

Srivastava RK, Ernst RE (2013) Global intracratonic boninitic–norite magmatism during the Neoarchean–Paleoproterozoic: revisited. LIP of the Month, July 2013. http://www.largeigneousprovinces.org/13jul

Stanistreet IG, McCarthy TS (1991) Changing tectono-sedimentary scenarios relevant to the development of the Late Archaean Witwatersrand Basin. J Afr Eart Sci 13:65–81

Sumner DY, Beukes NJ (2006) Sequence stratigraphic development of the Neoarchean Transvaal carbonate platform Kaapvaal Craton, South Africa. S Afr J Geol 109:1–22

Sumner DY, Bowring SA (1996) U–Pb geochronologic constraints on deposition of the Campbellrand Subgroup Transvaal Supergroup South Africa. Precambrian Res 79:25–35

Sumner DY, Grotzinger J (2004) Implications for Neoarchean ocean chemistry from primary carbonate mineralogy of the Campbellrand-Malmani Platform, South Africa. Sedimentology 51:1273–1299

Sun SS, Nesbitt RW, McCulloch MT (1989) Geochemistry and petrogenesis of Archaean and early Proterozoic siliceous high-magnesian basalts In: Crawford AJ (ed) Boninites and Related Rocks. Unwin Hyman, London, p 465

Van der Westhuizen WA, De Bruiyn H, Meinties PG (1991) The Ventersdorp Supergroup; an overview. J Afr Earth Sci 13:83–105

Van der Westhuizen WA, De Bruiyn H, Meintjies PG (2006) The Ventersdorp Supergroup. In: Johnson MR, Anhaeusser C (eds) The Geology of South Africa, Geological Society of South Africa. Council for Geoscience, Johannesburg, Pretoria, pp 187–208

Visser JNJ (1971) The deposition of the Griquatown glacial member in the Transvaal Supergroup. Trans Geol Soc S Afr 74:187–199

Visser JNJ (1999) Lithostratigraphy of the Makganyene Formation (Postmasburg Group). South African Committee for Stratigraphy Lithostratigraphic Series 34:1–7

Visser JNJ, Grobler NJ, Joubert CW, Potgieter CD, Potgieter GJA, McLaren CH, Liebenberg J (1976) The Ventersdorp Group between Taung and Britstown northern Cape Province. Ann Geol Surv Pretoria 11:15–28

Wabo H, Humbert F, de Kock MO, Belyanin G, Söderlund U, Maré LP, Beukes NJ (2019) Constraining the chronology of the Mashishing dykes from the eastern Kaapvaal craton in South Africa. In: Srivastava RK, Ernst RE, Peng P (eds) Dyke Swarms of the world—A modern perspective. Springer Geology, pp 215–261. https://doi.org/10.1007/978-981-13-1666-1_6

Walraven F, Martini J (1995) Zircon Pb-evaporation age determinations for the Oak Tree Formation Chuniespoort Group Transvaal Sequence; implications for Transvaal-Griqualand West basin correlations. S Afr J Geol 98:58–67

Walraven F, Smith CB, Kruger FJ (1991) Age determinations of the Zoetlief Group—a Ventersdorp Supergroup correlative. S Afr J Geol 94:220–227

Walraven F, Retief EA, Moen HFG (1994) Single-zircon Pb-evaporation evidence for 277 Ga magmatism in northwestern Transvaal, South Africa. S Afr J Geol 97:107–113

Walraven F, Grobler DF, Key RM (1996) Age equivalents of the Plantation Porphyry and the Kanye Volcanic Formation southeastern Botswana. S Afr J Geol 99:23–31

White JDL, Bryan SE, Ross PS, Self S, Thordarson T (2009) Physical volcanology of continental large igneous provinces: update and review. In: Thordarson T, Self S, Larsen G, Rowland SK, Hoskuldsson A (eds) Studies in volcanology: the legacy of George Walker, Special Publication of IAVCEI 2:291–321

Winchester JA, Floyd PA (1977) Geochemical discrimination of different magma series and their differentiation products using immobile elements. Chem Geol 20:325–343

Wingate MTD (1998) A palaeomagnetic test of the Kaapvaal—Pilbara (Vaalbara) connection at 278 Ga. S Afr J Geol 101:257–274

Zeh A, Gerdes A, Barton JM Jr (2009) Archean accretion and crustal evolution of the Kalahari Craton—the zircon age and Hf isotope record of granitic rocks from Barberton/Swaziland to the Francistown Arc. J Petrol 50:933–966

Zeh A, Ovtcharova M, Wilson AH, Schaltegger U (2015) The Bushveld Complex was emplaced and cooled in less than one million years—results of zirconology and geotectonic implications. Earth Planet Sci Lett 418:103–114

Zirakparvar NA, Mathez EA, Scoates JS, Wall CJ (2014) Zircon Hf isotope evidence for an enriched mantle source for the Bushveld Igneous Complex. Contrib Mineral Petrol 168:1050. https://doi.org/10.1007/s00410-014-1050-2

MIX
Papier aus verantwortungsvollen Quellen
Paper from responsible sources
FSC® C105338

If you have any concerns about our products,
you can contact us on
ProductSafety@springernature.com

In case Publisher is established outside the EU,
the EU authorized representative is:
Springer Nature Customer Service Center GmbH
Europaplatz 3, 69115 Heidelberg, Germany

Printed by Libri Plureos GmbH
in Hamburg, Germany